Design and Applications
of
Analog Integrated Circuits

Design and Applications
of
Analog Integrated Circuits

SIDNEY SOCLOF
California State University
Los Angeles

PRENTICE HALL, Englewood Cliffs, New Jersey 07632

Library of Congress Cataloging-in-Publication Data

Soclof, Sidney.
    Design and applications of analog integrated circuits / Sidney
Soclof.
      p.   cm. — (Prentice-Hall series in solid state physical
electronics)
    Includes bibliographical references and index.
    ISBN 0-13-026030-4
    1. Linear integrated circuits.   I. Title.   II. Series.
TK7874.S643   1991
621.381′5—dc20   90-47162
                                                         CIP

Prentice Hall Series in Solid State Physical Electronics
Nick Holonyak, Jr., Series Editor

Editorial/production supervision: *Raeia Maes*
Manufacturing buyer: *Lori Bulwin*
Cover design: *Joseph DiDomenico*

© 1991 by Prentice-Hall, Inc.
A division of Simon & Schuster
Englewood Cliffs, New Jersey 07632

Printed in the United States of America

10 9 8 7 6 5 4 3 2 1

ISBN   0-13-026030-4

Prentice-Hall International (UK) Limited, *London*
Prentice-Hall of Australia Pty. Limited, *Sydney*
Prentice-Hall Canada Inc., *Toronto*
Prentice-Hall Hispanoamericana, S.A., *Mexico*
Prentice-Hall of India Private Limited, *New Delhi*
Prentice-Hall of Japan, Inc., *Tokyo*
Simon & Schuster Asia Pte Ltd., *Singapore*
Editora Prentice-Hall do Brasil, Ltda., *Rio de Janeiro*

Contents

14 MODULATORS, DEMODULATORS, AND PHASE DETECTORS

15 VOLTAGE-CONTROLLED OSCILLATORS AND WAVEFORM GENERATORS   698

16 PHASE-LOCKED LOOPS   729

Preface

This book is a comprehensive treatment of the analysis, design, applications, and fabrication of analog integrated circuits. There is enough material in this book for a three- or four-semester senior- or graduate-level sequence of courses. The various chapters in this book are designed to be relatively self-sufficient, in that they may be used without much reference to or study of the other chapters. Therefore, a one-semester course covering a particular aspect of analog integrated circuits can be taught by selection of the appropriate chapters. For example, a one-semester course with primary emphasis on operational amplifiers can be built out of Chapters 5 through 9, and selected parts of Chapters 3 and 4 can also be included. A course with principal emphasis on integrated-circuit fabrication, devices, and the internal circuit design can be designed by using Chapters 1 through 3, with selections made from various other chapters. A one- or two-semester course with primary emphasis on applications of analog integrated circuits other than operational amplifiers can be constructed by choosing from Chapters 10 through 17. Thus there is enough material in this book for any advanced analog electronics course or sequence of courses.

Homework problems and references for further reading are given at the end of each chapter.

The material in this book is of an advanced nature and the reader should have the background of a two-semester introductory electronics course sequence.

Chapter 1 presents the basic processes that are used to manufacture various semiconductor devices, with particular emphasis on the processes used for integrated circuits. This chapter includes material on crystal growth, wafer preparation, diffusion, ion implantation, oxidation, photolithography, metallization, and chemical-vapor deposition processes, including epitaxy.

In Chapter 2, the basic processes described in Chapter 1 are combined to produce various semiconductor devices and, in particular, devices for integrated circuits. This chapter discusses the basic characteristics of PN junctions and the advantages of the use of epitaxial layers. The processing sequences used for diodes, bipolar transistors, junction field-effect transistors, and MOSFETs are described, followed by the various techniques used for the isolation of devices on an integrated-circuit circuit chip. The fabrication and characteristics of various integrated-circuit devices are presented, including NPN and PNP bipolar transistors, JFETs and MOSFETs, diodes, resistors, and capacitors. Other material in Chapter 2 deals with considerations of chip size, circuit complexity, and thermal design.

Chapters 3 and 4 present various electronic circuit configurations common to many types of analog integrated circuits. The first part of Chapter 3 deals with the various types of constant-current source circuits that are an important part of many integrated circuits. The second part of this chapter is concerned with low-impedance voltage source circuits, and the last part is a study of temperature-compensated voltage reference circuits.

The subject matter of Chapter 4 is differential amplifiers. In the first part of this chapter, the design and characteristics of bipolar transistor differential amplifiers are considered. This is followed by a presentation of JFET and MOSFET differential amplifiers. Closely associated with the subject of differential amplifiers is that of active-load circuits, which is discussed in the last part of the chapter, along with bipolar transistor and FET active loads.

The most widely used type of analog integrated circuit is the operational amplifier, and this is the subject matter of Chapters 5 through 9. The discussion commences in Chapter 5 with an analysis of closed-loop operational amplifier circuits based on the case of an ideal device. The various nonideal characteristics of operational amplifiers are discussed, including the effects of the input offset voltage, input bias current, input and output impedances, common-mode gain, and the noise generated both within the operational amplifier and the external resistances. As in any feedback system, the question of system stability is an important one. This is considered in Chapter 5, and the various compensation techniques to ensure an adequate margin of stability are also discussed. Later in the chapter, examples of a large number of applications of operational amplifiers are presented. An important category of operational amplifier applications is that of active filters. These applications are discussed in Chapter 6, together with material on switched-capacitor filters.

Whereas Chapters 5 and 6 deal mostly with the external circuitry and applications of operational amplifiers, attention in Chapter 7 is turned principally to the internal circuit design of operational amplifiers. The chapter begins with a discussion of the basic configuration of operational amplifiers and a general analysis of the open-loop frequency response, including a discussion of the unity gain frequency. This is followed by a discussion of large-signal response characteristics, including the slewing rate and the full-power bandwidth. A representative example of an operational amplifier circuit is then presented and analyzed. This analysis includes evaluation of dc biasing, the ac small-signal voltage gain of each stage and the entire amplifier, and various other amplifier characteristics, including the common-mode rejection ratio, the power supply rejection ratio, the open-loop frequency response, and the equivalent input noise voltage and current. Operational amplifiers that use field-effect transistors in just the input stage or throughout the entire amplifier are important devices and are considered in Chapter 8. A number of representative JFET and MOSFET operational-amplifier circuits are presented. In Chapter 9, current-feedback operational amplifiers, Norton operational amplifiers, and transconductance operational amplifiers are discussed. The current-feedback operational amplifiers are of particular interest due to their very wide bandwidth and fast slewing rate characteristics.

The voltage comparator is a device closely related to the operational amplifier, and a full discussion of this type of integrated circuit is given in Chapter 10. A consideration of some of the basic characteristics of voltage comparators and how these devices compare to operational amplifiers are thoroughly discussed. Examples of voltage comparators are then presented. One of the most important characteristics of the voltage comparator is its response time, and this is considered in detail. Various techniques that are used to reduce the response time are also presented. Complementary-symmetry MOSFETs can be used for comparators, especially for high-density, low-power applications. The CMOS inverter and CMOS voltage comparators are presented at the end of this chapter.

Integrated-circuit voltage regulators are presented in Chapter 11. In addition to a presentation of the basic theory, characteristics, and the protective circuitry of voltage regulators, examples of various types of voltage regulators are given, including adjustable positive and negative regulators, three-terminal fixed regulators, and switching-mode regulators.

In Chapter 12, integrated-circuit power amplifiers are investigated. Power conversion efficiency and distortion of power amplifiers are considered, and examples are given of integrated-circuit audio power amplifiers. Examples of power operational amplifiers are also presented.

In Chapter 13, attention is turned to the high-frequency performance of integrated circuits with a discussion of wide-bandwidth or video amplifiers. First, the performance of common-emitter, cascode, emitter-follower, and FET circuits at high frequencies is discussed. This is followed by a presentation of examples of video amplifier integrated cicuits. Then examples of operational amplifiers that have very wide bandwidths and high slewing rates are given.

Modulators, demodulators, and phase detector integrated circuits are discussed in Chapter 14. These are very closely related topics since the same basic circuit can be used for all three functions. The chapter opens with an analysis of the basic circuit configuration that is used for these three functions. Then application examples are given, including amplitude modulation and demodulation, frequency modulation and FM detection, frequency doubling, and phase detection.

Integrated circuits that are used to generate various types of waveforms are presented in Chapter 15. Examples of voltage-controlled oscillators in which the frequency can be varied by an input voltage are considered first, followed by a discussion of waveform generators for the production of square, triangular, pulse, and sinusoidal waveforms.

The phase-locked loops of Chapter 16 are an important element of many communications and signal-processing systems. The basic operation of these devices is discussed first, followed by examples of various applications, such as AM and FM detection, frequency synthesis, and stereo demodulation.

Analog-to-digital and digital-to-analog converters are essential elements for communication between analog and digital systems. These integrated circuits are discussed in Chapter 17 and various examples are given.

This book is intended to be of a practical nature, and the emphasis on the integrated circuits is mainly on what is, rather than what has been, or what might be. That is, the emphasis is on devices that are currently being used and are commercially available. Many circuits and devices presently being reported on in the research and development stage are not treated in this book because of size limitations.

I wish to acknowledge and thank the many reviewers for their time and helpful comments regarding this book, including Raymond J. Black, Jr., of New Mexico State University, Las Cruces; C. W. Bray of Memphis State University; Stanley G. Burns of Iowa State University; Robert A. Curtis of Ohio University; Frank H. Hielscher of Lehigh University; Richard Kwor of the University of Colorado, Colorado Springs; Anton Mavretic of Boston University; James E. Morris of the State University of New York, Binghamton; Andrzej Rusek of Oakland University; and Paul Van Halen of Portland State University.

Sidney Soclof

Design and Applications
of
Analog Integrated Circuits

1

Integrated-circuit Fabrication

1.1 SILICON DEVICE PROCESSES

The basic processing steps that are used to fabricate various silicon devices, such as diodes, transistors, and integrated circuits (ICs), can be categorized as follows:

1. Ion implantation
2. Diffusion
3. Oxidation
4. Photolithography
5. Chemical-vapor deposition (including epitaxy)
6. Metallization

Starting with single-crystal silicon wafers, the processes listed above can be used to produce functioning discrete devices (that is, individual diodes and transistors) and ICs. These devices or ICs will be in wafer form, with tens, hundreds, or even thousands of devices or ICs on the same silicon wafers. The wafer must be then divided up to obtain the individual *dice* or *chips*.

These chips are then encapsulated or *packaged*, with a wide variety of packages and packaging methods being possible. There are three basic purposes of packaging: (1) to provide encapsulation of the chip for protection of the chip from environmental

1

effects, (2) to provide easy access to the various parts of the chip by means of a lead or pin structure such that the device may be conveniently plugged into or attached to the rest of the system, and (3) to facilitate heat transfer out from the chip to the ambient.

The basic *wafer fabrication* processes listed above are generally applied a number of times in succession, especially in the case of ICs, where as many as 20 repetitions of the photolithography, oxidation, ion implantation, and diffusion steps may be used.

1.2 SILICON WAFER PREPARATION

The starting materials for silicon device processing are single-crystal silicon wafers of suitable conductivity type and doping. The silicon wafer preparation sequence involves the following basic steps:

1. Crystal growth and doping
2. Ingot trimming and grinding
3. Ingot slicing
4. Wafer polishing and etching
5. Wafer cleaning

We will now discuss these steps briefly.

1.2.1 Crystal Growth

In the crystal growth process, single-crystal silicon ingots of the appropriate doping level and type are produced. The starting material for crystal growth is highly purified polycrystalline silicon called *semiconductor-grade* silicon. The purity of this material is generally down in the range of more than 99.9999999% or "9-nines" purity. This corresponds to an impurity concentration of less than 1 part per billion atoms (<1 ppba), that is, less than one impurity atom for every billion (10^9) silicon atoms. The number of silicon atoms per unit volume is 5.0×10^{22} cm^{-3}, so an impurity concentration of 1 ppba corresponds to an impurity density of 5×10^{13} cm^{-3}. Much of this residual impurity density will be acceptor impurities such as boron, and the resistivity corresponding to the impurity density above is approximately 300 Ω-cm. Polycrystalline silicon with impurity concentrations of less than 0.1 ppba is available.

The *Czochralski* crystal growth process is the one most often used for producing single-crystal silicon ingots. The polycrystalline silicon together with an appropriate amount of dopant or doped silicon is put into a quartz crucible, which is then placed inside a crystal growth furnace. The material is then heated to a temperature that is slightly in excess of the silicon melting point of 1420°C. A small single-crystal rod of silicon called a *seed crystal* is then dipped into the silicon melt and slowly withdrawn, as shown in Figure 1.1. The conduction of heat up the seed crystal will produce a reduction in the temperature of the melt in contact with the seed crystal to slightly below the silicon melting point. This silicon will therefore freeze onto the end of the seed crystal, and as the seed crystal is slowly pulled up out of the melt it will pull

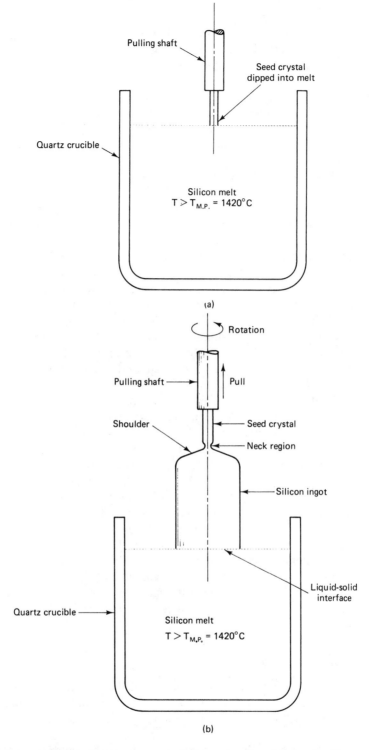

Figure 1.1 Czochralski crystal growth: (a) seed setting; (b) crystal pulling.

up with it a solidified mass of silicon that will be a crystallographic continuation of the seed crystal. Both the seed crystal and the crucible are rotated, but in opposite directions, during the crystal pulling process in order to produce crystalline ingots of circular cross section.

The crystal pulling is done in an inert-gas atmosphere, usually argon or helium, and sometimes a vacuum is used. The diameter of the ingot is controlled by the pulling rate and the melt temperature, with ingot diameters of about 100 to 150 mm (4 to 6 in.) being the most common, with some ingots being as large as 200 mm (8 in.) in diameter. The ingot length is generally on the order of 100 cm, and several hours are required for the pulling of a complete ingot.

The *float-zone* process is often used for the production of single-crystal silicon ingots, especially where very high resistivity material (>100 Ω-cm) is required. In this crystal growth process, a rod of polycrystalline silicon about 100 mm in diameter is placed in a quartz tube and a high vacuum is produced. A small single-crystal seed section of silicon is placed in contact with one end of the polycrystalline silicon rod. As shown in Figure 1.2, a water-cooled radio-frequency (RF) induction heating coil induces eddy currents in the silicon and raises the temperature of the region of the rod near the induction heating coil to above the melting point. This produces a molten zone held in place by surface-tension effects. The molten zone is started out at the seed end of the polycrystalline silicon rod, and by moving the induction heating coil the molten zone can be slowly moved through the length of the rod. As the zone moves through the rod, the silicon melts and then resolidifies as a crystallographic continuation of the seed crystal.

The float-zone process can also be used for the purification of silicon; this process is known as *zone refining*. Virtually all impurities in silicon tend to stay in the liquid phase rather than be incorporated into the solid phase at a moving liquid–solid interface. If a molten zone is passed through a silicon rod, the impurities will therefore tend to become segregated in the molten zone and thereby be swept to the far end of the rod. If a molten zone is passed along the length of the silicon rod to one end and then started again at the opposite end, and the process is repeated a number of times, a very high degree of purification can be achieved.

Float-zone crystals are grown in a vacuum atmosphere, and there is no contact of the ingot with any crucible. As a result of this, very high resistivity crystals with a very low content of oxygen and other impurities are possible with the float-zone process. In the Czochralski process, the contact of the molten silicon with the quartz crucible results in contamination of the melt with oxygen and other impurities from the crucible. In Table 1.1, a general comparison of the Czochralski and float-zone silicon processes is presented.

1.2.2 Ingot Trimming and Slicing

After the crystal growth process is completed, the extreme top and bottom portions of the ingot are cut off and the ingot surface is ground to produce a constant and exact diameter. A crystallographic orientation flat is also ground along the length of the ingot.

The ingot is then sliced using a large-diameter stainless steel saw blade with

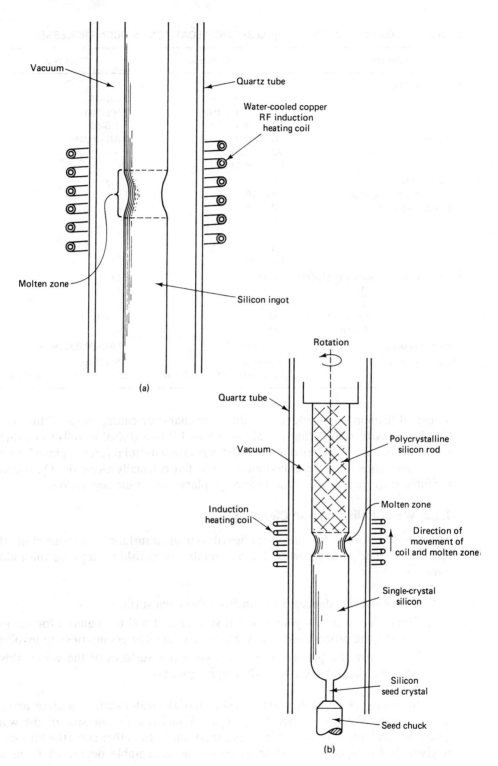

Figure 1.2 Vacuum float-zone refining and crystal growth: (a) float-zone refining; (b) crystal growth.

TABLE 1.1 COMPARISON OF CZOCHRALSKI AND FLOAT-ZONE SILICON PROCESSES

Parameter		Czochralski	Float zone
Resistivity (Ω-cm)			
N-type	As:	0.001–25	0.001–25
	P:	0.001–100	0.001–500
	Sb:	0.005–20	1.0–50
P-type	B:	0.0003–100	0.001–10,000
	Al:	0.060–15	
	Ga:	0.10–20	
Radial resistivity			
Gradient (from center	As:	20	15
to half-radius) (%)	P:	15	10
	Sb:	10	15
	B:	10	10
	Al:	10	
	Ga:	10	
Lifetime (μs) Resistivity (Ω-cm)	τ (μs)		τ (μs)
1–5	10		10
5–10	25		25
10–50	50		50–100
Above 50	100		75–200
Dislocation density		0–1500 cm^{-2}	10,000–30,000 cm^{-2}
Oxygen content		5–10 ppm	0.5–1.0 ppm
		2.5×10^{17} to 5×10^{17} cm^{-3}	2.5×10^{16} to 5×10^{16} cm^{-3}

industrial diamonds embedded into the inner-diameter cutting edge. This produces circular slices or wafers that are about 0.5 to 1.0 mm (0.020 to 0.040 in.) thick, as shown in Figure 1.3. The orientation flat serves as a useful reference plane for various device processes to be described later. The flat is usually along the (110) plane to facilitate fracture along a natural cleavage plane during die separation.

1.2.3 Wafer Polishing and Cleaning

The raw cut silicon wafers have very heavily damaged surfaces as a result of the slicing operation. The wafers now undergo a number of polishing steps for the following purposes:

1. To remove the damaged silicon from the sawn surfaces
2. To produce a highly planar or flat surface that will be required for the photolithographic process, especially when very fine line geometries are involved
3. To improve the parallelism of the two major surfaces of the wafer, this also being needed for the photolithographic process

In Figure 1.4, some diagrams are shown of the wafer surface contour and thickness following the various polishing steps. Usually, only one side of the wafer is given the final mirror-smooth highly polished finish; the other side (the "back side") is given just a lapping operation to ensure an acceptable degree of flatness and

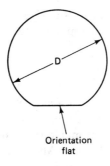

Orientation
flat

Figure 1.3 Silicon wafer. $d = 100$, 125, or 150 mm.

parallelism. The final polishing operation is often a very light chemical etching to remove the last vestiges of mechanical damage arising from the previous polishing operations, and sometimes a combined chemical–mechanical polishing operation is used.

After the wafer polishing operations are completed, the wafers are thoroughly cleaned, rinsed, and dried, and they are now ready to be used for the various device processing steps to be discussed next.

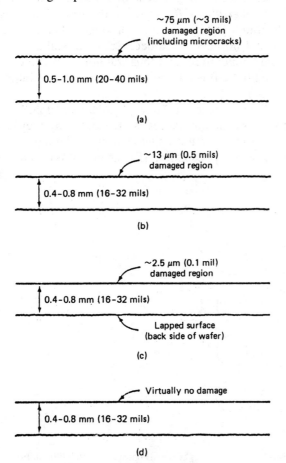

~75 μm (~3 mils)
damaged region
(including microcracks)

0.5–1.0 mm (20–40 mils)

(a)

~13 μm (0.5 mils)
damaged region

0.4–0.8 mm (16–32 mils)

(b)

~2.5 μm (0.1 mil)
damaged region

0.4–0.8 mm (16–32 mils)

Lapped surface
(back side of wafer)

(c)

Virtually no damage

0.4–0.8 mm (16–32 mils)

(d)

Figure 1.4 Wafer lapping and polishing: (a) raw cut slice; (b) lapped slice; (c) polished wafer; (d) etched slice.

We will now discuss the various device processes separately, and then we will see how they all fit together in the manufacturing of semiconductor devices.

1.3 DIFFUSION

The diffusion process under consideration here is *solid-state impurity diffusion* in which various dopant impurities can be introduced into the surface region of a silicon wafer. For this process to occur at a reasonable rate, the temperature of the wafer must be in the range of about 900° to 1200°C. Although these are rather high temperatures, they are still well below the melting point of silicon, which is at 1420°C. The rate at which the various impurities diffuse into silicon is on the order of 1 μm/h at the temperature range above, and the penetration depths that are involved in most diffusion processes are on the order of 0.3 to 30 μm. At room temperature, the diffusion process is so extremely slow that the impurities can be considered to be essentially "frozen" in place.

Dopants of very small atomic or ionic radii, such as lithium (Li$^+$), can fit in the gaps or voids (*interstices*) between silicon atoms and can therefore diffuse very rapidly. These small-size dopants will be *interstitial diffusants*; the diffusion process is called *interstitial diffusion* and is illustrated in Figure 1.5. Although lithium will act as a donor impurity in silicon, it is not normally used because it will still move around, even at temperatures near room temperature, and thus will not be "frozen" in place. This is true of most other interstitial diffusions, so long-term device stability cannot be assured with this type of impurity.

The more useful diffusants in silicon are the larger *substitutional* dopants. These dopants atoms are too big to fit into the interstices, so the only way they can enter the silicon crystal structure is to substitute for a silicon atom. The most commonly used substitutional donor dopants are phosphorus (P), arsenic (As), and antimony

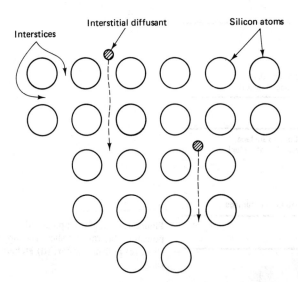

Figure 1.5 Interstitial diffusion.

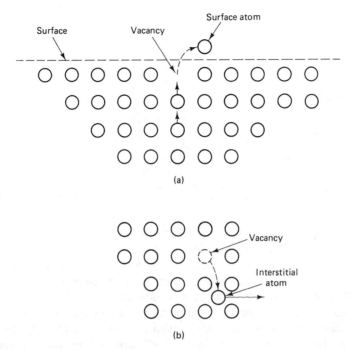

Figure 1.6 Vacancy generation processes: (a) Schottky defect (surface atom–vacancy pair); (b) Frenkel defect (interstitial–vacancy pair).

(Sb), with phosphrous being used most often. The only acceptor dopant that is used to any extent is boron (B).

For the substitutional diffusion process, the presence of vacancies in the crystal structure is of primary importance. A *vacancy* is the absence of an atom from a point in the crystal structure where an atom would ordinarily be present. Although the ideal crystal structure has no vacancies, some vacancies are naturally present in any real crystal. The vacancy density or population can be vastly increased by raising the temperature of the crystal, since the vacancy concentration increases exponentially with temperature. Vacancies can be generated at the surface of the crystal (*Schottky defects*), as shown in Figure 1.6a, or throughout the interior of the crystal (*Frenkel defects*), as shown in Figure 1.6b. The vacancies generated at the surface can diffuse into the interior of the crystal and become uniformly distributed throughout the volume of the crystal.

The presence of vacancies permits impurity atoms to enter the surface layer of the crystal and then to diffuse slowly into the interior, as shown in Figure 1.7. This impurity diffusion is a very slow, step-by-step process, since a vacancy must be present in a suitable position adjacent to the impurity atom for it to move down one atomic layer at a time.

1.3.1 Diffusion Equations

The diffusion process involves the flow of atoms or other particles under the influence of a concentration gradient, as shown in Figure 1.8. Since there are more particles

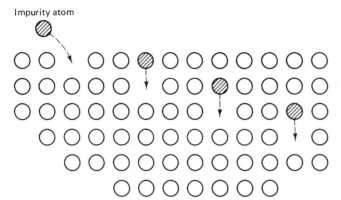

Impurity atom

Figure 1.7 Solid-state impurity diffusion.

to the left of the line at $x = x_1$ than to the right of this line, there is a net flow of particles from left to right. Note that in Figure 1.8 the concentration gradient dN/dx is negative and the particle flow is in the $+x$ direction. The particle flow or flux is proportional to the concentration gradient given by $F \propto -dN/dx$, where F is the diffusant flux, which is the rate of flow of the particles or atoms per unit area. To make this into an equation, we introduce a constant of proportionality to obtain an equation called Fick's first law, given by $F = -D(dN/dx)$. In this equation, D is the *diffusion constant* or *diffusion coefficient* and has units of cm²/s. Looking at the overall balance of units, we have

$$F(1/\text{cm}^2\text{-s}) = -D(\text{cm}^2/\text{s}) \times \frac{dN}{dx} (\text{cm}^{-3}/\text{cm}) \qquad (1.1)$$

The diffusion constant D is a function of the type of diffusant and the material into which the diffusion takes place and increases exponentially with temperature. An equation for D can be written as $D = D_0 \exp(-qE_A/kT)$, where q is the electronic charge (1.6×10^{-19} C), k is Boltzmann's constant (1.38×10^{-23} J/K), T is the absolute temperature (K), and D_0 is a preexponential constant with units of cm²/s. The quantity E_A is an activation energy with units of electron volts (eV) and is in the

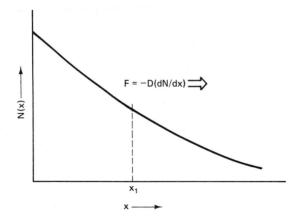

Figure 1.8 Diffusion: Fick's first law.

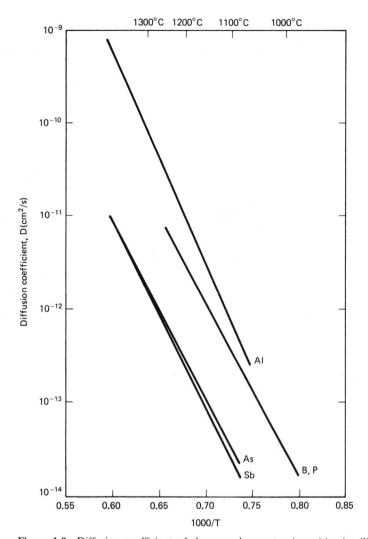

Figure 1.9 Diffusion coefficient of donor and acceptor impurities in silicon. (Adapted from C. S. Fuller and J. A. Ditzenbereger, "Diffusion of Donor and Acceptor Elements in Silicon." *Journal of Applied Physics*, Vol. 27, pp. 544–553, May 1956.)

range from 4 to 5 eV for most substitutional dopant diffusants of interest in silicon, such as P, B, As, and Sb. Figure 1.9 is a graph of the variation of the diffusion constant with temperature for these diffusants. Note the straight-line relationship that is obtained by having D on a logarithmic scale and the temperature on a reciprocal temperature, $1/T$, scale.

A second relationship involved in the diffusion process is the *continuity equation*, given by

$$\frac{\partial N}{\partial t} = \frac{F(x) - F(x + dx)}{dx} = \frac{-\partial F}{\partial x} \tag{1.2}$$

This equation is basically a statement of the conservation of matter and states that the rate of change of atoms or particles $\partial N/\partial t$ in a given volume is equal to the rate of flow of the particles into the volume minus the rate of flow of particles out of the volume.

If we now combine Fick's first law and the continuity equation, we obtain *Fick's second law*, given by

$$\frac{\partial N}{\partial t} = \frac{-\partial F}{\partial x} = \frac{-\partial}{\partial x}\left(-D\,\frac{\partial N}{\partial t}\right) = D\frac{\partial^2 N}{\partial x^2} \tag{1.3}$$

The solution of this partial differential equation, subject to the appropriate boundary conditions, will give us the distribution or profile of the diffusant atoms $N(x, t)$. The two sets of boundary conditions that are of principal interest for solid-state impurity diffusion are (1) constant surface concentration and (2) limited source conditions.

1.3.2 Constant Surface Concentration Diffusion

We will first consider diffusion under constant surface concentration or "infinite source" conditions. This diffusion process is commonly known as a *predeposition diffusion* or a *deposition diffusion*. For this case, the diffusant concentration at the surface ($x = 0$) will be assumed to be constant, so we will have that $N(0, t) = N_0 =$

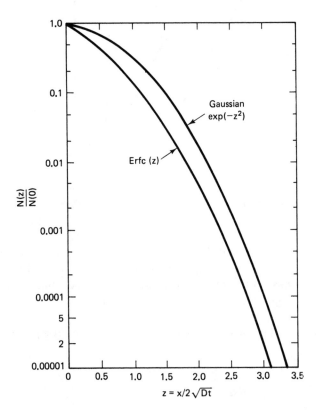

Figure 1.10 Complementary error function and Gaussian functions.

constant. The other boundary condition for this case is that as $x \to \infty$ the diffusant concentration decays to zero. Subject to these boundary conditions, the solution to Fick's second law is

$$N(x, t) = N_0 \, \text{erfc}\left(\frac{x}{2\sqrt{Dt}}\right) \qquad (1.4)$$

where N_0 is the constant surface concentration, x the distance into the material from the surface ($x = 0$), t the diffusion time, and erfc is a mathematical function called the *complementary error function*.

In Figure 1.10, a graph of the complementary error function erfc (z) is presented. At $z = 0$, we have that erfc (0) = 1, and as $z \to \infty$, erfc (z) $\to$ 0, and we note that erfc (z) decreases very rapidly with increasing z.

In Figure 1.11, a curve of an erfc type of impurity profile is shown for the example of a P-type boron diffusion into an N-type (phosphorus-doped) substrate. At the position $x = x_J$, the boron diffusant concentration is equal to the phosphorus substrate concentration. In the region of $x < x_J$, the boron (acceptor) concentration is greater than the phosphorus (donor) substrate concentration, so this region is therefore P type. In the region of $x > x_J$, the donor concentration is greater than the acceptor concentration, so this region is N type. At $x = x_J$, there is therefore

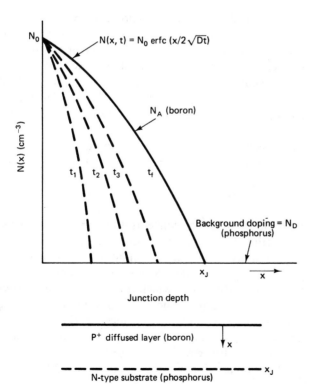

Figure 1.11 Diffusion profile: constant surface concentration diffusion (deposition diffusion).

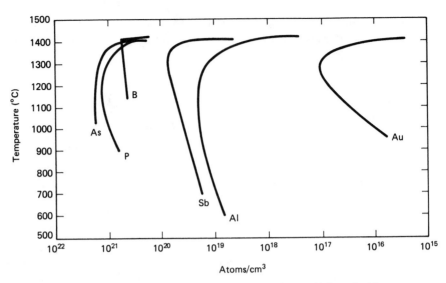

Figure 1.12 Solid-state solubility of impurities in silicon. (Adapted with permission from F. A. Trumbore, "Solid Solubilities of Impurity Elements in Germanium and Silicon," *Bell Systems Technical Journal*, Vol. 39, pp. 205–233. January 1960, AT&T.)

the transition from a net P-type doping to a net N-type doping, so x_J marks the position of the PN junction.

Therefore, a PN junction is produced by this diffusion process. Since the doping level in the P-type diffused layer is relatively heavy, especially near the surface where concentrations of about 10^{20} cm^{-3} can be expected, the diffused layer can be designated as a P$^+$ region. The surface concentration will usually be fixed by the solid-state solubility of the impurity in silicon. In Figure 1.12, a graph of the solid-state solubilities of various dopants in silicon versus temperature is given. For boron in silicon, the solubility over the range of usual diffusion temperatures (900° to 1200°C) will be in the range 1×10^{20} to 2×10^{20} cm^{-3}.

In Figure 1.11, a series of impurity profiles is shown for a number of different diffusion times. The PN junction occurs at $x = x_J$, at which point we have that

$$N_A(x_J,\ t) = N_0 \ \mathrm{erfc} \ \frac{x_J}{2\sqrt{Dt}} = N_B \tag{1.5}$$

where N_B represents the background or substrate concentration. From this we see that the junction depth x_J is proportional to the square root of the diffusion time.

Let us now consider as a representative example a constant surface concentration boron diffusion at 1150°C into a phosphorus-doped N-type silicon substrate. The substrate doping level is chosen to be 6×10^{15} cm^{-3}, which corresponds to a resistivity of 0.9 Ω-cm. At 1150°C, the solid-state solubility of boron in silicon is approximately 3×10^{20} cm$^{1-3}$, and the diffusion coefficients of boron in silicon is $D = 1 \times 10^{-12}$ cm^2/s. From the normalized erfc (z) graph of Figure 1.10, we see that at the junction we have that erfc $(z_J) = N_B/N_0 = (6 \times 10^{15}$ cm$^{-3})/(3 \times 10^{20}$ cm$^{-3}) = 2 \times 10^{-5}$, so that $z_J = x_J/2 \sqrt{Dt} = 3.0$. Solving for x_J, this gives $x_J = 3.0 \times 2 \times$

$\sqrt{1 \times 10^{-12} \text{ cm}^2/\text{s} \times t} = 6 \times 10^{-6} \text{ cm } \sqrt{t/s}$. Since 1 μm = 10^{-4} cm, this can be reexpressed as $x_J = 0.06 \text{ μm } \sqrt{t/s} = 3.6 \text{ μm } \sqrt{t/h}$. Therefore, a 1.0-h diffusion at 1150°C will result in a junction depth of 3.6 μm. For a 1.0-μm junction depth, the required diffusion time is 0.077 h = 4.63 min. At the other extreme, a 10-μm junction depth requires a total diffusion time of 7.72 h, so we see that deep junctions can require some very long diffusion times.

1.3.3 Constant Surface Concentration Diffusion Processes

For the various types of diffusion and oxidation processes, a resistance-heated tube furnace is usually used. A tube furnace has a long (about 2 to 3 m) hollow opening into which a quartz tube about 100 to 150 mm (4 to 6 in.) in diameter is placed, as shown in Figure 1.13. The temperature within the quartz furnace tube can be controlled very accurately such that a temperature within $\frac{1}{2}$°C of the set-point temperature can be maintained uniformly over a "hot zone" about 1 m in length. The silicon wafers to be processed are stacked up vertically into slots in a quartz carrier or "boat" and inserted into the furnace tube.

A number of gases are metered into the furnace. The principal gas flow in the furnace is nitrogen (N_2), which acts as a relatively inert gas and is used as a carrier gas to be a dilutent for the other more reactive gases. The N_2 carrier gas generally makes up some 90% to 99% of the total gas flow. A small amount of oxygen and a very small amount of a compound of the dopant make up the rest of the gas flow. For boron diffusions, such compounds as B_2H_6 (diborane) and BBr_3 (boron tribromide) are commonly used.

The following reactions will be occurring simultaneously at the surface of the silicon wafers:

$$\text{(1)} \quad Si + O_2 \longrightarrow SiO_2 \text{ (silica glass)} \tag{1.6}$$

$$\text{(2)} \quad 4BBr_3 + 3O_2 \longrightarrow 2B_2O_3 \text{ (boron glass)} + 6Br_2, \text{ or} \tag{1.7}$$

$$2B_2H_6 + 3O_2 \longrightarrow 2B_2O_3 \text{ (boron glass)} + 6H_2 \tag{1.8}$$

This process is the chemical-vapor deposition (CVD) of a glassy layer on the silicon

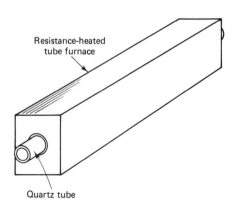

Resistance-heated
tube furnace

Quartz tube

Figure 1.13 Diffusion furnace.

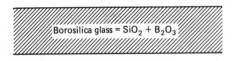

Figure 1.14 Boron diffusion glass layer on silicon wafer.

Silicon wafer

surface, which is a mixture of silica glass (SiO$_2$) and boron glass (B$_2$O$_3$) called *borosilica glass* (BSG). The BSG glassy layer, shown in Figure 1.14, is a viscous liquid at the diffusion temperatures and the boron atoms can move around relatively easily. Furthermore, the boron concentration in the BSG is such that the silicon surface will be saturated with boron at the solid-state solubility limit throughout the time of the diffusion process as long as the BSG remains present.

For phosphorus diffusions, such compounds as PH$_3$ (phosphine) and POCl$_3$ (phosphorus oxychloride) can be used. In the case of a diffusion using POCl$_3$, the reactions occurring at the silicon wafer surfaces will be

$$(1) \ Si + O_2 \ \longrightarrow \ SiO_2 \text{ (silica glass)} \tag{1.9}$$

$$(2) \ 4POCl_3 + 3O_2 \ \longrightarrow \ 2P_2O_5 \text{ (phosphorus glass)} + 6Cl_2 \tag{1.10}$$

This results in the production of a glassy layer on the silicon wafers that is a mixture of phosphorus glass and silica glass called *phosphorosilica glass* (PSG), which is a viscous liquid at the diffusion temperatures. The mobility of the phosphorus atoms in this glassy layer and the phosphorus concentration are such that the phosphorus concentration at the silicon surface will be maintained at the solid-state solubility limit throughout the time of the diffusion process. Similar processes occur with other dopants, such as arsenic, in which arsenosilica glass is formed on the silicon surface.

The constant surface concentration diffusion results in the deposition of dopant atoms into the surface regions of the silicon wafers, in most cases to depths in the range from 0.3 to 3 μm. This type of diffusion is often called a *deposition diffusion*.

1.3.4 Drive-in Diffusion

The deposition diffusion is often followed by a second diffusion process in which the external dopant source (the diffusion glass) is removed such that *no* additional dopants enter the silicon. During this diffusion process, the dopants that are already in the silicon move farther in and are thus redistributed. The junction depth increases, and at the same time the surface concentration decreases. This type of diffusion is called *drive-in*, or *redistribution*, or *limited-source* diffusion.

The impurity profile for the drive-in diffusion can be obtained by solving Fick's second law subject to the boundary condition that the total diffusant density (diffusant atoms per unit area) remains constant. This condition can be stated mathematically as $Q = \int_0^\infty N(x, t) \, dx$ = constant, where Q is the diffusant density. An alternative way of expressing this boundary condition is to note that under the stated conditions there is no net flow of diffusant atoms into or out of the silicon wafers at the surface $(x = 0)$, so in terms of Fick's first law we have $D(dN/dx)|_{x=0} = 0$. If Fick's second law is solved subject to this boundary condition, the resulting impurity distribution is given by

$$N(x, t) = \frac{Q}{\sqrt{\pi Dt}} \exp\left(-\frac{x^2}{4Dt}\right) \tag{1.11}$$

This is a Gaussian distribution, and a normalized curve of $\exp(-z^2)$ is shown in Figure 1.10. Note that the surface diffusant concentration will be $N(0, t) = Q/\sqrt{\pi Dt}$ and thus will not be constant, but rather will decrease with increasing diffusion time. Note also that the slope of the impurity profile, dN/dx, at the surface ($x = 0$) will be zero, so the Gaussian profile will indeed satisfy the boundary conditions for the drive-in diffusion.

The foregoing equation for the Gaussian impurity profile is based on the assumption that at $t = 0$ there is a diffusant density equal to Q right at the silicon surface ($x = 0$). However, if the drive-in diffusion is preceded by a deposition diffusion such that the junction depth resulting from this first diffusion x_{J_1} is considerably smaller than the junction depth x_{J_2} after the completion of the drive-in diffusion, such that $x_{J_2}^2 \gg x_{J_1}^2$, the final impurity profile will, to a close approximation, be given by the Gaussian equation above. This two-step diffusion process is illustrated in Figure 1.15.

The diffusant density produced as a result of a deposition diffusion is given by $Q = 2N_0\sqrt{D_1 t_1/\pi}$, where D_1 and t_1 are the diffusion constant and diffusion time, respectively, for the deposition diffusion. If a short deposition diffusion is followed by a drive-in diffusion, the resulting impurity profile is approximately a Gaussian profile as given by $N(x, t_2) = Q/\sqrt{\pi D_2 t_2} \exp(-x^2/4D_2 t_2)$, where D_2 and t_2 refer to the diffusion constant and time of the drive-in diffusion, respectively. The surface concentration at the end of the drive-in diffusion is

$$N(0, t_2) = \frac{Q}{\sqrt{\pi D_2 t_2}} = \frac{2}{\pi}\sqrt{\frac{D_1 t_1}{D_2 t_2}}\, N_0 \tag{1.12}$$

Thus, if the drive-in diffusion is a relatively long time, high temperature diffusion compared to the deposition diffusion such that $D_2 t_2 \gg D_1 t_1$, the surface concentration can be reduced substantially below the value of N_0 as imposed by the solid-state solubility limit. In Figure 1.15, some impurity profiles for a number of drive-in times are presented. Note that the total diffusant density Q does not change but is only redistributed or "driven in" such that the junction depth increases and at the same time the surface concentration decreases. This two-step combination of a deposition diffusion followed by a drive-in diffusion is often used to produce the base region of transistors.

In Figure 1.16, a diagram is presented of a typical impurity profile of a double-diffused NPN transistor. Starting with the N-type silicon substrate, a boron deposition diffusion is performed. The boron diffusion glass is then removed, and a drive-in diffusion is performed to reduce the boron surface concentration and to increase the junction depth. This produces the P-type base region of the transistor. The silicon wafer is then subjected to a high-concentration phosphorus deposition diffusion that

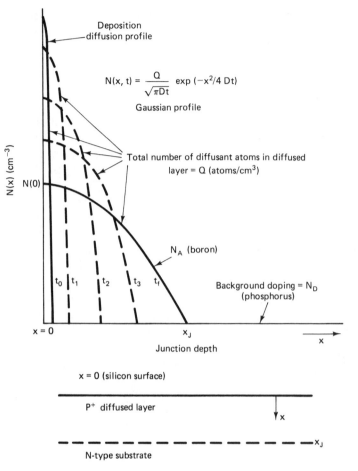

Figure 1.15 Diffusion profile: drive-in diffusion.

converts part of the P-type diffused layer back to N type. This diffusion results in a heavily doped N^+ region that will be the emitter of the transistor.

After the diffusions have been completed, the collector–base junction depth will be $x_{J(CB)}$ and will generally be about 3 μm. The emitter–base junction depth $x_{J(EB)}$ will be around 2 to 2.5 μm. The base width of the transistor will be $W_B = x_{J(CB)} - x_{J(EB)}$ and will generally be in the range from 0.3 to 1.0 μm, although some "supergain" or "super-beta" transistors have base widths as small as in the range from 0.1 to 0.2 μm.

The emitter surface concentration is usually up in the range of 10^{21} cm^{-3}, and the base surface concentration is reduced by the drive-in diffusion down to the range of around 3×10^{18} cm^{-3}. The doping density at the emitter–base junction generally falls in the range of 3×10^{17} cm^{-3}, and the collector doping typically falls in the range 10^{15} to 10^{16} cm^{-3}, corresponding to resistivities of from 5 Ω-cm down to about 0.5 Ω-cm. The double-diffused transistor structure with a diffused base region fol-

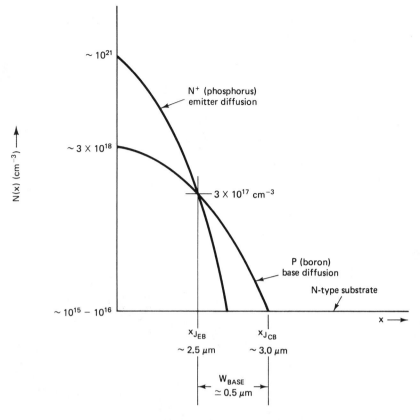

Figure 1.16 Impurity profile of a double-diffused NPN transistor.

lowed by a diffused emitter region is the technique used for the manufacture of almost all silicon transistors, including virtually all bipolar IC transistors.

1.3.5 Sheet Resistance

The two principal parameters that are used for the evaluation and characterization of diffused layers are the junction depth x_J and the sheet resistance R_S. The resistance of a layer of material, as shown in Figure 1.17a, is given by the familiar equation $R = \rho L/A = \rho L/tW$, where ρ is the average resistivity of the layer, L the length, t the layer thickness, and W the width of the layer.

If the shape is such that $L = W$, such that the material is a square, the resistance equation will become

$$R = \frac{\rho}{t} \tag{1.13}$$

Notice that the resistance of a "square" is independent of the lateral dimensions, L

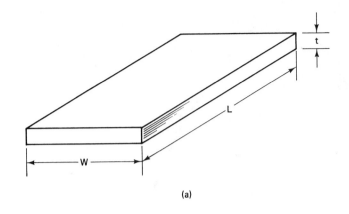

(a)

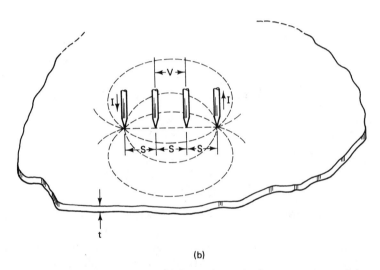

(b)

Figure 1.17 (a) Sheet resistance; (b) four-point probe for measurement of sheet resistance.

and W. This is called the *sheet resistance*, R_S, and is often expressed in such units as Ω/square (or $\Omega/\square$). The resistance of a layer of material can thus be expressed in terms of the sheet resistance as

$$R = R_S \frac{L}{W} \tag{1.14}$$

The sheet resistance of thin layers, including diffused layers, can be conveniently measured using a four-point probe apparatus, as shown in Figure 1.17b. If the conditions are satisfied that the layer thickness t is small compared to the probe spacing, such that $t \ll s$, and that the edge of the layer is relatively remote from the probe array, the sheet resistance is given approximately by $R_S = 4.5324(V/I)$. The current I is supplied by the two outer probes, and the voltage V is measured by a high-impedance voltmeter across the two inner probes. As a result of having separate pairs of probes for supplying the current and measuring the voltage drop, the probe contact resistance will not influence the measurement of the sheet resistance. The four-point probe can be used to measure the sheet resistance of various types of diffused layers, as well as that of silicon wafers for the measurement of the resistivity.

Sheet resistance values of diffused layers generally fall in the range from 1 Ω/square up to about 1000 Ω/square. The transistor base diffused layer will have a sheet resistance of typically about 200 Ω/square, and the N^+ emitter diffused layer will be down in the range of around 2 Ω/square. In Figure 1.18a and b, graphs are

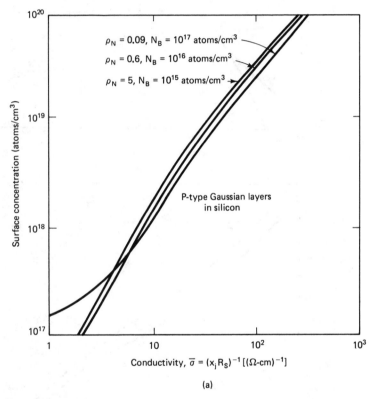

$\rho_N = 0.09$, $N_B = 10^{17}$ atoms/cm^3
$\rho_N = 0.6$, $N_B = 10^{16}$ atoms/cm^3
$\rho_N = 5$, $N_B = 10^{15}$ atoms/cm^3

P-type Gaussian layers in silicon

Surface concentration (atoms/cm^3)

Conductivity, $\bar{\sigma} = (x_j R_s)^{-1} [(\Omega\text{-cm})^{-1}]$

(a)

Figure 1.18 Average conductivity of diffused layer in silicon: (a) average conductivity versus surface concentration for P-type Gaussian layers; (b) average conductivity versus surface concentration for N-type erfc layers; (c) resistivity versus impurity concentration for silicon. [(a) and (b) adapted with permission from J. C. Irvin, "Resistivity of Bulk Silicon and of Diffused Layers," *Bell System Technical Journal*, Vol. 41, pp. 387–410, 1962, copyright 1962, AT&T; and S. K. Ghandi, *Theory and Practice of Microelectronics*, Wiley, New York, 1968; (c) adapted with permission from J. C. Irvin, "Resistivity of Bulk Silicon and of Diffused Layers," *Bell System Technical Journal*, Vol. 41, pp. 387–410, 1962, copyright 1962, AT&T.]

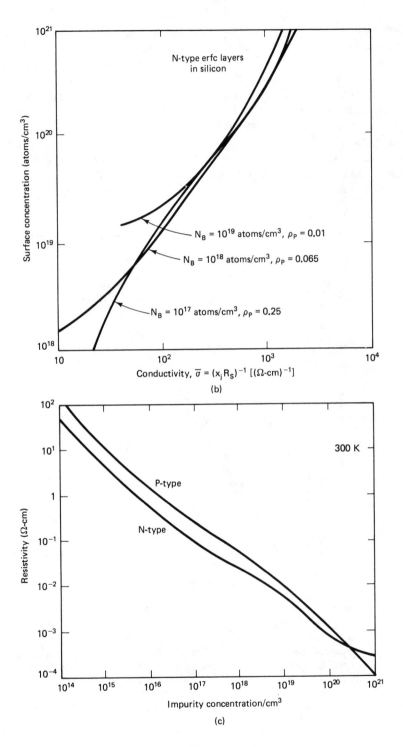

Figure 1.18 (continued)

Integrated-circuit Fabrication   Chap. 1

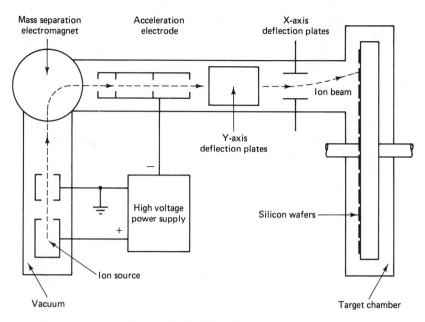

Figure 1.19 Ion-implantation system.

given that can be used to determine the sheet resistance of P-type Gaussian and N-type erfc diffused layers in silicon. In Figure 1.18c, a graph of resistivity versus impurity concentration for bulk silicon is presented.

1.3.6 Ion Implantation

Ion implantation can be used to produce a shallow surface region of dopant atoms deposited into a silicon wafer, and thus can be used as an alternative to a deposition diffusion. Ion implantation allows for much more precise control over the sheet resistance than does deposition diffusion and as a result is now used more often than deposition diffusion. In the ion implantation process, silicon wafers are placed in a vacuum chamber and are scanned by a beam of high-energy dopant ions, as shown in Figure 1.19. These ions have been accelerated through potential differences of from 20 kV to as much as 250 kV, and as the ions strike the silicon wafers they will penetrate some small distance into the wafer. The depth of penetration of any particular type of ion will increase with increasing accelerating voltage, as shown by Figure 1.20.

The penetration depth will generally be in the range from 0.1 to 1.0 μm. For boron ions (B^+) in silicon, for example, the average penetration depth, called the *projected range*, will be only 0.067 μm for an accelerating voltage of 20 kV. At 100 kV, the projected range increases to 0.30 μm, and at 200 kV it will be 0.52 μm. Even at a large accelerating potential of 300 kV, the projected range is still only 0.70 μm. For phosphorus ions in silicon, the projected ranges are even smaller: 0.026 μm at 20 kV, 0.123 μm at 100 kV, 0.254 μm at 200 kV, and 0.386 μm at 300 kV, as given in Table 1.2.

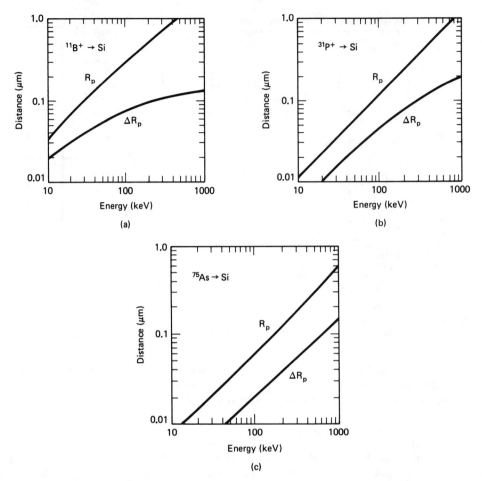

Figure 1.20 Ion-implantation range and straggle of boron, phosphorus, and arsenic in silicon. (Adapted from D. H. Lee and J. W. Mayer, "Ion Implanted Semiconductor Devices," *Proceedings of the IEEE*, pp. 1241–1255, September 1974, © 1974 IEEE.)

The distribution of the implanted ions as a function of distance x from the surface will be a Gaussian distribution, given by

$$N(x) = N_p \exp\left[-\frac{(x - R_p)^2}{2\Delta R_p^2} \right] \tag{1.15}$$

where x = distance into substrate from surface
R_p = projected range
ΔR_p = straggle (standard deviation) of the projected range
N_p = peak concentration of implanted ions

An ion implantation impurity profile is shown in Figure 1.21. The peak implanted ion concentration is related to the *implantation dosage* Q by $N_p = Q/(\sqrt{2\pi}\ \Delta R_p)$

TABLE 1.2 PROJECTED RANGES AND STRAGGLE FOR BORON, PHOSPHORUS, AND ARSENIC IN SILICON

Energy (kV)	Boron R_p (μm)	ΔR_p (μm)	Phosphorus R_p (μm)	ΔR_p (μm)	Arsenic R_p (μm)	ΔR_p (μm)
10	0.0344	0.0156	0.0144	0.0070	0.0096	0.0037
20	0.0674	0.0253	0.0260	0.0122	0.0159	0.0060
30	0.0999	0.0331	0.0375	0.0169	0.0216	0.0081
40	0.1311	0.0392	0.0490	0.0214	0.0271	0.0101
50	0.1616	0.0445	0.0610	0.0259	0.0324	0.0120
60	0.1914	0.0491	0.0732	0.0303	0.0377	0.0139
70	0.2202	0.0530	0.0855	0.0345	0.0429	0.0158
80	0.2478	0.0564	0.0980	0.0386	0.0481	0.0176
90	0.2739	0.0593	0.1106	0.0425	0.0533	0.0194
100	0.2992	0.0618	0.1233	0.0463	0.0584	0.0211
110	0.3238	0.0641	0.1362	0.0500	0.0635	0.0229
120	0.3479	0.0663	0.1491	0.0535	0.0686	0.0246
130	0.3714	0.0682	0.1621	0.0570	0.0738	0.0263
140	0.3942	0.0700	0.1752	0.0604	0.0789	0.0279
150	0.4165	0.0716	0.1883	0.0636	0.0840	0.0296
160	0.4383	0.0731	0.2014	0.0668	0.0891	0.0312
170	0.4594	0.0744	0.2146	0.0698	0.0943	0.0329
180	0.4799	0.0757	0.2277	0.0726	0.0995	0.0345
190	0.500	0.0769	0.2407	0.0753	0.1048	0.0362
200	0.5198	0.0780	0.2538	0.0780	0.1101	0.0378
220	0.5582	0.0800	0.2801	0.0831	0.1208	0.0412
240	0.5954	0.0818	0.3065	0.0880	0.1316	0.0445
260	0.6314	0.0834	0.3329	0.0927	0.1425	0.0478
280	0.6663	0.0849	0.3594	0.0972	0.1535	0.0510
300	0.7001	0.0862	0.3858	0.1015	0.1645	0.0543

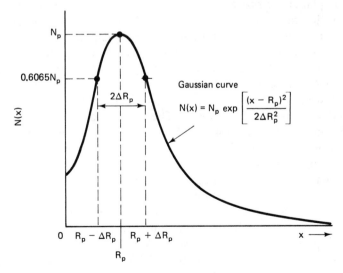

Figure 1.21 Ion-implantation impurity profile.

$$N(x) = N_p \exp\left[\frac{(x - R_p)^2}{2\Delta R_p^2}\right]$$

$= 0.4(Q/\Delta R_p)$. The implantation dosage Q is the number of implanted ions per unit of surface area as given by such units as ions/cm². The ion density drops off rapidly from the peak value with distance as measured from R_p in either direction, as indicated by the following values of the Gaussian function:

x	$N(x)/N_p$
R_p	1.0
$R_p \pm 1.2\Delta R_p$	0.5
$R_p \pm 2\Delta R_p$	0.1
$R_p \pm 3\Delta R_p$	0.01
$R_p \pm 3.7\Delta R_p$	0.001
$R_p \pm 4.3\Delta R_p$	0.0001

Note that the Gaussian implanted ion profile will be truncated at $x = 0$.

After the ions have been implanted, they are lodged principally in interstitial positions in the silicon crystal structure, and the surface region into which the implantation has taken place will be heavily damaged by the impact of the high-energy ions. The disarray of silicon atoms in the surface region is often to the extent that this region is no longer crystalline in structure, but rather amorphous. To restore this surface region back to a well-ordered crystalline state and to allow the implanted ions to go into substitutional sites in the crystal structure, the wafer must be subjected to an *annealing* process. The annealing process usually involves the heating of the wafers to some elevated temperature, often in the range from 800° to 1000°C for a suitable length of time, such as 30 min. Since the damage caused by the ion implantation to the surface layers of silicon atoms is so severe that it cannot be annealed, implantations are usually done through a thin (~300 Å) sacrificial oxide layer.

There are other annealing techniques, such as laser beam and electron-beam annealing, in which only the surface region of the wafer is heated and recrystallized. An ion implantation process is often followed by a conventional-type drive-in diffusion, in which case the annealing process will occur as part of the drive-in diffusion.

Ion implantation is a substantially more expensive process than an ordinary deposition diffusion, both in terms of the cost of the equipment and the throughput. It does, however, offer the advantage of much more precise control over the density of dopants Q deposited into the wafer, and hence the sheet resistance. It also allows for very low values of dosage Q ($<10^{14}$ cm⁻³), so very large values of sheet resistance (>1000 Ω/square) can be obtained. These high sheet resistance values are useful for obtaining large-value resistors ($\gtrsim 50$ kΩ) for ICs. Very low dosage, low energy implantations are also used for the adjustment of the threshold voltage of MOSFETs and other applications.

To define the ion-implanted pattern on a silicon wafer, a masking layer is needed on the surface of the wafer to stop the ions from penetrating the silicon in the various regions where this is needed. Commonly used materials for the masking of lower-energy (less than 100 kV) ion implantations are SiO_2, Si_3N_4, and photoresist. The relative stopping power of these materials, with silicon as a reference for both boron and phosphorus implantations, are SiO_2: 1.25, Si_3N_4: 1.62, and photoresist: 0.75. For higher-energy implantations, heavy metals such as thin films of gold, platinum, tungsten, or tantalum deposited on top of the SiO_2 layer can be used. The relative stopping power for these materials for boron or phosphorus implantations at 1000 kV is in the range from 2 to 4.

To consider an example of ion implantation masking, take the case of a boron implantation into silicon at an energy of 100 kV. For this implantation the projected range R_p is 0.30 μm, and the straggle ΔR_p is 0.062 μm. The distance into the silicon to the point where the implanted ion concentration is down to less than 0.1% of the peak concentration is $x = R_p + 3.7\Delta R_p = 0.30\,\mu m + (3.7 \times 0.062\,\mu m) = 0.53\,\mu m$. To mask against this ion implantation to the extent that the implanted ion concentration in the silicon will not exceed 0.1% of the peak concentration, the thickness of the masking layer must be 0.53 μm/stopping power. If an SiO_2 layer is to be used for the patterning of the ion implantation, as is very often the case, the required thickness of the oxide layer will be 0.53 μm/1.25 = 0.424 μm = 4240 Å. This thickness of oxide can easily be obtained by the thermal oxidation process.

1.4 OXIDATION

If a bare silicon surface is first exposed to air, a very thin oxide layer will almost immediately be formed by the reaction $Si + O_2 \rightarrow SiO_2$. This thin oxide layer is called the *native oxide* and it will be only some 20 to 30 Å thick. For the oxidation process to take place, silicon and oxygen atoms must react to form SiO_2. The resultant SiO_2 layer that is produced will be a dense, continuous protective layer on the silicon surface and will thereby act to inhibit the further growth of the oxide layer, so the oxidation will be a self-limiting process.

Silicon dioxide (SiO_2) layers are used for two principal purposes:

1. To act as a diffusion or implantation mask to produce a desired pattern of the diffused or implanted layers
2. To serve as a protective or *passivating* layer on the silicon surface for the protection of the semiconductor devices

For these two purposes, the thickness of the oxide layer that is required will generally be in the range from about 5000 to 10,000 Å. To produce these thick layers of oxide, a process called *thermal oxidation* can be used.

The silicon wafers are stacked up in a quartz boat, which is then inserted into a quartz furnace tube inside a tube furnace, very similar to a diffusion furnace. The silicon wafers are raised to a high temperature, generally in the range from 950° to 1150°C, and at the same time the wafers are exposed to a gas containing O_2 or H_2O, or both.

The diffusion rate of O_2 and H_2O in SiO_2 increases exponentially with temperature, so at these elevated temperatures the diffusion of these oxidants through the oxide layer will continue such that the required thickness of oxide can be produced within an acceptable period of time, as shown by the graphs of Figure 1.22. The chemical reactions taking place at the silicon surface are

$$(1) \quad \text{For } O_2: Si + O_2 \rightarrow SiO_2 \tag{1.16}$$

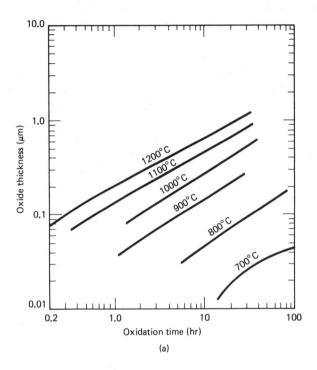

(a)

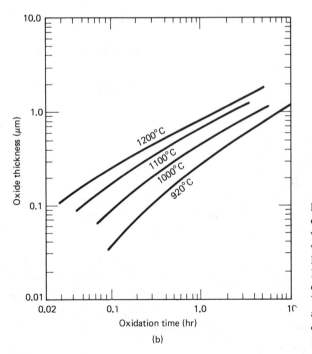

Figure 1.22 Oxide thickness versus oxidation time: (a) thermal oxidation with dry oxygen; (b) thermal oxidation with wet oxygen (95°C H$_2$O). (From B. E. Deal and A. S. Grove, "General Relationship for the Thermal Oxidation of Silicon," *Journal of Applied Physics*, Vol. 36, No. 12, pp. 3370–78, 1965, and S. K. Ghandi, *Theory and Practice of Microelectronics*, Wiley, New York, 1968.)

(b)

$$(2) \quad \text{For } H_2O: \quad Si + 2H_2O \rightarrow SiO_2 + 2H_2 \qquad (1.17)$$

This oxidation process is called thermal oxidation because of the use of high temperatures to promote the growth of the oxide layer.

The initial growth of the oxide is limited by the rate at which the chemical reaction takes place. After the first 100 to 300 Å of oxide has been produced, the growth rate of the oxide layer will be limited principally by the rate of diffusion of the oxidant (O_2 or H_2O) through the oxide layer, as illustrated in Figure 1.23. The rate of diffusion of O_2 or H_2O through the oxide layer will be inversely proportional to the thickness of the layer, so we will have that $dx/dt = C/x$, where x is the oxide thickness and C is a constant of proportionality. Rearranging this equation gives x $dx = C\,dt$, and then integrating both sides gives $x^2/2 = Ct$. Solving for the final oxide thickness x_f gives $x_f^2 = 2Ct + x_i^2$, where x_i is the initial thickness of the oxide (at $t = 0$). We see that after an initial reaction rate-limited linear growth phase, the oxide growth becomes diffusion rate limited, with the oxide thickness increasing as the square root of the growth time.

The oxide growth rate that is obtained when using H_2O as the oxidant is about four times faster than the rate obtained with O_2. This is primarily the result of the

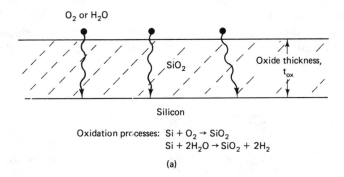

(a)

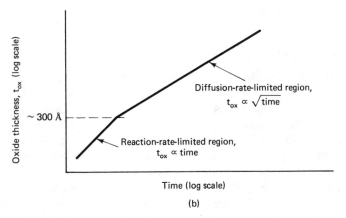

(b)

Figure 1.23 Oxidation of silicon: (b) oxide thickness versus time for thermal oxidation of silicon.

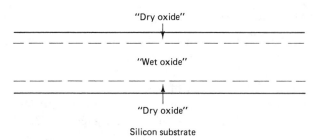

"Dry oxide"

"Wet oxide"

"Dry oxide"

Silicon substrate

Figure 1.24 Composite "dry–wet–dry" oxide.

H_2O molecule's being about one-half the size of the O_2 molecule, so the rate of diffusion of H_2O through the SiO_2 layer is much greater than the O_2 diffusion rate. Although the oxide growth rate with H_2O is much faster than with O_2, the "dry" (O_2) oxide is a slightly denser oxide with a higher dielectric strength than the "wet" (H_2O) oxide. In many cases, a "dry–wet–dry" oxidation process is used, starting off the initial oxide growth using O_2. This is followed by a H_2O oxide growth phase to produce the bulk of the oxide thickness and then completed by a final "dry" oxidation. This will produce a composite oxide layer as shown in Figure 1.24, with the denser "dry" oxide regions being adjacent to the silicon surface and serving as a protective cap on top and the less dense "wet" oxide being sandwiched in between.

In the thermal growth process of an oxide layer, some silicon from the substrate is consumed. If the resulting thickness of the SiO_2 layer is designated as t_{ox}, the thickness of the silicon consumed will be $0.44t_{ox}$, as shown in Figure 1.25.

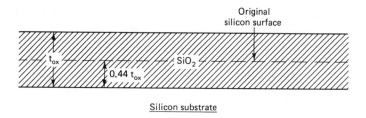

Original silicon surface

t_{ox}

$0.44\, t_{ox}$

SiO_2

Silicon substrate

Figure 1.25 Consumption of silicon during oxide growth.

1.4.1 Oxide Masking

An oxide layer can be used to mask an underlying silicon surface against a diffusion or ion implantation process. If an oxide layer is patterned, as by the photolithographic process, to produce regions where there are openings or "windows" where the oxide has been removed to expose the underlying silicon, these exposed silicon regions will be subjected to the diffusion or implantation of dopants, whereas the unexposed silicon regions will be protected. The pattern of dopant deposited into the silicon is thus a replication of the pattern of openings in the oxide layer. The replication is a key factor in the production of microscopic-size solid-state electronic devices.

The thickness of oxide needed for diffusion masking is a function of the type of diffusant and the diffusion time and temperature conditions, as shown by Figure 1.26. In particular, we see that an oxide thickness of some 5000 Å is sufficient to mask

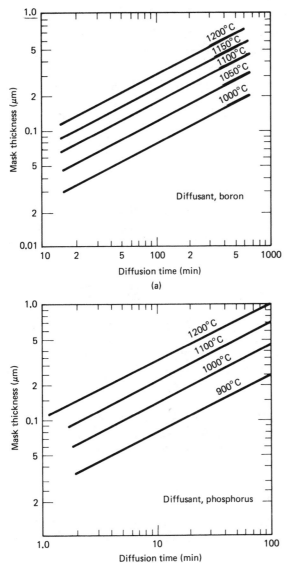

Figure 1.26 Oxide thickness needed for diffusion masking: (a) mask thickness for boron diffusions; (b) mask thickness for phosphorus diffusions. (From S. K. Ghandi, *Theory and Practice of Microelectronics*, Wiley, New York, 1968.)

against almost all diffusions. This oxide thickness is also sufficient to block almost all but the highest-energy ion implantations.

1.4.2 Planar Devices: Oxide Passivation

In Figure 1.27 a cross-sectional view is presented of a PN junction produced by diffusion through an oxide window. Since diffusion is an isotropic process, the diffusion will not only proceed in the downward direction, but also sideways as well. The junction depth in the vertical direction is indicated as x_J. The distance from the

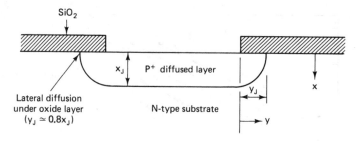

Figure 1.27 Diffusion masking using an oxide layer.

edge of the oxide window to the junction in the lateral direction underneath the oxide is indicated as y_J. The relationship between y_J and x_J is that $y_J \simeq 0.8x_J$, so the curvature of the junction in the regions underneath the edge of the oxide window is approximately that of a quarter-circle with a radius of curvature that is approximately equal to the junction depth.

We note that the junction intersects the silicon surface well underneath the protective thermally grown oxide layer. This oxide layer will protect the junction against various environmental effects and is therefore called a *passivated junction*. We also note that the locus of intersection of the junction with the silicon surface is entirely within a single geometric plane, and for that reason this type of junction is called a *planar junction*. Note that the junction itself is not flat or plane, but rather there is a curvature of the junction in the regions underneath the edges of the oxide window. This junction curvature will result in an increase in the electric field intensity in these regions, which will cause the breakdown voltage of the junction to be lower than that of a corresponding plane junction with the same doping levels.

The production of the pattern of oxide openings or windows involves a number of processing steps that are collectively called the *photolithographic process*. This process will be discussed in a following section.

1.4.3 High-pressure Oxidation

The oxidation rate can be greatly accelerated by carrying out the thermal oxidation process at pressures that are much above atmospheric pressure. The rate of diffusion of the oxidant molecules through an oxide layer is proportional to the ambient pressure. For example, at a pressure of 10 atm the diffusion rate will be increased by a factor of 10, and the corresponding oxidation time can be reduced by approximately the same factor. Alternatively, the oxidation can be done for the same length of time, but the temperature required will be substantially lower. For example, a steam (H_2O) oxidation at 1200°C and a 1.0-atm ambient pressure will produce a oxide layer that is 6000 Å (0.6 μm) thick, for an oxidation time of 36 min. This same thickness of oxide can be produced in the same period of time at a temperature of only 920°C if the ambient pressure is increased to 10 atm. As a second example, a 1.0-h steam oxidation at 920°C and at 1.0 atm ambient pressure will result in a 2000-Å-thick oxide. If the ambient pressure is increased to 10 atm, the temperature can be reduced to only 795°C for the same oxide thickness produced in the same period of time.

The possibility of lower-temperature processing is one of the principal advantages of the high-pressure oxidation process. The lower processing temperature reduces the formation of crystalline defects and produces less effect on previous diffusions and other processes. The shorter oxidation time is also advantageous in increasing the system throughput. The major disadvantage of this process is the high initial cost of the system.

1.5 PHOTOLITHOGRAPHY

The photolithographic process is the means by which microscopically small circuit and device patterns can be produced on silicon wafers, resulting in as many as 10,000 transistors on a 1 cm × 1 cm chip. With the conventional photolithographic process using ultraviolet light exposure, device dimensions or line widths as small as 1.5 μm can be obtained. By the use of *electron-beam* or *x-ray* exposure, device details down into the submicron (<1 μm) range can be obtained, with device dimensions as small as 0.2 μm having been achieved.

The photolithographic process encompasses a number of processing steps, which will now be described.

1. Photoresist application (spinning). A drop of light-sensitive liquid called *photoresist* is applied to the center of a silicon wafer that is held down by a vacuum chuck. The wafer is then accelerated rapidly to a rotational velocity in the range from 3000 to 7000 r/min for some 30 to 60 s. This flings off most of the photoresist, leaving a thin photoresist film on the silicon wafer that is in the range from 5000 to 10,000 Å thick, as illustrated in Figure 1.28a. The thickness of the photoresist layer will be approximately inversely proportional to the square root of the rotational velocity.

Sometimes, prior to the application of the photoresist, the silicon wafers are given a "bake-out" at a temperature of at least 100°C to drive off moisture from the wafer surfaces so as to obtain better adhesion of the photoresist.

2. Prebake. The silicon wafers are now put into an oven at about 80°C for about 30 to 60 min to drive off solvents in the photoresist and to harden it into a semisolid film.

3. Alignment and exposure. The photoresist-coated wafer is now placed in an apparatus called a *mask aligner* in very close proximity (about 25 to 125 μm) to a *photomask*. The relative positions of the wafer and the photomask are adjusted such that the photomask is correctly lined up with reference marks or a preexisting pattern on the wafer.

The photomask is a glass plate, typically about 125 mm (5 in.) square and about 2 mm thick. The photomask has a photographic emulsion or thin film metal (generally chromium) pattern on one side. The pattern has just clear or opaque areas, with no shades of gray.

The alignment of the photomask to the wafer is often required to be accurate

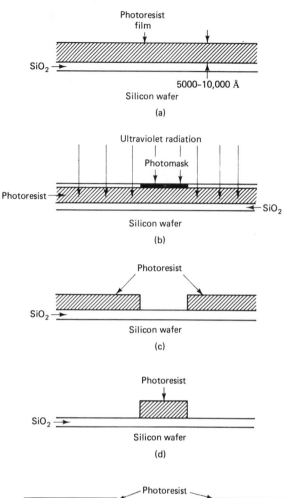

Figure 1.28 Photolithographic process:
(a) photoresis spinning; (b) exposure;
(c) development: negative photoresist;
(d) development: positive photoresist;
(e) oxide etching.

to within less than 1 μm, and in some cases to within 0.5 μm. After proper alignment has been achieved, the wafer is brought into direct contact with the photomask. The areas of the silicon wafer that are not covered by the opaque areas of the photomask are exposed to a highly collimated ultraviolet light, as shown in Figure 1.28b. The exposure time will generally be in the range from 3 to 30 s and is carefully controlled such that the total ultraviolet radiation dosage in watt-seconds or joules is the required amount.

The photomask can be obtained from an original large-size drawing of the desired mask pattern that is then photographically reduced. Generally, two consecutive reductions are used, each being on the order of 10 to 30 times for a total reduction of about 100 to 1000. The second reduction may also include a step-and-repeat exposure process for making a master photomask. This process reproduces the original mask drawing in reduced form on the photomask in an array containing tens, hundreds, or even thousands of replications of the original drawing. More commonly, however, CAD tools are used to "draw" the mask with a computer interface to the mask-making machine. The resulting circuit or device pattern on the wafer will be such that the wafer can be correspondingly divided up into many chips, each with a complete device or with a complete IC. The master photomask is then used to make many copies that are actually used in the alignment and exposure process.

4. Development. Two basic types of photoresist are used, negative photoresist and positive photoresist. If a negative photoresist is used, the areas of the photoresist that are exposed to the ultraviolet radiation become polymerized. This polymerization process increases the length of the organic chain molecules that make up the photoresist (that is, cross-linked polymers). This makes the resist tougher and makes it essentially insoluble in the developer solution. The areas of the resist that are unexposed to the ultraviolet radiation, however, remain unpolymerized and are readily dissolved by the developer solution. The resisting photoresist pattern after the development process will therefore be a replication of the photomask pattern, with the clear areas on the photomask corresponding to the areas where the photoresist remains on the wafers, as shown in Figure 1.28c.

With a *positive* photoresist, the opposite type of process occurs. Exposure to ıltraviolet radiation results in *depolymerization* of the photoresist (that is, breaks cross-links between polymers). This makes these exposed areas of the photoresist readily soluble in the developer solution, whereas the unexposed areas are essentially insoluble. The developer solution will thus remove the exposed or depolymerized regions of the photoresist, whereas the unexposed areas will remain on the wafer, as shown in Figure 1.28d. Thus again there is a replication of the photomask pattern, but this time the clear areas of the photomask produce the areas on the wafer from which the photoresist has been removed.

5. Postbake. After development and rinsing, the wafers are usually given a postbake in an oven at a temperature of about 150°C for about 30 to 60 min to toughen further the remaining resist on the wafer. This is to make it adhere better to the wafer and to make it more resistant to the hydrofluoric acid (HF) solution used for etching of the SiO_2.

6. Oxide etching. (*a*) *Wet Etching.* The silicon wafers are now immersed in or sprayed with a hydrofluoric (HF) acid solution. This solution is usually a diluted solution of typically 10:1 H_2O:HF or, more often, a H_2O + NH_4F (ammonium fluoride) + HF solution. The HF solution will etch the SiO_2 but will not attack the underlying silicon, nor will it attack the photoresist layer to any appreciable extent. The wafers are exposed to the etching solution long enough to remove the SiO_2 completely in the areas of the wafer that are not covered by the photoresist, as shown

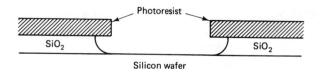

Photoresist

SiO₂   SiO₂

Silicon wafer

Figure 1.29 Effect of oxide overetching.

in Figure 1.28e. For the 10:1 buffered HF solution (NH_4F:HF), the SiO_2 etching rate is about 1000 Å/min at 25°C, so only about 5 min will be required to remove a typical oxide layer of 5000 Å thickness.

The result of the oxide etching process is a pattern of openings or windows in the SiO_2 layer that replicates the photoresist pattern and therefore is a replication of the pattern on the photomask.

The oxide etching time is very carefully controlled so that all the oxide in the photoresist windows is removed. The etching time, however, should not be overly prolonged beyond this point because it will result in more undercutting underneath the photoresist and thus to a widening of the oxide opening beyond what is desired, as shown in Figure 1.29.

(b) *Dry (Plasma) Etching.*   The oxide etching process just described is a wet etching process and the chemical reagents used are in liquid form. A newer process for oxide etching is a dry etching process called plasma etching. In the plasma etching process, the wafers to be etched are placed in a vacuum chamber or enclosure at reduced pressures, often in the range from 1 to 10 torr. Suitable reagent gases such as CF_4 (Freon 14) or C_2F_6 are then metered into the reaction chamber. A radio-frequency (RF) field is then applied, which results in the ionization of the gas molecules and the breaking up of some of the gas molecules into highly reactive free radicals. The density of positive and negative ions and electrons is such that there will be approximately overall charge neutrality, so this is called a gaseous plasma.

The free radicals react with the SiO_2 layer to produce O_2 and various vapor-phase silicon compounds, which are then vented out of the system by the vacuum pump. By means of this dry etching process, the SiO_2 in the areas is exposed by openings in the photoresist. The plasma etching also results in some attack on the photoresist and any exposed silicon, but not at the same rate as the SiO_2. In any case, the etching process is carefully monitored with some end-point detection method to stop the process shortly after the oxide windows have been cleared. With CF_4, the silicon etching rate will actually be larger than the SiO_2 rate, but with C_2F_6 an SiO_2/Si etch ratio of 15:1 is possible.

A major advantage of the dry etching process is that smaller line openings can be achieved ($\lesssim 1$ μm) than with the wet etching process. With wet etching, surface-tension effects can inhibit the flow of the etchant into very narrow photoresist openings. Such is not the case with dry etching. The wet etching process is isotropic, so the SiO_2 is attacked equally well in all directions. As a result, there is some undercutting of the photoresist layer, as shown in Figure 1.30. This undercutting restricts the minimum line width that can reasonably be obtained to a little more than twice the oxide thickness. With an oxide thickness of generally about 5000 Å being used, this limits the minimum line width to something in excess of 1 μm.

The dry etching process can be designed to produce some degree of etching

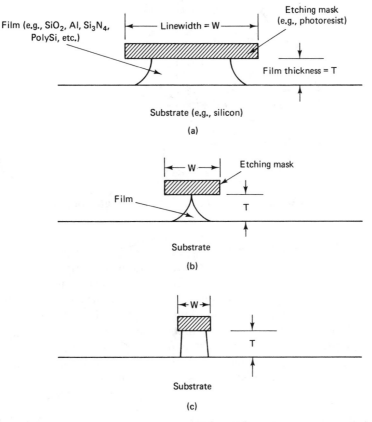

Figure 1.30 Advantages of etching anisotropy: (a) isotropic etching: line width $\gg$ film thickness, $W \gg T$; (b) isotropic etching: line width $= 2 \times$ film thickness, $W = 2T$; (c) anisotropic etching (large etching ratio). In (c) line width can be comparable to or even less than the film thickness.

anisotropy, as illustrated in Figure 1.30c. With a high degree of etching anisotropy, the etching can proceed in the downward direction at a rate substantially faster than sideways etching. This anisotropy in the etching process helps to make possible the attainment of submicron dimensions in ICs.

A dry etching process that offers a relatively high degree of etching anisotropy is *planar plasma etching*. In this process, an electric field is oriented perpendicular to the wafer surface, as shown in Figure 1.31. This electric field accelerates the ions and free radicals of the plasma toward the wafer surface. This directed flow of the ions and free radicals results in a much greater etching rate in the downward direction than in the sideways direction, so a large *etch ratio* can be achieved.

Another dry etching process that can produce a large etch ratio is called *ion milling*. The wafers are placed in a vacuum chamber and a beam of inert-gas ions, usually argon (Ar^+), is accelerated through a large potential difference of generally 400 to 1000 V. The beam of Ar^+ ions then strikes the wafer surface, and the kinetic energy of the ions is such that some of the atoms at or near the surface are knocked

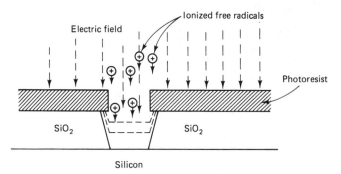

Figure 1.31 Planar plasma etching.

off; this process is called *sputtering*. Sputtering rates of about 100 to 1000 Å/min are commonly obtained.

7. Photoresist stripping. The last step in the photolithographic process is the removal of the photoresist, known as photoresist stripping. Positive photoresists can usually be easily removed in organic solvents such as acetone. Negative photoresists are more difficult to remove. A hot sulfuric acid immersion together with mechanical scrubbing is often used with negative photoresists.

Plasma reactors of the type that are used for plasma etching can also be used for photoresist stripping. Oxygen is used as the reagent gas, and the organic resist material is oxidized to produce various gases, such as CO, CO_2, and H_2O, which are then vented out of the system through the vacuum pump.

1.6 CHEMICAL-VAPOR DEPOSITION

The chemical-vapor deposition (CVD) process is a means for the deposition of thin films on a substrate. The materials to be deposited enter a reaction chamber in the gaseous or vapor phase and react on or near the surface of the substrates, which are at some elevated temperature. The chemical reaction that occurs produces the atoms or molecules that are deposited on the substrate surface. A number of different materials can be deposited by the CVD process. In the sections to follow, some of the techniques used to deposit CVD films that are of importance in semiconductor device fabrication are discussed.

1.6.1 Silicon Dioxide Deposition

An important example of the CVD process is the deposition of SiO_2 layers on silicon substrates. A number of chemical reactions can be used for this process, and one that is very commonly used is

$$SiH_4 \text{ (silane)} + O_2 \longrightarrow SiO_2 + 2H_2 \qquad (1.18)$$

Silane is a gas that is metered into the reaction chamber together with oxygen. The silicon wafers are placed on a heated substrate and raised to a temperature of about 450° to 600°C, and the reaction above takes place. This results in a *deposited* (CVD) oxide layer (called *silox*), as compared to the thermally grown oxide layer discussed previously.

The CVD oxide can be deposited in a matter of only a few minutes, but it will not be as dense as the thermally grown oxide. The thermally grown oxide will generally be a higher-quality oxide, with higher dielectric strength, and serves as a better passivating layer than the CVD oxide. However, the temperature required for the CVD oxide is much lower than that for the thermal oxide, and the time required is much shorter. As a result, the CVD oxide can often be deposited on top of the wafer after previous processing steps such as metallization have been completed without seriously affecting the prior processes. The CVD oxide layer can thus be used for such purposes as "postmetallization passivation" to serve as a protective layer covering the device or IC after all processes, including metallization, have been completed. It can also be used as the insulator between metallization levels.

1.6.2 Silicon Nitride Deposition

Silicon nitride (Si_3N_4) thin-film layers can be deposited by the CVD process. These layers can be used for device protection and passivation. In particular, Si_3N_4 serves as a very good barrier against the penetration of such contaminants as Na^+, K^+, and other ions, against which SiO_2 is much less effective. Silicon nitride can also serve as a diffusion or ion implantation mask. An interesting application of silicon nitride is as an oxidation mask. The diffusion rate of the various oxidants, such as O_2 and H_2O, is very small in silicon nitride, such that a very thin nitride layer (~ 1000 Å) will be sufficient to prevent the oxidation of the underlying silicon. This is the basis of the *local or selective oxidation of silicon* process used in the fabrication of some types of high-density ICs. In this process, a silicon nitride layer is deposited on a silicon wafer and patterned by a photolithographic process. The wafer is then subjected to a thermal oxidation process, but the thermal oxide is grown only in those areas that are not covered by the nitride film, as shown in Figure 1.32.

The most commonly used process for the deposition of silicon nitride layers uses the CVD reaction given by

$$3SiH_4 \text{ (silane)} + 4NH_3 \text{ (ammonia)} \rightarrow Si_3N_4 + 12H_2 \qquad (1.19)$$

at temperatures in the range from 600° to 800°C.

Silicon nitride films can be etched with a hot (~ 180°C) phosphoric acid (H_3PO_4) solution. Unfortunately, photoresists will not stand up to this etching solution, so a CVD oxide layer is used as the etching mask, as shown in Figure 1.33. The nitride layer is deposited first and then the CVD oxide is deposited on top of the nitride. Then photoresist is spun on the wafer and, using the standard photolithographic process, the CVD oxide is patterned. The HF solution that is used for the etching of the CVD oxide will not attack the nitride layer very rapidly, so openings or windows are etching in only the CVD oxide at this point. The photoresist is now completely removed and the wafers are immersed in the hot phosphoric acid solution, which etches the nitride layer but does not attack the CVD oxide. The photoresist thus serves as an etching mask for the CVD oxide, which in turn serves as an etching mask for the nitride layer.

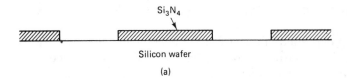

(a)

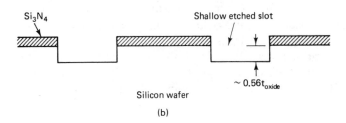

(b)

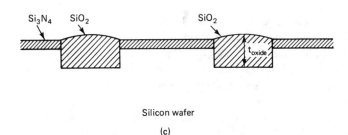

Figure 1.32 Selective oxidation of silicon: (a) deposition and patterning of a silicon nitride film; (b) etching slots in silicon using silicon nitride as an etching mask; (c) growth of thermal oxide layer using silicon nitride as an oxidation mask.

A more convenient etching process for silicon nitride is the use of plasma etching, in which case photoresist can be used. Silicon nitride can be etched with a SiF_4/O_2 gas mixture with an etch ratio of 5:1 compared to silicon and 50:1 with respect to SiO_2.

1.6.3 Silicon Epitaxial Layers

A special case of chemical-vapor deposition is called *epitaxy* or *epitaxial layer deposition*, in which case the deposited layer is in single-crystal form. This epitaxial process will occur only for certain combinations of substrate and layer materials and under certain deposition conditions.

The most common example of epitaxy is the deposition of a silicon epitaxial layer on a single-crystal silicon substrate. In this case, the substrate and layer materials are the same, and this is called *homoepitaxy*. The epitaxial layer becomes a crystallographic continuation of the substrate.

The doping and conductivity type of the epitaxial layer are unrestricted by the substrate, so such epitaxial layer/substrate combinations as N/N^+, N/P, P/N, and P/P^+ are possible. With epitaxy, a lightly doped epitaxial layer can be deposited on a heavily doped substrate. This is in contrast with the situation with a diffused layer, in which case the diffused layer doping is almost always very much heavier than the substrate doping. Although very thick and very thin epitaxial layers can be deposited, most epitaxial layers are in the thickness range from 3 to 30 μm.

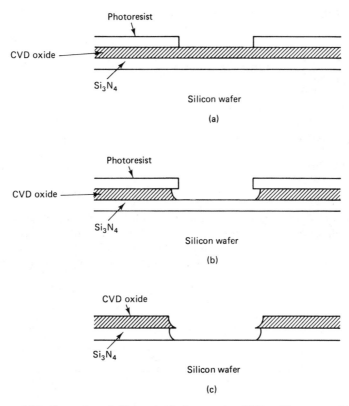

Figure 1.33 Patterning of silicon nitride layers using CVD oxide as an etching mask: (a) photolithography to produce a patterned photoresist film; (b) etching of CVD oxide using HF solution; (c) removal of photoresist and etching of silicon nitride using hot phosphoric acid.

A number of different chemical reactions can be used for the deposition of epitaxial layers; the most important reactions are shown in Table 1.3. The epitaxial layer deposition takes place in a chamber or apparatus called an *epitaxial reactor*. The three basic types of epitaxial reactors are the horizontal reactor, the vertical reactor, and the cylindrical reactor, shown in Figure 1.34. In most cases, the means for heating the silicon wafers to the required temperature is radio-frequency (RF) induction heating, although radiant heating using an array of focused high-intensity quartz lamps and resistance heating can also be used. For RF induction heating, the silicon wafers rest on a silicon carbide-coated graphite susceptor. A water-cooled copper induction coil serves as the primary winding of a transformer. The graphite susceptor serves, in effect, as a single-turn secondary winding. The voltage induced in the susceptor produces a circulating eddy current, which, as a result of heating produced by the I^2R power loss, raises the temperature to the required value.

Doped P- or N-type epitaxial layers can be grown to specified doping levels. A number of gases are metered into the reactor tube, including some very small amounts of doping gases, such as B_2H_6 (diborane) for boron doping and PH_3 (phosphine) for phosphorus doping of the epitaxial layer. During the epitaxial layer dep-

TABLE 1.3 CHEMICAL REACTION USED TO DEPOSIT EPITAXIAL LAYERS

Reaction	Temperature (°C)	Deposition rate (μm/min)
1. $SiCl_4 + 2H_2 \longrightarrow Si + 4HCl$ (silicon tetrachloride)	1150–1250	0.4–1.5
2. $SiHCl_3 + H_2 \longrightarrow Si + 3HCl$ (trichlorosilane)	1100–1200	0.4–2.0
3. $SiH_2Cl_2 \longrightarrow Si + 2HCl$ (dichlorosilane)	1050–1150	0.4–3.0
4. $SiH_4 \longrightarrow Si + 2H_2$ (silane)	950–1050	0.2–0.3

osition, the dopant gas molecules react and become decomposed, and the dopant atoms thus produced become incorporated into the epitaxial layer.

Before the start of the epitaxial layer deposition, H_2 gas is used to remove the native SiO_2 and then anhydrous HCl gas is fed into the reactor. This HCl gas reacts

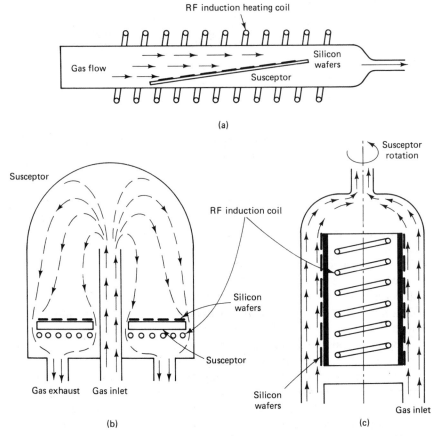

Figure 1.34 Epitaxial reactors: (a) horizontal reactor; (b) vertical reactor; (c) cylindrical reactor.

with the silicon at the surface of the wafers in reactions that are the reverse of those listed above for epitaxial layer deposition. These reverse reactions result in *vapor-phase etching* of the silicon surface. Vapor-phase etching is done immediately before the deposition process to remove a small amount of silicon and other contaminants from the wafer surfaces to ensure that a clean, freshly etched silicon surface will be available for epitaxial layer deposition.

Epitaxial layer deposition takes place at temperatures in the range from 950° to 1250°C, and, as a result, during the deposition as well as during all subsequent high-temperature processing steps there will be diffusion of impurities across the epitaxial layer–substrate interface. This will cause a blurring of the impurity profile in the region of this interface. The most serious problem in this regard will be in the case of a very thin and very lightly doped epitaxial layer that is deposited on a very heavily doped substrate. An example of this would be a 50-Ω-cm ($N_D \simeq 1 \times 10^{14}$ cm^{-3}) 5-μm N-type epitaxial layer that is deposited on a 0.005-Ω-cm ($N_D \simeq 1 \times 10^{19}$ cm^{-3}) N$^+$ substrate. The outdiffusion of impurities from the heavily doped substrate into the lightly doped epitaxial layer will obliterate the sharp N/N$^+$ transition that would otherwise be present at the layer–substrate interface, as shown in Figure 1.35. The influx of donor atoms from the substrate will reduce the effective thickness of the lightly doped epitaxial layer by 1 or 2 μm. To minimize this problem of outdiffusion

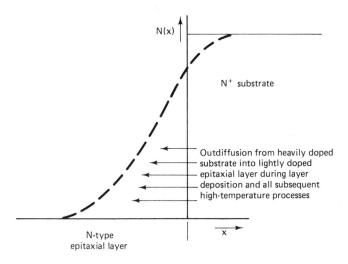

Epitaxial layer/substrate interface

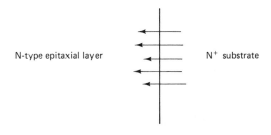

Figure 1.35 Epitaxial layer substrate outdiffusion.

from heavily doped N^+ substrates, slow donor diffusants such as antimony (Sb) and arsenic (As) are often used for the doping of substrate in preference to phosphorus.

1.6.4 Silicon Heteroepitaxy

The crystal structure of sapphire, which is crystalline Al_2O_3, is such that under carefully controlled deposition conditions an epitaxial silicon layer can be deposited on a sapphire substrate. This is an example of *heteroepitaxy*, since the substrate and epitaxial layer are different materials. For the heteroepitaxial process, the crystal structure and atomic spacing of the substrate must be a very close match to that of the layer to be deposited. If this is not the case, the CVD layer that is deposited will be polycrystalline or in some cases amorphous.

The *silicon-on-sapphire* (SOS) process uses a very thin (~ 1 μm) silicon epitaxial layer on an insulating sapphire substrate. This SOS structure is often used for CMOS ICs (CMOS/SOS). The insulating substrate results in a lower parasitic capacitance, which in turn results in high-speed performance and very low power consumption CMOS circuits.

1.7 METALLIZATION

The final step in the wafer processing sequence is that of metallization. The purpose of this process is to produce a thin-film metal layer that will serve as the required conductor pattern for the interconnection of the various devices and circuit elements on the chip. The metallization pattern is also used to produce metallized areas called *bonding pads* around the periphery of the chip to provide areas for the bonding of wire leads from the package to the chip. The bonding wires are typically 25-μm (0.001-in.) diameter gold wires, and the bonding pads are usually made to be around 100 μm $\times$ 100 μm (0.004 in. $\times$ 0.004 in.) square to accommodate fully the flattened ends of the bonding wires and to allow for some registration errors in the placement of the wires on the pads.

The material used for the metallization of most ICs and discrete diodes and transistors is aluminum (Al). This film thickness is about 1 μm and conductor widths of about 2 to 25 μm are commonly used. The use of aluminum offers the following advantages:

1. It is a relatively good conductor.
2. It is easy to deposit thin films of aluminum by vacuum evaporation.
3. Aluminum forms good mechanical bonds with silicon by sintering at about 500°C or by alloying at the eutectic temperature of 577°C.
4. Aluminum forms low-resistance, nonrectifying (ohmic) contacts with P-type silicon and with heavily doped ($\gtrsim 10^{19}$ cm^{-3}) N-type silicon.

The aluminum usually contains 1% silicon to improve reliability and prevent the formation of voids under the contacts. For the metallization process the first step is the deposition of a thin film (~ 1 μm) of aluminum on the silicon wafers. This is done by placing the wafers in a vacuum evaporation chamber. The pressure in the

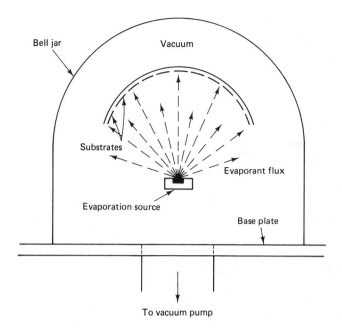

Bell jar

Vacuum

Substrates

Evaporant flux

Evaporation source

Base plate

To vacuum pump

Figure 1.36   Vacuum evaporation.

chamber is reduced to the range of about 10^{-6} to 10^{-7} torr (1 atm = 760 torr = 760 mm Hg).   The material to be evaporated is placed in a resistance-heated tungsten coil or basket.   Electron-beam heating is also very commonly used for vacuum evaporation.   The material to be evaporated is placed in a water-cooled crucible.   A focused electron beam of very high power density is directed at the surface of the material to be evaporated.   This causes a small region of the material to heat up to very high temperatures and to start to vaporize.   The molecules that are evaporated from the surface then travel in straight-line paths.   The evaporated molecules that hit the substrates will condense there and form a thin-film coating, as shown in Figure 1.36.

A high vacuum is required for the vacuum thin-film deposition process to permit the evaporated molecules to travel to the substrate unimpeded by collisions with gas molecules.   The mean free path of gas molecules in a vacuum chamber is related to the pressure by mean free path (mfp) = 4.5×10^{-3} cm/pressure (torr).   At a pressure of 10^{-3} torr, the mfp will be 4.5 cm, increasing to 45 cm at 10^{-4} torr and 450 cm at 10^{-5} torr.   For a source-to-substrate distance of typically 50 cm, a pressure of 10^{-6} torr will thus ensure that virtually all the evaporant flux will reach the substrate unimpeded.   The high vacuum is also needed to prevent undesirable chemical reactions between the evaporant molecules and the residual gases in the vacuum chamber.

The thickness of the deposited film can be monitored during the evaporation process by a quartz crystal film thickness monitor.   Part of the evaporant flux is deposited on the quartz crystal.   As the film thickness builds up on the exposed face of the quartz crystal, it increases the net mass of the crystal.   This will result in a decrease in the resonant frequency of the crystal that is proportional to the net mass of the crystal.   Consequently, the shift in the oscillator frequency will be directly related to the film thickness that has been deposited on the crystal.   When the required

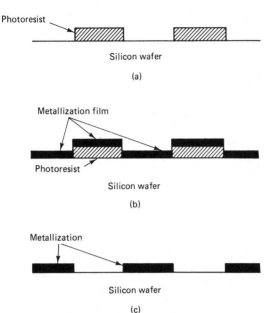

Figure 1.37 Lift-off process: (a) photolithographic process to pattern photoresist layer; (b) deposition of metallization thin film; (c) photoresist lift-off.

film thickness has been reached, the evaporant flux can be shut off automatically by means of a shutter that is placed in front of the evaporation source.

After the thin-film metallization has been done, the film must be patterned to produce the required interconnection and bonding pad configuration. This is done by a photolithographic process of the same type that is used for producing patterns in SiO_2 layers. Aluminum can be etched by a number of acid and base solutions, including HCl, H_3PO_4, KOH, and NaOH. The most commonly used aluminum etchant is phosphoric acid (H_3PO_4), often with the addition of small amounts of HNO_3 (nitric acid) and acetic acid, to result in a moderate etch rate of about 1 μm/min at 50°C. Plasma etching can also be used with aluminum. Using a CCl_4/He plasma at a pressure of 3×10^{-4} torr, an aluminum etch rate of 0.18 μm/min can be produced. Highly anisotropic aluminum etching with 10 : 1 (vertical-to-horizontal) etch ratios is possible. As a result, very fine line (~1 μm) aluminum patterns are possible with aluminum film thicknesses that are substantially greater than the line widths.

An alternative metallization patterning technique is the *lift-off* process. In this process, a positive photoresist is spun on the wafer and patterned using the standard photolithographic process. Then the metallization thin film is deposited on *top* of the remaining photoresist. The wafers are then immersed in a suitable solvent such as acetone and at the same time subjected to ultrasonic agitation. This causes swelling and dissolution of the photoresist. As the photoresist comes off, it lifts off the metallization on top of it, as shown in Figure 1.37. For this lift-off process to work, the metallization film thickness must generally be somewhat less than the photoresist thickness. The process can, however, produce a very fine line width (~1 μm) metallization pattern, even with metallization film thicknesses that are greater than the line width.

With very narrow line widths, it is desirable to keep up the film thickness to a

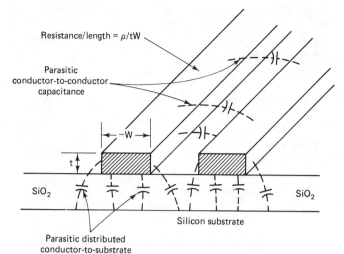

Resistance/length = ρ/tW

Parasitic conductor-to-conductor capacitance

$\leftarrow W \rightarrow$

t

SiO$_2$

SiO$_2$

Silicon substrate

Parasitic distributed conductor-to-substrate capacitance

Figure 1.38 Interconnection RC time constant.

value such that the resistance of the interconnections is maintained at an acceptably small value. A large value of interconnection resistance when combined with the parasitic capacitance, as shown in Figure 1.38, results in an RC time constant that limits the overall speed of the circuit. Indeed, with ICs containing ultrafast small-geometry transistors, the principal limitation on the speed of operation can be the interconnection RC time constant rather than the devices themselves.

PROBLEMS

Diffusion and Ion Implantation Problems

1.1. Given: Constant surface concentration diffusion of phosphorus at 1150°C for 60 min into a 1.0-Ω-cm P-type (boron-doped) silicon substrate. The phosphorus surface concentration will be solubility limited. Find:

    (a) Phosphorus surface concentration, N_0.   (*Ans.*: ~1.5 × 10²¹ cm⁻³)

    (b) Phosphorus diffusion constant, D.   (*Ans.*: ~1 × 10⁻¹² cm²/s)

    (c) Substrate doping level, N_B.   (*Ans.*: ~1.8 × 10¹⁶ cm⁻³)

    (d) Junction depth, x_J.   (*Ans.*: 3.7 μm)

    (e) Junction depth if diffusion time is 15 min.   (*Ans.*: 1.9 μm)

    (f) Junction depth for a 2.0-h diffusion.   (*Ans.*: 5.2 μm)

    (g) Time required for $x_J = 10$ μm.   (*Ans.*: 7.3 h)

1.2. Given: Limited source boron diffusion at 1150°C for 90 min into a 10-Ω-cm N-type silicon substrate. The boron deposited into the silicon surface region has a density of $Q = 1 \times 10^{15}$ cm⁻². Find:

    (a) Boron diffusion constant, D.   (*Ans.*: ~1 × 10⁻¹² cm²/s)

    (b) Substrate doping.   (*Ans.*: ~4.7 × 10¹⁴ cm⁻³)

    (c) Junction depth, x_J.   (*Ans.*: 4.6 μm)

    (d) Boron surface concentration after the diffusion.   (*Ans.*: $N(0) = 7.7 \times 10^{18}$ cm⁻³)

    (e) Average conductivity of diffused layer, $\bar{\sigma}$.   (*Ans.*: ~24 S/cm)

    (f) Sheet resistance of diffused layer, R_S.   (*Ans.*: ~93 Ω/square)

(g) End-to-end resistance of the diffused resistor, if the lateral dimensions of this diffused layer are 0.5 mil $\times$ 100 mil.   (*Ans.*: ~18.5 kΩ)

1.3. To deposit a boron surface density of $Q = 1 \times 10^{15}$ cm^{-2} into the silicon surface region, a short, low-temperature deposition diffusion step is performed at 950°C.
   (a) Find the boron diffusion constant for the deposition diffusion.   (*Ans.*: ~7.6 $\times$ 10^{-15} cm^2/s)
   (b) Find the surface concentration during deposition diffusion, N_0.   (*Ans.*: ~5 $\times$ 10^{20} cm^{-3})
   (c) Find the time required for deposition diffusion.   (*Ans.*: 6.9 min)
   (d) Would it be preferable to have this deposition diffusion done at a somewhat lower temperature, such as 900°C? Explain.

1.4. The variation of diffusion constant with temperature can be described by the relationship $D = D_0 \exp(-qE_A/kT)$, where D_0 is the preexponential constant, q is the electronic charge, k is Boltzmann's constant (1.384 $\times$ 10^{-23} J/K), and E_A is the activation energy for the diffusion process (in electron volts).
   (a) From the graph of the diffusion constant versus temperature, find the activation energy E_A for the diffusion of boron and phosphorus into silicon.   (*Ans.*: $E_A = 3.61$ eV)
   (b) Show that the fractional change of the diffusion constant with temperature $(1/D)(dD/dT)$ is given by $(1/D)(dD/dT) = qE_A/kT^2$. Find the value of $(1/D)(dD/dT)$ for the diffusion of boron or phosphorus into silicon for a diffusion temperature of 1100°C. Express the answer in percentage change as well as in fractional change.   (*Ans.*: 0.022/°C or 2.2%/°C)
   (c) Given that the junction depth for either a deposition or drive-in type of diffusion is approximately proportional to $\sqrt{Dt}$, show that the fractional change in the junction depth will be related to the change in the diffusion temperature by $(1/x_J)(dx_J/dT) = \frac{1}{2}(1/D)(dD/dT) = \frac{1}{2}qE_A/kT^2$. Find the fractional and percentage change in the junction depth with temperature for the diffusion of boron or phosphorus into silicon at 1100°C.   (*Ans.*: 0.011/°C or 1.1%/°C)
   (d) Given that the current gain h_{FE} of a transistor is inversely proportional to the base width, show that the fractional change in h_{FE} with diffusion temperature will be given by

$$\frac{1}{h_{FE}} \frac{dh_{FE}}{dT} \simeq \frac{1}{2} \frac{x_J}{W} \frac{qE_A}{kT^2}$$

where W is the base width and x_J can refer to either the emitter–base or the collector–base junction depth of the transistor. (*Hint:* Assume that W is small compared to $x_{J(EB)}$ and $x_{J(CB)}$.)
   (e) For a transistor with a base width of 0.4 μm and an emitter–base junction depth of 2.4 μm, find the fractional and percentage variation in h_{FE} due to changes in either the emitter or the base diffusion temperature. Use a diffusion temperature of 1100°C.   (*Ans.*: 0.066/°C or 6.6%/°C. Note that, since a 1°C change in the diffusion temperature is only about a 0.1% change, the value of h_{FE} will change by about 70% for only a 1% change in the diffusion temperature.)

1.5. (*Boron ion implantation and drive-in diffusion*) A boron ion implantation followed by a drive-in diffusion at 1150°C produces a diffused layer with a boron surface concentration of 1 $\times$ 10^{18} cm^{-3} and a junction depth of 3.0 μm. The substrate phosphorus doping is 1 $\times$ 10^{15} cm^{-3}. Find:
   (a) Diffusion time.   (*Ans.*: 54.3 min)

(b) Ion implantation dosage.   (*Ans.*: 1.0×10^{14} cm^{-2})

(c) Average resistivity of diffused layer.   (*Ans.*: ~0.15 Ω-cm)

(d) Sheet resistance of the diffused layer.   (*Ans.*: ~500 Ω/square)

(e) Average hole mobility in the diffused layer.   (*Ans.*: ~123 cm^2/V-s)

(f) Implantation time per wafer if the implantation beam current is 50 μA, the wafer diameter is 100 mm, and the wafers are scanned by the implantation ion beam in a rectangular raster.   (*Ans.*: 32 s)

1.6. (*Impurity profile evaluation*)   A P$^+$ diffused layer is produced by a boron ion implantation followed by a drive-in diffusion at 1150°C to produce a junction depth of 2.5 μm in a 1×10^{15} cm^{-3} phosphorus-doped silicon substrate.   The impurity profile of this diffused layer is to be evaluated by etching off successive 0.2-μm-thick layers and measuring the sheet resistance after each etching step.

(a) Show that the sheet resistance of the removed layer will be given by

$$\frac{1}{\Delta R_S} = \frac{1}{R_{S_b}} - \frac{1}{R_{S_a}}$$

where ΔR_S is the sheet resistance of the removed layer, R_{S_b} the sheet resistance that is measured *before* the removal of the layer, and R_{S_a} the sheet resistance that is measured *after* the removal of the layer.   The initial sheet resistance of 934 Ω/square, and after the removal of the first 0.2-μm layer, the sheet resistance has increased to 1148.6 Ω/square.

(b) Find the sheet resistance of the removed layer.   (*Ans.*: 5000 Ω/square)

(c) Find the average resistivity of the removed layer.   (*Ans.*: 0.1 Ω-cm)

(d) Find the surface concentration of the diffused layer, $N(0)$.   (*Ans.*: 5×10^{17} cm^{-3})

(e) Find the drive-in diffusion time.   (*Ans.*: 42 min)

(f) Find the ion implantation dosage.   (*Ans.*: 4.4×10^{13} cm^{-2})

(g) By how much will the boron concentration vary over the thickness of the first, second, third, fourth, fifth, and sixth layers that are removed?   Does the choice of 0.2 μm as the thickness increment appear to be a good one?   Explain.   (*Ans.*: first: 4.1%; second: 12.8%; third: 22.1%; fourth: 32.3%; fifth: 44.3%; sixth: 55.3%)

(h) Find the average boron concentration in the second, fourth, sixth, eighth, and tenth layers that are removed.   (*Ans.*: second: 4.6×10^{17} cm^{-3}; fourth: 3.1×10^{17} cm^{-3}; sixth: 1.5×10^{17} cm^{-3}; eighth: 5.3×10^{16} cm^{-3}; tenth: 1.4×10^{16} cm^{-3})

Oxidation Problems

1.7. Find the time required to grow a 5000-Å SiO$_2$ layer on silicon at a temperature of 1100°C using **(a)** dry oxygen and **(b)** wet oxygen (H$_2$O).   (*Ans.*: 10 h, 40 min)

1.8. Show that if a silicon wafer is subjected to a number of successive thermal oxidation processes the resulting final oxide thickness x_f will be given by $x_f^2 = x_1^2 + x_2^2 + x_3^2 + \cdots$, where $x_1, x_2, x_3, \ldots$ are the oxide thicknesses that would result from the application of each oxidation process to an unoxidized wafer.

1.9. A diffusion window is etched in a field oxide layer of thickness x_{f1} as shown in Figure P1.9a.   There is then a regrowth of oxide in the window to a thickness of x_w, during which time the field oxide thickness increases to x_{f2}, as shown in Figure P1.9b.

(a) Show that the difference in the oxide thicknesses is given by

$$x_{f2} - x_w = \sqrt{x_{f1}^2 + x_w^2} - x_w$$

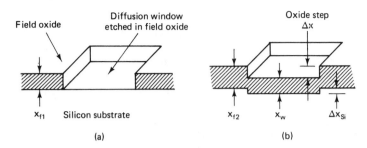

(a)

(b)

Figure P1.9

(b) Show that the height of the step in the silicon surface is given by

$$\Delta x_{S1} = 0.44(x_w - x_{f2} + x_{f1})$$

(c) Show that the oxide step height will be given by

$$\Delta x = x_{f2} - x_w + \Delta x_{S1}$$

(d) If $x_{f1} = 5000$ Å and $x_w = 5000$ Å, find the heights of the silicon step and the oxide step.   (*Ans.*: 1289 Å, 3360 Å)

1.10. A silicon wafer has a field oxide of 5000 Å thickness.   A window is etched in the oxide for a base diffusion and then the oxide is regrown to a thickness of 5000 Å.   A second window is then etched within the confines of the first window for the emitter diffusion, and then the oxide in this area is regrown to a thickness of 5000 Å, producing the result shown in Figure P1.10.   Find x_2, x_1, Δx_2, Δx_1, and the heights of the two silicon steps.   (*Ans.*: 8660 Å, 7071 Å, 3090 Å, 3360 Å, 1501 Å, 1289 Å)

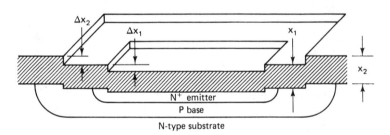

Figure P1.10

1.11. An oxidized silicon wafer is subjected to a measurement of the optical reflectance as a function of wavelength.
   (a) Show that the thickness t of the oxide layer is related to the distance between the adjacent reflectance maxima by

$$\frac{1}{t} = 2n \left(\frac{1}{\lambda_m} - \frac{1}{\lambda_{m+1}} \right)$$

where n is the index of refraction of the oxide layer and λ_m and λ_{m+1} are the wavelengths of adjacent reflection peaks.
   (b) Show that the relationship above will also hold true for the reflection minima, where λ_m and λ_{m+1} are the wavelengths of the adjacent minima.
   (c) For an SiO_2 layer on silicon, adjacent reflection peaks are observed at wavelengths

of 5000 and 6000 Å. If the index of refraction of SiO_2 is 1.5, find the thickness of the oxide layer. (*Ans.:* 10,000 Å)

(d) For an SiO_2 layer, adjacent reflection minima are observed at 5454 and 6667 Å. Find the thickness of the layer. (*Ans.:* 10,000 Å)

(e) An SiO_2 layer is 6000 Å thick. Find the wavelengths of the reflection maxima and minima. (*Ans.:* maxima at 18 kÅ, 9 kÅ, 6 kÅ, 4.5 kÅ, 3.6 kÅ, 3.0 kÅ, . . . ; minima at 36 kÅ, 12 kÅ, 7.2 kÅ, 5.14 kÅ, 4.0 kÅ, . . .)

1.12. (*Charge carrier mobility*) The mobility of charge carriers (free electrons and holes) in a semiconductor will decrease with increasing doping level due to the scattering of the charge carriers by the ionized dopant impurities. The mobility can be expressed approximately as $1/\mu = 1/\mu_L + 1/\mu_i$, where μ is the charge carrier mobility, μ_L is the mobility value that is obtained when there is no ionized impurity scattering so that the only scattering process is lattice scattering, and μ_i is the mobility value that would be obtained if there were only ionized impurity scattering.

An approximate relationship for μ_i is given by $\mu_i = KN_i^\alpha$, where N_i is the ionized impurity (dopant) concentration, and K and α are constants. Given the mobility values listed below for silicon at 25°C, find K and α such that the mobility equation will be an exact fit to the mobility versus doping curve at $N_i = 1 \times 10^{16}$ cm^{-3} and at $N_i = 1 \times 10^{18}$ cm^{-3}. Then find the percentage error at $N_i = 1 \times 10^{17}$ cm^{-3} and at $N_i = 1 \times 10^{19}$ cm^{-3} for the hole mobility (P type) and the electron mobility (N type).

Dopant density, N_i (cm^{-3})	Hole mobility, μ_p (cm^2/V-s)	Electron mobility, μ_n (cm^2/V-s)
1×10^{16}	400	1000
1×10^{17}	250	500
1×10^{18}	100	280
1×10^{19}	28	115
Lattice mobility, μ_L	500	1500

(*Ans.:* P type: $K = 1.164 \times 10^{-13}$, $\alpha = 0.6021$, no error at 1×10^{17} cm^{-3}, 5.05% error at 1×10^{19} cm^{-3}; N type: $K = 1.0 \times 10^{-11}$, $\alpha = 0.470$, +21% error at 1×10^{17} cm^{-3}, -5.9% error at 1×10^{19} cm^{-3})

*1.13. (*Resistivity versus impurity concentration*) Write a computer program to obtain the resistivity values as a function of doping level for P- and N-type silicon at 25°C. Use the mobility results of Problem 1.12. Compare the results obtained with the graph of Figure 1.18c.

*1.14. (*Sheet resistance of diffused layers*) Write a computer program to obtain the sheet resistance of a diffused layer with a Gaussian impurity profile as a function of surface concentration $N(O)$ and junction depth x_J. The results of Problems 1.12 and 1.13 can be used in the computer program. Use the computer program to obtain the sheet resistance for the following P-type diffused layers.

(a) $N(O) = 1 \times 10^{18}$ cm^{-3}, $x_J = 3.0$ μm, $N_{substrate} = 1 \times 10^{15}$ cm^{-3}
(b) $N(O) = 1 \times 10^{18}$ cm^{-3}, $x_J = 3.0$ μm, $N_{substrate} = 1 \times 10^{17}$ cm^{-3}
(c) $N(O) = 1 \times 10^{17}$ cm^{-3}, $x_J = 3.0$ μm, $N_{substrate} = 1 \times 10^{15}$ cm^{-3}
(d) $N(O) = 1 \times 10^{19}$ cm^{-3}, $x_J = 3.0$ μm, $N_{substrate} = 1 \times 10^{15}$ cm^{-3}
(e) $N(O) = 1 \times 10^{19}$ cm^{-3}, $x_J = 2.0$ μm, $N_{substrate} = 1 \times 10^{15}$ cm^{-3}

Compare the results obtained with those obtained from the graph of Figure 1.18a.

*1.15. (*Sheet resistance of ion-implanted diffused layers*) Write a computer program to obtain the sheet resistance of diffused layers that is produced by a low-energy ion implantation followed by a drive-in diffusion. The sheet resistance should be obtained as a function of ion-implantation dosage, junction depth, and background or substrate doping. Find the sheet resistance for a boron-doped layer with a junction depth of 2.5 μm, a background N-type doping of 1×10^{15} cm^{-3}, and the following boron ion-implantation dosages (cm^{-2}): $1 \times 10^{10}, 1 \times 10^{11}, 1 \times 10^{12}, 1 \times 10^{13}, 1 \times 10^{14}, 1 \times 10^{15},$ and 1×10^{16}. Compare the results obtained with those of Figure 2.54.

REFERENCES

BAR-LEV, A., *Semiconductors and Electronic Devices*, Prentice-Hall, Englewood Cliffs, N.J., 1984.

BEADLE, W. E., J. C. C. TSAI, and R. D. PLUMMER (eds.), *Quick Reference Manual for Silicon Integrated Circuit Technology*, Wiley, New York, 1985.

BRODIE, I., *The Physics of Microfabrication*, Plenum Press, New York, 1982.

BURGER, R. M., and R. P. DONOVAN, *Fundamentals of Silicon Integrated Circuit Device Technology*, Vol. 1, Prentice-Hall, Englewood Cliffs, N.J., 1967.

CAMENZIND, H. R., *Electronic Integrated Systems Design*, Van Nostrand Reinhold, New York, 1972.

CHEN, J., *CMOS Devices and Technology for VLSI*, Prentice-Hall, Englewood Cliffs, N.J., 1990.

COLCLASER, R. A., *Microelectronics: Processing and Device Design*, Wiley, New York, 1980.

CONNELLY, J. A., *Analog Integrated Circuits*, Wiley, New York, 1975.

DeFOREST, W. S., *Photoresist: Materials and Processes*, McGraw-Hill, New York, 1975.

DOYLE, J. M., *Thin Film and Semiconductor Integrated Circuitry*, McGraw-Hill, New York, 1966.

EIMBINDER, J., *Application Considerations for Linear Integrated Circuits*, Wiley, New York, 1970.

EINSPRUCH, N. G., *VLSI Electronics: Microstructure Science*, Vols. 1–6, Academic Press, New York, 1982.

ELLIOTT, D. J., *Integrated Circuit Fabrication Technology*, McGraw-Hill, New York, 1982.

———, *Integrated Circuit Mask Technology*, McGraw-Hill, New York, 1985.

———, *Microlithography*, McGraw-Hill, New York, 1986.

FOGIEL, M., *Microelectronics: Basic Principles, Circuit Design, Fabrication Technology*, Research & Education Associates, Piscataway, N.J., 1972.

GHANDI, S. K., *The Theory and Practice of Microelectronics*, Wiley, New York, 1968.

———, *VLSI Fabrication Principles*, Wiley, New York, 1983.

GLASER, A. B., and G. E. SUBAK-SHARPE, *Integrated Circuit Engineering*, Addison-Wesley, Reading, Mass., 1977.

GREBENE, A. B., *Analog Integrated Circuit Design*, Van Nostrand Reinhold, New York, 1972.

GROVE, A. S., *Physics and Technology of Semiconductor Devices*, Wiley, New York, 1967.

HAMILTON, D. J., and W. G. HOWARD, *Basic Integrated Circuit Engineering*, McGraw-Hill, New York, 1975.

HNATEK, E. R., *A User's Handbook of Integrated Circuits*, Wiley, New York, 1973.

JAEGER, R. C., *Introduction to Microelectronic Fabrication*, Addison-Wesley, Reading, Mass., 1988.

MILLMAN, J., *Microelectronics*, McGraw-Hill, New York, 1979.

MOREAU, W. M., *Semiconductor Lithography*, Plenum Press, New York, 1988.

MORGAN, R. A., *Plasma Etching in Semiconductor Fabrication*, Elsevier, New York, 1985.

MOTOROLA, INC., *Integrated Circuits: Design Principles and Fabrication*, Vol. 1, McGraw-Hill, New York, 1967.

MULLER, R. S., and T. I. KAMINS, *Device Electronics for Integrated Circuits*, 2d ed., Wiley, New York, 1986.

RUNYAN, W. R., *Silicon Semiconductor Technology*, McGraw-Hill, New York, 1965.

RUSKA, W. S., *Microelectronic Processing*, McGraw-Hill, New York, 1987.

RYSSEL, H., and I. RUGE, *Ion Implantation*, Wiley, New York, 1986.

SHIMURA, F., *Semiconductor Silicon Crystal Technology*, Academic Press, New York, 1989.

STREETMAN, B. G., *Solid State Electronic Devices*, Prentice-Hall, Englewood Cliffs, N.J., 1980.

SZE, S. M., *VLSI Technology*, 2d ed., McGraw-Hill, New York, 1988.

TILL, W. C., and J. T. LUXON, *Integrated Circuits: Materials, Devices, and Fabrication*, Prentice-Hall, Englewood Cliffs, N.J., 1982.

TSIVIDIS, Y., and P. ANTOGNETTI (eds.), *Design of MOS VLSI Circuits for Telecommunications*, Chapter 1, Fabrication Technology of MOS IC's for Telecommunications by E. DEMOULIN, Prentice-Hall, Englewood Cliffs, N.J., 1985.

VERONIS, A., *Integrated Circuit Fabrication Technology*, Prentice-Hall, Englewood Cliffs, N.J., 1979.

VOSSEN, J. L., and W. KERN, *Thin Film Processes*, Academic Press, New York, 1978.

WOLF, S., and R. N. TAUBER, *Silicon Processing for the VLSI Era, Vol. 1: Process Technology*, Lattice Press, Sunset Beach, Calif., 1986.

2

Integrated-circuit Devices

2.1 JUNCTION CHARACTERISTICS

Before continuing with a discussion of the fabrication sequence for diodes, transistors, and ICs, it is of importance to look at some characteristics of PN junctions, particularly the junction capacitance C_J, the breakdown voltage, and the series resistance R_S. The results of this discussion will provide insight into the use of epitaxial structures for devices.

2.1.1 Junction Capacitance

A reverse-biased PN junction can be considered to be a parallel-plate capacitor with the depletion region being the insulator or dielectric, as shown in Figure 2.1. The depletion or space-charge region is the region adjacent to the PN junction that is essentially depleted or devoid of all mobile charge carriers (free electrons and holes), so it indeed acts like an insulator.

The junction capacitance is given by the parallel-plate capacitance equation as $C_J = \varepsilon A/W$, where $\varepsilon = \varepsilon_r \times \varepsilon_0 = 11.8 \times 8.85 \times 10^{-14}$ F/cm = 1.0443×10^{-12} F/cm = permittivity of silicon, A is the junction area, and W is the depletion region width. For the case in which there is uniform doping on both sides of the junction,

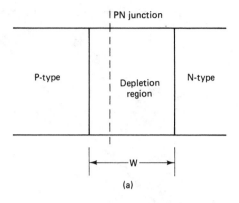

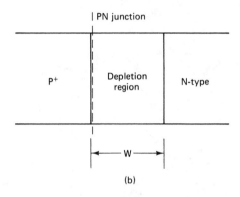

Figure 2.1 PN junction depletion region width: (a) general case; (b) P^+N "one-sided" junction.

the depletion region width is given by

$$W = \sqrt{\frac{2\varepsilon V_J}{qN}} \qquad (2.1)$$

where V_J = junction voltage = $\phi + V_R$ = contact potential (~ 0.8 V) + applied reverse-bias voltage, and $N = N_A N_D/(N_A + N_D)$. The corresponding equation for junction capacitance is

$$C_J = A \sqrt{\frac{q\varepsilon N}{2V_J}} \qquad (2.2)$$

At $N = 1 \times 10^{16}$ cm^{-3} and $V_J = 1.0$ V, these equations give $W = 0.361$ μm and $C_J = 2.89 \times 10^{-8}$ F/cm^2 = 289 pF/mm^2. In Figure 2.2, a graph of C_J versus N is given for several different values of V_J.

If the junction area for the preceding case is 500 μm $\times$ 500 μm (0.020 in. $\times$ 0.020 in.), the junction capacitance will be C_J = 289 pF/mm^2 $\times$ (0.5 mm $\times$ 0.5 mm) = 72.3

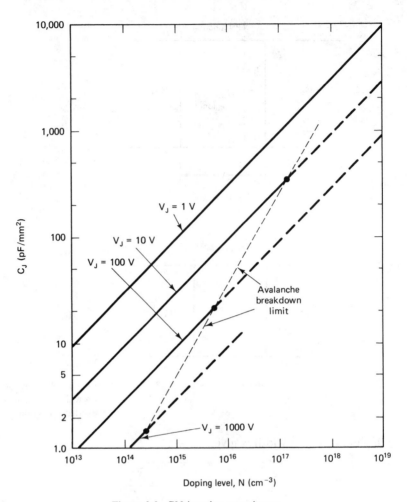

Figure 2.2 PN junction capacitance.

pF at $V_J = 1.0$ V, which corresponds to a reverse-bias voltage of only about 0.2 V. If V_J increases to 25 V, the capacitance will decrease by a factor of 5 to become 14.5 pF.

For almost all diffused PN junctions, the doping on the diffused layer side of the junction will be very much heavier than the doping on the other (substrate) side of the junction, so most diffused junctions can be considered to be *one-sided* junctions. In Figure 2.1b, a one-sided P^+N junction is shown. Notice that the depletion region is almost entirely on the lightly doped (substrate) side of the junction and extends very little into the diffused layer side. For a P^+N one-sided junction, we have that $N = N_A N_D/(N_A + N_D) \simeq N_D$, and for an N^+P one-sided junction, we have $N \simeq N_A$.

For many applications, it is desirable to have a very small junction capacitance,

TABLE 2.1 BREAKDOWN VOLTAGE VERSUS DOPING LEVEL

Doping level	Resistivity (Ω-cm)		Breakdown voltage (V)	
	N type	P type	Equation	Experimental
1×10^{14}	50	135	1250	1250
3×10^{14}	16	45	600	600
1×10^{15}	5	13.5	270	300
3×10^{15}	1.7	4.5	130	150
1×10^{16}	0.60	1.5	58	65
3×10^{16}	0.23	0.63	28	30
1×10^{17}	0.09	0.28	12.5	15
3×10^{17}	0.05	0.14	6.0	8.0
1×10^{18}	0.025	0.065	2.7	5.3
2×10^{18}	0.017	0.040	1.7	5.0

so we see that a lightly doped substrate will be needed. This will be especially true in the case of high-speed and high-frequency device applications.

2.1.2 Breakdown Voltage

The breakdown voltage of a PN junction is in general a function of the doping levels on both sides of the junction. For a one-sided junction, the breakdown voltage will be a function principally of the doping level on the more lightly doped side of the junction. An approximate equation for the breakdown voltage of a one-sided, flat (that is, no junction curvature) junction is

$$\text{breakdown voltage} = 2.7 \times 10^{12} \text{ V}/N^{2/3}$$

where N is the doping level (cm^{-3}). This equation is a good approximation up to doping levels of about 1×10^{17} cm^{-3}. Table 2.1 demonstrates the results obtained by using the breakdown voltage equation.

The breakdown voltage for planar junctions will be somewhat lower than these values, especially at the lighter doping levels, where the breakdown voltage can be very substantially less. This is due to the effect of the junction curvature in the regions underneath the edges of the oxide window, which results in an increase in the electric field intensity. The effect of junction curvature on breakdown voltage is shown in Figure 2.3. Looking at an extreme case, for $N = 1 \times 10^{14}$ cm^{-3}, the breakdown voltage for a plane (flat) junction will be around 1250 V. A junction curvature corresponding to a junction depth of $x_J = 1$ μm will drop the breakdown voltage all the way down to about 50 V. Increasing x_J to 3 μm will increase the breakdown voltage back up to around 100 V, and at $x_J = 10$ μm, the breakdown voltage will reach around 250 V. At heavier dopings, however, we see that the influence of junction curvature on breakdown voltage becomes less.

From the preceding discussion we see that for a high breakdown voltage a light doping is required. The junction depth should also not be too small, especially for the cases in which very high breakdown voltages ($\gtrsim 100$ V) are required.

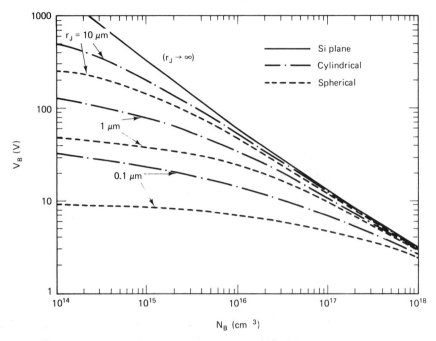

Figure 2.3 Breakdown voltage as a function of substrate doping and junction curvature for abrupt one-sided junctions in silicon. (Reprinted with permission from S. M. Sze and G. Gibbons, "Effect of Junction Curvature on Breakdown Voltage in Semiconductors," *Solid State Electronics*, Vol. 9, p. 831, 1966. Copyright 1966, Pergamon Press Ltd.)

2.1.3 Series Resistance

The bulk series resistance of a PN junction is due to the finite resistivity of the P- and N-type regions of the junction, outside the depletion region. For a P^+N diffused junction, the series resistance due to the P^+ diffused layer will be negligible compared to the resistance due to the N-type side of the junction. Thus, for a small value of series resistance, a low-resistivity (that is, a heavily doped) substrate is needed.

2.2 EPITAXIAL STRUCTURE

From the previous discussion we have the following requirements:

1. For low junction capacitance, C_J: low doping (lightly doped substrate)
2. For high breakdown voltage: low doping (lightly doped substrate)
3. For low series resistance, R_S: heavy doping (low resistivity substrate)

We see that the series resistance requirement will be incompatible with the capacitance and breakdown voltage requirements.

The epitaxial structure shown by the diode example of Figure 2.4 offers a good

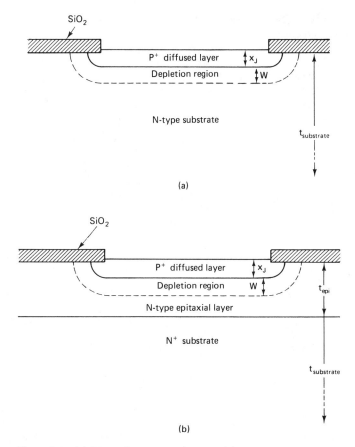

Figure 2.4 (a) Planar (nonepitaxial) diode; (b) planar epitaxial diode.

way of resolving this incompatibility and simultaneously satisfying the capacitance, breakdown voltage, and series resistance requirements. As long as the depletion region remains entirely within the lightly doped epitaxial layer and does not reach the heavily doped N^+ substrate, the capacitance and breakdown voltage will be a function only of the epitaxial layer doping and will be independent of the substrate doping. The series resistance will, however, be determined to a major extent by the N^+ substrate doping since the epitaxial layer is very thin (~ 10 μm) compared to the substrate thickness of some 250 to 400 μm.

The series resistance of the epitaxial diode is given by

$$R_S = R_{\text{epi}} + R_{\text{substrate}} = \frac{\rho_{\text{epi}}(t_{\text{epi}} - x_J - W)}{A} + \frac{\rho_{\text{substrate}}t_{\text{substrate}}}{A} \qquad (2.3)$$

Note that $t_{\text{epi}} - x_J - W$ is the thickness of the *undepleted* portion of the epitaxial layer, which is the distance from the edge of the depletion region to the substrate. The heavily doped, low-resistivity substrate that constitutes the major part of the overall device thickness can lead to a very substantial reduction in the series resistance.

Sec. 2.2   Epitaxial Structure

59

2.2.1 Epitaxial Diode Example: Voltage-variable Capacitance Diode

As an example of the efficacy of the epitaxial structure in reducing the series resistance, let us consider a planar epitaxial P^+NN^+ diode that will be used as a *voltage-variable capacitance* (VCC) or *varactor* (variable-reactance) diode. A VVC diode makes specific use of the dependence of the diode capacitance on the bias voltage. This device can be used in a variety of applications, including voltage-controlled oscillators (VCOs) and automatic frequency control (AFC) circuits, in which case the VVC diode is part of an LC tuned circuit.

Let us consider a P^+NN^+ VVC diode of the following design:

Junction diameter = 500 μm = 0.50 mm (0.020 in.)

Junction depth, x_J = 2.0 μm

Epitaxial layer thickness, t_{epi} = 10 μm

Epitaxial layer doping, N_{epi} = 1 × 10^{15} cm^{-3} (ρ_{epi} = 5 Ω-cm)

Substrate resistivity, $\rho_{substrate}$ = 0.005 Ω-cm (Sb-doped)

Substrate thickness, $t_{substrate}$ = 300 μm = 0.3 mm (0.012 in.)

$$(2.4)$$

The junction area corresponding to the diameter of 500 μm is

$$A = \frac{\pi d^2}{4} = \frac{\pi}{4}(0.5 \text{ mm})^2 = 0.196 \text{ mm}^2 = 0.196 \times 10^{-2} \text{ cm}^2 \quad (2.5)$$

Under zero bias conditions, we have that $V_J = \phi = 0.8$ V, so $C_J(0) = 289/\sqrt{10 \times 0.8}$ pF/mm^2 × 0.196 mm^2 = 20.1 pF, and the zero-bias depletion layer width is $W(0) = 0.361 \times \sqrt{10 \times 0.8} = 1.02$ μm. The series resistance under these conditions will be given by $R_S(0) = R_{epi}(0) + R_{sub} = [5 \ \Omega\text{-cm} \times (10 - 2 - 1.02) \times 10^{-4} \text{ cm}/0.196 \times 10^{-2} \text{ cm}^2] + [0.005 \ \Omega\text{-cm} \times 0.03 \text{ cm}/0.196 \times 10^{-2} \text{ cm}^2] = 1.78 \ \Omega + 0.0765 \ \Omega = 1.86 \ \Omega$. If a nonepitaxial structure were used, the series resistance would be $R_S = 5 \ \Omega\text{-cm} \times 0.03 \text{ cm}/0.196 \times 10^{-2} \text{ cm}^2 = 76.5 \ \Omega$. The quality factor Q of the diode under zero-bias conditions at a frequency of 50 MHz will be $Q(0) = (1/\omega C_J)/R_S = 1/(2\pi \times 50 \text{ MHz} \times 20.1 \text{ pF} \times 1.86 \ \Omega) = 85.5$. This is to be compared to a value of $Q(0) = 2.0$ for a nonepitaxial structure.

At a reverse-bias voltage of 3.0 V, we have that $V_J = 3.8$ V, so the capacitance is now $C_J(-3 \text{ V}) = 20.1 \times \sqrt{0.8/3.8} = 9.20$ pF and the depletion layer width is $W(-3 \text{ V}) = 1.02 \ \mu\text{m} \times \sqrt{3.8/0.8} = 2.22 \ \mu\text{m}$. The series resistance is now given by $R_S = R_{epi}(-3 \text{ V}) + R_{sub} = 5 \ \Omega\text{-cm} \times (10 - 2 - 2.22) \times 10^{-4}/0.196 \times 10^{-2} + 0.0765 = 1.47 + 0.0765 \ \Omega = 1.551 \ \Omega$. The Q-value at 3-V bias is $Q(-3 \text{ V}) = 223$ and has increased from the zero-bias value, due to the decrease in the both the capacitance and the series resistance. If a nonepitaxial structure were to be used, the Q value at this voltage would be only 4.5.

The voltage at which the depletion region will extend all of the way across the epitaxial layer from the junction to the N^+ substrate (that is, "full depletion") is given by 1.02 μm × $\sqrt{V_J/0.8 \text{ V}} = t_{epi} - x_J = 8 \ \mu$m, so $V_J = 49.2$ V, and thus the required reverse-bias voltage will be $V_R = V_J - \phi = 48.4$ V.

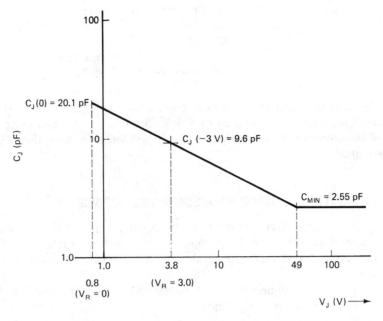

Figure 2.5 Capacitance versus voltage characteristic for VVC diode example.

When the epitaxial layer is fully depleted between the junction and the N^+ substrate, the capacitance levels off at a minimum value as given by

$$C_{\text{MIN}} = \frac{\varepsilon A}{W_{\text{MAX}}} = \frac{\varepsilon A}{t_{\text{epi}} - x_J} = \frac{1.04 \times 10^{-12}\,\text{F/cm} \times 0.196 \times 10^{-2}\,\text{cm}^2}{8 \times 10^{-4}\,\text{cm}} \tag{2.6}$$
$$= \underline{2.55\,\text{pF}}$$

The series resistance will now be due to just the substrate resistance, since the epitaxial layer is now fully depleted, and so will now be $R_{S(\text{MIN})} = R_{\text{substrate}} = 0.0765\ \Omega$. The Q value at 50 MHz will now be at its maximum value, since both the capacitance and resistance have now reached their minimum values, and will be $Q_{\text{MAX}} = 16{,}300$. This is to be compared to a Q value of about 136 for the nonepitaxial structure at the same reverse-bias voltage.

From this example, we see the clear advantage of the epitaxial structure. In Figure 2.5, a graph of the variation of capacitance with junction voltage is presented for this P^+NN^+ epitaxial diode. The capacitance ratio is of interest, and we have that $C_j(\text{O})/C_{\text{MIN}} = \underline{7.87}$ and $C_J(-3\,\text{V})/C_{\text{MIN}} = \underline{3.61}$, so quite a large capacitance variation can be obtained.

If this VVC diode is part of an LC tuned circuit, as shown in Figure 2.6, with a fixed capacitance of 2.0 pF, the frequency or tuning ratio will be given by

$$\frac{f_{\text{max}}}{f_{\text{min}}} = \frac{\sqrt{1/LC_{\text{min}}}}{\sqrt{1/LC_{\text{max}}}} = \sqrt{\frac{(20.1 + 2)\,\text{pF}}{(2.55 + 2)\,\text{pF}}} = \underline{2.20} \tag{2.7}$$

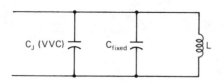

Figure 2.6 *LC* tuned circuit with VVC diode.

for the case in which the minimum reverse-bias voltage is 0 V. If the minimum reverse-bias voltage is restricted to -3 V, the tuning ratio will be reduced to 1.57. Thus an appreciable frequency swing will be available using this VVC diode in a tuned circuit.

2.3 PLANAR EPITAXIAL DIODE PROCESSING SEQUENCE

Now that we have discussed the basic processing steps for the fabrication of silicon devices, and we have seen the purpose of the epitaxial structure, we will summarize the processing sequence for a number of devices, starting with a planar P^+N/N^+ diode.

A processing sequence for a P^+N/N^+ epitaxial planar diode of the type shown in Figure 2.4b is summarized below.

1. *Starting material:* N/N^+ epitaxial wafers with a 0.005 Ω-cm (Sb-doped) substrate and an epitaxial layer of anywhere from 5 to 25 μm thick and phosphorus doped to resistivities in the range from 5 to 50 Ω-cm.
2. *Oxidation:* An oxide layer about 5000 to 8000 Å thick is grown.
3. *First photolithography:* Window openings in the oxide layer for the P^+ diffusion are produced.
4. *Boron diffusion:* A P^+ diffused layer about 1 to 3 μm thick is produced to be the anode regions of the diodes.
5. *Second photolithography:* Anode contact windows are produced.
6. *Metallization:* Aluminum deposition (~ 1 μm) produces the anode contacts.
7. *Third photolithography:* Metallization patterned for anode contacts.
8. *Contact sintering or alloying:* This is a heat treatment at about 500° to 600°C for sintering or alloying the metallization film to form a good mechanical bond to the silicon and to produce a low-resistance, nonrectifying ("ohmic") contact.
9. *Back-side metallization:* A thin film of gold is evaporated onto the lapped back sides of the wafers. This is for the eutectic die (chip) bonding of the chips to gold-plated headers or substrates at temperatures in the range from 400° to 420°C, the gold–silicon eutectic temperature being 370°C.

2.4 PLANAR EPITAXIAL TRANSISTOR

We will now enumerate the processing steps for a representative NPN planar epitaxial transistor of the type shown in Figure 2.7, where a cross-sectional view of the transistor is presented together with a graph showing the impurity profiles.

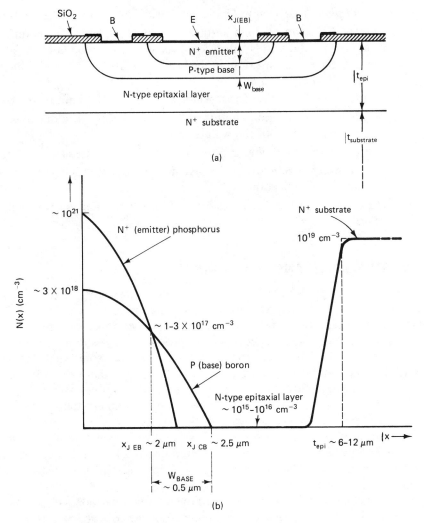

Figure 2.7 NPN double-diffused planar epitaxial transistor: (a) cross-sectional view; (b) impurity profiles.

1. *Starting material:* N/N$^+$ epitaxial wafer with 0.005 Sb-doped substrate and N-type epitaxial layer of about 6 to 12 μm thickness and 0.3 to 3 Ω-cm resistivity.

2. *Oxidation:* An oxide layer of about 5000 to 8000 Å thickness is grown.

3. *First photolithography:* Oxide windows are etched for the base diffusion.

4. *Base diffusion:* A two-step process, a boron ion implantation, or a boron deposition diffusion, followed by a drive-in diffusion to produce a P-type diffused layer with a junction depth of about 2 to 3 μm, a surface concentration of around 3 × 10^{18} cm^{-3}, and a sheet resistance of about 200 Ω/square. The drive-in diffusion is performed in an oxidizing ambient (O$_2$) so that oxide is regrown in the windows that were produced in the preceding step.

5. *Second photolithography:* Oxide windows are etched for the emitter diffusion.

6. *Emitter diffussion:* A high surface concentration phosphorus diffusion is performed to produce an N^+ diffused layer with a junction depth of about 2 to 2.5 μm, a surface concentration in the range 1×10^{21} cm^{-3}, and a sheet resistance of about 2 to 2.5 Ω/square.

7. *Third photolithography:* Oxide windows are etched for emitter and base contacts.

8. *Metallization:* An aluminum thin film of about 0.5 to 1 μm thickness is deposited on the front surface of the wafers.

9. *Fourth photolithography:* Emitter and base contact areas are defined.

10. *Contact sintering or alloying:* This is a heat treatment at 500° to 600°C for sintering or alloying metallization.

11. *Back-side metallization:* A gold thin film is deposited on the back side of the wafers.

2.5 CHIP SEPARATION AND CHIP BONDING

After the wafer processing sequence for diodes, transistors, or ICs is completed, the wafers must be divided into individual dice or chips. This can be done by a "scribe-and-break" operation using a diamond-tipped scribe, a high-intensity laser beam (laser scribing), or a high-speed circular saw to produce grooves in the silicon. In the case of the diamond scribe, the grooves are very shallow, somewhat deeper with laser scribing, and may extend more than halfway through the wafer with the saw. The wafers will have a pattern of orthogonally oriented "scribing streets," which are kept clear of oxide and metal and are aligned along certain crystallographic directions to promote easy and smooth cleavage of the wafer.

The most popular process for chip separation is to use a wafer saw to cut entirely through the wafer. The wafer is mounted on an adhesive-coated tape prior to the sawing operation so that after sawing the chips will remain in matrix form for convenience in further operations.

The chips or dice are then bonded to either metal headers or ceramic substrates. The metal headers are usually gold-plated Kovar. Kovar is an iron–nickel–cobalt alloy whose thermal expansion coefficient is a close match to that of silicon. The headers are heated to temperatures in the range from 400° to 420°C in an inert-gas atmosphere (N_2 or a mixture of about 90% N_2 and 10% H_2). The chips are then bonded to the headers by means of the formation of a gold–silicon alloy, which results in a good mechanical bond and a low-resistance electrical contact. This contact will be to the collector of the transistors, the cathodes of the P^+N/N^+ diodes, and the substrates of ICs.

2.5.1 Lead Bonding and Encapsulation

Small-diameter ($\sim$20 to 40 μm or 0.8 to 1.6 mils) gold wires are now bonded from the package leads or terminal posts to the metallized contact areas or bonding pads on the chips. Aluminum bonding wire is also sometimes used, especially for high-current-power devices, where larger-diameter round or flat ribbon leads may be used.

The device is now encapsulated in a metal, ceramic, or plastic package. The plastic package is the lowest in cost, but the metal and ceramic packages offer the advantage of providing a hermetic seal and a higher operating temperature range.

2.6 JFET PROCESSING SEQUENCE

An N-channel JFET is shown in Figure 2.8a. The N-type channel of this device is formed by the N-type epitaxial layer region between the P$^+$ diffused layer (gate) and the P-type substrate. The processing sequence for this device closely follows that for the double-diffused transistor and is summarized below.

1. *Starting material:* N/P epitaxial wafers
2. *Oxidation*
3. *First photolithography:* Windows for P$^+$ boron top gate diffusion
4. *Boron diffusion:* P$^+$ top gate diffusion

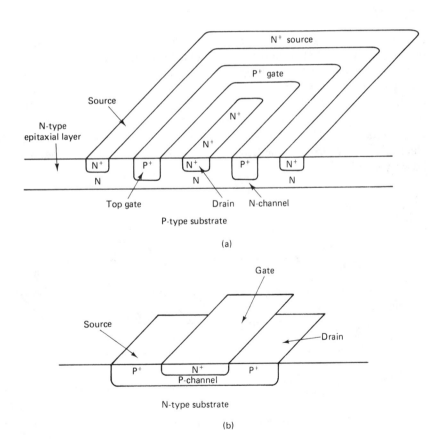

Figure 2.8 Junction-field-effect transistor structures: (a) N-channel epitaxial JFET; (b) double-diffused P-channel JFET.

5. *Second photolithography:* Windows for N$^+$ source and drain diffusion
6. *Phosphorus diffusion:* N$^+$ diffusion to produce the source and drain regions of the JFET
7. *Third photolithography:* Contact windows
8. *Metallization*
9. *Fourth photolithography:* Metallization patterning for source, drain, and gate contact areas
10. *Contact sintering or alloying*
11. *Back-side metallization*

In Figure 2.8b, a P-channel double-diffused JFET is shown. The P-type channel of this device is formed by the P-type diffused region between the N$^+$ diffused layer and the N-type substrate. This device is structurally similar to a double-diffused NPN transistor, with the P-type channel corresponding to the P-type base region of the bipolar transistor, and the processing sequence is basically the same.

2.6.1 MESFET

In Figure 2.9, a diagram of a gallium arsenide (GaAs) MESFT (metal-semiconductor field-effect transistor) is shown. This device operates in essentially the same way as does a junction-gate FET, except that instead of a gate–channel PN junction there is a gate–channel Schottky barrier. The depletion region associated with this Schottky barrier controls the effective height of the conducting channel and can thereby control the drain-to-source current of the device. The width of this depletion region increases with increasing gate voltage so that we see again that the gate will be the control electrode, and as long as the Schottky barrier is reverse biased, the gate current will be very small.

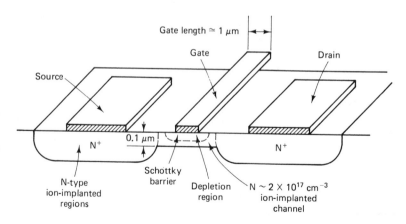

Figure 2.9 Gallium arsenide MESFET structure.

N-type GaAs offers the advantage of an electron mobility that is much higher than that of silicon (8500 cm^2/V-s for GaAs versus 1400 cm^2/V-s for lightly doped silicon). This higher electron mobility together with a very short channel length ($\sim$1 μm) results in very short channel transit times for electrons and therefore leads to very high speed devices that can operate well in the gigahertz range.

2.7 MOSFET PROCESSING SEQUENCE

A cross-sectional view of a simple N-channel aluminum gate MOSFET is shown in Figure 2.10a. The processing steps for this device are listed below.

1. *Starting material:* P-type silicon, $\sim$10 Ω-cm resistivity.
2. *Oxidation:* Thermally grown oxide of about 10,000 Å thickness.
3. *First photolithography:* Oxide windows for source and drain diffusion.
4. *Phosphorus diffusion:* N$^+$ diffusion to produce the source and drain regions.
5. *Second photolithography:* Oxide removed from channel region between source and drain.
6. *Oxidation:* Growth of a very thin oxide layer over the channel region. This gate oxide will generally be in the range from 300 to 800 Å (Figure 2.11) and is grown under very carefully controlled conditions to minimize contamination of the oxide by such impurities as various alkali (Na$^+$, K$^+$, etc.) ions.
7. *Third photolithography:* Contact windows.
8. *Metallization:* Aluminum thin film.
9. *Fourth photolithography:* Metallization patterning to produce the gate electrode and the source and drain contact areas.
10. *Contact sintering*
11. *Back-side metallization*

2.7.1 Self-aligned Gate MOSFETs: Overlap Capacitance

For a MOSFET to be turned on, a conducting channel in the form of a surface inversion layer must be produced over the entire distance between the source and drain areas. Therefore, the gate electrode must extend all the way between the source and the drain. To allow for possible mask registration errors, the gate is designed to overlap the edges of the source and drain regions by a small amount, often in the range of about 5 μm. This condition will result in a small *overlap capacitance* between the gate and the source (C_{gs}) and the source and the drain (C_{gd}). These capacitances will generally be on the order of 1 to 3 pF. The gate-to-drain capacitance is of particular interest since it represents a feedback capacitance from output (drain) to input (gate), and its effect on the input capacitance of the MOSFET is increased by the Miller effect.

In Figure 2.12, a MOSFET structure that features a self-aligned gate is shown. The processing steps for this device are similar to those of a conventional P-channel

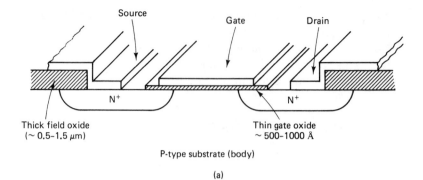

Figure 2.10 MOSFET: (a) cross-sectional perspective view: (b) top view.

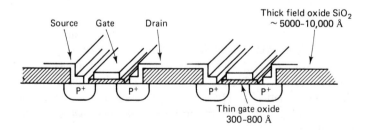

Figure 2.11 Self-isolating characteristics of MOS transistors.

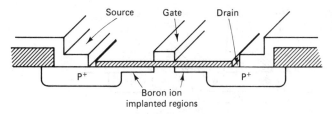

Figure 2.12 Self-aligned gate MOSFET using ion implantation.

MOSFET, except that the diffused source and drain regions do not extend all the way to the gate. A boron ion implantation is used to produce P-type extensions of the source and drain regions right up to the edge of the gate. The boron ions, with about 100 kV of energy, are able to penetrate the thin gate oxide, but are blocked by the much thicker gate and by the thick field oxide. The gate electrode thus serves as an implantation mask such that the source and drain regions effectively terminate right under the edges of the gate so that the overlap capacitance is thereby minimized. The annealing of the ion implantation takes place at a relatively low temperature in the range from 400° to 500°C such that the lateral diffusion of the implanted boron ions underneath the gate is negligible.

Ion implantation is also useful in MOSFETs for adjusting the threshold voltage V_T. For this purpose, a very low dosage ($\sim 10^{11}$ to 10^{12} cm^{-2}), low energy (~ 30 kV) implantation is used to modify the doping of the silicon surface in the channel region. This technique is especially useful for N-channel MOSFETs, where the threshold voltage would otherwise be too low ($\lesssim 1$ V); this is true for many applications, including most digital circuits. A low-dosage boron implantation can be used here to increase the effective doping of the P-type substrate surface region and thus raise the threshold voltage to an acceptable level.

Another self-aligned gate structure is shown in Figure 2.13. This device uses a gate of polycrystalline silicon (polysilicon) or a refractory metal (Mo, Ta, or W) or metal silicide (MoSi$_2$, TaSi$_2$, WSi$_2$). These gate materials can withstand the high temperatures used in the diffusion process, so the gate can now serve as a diffusion mask. There will be some lateral diffusion underneath the gate, but the overlap capacitance will still be considerably smaller than in the conventional MOSFET structure.

The polysilicon gate also has the advantage of producing a small, favorable shift in the threshold voltages, reducing the P-channel threshold voltage and increasing the threshold voltage of the N-channel devices; this gate is very widely used.

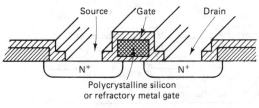

Polycrystalline silicon
or refractory metal gate

P-type silicon substrate

Figure 2.13 Polysilicon or refractory metal gate MOSFET.

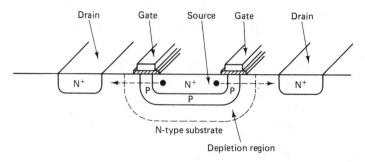

Figure 2.14 Diffused channel MOS (DMOS) transistor.

2.7.2 Short-channel MOSFETs

A short channel length is advantageous in a MOSFET since it results in a higher transfer conductance. The higher transfer conductance, in turn, leads to a larger voltage gain and gain–bandwidth product. Also, the drain-to-source current I_{DS} at any given gate voltage will be larger, so the current-handling capability of the device is increased. Indeed, current ratings of up to 10 A are available with some VMOS and vertical DMOS devices.

In Figure 2.14 a diagram of a double-diffused MOSFET (DMOS) is shown. The channel length underneath the gate oxide is the lateral distance between the N^+P junction and the PN substrate junction and is thus controlled by the junction depths produced by the N^+- and P-type diffusions. This is similar to the situation with respect to the base width of a double-diffused bipolar transistor, and the channel length can be made to be as small as 0.5 μm.

The application of a suitably large positive voltage to the gate ($V_{GS} > V_t$) will invert the P-type surface region underneath the gate to N-type, and the resulting N-type surface inversion region will serve as a conducting channel for the flow of electrons from source to drain.

The lightly doped N-type substrate and the room available for the expansion of the depletion region between the P-type diffused region and the N^+ drain contact region will make possible a relatively high breakdown voltage between drain and source (BV_{DS}).

A vertical DMOS structure is shown in Figure 2.15. In this case the drain

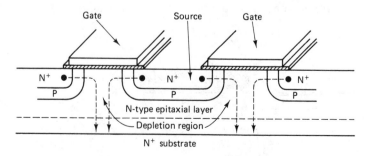

Figure 2.15 Vertical PMOS (V-DMOS) transistor.

contact region is the N$^+$ substrate. The removal of the N$^+$ drain contact regions from the top surface allows for more parallel-connected channels to be formed so that the transfer conductances and the drain-to-source current capability of the device can be correspondingly increased. With a large, high-density array of very many gate electrodes on the top surface, vertical DMOS devices with current ratings of up to 10 A are possible.

Figure 2.16 shows a VMOS transistor. This is again a double-diffused device in which the channel length is set by the difference between the N$^+$- and P-type diffusions. The lightly doped N-type epitaxial layer and the room allowed for the expansion of the depletion region between the P$^+$ diffused layer and the N$^+$ substrate leads to a high breakdown voltage ($BV_{DS} \gtrsim 50$ V) and a low drain capacitance. At the same time, the drain series resistance is kept to a small value as a result of the heavily doped N$^+$ substrate.

The V-grooves are produced by an anisotropic or orientation-dependent etching (ODE) process. The etchant that is used, such as KOH at 80° to 100°C, attacks silicon very rapidly in the [100] crystallographic direction, but very slowly in the [111] direction. For (100)-oriented silicon substrates, the result is the production of V-shaped grooves that have (111) sidewalls, as shown in Figure 2.17b. The angle of the (111) groove sidewalls with respect to the (100) silicon surface will be 54.74°. The width of the grooves, W, is controlled by the width of the opening in the oxide layer, which is used as an etching mask, since SiO$_2$ is attacked only very slowly by the etching solution. Figure 2.18 shows the relationships between the principal crystallographic planes. Note that with a (110) silicon substrate, the etching of vertical-walled slots is possible. The orientation-dependent etching process is useful for a number of silicon devices in addition to the VMOS transistors. VMOS transistors with large arrays of very many V-groove gate structures are available with current ratings of up to several amperes.

Vertical DMOS transistors consisting of a very large number of parallel connected cells in a rectangular or hexagonal pattern on a common N/N$^+$ drain region

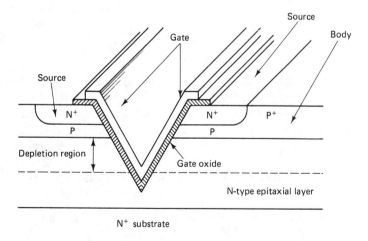

Figure 2.16 VMOS transistor.

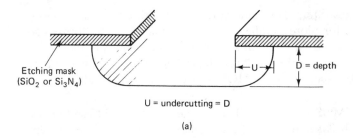

Etching mask
(SiO$_2$ or Si$_3$N$_4$)

U = undercutting = D

$\leftarrow$U$\rightarrow$ D = depth

(a)

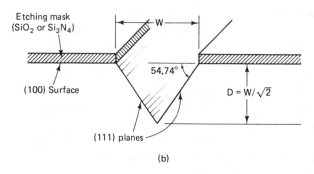

Etching mask
(SiO$_2$ or Si$_3$N$_4$)

W

54.74°

(100) Surface

(111) planes

D = W/$\sqrt{2}$

(b)

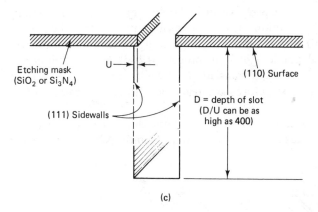

Etching mask
(SiO$_2$ or Si$_3$N$_4$)

U

(110) Surface

(111) Sidewalls

D = depth of slot
(D/U can be as
high as 400)

(c)

Figure 2.17 Isotropic and anisotropic etching of silicon: (a) isotropic etching (example: HF:HNO$_3$: acetic acid mixture); (b) anisotropic or orientation-dependent etching on (100) silicon wafer; (c) anisotropic or orientation-dependent etching on (111) silicon wafer.

are available with continuous current ratings in excess of 25 A at voltages of up to 500 V, which gives a power-handling capability of 12.5 kW. The very short channel length and very large total channel width that can be on the order of 1 million times greater than the channel length can also result in very small values for the drain-to-source resistance $r_{ds(ON)}$, with values as low as 0.12 Ω being achieved.

2.7.3 Examples of Ion Implantation for MOS Threshold Voltage Adjustment

It was mentioned previously that ion implantation can be used for the adjustment of the threshold voltage V_t for MOSFETs. As a result of positive-ion contamination in the gate oxide, surface charges at the oxide–silicon interface, and the gate electrode/

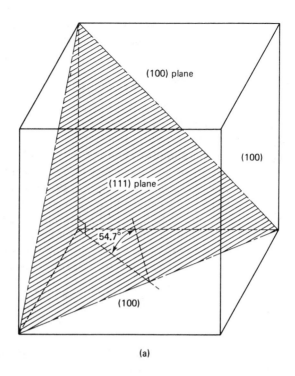

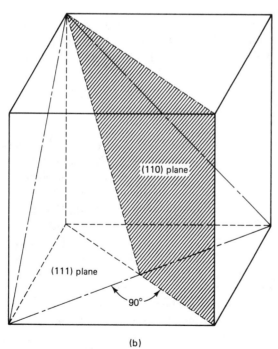

Figure 2.18 Principal crystallographic planes: (a) relationship between the (100) and (111) planes; (b) perpendicular set of (110) planes and (111) planes.

silicon contact potential, the threshold voltage for NMOS transistors will often be very low ($\lesssim 1$ V). In some cases the NMOS threshold voltage can actually be slightly negative, so there will be an N-type inversion layer present even in the absence of any applied gate voltage. At the same time, the gate voltage for PMOS devices will be relatively large (~ -5 V) in the negative direction due to these same effects.

A boron ion implantation can be used to shift the threshold voltage in the positive direction for both NMOS and PMOS devices. The boron ions implanted in the silicon form a shallow surface layer of negatively charged acceptor ions. The charge per unit area due to these implanted boron ions is $\Delta Q = q \times$ implantation dosage, where q is the electron charge. This surface charge density results in a shift in the threshold voltage of $\Delta Q = C_{oxide} \Delta V_t$, where C_{oxide} is the capacitance per unit area of the MOS gate capacitance. This capacitance is given by $C_{oxide} = \varepsilon_{oxide}/t_{oxide}$, where $\varepsilon_{oxide} = \varepsilon_r \times \varepsilon_0 = 3.8\varepsilon_0 = 3.8 \times 8.85 \times 10^{-14}$ F/cm $= 3.363 \times 10^{-13}$ F/cm.

Let us now consider the example of a PMOS transistor with a threshold voltage of -5.0 V and a 1000-Å gate oxide. The boron ion implantation dosage needed to shift the threshold voltage down to -2.5 V is given by

$$
\begin{aligned}
\Delta Q = q \times \text{dosage} &= C_{oxide} \Delta V_t = \frac{3.36 \times 10^{-13} \text{ F/cm}}{0.1 \times 10^{-4} \text{ cm}} \times 2.5 \text{ V} \\
&= 8.408 \times 10^{-8} C
\end{aligned}
\tag{2.8}
$$

so the required dosage will be 5.25×10^{11} boron ions/cm^2. If the ion implantation system has a beam current of 1.0 μA, the time required for the implantation of a 100-mm-diameter wafer is only 8.4 s.

For a second example, let us consider an NMOS transistor with a threshold voltage of only $+0.5$ V and a 1000-Å gate oxide. The boron ion dosage required to shift the threshold voltage from $+0.5$ to $+1.5$ V is 2.1×10^{11} boron ions/cm^2. The implantation time for a 100-mm wafer and a 1-μA beam current is only 3.4 s.

2.8 IC DEVICE ISOLATION

In monolithic ICs, many transistors and other devices share the same small single-crystal silicon chip, so there must be some means of providing electrical isolation between devices wherever needed. The most commonly used means of providing this isolation is called *junction isolation*, illustrated in Figure 2.19. Starting with a P-type substrate, an N-type epitaxial layer of about 10 μm thickness is deposited, as shown in Figure 2.19a. This is followed by a thermal oxidation process and photolithography to produce the line openings in the oxide, as shown in Figure 2.19b. A high-concentration deep boron P$^+$ diffusion is then performed. The boron diffusion proceeds long enough to penetrate all the way through the epitaxial layer, down into the P-type substrate, as shown in Figure 2.19c and d.

The N-type epitaxial layer has now been subdivided into separate N-type regions. Between every N-type and every other N-type region, there are two PN junctions. As long as all of these PN junctions are reverse biased, there will be electrical isolation between the various N-type regions and between the N-type regions and the P-type substrate. The P-type substrate is usually connected to the most negative voltage in

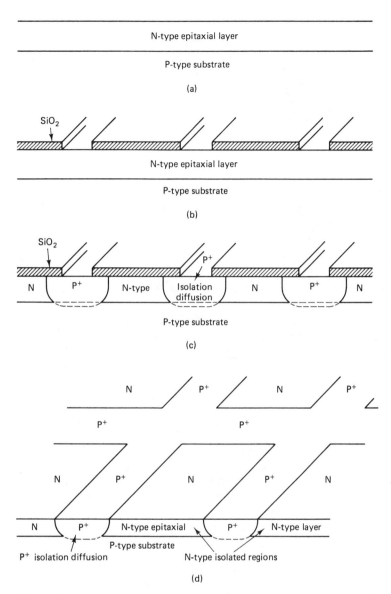

Figure 2.19 PN junction isolation for integrated circuits: (a) P-type substrate with N-type epitaxial layer; (b) oxide growth and patterning of oxide to produce openings for isolation diffusion; (c) P+ boron isolation diffused to produce isolated N-type regions; (d) N-type isolated regions.

the circuit, which is either the the negative supply voltage or the circuit ground in the case of single-supply operation. This substrate connection will automatically ensure that all the N-region to P-substrate junctions will be reverse biased, and there will therefore be electrical isolation between the N-type regions.

This electrical isolation is not perfect, however, and there will be a very small

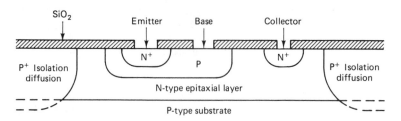

Figure 2.20 IC NPN transistor: cross-sectional view.

amount of leakage current between the N-type isolated regions and the P-type substrate that will be on the order of 1 nA/mm² at 25°C. Most IC transistors will have areas in the range from 0.001 to 0.01 mm², so the corresponding leakage current will be in the range from only 1 to 10 pA and will be negligible for most applications.

The principal isolation problem is the presence of the capacitance of the N-type isolated region/P-type substrate junction. This capacitance is a function of the doping levels and the bias voltage, but is generally in the range of 300 pF/mm² (0.19 pF/mil²) at zero bias, decreasing to 120 pF/mm² (0.08 pF/mil²) at 5-V reverse bias and about 95 pF/mm² (0.06 pF/mil²) at 10-V bias. For a typical bipolar transistor 100 μm × 100 μm = 0.01 mm² in area, the parasitic N-type region (collector) to substrate capacitance C_{CS} is therefore about 1 pF at bias voltages in the range from 5 to 15 V.

A cross-sectional view of an IC NPN transistor is shown in Figure 2.20. Note that the contacts to all three regions—emitter, base, and collector—of the transistor are on the front or top surface, unlike the case of an individual or discrete transistor, in which the collector contact is on the back side of the chip. As a result of the topside collector contact, there is a large collector series resistance $r_{cc'}$ present due to the high sheet resistance of the N-type collector region (about 1000 to 10,000 Ω/ square). This resistance can, however, be reduced very substantially by the use of an *N⁺ buried layer* or *subcollector diffusion*, as shown in Figure 2.21. Since the collector–base junction is still bounded by the more lightly doped N-type epitaxial region, the collector–base junction capacitance C_{CB} can be kept acceptably low, and the collector–base breakdown voltage can be maintained at a high value (~50 V). The basic processing sequence for monolithic ICs using this type of NPN transistor is given in Table 2.2.

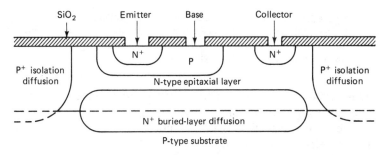

Figure 2.21 IC NPN transistor with buried layer.

TABLE 2.2 SILICON MONOLITHIC IC PROCESSING SEQUENCE

1. *Starting material:* semiconductor grade ($\lesssim 1$ ppba) polycrystalline silicon
2. *Crystal growth:* Czochralski (or float-zone) process to produce P-type single-crystal silicon ingots about 100 to 150 mm in diameter and boron doped to a resistivity in the range from 1 to 10 Ω-cm
3. *Mechanical preparation of silicon wafers:* ingot slicing, wafer lapping, and polishing to produce wafers about 0.5 to 1 mm (20 to 40 mils) thick
4. *Oxidation:* thermally grown SiO_2 layer about 5000 Å thick
5. *First photolithography:* windows etched in oxide for buried layer diffusion
6. *Buried layer diffusion:* diffusion of high-concentration Sb-doped N^+ layer about 3 μm deep in oxidizing ambient followed by oxide removal
7. *Epitaxial layer deposition:* deposition of phosphorus-doped N-type epitaxial layer of 0.1- to 1-Ω-cm resistivity and 5 to 15 μm thick
8. *Oxidation:* growth of thermal oxide about 5000 to 10,000 Å thick
9. *Second photolithography:* oxide windows etched for the P^+ isolation diffusion
10. *Isolation diffusion:* P^+ boron diffusion all the way through the epitaxial layer down to the P-type substrate, with the boron drive-in performed in an oxidizing ambient
11. *Third photolithography:* oxide window opening for base region (and resistor) diffusion
12. *Base and diffused resistor diffusion:* an ion implantation or deposition diffusion followed by a drive-in diffusion to produce a junction depth of about 2 to 3 μm and a sheet resistance of 200 Ω/square, with the drive-in done in an oxidizing ambient
13. *Fourth photolithography:* windows etched in oxide for emitter and collector contact N^+ diffusion
14. *Emitter and collector contact diffusion:* a phosphorus N^+ diffusion to a junction depth of about 2 μm and a sheet resistance of about 2.2 Ω/square, with the drive-in done in an oxidizing ambient
15. *Fifth photolithography:* contact windows etched in oxide
16. *Contact and interconnection metallization:* an aluminum thin film of about 0.5 to 1.0 μm deposited on the front surface of the IC wafers
17. *Sixth photolithography:* etching of metallization pattern
18. *Contact sintering or alloying:* heat treatment at 500° to 600°C to produce good mechanical bond and low-resistance electrical contact
19. *Back-side metallization:* gold thin film deposited on wafer backs
20. *Wafer probing:* testing of ICs with an automatic testing system that inks bad ICs for later identification
21. *Die separation:* wafer sawed or broken up into individual IC chips or dice
22. *Die mounting:* IC chips bonded to metallic headers, lead frames, or ceramic substrates using a eutectic alloy, solder preform, or epoxy
23. *Lead bonding:* small-diameter wire leads, usually ~1-mil gold wire, bonded from the package leads to the bonding pads on the IC chip
24. *Encapsulation:* IC sealed in a metal can (TO-5), ceramic, or plastic (DIP or dual-in-line) package
25. *Testing:* complete testing of ICs to determine if they meet the device specifications; for high-reliability and military-specification devices, an extensive testing program, including operation under full power condition at various temperature extremes for an extended period of time (*burn-in*) and temperature cycling, may be carried out.

The purpose of the N^+ collector contact regions that are formed at the same time as the N^+ emitter regions is to ensure a low-resistance, nonrectifying (ohmic) contact between the aluminum metallization and the collector region. In the absence of these N^+ regions, the alloying of the aluminum contacts results in a shallow region of silicon underneath the aluminum contact being saturated with aluminum to a concentration of around 10^{19} cm^{-3}. Since aluminum acts as an acceptor dopant in silicon, this region will be converted to P-type if the donor doping in this region is

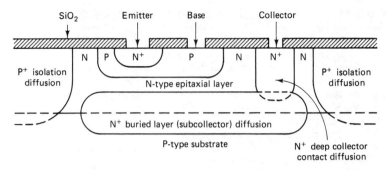

Figure 2.22 IC NPN transistor with buried layer and deep collector contact diffusion.

less than the aluminum concentration. A parasitic PN junction can thus be formed and will appear in series with the collector of the transistor.

To prevent the formation of this P-type region, the N-type donor doping in the collector region underneath the collector contact metallization should be well in excess of 10^{19} cm^{-3}. This can easily be done by producing an N$^+$ diffused layer in the collector contact regions by opening oxide windows at the same time that the emitter diffusion windows are opened. The resulting emitter diffusion that takes place will result in a surface concentration in the 10^{21} cm^{-3} range of phosphorus so that the contact regions will not be converted to P type.

For an even lower collector series resistance $r_{cc'}$, the transistor structure of Figure 2.22 can be used, in which there is a deep collector contact diffusion that extends all the way through the epitaxial layer to intersect the N$^+$ buried layer. This will produce a major reduction in $r_{cc'}$, but at the expense of requiring an additional photolithography and N$^+$ diffusion step.

2.9 TRANSISTOR LAYOUT AND AREA REQUIREMENTS

In Figure 2.23, a cross-sectional view and a top view of a single-base-stripe IC NPN transistor is shown. From the top view we see that the total chip area taken up by the transistor is very much greater than the active area of the transistor, which is the area of the N$^+$ emitter region, for it is in this area that the electron flow from emitter to collector takes place across the thin base region. In particular, the oxide opening for the P$^+$ isolation diffusion and the lateral spread of the diffusion underneath the edges of the oxide opening takes up an appreciable part of the total transistor area. If the epitaxial layer thickness is designated as t_{epi}, this lateral diffusion is approximately equal to t_{epi} if the P$^+$ isolation diffusion is to be deep enough to intersect the P-type substrate. To make absolutely certain that the P$^+$ isolation diffusion will indeed pass all the way down through the N-type epitaxial layer down to the P-type substrate in spite of variations in epitaxial layer thickness and diffusion conditions, the isolation diffusion is extended for a somewhat longer period of time. The resulting lateral diffusion under the oxide mask is nt_{epi}, where $n > 1$ and typically around 1.5.

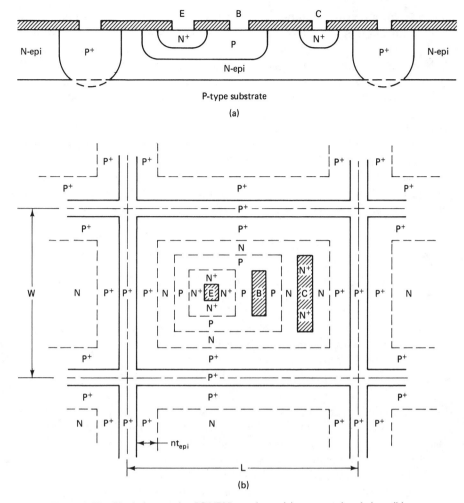

Figure 2.23 Single-base-stripe IC NPN transistor: (a) cross-sectional view; (b) top view.

A minimum clearance requirement is needed between the various diffused regions, contact openings, and metallized contacts to allow for mask registration errors and the minimum line resolution that is available with the photolithographic process. If this minimum clearance or *design rule* is specified as d for all the above, and if $t_{epi} = d$ and $n = 1.5$, for the transistor of Figure 2.23 we will have the overall dimensions $L = 15d$ and $W = 11d$. The transistor area is $A = LW = 165d^2$. Of this total transistor area, the active area or N^+ emitter region is only $9d^2$, which represents only 5.5% of the total transistor area.

If a 10-μm design rule is used, the transistor area is $A = LW = 150$ μm $\times$ 110 μm $= 0.0165$ mm^2 = <u>26 mil^2</u>, corresponding to a device density of <u>60 transistors/ mm^2</u>. If the photolithography resolution and mask alignment are such that a 5-μm design rule can be used, the transistor area can be reduced to about <u>0.004 mm^2</u> =

6.4 mil^2, for a density of about 250 transistors/mm^2. Thus a small IC chip measuring only 1 mm × 1 mm (40 mils × 40 mils) can accommodate a large number of transistors.

The base spreading resistance $r_{bb'}$ can be reduced by about a factor of 4 by using a double-base-stripe geometry in which there is a base contact strip on both sides of the emitter. This will, however, be at the expense of increasing dimension L by $2d$, so the transistor area will now be $A = 17d \times 11d = 187d^2$. For a 10-μm design rule, this gives an area of $\underline{0.019 \text{ mm}^2} = \underline{29 \text{ mil}^2}$, for a device density of $\underline{53 \text{ transistors/}}$ $\underline{\text{mm}^2}$.

The lowest base and collector series resistance is obtained by using a ring-type contact geometry for both the base and collector. This type of contact geometry increases L by $4d$ and W by $8d$, so $A = LW = 19d \times 19d = 361d^2$. For a 10-μm design rule, this gives an area of $\underline{0.036 \text{ mm}^3} = \underline{56 \text{ mil}^2}$, for a device density of $\underline{28}$ transistors/mm^2.

2.10 PNP TRANSISTORS

The favored type of bipolar transistor for ICs is the NPN transistor, for two basic reasons.

1. The electron mobility in silicon is about 2.5 times higher than the hole mobility, so the base transit time will generally be shorter in NPN transistors than in PNP transistors. This will lead to a somewhat higher current gain and to improved high-frequency performance for the NPN transistors.

2. The solid-state solubility of the donor dopants, phosphorus and arsenic, is substantially larger than that of the acceptor dopant, boron. It is desirable to make the emitter region much more heavily doped than the base region so that, when the emitter–base junction is forward biased, most of the current flow across the junction will be due to charge carriers emitted by the emitter into the base rather than the flow of charge carrier from base to emitter. It is only the charge carriers that are emitted by the emitter into the base that can contribute to the collector current. The opposite flow of carriers from base to emitter only adds to the base current. As a result of the higher solubility of the donor dopants compared to boron, a more efficient emitter–base structure can be obtained in the case of an NPN transistor than in the PNP case, thus leading to a higher current gain. As a result, the structure of most bipolar ICs is based on the use of NPN transistors.

For many applications, however, PNP transistors are needed at various places in the circuit. In Figure 2.24a, an IC PNP transistor that is compatible with the NPN transistor fabrication sequence is shown. No extra processing steps are needed to produce this PNP transistor than those needed for the NPN transistor. The transistor of Figure 2.24a is called a *vertical* or *substrate PNP* transistor. The collector of the transistor is the P-type substrate, and the current flow of holes emitted by the P$^+$ emitter is in the downward or vertical direction through the N-type epitaxial base region to the P-type collector.

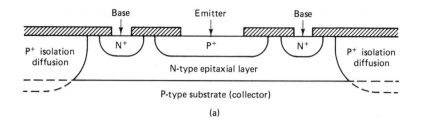

(a)

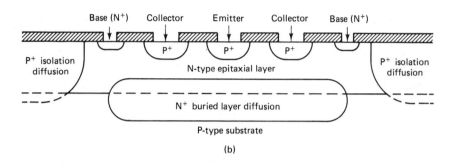

(b)

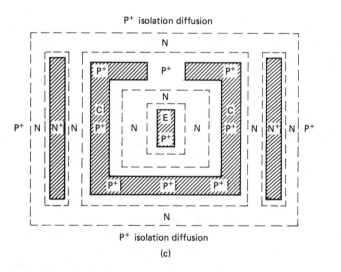

(c)

Figure 2.24 IC PNP transistors: (a) vertical (substrate) PNP transistor; (b) lateral PNP transistor (cross-sectional view); (c) lateral PNP transistor (top view).

The base width of the vertical PNP transistor is the epitaxial layer thickness between the P$^+$ diffused layer and the P-type substrate and is generally about 5 μm. This is to be compared to the base width of the NPN transistor of only about 0.5 μm. The large base width of the PNP transistor results in a long emitter-to-collector transit time for the holes traveling across the base region, which in turn produces a low value for the current gain, β (~5 to 30), and a poor high-frequency response with f_T values in the range from 10 to 30 MHz, compared to about 500 MHz for the NPN transistor.

TABLE 2.3 TYPICAL IC BIPOLAR TRANSISTOR PARAMETER VALUES

Transistor type	Current gain, β	f_T (MHz)	BV_{EBO} (V)	BV_{CBO} (V)
NPN	~50–200	~500	~6–8	~50
Vertical PNP	~5–30	~10–30	~50	~50
Lateral PNP	~5–20	~1–10	~50	~50

The collector of this transistor is the P-type substrate, which is connected to the negative supply voltage or to ground, so that in either case the substrate will be ac ground. This condition will restrict the operation of this PNP transistor to only the common-collector (or emitter-follower) configuration. In spite of this rather severe limitation, this transistor structure does find application in some ICs.

In Figure 2.24b, another PNP transistor structure is shown. This transistor is called a *lateral PNP* transistor and it is also compatible with the NPN transistor fabrication sequence. This PNP transistor is called a lateral PNP transistor because the direction of flow of the holes from the P$^+$ emitter to the P$^+$ collector region is parallel to the surface. The base width of this device is the spacing between the edges of the P$^+$ emitter and collector regions. Because of the limitations of the photolithographic process, this base width will be limited to a minimum of about 3 to 5 μm. This relatively large base width will again result in a long emitter-to-collector hole transit time across the base region. In addition, some of the holes emitted by the P$^+$ emitter will leave from the bottom area of the emitter and be lost by recombination with electrons in the N-type base region and will therefore not contribute to the collector current. Furthermore, many of the holes traveling across the base region near the surface will be lost to surface recombination and not reach the collector. All these facts will lead to a low current gain ($\beta \sim 5$ to 20) as compared to β values in the range from 50 to 200 for NPN transistors. The large base width will also be responsible for poor high-frequency response, with f_T values in the range from 1 to 10 MHz. The lateral PNP, however, unlike the vertical PNP transistor, can be used in any circuit configuration. Both types of PNP transistors suffer from high-level injection problems due to the low base doping concentration. This results in a severely limited current-carrying capacity and a rapid decrease of β at higher current levels.

Table 2.3 summarizes the characteristics of IC NPN and PNP transistors. It is possible to produce IC PNP transistors with characteristics that are comparable to those of NPN transistors. These higher-performance PNP transistors will, however, require extra processing steps beyond those required for NPN transistors, and so add to cost of the IC.

2.11 IC JUNCTION-FIELD-EFFECT TRANSISTORS

Figure 2.25 shows some IC JFET structures. The N-channel JFET structure of Figure 2.25a is compatible with the NPN transistor fabrication sequence. Another view of this N-channel JFET is shown in Figure 2.26a, where it is to be noted that the top

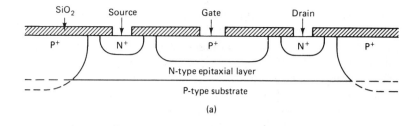

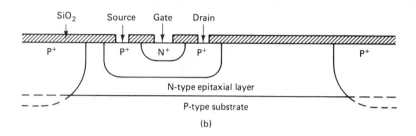

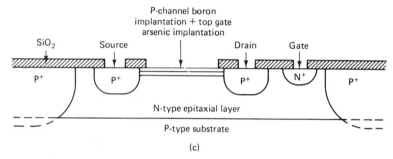

Figure 2.25 Integrated circuit JFET structures: (a) N-channel JFET; (b) P-channel JFET; (c) ion-implanted P-channel JFET.

P$^+$ gate region extends beyond the N-type epitaxial layer region to make contact with the P-type substrate bottom gate. The N-type channel is thus completely encircled by the gate structure, and the application of a suitably large negative voltage to the gate can pinch the channel off and reduce the drain-to-source current to essentially zero. If the P$^+$ top gate did not extend out to overlap the P-type substrate, the N-type channel would not be completely encircled by the gate structure and it would not be possible to cut off the drain-to-source current. The major disadvantage, however, of the JFET structure of Figure 2.26a is that the gate is connected to the P-type substrate, which is at ac ground potential. This can severely restrict the use of this JFET to only the common-gate configuration.

Another N-channel JFET is shown in Figure 2.26b. The fabrication of this JFET is similar to the one just considered, the principal difference being in the top surface geometry. In this JFET the P$^+$ top gate is in the form of an annular ring that completely encloses the drain region of the JFET. The only current path source to drain will be underneath the P$^+$ top gate. Therefore, the application of a suitably large negative voltage to the top gate can pinch off the channel and reduce the drain-to-source current to essentially zero.

A P-channel JFET is shown in Figure 2.25b. The N^+ gate region of this device extends out beyond the P^+ source/drain/channel region to overlap the N-type epitaxial layer so that the gate completely encircles the channel. The same processing sequence can be used for this JFET as for the NPN transistor, but if this is done, the gate-to-channel breakdown voltage (corresponding to BV_{EBO}) will be down in the range from 6 to 8 V, and the full pinch-off of the channel may not be possible. As a result, a specially tailored low-concentration boron diffusion will be required to produce a higher gate–channel breakdown voltage and lower channel doping so that the channel can be pinched off at a voltage that is conveniently less than the breakdown voltage. This, however, involves some extra processing steps and adds to the cost of the IC.

Figure 2.25c shows a P-channel JFET that features a boron-ion-implanted channel region. Since the ion-implantation dosage can be very precisely controlled, the JFET parameters, such as the pinch-off voltage V_P and the value of I_{DSS} (I_{DS} at $V_{GS} = 0$), can be closely set to the values desired. This JFET uses the same processing

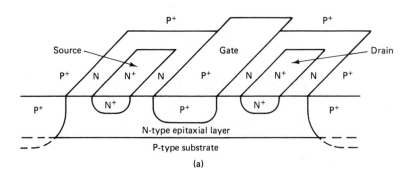

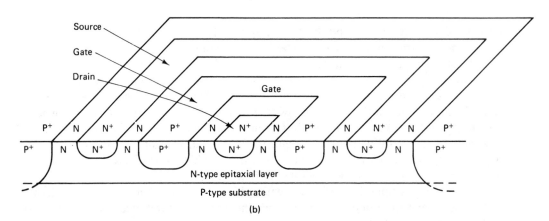

Figure 2.26 IC N-channel JFET geometries: (a) common top and bottom gate structure; (b) annular ring gate structure.

steps as the NPN transistor, with the addition of photolithography, boron ion implantation, arsenic ion implantation, and annealing steps. The boron ion implantation produces the P-type channel region that extends about 0.7 μm down from the surface. This is followed by a very shallow N^+ arsenic ion implantation for the top gate. This produces a gate–channel PN junction at a distance of 0.2 μm from the surface. The resulting channel height is only about 0.5 μm, and the pinch-off voltage is around 1 V.

2.12 IC MOSFETS

In Figure 2.27, some P-channel MOSFETs sharing a common N-type silicon substrate are shown to illustrate the self-isolating characteristics of these devices. The oxide underneath the gate electrode is only about 500 to 1000 Å thick, compared to the field oxide, which is about 1.0 to 1.5 μm thick. The application to the MOSFET gate of a negative voltage that is larger than the threshold voltage V_T produces a P-type surface inversion layer underneath the gate oxide. This P-type inversion layer serves as a conducting channel between the P^+ source and drain regions. The voltage required to invert the N-type silicon under the thick field oxide is very much greater than the gate threshold voltage V_t, which is generally in the range from -2 to -10 V. This is a direct result of the greater thickness of the field oxide compared to the very thin gate oxide. The voltage required to invert the N-type silicon underneath the field oxide is in excess of maximum negative voltage in the circuit, so the metallization on top of the field oxide is not able to produce an inversion layer. Thus no P-type conducting channels will form between adjacent MOSFETs. Therefore, the MOSFETs are self-isolating and no special isolation diffusions or isolated regions are necessary. As a result of this and the simple geometry of the MOSFETs, the area required for a MOSFET on the IC chip will be much smaller than that for a bipolar transistor.

Figure 2.28 shows a cross-sectional view and a top view of a minimum-area P-channel MOSFET (PMOS). If a design rule dimension of d is assumed for all clearances and spacings, the overall PMOS transistor dimensions are given approximately by $L = 7.5d$ and $W = 3d$ for an area of $A = LW = 22.5d^2$ for each transistor.

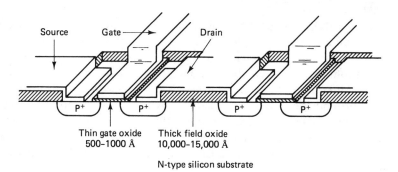

Figure 2.27 Self-isolating characteristic of MOSFETs.

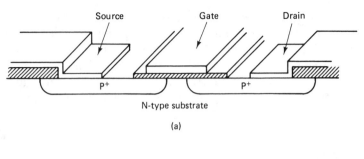

(a)

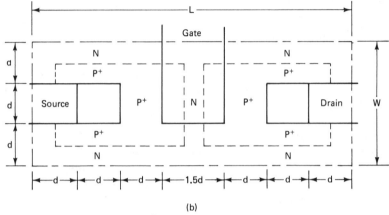

(b)

Figure 2.28 (a) P-channel IC MOSFET; (b) top surface layout of P-channel MOSFET.

This is to be compared to the area of $165d^2$ for the minimum size (single base strip–single collector stripe) bipolar transistor, so we see that a much higher transistor density will be available with MOSFETs.

For a 10-μm design rule, the PMOS transistor area will be $A = 0.0023 \text{ mm}^2 = 3.5 \text{ mil}^2$, for a density of 444 transistors/mm². If a 5-μm design rule can be used, the transistor area can be reduced to about $0.0006 \text{ mm}^2 = 0.9 \text{ mil}^2$, for a device density of 1800 transistors/mm². A 1 cm × cm IC chip therefore can contain as many as 180,000 transistors.

With PMOS devices there will usually be no problem with device isolation, but there can be a problem with NMOS transistors, especially on the more lightly doped substrates. Positive ions (such as Na^+ and K^+) trapped in the oxide can act to reduce the threshold voltage such that conducting channels may be formed between the various NMOS transistors. To prevent this, P^+ guard rings or channel stoppers may be used around the NMOS transistors. Another technique is to use a boron ion implantation to increase the net acceptor doping level in the surface region of the P-type substrate between the NMOS transistors so that the surface is prevented from becoming inverted. Some guard ring structures for IC NMOS transistors are shown in Figure 2.29.

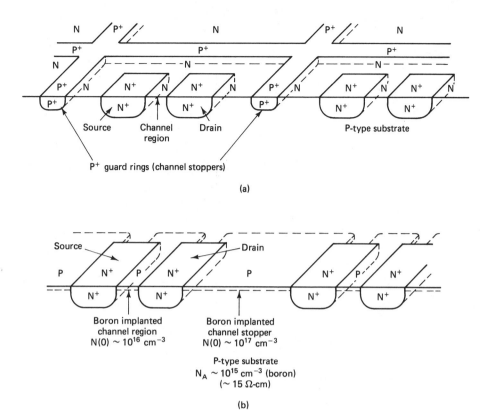

Figure 2.29 Guard ring structures for NMOS transistors: (a) NMOS structure using diffused P⁺ channel stoppers; (b) NMOS structure using a boron ion implanted channel stopper.

2.12.1 Depletion-mode MOSFETs–FET Transfer Characteristics

The type of MOSFET discussed up to this point is the enhancement-mode MOSFET in which the conducting channel between source and drain is produced by the action of the electric field that results from the application of a gate voltage. If the gate voltage is greater than the threshold voltage V_t, then a surface inversion layer will be formed at the silicon surface adjacent to the silicon–gate oxide interface. This surface inversion layer will be of opposite conductivity type to that of the silicon substrate (or body) and constitutes a conducting channel between the source and drain regions.

In contrast to the enhancement-mode MOSFET, a *depletion-mode* MOSFET has a built-in channel between source and drain. This built-in channel is not a field-induced channel, but rather is formed by producing a shallow, lightly doped region at the surface, connecting the source and drain, and is of the same doping type as the source and drain regions. This channel can conveniently be formed by means of a shallow, low-dosage ion implantation. The depletion-mode MOSFET (or D-MOS-

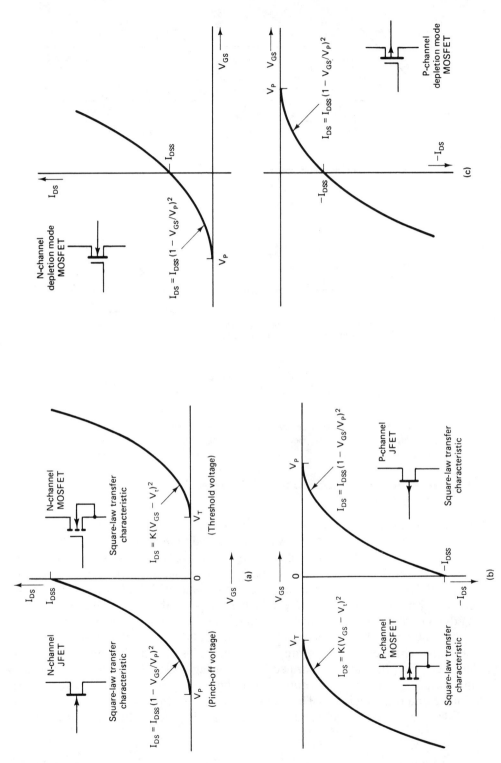

Figure 2.30 Field-effect transistor transfer characteristics: (a) N-channel FETs; (b) P-channel FETs; (c) depletion-mode MOSFETs.

FET) combines some of the features of JFETS and enhancement-mode MOSFETs. Taking an N-channel D-MOSFET as an example, the application of a gate-to-source voltage of positive polarity will draw additional electrons into the already existing N-type channel. This increases the conductance of the channel and thereby increases the drain-to-source current. Application of a negative voltage produces the opposite effect. Electrons will be repelled from the channel and the channel will become thinner. This decreases the conductance of the channel and decreases the drain current. The gate voltage at which the channel becomes fully depleted of electrons is called the *pinch-off voltage* (V_P or $V_{GS(OFF)}$), just as in the case of the JFETs. Depletion-mode MOSFETs can be used as active loads and as current-regulator diodes in various MOSFET circuits.

Figure 2.30 presents graphs of the I_{DS} versus V_{GS} transfer characteristics of the various types of FETs. Note that the D-MOSFETs have a transfer characteristic very similar to that of JFETs, except that both polarities of V_{GS} are permitted and, as a result, currents that are in excess of I_{DSS} can be obtained. Note also the symbols used for the enhancement- and depletion-mode MOSFETs. The solid line between the source and drain regions in the depletion-mode MOSFET is symbolic of the preexisting channel, whereas the dashed line for the enhancement-mode devices indicates a channel that is produced only by means of the application of a gate voltage in excess of the threshold voltage.

From Figure 2.30, we note that both the JFETs and the MOSFETs exhibit a square-law transfer characteristic. An exception to this general rule, however, is that of the short-channel ($L \lesssim 1$ μm) MOSFETs for which an approximately linear transfer characteristic is obtained.

2.13 COMPLEMENTARY-SYMMETRY MOSFETS

Figure 2.31 shows a complementary-symmetry pair of MOSFETs (CMOS). The CMOS pair consists of an NMOS transistor and a PMOS transistor, both sharing the same N-type silicon substrate, together with other CMOS transistor pairs.

The NMOS transistor is made by first producing a deep P-type boron-doped diffused layer with a low surface concentration called the *P-well*. The source and

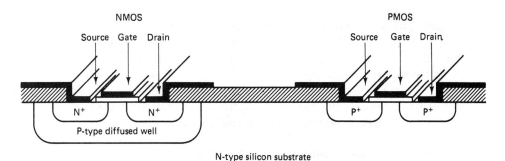

Figure 2.31 Complementary-symmetry MOSFETs (CMOS).

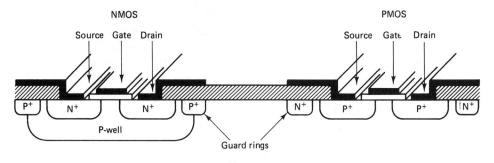

Figure 2.32 CMOS with guard rings.

drain regions are then formed by a N$^+$ phosphorus diffusion. The PMOS transistor is formed by the conventional processing sequence.

Figure 2.32 shows a CMOS structure in which there are P$^+$ and N$^+$ guard rings surrounding the NMOS and PMOS devices, respectively. This is to prevent a *latch-up* condition that could otherwise occur as a result of a four-layer N$^+$–P–N–P$^+$ switching action.

Figure 2.33 is a diagram of the basic CMOS transistor circuit, together with the individual I_{DS} versus V_{DS} output characteristics for the NMOS and PMOS devices. In Figure 2.33c, the characteristics of the NMOS and PMOS devices are superimposed with a family of curves for various values of the input voltage. In Figure 2.33d, the corresponding CMOS transfer characteristic is given.

When the input voltage V_{IN} is "low" such that the NMOS transistor is off, the PMOS device will be on and the output voltage will be pulled up to the "high" state near the positive supply voltage V_{DD}. When V_{IN} goes "high" the NMOS is now turned on and the PMOS goes off, so the output voltage will now drop down to the low state near ground potential.

In either of the foregoing conditions, one of the transistors of the CMOS pair is off, so the current through the CMOS pair is negligible. If the load driven by this CMOS circuit is CMOS devices or MOSFETs, the only significant power consumption will occur during the switching transitions. This very low power dissipation of CMOS devices is a principal advantage of this transistor configuration, especially for large digital systems. Another favorable feature of the CMOS circuit is the large logic voltage swing, with the high-state output voltage being very close to $+V_{DD}$ and the low-state output voltage dropping to very close to either ground potential or the negative supply voltage $-V_{SS}$.

Figure 2.34 shows a basic CMOS inverter, and the corresponding transfer characteristic is given in Figure 2.35. Figure 2.36 shows a CMOS NAND gate, and a CMOS NOR gate is presented in Figure 2.37.

2.13.1 CMOS Silicon-on-Sapphire (CMOS/SOS)

Figure 2.38 shows a heteroepitaxial CMOS structure. A thin ($\sim$1 μm) single-crystal silicon epitaxial layer is deposited on a highly polished single-crystal sapphire sub-

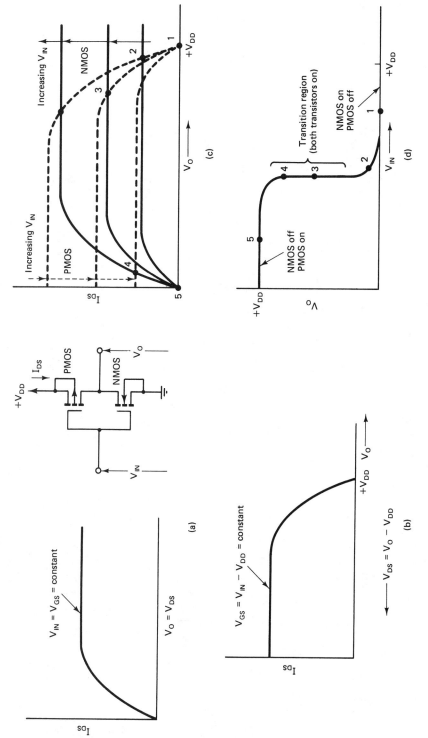

Figure 2.33 CMOS I_{DS} versus V_{DS} characteristics: (a) NMOS I_{DS} versus V_{DS}; (b) PMOS I_{DS} versus V_{DS}; (c) composite NMOS and PMOS characteristics; (d) CMOS transfer characteristics.

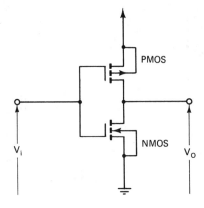

Figure 2.34 CMOS (complementary MOS) inverter.

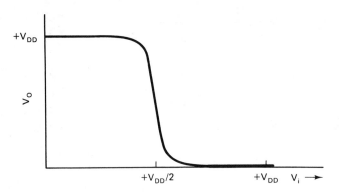

Figure 2.35 CMOS transfer curve.

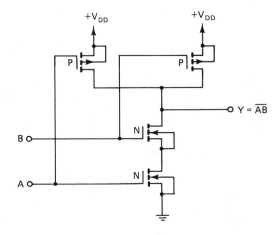

Figure 2.36 CMOS NAND gate.

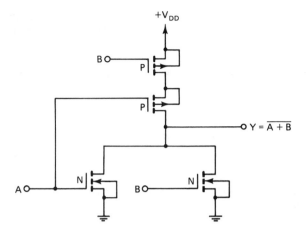

Figure 2.37 CMOS NOR gate.

strate. The silicon film is doped N^+ and P^+ by phosphorus and boron ion implantations, followed by an annealing or drive-in diffusion step, as required. The thin silicon film is then etched into many separate NMOS and PMOS devices and interconnected by the metallization pattern.

This CMOS/SOS structure is an example of a dielectrically isolated IC, and the use of the insulating substrate results in a greatly reduced parasitic capacitance. This leads to very high speed device performance and very low power dissipation. These features are especially useful for very high speed, high-density ICs.

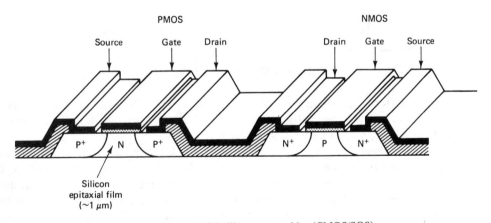

Figure 2.38 CMOS silicon-on-sapphire (CMOS/SOS).

2.14 IC DIODES

A number of IC diode structures are compatible with the NPN transistors; that is, no additional processing steps are required beyond those for the NPN transistor. Six basic diode configurations are shown in Figure 2.39, together with some of the basic characteristics such as series resistance, forward voltage drop, breakdown voltage, and storage time. The base spreading resistance $r_{bb'}$ will generally be about 100 Ω

Diode configuration		Series resistance	V_F at 10 mA (mV)	Breakdown voltage	Storage time (ns)
1.	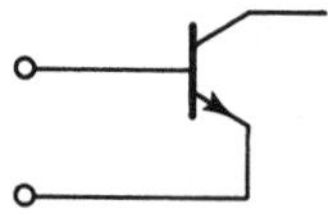 $I_C = 0$	$r_{bb'}$	~960	BV_{EBO}	~70
2.	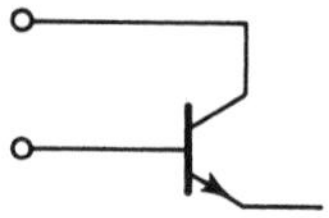 $I_E = 0$	$r_{bb'} + r_{cc'}$	~950	BV_{CBO}	~130
3.	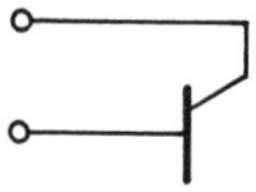 $I_E = 0$ (no emitter N$^+$ diffusion)	$r_{bb'} + r_{cc'}$	~950	BV_{CBO}	~80
4.	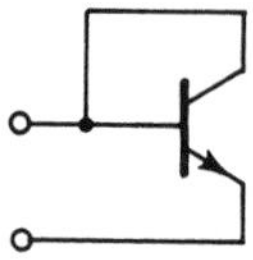 $V_{CB} = 0$	$r_{bb'}/\beta$	~850	BV_{EBO}	~6
5.	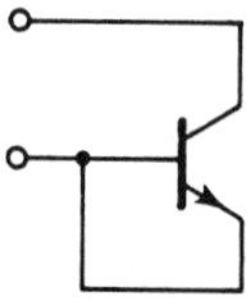 $V_{EB} = 0$	$r_{bb'}/\beta_R + r_{cc'}$	~940	BV_{CBO}	~90
6.	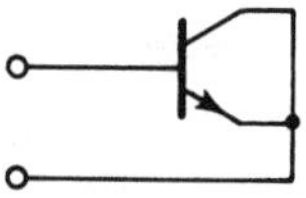 $V_{CE} = 0$	$r_{bb'}$	~920	BV_{EBO}	~150

Figure 2.39 Diodes for integrated circuits.

and the collector resistance $r_{cc'}$ will be about 30 Ω. The emitter–base breakdown voltage BV_{EBO} will be in the range 6 to 8 V, and the collector–base breakdown voltage BV_{CBO} will usually be about 50 V.

From Figure 2.39 we see that the $V_{CB} = 0$ diode configuration 4 will offer the lowest series resistance and forward voltage drop and also the lowest storage time. The small value of minority carrier storage time will mean a fast switching transition from forward-bias conditions to reverse-bias conditions. The breakdown voltage is, however, limited to BV_{EBO} or only 6 to 8 V. Thus for all applications in which the diode will not be subjected to reverse-bias voltages in excess of about 6 V, this will be the preferred diode configuration.

In this diode circuit ($V_{CB} = 0$), most of the diode forward current I_F is the collector current; the base current I_B is smaller than the collector current by a factor of β, the transistor current gain. As a result, the diode forward voltage drop due to the base spreading resistance $r_{bb'}$ is given by

$$I_B r_{bb'} = (I_C/\beta)r_{bb'} \cong I_F(r_{bb'}/\beta)$$

Since the collector current remains relatively independent of the collector voltage as

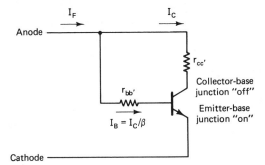

Figure 2.40 IC diode ($V_{CB} = 0$) series resistance.

long as the transistor operates in the active mode of operation, the voltage drop due to the collector series resistance $I_C r_{cc'}$ will not affect the diode voltage drop. Therefore, the effective series resistance of the diode is approximately $r_{bb'}/\beta$, which usually is only a few ohms. The I_B and I_C components of the total diode forward-bias current flow I_F are illustrated in Figure 2.40, in which the resistances $r_{bb'}$ and $r_{cc'}$ are shown as elements external to the transistor, although they are really inside the transistor structure.

2.14.1 Storage Time

A major component of the switching speed of diodes is the *reverse recovery time*, which is limited by the storage time of the minority carriers. When a PN junction is forward biased, electrons will be injected from the N-type side of the junction into the P-type side, where they become minority carriers, and similarly holes will be emitted by the P-type side into the N-type side, where they also become minority carriers. As the injected minority carriers travel deeper into each side, the population of these injected minority carriers steadily drops off with increasing distance as a result of recombination with the majority carriers. The average lifetime of these injected minority carriers ranges from much less than 1 μs for heavily doped (N^+ or P^+) silicon to about 10 μs for light to moderately doped silicon.

If the bias voltage across the PN junction is now suddenly switched from forward bias to reverse bias, many of the injected minority carriers can turn around and start flowing back to the opposite side of the junction, from which they first were injected. Thus a large reverse current can flow through the diode for a period of time called the *storage time*, which will often constitute the major portion of the reverse recovery time of the diode. The excess minority carrier density that is responsible for this large reverse current flow dissipates after a short period of time due to the withdrawal of carriers back into their region of origin and due also to the recombination of the minority carriers with the majority carriers.

The short storage time of the $V_{CB} = 0$ diode configuration is due to the fact that when the emitter–base junction is forward biased most of the current flow across the emitter–base junction is the result of electrons emitted from the N^+ emitter into the P-type base region, rather than vice versa. The electrons emitted into the base will then very rapidly travel across the very short base region and be collected by the collector. The transit time of the electrons across the base region is very short

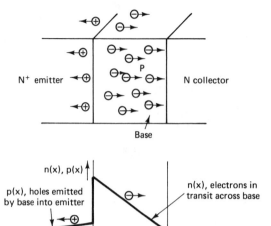

Figure 2.41 Minority carrier storage for transistor in active mode of operation.

($\lesssim$ 100 ps), so that there will not be many electrons at any one time in transit across the base region. It is because of this very small density of minority carriers (electrons) in the base region that this diode configuration has a very short storage time ($\sim$6 ns). This is to be compared to the storage time of some 70 ns for the $I_C = 0$ diode configuration. In this case, there is no net flow of electrons out of the collector; consequently, a buildup of minority carrier (electrons) density in the base region occurs. Figure 2.41 presents a diagram of the minority carrier distribution in the transistor operating in the active mode, as is the case for the $V_{CB} = 0$ diode configuration.

2.14.2 Schottky-barrier Diodes

A Schottky-barrier diode is shown in Figure 2.42a. In this type of diode, when the metal–semiconductor contact (Schottky barrier) is forward biased, there is only the injection of electrons from the N-type semiconductor into the metal. For every electron injected from the semiconductor into the metal, there is a corresponding electron flowing out of the metal, around the external circuit, and back into the semiconductor. There will therefore be no minority carrier injection or storage, and the storage time for the Schottky-barrier diodes will be essentially zero, although a small space charge capacitance is still present. These diodes will thus exhibit a very fast switching action.

In addition to the fast switching characteristics of Schottky-barrier diodes, the forward voltage drop is about one-half of that of the PN junction diodes, typically about 250 to 350 mV for currents in the range from 1 to 10 mA, as shown in Figure 2.42b.

In the Schottky-barrier diode of Figure 2.43, there is a large localized intensification of the electric field strength at the edges of the oxide window due to the sharp curvature of the metal contact in this region. This can lead to a decreased breakdown voltage. The Schottky-barrier diode of Figure 2.44 has a P$^+$ guard ring

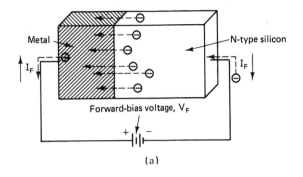

(a)

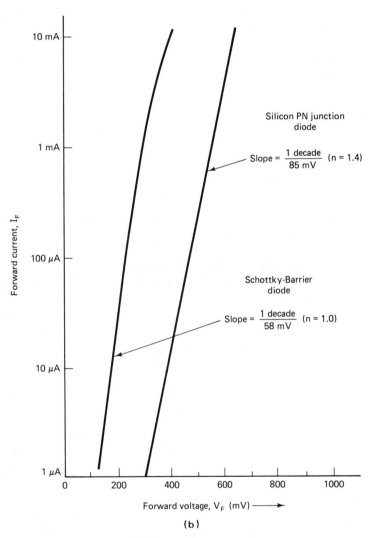

(b)

Figure 2.42 (a) Forward-biased Schottky barrier diode; (b) comparison of Schottky barrier and PN junction silicon diode characteristics.

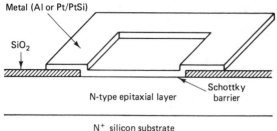

Metal (Al or Pt/PtSi)

SiO₂

N-type epitaxial layer

Schottky barrier

N⁺ silicon substrate

Figure 2.43 Schottky barrier diode.

around the edge of the oxide window.   The purpose of this guard ring is to spread out and decrease the electric field intensity in the region of the oxide window edge and thereby increase the breakdown voltage and reduce the junction leakage current. An IC Schottky-barrier diode is shown in Figure 2.45.

Schottky-barrier diodes are often placed across the collector–base junction of transistors as a clamping diode to increase the transistor switching speed in going from

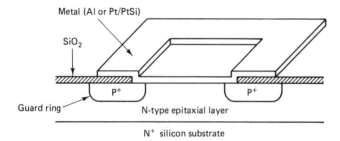

Metal (Al or Pt/PtSi)

SiO₂

Guard ring

P⁺   P⁺

N-type epitaxial layer

N⁺ silicon substrate

Figure 2.44 Schottky barrier diode with guard ring.

the saturation region to the active region.   Since the voltage required to turn on a Schottky-barrier diode is only around 0.25 V, compared to 0.5 V for the collector–base PN junction, the collector–base junction is prevented from turning on when the transistor–Schottky-barrier clamping diode combination goes into saturation.   As a result, there is no injection of electrons (for the NPN case) from the collector back into the base, and therefore the base is not inundated with a great increase in the

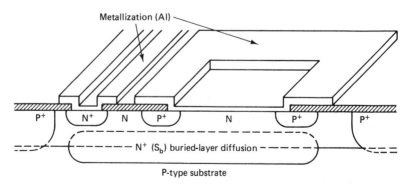

Metallization (Al)

P⁺   N⁺   N   P⁺   N   P⁺   P⁺

N⁺ (S_b) buried-layer diffusion

P-type substrate

Figure 2.45 Integrated-circuit Schottky barrier diode.

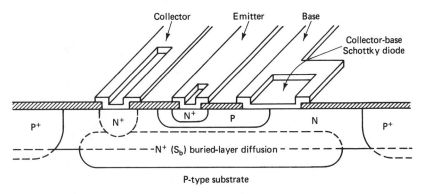

Figure 2.46 Schottky diode clamped IC transistor.

minority carrier (electrons) density as would otherwise be the case. Consequently, the transistor storage time will be kept small and it will be possible to bring the transistor rapidly out of the saturation region.

The Schottky-barrier diode is usually incorporated as an integral part of the transistor structure, as is illustrated in Figure 2.46. Schottky diode clamped transistors are used in many switching-type circuits, including voltage comparators and TTL logic gates, such as the 54LS, 74LS, 54S, and 74S series of TTL gates.

2.14.3 Zener Diodes

There are three PN junctions associated with the IC NPN transistor structure: the emitter–base, collector–base, and collector–substrate junctions. The collector–base and collector–substrate junctions will have breakdown voltages (V_Z) of around 50 V. Since this is in excess of the voltages available in the circuit, these two PN junctions cannot be used as a zener or voltage regulator diode in ICs. The emitter–base junction, however, will have a breakdown voltage in the range 6 to 8 V, and it therefore can be used as a zener diode, as shown by Figure 2.47.

The *temperature coefficient of the zener voltage* or TCV_Z, given by $TCV_Z = dV_Z/dT$, increases with increasing values of V_Z and will be in the range of +2 mV/°C to +4 mV/°C for the emitter–base V_Z values in the 6- to 8-V range. The temperature coefficient of the forward voltage drop across the emitter–base junction, $TCV_{BE} = dV_{BE}/dT$, will be negative and approximately −2.2 mV/°C. The negative temperature coefficient of the forward-biased emitter–base junction can therefore be used to compensate partially for the positive temperature coefficient of the zener

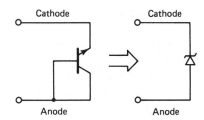

Figure 2.47 IC zener diode.

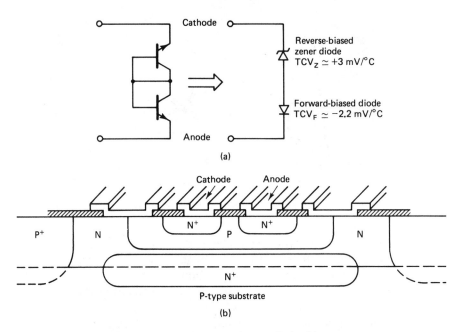

Figure 2.48 IC temperature-compensated zener diode: (a) temperature-compensated zener diode circuit; (b) IC structure for temperature-compensated zener diode.

diode by the use of a series combination of a zener diode and a forward-biased diode, as shown in Figure 2.48a. In Figure 2.48b, the IC structure of this temperature-compensated zener diode is shown. Note that the two diodes share a common N-type collector region and a common P-type base region.

2.15 IC RESISTORS

In Figure 2.49 a number of different IC resistor configurations are shown together with some typical sheet resistance values. The most widely used resistor type is the diffused resistor, which uses the sheet resistance of the transistor base diffused layer, as shown in Figure 2.49a. The sheet resistance is typically about 200 Ω/square and the *temperature coefficient of resistance TCR*, given by $TCR = (1/R) \, dR/dT$, is about $+2000 \, \text{ppm/}°\text{C} = 2 \times 10^{-3}/°\text{C} = 0.2\%/°\text{C}$. In Figures 2.50 and 2.51, the top surface layout of a diffused resistor is shown.

The N-type epitaxial layer isolated region is normally tied to the positive supply voltage, and the P-type substrate is connected to the negative supply or ground, so that not only will the N-layer/P-substrate junction be reverse biased, but also the P-resistor/N-layer junction. These two reverse-biased junctions eliminate the possibility of the parasitic PNP (base–collector–substrate) transistor being turned on and also minimize the parasitic junction capacitance of the collector–base and the collector–substrate junctions. In addition, many resistors can share the common N-type isolated region as shown in Figure 2.50b.

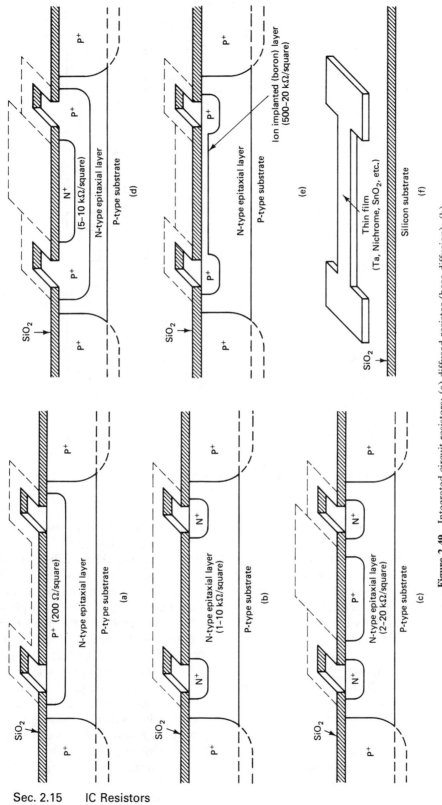

Figure 2.49 Integrated circuit resistors: (a) diffused resistor (base diffusion); (b) epitaxial layer resistor; (c) pinched epitaxial layer resistor; (d) pinched base diffusion resistor; (e) ion-implanted resistor; (f) thin-film resistor.

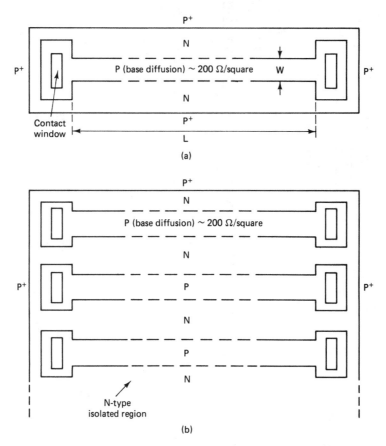

Figure 2.50 IC diffused resistor top surface geometry: (a) small-value resistor; (b) several resistors sharing the same N-type isolated region.

The resistance value of a resistor can be expressed in terms of the sheet resistance R_S as $R = R_S(L/W)$, where L is the total length of the resistor and W is the width. The L/W ratio is, in effect, the number of squares in series. There will be statistical deviations in the resistor value from the mean or design center value due in part to variations in the sheet resistance from wafer to wafer, and even over the surface of any given wafer. For resistors of widths less than about 25 μm, however, the principal cause of resistor value deviations will be statistical fluctuations in the line width W. These width fluctuations are the result of variations in the amount of oxide undercutting, the extent of the lateral diffusion underneath the oxide, and other variations in the photolithographic process. These statistical deviations are generally in the range ±1 or 2 μm. Thus, if a resistor line width of only 5 μm is chosen, these variations in the line width can result in a resistor tolerance or standard deviation of as much as ±30%. An increase in the line width will result in a reduced standard deviation, but for widths beyond about 25 μm the standard deviation will level out at about ±5% due to the fluctuations in the sheet resistance.

The great resistor line widths, however, will be at the price of a larger chip area

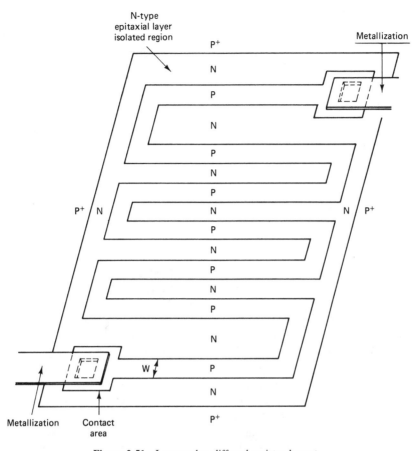

Figure 2.51 Large-value diffused resistor layout.

that is occupied by the resistor. The length of the resistor will be given by $L = (R/R_S)W$, so the chip area occupied by the resistor will be proportional to W^2. In Table 2.4 some of the important characteristics of the various IC resistors are listed.

The *ratio tolerance* of an IC resistor is the statistical deviation in the ratio of two similar-type IC resistors on the same chip from the mean or design center value. Since the resistors on the same chip that are thus being compared are within a very small distance from each other and have undergone identical processing, it is to be expected that the resistor characteristics will indeed closely match each other. Thus the ratio tolerance is usually very much smaller than the absolute value tolerance. Considering a pair of IC P-type diffused resistors of equal nominal or design center value, the 1:1 ratio tolerance for wide-line-width (≥ 25 μm) resistors is typically in the range of $\pm 0.5\%$. If the line width is decreased to 7 μm, the ratio tolerance will increase to about $\pm 2\%$. Also, as the resistance ratio increases, the ratio tolerance will increase, from 0.5% for a 1:1 ratio to 1.5% for a 5:1 ratio for wide-line-width resistors. For very low value resistors ($\leq 10\ \Omega$) and for diffused crossovers, as shown in Figure 2.57, the N^+ emitter diffusion with a sheet resistance of about 2.2 Ω/square can be used.

TABLE 2.4 IC RESISTOR CHARACTERISTICS

Resistor type	Sheet resistance (Ω/square)	Temperature coefficient, TCR (ppm/C)	Tolerance	
			Absolute value (%)	Ratio (%) (1:1)
1. Diffused (base)	100–300	+1000 to +3000	±10	±1
2. Diffused (emitter)	2–3	+100	±10	±1
3. Epitaxial layer	1–10 kΩ	+3500 to +5000	±30	±5
4. "Pinched" epitaxial	2–20 kΩ	+4000	±50	±10
5. "Pinched" base diffusion	5–10 kΩ	+3000 to +5000	±40	±6
6. Ion implanted	500 Ω–20 kΩ	+200 to +1000	±6	±2
7. Thin film				
Tantalum (Ta)	200 Ω–5 kΩ	100	±5	±1
Nichrome (ni–Cr)	40–400	100	±5	±1
Tin oxide (SnO_2)	80 Ω–4 kΩ	0 to −1500	±8	±2

The epitaxial layer resistor of Figure 2.49b offers a high sheet resistance in the range from 1000 to 10,000 Ω/square, but it has a high temperature coefficient and exhibits a large standard deviation. This large standard deviation or tolerance range is the result of variations in the epitaxial layer thickness and doping level. An even higher sheet resistance is available with the "pinched" epitaxial layer resistor of Figure 2.49c. In this case the epitaxial layer resistor cross-sectional area is reduced by the intrusion of the P-type (base) diffused layer. Unfortunately, the resistance standard deviation will be even greater in this case as a result of the greater effect of variations in the epitaxial layer thickness and the additional effect of the variations in the junction depth of the P-type diffused layer.

The "pinched" P-type base diffused layer of Figure 2.49d offers the highest sheet resistance of all the diffused resistor structures, but at the expense of a very large standard deviation in the resistance value due to the combined effects of variations in the N^+ emitter and P-type base diffusion depths.

The ion-implanted resistor of Figure 2.49e is a type that is often used and offers the combination of a high sheet resistance and a relatively low standard deviation of the resistance value. This is the result of the very close control over the implanted-ion dosage that is possible. This is at the expense, however, of an extra photolithographic and ion-implantation processing step.

The thin-film resistors of the type shown in Figure 2.49f are deposited on *top* of the SiO_2 layer. Relatively high sheet resistances are available with a variety of thin-film materials, together with a low standard deviation of the resistance value. One particularly attractive feature of these nonsilicon resistors is the low temperature coefficients, only about 100 ppm/°C or 0.01%/°C. This is at the expense again of extra processing steps. As a result, these thin-film resistors are not used much on silicon monolithic IC chips, but are used extensively for thin-film hybrid integrated circuits.

An interesting feature of thin-film resistors when used with either silicon monolithic ICs or with hybrid ICs is their ability to be trimmed or adjusted in value. The resistance value can be increased by cutting a slot partway across the width of the

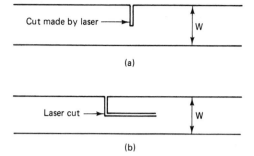

Figure 2.52 Laser trimming of thin-film resistors: (a) straight cut for small resistance increase; (b) right-angle cut for larger resistance increase.

resistor, as shown in Figure 2.52, usually by means of a high-energy laser beam. The resistance value can thus be increased to some desired value, usually by means of a closed-loop laser trimming system in which the resistance value is monitored during the trimming operation. In many IC circuits, a pair of resistors is involved, and the value of one of the resistors is increased to balance the circuit. This technique is often used for offset voltage nulling in high-precision operational amplifiers.

2.15.1 Resistor Area

Large-value IC resistors will generally take the form of a meandering pattern, as shown in Figure 2.51. The total resistor length L is given by $L = (R/R_S)W$, and each corner counts as 0.65 square. If the spacing between the P-type diffused regions is chosen to be equal to the resistor line width W, the area required by a large-value ($\geqslant 10$ kΩ) resistor will be given approximately by $A = L \times 2W = 2(R/R_S)W^2$. The resistor line width will generally be in the range from 7 to 25 μm depending on the resistance tolerance and power density requirements.

For example, a 50-kΩ diffused resistor with $R_S = 200$ Ω/square and a line width of $W = 15$ μm (0.6 mil) will require an area on the IC chip of about $A = 2(50 \text{ k}\Omega/200 \text{ }\Omega) \times (15 \text{ }\mu\text{m})^2 = \underline{0.1125 \text{ mm}^2} = \underline{174 \text{ mil}^2}$. This is to be compared to the area required by a single-base-stripe $\overline{\text{NPN}}$ transistor with a 10-μm design rule, which is 0.0165 mm^2, so the resistor occupies an area equivalent to that required by seven bipolar transistors. A MOSFET with a 10-μm design rule will need an area of only 0.00225 mm^2, so the 50-kΩ resistor occupies an area that can accommodate about 50 MOSFETs. This area requirement for resistors is one of the reasons for the vast preponderance of transistors over resistors in the design of both analog and digital ICs.

2.15.2 Parasitic Capacitance

For the various IC resistors, there will be a significant parasitic capacitance distributed along the length of the resistor that will have an important effect on the circuit behavior of the resistor. For the P-type diffused resistor (base region diffusion) and for the P-type ion-implanted resistor, there will be a distributed parasitic capacitance along the length of the resistor resulting from the capacitance of the PN junction formed between the P-type layer and the N-type epitaxial layer. Figure 2.53 shows an equivalent circuit for the resulting distributed RC structure. The distributed RC trans-

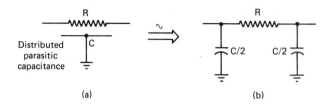

(a)

(b)

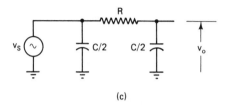

(c)

Figure 2.53 IC resistors distributed RC network: (a) distributed RC network; (b) approximate equivalent circuit; (c) approximate equivalent circuit used for evaluation of transfer function.

mission line of Figure 2.53a can be approximated by the lumped circuit of Figure 2.53b, in which the total parasitic capacitance is divided into two equal parts and concentrated at the two extreme ends of the resistor. This approximate circuit is useful for the magnitude of the transfer function, but generally gives very inaccurate results for the phase shift.

If the approximate equivalent circuit is connected as shown in Figure 2.53c, we see that it takes the form of a simple RC low-pass network, and the time constant τ will be $\tau = RC/2$, where C is the total parasitic capacitance. Since $R = R_S(L/W)$ and $C = (C/A)LW$, where C/A is the capacitance per unit area, we have that

$$\tau = \frac{RC}{2} = \frac{R_S(C/A)L^2}{2} = \frac{R_S(C/A)(R/R_S)^2 W^2}{2} = \frac{(R^2/R_S)(C/A) W^2}{2} \quad (2.9)$$

The 10% to 90% rise time is 2.2τ and the 3-dB bandwidth or breakpoint frequency is $f_1 = 1/(2\pi\tau)$.

To consider an example, let us use as representative values $R_S = 200$ Ω/square, $C/A = 100$ pF/mm² (at about 10 V bias), and a resistor width of 15 μm (0.6 mil). In Table 2.5, some values of τ, rise time, and 3-dB bandwidth are listed for various values of total resistance. We see that the parasitic capacitance can be an important factor in limiting the usefulness of large-value resistors. This is an additional reason for the emphasis on using mostly transistors and very few resistors in the design of integrated circuits. In many ICs, very few resistors are used, and in some MOSFET ICs, no resistors at all are used. We also take note of the important role of the resistor width W and the trade-offs involved between resistor area and high-frequency performance, on the one hand, and the absolute value and ratio tolerances, on the other.

2.15.3 Ion-implantation Dosage for Resistors and Base Region

It is of interest to do some calculations of the ion-implantation dosage required for various P-type resistors and for the base region of an NPN transistor. The *sheet*

TABLE 2.5 RESISTOR FREQUENCY RESPONSE
CHARACTERISTICS

$$R_S = 200 \ \Omega/\text{square}$$
$$W = 15 \ \mu\text{m} \ (0.6 \ \text{mil})$$
$$C/A = 100 \ \text{pF/mm}^2$$

Resistance value (kΩ)	Time constant		Rise time		Bandwidth	
1	113	ps	248	ps	1.41	GHz
10	11.3	ns	24.8	ns	14.1	MHz
20	45	ns	99	ns	3.5	MHz
30	101	ns	223	ns	1.6	MHz
50	281	ns	619	ns	566	kHz
70	551	ns	1.21	μs	289	kHz
100	1.13	μs	2.48	μs	141	kHz
150	2.53	μs	5.57	μs	62	kHz
200	4.50	μs	9.90	μs	35	kHz
300	10.13	μs	22.3	μs	16	kHz

conductance G_S of a P-type ion implanted layer of thickness x_J is given by

$$G_S = \frac{1}{R_S} = \int_0^{x_J} \sigma(x) \ dx = \int_0^{x_J} \mu_p qp(x) \ dx \simeq \int_0^{x_J} \mu_p qN_A(x) \ dx \qquad (2.10)$$

where $N_A(x)$ is the acceptor (boron) density in the P-type implanted layer, $p(x)$ is the corresponding hole density, μ_p is the hole mobility, and e is the electronic charge. The hole mobility μ_p is a function of the doping level, decreasing at high doping concentrations due to the scattering of holes by the ionized dopant impurities. The hole mobility starts off at around 450 cm²/V-s for light to moderately doped silicon ($N_A \lesssim 10^{16}$ cm^{-3}), and decreases to around 250 cm²/V-s at a doping concentration of 1×10^{17} cm^{-3}, and about 100 cm²/V-s at doping levels of 1×10^{18} cm^{-3}.

If we can use an appropriately weighted average value for the hole mobility μ_p, the equation for the sheet conductance can be written as

$$G_S = \overline{\mu}_p q \int_0^{x_J} N_A(x) \ dx = \overline{\mu}_p q\phi \qquad (2.11)$$

where ϕ is the ion-implantation dosage. For the sheet resistance, we therefore have that $R_S = 1/G_S = 1/(\overline{\mu}_p q\phi)$.

Let us now find the dosage required for some high-value resistors. Since the doping levels involved will be in the moderate range, we will use an average mobility value of $\overline{\mu}_p = 200$ cm²/V-s for this calculation. For an ion-implanted P-type resistor with a sheet resistance of 10 kΩ/square, we have that

$$10 \ \text{k}\Omega = \frac{1}{200 \ \text{cm}^2/\text{V-s} \times 1.6 \times 10^{-19} \ \text{C} \times \phi} \qquad (2.12)$$

so the required dosage is $\phi = 3 \times 10^{12}$ cm^{-2}. For a sheet resistance of 1 kΩ/square, the required dosage is somewhat more than $\phi = 3 \times 10^{13}$ cm^{-2} due to the decrease in the mobility at the higher doping levels.

Let us now consider the case of the base region of an NPN transistor that is produced by a low-energy boron ion implantation followed by a drive-in diffusion for a resulting sheet resistance of 200 Ω/square. As a result of the higher doping, the average mobility in the layer is now substantially lower than in the previous cases, so a choice of $\overline{\mu_p} = 100$ cm^2/V-s will be appropriate. From this we obtain an implantation dosage of $\phi = 4 \times 10^{14}$ cm^{-2}.

In Figure 2.54, curves are given of the sheet resistance versus implantation dosage for P- and N-type layers that are produced by an ion implantation followed by a drive-in diffusion/annealing step. The drive-in diffusion results in a junction depth x_J in silicon with a background doping of $N_B = 1 \times 10^{15}$ cm^{-3}. A background doping that is not 1×10^{15} cm^{-3} will generally result in only a small change in the sheet resistance. The effect of a different background doping can be approximately taken into account by changing the ion-implantation dosage that is required for a given sheet resistance by $\Delta\phi = \Delta N_B\, x_J$, where $\Delta N_B = N_B - (1 \times 10^{15}$ cm$^{-3})$ and N_B is the background doping.

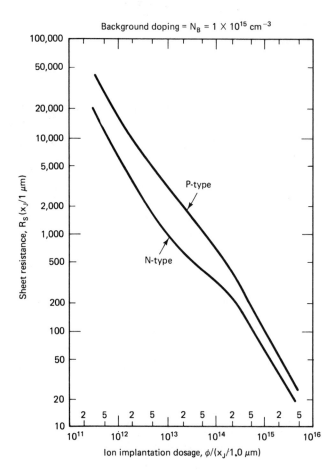

Figure 2.54 Sheet resistance versus ion-implantation dosage in silicon.

We will now compare the approximate values of the implantation dosage obtained in the preceding examples with the more exact values obtained from Figure 2.54, based on a junction depth of 2.0 μm. For $R_S = 10 \, \text{k}\Omega/\text{square}$, we obtain $\phi = 1.6 \times 10^{12}$ cm^{-2}; for $R_S = 1.0 \, \text{k}\Omega/\text{square}$, the dosage is $\phi = 4.5 \times 10^{13}$ cm^{-2}; and for $R_S = 200$ Ω/square, the required dosage is $\phi = 4 \times 10^{14}$ cm^{-2}.

2.16 IC CAPACITORS

A number of capacitor configurations are available for ICs, as shown in Figure 2.55. In Table 2.6 the approximate values for the capacitance per unit area and breakdown voltage for these capacitors are listed. The emitter–base junction capacitor, C_{BE}, offers the highest capacitance density, about 1000 pF/mm^2, but the breakdown voltage is only about 7 V, which severely limits the use of this capacitor configuration.

The *metal oxide–silicon* (MOS) capacitor offers the next highest capacitance density, about 350 pF/mm^2 for a 1000 Å $= 0.1$ μm oxide thickness. The breakdown electric field strength of SiO$_2$ is about 600 V/μm, so a 1000-Å oxide layer will have a breakdown voltage of about 60 V. As a result of the low resistance of the N$^+$ diffused layer (the emitter region of transistors), the parasitic series resistance of this capacitor is very low compared to that of the other capacitor configurations. For a typical 50-pF MOS capacitor with $L = W$ and a sheet resistance of the N$^+$ diffused layer of 2 Ω/square, the effective series resistance of the capacitor is about 1 Ω, and at a frequency of 1 MHz the Q-factor will be about 3000.

The area occupied by this 50-pF MOS capacitor is 0.15 mm^2, which is equivalent to the area needed by about 10 bipolar transistors and about 100 MOSFETs. We therefore see that large capacitances, especially above about 100 pF, are undesirable components for ICs, due to the large chip area required.

Associated with the various IC capacitors are both a parasitic series resistance and a parasitic capacitance, as shown in Figure 2.56. When the capacitors are used as bypass capacitors with one end tied to ground, the parasitic capacitance will pose no problem since it will either be shorted out or will just add to the total bypass capacitance. For other circuit applications, however, the parasitic capacitance can result in a substantial signal loss. For example, in the case of the collector–base

TABLE 2.6 IC CAPACITORS

Capacitor type	Capacitance per unit area (5-V reverse bias)		Breakdown voltage (V)
	pF/mm^2	pF/mil^2	
Collector–base junction, C_{CB}	125	0.08	50
Emitter–base junction, C_{BE}	1000	0.65	7
Collector–substrate junction, C_{CS}			
With N$^+$ buried layer	90	0.06	50
No N$^+$ buried layer	60	0.04	50
MOS capacitor (0.1 μm oxide thickness)	350	0.22	60

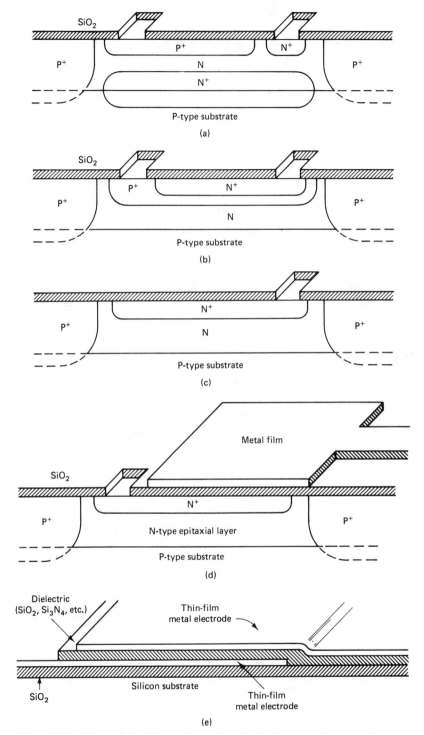

Figure 2.55 Integrated-circuit capacitors: (a) collector–base junction capacitor; (b) emitter–base junction capacitor; (c) collector–substrate capacitance; (d) metal oxide–silicon (MOS) capacitor; (e) thin-film capacitor.

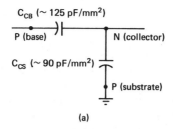

(a)

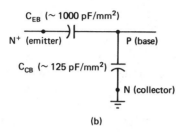

(b)

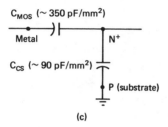

(c)

Figure 2.56 IC capacitor parasitic capacitance: (a) collector–base junction capacitor; (b) emitter–base junction capacitor; (c) MOS capacitor.

junction capacitance C_{CB} of Figure 2.56a, the parasitic collector–substrate junction capacitance is almost equal to the collector–base capacitance, so the voltage transfer ratio of this capacitor configuration is only a little better than one-half. For the MOS capacitor of Figure 2.56c, the ratio of the MOS capacitance to the collector–substrate capacitance is about 350 pF/90 pF $\simeq$ 4, so a voltage transfer ratio of about 0.8 is obtained.

A further advantage of the MOS capacitor is that its capacitance is substantially independent of the voltage across the capacitor. This is unlike the situation with the junction capacitors (C_{CB}, C_{BE}, and C_{CS}), for which the capacitance is a function of the bias voltage. Furthermore, the MOS capacitor is a *nonpolarized* capacitor that can operate with either polarity of applied voltage, whereas the junction capacitors are polarized in that the PN junction must not be allowed to become forward biased. The MOS capacitor thus offers a number of advantages over the IC junction capacitors. It does, however, require some extra processing steps. These include a photolithographic step to produce an open area in the thick passivating oxide and an extra oxidation process to grow the thin oxide layer, about 1000 Å thick, that serves as the dielectric of the MOS capacitor. Nevertheless, the MOS capacitors are generally the most widely used type of capacitor for monolithic ICs.

2.17 IC INDUCTORS

Inductors pose a special problem for ICs. IC devices are essentially two-dimensional, in that the depth dimension is usually very small (~1 to 10 μm) compared to the lateral dimensions. IC inductors can be made in the form of a flat, metallic, thin-film spiral, but the inductance will be only a few nanohenries. The combination of the very low inductance value and the appreciable series resistance of the thin-film metallization spiral will result in very low Q-factors, so this type of inductor is of very limited usefulness. For any reasonable inductance value, a three-dimensional coil structure is needed to obtain the large number of turns needed to produce a large magnetic flux density and a very large number of flux linkages.

In most cases the necessity for inductances is eliminated by the design of the IC system. In many cases, feedback circuits using RC networks can take the place of LC tuned circuits or produce a net input admittance that looks inductive. For other applications, such as RF and IF circuits, where inductors are an absolute necessity, inductors external to the IC package must be used. An exception to this is in the case of thin-film hybrid microwave integrated circuits (MICs), where thin-film inductor spirals may be of use.

2.18 IC CROSSOVERS

In the conventional IC, all the metallization for device interconnection is in the same geometric plane. In almost all ICs, there will be various places in the circuit where there must be crossovers, although the number of crossovers can be minimized by the careful design layout of the IC.

Although all the metallization is in the same plane, a diffused region in the silicon can be used as a connecting link in one of the metallization runs that would otherwise intersect. Such a crossover is shown in Figure 2.57, in which an N$^+$ layer is used to tunnel under the conductor to form a connecting link for another conductor. This N$^+$ diffused layer can be produced at the same time as the emitter diffusion and thus does not require any additional processing steps. It does, however, occupy a substantial amount of chip area and adds a significant amount of parasitic series resistance and shunt capacitance to the circuit. For an N$^+$ crossover with a sheet resistance of 2 Ω/square, the added series resistance is about 10 Ω.

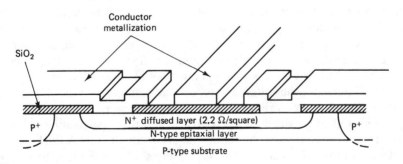

Figure 2.57 Diffused crossover.

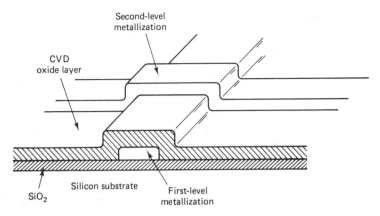

Figure 2.58 Crossover using two-level metallization.

A diffused crossover that takes up no additional chip area and adds no extra series resistance or shunt capacitance can be obtained by using a P-type resistor that is already part of the circuit. In a high-denisty IC, there is the need for a great many crossovers. The use of the N^+ type of "crossunder" is very undesirable because of the area requirements and the added series resistance, and there may not be a sufficient number of suitably located P-type resistors in the circuit to be used for crossovers. For this case, a multiple-level metallization structure may be the best solution.

A two-level metallization IC is shown in Figure 2.58. The first metallization level is produced by conventional means. Then a low-temperature ($\sim$400°C) CVD SiO_2 layer is deposited and contact windows or "vias" to the first metallization level are produced by a photolithographic process. The second metallization layer is then deposited and patterned. Some ICs have as many as three levels of metallization.

2.19 DIELECTRIC ISOLATION PROCESSES

In the dielectrically isolated IC, the various devices on the chip are electrically isolated from the substrate and from each other by an insulating or dielectric layer. This is to be compared to the junction isolated IC, in which the isolation is by means of reverse-biased PN junction.

A method for producing a dielectrically isolated IC is illustrated in Figure 2.59. Starting with an N-type substrate, an N^+ diffusion is performed. This is followed by the growth of an SiO_2 layer, which is then patterned to form a grid of intersecting line openings in the oxide. A cross-sectional view is presented in Figure 2.59a. The wafer is then subjected to an orientation-dependent etching (ODE) process using the patterned oxide layer as the etching mask. This results in the V-shaped grooves shown in Figure 2.59b, in which the (111) plane sidewalls are at an angle of 54.74° with respect to the (100) top surface of the silicon wafer. The depth of the V-groove will thus be related to the width of the oxide opening by $D = W/\sqrt{2}$.

The wafer now undergoes a thermal oxidation process to cover the sidewalls of the V-groove with an oxide layer. This is followed by the deposition of a very thick layer of polycrystalline silicon, as shown in Figure 2.59c. This CVD silicon layer will not be single-crystal because it is not deposited directly on the silicon substrate, but on the SiO_2 film, which has an amorphous structure, so a polycrystalline silicon layer results.

The next step is the most critical and difficult one. The silicon wafers are mounted on a lapping plate with the polycrystalline side of the wafers down, and the N-type silicon substrate is then carefully lapped down to the level at which the vertices of the V-grooves become exposed, producing the result shown in Figure 2.59d. This results in an array of N-type single-crystal silicon regions that are isolated from the

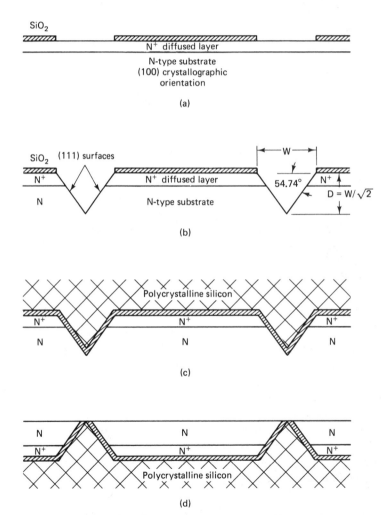

Figure 2.59 Dielectric isolation for integrated circuits: (a) N^+/N wafers with openings in the oxide layer; (b) orientation-dependent etching to produce V-shaped slots; (c) thermal oxidation followed by deposition (CVD) polycrystalline silicon layer; (d) lapping of N-type silicon substrate from backside to expose vertices of the V-grooves.

polycrystalline silicon substrate and from each other by the thermally grown oxide layer. The polycrystalline silicon now becomes the substrate and provides the mechanical support for the IC. Although the polycrystalline silicon serves no electrical function, it is a very suitable material for this application because its thermal expansion coefficient is a very good match to that of single-crystal silicon and it can withstand the high processing temperatures.

Note that it is very important that the lapping operation remove the N-type silicon all the way down to the vertices of the V-grooves. If it falls short of this, the N-type regions will be connected and device isolation will not be achieved. If, on the other hand, the lapping operation proceeds too far, the N-type regions will be too thin or possibly even removed completely. This will require very close control of the lapping process over the entire surface of the wafer, since the wafer diameter will be about 100 mm and the V-groove depth will be only about 10 μm.

The N^+ diffused layer that started off on the top surface of the N-type silicon wafer now ends up at the bottom of the N-type isolated regions. The N^+ layer serves as a buried layer to reduce the collector series resistance of the NPN transistors. The rest of the processing sequence for the dielectrically isolated ICs follows along the same line as for the conventional junction isolated IC.

The processing of dielectrically isolated ICs is much more expensive than for junction isolated ICs. The dielectrically isolated ICs are, however, very useful for such applications as high-voltage and radiation-resistance ICs. The dielectric strength of SiO_2 is about 600 V/μm, so a 5000-Å layer of oxide will result in an N-type isolated region-to-substrate breakdown voltage in the region of 300 V. In comparison, the N-region-to-substrate breakdown voltage of the junction isolated IC is only about 50 V. As a result, with dielectrically isolated ICs, high-voltage transistors and diodes are now possible.

A burst of high-energy ionizing radiation of x-rays or gamma rays can produce very large numbers of excess free electrons and holes in silicon as a result of the photogeneration process. X-rays and gamma rays consist of high-energy photons, above 100 eV for x-rays and in excess of 100 keV for gamma rays. To remove an electron from a covalent bond in silicon and thereby produce a free electron and a hole requires an energy E_G equal to 1.1 eV. Therefore, each x-ray or gamma-ray photon can result in the photogeneration of a large number of free electrons and holes. These excess photogenerated electrons and holes will result in large increases in the leakage current of the PN junctions in the IC. This can produce severe problems in junction isolated ICs, where a burst of ionizing radiation can result in large current transients or spikes. The dielectrically isolated IC is more resistant to the effects of ionizing radiation because of the presence of the insulating oxide layer between the N-type isolated regions and the polycrystalline silicon substrate.

Another type of dielectrically isolated IC is the *silicon-on-sapphire* (SOS) structure discussed earlier in conjunction with the CMOS/SOS device structure. The crystallographic mismatch that exists at and near the silicon/sapphire interface will result in severe stresses in the silicon and crystallographic defects in the thin (~1 μm) silicon layer. This will produce a very low minority carrier lifetime in the silicon film, especially in the region close to the sapphire substrate. This low minority carrier lifetime limits the usefulness of the SOS structure for bipolar devices, but this structure is still very useful for MOS devices, as in the case of the CMOS/SOS configuration. In this case, a principal feature provided by the insulating substrate is a large reduction in the parasitic capacitance, thus increasing the high-speed performance of the circuit.

2.19.1 Silicon-on-Insulator Technology

A more recently developed technique for dielectrically isolated devices is the *silicon-on-insulator* (SOI) process, in which a thin layer of single-crystal silicon can be produced on top of a thermal SiO_2 layer on a silicon wafer. The oxide layer is photolithographically patterned to produce islands or strips of oxide. A thin CVD layer of silicon is then deposited on the wafer. In the regions where the deposited silicon layer overlays the oxide, it will be polycrystalline, but it will be single-crystal in the regions where it is in direct contact with the silicon substrate. The silicon layer is then directionally recrystallized using a scanned laser, electron beam, or resistance-heated strip heater. The silicon film that is in direct contact with the substrate recrystallizes, with the silicon substrate serving as the nucleation center. As the heated zone is scanned across the wafer the crystal growth propagates from these nucleation regions to the regions of the silicon film on top of the oxide islands or strips. The result is a complete single-crystal layer of silicon.

A related SOI process is the *epitaxial lateral overgrowth* (ELO) process. The starting material is again a thermally oxidized silicon wafer in which the oxide layer has been photolithographically patterned into islands or strips of oxide. A repeated sequence of carefully controlled CVD silicon deposition and vapor-phase etching cycles is then carried out to produce a single-crystal silicon film on the silicon substrate in the exposed regions between the oxide islands. The vapor-phase etching process preferentially removes any polycrystalline silicon that is deposited on top of the SiO_2. As successive cycles of the CVD deposition and vapor-phase etching processes continue, the single-crystal silicon that is formed in the oxide windows starts to extend over the adjoining oxide regions, and at the same time any polycrystalline silicon that is deposited on top of the oxide is removed by the vapor-phase etching process. The end result is the formation of a complete single-crystal layer of silicon. This heteroepitaxial silicon layer can have mobility and minority carrier lifetime values that are comparable to those obtained in homoepitaxial layers and are suitable for the fabrication of good-quality field-effect and bipolar transistors.

2.20 ISOPLANAR AND OTHER IC STRUCTURES

An isoplanar IC structure is shown in Figure 2.60. It is made by using the selective oxidation of silicon process, which was discussed earlier. This process produces oxidized isolation moats between the N-type silicon islands. This is not a dielectrically isolated IC, however, since there is a PN junction between each N-type island and the common P-type substrate. The oxide isolation moat can, however, result in a reduced collector-to-substrate capacitance and an increased breakdown voltage due to the absence of the heavily doped P^+ regions. A main advantage of the isoplanar process is in the increased device density, since the N^+ emitter and P^+ base regions can be put directly up against the oxide isolation regions as shown in Figure 2.60.

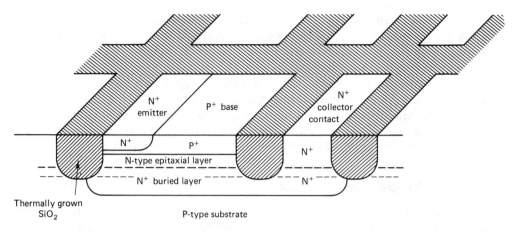

Figure 2.60 Isoplanar integrated-circuit processing using selective oxidation of silicon technique.

2.20.1 Double-diffused Isolation

In the conventional junction isolated IC, the P^+ isolation diffusion must extend downward all the way through the N-type epitaxial layer to the P-type substrate. As a result of this deep diffusion, there will be a large lateral diffusion underneath the oxide so that the P^+ isolation diffusion will occupy a rather large strip and therefore take up a large amount of chip area. For a 10-μm epitaxial layer thickness and a 10-μm-wide oxide opening, the total width of the P^+ region is some 30 to 50 μm.

To reduce the chip area taken up by the isolation diffusion, the structure shown in Figure 2.61 can be used. Prior to the deposition of the N-type epitaxial layer, there is the usual N^+ (Sb) buried layer diffusion and, in this case, also a P^+ (B) diffusion performed into the P-type substrate. During the subsequent epitaxial layer

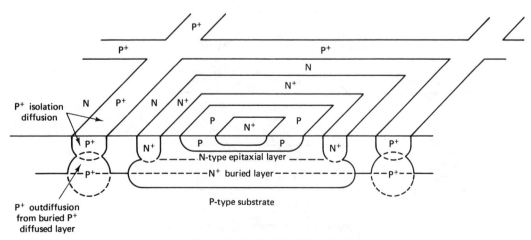

Figure 2.61 Double-diffused isolation.

deposition and high-temperature diffusion and oxidation processes, outdiffusion occurs from the buried N^+ and P^+ regions into the epitaxial layer. As a result of the P^+ outdiffusion, the P^+ isolation diffusion that comes in from the top surface will have to proceed only less than halfway through the epitaxial layer before it merges with the P^+ outdiffusion from the substrate and thereby completes the isolation of the N-type regions. Although this process can substantially reduce the area occupied by the P^+ isolation regions, it does require some extra processing steps and also requires a critical pattern alignment such that the P^+ outdiffusion from the substrate will be lined up correctly with the P^+ isolation diffusion from the top surface.

2.20.2 Collector-diffused Isolation

A very effective way of increasing the device density in ICs is to use the collector-diffused isolation process in which the P^+ isolation is eliminated altogether, as shown in Figure 2.62. The P-type substrate has an N^+ buried layer diffusion as in the conventional case, but now a very thin P-type epitaxial layer only about 2 μm in thickness is used. An N^+ diffusion is performed to penetrate all the way through the P-type epitaxial layer to merge with the N^+ outdiffusion from the substrate. This N^+ diffusion from the top surface serves the dual function of providing device isolation and being a deep collector contact diffusion to minimize the collector series resistance. The devices are isolated by means of the reverse-biased PN junctions formed between the N^+ diffused layer and the P-type epitaxial layer.

The resulting NPN transistors have an epitaxial base region rather than a diffused base. The base width is determined by a combination of the epitaxial layer thickness, the N^+ outdiffusion from the substrate, and the N^+ emitter diffusion from the top surface. As a result, there is generally a wider spread in the base widths for this type of transistor than in the case of the double-diffused devices. This leads to a larger spread in the current gains and poorer transistor matching. In addition, the heavily doped N^+ collector region that is directly adjacent to the P-type base region will lead to a lower collector-to-emitter breakdown voltage.

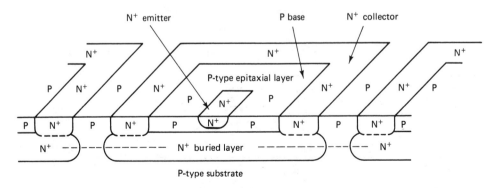

Figure 2.62 Collector-diffused isolation.

2.21 BONDING PADS AND CHIP EDGE CLEARANCE

Around the periphery of any IC chip is an array of bonding pads, which are metallized regions to which the small-diameter wires from the package are bonded. These wires are typically 1-mil (25-μm) diameter gold wires and are bonded to the bonding pads by a process known as *thermocompression bonding,* which uses a combination of heat ($\sim$250°C) and pressure to obtain a good mechanical bond and a very low resistance electrical contact. In the thermocompression and similar types of bonding processes, the end of the wire becomes expanded to about two to three times its original wire diameter on the bonding pad; so the bonding pad must be made large enough to accommodate this, as well as to allow for alignment errors during the lead bonding operation. As a result, the bonding pads are generally in the range from 100 to 150 μm (4 to 6 mils) square, as shown in Figure 2.63.

For the wafer sawing or scribing operation, it is generally desirable to keep the areas where the sawing or scribing will take place clear of oxide or metal. To do this the IC photomask set is designed to provide for a region about 75 to 100 μm (3

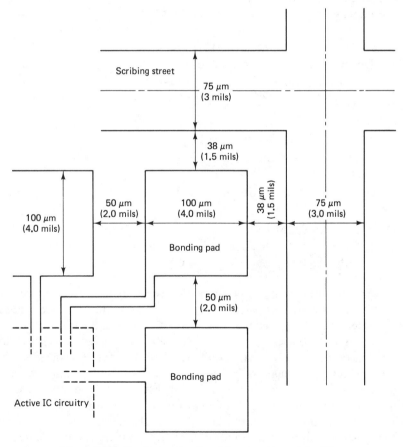

Figure 2.63 Bonding pad and scribing street layout for ICs.

to 4 mils) in width between the adjacent ICs, and the bare silicon is kept exposed. This area is often referred to as a *scribing street*.

A substantial amount of mechanical damage occurs to the edge of the chips during the wafer sawing or scribing operation, including the generation of microcracks that extend a short distance into the chip from the edge. To keep the active circuitry of the IC well away from this region, as well as to accommodate the bonding pads, scribing streets, and the required clearances, there is a strip about 250 µm (10 mils) wide around the periphery of the chip that does not have any active circuitry on it, as shown in Figure 2.63. For small ICs in which the chip size is only 1 mm × 1 mm (40 mils × 40 mils), this can amount to about one-half of the total chip area.

2.22 IC CHIP SIZE AND CIRCUIT COMPLEXITY

The first transistor was developed in 1948 and was a germanium-alloy junction transistor. Silicon devices were developed in the mid-to-late 1950s, and integrated circuits were first made in the early 1960s. Since that time, the size and complexity of ICs have increased very rapidly, as shown by the following brief chronology.

Invention of the transistor (germanium)	1948
Development of silicon transistors	1955–1959
Planar processing	1959
First ICs, small-scale integration (SSI), 3 to 30 gates/chip	~1960
Medium-scale integration (MSI), 30 to 300 gates/chip	~1965–1970
Large-scale integration (LSI), 300 to 3000 gates/chip	~1970–1975
Very large-scale integration (VLSI), more than 3000 gates/chip	~1975
First commercial VLSI chip: 64K RAM	Late 1970s
256K RAMs	Early 1980s
512K RAMs, 1M ROMs, very high-speed GaAs ICs, "three-dimensional" (multilayer) ICs, silicon-on-insulator technology	Middle 1980s

The evolution of ICs over the years has been in terms of some very large increases in the device density together with some increases in the chip area. Figure 2.64 shows the dramatic and approximately exponential increase with time of the number of devices on an IC chip. The graph essentially follows Moore's law, which states that there has been an approximate doubling of the number of devices per chip every year. Some of this rapid increase is due to an increase in the chip size, but most of it is due to improvements in the resolution of the photolithographic process, with device dimensions and line widths shrinking from the 10- to 15-µm range in the 1960s down to the 1- to 2-µm range in the late 1970s and early 1980s. Advances in x-ray and electron-beam lithography can result in continued increases in IC complexity, although probably not at the same rate as in the 1960s and 1970s.

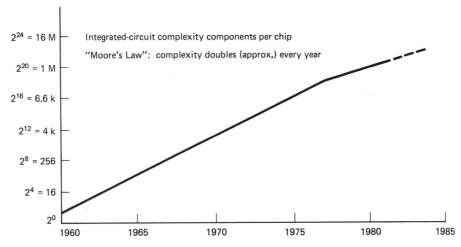

Figure 2.64 Increases in the number of devices on an IC chip. (Adapted from G. E. Moore, "Progress in Digital Integrated Electronics," *IEEE, Intl. Electronic Devices Meeting Tech. Digest*, pp. 11–13, Washington, D.C., 1975 © 1975 IEEE.)

2.22.1 X-Ray and Electron-Beam Lithography

With the conventional ultraviolet (UV) photolithography process in which the UV wavelengths used are in the range from 0.3 to 0.4 μm, the minimum device dimensions or line widths are limited by diffraction effects to around five wavelengths or about 1.75 μm. This is what puts an upper limit on the IC device density using UV photolithography.

In the mid-to-late 1970s, x-ray and electron-beam lithography techniques came into use to produce device dimensions down into the submicron (<1 μm) range. This is due to the much shorter wavelengths involved. In the case of x-rays, wavelengths on the order of 10 to 100 Å (0.001 to 0.01 μm) are available, and the wavelengths for electron-beam exposure are even less. With these techniques, MOSFETs with gate lengths as small as 0.25 μm have been made.

The cost of the x-ray or electron-beam exposure equipment is very high, and the exposure times are very much longer than with UV photolithography; so this is a much more expensive process and is used only when the very small device dimensions (≤1 μm) are needed.

The use of shorter-wavelength UV radiation, called *deep UV*, can result in device dimensions down to about 1.25 μm. The deep UV sources that are used for photoresist exposure have wavelengths in the range from 2000 to 3000 Å.

In the case of x-ray lithography, the mask often used consists of a very thin transparent plastic (Mylar) membrane only about 3 μm (0.1 mil) in thickness. The Mylar membrane has a thin film of gold, about 0.5 μm thick, deposited on it and then patterned to produce the desired x-ray exposure mask. Gold is used because it is a good absorber of x-rays.

For electron-beam lithography, the electron beam that strikes the photoresist produces the same type of polymerization or depolymerization effects that UV or x-ray photon irradiation produces. For electron-beam exposure, an electron beam that is focused down to a spot size of about 0.1 to 0.2 μm is scanned across the wafer surface. The beam can be scanned across the wafer in a raster pattern and is turned on only when passing over the areas to be exposed. This is similar to the raster type of pattern that is used in the production of a TV picture on a CRT, except that there is no gray scale, the electron beam being turned completely on or completely off.

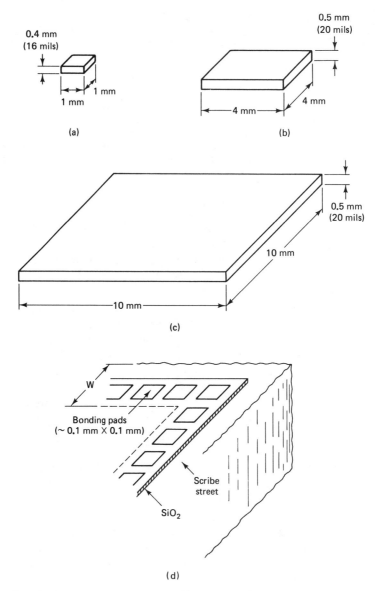

Figure 2.65 Integrated circuit chips: (a) SSI chip; (b) MSI chip; (c) LSI or VLSI chip; (d) detail of chip edge region.

An alternative technique is to use a vector scan in which the electron beam is kept turned on all the time, but is directed only to those areas on the wafer that are to be exposed.

With electron-beam lithography, an exposure mask is not required and the desired circuit patterns can be "written" directly on the wafer by a computer-controlled electron-beam exposure system. This is without the need for any set of masks or even any artwork of the circuit configuration. The equipment cost is, however, very high and the exposure times are generally very long, up to as much as several hours for one wafer. The electron-beam lithography process is very useful for producing photomasks that can be used in the x-ray and UV photolithography processes.

2.22.2 IC Chip Sizes

Figure 2.65 shows representative small (SSI), medium (MSI), and large (LSI or VLSI) IC chips. The chip areas range from 1 mm^2 (1600 mil^2) for the SSI chip to 1 cm^2 (160,000 mil^2) for the LSI chip. In each case, room must be allowed on the chip for the bonding pads, scribe streets, and other clearances. In addition, the active devices on the chip must be set back some safe distance from the edge of the chip due to the mechanical damage that occurs in this region during the chip separation process.

For the SSI chip of Figure 2.65a, a peripheral strip of about 150 μm (6 mils) in width is not available for the active circuitry of the IC. The chip area left of this circuitry is approximately $(1 \text{ mm} - 0.3 \text{ mm})^2 = 0.5 \text{ mm}^2$. Assuming a 10-$\mu$m design rule, this chip can accommodate about 30 bipolar transistors or over 200 MOSFETs. Approximately 7000 of these small IC chips can be obtained from a 100-mm wafer.

For the large 1-cm^2 LSI or VLSI chip, about 70 ICs can be obtained from a 100-mm wafer. If a 200-μm (8-mil) peripheral strip is used for the bonding pads, scribing streets, and other clearances, the remaining chip area for the active circuitry is approximately $(10 \text{ mm} - 0.4 \text{ mm})^2 = 92 \text{ mm}^2$. This area accommodates as many as 6000 bipolar transistors or 40,000 MOSFETs, assuming a 10-μm design rule. With a 4-μm design rule, the number of MOSFETs can reach 250,000.

2.23 THERMAL DESIGN CONSIDERATIONS

The dissipation of power in semiconductor devices results in a temperature rise of the device above the ambient or surrounding temperature. An excessive device temperature can result in the device operating beyond the specified temperature range so that the device will not operate within the manufacturer's specifications, and in particular there will be an excessive junction leakage current. Another more serious consequence of an excessive temperature rise can be the permanent irreversible damage of the device.

The *net* electrical power delivered to the device must under equilibrium conditions be equal to the heat power dissipated by the device, so we have $P_d = P_{\text{in}}$, where P_{in} is the *net* electrical power delivered to the device and P_d is the heat power that is dissipated by the device, as shown in Figure 2.66. The heat flows out of the device to its surroundings by means of the heat conduction and convection processes.

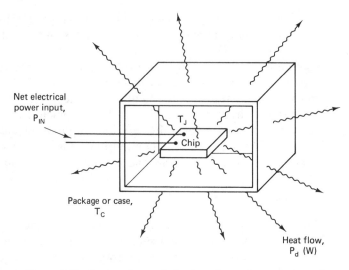

Figure 2.66 Semiconductor device heat flow and temperature rise.

Normally, heat radiation is not a significant factor, due to the relatively low temperature rises involved.

The temperature rise is therefore directly proportional to the heat flow so that $\Delta T = \Theta P_d$, where Θ is the *thermal impedance* and $\Delta T = T_J - T_A$ is the temperature rise of the device junction above the ambient temperature T_A. The critical point is at the PN junctions of the device; hence the reference that is made to the junction temperature T_J. However, as a result of the small size of the silicon chip and the high thermal conductivity of silicon, the average temperature of the chip is generally within a few degrees of the junction temperature.

In Figure 2.67, an electrical analog thermal equivalent circuit is shown. The heat flow P_d through the thermal impedances from the chip (or junction) to the case (or package) Θ_{JC} and from the case to the ambient Θ_{CA} produces temperature drops given by $\Delta T_{JC} = T_J - T_C = \Theta_{JC} P_d$ and $\Delta T_{CA} = T_C - T_A = \Theta CA P_d$. The total temperature drop from junction to ambient is

$$\Delta T_{JA} = T_J - T_A = \Delta T_{JC} + \Delta T_{CA} = (\Theta_{JC} + \Theta_{CA})P_d = \Theta_{JA} P_d \qquad (2.13)$$

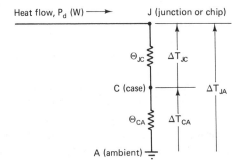

Figure 2.67 Thermal equivalent circuit.

The thermal impedances have units of °C/W, although the thermal impedance is sometimes given in units of °C/mW.

In Figure 2.68 a variety of IC packages are shown together with values of the thermal impedance. The most popular package type is the *dual-in-line package* or DIP. For the lowest-cost plastic DIP packages, the thermal impedance ranges from 200°C/W for the 8-pin DIP (or mini-DIP) to 160°C/W for the 14-pin DIP and 150°C/W for the 16-pin DIP. The ceramic DIP, which is more expensive than the plastic DIP, has a substantially lower thermal impedance: 100°C/W for the 14-pin DIP and 90°C/W for the 16-pin DIP. The flat-pack has as a principal feature a low profile, being a maximum of only 0.07 in. (1.8 mm) high. However, together with the small package volume, there is a relatively high thermal impedance, about 300°C/W for both the 14-pin and 16-pin packages. The metal can packages, including the 8-pin TO-99 and the 10-pin TO-100 packages, have a thermal impedance of about 160°C/W. These impedance values are the junction-to-ambient thermal impedance Θ_{JA} under free-air conditions; that is, the device is cooled by natural convection and no heat sink is attached to the device. The junction-to-case thermal impedance for most of the IC packages is in the range from 20° to 40°C/W.

2.23.1 Surface-Mount Packages

In Figure 2.69, some of the newer *small-outline* (SO) surface-mount packages are shown. Unlike the dual-in-line package (DIP), the surface-mount packages do not have any pins that fit into sockets or go into holes in a printed circuit (PC) board. The pins of a surface-mount package are soldered directly to the copper traces (metallization) on the surface of the PC board. The SO packages are about half the width of the DIP packages, and the pin center-to-center spacing is 0.050 in. for the SO package as compared to 0.10 in. for the DIP. As a result, the SO package is about one-fourth the size of a DIP package with the same number of pins, resulting in a much higher device density on the PC board. The smaller size of the SO package also results in shorter lead lengths and thus a reduced parasitic resistance and inductance, giving improved high-frequency performance.

The SO devices do not have pins that pass through holes on the PC board, so SO devices can be mounted right opposite each other on both sides of the PC board. This contributes to high device density. In addition to active devices such as diodes, transistors, and ICs, passive components such as resistors and capacitors are available in surface-mount packages.

2.23.2 Examples

Let us consider now as a representative example a 14-pin plastic DIP package that has a thermal impedance of 160°C/W. The usual upper temperature limit for plastic package devices is 150°C, due in part to the softening of the plastic package material. If the ambient temperature is 25°C, the maximum allowable power dissipation rating $P_{d(\mathrm{MAX})}$ will be obtained from $\Delta T_{\mathrm{MAX}} = T_{J(\mathrm{MAX})} - T_A = \Theta_{JA} P_{d(\mathrm{MAX})}$, so

$$P_{d(\mathrm{MAX})} = (T_{J(\mathrm{MAX})} - T_A)/\Theta_{JA} \qquad (2.14)$$

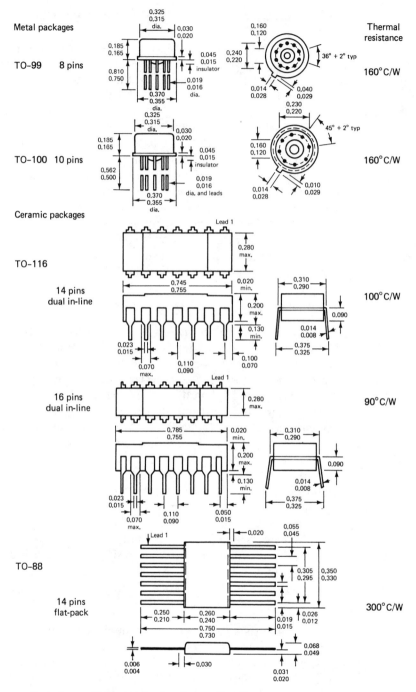

Most often used integrated circuit package configurations.

Figure 2.68 Integrated circuit packages. (From Hans R. Camenzind, *Electronic Integrated Systems Design*, Van Nostrand Reinhold, New York, 1972.)

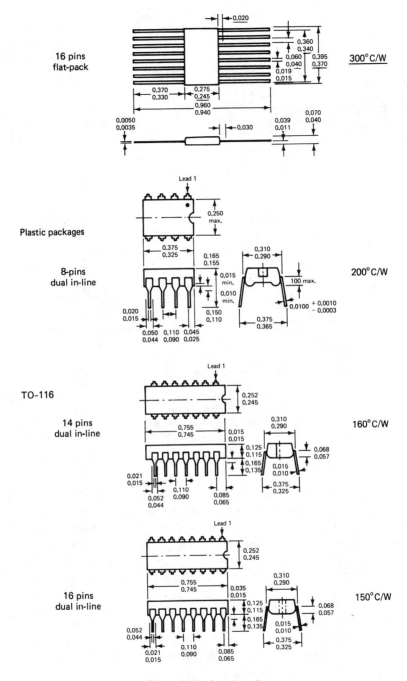

Figure 2.68 (continued)

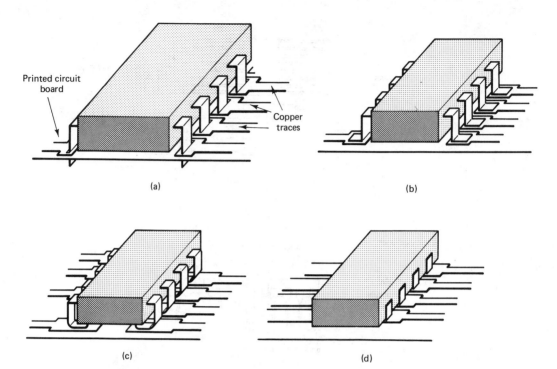

Figure 2.69 Comparison of through-hole DIP and surface-mount packages: (a) DIP package (pin spacing = 0.10 in. = 2.5 mm); (b) small-outline IC (SOIC) surface-mount package (pin spacing = 0.050 in. = 1.25 mm); (c) plastic leaded chip carrier (PLCC) package (pin spacing = 0.050 in. = 1.25 mm); (d) leadless ceramic chip carrier (LCCC) package (pin spacing = 0.050 in. = 1.25 mm).

and thus
$$P_{d(\text{MAX})} = \frac{(150 - 25)°\text{C}}{160°\text{C/W}} = \underline{0.78\text{ W}} = \underline{780\text{ mW}} \tag{2.15}$$

If the ambient temperature is higher than 25°C, the maximum power dissipation rating is reduced, and as the ambient temperature T_A approaches the maximum allowable junction temperature $T_{J(\text{MAX})}$, the value of $P_{d(\text{MAX})}$ approaches zero. In Figure 2.70, a *derating curve* is shown for this example, showing the decrease in the maximum power dissipation rating of the device as the ambient temperature increases. The slope of the derating line is equal to the negative of the reciprocal of the thermal impedance, $-1/\Theta_{JA}$.

For devices in the metal or ceramic packages, the usual upper temperature limit for the junction temperature, $T_{J(\text{MAX})}$, is about 175°C. Thus, for the metal TO-99 (8-pin) and TO-100 (10-pin) packages with a thermal impedance of 160°C/W and an ambient temperature of 25°C, the maximum power dissipation rating is

$$P_{d(\text{MAX})} = \frac{(175 - 25)°\text{C}}{160°\text{C/W}} = \underline{0.94\text{ W}} = \underline{940\text{ mW}} \tag{2.16}$$

A further advantage of the metal package is the possibility of obtaining a hermetic gastight seal so that better long-term device protection can be obtained. The metal

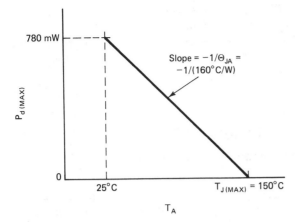

Figure 2.70 Maximum power dissipation derating characteristic.

package is, however, much more costly than the plastic package. Indeed, a major part of the cost of many semiconductor devices is the packaging cost.

2.23.3 Heat Sinks

The maximum power dissipation rating of semiconductor devices can be increased substantially by the use of a *heat sink*. A heat sink is a metal part, usually some type of finned structure, that can be clipped, clamped, or bonded to the semiconductor device package and aids the convective heat transfer process. A wide variety of heat sinks are available for various device package configurations.

The use of a heat sink can reduce the case-to-ambient thermal impedance Θ_{CA} by a very large factor so that the maximum power dissipation rating of the device can be greatly increased, often by a factor as large as 3 or 4. With a heat sink present, the case-to-ambient thermal impedance is reduced from the free-air value to a value determined by the heat sink so that Θ_{CA} is now equal to Θ_{HS}, which is the heat sink thermal impedance. Depending on the design of the heat sink, Θ_{HS} values ranging from 100°C/W for very small clip-on heat sinks down to around 10°C/W for larger area and more massive heat sinks are available. Simple clip-on heat sinks for DIPs are available that can provide thermal impedances in the range 20° to 30° C/W. In addition to the clip-on heat sinks, there are heat sinks that can be bonded to the semiconductor device package by a thermally conductive epoxy adhesive. The heat sink adhesives have thermal impedances in the range of 70°C-cm/W. For an area of 1 cm², the thermal impedance of an adhesive film of thickness t(mm) is $\Theta = 7$°C/W $\times t$(mm). Thus an adhesive film of $t = 0.1$ mm $= 0.004$ in. in thickness and 1 cm² in area will have a thermal impedance of only 0.7°C/W.

For a representative example of the advantage to be gained by the use of a heat sink, let us consider a device with $\Theta_{JA} = 150$°C (free air), $\Theta_{JC} = 40$°C/W, and $T_{J(MAX)} = 175$°C. For an ambient temperature of 25°C, the free-air maximum power dissipation rating is $P_{d(MAX)} = (175 - 25)$°C/(150°C/W) $= \underline{1.0 \text{ W}}$. If a heat sink is now used that has a heat sink-to-ambient thermal impedance of Θ_{HS} of 20°C/W, the case-to-ambient thermal impedance is reduced from 110°C/W to the 20°C/W impedance of the heat sink. The total thermal impedance is now $\Theta_{JA} = \Theta_{JC} +$

$\Theta_{HS} = (40 + 20)°C/W = 60°C/W$, and the corresponding value of $P_{d(MAX)}$ is increased to $P_{d(MAX)} = 150°C/(60°C/W) = \underline{2.5\ W}$. In the limiting case of an ideal infinite heat sink that has a zero thermal impedance, the case-to-ambient thermal impedance is reduced to zero, so $\Theta_{JA} = \Theta_{JC} = 40°C/W$. The maximum power dissipation rating reaches a value of $P_{d(MAX)} = 150°C/(40°C/W) = \underline{3.75\ W}$.

PROBLEMS

Voltage-variable Capacitance Diode Problems: P⁺NN⁺
Epitaxial Diode (Problems 2.1–2.14)

2.1. Given that the VHF TV bands are 54 to 88 MHz (channels 2 to 6) and 174 to 216 MHz (channels 7 to 13) and each channel is 6.0 MHz wide, can one VVC diode be used for tuning over the entire VHF TV band? Assume that the minimum allowable reverse-bias voltage is 3.0 V and the maximum voltage available in the circuit is 30 V. Explain your answer.

2.2. Can one VVC diode be used for tuning over the TV band above (channels 2 to 13) if the inductor is switched, with one value inductance L_1 being used for the VHF low band (channels 2 to 6) and a smaller inductance L_2 being used for the VHF high band (channels 7 to 13)? Explain.

2.3. If the switched inductance technique described above is used, and there is a fixed capacitance of 4.0 pF in the circuit, find the required minimum and maximum diode capacitance values required.   [*Ans.*:

$$C_{J(MIN)} = 34.62\ \text{pF} \quad \text{(full depletion of epitaxial layer)}$$

$$C_{J(MAX)} = C_J(-3\ \text{V}) = 98.56\ \text{pF}$$

2.4. If the epitaxial layer doping is $N_{epi} = 1 \times 10^{15}\ \text{cm}^{-3}$ (5 Ω-cm), find the junction area required.   (*Ans.*: $A = 2.10\ \text{mm}^2$)

2.5. If the junction depth is $x_J = 2.0\ \mu\text{m}$, find the required thickness of the epitaxial layer.   (*Ans.*: $t_{epi} = 8.33\ \mu\text{m}$)

2.6. Find the maximum value of the series resistance, $R_S(3\ \text{V})$.   [*Ans.*: $R_S(3\ \text{V}) = 0.105\ \Omega$]

2.7. Find the minimum value of Q at 3 V and 174 MHz.   (*Ans.*: $Q = 88.4$)

2.8. What value of Q will be needed (at 174 MHz) for a 6-MHz bandwidth? Will the VVC Q-value be satisfactory for this?   (*Ans.*: 29, yes)

2.9. If the junction area is increased to twice the minimum value, by what factor will these parameters change?
  (a) $C_{J(MAX)}$.   (*Ans.*: 2 : 1)
  (b) $C_{J(MIN)}$.   (*Ans.*: 2 : 1)
  (c) Diode capacitance ratio.   (*Ans.*: unaffected by an area change)
  (d) Series resistance, R_S.   (*Ans.*: 1 : 2)
  (e) Q.   (*Ans.*: unaffected by an area change)
  (f) $C_{\text{TOTAL MAX}}/C_{\text{TOTAL MIN}}$.   (*Ans.*: 2.746 : 2.656)
  (g) Tuning ratio, f_{MAX}/f_{MIN}.   (*Ans.*: 1.657 : 1.630)

2.10. If a nonepitaxial 5-Ω-cm N-type wafer were used instead of the N/N⁺ wafer, what will the series resistance be at 3-V bias? What will the Q be at 3 V and 174 MHz? Will this Q-value be satisfactory? [*Ans.*: $R_s = 7.1\ \Omega$, $Q(-3\ \text{V}, 174\ \text{MHz}) = 1.30$, no]

2.11. Will the use of the complementary N^+PP^+ epitaxial diode structure offer any advantages over the P^+NN^+ structure? Explain. How will the Q-values compare for the two structures, assuming equal doping levels and dimensions? (*Ans.*: no, the Q will be lower by a factor of about 2.5)

2.12. The UHF TV band extends from 470 to 890 MHz (channels 14 to 83). If the minimum allowable reverse-bias voltage is 3.0 V and the maximum voltage available is 30 V, can one VVC diode be used for tuning over the UHF TV band?

2.13. To what value will the minimum diode bias voltage be reduced to allow for tuning over the entire UHF TV band? (*Ans.*: $V_{R(MIN)} = 1.6$ V)

2.14. The junction capacitance of a VVC diode will exhibit a temperature dependence. The rate of change of capacitance with temperature can be designated as $TCC_J = (1/C_J)(dC_J/dT)$. A representative value for TCC_J is 400 ppm/°C = 0.04%/°C.

  (a) What will the corresponding temperature coefficient of the frequency of the tuned circuit be? [*Ans.*: $TCF = (1/f)(df/dT) = 200$ ppm/°C = 0.02%/°C]

  (b) Will this frequency shift cause any serious problems for a UHF tuner, considering channel 83 (890 MHz) as the worst-case condition? (*Ans.*: $\Delta f = 0.18$ MHz/°C, or only 3%/°C of the 6.0-MHz channel width, so a temperature change limited to a few °C should not cause any serious problem, although a larger temperature change can be a problem)

  (c) Repeat part (b) for the case of an FM tuner (88 to 108 MHz) with a 200-kHz bandwidth per FM channel. (*Ans.*: $\Delta f = 22$ kHz/°C or about 11%/°C of the channel bandwidth, so a temperature change of even a few degrees can produce a substantial mistuning if not compensated for)

2.15. (*PIN switching diode*) Given: P^+NN^+ (PIN) planar epitaxial VHF switching diode as shown in Figure P2.15. The package capacitance is 0.05 pF.

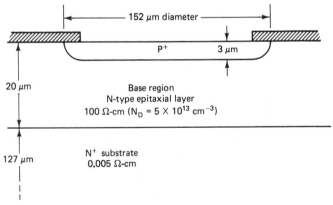

Figure P2.15

  (a) Find the minimum capacitance, C_{min}. (*Ans.*: 0.162 pF)

  (b) Find the value of the base (epitaxial layer) region component of the series resistance, R_o, for very small forward currents. (*Ans.*: ~936 Ω)

  (c) Find the substrate component of the series resistance. (*Ans.*: 0.35 Ω)

(d) As a result of the injection of holes from the P$^+$ region into the very lightly doped N-type epitaxial layer under forward-bias conditions, additional electrons will be drawn into the epitaxial layer from the N$^+$ substrate. These electrons are drawn in to neutralize the charge density produced by the holes in transit across the epitaxial layer or base region. This flow of electrons and holes into the base region results in an increased conductivity and therefore a decreased series resistance for the base region. This effect is called *conductivity modulation*. As a result of this conductivity modulation, the base-region series resistance is given by

$$R_{\text{base}} = \frac{V_T}{I(1 + b)} \ln \left[1 + \frac{IR_o \, (1 + b)}{V_T} \right]$$

where $b = \mu_n/\mu_p$ is the mobility ratio (2.5 for silicon) and R_o is the unmodulated value of the base resistance (R_{base} at $I = 0$). Using $b = 2.5$ and $V_T = 26$ mV (27°C), find the base-region resistance at the following forward current values: 100 nA, 1.0 µA, 10 µA, 100 µA, 1 mA, 10 mA, 30 mA, 50 mA, 100 mA, and 200 mA. (*Ans.*: 926, 878, 604, 194, 36, 5.3, 2.04, 1.30, 0.70, 0.376 Ω)

(e) The total dynamic forward resistance of this diode is given by

$$r_d = \frac{nV_T}{I} + R_{\text{series}} = \frac{nV_T}{I} + R_{\text{base}} + R_{\text{substrate}}$$

Find the diode dynamic resistance for the current values given in part (d). Use $n = 1.6$. (*Ans.*: 417 kΩ, 42.5 kΩ, 4.76 kΩ, and 610, 78, 9.75, 3.72, 2.42, 1.41, 0.88 Ω)

(f) What reverse-bias voltage is required for the full depletion of the base region such that the junction capacitance is at its minimum value? Assume 0.8 V for the contact potential. (*Ans.*: 10.3 V)

(g) If the complementary structure (that is, N$^+$ PP$^+$) were used for this switching diode, but with the same resistivity values and dimensions as before, how would the equation for R_{base} be changed? Calculate R_{base} and r_d for the N$^+$PP$^+$ diode at a forward current of 100 mA. Find the value of reverse-bias voltage required for the full depletion of the base region. (*Ans.*: $R_{\text{base}} = 1.58$ Ω, $r_d = 2.35$ Ω, $V_R = 27.7$ V)

(h) The P$^+$NN$^+$ diode is to be used as a series switch in a 50-Ω system (50-Ω source, 50-Ω load) at 100 MHz. Find the *isolation* (in dB) with a 15-V reverse bias, and the *insertion loss* (in dB) when there is a forward current of 50 mA through the diode. (*Ans.*: 39.85 dB, 0.208 dB)

(i) Repeat part (h) for the case of the PIN diode being used as a *shunt* switch. (*Ans.*: 21.1 dB, ~0 dB)

(j) Repeat part (h) for the case of a *series-shunt* switch in which two PIN diodes are used, one in the series position and one in the shunt position. (*Ans.*: 66.6 dB, 0.21 dB)

2.16. (*Short-channel MOSFET*) In a short-channel MOSFET the drain-to-source electric field strength in the channel region is such that the charge carriers travel at drift velocities close to the saturation velocity v_{sat}. The total mobile charge in the channel is $Q_{\text{channel}} = C_{\text{ox}}(V_{GS} - V_{\text{threshold}})$, where C_{ox} is the gate-to-channel capacitance. The drain-to-source current I_{DS} will be given by $I_{DS} = Q_{\text{channel}}/t_{\text{transit}}$, where t_{transit} is the source-to-drain transit time and is given approximately by $t_{\text{transit}} = L/v_{\text{sat}}$, with L being the channel length.

(a) Show that I_{DS} is given by

$$I_{DS} = g_{fs}(V_{GS} - V_{\text{threshold}})$$

where $g_{fs} = (\varepsilon_{\text{ox}}/t_{\text{ox}})Wv_{\text{sat}}$ and W is the channel width.

(b) A short-channel MOSFET has an oxide thickness of $t_{\text{ox}} = 1000$ Å. Find g_{fs} per centimeter of channel width. Use $\varepsilon_{r(\text{oxide})} = 3.8$ and $v_{\text{sat}} = 1 \times 10^6$ cm/s for electrons in the N-type channel. (*Ans.*: 33.6 mS/cm)

(c) A VMOS short-channel transistor is on a 1 mm × 1 mm chip. The V-grooves have a center-to-center spacing of 40 μm. Using the same parameters as given above, find the g_{fs} of the transistor. (*Ans.*: 168 mS)

(d) For the above VMOS transistor, the threshold voltage is +2.0 V. Find I_{DS} at $V_{GS} = +12$ V. (*Ans.*: $I_{DS} = 1.68$ A)

(e) For the MOSFET described above, the junction depth of the N^+ source region is 2.0 μm. Find the gate-to-source capacitance C_{gs}. (*Ans.*: 41 pF)

2.17. (*Ion-implanted P-channel JFET*) The P-type channel of a JFET is formed by a boron ion implantation into an N-type epitaxial layer as shown in Figure 2.25c.

(a) Show that the pinch-off voltage V_P of the JFET is given approximately by $V_P = q\phi^2/2\varepsilon N_D - V_{CP}$, where ϕ is the ion implantation dosage, N_D is the epitaxial layer doping, and V_{CP} is the contact potential.

(b) Find the ion implantation dosage required to produce a JFET with a pinch-off voltage of +4.0 V if the epitaxial layer doping is $N_D = 1 \times 10^{16}$ cm^{-3} (0.6 Ω-cm) and $V_{CP} = 0.8$ V. (*Ans.*: $\phi = 7.91 \times 10^{11}$ cm^{-2})

(c) If the average hole mobility in the implanted layer is 200 cm²/V-s, find the sheet resistance of the layer. Assume bias conditions such that the channel resistance will be at its minimum value (that is, essentially zero bias or a slight forward bias at the gate-to-channel junction). (*Ans.*: 39,454 Ω/square)

(d) If the JFET channel length is 8 μm and the channel width is 160 μm, find $r_{ds(\text{ON})}$, I_{DSS}, and g_{fsO}. The following approximate relationship can be used for I_{DSS}: $I_{DSS} = -V_P/[3r_{ds(\text{ON})}]$. (*Ans.*: 1973 Ω, −0.676 mA, 0.338 mS)

(e) Repeat the preceding problems for an epitaxial layer doping of $N_D = 1 \times 10^{17}$ cm^{-3} (0.09 Ω-cm). (*Ans.*: $\phi = 2.502 \times 10^{12}$ cm^{-2}, $R_S = 12,476$ Ω/square, $r_{ds(\text{ON})} = 624$ Ω, $I_{DSS} = -2.14$ mA, $g_{fsO} = 1.069$ mS)

(f) Design a JFET with an ion-implanted P-type channel to meet the following specifications: $V_P = +2.2$ V and $I_{DSS} = -100$ μA. The epitaxial layer doping is 1×10^{16} cm^{-3} (0.6 Ω-cm) and the channel length is 10 μm. Specify the ion-implantation dosage ϕ and the channel width W. Assume an average hole mobility of 200 cm²/V-s in the P-type channel and a contact potential of 0.8 V. (*Ans.*: $\phi = 6.25 \times 10^{11}$ cm^{-2}, $W = 50$ μm)

2.18. (*Ion-implanted depletion-mode MOSFET*) The device structure of Figure 2.25c is modified into a depletion-mode MOSFET by the formation of a thin oxide layer over the P-type implanted channel and the deposition of a gate electrode on top of the oxide layer. The P-type channel is formed by a low-energy, low-dosage boron ion implantation followed by a low-temperature ($\sim 700°C$) annealing step to activate the implanted ions by causing them to go into substitutional sites in the silicon crystal lattice.

(a) Show that the pinch-off voltage is given approximately by $V_P = (q\phi t_{\text{ox}}/\varepsilon_{\text{ox}}) - V_{CP}$, where ϕ is the ion-implantation dosage and V_{CP} is the gate-to-silicon contact potential (generally in the range from 0 to 0.5 V).

(b) Find the ion-implantation dosage ϕ required for a pinch-off voltage of +4.0 V if the oxide thickness t_{ox} is 800 Å. Assume that $V_{CP} \simeq 0$, as for the case of a silicon gate device. (*Ans.*: 1.05×10^{12} cm^{-2})

(c) Find the minimum value of the sheet resistance of the P-type channel region ($V_{GS} = 0$). Assume that the average hole mobility in the channel is 200 cm^2/V-s. (*Ans.*: 29.74 kΩ/square)

(d) If the channel length is $L = 8.0$ μm and the width is $W = 40$ μm, find $r_{ds(ON)}$, I_{DSS}, and g_{fsO}. (*Ans.*: 5.55 kΩ, 224 μA, 112 μS)

(e) Show that I_{DSS} is given approximately by $I_{DSS} \simeq (\varepsilon_{ox}\overline{\mu}_p W/t_{ox}L)V_P^2$ and g_{fsO} is given approximately by $g_{fsO} \simeq \frac{2}{3}(\varepsilon_{ox}\overline{\mu}_p W/t_{ox}L)V_P$.

(f) Compare this ion-implanted depletion-mode MOFSET with the ion-implanted JFET of Problem 2.17. What are the relative advantages and disadvantages of the two devices?

2.19. (*Ion-implanted IC resistor*) A boron ion implantation is followed by a drive-in diffusion at 1150°C to produce a P-type layer with a sheet resistance of 5000 Ω/square and a junction depth of 2.0 μm. The substrate is doped with 1×10^{15} cm^{-3} phosphorus. Find:

(a) Average resistivity of the layer. (*Ans.*: 1.0 Ω-cm)

(b) Boron surface concentration. (*Ans.*: 4×10^{16} cm^{-3})

(c) Drive-in diffusion time. (*Ans.*: 45.2 min)

(d) Ion-implantation dosage. (*Ans.*: 3.7×10^{12} cm^{-2})

(e) Average hole mobility in the layer. (*Ans.*: 338 cm^2/V-s)

(f) Ion implantation time per 100-mm-diameter wafer if the beam current is 5 μA with a rectangular raster scan. (*Ans.*: 11.8 s)

(g) Chip area required for a 1.0-MΩ resistor if the resistor line width is to be 15 μm and the line-to-line spacing is to be 15 μm. (*Ans.*: ~0.1 mm^2)

(h) Capacitance per unit area and the total parasitic capacitance for the 1.0-MΩ resistor at a reverse-bias voltage of 10 V. (*Ans.*: 27.8 pF/mm^2, ~1.25 pF)

(i) Cutoff frequency (3-dB bandwidth) for the resistor. (*Ans.*: ~254 kHz)

2.20. (*Gallium arsenide MESFET*) A GaAs MESFET similar to that of Figure 2.9 has a channel region formed by a low-energy, low-dosage ion implantation followed by a short drive-in diffusion/annealing step.

(a) If the Gaussian impurity profile of the channel can be approximated by a rectangular-shaped profile, show that the pinch-off voltage is given by $V_P = q\phi t_c/\varepsilon - V_{cp}$, where ϕ is the implantation dosage, t_c is the channel thickness, and V_{cp} is the metal–semiconductor contact potential (typically ~0.8 V for metal–GaAs Schottky barriers).

(b) If the implanted channel has a surface concentration of $N(0) = 2 \times 10^{17}$ cm^{-3} and at a distance of 0.25 μm from the surface the concentration is down to 2×10^{15} cm^{-3}, find:

(1) The ion implantation dosage. (*Ans.*: 2.065×10^{12} cm^{-2})

(2) The pinch-off voltage using the approximate relationship of part (a) with $V_{cp} = 0.8$ V and $t_c = 0.25$ μm ($\varepsilon_r = 10.9$ for GaAs). (*Ans.*: -7.77 V)

(3) $r_{ds(ON)}$, I_{DSS}, and g_{fso} for a channel length of $L = 2.0$ μm and a width of $W = 10$ μm. Use an average electron mobility of 5000 cm^2/V-s for the channel. (*Ans.*: 121 Ω, 21 mA, 5.5 mS)

(4) The minimum source-to-drain transit time and the corresponding frequency. The electron saturation velocity in GaAs is ~10^7 cm/s. (*Ans.*: 20 ps, 8.0 GHz)

*(c) Using the equation for the Gaussian impurity profile and Posson's equation, write

a computer program to obtain the pinch-off voltage. Find the pinch-off voltage and compare the value obtained with the approximate result obtained in part (b,2).

2.21. (*Thermal design*) A power IC has $\Theta_{JC} = 10°C/W$ and $T_{J(MAX)} = 150°C$. Find the maximum allowable value for the heat sink thermal resistance (Θ_{CA}) for a $P_{d(MAX)}$ rating of 10 W at an ambient temperature of 25°C.

REFERENCES

BAR-LEV, A., *Semiconductors and Electronic Devices*, Prentice-Hall, Englewood Cliffs, N.J., 1984.

BURGER, R. M., and R. P. DONOVAN, *Fundamentals of Silicon Integrated Circuit Device Technology*, Vol. 1, Prentice-Hall, Englewood Cliffs, N.J., 1967.

CAMENZIND, H. R., *Electronic Integrated Systems Design*, Van Nostrand Reinhold, New York, 1972.

COLCLASER, R. A., *Microelectronics: Processing and Device Design*, Wiley, New York, 1980.

CONNELLY, J. A., *Analog Integrated Circuits*, Wiley, New York, 1975.

EIMBINDER, J., *Application Considerations for Linear Integrated Circuits*, Wiley, New York, 1970.

EINSPRUCH, N. G., *VLSI Electronics: Microstructure Science*, Academic Press, New York, 1982.

FOGIEL, M., *Microelectronics: Basic Principles, Circuit Design, Fabrication Technology*, Research & Education Association, Piscataway, N.J., 1972.

FURUKOW, S. (ed.), *Silicon-on-Insulator*, D. Reidel Publishing Co., 1985.

GHANDI, S. K., *The Theory and Practice of Microelectronics*, Wiley, New York, 1968.

GLASER, A. B., and G. E. SUBAK-SHARPE, *Integrated Circuit Engineering*, Reading, Mass., Addison-Wesley, 1977.

GRAY, P. R., and R. G. MEYER, *Analysis and Design of Analog Integrated Circuits*, Wiley, New York, 1984.

GREBENE, A. B., *Analog Integrated Circuit Design*, Van Nostrand Reinhold, New York, 1972.

GROVE, A. S., *Physics and Technology of Semiconductor Devices*, Wiley, New York, 1967.

HAMILTON, D. J., and W. G. HOWARD, *Basic Integrated Circuit Engineering*, McGraw-Hill, New York, 1975.

HINH, S. W., *Handbook of Surface Mount Technology*, Wiley, New York, 1988.

HNATEK, E. R., *A User's Handbook of Integrated Circuits*, Wiley, New York, 1973.

———, *Integrated Circuit Quality and Reliability*, Marcel Dekker, New York, 1987.

HOWES, M. J., and D. V. MORGAN (eds.), *Gallium Arsenide—Materials, Devices, and Circuits*, Chapter 10, GaAs MESFETs by B. TURNER, Wiley, New York, 1985.

JASTRZEBSKI, L., A. C. IPRI, and J. F. CORBOY, "Device Characterization on Monocrystalline Silicon Growth over SiO_2 by the ELO (Epitaxial Lateral Overgrowth) Process," *IEEE Electron Device Letters*, Vol. EDL-4, No. 2, February 1983.

MANGIN, C. H., and S. McCLELLAND, *Surface Mount Technology*, IFS Publications, 1987.

MATISOFF, B. S., *Handbook of Electronics Packaging Design and Engineering*, Van Nostrand Reinhold, New York, 1982.

MILLMAN, J., *Microelectronics*, McGraw-Hill, New York, 1979.

MORTENSON, K. E., *Variable Capacitance Diodes*, Artech House, Norwood, Mass., 1974.

MOTOROLA, INC., *Analysis and Design of Integrated Circuits*, McGraw-Hill, New York, 1967.

MULLER, R. C., *Device Electronics for Integrated Circuits*, Wiley, New York, 1977.

MUN, J. (ed.), *GaAs Integrated Circuits*, McGraw-Hill, New York, 1988.

OXNER, E. S., *Power FETs and Their Applications*, Prentice-Hall, Englewood Cliffs, N.J., 1982.

RICHMAN, P., *MOS Field-Effect Transistors and Integrated Circuits*, Wiley, New York, 1973.

SARTELL, J. A. (ed.), *Electronic Packaging—Materials and Processes*, American Society for Metals, Metals Park, Ohio, 1986.

SERAPHIM, D. P., R. C. LASKY, and C. LI (eds.), *Principles of Electronic Packaging*, McGraw-Hill, New York, 1989.

SEVIN, L. J., *Field-Effect Transistors*, McGraw-Hill, New York, 1965.

SHAH, R. R., S. A. EVANS, D. L. CROSTHWAIT, and R. L. YEAKLEY, "Laser Recrystallized Polysilicon for High Performance I²L," *IEEE Trans. Electron Devices*, Vol. ED-28, No. 12, December 1981.

SOARES, R., J. GRAFFEUIL, and J. OBREGON (eds.), *Applications of GaAs MESFETs*, Artech House, Norwood, Mass., 1983.

STREETMAN, B. G., *Solid State Electronic Devices*, Prentice-Hall, Englewood Cliffs, N.J., 1980.

SZE, S. M., *Physics of Semiconductor Devices*, Wiley, New York, 1969.

TILL, W. C., and J. T. LUXON, *Integrated Circuits: Materials, Devices, and Fabrication*, Prentice-Hall, Englewood Cliffs, N.J., 1982.

TSIVIDIS, Y. P., and P. ANTOGNETTI, *Design of MOS VLSI Circuits for Telecommunications*, Chapter 1, Fabrication Technology for MOS IC's for Telecommunications by E. DEMOULIN, Prentice-Hall, Englewood Cliffs, N.J., 1985.

———, *Operation and Modeling of the MOS Transistor*, McGraw-Hill, New York, 1987.

TUMMALA, R. R., and E. J. RYMASZEWSKI, *Microelectronics Packaging Handbook*, Van Nostrand Reinhold, New York, 1989.

VERONIS, A., *Integrated Circuit Fabrication Technology*, Prentice-Hall, Englewood Cliffs, N.J., 1979.

WALLMARK, J. T., and H. JOHNSON, *Field-Effect Transistors: Physics, Technology, and Applications*, Prentice-Hall, Englewood Cliffs, N.J., 1966.

WOLF, H. F., *Semiconductors*, Wiley, New York, 1971.

YANG, E. S., *Microelectronic Devices*, McGraw-Hill, New York, 1988.

3

Constant-current Sources, Voltage Sources, and Voltage References

3.1 CONSTANT-CURRENT SOURCES

The ideal constant-current source is an electric circuit element that provides a current to a load that is independent of the voltage across the load, or equivalently the load impedance. Note that the adjective "constant" in the term "constant-current source" refers to the fact that the current is independent of the load conditions. Therefore, under this definition a constant-current source can produce a current that is time varying, that is, an ac current. The constant-current source can also be a controlled source such that the strength of the current source is a function of some other voltage or current in the system, but not of the voltage across the load supplied by the constant-current source under consideration.

In electronic circuits, especially for integrated circuits, there are many applications for constant-current sources, and in particular dc constant-current sources. Although it is never possible to have an ideal constant-current source in a real electronic circuit, there are ways to produce circuits that provide a very close approximation to the ideal constant-current source.

For integrated-circuit applications, these constant-current sources generally make use of the fact that for a transistor in the active mode of operation, the collector or drain current is relatively independent of the collector or drain voltage. For a bipolar transistor in the active mode or region of operation (emitter–base junction "on," collector–base junction "off"), the voltage across the transistor from collector to emitter, V_{CE}, should be greater than approximately 0.2 V but less than the collector-to-emitter breakdown voltage, BV_{CEO}, which is typically at least 50 V for IC transistors. Over this range of voltages, the collector current, I_C, is relatively independent of the collector-to-emitter voltage, V_{CE}, of the transistor. FETs operating in the active mode in which the drain current is relatively independent of the drain-to-source voltage can also be used for current sources. Both JFET and MOSFET current sources will be presented in this chapter.

As a preface to looking at some constant-current sources, let us first consider the basic circuit shown in Figure 3.1. We will first consider the case of the two transistors being identical in every respect. Noting that the bases of the two transistors are tied together and that the two emitters go to the same point, we have that V_{B_1} = V_{B_2}, and $V_{E_1} = V_{E_2}$, so $V_{BE_1} = V_{BE_2}$. As a result, the two transistors will have *exactly* the same base-to-emitter voltages. Transistor Q_1 is a *diode-connected* transistor with the collector shorted to the base so that $V_{CB} = 0$. The base–emitter junction of Q_1 will be biased "on" as a result of the current I_1 that is forced through this transistor. As a result of the fact that $V_{CB} = 0$, the collector–base junction will be "off." Therefore, transistor Q_1 will be operating in the active region.

Transistor Q_2 will also be in the active region as long as the voltage across Q_2, V_{CE_2}, is greater than 0.2 V but less than the breakdown voltage, BV_{CEO}. Since we have two identical transistors, both in the active region with identical base-to-emitter voltages, the collector currents of the two transistors are approximately equal, so $I_{C_2} = I_{C_1}$. Since $I_1 = I_{C_1} + I_{B_1} + I_{B_2} = I_C + 2I_C/\beta = I_C (1 + 2/\beta)$, we have that $I_{C_2} = I_{C_1} = I_1/(1 + 2/\beta)$. Since the current gain β (or h_{fe}) values of IC transistors are typically much greater than unity, we can say that $I_{C_2} = I_{C_1} \simeq I_1$. For a typical current gain value of 100, the effect of the base current will be to produce only a 2% difference between I_C and I_1. Even a current gain as low as 50 will result in only 4% difference between I_C and I_1. Therefore, for most applications we can safely neglect the effects of the base current and say that $I_{C_2} = I_{C_1} \simeq I_1$.

The circuit that we have been considering so far can be called a *current mirror*, since the current that flows through the left side of the circuit produces essentially a mirror image on the right side. This circuit will be the basis of most of the current source circuits that we will be looking at, and it will also be the basis of most of the differential amplifier active load circuits to be considered later.

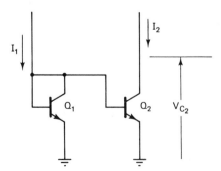

Figure 3.1 Current mirror circuit.

The foregoing analysis of the current mirror transistor pair was made on the basis of the two transistors being identical in every respect. We will now consider what happens in an actual situation in which this is not the case, for even in the case of two integrated-circuit transistors of identical design right next to each other on the same IC chip, there will still be small differences in the electrical characteristics of the two transistors.

The most important difference that exists between two nominally identical transistors is the base width, W. Differences in the base widths of two otherwise identical transistors result in differences in the current gain values and produce an offset voltage, V_{OS}.

For the current mirror circuit, the differences in the current gain values are not important since the base current is small anyway. The offset voltage, however, can be of importance. The concept of the offset voltage of a pair of transistors will also be important in the discussion of the differential amplifier. For two transistors that are characterized in the active region by the relationships

$$I_{C_1} = I_{CO_1} \exp \left(\frac{V_{BE_1}}{V_T} \right), \qquad \text{for } Q_1 \tag{3.1}$$

$$I_{C_2} = I_{CO_2} \exp \left(\frac{V_{BE_2}}{V_T} \right), \qquad \text{for } Q_2 \tag{3.2}$$

the offset voltage, V_{OS}, is given by the equation $\exp(V_{OS}/V_T) = I_{CO_1}/I_{CO_2}$. Thus, for a current mirror pair ($V_{BE_1} = V_{BE_2}$), the two collector currents are not exactly equal, but differ by the ratio $I_{C_1}/I_{C_2} = I_{CO_1}/I_{CO_2} = \exp(V_{OS}/V_T)$. For nominally identical IC transistors, offset voltages of about ± 1 mV are typical. This corresponds to a current ratio of $I_{C_1}/I_{C_2} = \exp(V_{OS}/V_T) = \exp(\pm 1 \text{ mV}/25 \text{ mV}) = 1 \pm \frac{1}{25} = 1 \pm 0.04$, or a $\pm 4\%$ difference in the two collector currents of the pair of transistors.

Let us now consider the simple constant-current source circuit shown in Figure 3.2. Current I_1 is given by

$$I_1 = \frac{V^+ - V_{BE}}{R_1} \tag{3.3}$$

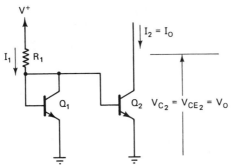

Figure 3.2 Basic current source circuit.

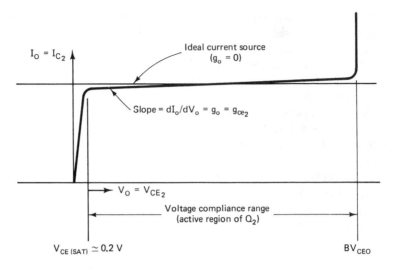

Figure 3.3 Current source output characteristics.

where V^+ is the positive dc supply voltage for the circuit. Since $I_2 = I_{C_2} = I_{C_1} \simeq I_1$, we see that the design of this circuit is relatively simple. If, for example, we want this current source to supply a current of $I_2 = 1.0$ mA, we need $I_1 = 1.0$ mA. If $V^+ = 15$ V, we have that $R_1 = (V^+ - V_{BE})/I_1 = (15 - 0.7)$ V/1.0 mA $= 14.3$ kΩ.

It is important to note that I_2 will stay approximately constant at 1.0 mA only over the range of V_{C_2} values such that transistor Q_2 remains in the active region. If $BV_{CEO} = 50$ V, the range of V_{C_2} voltages for constant-current source operation will be from $+0.2$ to $+50$ V. Outside of this range of voltages, the circuit will no longer act, even approximately, as a constant-current source.

The voltage range over which a circuit acts as an approximate constant-current source is called the *voltage compliance range*. For the circuit under consideration, the voltage compliance range is thus from $+0.2$ to $+50$ V.

Even within the voltage compliance range, however, this circuit is only a good approximation to an ideal current source. In Figure 3.3 the output characteristics of this current source circuit are compared to those of an ideal constant-current source. Note that within the voltage compliance range the output current of this current source does increase slightly with increasing voltage across the current source, $V_O = V_{C_2}$. The I_O versus V_O characteristic curve of the current source circuit has a relatively constant slope over most of the voltage compliance range. This slope is given mathematically by *slope* $= dI_O/dV_O = g_o = $ *dynamic output conductance of the current source*. The reciprocal of g_o is r_o as given by $r_o = 1/g_o$ and is the *dynamic output resistance* of the current source. We note that the ideal current source has a dynamic output conductance, g_o, of zero ($g_o = 0$) and, correspondingly, an infinite dynamic output resistance.

In Figure 3.4 a circuit representation of the real constant-current source is shown and contrasted with the ideal current source for which $g_o = 0$. Note that this representation holds only within the voltage compliance range of the current source.

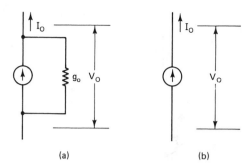

Figure 3.4 Circuit representations of current sources: (a) real current source; (b) ideal current source ($g_{ds} = 0$).

We will now proceed to determine some typical values for the output conductance, g_o. For g_o, we have

$$g_o = \frac{dI_o}{dV_o} = \frac{dI_{C_2}}{dV_{C_2}} = \frac{dI_{C_2}}{dV_{CE2}} = g_{ce2} \doteq \frac{I_C}{V_A} \tag{3.4}$$

where g_{ce2} is the dynamic collector-to-emitter conductance of Q_2, and V_A is the *Early voltage* or *base-width modulation coefficient* of Q_2 as discussed in Appendix B. Typical values of V_A for IC transistors are in the range from 100 to 300 V, with 250 V being selected here as a representative value. Since $I_{C_2} = 1.0$ mA, we have that $g_o = g_{ce2} = I_C/V_A = 1.0$ mA/250 V $= 1000$ μA/250 V $= 4$ μA/V $= 4$ μS, and correspondingly $r_o = 250$ V/1.0 mA $= 250$ kΩ. Therefore, we see that $I_O = I_{C_2}$ changes by 4 μA for every 1-V change in V_O. On a percentage-wise basis, the percentage of change in I_O per 1-V change in V_O is given by $(1/I_C)(dI_C/dV_C) \times 100\% = 4$ μA/V/1000 μA $\times 100\% = \underline{0.4 \%/V}$. Therefore, we see that the output current of this current source increases by 0.4% for every 1-V increase in the voltage across the current source.

On a more general level, we note that since $g_o = I_O/V_A$, the *fractional* change in the output current per 1-V change in the output voltage is given by

$$\frac{1}{I_O}\frac{dI_O}{dV_O} = \frac{1}{I_O}g_o = \frac{1}{I_O}\frac{I_O}{V_A} = \frac{1}{V_A} \tag{3.5}$$

Correspondingly, the percentage of change in I_O per 1-V change in V_O is given by

$$\frac{1}{I_O}\frac{dI_O}{dV_O} \times 100\% = \frac{100\%}{V_A} \tag{3.6}$$

Note that this is *independent* of the output current level.

To summarize the characteristics of the basic current source circuit, it has a basic current output of 1.0 mA with a voltage compliance range of $+0.2$ to $+50$ V. The dynamic output conductance is 4 μS or, on a percentage basis, 0.4%/V. Note that the direction of current flow (at the top or ungrounded terminal) is *into* the current source circuit. As a result of this direction of current flow, this type of constant-current circuit is often referred to as a current *sink*, as contrasted to a current *source* in which the current flows *out* of the top (ungrounded) terminal.

To obtain a current "source" equivalent to this current "sink," the same basic

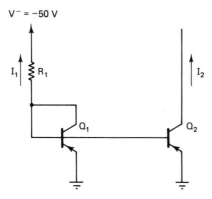

Figure 3.5 Current source circuit.

circuit configuration can be used with the supply voltage V^+ being replaced by a negative dc voltage V^- and the NPN transistors being replaced by PNP transistors, giving the result shown in Figure 3.5. This current source will have a voltage compliance range of from -0.2 to -50 V.

3.1.1 Current Source with Bipolar Voltage Compliance Range

For many IC applications, it is desirable to have a current source that has a *bipolar* compliance range, that is, a compliance range that extends over both positive and negative voltages. The circuit of Figure 3.6 is a simple current sink circuit that has a bipolar voltage compliance range. The basic principles of operation of this circuit are the same as the circuit previously considered. Since the emitter of Q_2 now goes V^- rather than ground, the voltage compliance range of this current sink is shifted in the negative direction by V^- volts.

To consider an example, if $V^+ = 15$ V and $V^- = -15$ V, the voltage compliance range is from -14.8 to $+35$ V. Thus the voltage compliance range extends through ground potential to negative voltage values.

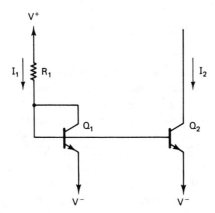

Figure 3.6 Current source with bipolar voltage compliance range.

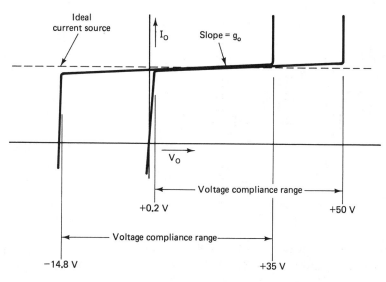

Figure 3.7 Comparison of voltage compliance ranges.

For R_1 we now have the relationship that $I_1 = (V^+ - V^- - V_{BE})/R_1$. Thus, for $I_1 = 1.0$ mA, the corresponding value that is needed for R_1 is $R_1 = (30 - 0.7)$ V/1.0 mA = 29.3 kΩ. If, on the other hand, the top end of R_1 is returned directly to ground rather than to V^+, the required value of R_1 can be reduced to 14.3 kΩ, as before.

The dynamic output conductance of this circuit is the same as the circuit previously considered: 4 μS (or 0.4%/V). In Figure 3.7, the output characteristics of this current source are compared to those of the first current source and to that of an ideal current source.

3.1.2 Current Source for Low Current Levels

In the current source circuits just considered, a resistance R_1 of 29.3 kΩ was required for a current level of 1.0 mA. For smaller current levels, the value of R_1 will have to be increased proportionally. For many IC applications, currents in the microampere range or less are required. For a current of 1.0 μA, a resistance value of $R_1 = 29.3$ MΩ would be needed. If R_1 were to be an ordinary (discrete) resistor, this would pose no major problem since a 29-MΩ resistor costs about the same, and is about the same size, volume, and weight as a 29-kΩ resistor. Integrated-circuit resistors, however, take up an area on the silicon chip that is roughly proportional to the value of the resistor. Therefore, the prorated "cost" of the resistor in terms of the amount of "real estate" on the chip that it occupies increases with increasing resistance. Generally, anything over 50 kΩ is to be avoided for IC applications, if at all possible.

As a result, for current levels below about 1.0 mA, a modification of the current source previously analyzed will now be considered. This new current source is shown in Figure 3.8, and the modification consists of the insertion of a resistor R_2 in series

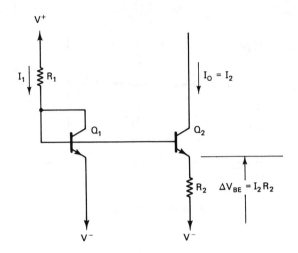

Figure 3.8 Current source for low current levels.

with the emitter of Q_2. With this change, we no longer have that $V_{BE_2} = V_{BE_1}$ due to the voltage drop across R_2. Indeed, we now have that

$$V_{BE_2} = V_{BE_1} - I_2 R_2 \qquad (3.7)$$

To analyze this new situation, we must go back to the basic exponential relationship for a transistor in the active region between the collector current and the base-to-emitter voltages as given by

$$I_C = I_{CO} \exp\left(\frac{V_{BE}}{V_T}\right) \qquad (3.8)$$

For two identical transistors, the ratio of the collector currents is given by

$$\frac{I_{C_1}}{I_{C_2}} = \frac{\exp\left(V_{BE_1}/V_T\right)}{\exp\left(V_{BE_2}/V_T\right)} = \exp\left(\frac{\Delta V_{BE}}{V_T}\right) \qquad (3.9)$$

where $\Delta V_{BE} = V_{BE_1} - V_{BE_2}$. For the circuit under consideration here, we have that $\Delta V_{BE} = I_2 R_2$, so $I_1/I_2 = I_{C_1}/I_{C_2} = \exp\left(I_2 R_2/V_T\right)$. We now see that, as a result of the introduction of R_2 into the circuit, I_2 is no longer equal to I_1, and indeed we can make I_2 much smaller than I_1.

To consider a specific example, let us require that $I_1 = 1.0$ mA, as before, but now let us design for an $I_O = I_2 = 10$ μA $= 0.01$ mA. Since the current ratio is $I_1/I_2 = 1.0$ mA/0.01 mA $= 100 = \exp\left(I_2 R_2/V_T\right)$, we have that $I_2 R_2/V_T = \ln 100 = 4.6$, so $R_2 = 4.6 \times 25$ mV/10 μA $= 11.5$ kΩ. Since R_1 is 29.3 kΩ, as before, for a total resistance $R_1 + R_2 = 40.8$ kΩ, a current level for $I_O = 10$ μA can be obtained, whereas if R_2 was not present, R_1 would have to be increased to 29.3 V/10 μA $= 2.93$ MΩ. Thus, with this circuit, current levels down in the microampere range with acceptable total circuit resistance values (less than 50 kΩ) are possible.

We will now consider the effect that this circuit modification has on other current source parameters, such as voltage compliance range and output conductance. As a result of the voltage drop across R_2, the total voltage compliance span is shifted by $I_2 R_2$. This generally represents only a very small shift in the compliance range. For example, in the case just considered, $I_2 R_2 = 4.6 \times 25$ mV $= 115$ mV. Thus for

$V^+ = 15$ V and $V^- = -15$ V, the compliance range is shifted from -14.8 to $+35$ V to -14.7 to $+34.9$ V, an almost negligible change.

The addition of R_2 to the circuit will, on the other hand, produce a very major change in the output conductance, and a beneficial one at that. For the case of the circuit with R_2, we must consider the full equation for the transistor output conductance as derived in Appendix B and given by

$$g_o = g_c = \frac{I_C}{V_A} \frac{1 + (I_C/V_T)(Z_E + Z_B)/\beta}{1 + (I_C/V_T)[Z_E + (Z_E + Z_B)/\beta]} \tag{3.10}$$

For the circuit under consideration, $Z_E = R_2$, and Z_B is the dynamic impedance that is seen looking out from the base of Q_2. This dynamic impedance consists of the parallel combination of R_1 and the dynamic resistance of the diode-connected transistor Q_1. The dynamic resistance of the diode-connected transistor is given by $r_d = V_T/I_1 = 25\,\text{mV}/1.0\,\text{mA} = 25\,\Omega$. Therefore, $Z_B = R_1 \parallel r_d = 14.3\,\text{k}\Omega \parallel 25\,\Omega \simeq 25$ Ω. Since $I_C = I_{C_2} = I_O = 10\,\mu\text{A}$, and again using $V_A = 250$ V, we obtain

$$g_o = g_c = \frac{10\,\mu\text{A}}{250\,\text{V}} \frac{1 + (10\,\mu\text{A}/25\,\text{mV})(11.5\,\text{k}\Omega)/\beta}{1 + (10\,\mu\text{A}/25\,\text{mV})(11.5\,\text{k}\Omega + 11.5\,\text{k}\Omega/\beta)} \tag{3.11}$$

Since $25\,\text{mV}/10\,\mu\text{A} = 2.5\,\text{k}\Omega$, for $\beta \geq 100$, $(11.5\,\text{k}\Omega/2.5\,\text{k}\Omega)/\beta \ll 1$, so that

$$g_o = g_c \simeq 0.04\,\mu\text{S} \frac{1}{1 + (11.5/2.5)} = \frac{40\,\text{nS}}{5.6} = 7.14\,\text{nS} \tag{3.12}$$

Thus $g_o = 7.14$, nS $= 7.14$ nA/V, and $r_o = 1/g_o = 140\,\text{M}\Omega$. On a percentage basis, this becomes

$$\frac{1}{I_O} \frac{dI_O}{dV_O} = \frac{7.14\,\text{nA/V}}{10\,\mu\text{A}} \times 100\% = \underline{0.0714\%/\text{V}} \tag{3.13}$$

Thus we see that the output current increases by only 0.07% for every volt increase in the voltage across the current source. For a 10-V change in V_o, the change in I_o will still be less than 1%, so this current source represents a very close approximation indeed to an ideal current source. In comparison with the circuit with $R_2 = 0$, we note that for that case the normalized change in the output current was 0.4%/V, so this circuit with R_2 represents an improvement by a factor of 5.6.

3.1.3 Dependence of Current Source Current on Supply Voltage

It is desirable that the strength of the current source be as independent of the dc supply voltage as possible for most applications. We will now consider the dependence of the strength of the current source, I_O, on the supply voltage using the circuit just considered for the "current source for low current levels" as presented in Figure 3.8. Using the exponential relationship between the collector currents and the base-to-emitter voltages, we have previously obtained the relationship

$$\frac{I_1}{I_2} = \exp\left(\frac{\Delta V_{BE}}{V_T}\right) = \exp\left(\frac{I_2 R_2}{V_T}\right) \tag{3.14}$$

so $I_1 = I_2 \exp (I_2 R_2/V_T)$. Taking the derivative dI_1/dI_2 gives

$$\frac{dI_1}{dI_2} = \exp \left(\frac{I_2 R_2}{V_T} \right) + \frac{R_2}{V_T} I_2 \exp \left(\frac{I_2 R_2}{V_T} \right)$$

$$= \frac{I_1}{I_2} + \frac{I_1}{I_2} \ln \frac{I_1}{I_2} \tag{3.15}$$

$$= \frac{I_1}{I_2} \left(1 + \ln \frac{I_1}{I_2} \right)$$

The reciprocal quantity becomes

$$\frac{dI_2}{dI_1} = \frac{I_2}{I_1} \frac{1}{1 + \ln (I_1/I_2)} \tag{3.16}$$

In terms of the fractional change in the currents dI_2/I_2 and dI_1/I_1, we now have

$$\frac{dI_2}{I_2} = \frac{dI_1}{I_1} \frac{1}{1 + \ln (I_1/I_2)} \tag{3.17}$$

For example, if $I_1 = 1.0$ mA and $I_2 = 10$ μA $= 0.01$ mA as considered before, ln $(I_1/I_2) = \ln 100 = 4.6$, so

$$\frac{dI_2}{I_2} = \frac{dI_1}{I_1} \frac{1}{5.6} \tag{3.18}$$

and thus we see that the fractional or percentage of change in the current source output current $I_o = I_2$ is 5.6 times *smaller* than the change in I_1.

In contrast with this, for the case in which $R_2 = 0$, we have that $I_2 = I_1$, so ln $(I_1/I_2) = 0$, and the fractional change in I_2 is thus equal to the change in I_1.

The current I_1 is related to the supply voltage by

$$I_1 = \frac{V^+ - V^- - V_{BE}}{R_1} \simeq \frac{V^+ - V^-}{R_1} = \frac{V_{supply}}{R_1} \tag{3.19}$$

From this we can easily see that

$$\frac{dI_1}{I_1} \simeq \frac{dV_{supply}}{V_{supply}} \tag{3.20}$$

That is, the fractional change in I_1 is equal to the fractional change in the supply voltage. Therefore, for the fractional change in I_2 we have

$$\frac{dI_2}{I_2} \simeq \frac{dV_{supply}}{V_{supply}} \frac{1}{1 + \ln (I_1/I_2)} \tag{3.21}$$

Thus for a 100 : 1 current ratio the fractional (or percentage) change in I_2 is approximately 5.6 times *smaller* than the fractional (or percentage) change in the total supply voltage.

As an example, if $V^+ = +15$ V and $V^- = -15$ V, we have that $V_{supply} = 30$ V. For every 1-V change in the total supply voltage, the percentage of change in the current source current, $I_O = I_2$, is

$$\frac{dI_o}{I_o} \times 100\% = \frac{1 \text{ V}}{30 \text{ V}} \times \frac{1}{5.6} \times 100\% = \underline{0.6\%/\text{V}} \quad (3.22)$$

Thus the current source current changes by approximately 0.6% for every 1-V change in the total supply voltage. This dependence of the strength of the current source on the supply voltage is important when such things as the dependence of amplifier gain on the supply voltage and the power supply rejection ratio (PSRR) are considered.

3.1.4 Temperature Coefficient of the Current Source

We will now consider the effect of temperature on the current source. We will start with the basic interrelationship between the two transistor currents given by $I_2 = I_1$ exp $(-I_2 R_2/V_T)$. Noting that I_1 itself will have a temperature dependence and that $V_T = kT/q$, taking the derivative of I_2 with respect to temperature gives

$$\frac{dI_2}{dT} = \frac{dI_1}{dT} \frac{I_2}{I_1} + I_2 \left[\frac{-I_2 R_2}{V_T} \frac{1}{I_2} \frac{dI_2}{dT} + \frac{I_2 R_2}{TV_T} - \frac{I_2}{V_T} \frac{dR_2}{dT} \right] \quad (3.23)$$

so that dividing through by I_2 and collecting terms in I_2 on the left side gives

$$\frac{1}{I_2} \frac{dI_2}{dT} \left(1 + \ln \frac{I_1}{I_2} \right) = \frac{1}{I_1} \frac{dI_1}{dT} + \frac{1}{T} \ln \frac{I_1}{I_2} - \ln \frac{I_1}{I_2} \frac{1}{R_2} \frac{dR_2}{dT} \quad (3.24)$$

Since $I_1 = (V_{\text{supply}} - V_{BE})/R_1$, we have that

$$TCI_1 = \frac{1}{I_1} \frac{dI_1}{dT} = -\frac{1}{R_1} \frac{dR_1}{dT} - \frac{dV_{BE}/dT}{V_{\text{supply}} - V_{BE}} \simeq -TCR_1 - \frac{dV_{BE}/dT}{V_{\text{supply}}} \quad (3.25)$$

where $TCR_1 = (1/R_1)(dR_1/dT)$ = temperature coefficient of R_1. Substituting the expression for the temperature coefficient of I_1, TCI_1, into the expression for the temperature coefficient of I_2 gives

$$TCI_2 = \frac{1}{I_2} \frac{dI_2}{dT}$$
$$\simeq \frac{-TCR_1 - [(dV_{BE}/dT)/V_{\text{supply}}] + (1/T) \ln(I_1/I_2) - \ln(I_1/I_2)TCR_2}{1 + \ln(I_1/I_2)} \quad (3.26)$$

We will now consider a typical example. For IC resistors, the temperature coefficient is typically in the range of $+2000 \text{ ppm/°C} = +2 \times 10^{-3}/\text{°C}$. For dV_{BE}/dT, a value of -2.2 mV/°C is appropriate. We will again choose $I_1/I_2 = 100$ so that $\ln(I_1/I_2) = 4.6$, and $V_{\text{supply}} = 30 \text{ V}$. Substituting values into the equation for TCI_2 gives

$$TCI_2 = \frac{-2 \times 10^{-3} + (2.2 \text{ mV}/30 \text{ V}) + (1/300)(4.6) - (4.6)(2 \times 10^{-3})}{1 + 4.6}$$
$$= 0.751 \times 10^{-3}/\text{°C} = 751 \text{ ppm/°C} = \underline{0.0751\%/\text{°C}} \quad (3.27)$$

Note that the most important terms in the expression for TCI_2 are the TCR_1,

TCR_2, and the $(1/T) \ln (I_1/I_2)$ terms. A calculation of TCI_2 based on just these three terms gives a result of $0.074\%/°C$.

As a result of the calculation of the temperature coefficient of the current source output current, we see that for the preceding example the current will increase by only $0.075\%/°C$. This variation of the current source current with temperature will contribute to the variation of the amplifier gain and to a small extent to the variation of offset voltage with temperature.

3.1.5 Wilson Current Mirror

We will now consider a compound type of current mirror known as the Wilson current mirror, as shown in Figure 3.9. It will be demonstrated that this type of current mirror can offer some significant advantages over the simple type of current mirror considered earlier.

Base current cancellation. For this analysis, we will consider all transistors to be essentially identical. Since Q_1 and Q_2 have identical base-to-emitter voltages, then $I_{C_1} = I_{C_2}$. Noting that the base currents are very small compared to the collector currents, we can say that $I_3 \simeq I_2 \simeq I_{C_2}$, so all the base currents will be approximately equal to each other. We can therefore write the following nodal equations:

$$(1) \quad I_1 = I_{C_1} + I_B \qquad (3.28)$$

$$(2) \quad I_2 = I_{C_2} + 2I_B \qquad (3.29)$$

$$(3) \quad I_2 = I_3 + I_B \qquad (3.30)$$

and the relationship given above that

$$(4) \quad I_{C_1} = I_{C_2} \qquad (3.31)$$

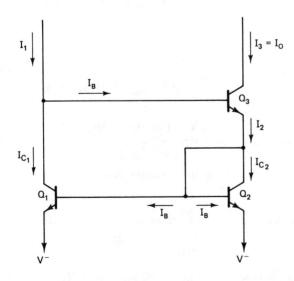

Figure 3.9 Wilson current mirror.

Combining equations (2) and (3) gives

$$I_3 = I_2 - I_B = I_{C_2} + 2I_B - I_B = I_{C_2} + I_B = I_{C_1} + I_B \qquad (3.32)$$

Identifying this last result with equation (1) gives $I_3 = I_1$. We note that there has been a cancellation of the effect of the finite base currents. In actuality, the base current cancellation will not be exact due to mismatches in the transistors, but the resulting difference between I_3 and I_1 will be extremely small.

Voltage compliance range. For this circuit to operate properly, all three transistors must be operating in the active region. Since the voltage drop across Q_2 is V_{BE} or about 0.6 V, and the voltage required to keep Q_3 out of saturation is about $+0.2$ V, the total voltage required across Q_2 and Q_3 is approximately 0.8 V. For example, if $V^- = -15$ V, the lower limit of the voltage compliance range will be -14.2 V.

Dynamic output conductance of Wilson current mirror. To determine the dynamic output conductance of this circuit, we will represent the dynamic collector-to-emitter conductance, g_{ce}, of Q_3 as a conductance external to the transistor, as shown in Figure 3.10. We will assume a change in the output voltage of ΔV_O and determine the corresponding change in output current, ΔI_O, that results. Then, taking the ratio of ΔI_O to ΔV_O will give us the output conductance as $g_O = \Delta I_O / \Delta V_O$.

The change in the output current ΔI_O in passing through Q_2 produces an equal change in the current through Q_1. If we assume that the supply current I_1 remains constant, the change in the base current of Q_3 will be $-\Delta I_O$. This change in the base current of Q_3 produces a change in the collector current of $-\beta \Delta I_O$.

The change in the output voltage, ΔV_O, produces a change in the current through

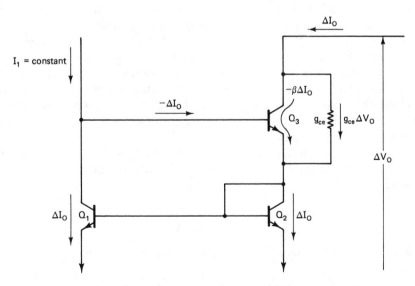

Figure 3.10 Dynamic output conductance of Wilson current mirror.

g_{ce} of $g_{ce} \Delta V_O$. If we now sum the currents at the collector or Q_3, we obtain

$$\Delta I_O = -\beta \Delta I_O + g_{ce} \Delta V_O \qquad (3.33)$$

After collecting terms in I_O on the left side, we obtain

$$\Delta I_O(1 + \beta) = g_{ce} \Delta V_O \qquad (3.34)$$

so we now have

$$\text{dynamic output conductance} = g_O = \frac{\Delta I_O}{\Delta V_O} = \frac{g_{ce}}{1 + \beta} = \frac{I_O/V_A}{1 + \beta} \qquad (3.35)$$

If, for example, $I_O = I_{C_3} = 10 \ \mu A$ and $V_A = 250$ V (the Early voltage), and $\beta = 100$, we obtain for g_O

$$g_O = \frac{10 \ \mu A/250 \ V}{101} = 0.4 \ nA/V = 0.4 \ nS \qquad (3.36)$$

This is an extremely small output conductance. On a normalized or percentage basis, we obtain

$$\frac{1}{I_O} \frac{dI_O}{dV_O} = \frac{1}{V_A(1 + \beta)} = 4 \times 10^{-5}/V = \underline{4 \times 10^{-3}\%/V} \qquad (3.37)$$

Thus the output current changes by only $\underline{0.004\%}$ per 1-V change in the output voltage.

3.1.6 Compound Current Sink (Current-sink-biased Current Sink)

In Figure 3.11, another type of compound current source is illustrated. This circuit can be considered to be a current-sink-biased current sink in which Q_5 and Q_2 (together with R_1 and R_2) operate as a conventional current source circuit, and in turn the collector current of Q_2 acts to bias Q_1, which acts as a current source. Transistors Q_3 and Q_4 are operated as diodes, and the total voltage drop across these two diode-connected transistors, $2V_{BE} = 1.3$ V, is used to provide for the base-to-emitter voltage drop needed to keep Q_1 in the active region and the collector-to-base voltage required to keep Q_2 also in the active region. If Q_3 and Q_4 were not present (that is, replaced by a short-circuit), it would not be possible for Q_1 and Q_2 to be simultaneously in the active region of operation.

To consider a design example, let us choose $I_o = 10 \ \mu A$ and $I_1 = 1.0$ mA, with $V^- = -15$ V. Since $I_1 = (V^- - 3V_{BE})/R_1$ and assuming a V_{BE} of 0.7 V, we obtain for R_1

$$R_1 = \frac{(15 - 2.1) \ V}{1.0 \ mA} = 12.9 \ k\Omega$$

Since the current ratio, $I_1/I_2 = 100$, the voltage drop across R_2 is given by

$$I_2 R_2 = \Delta V_{BE} = V_T \ln 100 = 25 \ mV \times 4.6 = 115 \ mV$$

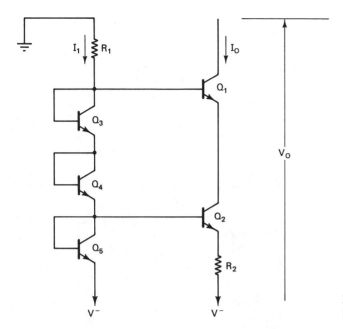

Figure 3.11 Compound current sink (current sink biased current sink).

Since $I_2 = I_o = 10$ μA, we have that $R_2 = 115$ mV/10 μA $= 11.5$ kΩ. Thus we see that the design of this circuit is very similar to that of the simple current source for low current levels considered earlier.

Voltage compliance range. For this current sink to operate within its voltage compliance range, all the transistors in the circuit must be in the active mode of operation. The lower limit of the compliance range is given by

$$V_{\text{lower limit}} = V^- + V_{BE5} + V_{BE4} + V_{BE3} + V_{CB(\text{SAT})1} \quad (3.38)$$
$$= -15 + 3(0.7) - 0.5 = -13.4 \text{ V}$$

so the compliance range extends to 1.6 V above the negative supply voltage. The upper limit of the voltage compliance range is determined by the collector-to-base breakdown voltage of Q_1 as given by

$$V_{\text{upper limit}} = BV_{CBO_1} + 3(V_{BE}) + V^- \quad (3.39)$$

For a 50-V breakdown voltage, this gives

$$V_{\text{upper limit}} = 50 + 2.1 - 15 = +37.1 \text{ V} \quad (3.40)$$

Dynamic output conductance of current-sink-biased current sink. The principal feature of the current-sink-biased current sink compared to the simpler current source circuits is its very low output conductance. This is due to the very high impedance in series with the emitter of Q_1, which is the output impedance of Q_2.

If we look at the equation for the output or collector conductance of a transistor as given by

$$g_c = \frac{I_C}{V_A} \frac{1 + (I_C/V_T)(Z_E + Z_B)/\beta}{1 + (I_C/V_T)[Z_E + (Z_E + Z_B)/\beta]} \quad (3.41)$$

we see that, in the limiting case as Z_E gets very large, g_c will approach $g_c = (I_C/V_A)/\beta$. Since that is approximately the case here, we see that this circuit will be characterized by a very small dynamic output conductance.

If we use $V_A = 250$ V as a representative value, and $\beta = 100$, we obtain

$$g_c \simeq \frac{I_C/V_A}{\beta} = \frac{10 \ \mu A/250 \ V}{100} = 0.4 \ nA/V = \underline{0.4 \ nS} \quad (3.42)$$

The corresponding value for r_o is $r_o = 1/g_o = 1/(0.4 \ nS) = 2.5 \ G\Omega$. On a percentage basis, the percent of change in I_o per 1-V change in V_o is given by

$$\frac{1}{I_o} \frac{dI_o}{dV_o} \times 100\% = \underline{0.004\%/V} \quad (3.43)$$

This circuit is, therefore, within its compliance range, a very good approximation to the ideal current source.

Dynamic output admittance. The very low value of the dynamic output conductance of this current sink is evident only at low frequencies. As we go to higher frequencies, we must consider the parasitic capacitances in the circuit, and in doing so the dynamic output *admittance* of this current sink is given by $y_o = g_o + j\omega C_{total}$. If this IC transistor has a collector–base junction capacitance of 1.0 pF and a collector-to-substrate capacitance of $C_{CS} = 2.0$ pF, the total capacitance seen looking into this current sink (looking into the collector of Q_1) is $C_{total} = 3.0$ pF. The output admittance is therefore given by

$$y_o = g_o + j\omega C_{total} = 0.4 \ nS + j\omega 3.09 \ pF \quad (3.44)$$

The breakpoint frequency for the admittance is given by

$$f_{breakpoint} = \frac{g_o}{2\pi C_{total}} = 21 \ Hz \quad (3.45)$$

Therefore, only for frequencies below about 10 Hz will the output admittance appear as approximately $g_o = 0.4$ nS. Above about 50 Hz, the dominant term in the output admittance is the susceptive term $j\omega C_{total}$. For example, at $f = 1.0$ kHz the output admittance will be $y_o = j19$ nS, and correspondingly the dynamic output impedance will be $-j53$ MΩ. At $f = 1.0$ MHz, the corresponding values will be $y_o = +j19$ μS and $z_o = -j53$ kΩ.

3.1.7 Multiple Current Sinks

Figure 3.12 shows a multiple, or ganged, group of current sinks. Transistor Q_2 and resistor R_2 serve as the reference for the current sink transistors Q_3 through Q_6. We

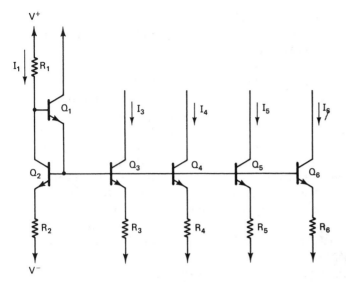

Figure 3.12 Multiple current sinks.

note that the bases of transistors Q_2 through Q_6 are tied together so that all these transistors share a common base voltage. Furthermore, the bottom end of resistors R_2 through R_6 all go to the same negative supply voltage terminal.

Now, if the emitter areas of transistors Q_2 through Q_6 are scaled so that the *current density* is the same in all these transistors, then the base-to-emitter voltage drops, V_{BE}, of all these transistors will be the same. As a result, the voltage across resistors R_2 through R_6 will be the same. Since the collector currents is approximately equal to the emitter currents, and the voltage drop across all the resistors is the same, we can say that $I_3 R_3 = I_4 R_4 = I_5 R_5 = I_6 R_6 = I_1 R_2$. Therefore, the currents are inversely proportional to the emitter resistances.

Let us now consider a design example. Let us specify the following current levels: $I_1 = 1.0$ mA, $I_3 = 1.0$ mA, $I_4 = 2.0$ mA, $I_5 = 4.0$ mA, and $I_6 = 8.0$ mA. We will also specify that the voltage drop across all the resistors, except R_1, be 4.0 V. Therefore, for the resistors we have the following values:

$$(1) \quad R_2 = R_3 = 4.0\ \text{V}/1.0\ \text{mA} = 4.0\ \text{k}\Omega$$

$$(2) \quad R_4 = 4.0\ \text{V}/2.0\ \text{mA} = 2.0\ \text{k}\Omega$$

$$(3) \quad R_5 = 4.0\ \text{V}/4.0\ \text{mA} = 1.0\ \text{k}\Omega$$

$$(4) \quad R_6 = 4.0\ \text{V}/8.0\ \text{mA} = 500\ \Omega$$

If $V^+ = +10$ V and $V^- = -10$ V, the total voltage drop across $R_1 + R_2$ will be $20 - 1.4 = 18.6$ V. Since $R_2 = 4.0$ kΩ, this gives $R_1 = 14.6$ kΩ.

Assuming a representative current gain value of 100, the total base current of transistors Q_2 through Q_6 will be $I_{B(\text{total})} = 16\ \text{mA}/100 = 0.16$ mA. If transistor Q_1 were not present, the current through R_1, I_1, would be 0.16 mA larger than the current through Q_2. To reduce the difference between these two currents, transistor Q_1 is used to reduce the total base current by a factor of its current gain. If the current gain of Q_1 is 100, the total base current of 0.16 mA is reduced to 0.16 mA/100 = 1.6 μA. Therefore, the difference between the current of Q_2 and I_1 is now only 1.6 μA or 0.16%.

To reduce the amount of area taken up on the IC chip by the transistors, it may be desirable not to scale the transistor emitter areas. This would be especially of interest when very large current ratios are needed. If the transistor areas are not scaled to produce equal current densities, the base-to-emitter voltage drops of the transistors will not be equal, and as a result the resistor values must be adjusted to account for this, especially if very accurate current ratios are desired.

To consider an example in which the areas are not scaled, let us return to the previous example. The current through Q_4 is twice that through Q_3 and Q_2, so the V_{BE} of Q_4 is greater than that of Q_3 and Q_2 by V_T ln (I_4/I_3) = 25 mV ln 2 = 17 mV. Therefore, the voltage drop across R_4 is 17 mV less than that across R_3 and R_2, so R_4 must be reduced by 17 mV/2 mA = 8.7 Ω, from 2000 to 1991 Ω. In a similar fashion, the V_{BE} of Q_5 is V_T ln 4 = 35 mV less than that of Q_3 and Q_2, so R_5 must be reduced by 35 mV/4 mA = 8.7 Ω, from 1000 to 991 Ω. Finally, the V_{BE} of Q_6 is V_T ln 8 = 52 mV less than that of Q_3 and Q_2, so R_6 should be reduced by 52 mV/8 mA = 6.5 Ω, from 500 to 493.5 Ω. These small adjustments in the resistance values will compensate for the differences in the V_{BE} values when the transistor areas are not scaled. Since the differences in the V_{BE} values will vary with temperature, the current ratios will have a slight temperature dependence.

3.1.8 Current Source Independent of Supply Voltage

For some applications it is desirable to have a current source whose strength is almost completely independent of the supply voltage. An example of such a circuit is shown in Figure 3.13. For this analysis, we will assume all transistors of the same type to be identical. We will also assume that $R_1 = R_3$ so that $I_1 = I_3$.

We note that I_3 and I_2 are related by I_2/I_3 = exp (I_3R_3/V_T). We also note that the ratio of I_5 to I_6 is determined by the ratio of R_6 to R_5 as given by $I_5/I_6 = R_6/R_5$. Since the base currents are small compared to the collector and emitter currents, we have that $I_2 = I_5$ and $I_3 = I_6$; so we can write

$$\frac{I_2}{I_3} = \exp\left(\frac{I_3R_3}{V_T}\right) = \frac{I_5}{I_6} = \frac{R_6}{R_5} \tag{3.46}$$

Solving this last equation for I_3 gives

$$I_1 = I_3 = \frac{V_T}{R_3} \ln \frac{R_6}{R_5} \tag{3.47}$$

Note that the output current of the current source will be independent of the supply voltage. To have the best accuracy for the I_5/I_6 current ratio, the emitter areas of Q_5 and Q_6 should be scaled such that the two transistors have equal current densities and therefore equal V_{BE} drops.

To consider an example, let us choose a current ratio of I_5/I_6 = 2 so that R_6/R_5

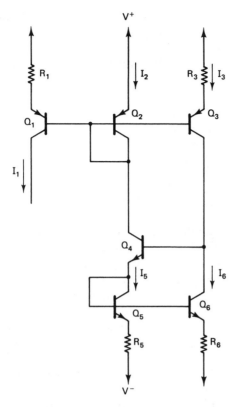

Figure 3.13 Current source that is independent of supply voltage.

= 2. For greatest accuracy, Q_5 should have twice the emitter active area as Q_6. For this case, we have that the current source output current is given by

$$I_{\text{output}} = I_1 = \frac{V_T}{R_3} \ln 2 \tag{3.48}$$

From an output current of 1.0 μA, the required value of $R_3 = R_1$ is

$$R_1 = R_3 = \frac{25 \text{ mV}}{1 \text{ μA}} \ln 2 = 17.3 \text{ k}\Omega$$

Note that a very small current has been obtained using only a very moderate resistance value. Note also that if the active areas of Q_5 and Q_6 are properly scaled it is possible to eliminate R_5 and R_6 from the circuit entirely. The current ratio I_5/I_6 is then determined entirely by the transistor active area ratio as given by $I_5/I_6 = A_5/A_6$. In this case, the current source output current is given by

$$I_{\text{output}} = I_1 = I_3 = \frac{V_T}{R_1} \ln \frac{A_5}{A_6} \tag{3.49}$$

Although the output current has been shown to be independent of the supply voltage, there is a minimum supply voltage level needed for the circuit to operate properly. For the case in which resistors R_5 and R_6 are not present, this minimum supply voltage is $2V_{BE} + V_{CE(SAT)}$, or about 1.6 V.

The voltage compliance range on the output ranges from an upper limit of $V^+ - I_1R_1 - V_{CE(SAT)1} = V^+ - 0.2$ V to a lower limit of $V^+ - BV_{CEO}$. For $V^+ = 10$ V and $BV_{CEO} = 50$ V, this will be a voltage compliance range of from $+9.8$ to -40 V.

This current source is characterized by a constant-current output over a very wide range of supply voltages, an ability to operate with total supply voltages as low as 1.6 V, and microampere current levels with only moderate resistance values. These characteristics make this type of current source especially suitable for micropower operational amplifier applications.

3.1.9 JFET Current Regulator Diodes

A diode-connected junction field-effect transistor (JFET) as shown in Figure 3.14 can be used as a current regulator diode or current source. The JFET is connected as a diode with the gate shorted to the source so that $V_{GS} = 0$. In Figure 3.15 the V–I characteristic of this diode is shown, and it is basically the I_{DS} versus V_{DS} drain characteristic of the JFET with $V_{GS} = 0$. Note that when the drain-to-source voltage V_{DS} goes above the pinch-off voltage V_P, the drain-to-source current I_{DS} saturates at a current level of I_{DSS}. This is not a complete saturation, for as V_{DS} continues to increase, the drain current increases slightly. However, from the point where $V_{DS} = V_P$ up to where $V_{DS} = BV_{DS}$, which is the drain-to-source breakdown voltage, the drain current stays approximately constant; so this region represents the voltage compliance range of this device. The dynamic conductance in this region is the drain-to-source conductance of the JFET g_{ds} and is the slope of the curve as given by $g_{ds} = dI_{DS}/dV_{DS}$. For an ideal current regulator diode or current source, this slope would be zero.

The current regulator diode can be considered the circuit *dual* of the voltage regulator or zener diode. In Figure 3.16, the V–I characteristics of a current regulator diode and a voltage regulator diode are compared. Whereas the voltage regulator diode when operated in the breakdown region acts to keep the *voltage* drop across

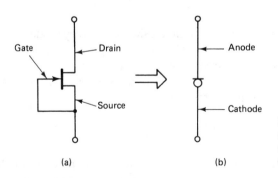

(a)

(b)

Figure 3.14 Diode-connected JFET as a current regulator diode: (a) diode-connected JFET ($V_{GS} = 0$); (b) current-regulator diode symbol.

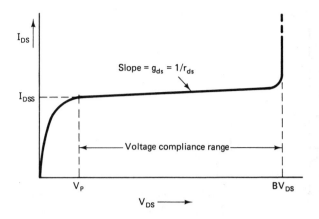

Figure 3.15 Current-regulator diode *V–I* characteristic.

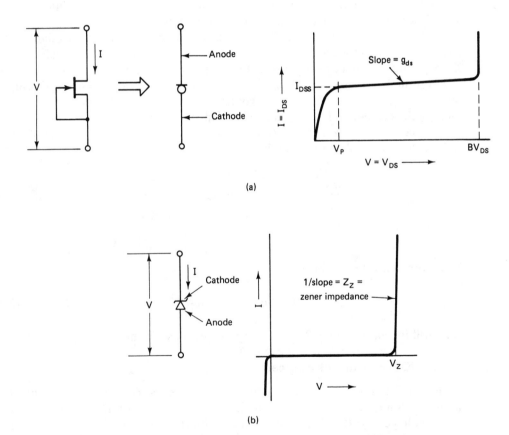

(a)

(b)

Figure 3.16 Current-regulator diode and voltage-regulator diode comparison: (a) current-regulator diode; (b) voltage-regulator diode (zener diode).

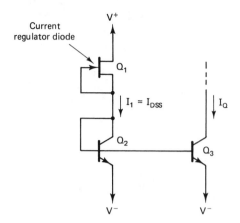

Figure 3.17 Current sink using a current-regulator diode.

it at a constant value, the current regulator diode when operated within its voltage compliance range acts to keep the current through it at a constant value.

In Figure 3.17, an example is presented of the use of a current regulator diode. The current regulator diode Q_1 regulates the current through Q_2 at a value of I_{DSS}. Since Q_2 and Q_3 are a current mirror, the collector current of Q_3, which is the current sink current I_Q, is also equal to I_{DSS}. As a consequence of the current through Q_1 being relatively independent of the voltage across Q_1, and therefore of the supply voltage, the current I_Q is also relatively independent of the supply voltage.

To consider an example, if $I_{DSS} = 1.0$ mA and $r_{ds} = 50$ kΩ, the rate of change of I_Q with supply voltage is given by

$$\frac{dI_Q}{dV_{supply}} = \frac{dI_Q}{dI_{DS}} \times \frac{dI_{DS}}{dV_{DS}} \times \frac{dV_{DS}}{dV_{supply}}$$

$$= 1 \times g_{ds} \times 1 \tag{3.50}$$

$$= 1/50 \text{ k}\Omega = \underline{20 \text{ μS}} = \underline{20 \text{ μA/V}}$$

On a normalized basis, this is

$$\frac{1}{I_Q} \frac{dI_Q}{dV_{supply}} = (1/1000 \text{ μA}) \times 20 \text{ μA/V} = \underline{0.02/V} \text{ or } \underline{2\%/V} \tag{3.51}$$

so I_Q will increase by only 2% for every 1-V increase in the total supply voltage.

3.1.10 MOSFET Current Sources

A MOSFET can also be used as a constant-current source. Figure 3.18 shows the output or drain characteristic of a MOSFET for the case of a fixed value of gate-to-source voltage V_{GS} that is greater than the threshold voltage V_t. As the drain-to-source voltage V_{DS} increases, the drain current I_{DS} increases. However, as V_{DS} increases, the voltage across the gate oxide at the drain end of the channel decreases. This results in a decrease in the mobile charge carrier population of the surface

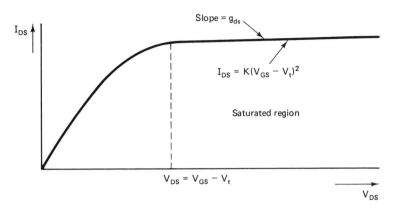

Figure 3.18 MOSFET output characteristic.

inversion layer at the drain end of the channel. This leads to a corresponding decrease in the channel conductance so that as V_{DS} increases the rate of increase of the drain current, dI_{DS}/dV_{DS}, which corresponds to the slope of the curve, decreases.

When V_{DS} reaches a value given by $V_{DS} = V_{GS} - V_t$, the voltage across the gate oxide at the drain end of the channel will be $V_{GD} = V_{GS} - V_{DS} = V_t$. Further increases in V_{DS} cause V_{GD} to drop below the threshold voltage V_t, and the channel becomes pinched off for a very short distance at the drain end. This results in the drain currents leveling off or saturating at a value given by $I_{DS} = K(V_{GS} - V_t)^2$. In this equation, $K = (\mu C_{ox}/2)(W/L)$, where μ is the carrier mobility in the surface inversion layer channel, C_{ox} is the capacitance per unit area of the gate oxide MOS capacitor, and W/L is the channel width to length ratio.

In the saturated region, the drain current increases only very slowly with increasing drain voltage. The slope of the I_{DS} versus V_{DS} curve in the saturated region is the dynamic drain-to-source conductance g_{ds}. This dynamic conductance is directly proportional to the drain current and can be expressed approximately as $g_{ds} = I_{DS}/V_A$, where V_A is a transistor parameter having units of volts and is analogous to the Early voltage of a bipolar transistor. The reciprocal quantity $1/V_A$ is the channel length modulation coefficient and is similarly closely related to the base width modulation coefficient of a bipolar transistor (see Appendix B). The V_A parameter for MOSFETs has about the same range of values as found for the Early voltage of bipolar transistors and is generally in the range of 30 to 150 V.

Many MOSFET current source configurations are similar to the bipolar transistor current source circuits considered previously. A simple example is the circuit of Figure 3.19, which uses a MOSFET current mirror. For current I_1, we have $I_1 = (V^+ - V^- - V_{GS})/R_1$. If transistors Q_1 and Q_2 are a matched pair, then $I_2 = I_1$, subject to the requirement that transistor Q_2 be operating in the saturated region. To satisfy this requirement, we must have the $V_{DS_2} \geq V_{GS} - V_t$, and this will also determine the lower limit of the voltage compliance range. For example, if $V_t =$

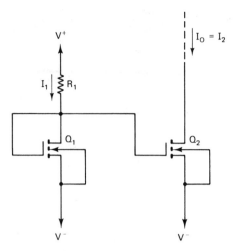

Figure 3.19 MOSFET current source.

+2 V, V_{GS} = +4 V, and V^- = −10 V, then for the operation of Q_2 in the saturated region $V_{DS_2} \geq$ +2 V, and the voltage compliance range extends down to −8 V.

The fractional change in the output current I_O per voltage change in the output voltage V_O is given by $(1/I_O)dI_O/dV_O = (1/I_O)g_o$. Since $g_o = g_{ds} = I_{DS_2}/V_A = I_O/V_A$, this becomes $(1/I_O)dI_O/dV_O = 1/V_A$. If V_A = 50 V, this gives $(1/I_O)dI_O/dV_O = 1/50$ V = 0.02/V = 2%/V.

If transistors Q_2 and Q_1 are identical in every respect except that the channel widths are in the ratio of W_2/W_1, then the current ratio is given by $I_2/I_1 = W_2/W_1$. This is similar to the case of a bipolar transistor current mirror circuit with the transistor active areas, A_2 and A_1, scaled to produce a current ratio given by $I_2/I_1 = A_2/A_1$.

Another example of a MOSFET current source is the compound current source of Figure 3.20. This is essentially the MOSFET counterpart of the bipolar transistor circuit of Figure 3.11. The principal advantage of this MOSFET current source over the simpler circuit just considered is a substantially lower dynamic output conductance and therefore much better current regulation. This, however, will be at the expense of a somewhat reduced voltage compliance range.

The dynamic output conductance g_o is given by $g_o = g_{ds_1}/(1 + g_{fs_1}Z_{S_1})$, where $Z_{S_1} = 1/g_{ds_2}$ (see Appendix B). Since $g_{ds_1} = g_{ds_2} = I_O/V_A$ and $g_{fs_1} = 2K(V_{GS} - V_t) = 2I_O/(V_{GS} - V_t)$, we have that $g_{fs_1}Z_{S_1} = g_{fs_1}/g_{ds_2} = [2I_O/(V_{GS} - V_t)] \times V_A/I_O = 2V_A/(V_{GS} - V_t)$. For g_o, we now have $g_o = dI_O/dV_O = (I_O/V_A)/[1 + 2V_A/(V_{GS} - V_t)]$. The fractional change in I_O per voltage change in V_O or current regulation is given by

$$current\ regulation = \frac{(1/I_O)dI_O}{dV_O} = \frac{g_o}{I_O} = \frac{1/V_A}{1 + 2V_A/(V_{GS} - V_t)} \qquad (3.52)$$

As a representative example, let us take $V_{GS} - V_t$ = 2 V and again assume V_A

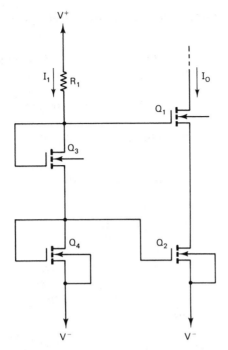

Figure 3.20 MOSFET compound current source.

$= 50$ V. The current regulation is given by

$$current\ regulation = \frac{(1/I_O)dI_O}{dV_O} = \frac{1/50\ V}{1 + 2 \times 50\ V/2\ V} = \frac{1/50\ V}{1 + 50}$$

$$= \frac{1}{2500\ V} = \underline{0.04\%/V} \tag{3.53}$$

This represents a substantial improvement over the current regulation value of 2%/V obtained for the first MOSFET current source circuit that was considered. At an output current level of $I_O = 100$ μA, the corresponding value of output conductance is $g_o = 100$ μA/2500 V = $\underline{40\ nS}$.

We will now determine the lower limit of the voltage compliance range. We will again use the values of $V_{GS} - V_t = +2$ V and $V_t = +2$ V, and we will assume $V^- = -10$ V. We now obtain

$$V_{D_4} = V_{G_2} = V^- + V_{GS_2} = -10 + 4\ V = -6\ V \tag{3.54}$$

and $V_{G_1} = V_{D_3} = V_{G_3} = V_{D_4} + V_{GS_3} = -6 + 4\ V = -2\ V \tag{3.55}$

For Q_1 to be in the saturated region, we must have that $V_{GD_1} = V_{G_1} - V_{D_1} \le V_t = +2$ V. Since $V_{G_1} = -2$ V, then $V_O = V_{D_1} \ge -4$ V, so the voltage compliance range will extend down to -4 V.

3.1.11 MOSFET Current Source Relatively Independent of Supply Voltage

Figure 3.21 shows a MOSFET current source that produces an output current that is relatively independent of the supply voltage. The output current I_O is given by $I_O = I_{DS2} = V_{GS1}/R_2$, and I_1 is given by $I_1 = (V^+ - V^- - V_{GS2} - V_{GS1})/R_1$. For the usual case in which $V^+ - V^- \gg V_{GS}$, this can be simplified to $I_1 \simeq (V^+ - V^-)/R_1$.

The change in I_O with respect to changes in the supply voltage (V^+ or V^-) is

$$\frac{dI_O}{dV^+} = \frac{dI_O}{dV_{GS1}} \times \frac{dV_{GS1}}{dI_1} \times \frac{dI_1}{dV^+} \tag{3.56}$$

$$\simeq \frac{1}{R_2} \times \frac{1}{g_{fs1}} \times \frac{1}{R_1} \tag{3.57}$$

For g_{fs1}, we have

$$g_{fs1} = 2K(V_{GS1} - V_t) = \frac{2I_1}{V_{GS1} - V_t} \simeq \frac{2(V^+ - V^-)}{(V_{GS1} - V_t)R_1} \tag{3.58}$$

so dI_O/dV^+ can now be expressed as

$$\frac{dI_O}{dV^+} \simeq \frac{1}{R_2} \times \frac{(V_{GS} - V_t)R_1}{2(V^+ - V^-)} \times \frac{1}{R_1} = \frac{V_{GS} - V_t}{2R_2(V^+ - V^-)} \tag{3.59}$$

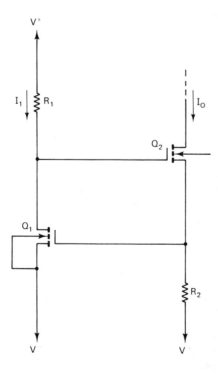

Figure 3.21 MOSFET current source that is relatively independent of the supply voltage.

The fractional change in I_O can now be related to the fractional change in the supply voltage by

$$\frac{dI_O}{I_O} \simeq \left[\frac{V_{GS} - V_t}{2R_2}\right]\left[\frac{dV^+}{V^+ - V^-}\right]\frac{1}{I_O} = \left[\frac{V_{GS1} - V_t}{2V_{GS1}}\right]\frac{dV^+}{V^+ - V^-} \qquad (3.60)$$

For example, let us take $V_t = 1$ V and $V_{GS} = 2$ V, for which we obtain $dI_O/I_O \simeq (1/4)\,dV^+/(V^+ - V^-)$. If $V^+ = +10$ V and $V^- = -10$ V, then a 1-V change in V^+ (or a -1-V change in V^-) will result in a change in I_O of only 1.25%.

Depletion-mode MOSFETs as current sources. Depletion-mode MOSFETs can be used as current regulator diodes in a manner similar to the JFET current regulator diodes. In Figure 3.22a, an example of a depletion MOSFET current sink is shown. The output conductance g_o of this current sink is equal to the dynamic drain–source conductance g_{ds} of the transistor. For a representative example, we use $I_{DSS} = 100$ μA, $V_P = -4.0$ V, $V_A = 80$ V, $BV_{DS} = 40$ V, and $V^- = -10$ V. The output current I_O is $I_O = I_{DSS} = 100$ μA. The dynamic output conductance is given by

$$g_o = \frac{dI_O}{dV_O} = \frac{dI_{DS}}{dV_{DS}} = g_{ds} = \frac{I_O}{V_A} = \frac{100\ \mu A}{80\ V} = \underline{1.25\ \mu S} \qquad (3.61)$$

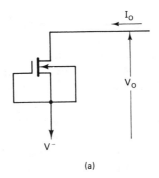

(a)

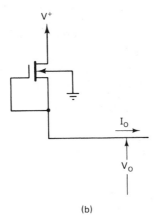

(b)

Figure 3.22 Depletion-mode MOSFETs used as current regulator diodes: (a) depletion-mode MOSFET current sink; (b) depletion-mode MOSFET current source.

The current regulation is

$$current\ regulation = \frac{1}{I_O}\frac{dI_O}{dV_O} \times 100\% = \frac{100\%}{V_A} = \underline{1.25\%/V} \tag{3.62}$$

The voltage compliance range extends from -6 to $+40$ V.

In Figure 3.22b, the same transistor is shown used as a current source. In many IC situations the body of the N-channel MOSFETs will be the common P-type substrate of the IC and at ac ground potential, and it is shown as such in Figure 3.16b. The resulting body-to-source voltage V_{BS} of the transistor will have a major effect on the dynamic output conductance. In Appendix B, there is a discussion of the *substrate bias effect* or *body effect,* and an equation for the output conductance for this case is given as $g_o = g_{ds} + g_{fs}K_{BS}$. If we assume the same parameters as in the foregoing discussion and use a representative value for the body effect coefficient of $K_{BS} = 0.08$, we obtain

$$g_o = \frac{100\,\mu A}{80\,V} + \frac{2 \times 100\,\mu A}{4\,V} \times 0.08 = 1.25\,\mu S + 4.0\,\mu S = \underline{5.25\,\mu S} \tag{3.63}$$

The corresponding current regulation is 5.25%/V. We see that this type of current source has a relatively high output conductance and poor current regulation as a result of the body effect.

Current sources used as active loads. A very important application of current sources for both analog and digital ICs is as an active load for various amplifier and digital logic circuits. Examples of the most commonly used active load configurations are shown in Figure 3.23. The current mirror active load circuit that is widely used for both bipolar transistor and FET differential amplifiers is discussed in Chapter 4.

Equations for the small-signal ac voltage gain for each of the active load circuits of Figure 3.23 will now be given. The voltage gain expression is valid for the case in which the driving transistor is operating in the active region. The external load that is driven by each circuit is represented by a conductance of value g_L. The dynamic forward transfer conductance of the driving transistors is represented by g_{m_1} for the bipolar transistor case and g_{fs_1} for the MOSFET circuits. The output conductances of the driving and load transistors are given as g_{o_1} and g_{o_2}, respectively. The parameters V_{A_1} and V_{A_2} represent the Early voltage, which is the base width modulation coefficient for bipolar transistors and the channel length modulation coefficient for MOSFETS (see Appendix B) for the driving and load transistors, respectively.

1. NPN driver: PNP current source active load

$$A_V = -\frac{g_m}{g_{o_1} + g_{o_2} + g_L} = -\frac{(I_C/V_T)}{(I_C/V_{A_1}) + (I_C/V_{A_2}) + g_L} \tag{3.64}$$

For $g_L \simeq 0$,

$$A_V \simeq -\frac{1/V_T}{(1/V_{A_1}) + (1/V_{A_2})} \tag{3.65}$$

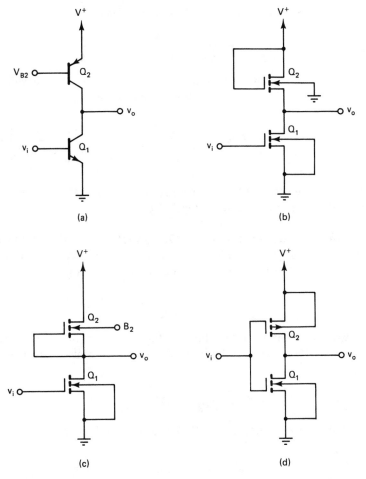

Figure 3.23 Active load circuits: (a) NPN driver–PNP current source active load; (b) NMOS driver–NMOS load; (c) NMOS driver–depletion-mode NMOS current source active load; (d) complementary symmetry (CMOS).

2. NMOS driver: NMOS load

$$A_V = -\frac{g_{fs1}}{g_{o1} + g_{fs2} + g_L} \simeq -\frac{g_{fs1}}{g_{fs2}} \tag{3.66}$$

If the driver and load transistors are identical, except for different channel width W to length L ratios, this can be written as

$$A_V \simeq -\frac{g_{fs1}}{g_{fs2}} = -\frac{2\sqrt{K_1 I_{DS}}}{2\sqrt{K_2 I_{DS}}} = -\sqrt{\frac{K_1}{K_2}} = -\sqrt{\frac{(W/L)_1}{(W/L)_2}} \tag{3.67}$$

3. NMOS driver: depletion NMOS active load

$$A_V = -\frac{g_{fs1}}{g_{o1} + g_{o2} + g_L} = -\frac{2I_{DS}/(V_{GS_1} - V_t)}{(I_{DS}/V_{A1}) + (I_{DS}/V_{A2}) + g_L}, \quad (\text{if } V_{B_2} = V_{S_2}) \qquad (3.68)$$

If the body of the load transistor is the IC substrate, which will be at ac ground potential, then as a result of the body effect $g_{o2} = g_{ds2} + g_{fs2}K_{BS}$ the voltage gain expression becomes

$$A_V = -\frac{g_{fs1}}{g_{o1} + g_{o2} + g_L} = -\frac{2I_{DSS}/(V_{GS_1} - V_t)}{\dfrac{I_{DSS}}{V_{A1}} + \dfrac{I_{DSS}}{V_{A2}} + \dfrac{2I_{DSS}K_{BS}}{-V_P} + g_L} \qquad (3.69)$$

where I_{DSS} is the saturated drain current ($V_{GS} = 0$) of Q_2, V_P is the pinch-off voltage of Q_2, and V_t is the threshold voltage of Q_1.

Complementary symmetry MOS (CMOS).   For the CMOS circuit transistors Q_1 (NMOS) and Q_2 (PMOS), both simultaneously perform the dual functions of driving transistor and active load.   The voltage gain is given by

$$A_V = -\frac{g_{fs1} + g_{fs2}}{g_{o1} + g_{o2} + g_L} \qquad (3.70)$$

If the two transistors have threshold voltages of equal magnitude and if the K values are the same, then at the midpoint of the active (or high-to-low transition) region we have that

$$g_{fs1} = g_{fs2} = \frac{2I_{DS}}{V_{GS} - V_t} = \frac{2I_{DS}}{\frac{1}{2}V^+ - V_t} \qquad (3.71)$$

If, in addition, we have that $g_L \simeq 0$, the expression for the voltage gain can be written as

$$A_V \simeq -\frac{2g_{fs}}{g_{o1} + g_{o2}} = -\frac{\dfrac{4I_{DS}}{\frac{1}{2}V^+ - V_t}}{\dfrac{I_{DS}}{V_{A1}} + \dfrac{I_{DS}}{V_{A2}}} = -\frac{\dfrac{4}{\frac{1}{2}V^+ - V_t}}{\dfrac{1}{V_{A1}} + \dfrac{1}{V_{A2}}} \qquad (3.72)$$

If the two transistors have equal channel length modulation coefficients such that $V_{A1} = V_{A2}$, then the voltage gain expression for the CMOS circuit reduces to simply

$$A_V \simeq -\frac{\dfrac{4}{\frac{1}{2}V^+ - V_t}}{(2/V_A)} = -\frac{2V_A}{\frac{1}{2}V^+ - V_t} \qquad (3.73)$$

If, for example, $V_A = 60$ V, $V^+ = 10$ V, and $V_t = +2$ V, we obtain $A_v = 40$.

3.1.12 Current Mirror ICs

Monolithic IC current mirrors are available such as the TLO11, TLO12, TLO14, and TLO21 (Texas Instruments) series.   These are fixed-ratio NPN Wilson current mirrors, as illustrated in Figure 3.24, where the number of emitters that are shown for

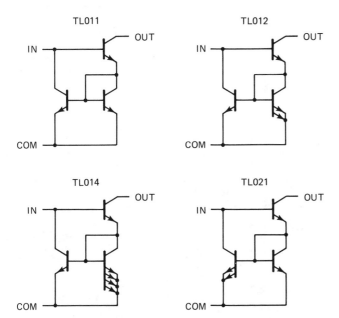

Figure 3.24 Fixed ratio current mirrors (Reprinted by permission of Texas Instruments).

each transistor is indicative of the relative active area. The nominal current ratios, $I_{OUT}:I_{IN}$, are 1:1 for the TLO11, 2:1 for the TLO12, 4:1 for the TLO14, and 1:2 for the TLO21. The current ratio tolerance is $\pm 8\%$ (MAX).

The recommended range of input currents is from 1 μA to 1 mA for all devices except the TLO21, for which the range isd 2 μA to 2 mA. The recommended range of output voltages is from 5 to 35 V for all devices. Since these are Wilson current mirrors, the dynamic output resistance will be very large. The output resistance r_o can be given by the general expression $r_o = 2 \times 10^3$ V/I_{OUT} (TYP), so at $I_{OUT} = 10$ μA the output resistance is 200 MΩ and is 2 MΩ at $I_O = 1$ mA. The dynamic output conductance g_o is $g_o = 1/r_o = I_{OUT}/(2 \times 10^3$ V), and thus the current regulation can be expressed as *current regulation* $= (1/I_{OUT})dI_{OUT}/dV_O = g_o/I_{OUT} = 1/(2 \times 10^3$ V) $= 0.5 \times 10^{-3}$/V $= 0.05\%$/V. Therefore, for every 1-V change in the output voltage, the output current changes by just 0.05%.

Another example of an IC current mirror is the TL010. This is an adjustable-ratio current mirror and a schematic diagram is shown in Figure 3.25, where again the relative active area of each transistor is proportional to the number of emitters shown in the diagram. For this current mirror, the $I_{OUT} : I_{IN}$ current ratio can be set by either connecting a transistor emitter to ground or leaving it open. If the base currents are neglected and all transistors are assumed to be identical, except for differences in the active areas, the current ratio is given by $I_{OUT}/I_{IN} = N/M$, where N is the total number of output emitters that are connected to ground and M is the number of input emitters that are connected to ground. In Figure 3.26, an example is shown for a current ratio of 2:13. If the base currents are taken into account, the equation for the current ratio becomes $I_{OUT}/I_{IN} = [N + (N + M)/\beta]/\{M + [(N + M)/\beta] + [(N + M)/\beta^2]\}$. If, for example, $N = 15$, $M = 1$, and $\beta = 250$, the current

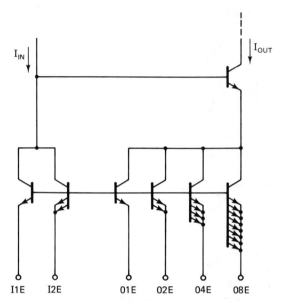

Figure 3.25 Adjustable ratio current mirror TL010 (Texas Instruments).

I1E I2E 01E 02E 04E 08E

ratio is $I_{OUT}/I_{IN} = 14.15$, as compared to the value of 15 that would be obtained if β goes to infinity. The dynamic output resistance of this current mirror is the same as for the TLO series of fixed-ratio current mirrors and is given by $r_o = 2 \times 10^3$ V/I_{OUT} (TYP), and the current regulation is again 0.05%/V (TYP).

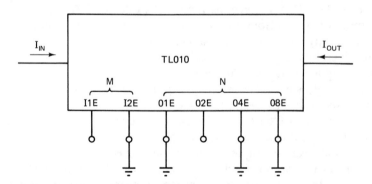

Figure 3.26 Current mirror set for a ratio of 2:13.

3.2 VOLTAGE SOURCES

A voltage source is an electric circuit element that produces an output voltage, V_o, that is independent of the load driven by the voltage source or, equivalently, of the output current. The voltage source is the circuit dual of the constant-current source. In Figure 3.27 the characteristics of the ideal voltage source and the ideal current are compared.

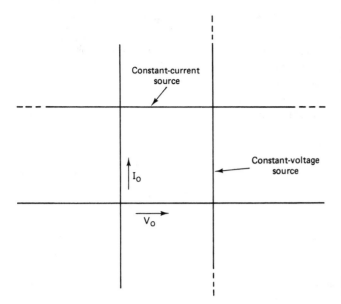

Figure 3.27 Characteristics of ideal constant current source and constant voltage source.

As is the case with the constant-current source, although it is not possible to produce an exact or ideal voltage source, it is possible to have electronic circuits that closely approximate the behavior of the ideal voltage source. Two principal electronic techniques can be used, individually or in combination, to produce a voltage source. One technique is to use the impedance-transforming properties of the transistor, which in turn is related to the current gain of the transistor. The other technique is the use of an amplifier with negative feedback.

3.2.1 Impedance Transformation

To investigate the impedance-transforming characteristics of the transistor, let us turn our attention to the circuit of Figure 3.28. The base of the transistor is driven by a source represented by a voltage source V_S and a series resistance R_S. We will now

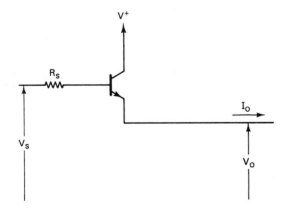

Figure 3.28 Voltage source using transistor impedance transformation.

investigate what happens to V_O as I_O increases. Let us assume an increase in I_O of amount dI_O. This produces an increase in the base current of amount $dI_B = dI_E/(\beta + 1) = dI_O/(\beta + 1)$. This increase in the base current increases the voltage drop across the source resistance by an amount $dI_B R_S = dI_O R_S/(\beta + 1)$. There will also be an increase in the base-to-emitter voltage drop of the transistor as given by $dV_{BE} = (dV_{BE}/dI_E) \, dI_E$. The quantity dV_{BE}/dI_E is the dynamic emitter-to-base resistance of the transistor. Its value can easily be determined from the following relationships. For the transistor in the active mode, we have that $I_E \simeq I_C = I_{CO} \exp (V_{BE}/V_T)$, so $dI_E/dV_{BE} \simeq dI_C/dV_{BE} = I_E/V_T$, and therefore $dV_{BE}/dI_E = r_{eb} = V_T/I_E$.

The total change in the output voltage is therefore given by

$$dV_O = -dI_B R_S - dV_{BE} = -\left[\frac{dI_O R_S}{\beta + 1} + dI_O r_{eb}\right] \qquad (3.74)$$

The effective output resistance of this circuit, as seen looking into the emitter of the transistor, is given by

$$\text{output resistance} = r_o = \frac{-dV_O}{dI_O} = \frac{R_S}{\beta + 1} + r_{eb} \qquad (3.75)$$

We see that as a result of the current gain of the transistor the effective value of R_S as seen looking into the emitter of the transistor is $R_S/(\beta + 1)$. Since the current gain, β, of the transistor is generally large, typically on the order of 100, this represents a very large impedance transformation.

To consider an example, let us assume that $R_S = 1000 \, \Omega$ and $I_O = 5$ mA, with $V_O = 10$ V and a current gain of 100 for the transistor. The effective output resistance is given by

$$r_o = \frac{1000}{\beta + 1} + \frac{25 \text{ mV}}{5 \text{ mA}} = 10 \, \Omega + 5 \, \Omega = 15 \, \Omega \qquad (3.76)$$

The change in V_O is therefore a drop of 15 mV per 1-mA increase in I_O. On a percentage basis, this is (15 mV/mA)/10 V $\times$ 100% = 0.15%/mA, so the output voltage decreases by 0.15% for every 1-mA increase in the output current.

The relationships above hold true only for small changes in the output current. We will now investigate the changes in V_O that result from large changes in the output current. Starting with the exponential relationship between current and base-to-emitter voltage for the transistor, we have that $I_E \simeq I_C = I_{CO} \exp (V_{BE}/V_T)$, so $V_{BE} = V_T \ln (I_E/I_{CO})$. The change in V_{BE} resulting from the output, or emitter current changing from I_{E_1} to I_{E_2}, is given by $\Delta V_{BE} = V_T \ln(I_{E_2}/I_{E_1}) = V_T \ln(I_{O_2}/I_{O_1})$. Therefore, the total change in the output voltage that results from the output current increasing from a value of I_{O_1} to I_{O_2} is given by

$$\Delta V_O = -\left[\frac{(I_{O_2} - I_{O_1})R_S}{\beta + 1} + V_T \ln \frac{I_{O_2}}{I_{O_1}}\right] \qquad (3.77)$$

Thus, if in the example above I_O changes from $I_{O_1} = 1.0$ mA up to a full-load

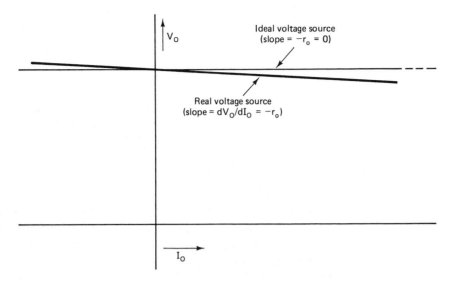

Figure 3.29 Voltage source graphical output characteristics.

value of $I_{O_2} = 5.0$ mA, the change in the output voltage is given by

$$\Delta V_O = -\left[\frac{4.0 \text{ mA} \times 1000 \text{ } \Omega}{101} + 25 \text{ mV (ln 5)} \right]$$

$$= -(40 \text{ mV} + 40 \text{ mV}) = -80 \text{ mV}$$

(3.78)

which represents a voltage decrease of 0.8%.

The *load regulation* of a voltage source of a voltage regulator is the change or decrease in the output voltage as the output current goes from some specified no-load current value to a full-load current. The load regulation is directly related to the output impedance of the voltage source or voltage regulator. In Figure 3.29, the output characteristics of the ideal voltage source are compared to those of a real voltage source. In Figure 3.30, the corresponding equivalent circuits are shown.

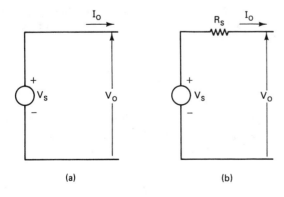

(a)

(b)

Figure 3.30 Voltage source equivalent circuits: (a) ideal voltage source; (b) real voltage source.

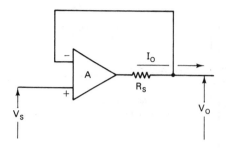

Figure 3.31 Use of amplifier with negative feedback to reduce output resistance.

3.2.2 Use of Negative Feedback to Reduce the Output Impedance

In Figure 3.31, a simple feedback circuit is presented to illustrate how an amplifier with negative feedback can be used to produce a very low value of output impedance and thus ensure a very good load regulation for the voltage source. In this circuit, A is the gain (open loop) of the amplifier and R_S represents the open-loop output impedance of the amplifier. The resistance R_S is actually internal to the amplifier, but for purposes of analysis is represented as a resistor external to the amplifier.

The output voltage of this circuit, V_O, is given by $V_O = (V_S - V_O)A - I_O R_S$, so collecting terms in V_O on the left side gives

$$V_O(1 + A) = V_S A - I_O R_S$$

and thus
$$V_O = V_S \frac{A}{1 + A} - I_O \frac{R_S}{1 + A}$$

$$= V_S \frac{A}{1 + A} - I_O R'_S$$

(3.79)

where $R'_S = R_S/(1 + A)$ is the *closed-loop* output resistance of this circuit. For the usual case of a very large open-loop gain such that $A \gg 1$, we have that $V_O \simeq V_S - R'_S I_O$, where $R'_S \simeq R_S/A$. For this case, we see that the closed-loop output resistance is very much smaller than the open-loop value.

Very often the output stage of the amplifier is an emitter-follower impedance-transforming stage of the type just studied. In addition, an emitter-follower stage is often added to the amplifier output to increase the current output range. An example of such a circuit is shown in Figure 3.32. The open-loop output resistance has been shown previously to be given by

$$R_o = \frac{R_S}{\beta + 1} + r_{eb} = \frac{R_S}{\beta + 1} + \frac{V_T}{I_O}$$

(3.80)

Under the feedback conditions of Figure 3.32, the corresponding closed-loop output resistance is given by $R'_O = R_O/(A + 1)$. For $R_S = 1000 \ \Omega$ and $I_O = 1.0$ mA, for example, we have that $R_O = (1000/101) + (256 \text{ mV}/1.0 \text{ mA}) = 35 \ \Omega$. If the amplifier open-loop gain at low frequencies is 10,000, the *closed-loop* output resistance is

$$R'_O = \frac{R_O}{A + 1} = \frac{35 \ \Omega}{10 \text{ k}\Omega} = 3.5 \text{ m}\Omega = 3.5 \ \mu\text{V/mA}$$

(3.81)

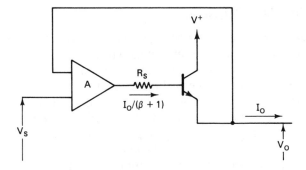

Figure 3.32 Feedback amplifier with emitter-follower output circuit.

We see that that output resistance of this circuit will have the very low value of 3.5 mΩ or 3.5 μV/mA. This means that the decrease in the output voltage for an increase in the output current of 1.0 mA is only 3.5 μV. This represents a very small change indeed, and thus it can be said that this circuit represents a very close approximation to an ideal voltage source.

The value of the dynamic output resistance obtained above is the dynamic output impedance of the voltage source circuit only at low frequencies. At higher frequencies the decreasing gain and phase angle of the voltage gain of the amplifier produces corresponding changes in both the magnitude and the angle of the output impedance. For most amplifiers, and especially for operational amplifiers, the voltage gain can be expressed in the form $A(f) = A(0)/[1 + j(f/f_1)]$, where $A(0)$ is the zero-frequency open-loop gain and f_1 is the breakpoint frequency. For frequencies that are more than one-half of a decade above this breakpoint frequency, the open-loop voltage gain can be expressed approximately as $A(f) \simeq A(0)f_1/jf = f_u/jf$, where f_u is the unity-gain frequency as given by $f_u = A(0)f_1$. The closed-loop output impedance as a result is given by

$$Z_S' = \frac{R_O}{A + 1} \simeq \frac{R_O}{f_u/jf} = jf\frac{R_O}{f_u}$$

We thus see that in the higher-frequency region the output impedance begins to look reactive and the closed-loop output impedance is given by

$$Z_S' \simeq jf\frac{R_O}{f_u} = j\omega\frac{R_O}{2\pi f_u} = j\omega L_S' \qquad (3.82)$$

where $L_S' = R_O/2\pi f_u = R_O/\omega_u$ is the effective output inductance of the circuit. To continue with the previous example, and assuming an amplifier unity gain frequency of 1.0 MHz, we have that $Z_s' \simeq jf(35\ \Omega/1\ \text{MHz})$. The breakpoint frequency f_1 is given by $f_1 = f_u/A(0) = 1.0\ \text{MHz}/10^4 = 100\ \text{Hz}$. Thus at a frequency of 1.0 kHz the dynamic output impedance will be $j35\ \text{m}\Omega = j35\ \mu\text{V/mA}$ as compared to the low-frequency value of 3.5 mΩ. At 10 kHz, the output impedance will have risen to $+j0.35\ \Omega$, and at 100 kHz it will be approximately $+j3.5\ \Omega$. The equivalent circuit of this voltage source can at these frequencies be represented as a dc voltage source, V_S, in series with a resistance of 3.5 mΩ and an inductance of 5.6 μH.

3.2.3 Power Supply Rejection (Line Regulation)

We have seen that one very desirable feature of a voltage source is that it have a very low dynamic output impedance so that the output voltage changes very little with changes in the output current. Another desirable feature for voltage sources or voltage regulators is that the output voltage be as independent as possible of the supply voltage. A simple example of a circuit in which this is accomplished is shown in Figure 3.33. In this circuit, the voltage-regulator diode or *zener diode* is biased by a current source. The characteristics of the voltage-regulator diode are such that, when it is biased in the breakdown region beyond the knee of the curve, the voltage drop across the diode is relatively independent of the current through the diode. The voltage drop across the diode is not, however, completely independent of the current. The change in the voltage drop across the diode divided by the change in the current is called the *zener impedance*, given by $Z_Z = dV_Z/dI_Z$, where V_Z is the voltage across the diode and I_Z is the current through the diode. Typical values of the zener impedance Z_Z range from a few ohms to a few tens of ohms.

In the circuit of Figure 3.33, a change in the supply voltage dV_{supply} results in a small change in the current through the current source of amount $dI_O = g_o\,dV_{supply}$, where g_o is the dynamic output conductance of the current source. This results in a change in the current through the voltage regulator diode of amount $dI_Z = dI_O$, which in turn changes the voltage drop across the voltage regulator diode by an amount $dV_Z = Z_Z\,dI_Z = Z_Z\,dI_O = g_oZ_Z\,dV_{supply}$. This ratio of the change in the voltage across the voltage regulator diode, and therefore of the output voltage, V_O, to the change in the supply voltage is given by

$$\frac{dV_O}{dV_{supply}} = \frac{dV_Z}{dV_{supply}} = g_oZ_Z \tag{3.83}$$

To consider a representative example, let us choose $Z_Z = 10\ \Omega$ and $g_o = 100$ nS, so that $dV_O/dV_{supply} = 100\,\text{nS} \times 10\,\Omega = 1 \times 10^{-6}$. Thus a change in the supply voltage of 1.0 V results in a change in the output voltage of only 1 μV.

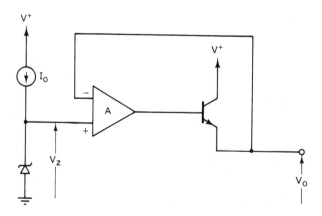

Figure 3.33 Voltage source with current source biasing for supply voltage rejection.

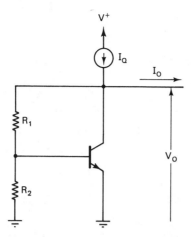

Figure 3.34 Voltage source using V_{BE} as the reference voltage.

3.2.4 Voltage Source Using V_{BE} as the Reference Voltage

Figure 3.34 shows a voltage source that uses the base-to-emitter voltage drop as the reference voltage. If we assume the base current to be small, the output voltage will be related to the V_{BE} of the transistor by the simple resistive voltage divider of R_1 and R_2 as given by the relationship

$$\frac{V_{BE}}{V_O} = \frac{R_2}{R_1 + R_2}$$

so that

$$V_O = \frac{V_{BE}(R_1 + R_2)}{R_2} = V_{BE}\left(1 + \frac{R_1}{R_2}\right) \tag{3.84}$$

The dynamic output resistance can be obtained by assuming a change in the output voltage V_O of amount dV_O, and finding the change in I_O that results. Taking the ratio of dV_O to dI_O gives $dV_O/dI_O = r_o =$ dynamic output resistance. For purposes of simplicity, we will again assume the base current to be small.

A change in the output voltage of amount dV_O changes the current through $R_1 + R_2$ by an amount $dV_O/(R_1 + R_2)$. This current change in turn produces a change in the V_{BE} of the transistor by an amount given by $dV_{BE} = [dV_O/(R_1 + R_2)]R_2$, which produces a change in the current flowing through the transistor by an amount $dI_C = g_m\, dV_{BE} = [g_m R_2/(R_1 + R_2)]\, dV_O$. The total change in the output current will be the sum of the change of current through $R_1 + R_2$ and the change in the transistor current, so we have

$$dI_O = dV_O\left(\frac{1}{R_1 + R_2} + \frac{g_m R_2}{R_1 + R_2}\right) = dV_O\, \frac{1 + g_m R_2}{R_1 + R_2} \tag{3.85}$$

The dynamic output resistance is therefore given by

$$r_o = \frac{dV_O}{dI_O} = \frac{R_1 + R_2}{1 + g_m R_2} \tag{3.86}$$

Since

$$\frac{V_O}{V_{BE}} = \frac{R_1 + R_2}{R_2} \tag{3.87}$$

we can rewrite r_o in the form

$$r_o = \frac{V_O}{V_{BE}} \frac{R_2}{1 + g_m R_2} \tag{3.88}$$

If $g_m R_2 \gg 1$, as is usually the case, then we can write r_o as

$$r_o \simeq \frac{V_O}{V_{BE}} \frac{1}{g_m} = \frac{V_O}{V_{BE}} \frac{V_T}{I_C} \tag{3.89}$$

If, for example, $I_C = 1.0$ mA and $V_O = 1.0$ V, and $V_{BE} = 650$ mV at this collector current, we have that

$$r_o \simeq \frac{1000 \text{ mV}}{650 \text{ mV}} \left(\frac{25 \text{ mV}}{1.0 \text{ mA}} \right) = \underline{38.5 \ \Omega} \tag{3.90}$$

To minimize the effect of the base current, we should make the current through R_1 and R_2 at least 10 times and preferably 20 times as large as the maximum expected base current. For a current gain of $\beta = 50$ (min.), this means that the current through R_1 and R_2 should be about $20(1.0 \text{ mA}/50) = 0.4$ mA. Since $V_o = 1.0$ V, we have that $R_1 + R_2 = 1.0 \text{ V}/0.4 \text{ mA} = 2.5 \text{ k}\Omega$. For R_2 we have that $R_2 = 650 \text{ mV}/0.4$ mA $= 1.625 \text{ k}\Omega$, and thus $R_1 = 2.5 \text{ k}\Omega - 1.625 \text{ k}\Omega = 875 \ \Omega$. The current source I_Q should be designed to supply the required amount of current through $R_1 + R_2$ plus the transistor collector and the maximum required load current. If the maximum required load current is, for example, 2.0 mA, the current source should be 3.4 mA.

The output voltage of this voltage source is relatively independent of the supply voltage by virtue of the small dynamic conductance that can be exhibited by the current source. A change dV_{supply} in the supply voltage will change the current source current by $dI_Q = g_o \, dV_{\text{suply}}$, where g_o is the dynamic output conductance of the current source. This current change will have the same effect on the output voltage, V_O, as will an equal change in I_O, so we will have that $dV_O = r_o \, dI_Q = r_o g_o \, dV_{\text{supply}}$, so that we end up with $dV_O/dV_{\text{supply}} = g_o r_o$. If g_o, for example, is 1.0 μS, we have

$$\frac{dV_O}{dV_{\text{supply}}} = 10^{-6} \text{ mS} \times 39 \ \Omega = 39 \times 10^{-6} = 39 \ \mu\text{V/V}$$

Thus a change in the supply voltage of 1 V results in a change in the output voltage of only 39 μV.

3.3 VOLTAGE REFERENCE

A voltage reference is an electric circuit designed to produce an output voltage that is independent of temperature. In practice, it is never possible to achieve this complete independence of temperature, especially over an extended temperature range.

The variation of the output voltage of the voltage reference circuit with temperature is expressed in terms of the temperature coefficient or *tempco*, given by

$$TC_{V(\text{REF})} = \frac{dV_{\text{REF}}}{dT} = \text{temperature coefficient of the reference voltage}$$

The most important characteristic of a voltage reference is the temperature coefficient of the output voltage, $TC_{V(\text{REF})}$. In most cases it is also desirable that the reference voltage be as independent of the supply voltage as possible, that is, that there be a good power supply rejection. Another desirable feature is that the output voltage be as independent of the loading or output current as possible; that is, the circuit should have a low output impedance. Indeed, in many cases a *voltage reference* circuit is used to bias a *voltage source* circuit, the combination for many applications being called a *voltage regulator*. A voltage regulator therefore combines the characteristics of a low temperature coefficient, a low output impedance (good load regulation), and good power supply rejection characteristics (good line regulation).

Since all electronic components exhibit temperature coefficients, the basic technique employed in voltage reference circuits is to design the circuit such that there are canceling effects so as to produce, at least nominally, a zero temperature coefficient at a given temperature. An example of a circuit in which such a cancellation technique is used is shown in Figure 3.35. In this circuit, transistors Q_1 through Q_3 are operated as diode-connected transistors. The current source, I_Q, drives a current through Q_1 in the reverse-bias direction such that Q_1 is operated in the reverse-bias breakdown region as a voltage regulator or zener diode. The voltage drop produced across Q_1

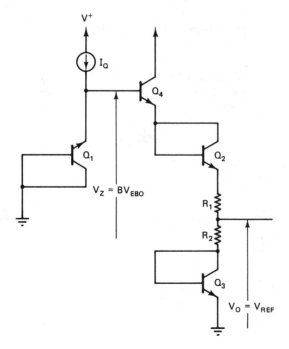

Figure 3.35 Temperature-compensated voltage reference.

is the emitter–base junction breakdown voltage, so we have $V_Z = BV_{EBO}$. This voltage is typically around 6 or 7 V.

The voltage at the emitter of Q_2 is $V_{E_2} = V_Z - V_{BE_4} - V_{BE_2}$. The voltage at the collector of Q_3 is $V_{C_3} = V_{BE_3}$. The output voltage of this circuit is therefore given by

$$V_{REF} = V_O = V_{E_2} \frac{R_2}{R_1 + R_2} + V_{C_3} \frac{R_1}{R_1 + R_2} \tag{3.91}$$

This result is obtained by the simple application of the voltage-division rule and the superposition theorem.

Since the currents through Q_4, Q_2, and Q_3 are essentially the same, and these transistors are IC transistors of similar construction, the V_{BE} voltage drop, and more important the temperature coefficient of the V_{BE} drop, dV_{BE}/dT, are approximately the same for these three transistors. We therefore can write V_O as

$$V_{REF} = V_O = \frac{(V_Z - 2V_{BE})R_2 + V_{BE}R_1}{R_1 + R_2} = \frac{V_Z R_2 - V_{BE}(2R_2 - R_1)}{R_1 + R_2} \tag{3.92}$$

The temperature coefficient of the reference voltage is therefore

$$TC_{V(REF)} = \frac{dV_{REF}}{dT} = \frac{R_2(dV_Z/dT) - (dV_{BE}/dT)(2R_2 - R_1)}{R_1 + R_2} \tag{3.93}$$

We must certainly note here that R_1 and R_2 are temperature dependent. Nevertheless, R_1 and R_2 are IC resistors of similar construction, so they will have identical temperature coefficients; so for any given temperature change both resistors will increase by the same fractional or percentage amount. Therefore, the ratio of R_1 to R_2 will not change with temperature, so in looking at $TC_{V(REF)}$ we do not have to consider the temperature variation of R_1 and R_2.

Looking at the equation for $TC_{V(REF)}$ we now note the possibility of producing a zero temperature coefficient by suitable choice of the resistance ratio such that the numerator becomes equal to zero. For this to be the case, we require that

$$R_2 \frac{dV_Z}{dT} = (2R_2 - R_1) \frac{dV_{BE}}{dT} \tag{3.94}$$

so that

$$\frac{2R_2 - R_1}{R_2} = \frac{dV_Z/dT}{dV_{BE}/dT} \tag{3.95}$$

Since dV_{BE}/dT is negative, this can be expressed in the most convenient form as

$$\frac{R_1}{R_2} - 2 = \frac{dV_Z/dT}{-dV_{BE}/dT}$$

or as

$$\boxed{\frac{R_1}{R_2} = 2 + \frac{dV_Z/dT}{-dV_{BE}/dT}} \tag{3.96}$$

To consider a representative example, we will use the following quantities:

$$V_Z = BV_{EBO} = 6.3 \text{ V}, \qquad \frac{dV_Z}{dT} = +3.0 \text{ mV/°C}$$

(3.97)

$$V_{BE} = 0.70 \text{ V}, \qquad \frac{dV_{BE}}{dT} = -2.3 \text{ mV/°C}$$

For a zero temperature coefficient for V_{REF}, the required resistor ratio is given by

$$\frac{R_1}{R_2} = 2 + \frac{dV_Z/dT}{-dV_{BE}/dT} = 2 + \frac{+3 \text{ mV/°C}}{+2.3 \text{ mV/°C}} = 3.30$$

(3.98)

The equation for V_{REF} given earlier can be rewritten in terms of the R_1/R_2 ratio by dividing through by R_2, giving

$$V_{REF} = V_Z \frac{1}{1 + R_1/R_2} - V_{BE} \frac{2 - R_1/R_2}{1 + R_1/R_2}$$

(3.99)

Insertion of the appropriate quantities gives for V_{REF} the value of 1.68 V. Note that the value of V_{REF} cannot be selected at will, for its value is a consequence of the requirement that $TC_{V(REF)} = 0$.

Using this circuit for a reference voltage source, it is possible to produce a $TC_{V(REF)}$ that is *nominally* zero. That is, the temperature coefficient will be zero if all the circuit parameters have values that correspond exactly to the values used in the design (that is, the "design center" values). If any of the circuit parameters do not in actuality correspond to the value used in the design, the temperature coefficient will not be zero, although it may still be very small.

To examine the possible effects of deviations of circuit parameters from their design center values on $TC_{V(REF)}$, let us take the equation for $TC_{V(REF)}$ given earlier and, by dividing through by R_2, rewrite it as

$$TC_{V(REF)} = \frac{(dV_Z/dT) - (dV_{BE}/dT)(2 - R_1/R_2)}{1 + R_1/R_2}$$

(3.100)

If all quantities have their design center values, the numerator will cancel out and $TC_{V(REF)}$ will be equal to zero. Now let us see what the resulting temperature coefficient will be if $TC_{V_Z} = dV_Z/dT$ differs from its design center value by ±5%. If this is the case, then for $TC_{V(REF)}$ we have

$$TC_{V(REF)} = \frac{\pm 0.05(dV_Z/dT)}{1 + 3.3} = \pm 0.0116 \times 3 \text{ mV/°C}$$

$$= \pm 0.035 \text{ mV/°C} = \pm 35 \text{ }\mu\text{V/°C}$$

(3.101)

Thus, instead of being zero, the actual value of $TC_{V(REF)}$ may be in the range of 35 μV/°C, or on a fractional basis 0.0021%/°C. Although the temperature coefficient of V_{REF} has not been reduced to zero, it still will be satisfactorily small for most applications.

To consider now a second example, let us see what temperature coefficient for V_{REF} will result from a deviation of ± 0.1 mV/°C in the value of dV_{BE}/dT for which a design center value of -2.3 mV/°C was chosen. After substitution into the equation for $TC_{V(REF)}$, we now have

$$TC_{V(REF)} = \frac{\pm 0.1 \text{ mV/°C}(2 - R_1/R_2)}{1 + R_1/R_2}$$

$$= \frac{\pm 0.1 \text{ mV/°C}(2 - 3.3)}{4.3} = \pm 0.030 \text{ mV/°C} \qquad (3.102)$$

$$= \pm 30 \text{ } \mu\text{V/°C}$$

and, on a percentage basis, $\pm 0.0018\%$/°C. This is, again, a small residual temperature coefficient and should prove satisfactory for most applications.

Finally, let us investigate the effect of a small deviation of the R_1/R_2 resistance ratio from the design center value. For integrated-circuit resistors, the ratio tolerance of similarly constructed resistors on the same IC chip is generally quite small, generally less than 5%. For this example, let us look at the effect of a $\pm 2\%$ difference between the actual resistance ratio and the design center value.

For this case, by suitable substitution into the equation for $TC_{V(REF)}$, we obtain

$$TC_{V(REF)} = \frac{(-dV_{BE}/dT)(R_1/R_2)(\pm 0.02)}{1 + R_1/R_2} = \frac{\pm 0.02 \times 2.3 \text{ mV/°C} \times 3.3}{4.3}$$

$$= \pm 0.035 \text{ mV/°C} = \pm 35 \text{ } \mu\text{V/°C} \qquad (3.103)$$

and, on a percentage basis, $\pm 0.0021\%$/°C. This, again, is a very small residual temperature coefficient, so this circuit should prove to perform adequately for most applications.

3.3.1 Band-gap Voltage Reference

In Figure 3.36 a very interesting and useful voltage reference circuit is presented. To proceed forthwith with the analysis of this circuit, we have that $V_{REF} = V_{BE3} + I_2R_2$. In this analysis we will assume all transistors to be identical, so that for the relationship between I_1 and I_2 we will have that $I_1 = I_2 \exp(I_2R_3/V_T)$. In this analysis, we will assume the base currents to be small enough to be neglected. From the equation just given, we can solve for I_2R_3 as $I_2R_3 = V_T \ln(I_1/I_2)$, so that

$$I_2R_2 = \frac{R_2}{R_3} I_2R_3 = \frac{R_2}{R_3} V_T \ln \frac{I_1}{I_2} \qquad (3.104)$$

Substituting this into the expression for V_{REF} gives

$$V_{REF} = V_{BE3} + \frac{R_2}{R_3} V_T \ln \frac{I_1}{I_2} \qquad (3.105)$$

We now will take note of the fact that V_{BE} will have a negative temperature coefficient (decreases with increasing temperature), whereas the last term in the equation will

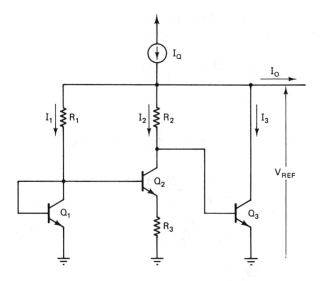

Figure 3.36 Band-gap voltage reference circuit.

have a positive temperature coefficient since $V_T = kT/q$. Therefore, by suitable choice of the resistance ratio and the current ratio, it should be possible to produce a cancellation of the two temperature coefficients and thereby end up with a zero net temperature coefficient.

To investigate this further, let us now obtain an expression for the temperature coefficient of V_{REF} as

$$TC_{V(REF)} = \frac{dV_{REF}}{dT} = \frac{dV_{BE}}{dT} + \frac{R_2}{R_3} \frac{k}{q} \ln \frac{I_1}{I_2} \qquad (3.106)$$

Note again that since R_1 and R_2 are two IC resistors of similar construction in close thermal contact on the same IC chip, the fractional change in both resistors will be the same, so the ratio of the two resistances is independent of temperature. To proceed further, we will now need to make a slight digression in order to obtain an expression for $TC_{V(BE)} = dV_{BE}/dT$.

Temperature coefficient of V_{BE}. For a transistor in the active region we have the now familiar exponential relationship between the collector current and the base-to-emitter voltage as given by

$$I_C = I_{CO} \exp\left(\frac{V_{BE}}{V_T}\right) \qquad (3.107)$$

where V_T is the thermal voltage (kT/q) and the preexponential constant, I_{CO}, is a strong function of temperature as given by $I_{CO} = CT^3 \exp(-qE_{GO}/kT)$. The quantity E_{GO} is the value of the energy band gap at absolute zero (0 K) as obtained by a linear extrapolation from room temperature (300 K) to absolute zero. The value of E_{GO} is 1.205 V = 1205 mV.

To obtain the temperature coefficient of V_{BE}, we first obtain an expression for

V_{BE} as given by $V_{BE} = V_T \ln (I_C/I_{CO})$, and then take the derivative of V_{BE} with respect to temperature for constant I_C. Doing this gives

$$\frac{dV_{BE}}{dT} = \frac{k}{q} \ln \frac{I_C}{I_{CO}} - V_T \frac{d(\ln I_{CO})}{dT}$$

$$= \frac{V_{BE}}{T} - V_T \frac{d(\ln I_{CO})}{dT} \tag{3.108}$$

Since $\ln I_{CO} = \ln C + 3 \ln T - qE_{GO}/kT$, we have that $d(\ln I_{CO}/dT = (3/T) + qE_{GO}/kT^2$, so we now have

$$\frac{dV_{BE}}{dT} = \frac{V_{BE}}{T} - V_T \left(\frac{3}{T} + \frac{qE_{GO}}{kT^2} \right)$$

$$= \frac{V_{BE}}{T} - 3 \left(\frac{k}{q} + \frac{E_{GO}}{T} \right) \tag{3.109}$$

$$= - \left(\frac{E_{GO} - V_{BE}}{T} + 3\frac{k}{q} \right)$$

Since

$$3\frac{k}{q} = 3(1.38 \times 10^{-23} \text{ J/K})(1.602 \times 10^{-19} \text{ C})$$

$$= 2.6 \times 10^{-4} \text{ V/K} = 0.26 \text{ mV/°C} \tag{3.110}$$

and

$$E_{GO} = 1205 \text{ mV}$$

we have

$$\boxed{TC_{V(BE)} = \frac{dV_{BE}}{dT} = - \left(\frac{1205 \text{ mV} - V_{BE}}{T} + 0.26 \text{ mV/°C} \right)} \tag{3.111}$$

For a representative value of $V_{BE} = 650$ mV, we obtain a temperature coefficient of $TC_{V(BE)} = -2.1$ mV/°C.

Conditions for a zero temperature coefficient. Now that we have obtained an analytical expression for $TC_{V(BE)}$ we can insert it into the expression for $TC_{V(REF)}$ to obtain

$$TC_{V(REF)} = \frac{dV_{REF}}{dT} = \frac{dV_{BE}}{dT} + \frac{R_2}{R_3} \frac{k}{q} \ln \frac{I_1}{I_2} \tag{3.112}$$

For $TC_{V(REF)} = 0$, we require that $(R_2/R_3)(k/q) \ln (I_1/I_2) = -dV_{BE}/dT$, so the corresponding value of V_{REF} is given by

$$V_{REF} = V_{BE} + \frac{R_2}{R_3} V_T \ln \frac{I_1}{I_2} = V_{BE} - T\frac{dV_{BE}}{dT}$$

$$= V_{BE} + (E_{GO} - V_{BE}) + 3\frac{k}{q} T$$

$$= E_{GO} + 3V_T = 1205 \text{ mV} + 78 \text{ mV}, \quad \text{at } 300 °C$$

$$= \underline{1283 \text{ mV}} = \underline{1.283 \text{ V}} \tag{3.113}$$

Thus this voltage reference circuit when operating under the condition that $TC_{V(\text{REF})} = 0$ will have an output voltage of 1.28 V. Notice that the reference voltage is determined principally by the energy band-gap value ($E_{GO} = 1205$ mV for silicon).

To produce the zero-temperature condition, the basic circuit requirement is given above as

$$\frac{R_2}{R_3} \frac{k}{q} \ln \frac{I_1}{I_2} = -\frac{dV_{BE}}{dT} = \frac{E_{GO} - V_{BE}}{T} + 3\frac{k}{q} \tag{3.114}$$

If we now multiply both sides of this equation by T, we obtain

$$\frac{R_2}{R_3} V_T \ln \frac{I_1}{I_2} = E_{GO} - V_{BE} + 3V_T = 1283 \text{ mV} - V_{BE} \tag{3.115}$$

Dividing through by $V_T = 25.9$ mV at 300 K gives

$$\frac{R_2}{R_3} \ln \frac{I_1}{I_2} = \frac{1283 \text{ mV} - V_{BE}}{25.9 \text{ mV}} \tag{3.116}$$

For a representative value of $V_{BE} = 650$ mV, this will yield the condition $(R_2/R_3) \ln (I_1/I_2) = 24.5$.

Design example. We will now consider a design example of the bandgap voltage reference. For this example we will make the following arbitrary, but reasonable choices: $I_1 = I_3 = 1.0$ mA, $I_1/I_2 = 5$. Since $I_2 R_3 = V_T \ln (I_1/I_2) = 41.6$ mV, we have that

$$R_3 = 41.6 \text{ mV}/0.2 \text{ mA} = \underline{208 \ \Omega}$$

For the resistance ratio, we have that $R_2/R_3 = 24.5/\ln (I_1/I_2) = 15.2$, so $R_2 = 15.2 \times 208$ $\Omega = \underline{3166 \ \Omega}$. For R_1, we have that

$$R_1 = \frac{V_{\text{REF}} - V_{BE}}{I_1} = \frac{1283 \text{ mV} - 650 \text{ mV}}{1.0 \text{ mA}} = \underline{633 \ \Omega} \tag{3.117}$$

All three resistance values have now been determined, and we note that they are all of reasonable and acceptable magnitude for high-accuracy IC resistors in not being excessively large or excessively small. The current values obtained are also within a reasonable range. The strength of the current source should be such that $I_Q = I_1 + I_2 + I_3 + I_{O(\text{MAX})} = 2.2 \text{ mA} + I_{O(\text{MAX})}$, where $I_{O(\text{MAX})}$ is the maximum output current that this voltage reference current will have to supply. Normally, for purposes of minimizing the effects of loading on the reference voltage, this output current will be very small.

Effect of loading on V_{REF}. To determine the effect of changes on the output current of the band-gap voltage reference circuit on V_{REF}, we note that changes in V_{REF} will result in changes in I_1, I_2, and I_3. However, due to the stabilizing effects of the resistors, the change in the current through Q_3 will be the predominant effect. Or, to put it another way, changes in I_O will be matched by almost equal but opposite changes in I_3. Corresponding to a change in I_3, there will be a change in the base-to-emitter voltage of Q_3, as given by the relationship $dI_3/dV_{BE3} = gm_3 = I_3/V_T$. As a result, we can write

$$\frac{dV_{\text{REF}}}{dI_O} = \frac{dV_{\text{REF}}}{-dI_3} = \frac{dV_{BE3}}{-dI_3} = \frac{-1}{g_{m_3}} \tag{3.118}$$

so therefore we have that

$$r_o \cong \frac{-dV_{\text{REF}}}{dI_O} = \frac{1}{g_{m_3}} = \frac{V_T}{I_3} \tag{3.119}$$

Thus, if $I_3 = 1.0 \text{ mA}$ as in the example above, then $r_o = 25 \text{ }\Omega$. If, for example, we wanted to limit the variations in V_{REF} due to variations in the output current to no more than 1.0 mV, we will have a corresponding limitation on variations in I_O as given by

$$\frac{\Delta V_{\text{REF}}}{\Delta I_O} = 25 \text{ }\Omega$$

so that

$$\Delta I_{O(\text{MAX})} = \frac{1 \text{ mV}}{25 \text{ }\Omega} = 40 \text{ }\mu\text{A} \tag{3.120}$$

As a result, we see that to obtain the full benefits available from this voltage reference we should generally insert a high-input-impedance buffer circuit between this voltage reference and the load that is to be driven.

We can also note, at this point, the desirability of using a current source to provide the overall biasing of this circuit to prevent variations in the supply voltage, V^+, from having an undue effect on V_{REF}. To illustrate the benefit afforded by the use of a current source with a low dynamic output conductance, let us consider the variation in the reference voltage with respect to the supply voltage as given by

$$\frac{dV_{\text{REF}}}{dV^+} = \frac{dV_{\text{REF}}}{dI_3} \frac{dI_3}{dI_Q} \frac{dI_Q}{dV^+} \tag{3.121}$$

wherein again we make use of the fact that changes in I_Q will be taken up principally by changes in the current through Q_3 so that, as a result, $dI_3/dI_Q \simeq 1$. Since $dV_{\text{REF}}/dI_3 = 1/g_{m_3} = V_T/I_3$, and $dI_Q/dV^+ = g_o$, we now have that

$$\frac{dV_{\text{REF}}}{dV^+} \simeq \frac{1}{g_{m_3}} g_o = \frac{V_T}{I_3} g_o \tag{3.122}$$

If, for example, $I_3 = 1.0 \text{ mA}$ as before and $g_o = 100 \text{ nS} = 1 \times 10^{-7} \text{ S}$, we have

$$\begin{aligned} \frac{dV_{\text{REF}}}{dV^+} &= 25 \text{ }\Omega \times 1 \times 10^{-7} \text{ S} = 2.5 \times 10^{-6} \\ &= 2.5 \text{ }\mu\text{V/V} \end{aligned} \tag{3.123}$$

so the change in V_{REF} is only $2.5 \text{ }\mu\text{V}$ for every 1-V change in the supply voltage.

This represents a very good degree of power supply rejection, indeed, so this circuit is well insulated from the two external effects of temperature variation and supply voltage variation.

Effect of error in parameter values on temperature coefficient. In this circuit, as was the case for the voltage reference circuit considered earlier, we must take note of the fact that although the nominal temperature coefficient will be zero, the actual temperature coefficient will not be zero, due to the inevitable deviation in the component values from their exact design center values. To illustrate this point again, let us consider the effect of a deviation in the R_2/R_3 resistance ratio from the design center value. For this example, we will choose a representative deviation of $\pm 2\%$.

The equation for $TC_{V(\text{REF})}$ has been given as

$$TC_{V(\text{REF})} = \frac{dV_{BE}}{dT} + \frac{R_2}{R_3}\frac{k}{q}\ln\frac{I_1}{I_2} \qquad (3.124)$$

When R_2/R_3 is at the design center value, we will assume that $TC_{V(\text{REF})} = 0$, so the two terms on the right side are equal in magnitude but opposite in algebraic sign. Therefore, a $\pm 2\%$ deviation in R_2/R_3 from the design center value will produce correspondingly a $\pm 2\%$ divergence in the magnitude of the entire quantity of which R_2/R_3 is a factor. Since at the design center point this quantity is equal in magnitude to dV_{BE}/dT, the $\pm 2\%$ variation in R_2/R_3 results in a variation equal to $\pm 2\%$ of dV_{BE}/dT in $TC_{V(\text{REF})}$. Since the design center value of $TC_{V(\text{REF})}$ is zero, the $\pm 2\%$ divergence of the resistance ratio results in a residual temperature coefficient for V_{REF} given by

$$TC_{V(\text{REF})} = \pm 0.02\,\frac{dV_{BE}}{dT} = \pm 0.02 \times 2.1 \text{ mV/}^\circ\text{C} = 42\ \mu\text{V/}^\circ\text{C} \qquad (3.125)$$

and, on a percentage basis, $\pm 0.0033\%/^\circ\text{C}$.

If a very small temperature coefficient is desired, a technique such as laser trimming can be utilized. If such a technique as laser trimming is to be employed with a band-gap regulator of the type under consideration here, resistors R_2 and R_3 would be in the form of thin-film resistors deposited by a vacuum deposition technique on the IC chip. A high-energy laser beam can then be used to cut a notch or slot in the thin-film resistor and thereby increase the resistance value as a consequence of the reduced effective width of the resistor. The laser is generally part of a feedback loop that senses the temperature coefficient of the reference voltage and controls the depth of the laser cut. If the temperature coefficient is too high in the positive direction, R_3 can be trimmed to increase its value and thereby bring the temperature coefficient down toward zero. If, on the other hand, the temperature coefficient is negative, R_2 would be trimmed to bring the temperature coefficient up toward zero. By this means, very small residual temperature coefficients can be achieved, even down to the range of a few microvolts per degree.

Band-gap voltage reference with larger reference voltage. The reference voltage produced by the band-gap voltage reference under consideration here is

$V_{REF} = E_{GO} + 3V_T = 1283$ mV. A larger reference voltage can be obtained by adding some diodes, in the form of diode-connected transistors, to the circuit at the lower end of R_2 in Figure 3.33. If n diodes are added to the circuit, the equation for V_{REF} becomes

$$V_{REF} = (n + 1)V_{BE} + \frac{R_2}{R_3} V_T \ln \frac{I_1}{I_2} \tag{3.126}$$

and correspondingly

$$TC_{V(REF)} = \frac{dV_{REF}}{DT} = (n + 1)\frac{dV_{BE}}{dT} + \frac{R_2}{R_3}\frac{k}{q} \ln \frac{I_1}{I_2} \tag{3.127}$$

For zero temperature coefficient, we thus have the requirement that

$$\frac{R_2}{R_3}\frac{k}{q} \ln \frac{I_1}{I_2} = -(n + 1)\frac{dV_{BE}}{dT} \tag{3.128}$$

When this condition is satisfied, the reference voltage is given by

$$V_{REF} = (n + 1)V_{BE} - (n + 1)T\frac{dV_{BE}}{dT} = (n + 1)\left(V_{BE} - \frac{T\,dV_{BE}}{dT}\right) \tag{3.129}$$

For the previous analysis, we have seen that $V_{BE} - T(dV_{BE}/dT) = E_{GO} + 3V_T$, so we now have that $V_{REF} = (n + 1)(E_{GO} + 3V_T) = (n + 1)(1.283$ V). Thus, if one additional diode (diode-connected transistor) is used, $V_{REF} = 2 \times 1.283$ V $= 2.566$ V, and for two additional diodes $(n = 2)$ we have that $V_{REF} = 3 \times 1.283 = 3.849$ V. Since

$$\frac{dV_{BE}}{dT} = -\left[\frac{E_{GO} + (3kT/q) - V_{BE}}{T}\right] \tag{3.130}$$

the design condition for the resistance ratio is

$$\frac{R_2}{R_3} \ln \frac{I_1}{I_2} = (n + 1)\frac{1283 \text{ mV} - V_{BE}}{25.9 \text{ mV}} \tag{3.131}$$

For a current ratio of 5 and for $n = 2$, the required resistance ratio is

$$\frac{R_2}{R_3} = (3)\frac{1283 \text{ mV} - 650 \text{ mV}}{25.9 \ln 5} = 45.6 \tag{3.132}$$

The large value of resistance ratio required in this case is not very desirable from the standpoint of IC circuit design, and the resistance ratio tolerance will probably be somewhat larger than in the previous example. As a result, the residual temperature coefficient on a percentage basis will likely be larger than in the case in which no additional diodes are used.

3.3.2 Voltage Reference with Feedback Amplifier

In Figure 3.37, the combination of a voltage reference and a feedback amplifier is shown. This circuit can provide for the step-up of the reference voltage to a value needed for the particular application and for the isolation of the voltage reference

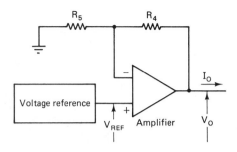

Figure 3.37 Voltage reference with feedback amplifier for load isolation and voltage step-up.

from the load so that changes in the output current will not change the reference voltage and produce only very minimal changes in the output voltage due to the very small value of the closed-loop output impedance of the feedback amplifier.

In this circuit a fraction of the output voltage given by the voltage division ratio $R_5/(R_4 + R_5)$ is fed back to the inverting input terminal of the amplifier. It is compared in the amplifier to the reference voltage, which is applied to the noninverting input terminal. The amplifier by means of the feedback loop acts to maintain the voltage at the output such that $V_O R_5/(R_4 + R_5) = V_{\text{REF}}$. As a result, we have that

$$V_O = V_{\text{REF}} \left(1 + \frac{R_4}{R_5} \right) \tag{3.133}$$

As a result, V_{REF} can be stepped up to any reasonable voltage to meet the circuit requirements. Note that whatever residual temperature coefficient there is for V_{REF} will be also stepped up, and by exactly the same factor as V_{REF} itself, so that we will have that $TC_{V_O} = dV_O/dT = (dV_{\text{REF}}/dT)(1 + R_4/R_5)$. Nevertheless, on a percentage or fractional basis the normalized temperature coefficient of V_O will be the same as that of V_{REF} as given by

$$\frac{1}{V_O} \frac{dV_O}{dT} = \frac{1}{V_{\text{REF}}} \frac{dV_{\text{REF}}}{dT} \tag{3.134}$$

This last result is based on the assumption that the resistance ratio R_4/R_5 does not change with temperature. To ensure compliance with this, these two resistors should be resistors with closely matched temperature coefficients and should be placed in the circuit such that they will be at the same temperature so that as a result the resistance variations of the two resistors will track each other.

3.3.3 Voltage Reference Diode

We will consider a two-terminal device based on the band-gap voltage reference circuit that produces a constant, temperature-compensated voltage drop across its terminals. This device thus is a diode characterized by a constant voltage drop and a very low temperature coefficient, and we will call it a *reference diode*.

The circuit of the reference diode is shown in Figure 3.38 and is a slight simplification of the circuit of the LM113 reference diode (National Semiconductor). In this circuit, transistors Q_1 through Q_3 and resistors R_1 through R_3 perform the same respective functions as in the basic band-gap voltage reference circuit of Figure 3.36. Therefore, the diode voltage drop V_D will be equal to 1.283 V.

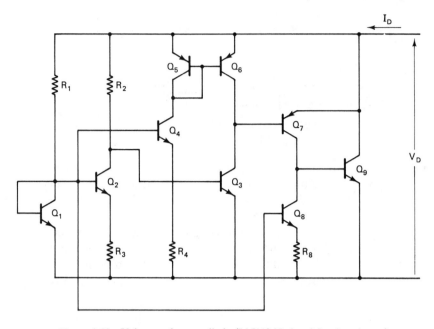

Figure 3.38 Voltage reference diode (LM113 National Semiconductor).

The basic function of the remainder of the circuit is to provide for a low output impedance and to allow for a relatively large diode current. Transistor Q_4 in conjunction with resistor R_4 acts as a current source and biases the current mirror comprised of Q_5 and Q_6. Transistor Q_6 acts as a current source and is the active load for Q_3. Transistor Q_3 is in the common-emitter configuration and drives transistor Q_7, which is also in the common-emitter configuration. Transistor Q_8 in conjunction with resistor R_8 is a current sink and provides an active load for Q_7. Transistor Q_9 in the common-emitter configuration is driven by Q_7 and provides most of the current passing through the diode.

The basic operation of this diode comprised of transistors Q_4 through Q_9 and the associated resistors can be understood as follows. An increase in V_D produces a corresponding increase in the base-to-emitter voltage of Q_3, since the voltage drop across R_2 remains relatively constant. The increase in the current through Q_3 is given by $\Delta I_3 = g_{m_3} \Delta V_{BE3} = g_{m_3} \Delta V_D$. Since Q_6 is a current source, current I_6 will not change, and thus the change in the current through Q_3 will produce an equal change in the base current of Q_7 as given by $\Delta I_{B_7} = \Delta I_3$. The change in the base current of Q_7 produces a change in the collector current of this transistor by an amount $\Delta I_7 = \beta_7 \Delta I_{B_7} = \beta_7 \Delta I_3$. Since Q_8 is a constant-current sink, the change in the collector current of Q_7 results in an equal change in the base current of Q_9, so $\Delta I_{B_9} = \beta_7 \Delta I_3$. The base current change produces a change in the collector current of amount $\Delta I_9 = \beta_9 \beta_7 \Delta I_3$. Since ΔI_3 has been related to ΔV_D by $\Delta I_3 = g_{m_3} \Delta V_D$, we now have that $\Delta I_9 = \beta_9 \beta_7 g_{m_3} \Delta V_D$. As a result of the current gains of Q_7 and Q_9, the current through Q_9 is much larger than that through Q_7 or Q_3, so the change in the diode current is due mostly to the change in the current through Q_9; so we have $\Delta I_D \simeq \Delta I_9 = \beta_9 \beta_7 g_{m_3} \Delta V_D$. As a result, the output resistance for this diode (the diode dynamic resistance) can be written as

$$r_d = \frac{\Delta V_D}{\Delta I_D} = \frac{1}{\beta_9 \beta_7 g_{m_3}} \qquad (3.135)$$

Since $g_{m_3} = I_3/V_T$, this can be rewritten as

$$\text{diode dynamic resistance} = r_d = \frac{V_T/I_3}{\beta_9 \beta_7} \qquad (3.136)$$

As a result of the current–gain product in the denominator of this expression, very low values of diode dynamic resistance can be obtained, as will be seen in the example to follow.

Voltage reference diode design example. For an example of the design of the voltage reference diode, we will require that the diode dynamic resistance be no greater than 0.1 Ω. We will assume that all the transistors have current gains of $\beta = 50$ minimum. We therefore have that

$$r_d = 0.1\ \Omega = \frac{25\ \text{mV}/I_3}{50 \times 50} \qquad (3.137)$$

so we have for I_3 the value of

$$I_3 = \frac{25\ \text{mV}}{0.1\ \Omega \times 50 \times 50} = 100\ \mu\text{A} \qquad (3.138)$$

In accordance with this value of I_3, a reasonable choice for I_1 is 200 μA; so, assuming a V_{BE} of 600 mV, we have

$$R_1 = \frac{(1.283 - 0.600)\ \text{V}}{0.2\ \text{mA}} = \underline{3.415\ \text{k}\Omega} \qquad (3.139)$$

We will select a current ratio of $I_1/I_2 = 5$, so for R_3 we have the relationship

$$\frac{I_2 R_3}{V_T} = \ln \frac{I_1}{I_2} = \ln 5$$

so that $\qquad R_3 = \dfrac{25\ \text{mV}}{40\ \mu\text{A}} \ln 5 = \underline{1.006\ \text{k}\Omega} \qquad (3.140)$

We can now easily solve for R_2 as

$$R_2 = \frac{V_D - V_{BE}}{I_2} = \frac{(1.283 - 0.600)\ \text{V}}{40\ \mu\text{A}} = \underline{17.1\ \text{k}\Omega} \qquad (3.141)$$

Since $I_3 = I_6 = I_5 = I_4 = 100\ \mu$A, we have for R_4 the relationship

$$\frac{I_4 R_4}{V_T} = \ln \frac{I_1}{I_4} = \ln 2 \qquad (3.142)$$

so that $\qquad R_4 = \dfrac{25\ \text{mV} \ln 2}{100\ \mu\text{A}} = \underline{173\ \Omega} \qquad (3.143)$

Based on the choices for the other currents, an acceptable choice for I_8 is 100 μA, so $R_8 = R_4 = \underline{173\ \Omega}$.

If we add up all the bias currents, we have a total quiescent current of $I_Q = I_1 + I_2 + I_5 + I_6 + I_8 = (200 + 40 + 100 + 100 + 100)\ \mu A = 540\ \mu A$. Therefore, it should be expected that, for this diode to be operating in its constant-voltage temperature-compensated regime, a current, I_D, of at least 540 μA should pass through the diode. If the current level is below this, the circuit will not function properly, giving rise to a larger dynamic resistance and a significant temperature coefficient.

3.3.4 Voltage References

Temperature-compensated zener diode voltage reference. A zener diode has a positive temperature coefficient that is typically in the +3-mV/°C range. The temperature coefficient does, however, depend on the breakdown voltage of the device, being smaller for zener diodes with lower breakdown voltages. A forward-biased PN junction has a negative temperature coefficient that is generally in the −2.0- to −2.5-mV/°C range, with the exact value being dependent on the current through the device. Since the zener diode and the forward-biased diode have temperature coefficients that are of opposite algebraic signs, but of comparable magnitudes, the combination of the two in series will result in a *temperature-compensated zener diode* that has a temperature coefficient that is much lower than that of just a zener diode alone. A diagram of a temperature-compensated zener diode is shown in Figure 3.39, and the physical structure of an IC temperature-compensated zener diode is discussed in Chapter 2. Indeed, by suitable design of the zener diode and suitable choice of the bias current, it is possible to achieve a temperature-compensated zener diode that has a nominal temperature coefficient of zero, with typical values being in the 1-ppm/°C range. The breakdown voltage of the zener diode portion of the temperature-compensated zener diode must be approximately 5.7 V in order to achieve temperature coefficient matching with the temperature coefficient of the forward-biased diode. Since the voltage drop across the forward-bias diode is approximately 0.6 V, the total voltage across the temperature-compensated zener diode is 6.3 V.

An example of an IC voltage reference based on the temperature-compensated zener diode is the REF10 (Burr-Brown). A simplified circuit diagram of this device is shown in Figure 3.40. The output voltage of this voltage reference circuit is given by $V_o = V_{TCZ}\,[1 + (R_2/R_1)]$, where V_{TCZ} is the voltage across the temperature-compensated zener diode. Using the nominal values given in the circuit of V_{TCZ} of

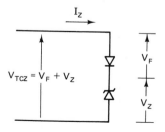

Figure 3.39 Temperature-compensated zener diode.

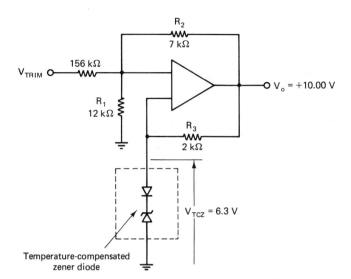

V_{TRIM} o— 156 kΩ

R_2 7 kΩ

R_1 12 kΩ

R_3 2 kΩ

$V_o = +10.00$ V

$V_{TCZ} = 6.3$ V

Temperature-compensated zener diode

Figure 3.40 REF10 voltage reference (Burr-Brown).

6.3 V, R_1 = 12 kΩ, and R_2 = 7 kΩ, the output voltage is calculated as +9.975 V. However, with the actual values of these components the output voltage will really be +10.000 V, with an untrimmed tolerance of just ±0.005 V. The output voltage can be adjusted over a −0.1- to +0.25-V range by use of an external 20-kΩ potentiometer connected across the output of the device, with wiper arm connected to the V_{TRIM} input terminal of the RF10.

The temperature coefficient of the output voltage is specified as a very impressive 1 ppm/°C (MAX) over a 0° to 70°C range for the REF10KM. The high accuracy of the output voltage and the extremely low temperature coefficient are due in large part to the laser trimming of on-chip resistors.

From the circuit we note that the temperature-compensated zener diode is biased by a current given by $I_z = (V_o - V_{TCZ})/R_3 = (10 - 6.3)$ V/2 kΩ = 1.85 mA. Note that this current is essentially independent of the supply voltage. As a result, V_{TCZ} and therefore the output voltage itself should exhibit very little dependence on the supply voltage. This is evidenced by the line regulation specification, which is just 0.001%/V (TYP) or 100 μV/V. Thus, for every 1-V change in the supply voltage, the output voltage changes by only 100 μV.

In addition to excellent line regulation, the voltage reference exhibits very good load regulation. The load regulation is specified as 0.001%/mA (TYP) over an output current range of from 0 to ±10 mA. This corresponds to a regulation of 0.1 mV/mA and an output resistance of 0.1 Ω. The no-load to full-load (±10 mA) regulation will therefore be 1 mV.

This IC can operate with a supply voltage from +15 up to +40 V, and with a +15-V supply draws a quiescent current of 4.5 mA (TYP), 6 mA (MAX). Although this device offers very impressive performance characteristics as a voltage reference from the standpoints of temperature coefficient, line regulation, and load regulation, the maximum output current of ±10 mA generally excludes it from being considered as a voltage regulator. However, it can be combined with other ICs and/or current

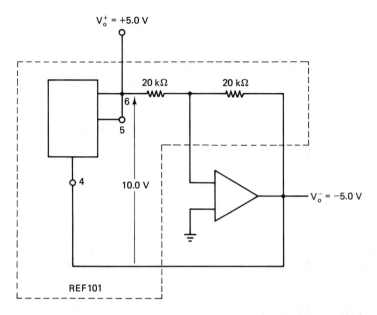

Figure 3.41 Using the REF 101 to produce at precision ±5.0 V reference.

boost transistors to provide much larger output currents and thereby serve as the basis of a very good voltage regulator.

The REF101 (Burr-Brown) is a +10-V reference that is very similar to the REF10, but it also contains a matched pair of user-accessible 20-kΩ resistors. This pair of resistors has a matching tolerance of ±0.01% (TYP), ±0.05% (MAX), and a temperature coefficient (TCR) of just 50 ppm/°C with a TCR tracking of 2 ppm/°C. This matched pair of resistors can be useful in a number of applications, an example of which is shown in Figure 3.41. In this circuit the REF101 together with its matched 20-kΩ resistor pair is used together with an operational amplifier to produce a precision ±5-V reference.

3.3.5 Voltage Reference Examples

Many voltage reference ICs are commercially available, many with temperature coefficients of 100 ppm/°C or less, and with nominal reference voltages of 1.2, 2.5, 5, and 10 V. Some examples of these voltage references, their voltage tolerances, and their maximum temperature coefficients are as follows.

1.2-volt reference ICs: AD589J (Analog Devices) (±3%) 100 ppm/°C; ICL8069D (Intersil) (±2%) 100 ppm/°C; AD589K (±3%) (Analog Devices) 50 ppm/°C; ICL8069C (Intersil) 50 ppm/°C; AD589L (Analog Devices) 25 ppm/°C; AD589M (Analog Devices) (±3%) 10 ppm/°C.

2.5-volt reference ICs: AD580J (Analog Devices) (±3%) 85 ppm/°C; AD580K (Analog Devices) (±1%) 40 ppm/°C; MC1403A (Motorola) (±1%) 25 ppm/

°C; AD580L (Analog Devices) ($\pm 0.4\%$) 25 ppm/°C; LM385-2.5 (National Semiconductor) ($\pm 3\%$) 25 ppm/°C (typ.); AD580M (Analog Devices) ($\pm 0.4\%$) 10 ppm/°C.

5-volt reference ICs: REF-02C (Precision Monolithics, Inc.) ($\pm 1\%$) 65 ppm/°C; LM336-5.0 (National Semiconductor) ($\pm 4\%$) 48 ppm/°C (typ.); MC1404U5 (Motorola) ($\pm 1\%$) 40 ppm/°C; AD584J (Analog Devices) ($\pm 0.3\%$) 30 ppm/°C; REF-02H (Precision Monolithics, Inc.) ($\pm 0.5\%$) 25 ppm/°C; AD584K (Analog Devices) ($\pm 0.12\%$) 15 ppm/°C; MC1400AG5 (Motorola) ($\pm 0.2\%$) 10 ppm/°C; REF-02E (Precision Monolithics, Inc.) ($\pm 0.3\%$) 8.5 ppm/°C; AD584L (Analog Devices) ($\pm 0.06\%$) 5 ppm/°C.

10-volt reference ICs: REF-01C (Precision Monolithics, Inc.) ($\pm 1\%$) 65 ppm/°C; MC1404U10 (Motorola) ($\pm 1\%$) 40 ppm/°C; AD581J (Analog Devices) ($\pm 0.3\%$) 30 ppm/°C; REF-01H (Precision Monolithics, Inc.) ($\pm 0.5\%$) 25 ppm/°C; AD584K (Analog Devices) ($\pm 0.1\%$) 15 ppm/°C; MC1400AG10 (Motorola) ($\pm 0.2\%$) 10 ppm/°C; REF-01E (Precision Monolithics, Inc.) ($\pm 0.3\%$) 8.5 ppm/°C; AD581L (Analog Devices) ($\pm 0.05\%$) 5 ppm/°C; AD2710K (Analog Devices) ($\pm 0.01\%$) 5 ppm/°C; AD2710L (Analog Devices) ($\pm 0.01\%$) 1 ppm/°C.

From this list we see that voltage reference ICs with temperature coefficients as low as 1 ppm/°C are available at all voltages from 1.2 to 10 V, and temperature coefficients as low as 5 ppm/°C are available for the 5- and 10-V references. We also note that another important voltage reference specification is the tolerance of the reference voltage with respect to the nominal value. We note that tolerances of around 1% are generally encountered, but that tolerances as small as 0.01% are available. The AD584 is an example of a pin-programmable voltage reference with reference voltages of 2.5, 5.0, 7.5, and 10 V being available.

Some examples of *ultrahigh precision* voltage references are the AD2710 and AD2712 series (Analog Devices). These ICs use a temperature-compensated zener diode and op amp together with laser trimmed resistors to produce a $+10.000$ V $\pm$ 1.0 mV output. Temperature coefficients as low as ± 1 ppm/°C over a $+25°$ to $+70°$C temperature range are available. The AD2712 also has available a -10-V output that tracks the $+10$-V output to within ± 1 ppm/°C.

3.3.6 Thermally Stabilized Voltage Reference

A thermally stabilized voltage reference is a voltage reference device whose temperature is maintained or stabilized at a constant value. As a result, the voltage reference output voltage is almost completely independent of the ambient temperature, with temperature coefficients of less than 1 ppm/°C being obtainable.

With any voltage reference circuit there will inevitably be some deviation of the device and component parameters away from the design center values such that there will be a small net temperature coefficient. One approach to reducing the temperature coefficient of a voltage reference is to thermally isolate it from the ambient temperature changes by maintaining the device temperature at some constant value. As the ambient temperature changes, the device temperature remains almost constant, and accordingly the reference voltage change will be extremely small. As a result, the temperature coefficient of the device, which is defined as the rate of change of reference voltage with respect to the ambient temperature, will correspondingly be extremely small.

To look at it mathematically, we have that the temperature coefficient of the reference voltage is given by

$$TC_{V(\text{REF})} = \frac{dV_{\text{REF}}}{dT_{\text{ambient}}} = \frac{dV_{\text{REF}}}{dT_{\text{chip}}} \frac{dT_{\text{chip}}}{dT_{\text{ambient}}} \tag{3.144}$$

As a result of the thermal stabilization of the chip temperature, the factor $dT_{\text{chip}}/dT_{\text{ambient}}$ is very much smaller than unity. Therefore, the temperature coefficient of the reference voltage (with respect to the ambient temperature) is much smaller than what would be obtained without stabilization of the chip temperature.

In all thermally stabilized voltage references, the chip temperature is maintained above the ambient temperature by an electrothermal feedback loop. The chip temperature is typically maintained at about 90° to 100°C by a feedback loop that controls the amount of electrical power dissipated on the chip and senses the resulting temperature rise. The temperature-stabilization circuit and the voltage reference circuit are on the same small silicon chip, so the two are in very good thermal contact, especially when the high thermal conductivity of the silicon and the small dimensions of the chip are considered.

Thermally stabilized voltage reference example. We will now consider a specific example of a thermally stabilized voltage reference. This voltage reference is the LM199/299/399 series (National Semiconductor). In Figure 3.42a, the thermal stabilization circuit is shown, and in Figure 3.36b the voltage reference circuit is given, with both circuits being located on the same chip to form a monolithic IC.

We will first consider the thermal-stabilization circuit. In this circuit, Q_4 acts as a thermal shutdown transistor that controls the power dissipation of the chip by turning the Darlington configuration of Q_1 and Q_2 off whenever the temperature rises above a certain value. Whenever the temperature is substantially below the stabilization temperature, Q_1 and Q_2 will be in full conduction, limited only by the current-limit transistor Q_3. As the temperature of the chip increases, transistor Q_4 starts to conduct and diverts base drive away from the base of Q_1. This causes a reduction in the current through Q_1 and thus a reduction in the power dissipation and of the rate of rise of chip temperature. As a result of this feedback action, the chip temperature will asymptotically approach the desired stabilization temperature.

The temperature-sensing circuitry for the thermostatic action of this circuit consists of transistors Q_4, Q_6, and Q_7 and zener diode D_1, which is a diode-connected transistor using the emitter–base breakdown voltage of 6.3 V. Assuming a V_{BE} of 0.65 V for Q_6 and Q_7, the net voltage across R_1 and R_2 is $V_Z - 2V_{BE} = 6.3 - 2(0.65) = 5.0$ V. The current through R_1 and R_2 resulting from this voltage is 5.0 V/12.2 kΩ = 0.41 mA. This current produces a voltage drop of 410 mV across R_2 at 25°C. Therefore, we see that at room temperature (25°C) the base-to-emitter voltage of Q_4 is insufficient to turn Q_4 on.

Transistor Q_7 is a multiple-collector lateral PNP transistor. The construction

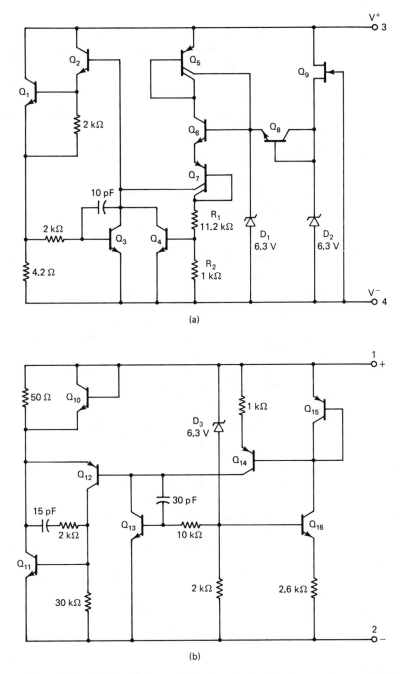

Figure 3.42 (a) Temperature stabilization circuit; (b) voltage reference circuit (National Semiconductor).

of Q_7 is such that the ratio of the collector currents is 0.3, so that when 410 μA flows out of the second collector 120 μA is the current supplied by the first collector. This current becomes the base drive of Q_2. This base drive is sufficient to produce a relatively large collector current for Q_1. Since this current can be quite large, as much as about 600 mA or even possibly 1 A, current limiting is incorporated into the design of the circuit. The current limiting is produced by the action of Q_3 in conjunction with the 4.2-Ω current-limit resistor. When the voltage drop across the current-limit resistor (4.2 Ω) approaches 600 mV, Q_3 is biased into full conduction and diverts an increasingly large proportion of the 120-μA collector current supplied by Q_7 away from the base of Q_2. This current-limiting action limits the current through Q_1 to about 140 mA.

If the circuit starts initially at 25°C, the initial current through Q_1 is 140 mA. With a terminal voltage of 15 V, the power dissipation is 2.1 W. The thermal impedance of this IC with a polysulfone thermal shield is about 220°C/W, so we see that if this current were to continue for any length of time there would be an excessive temperature rise.

As the temperature increases, the voltage across R_2 increases due to the positive temperature coefficient of the zener diode, about 3 mV/°C, and the negative temperature coefficient of the base-to-emitter voltage drops of Q_6 and Q_7, about −2.3 mV/°C each. After considering the voltage divider action of R_1 and R_2, we see that the voltage across R_2 increases at a rate of (3 mV/°C + 4.6 mV/°C)(1 kΩ/12.2 kΩ) = 0.62 mV/°C. At the same time that this is happening, the required base-to-emitter voltage to turn Q_4 on is decreasing at a rate of about 2.3 mV/°C. If we recall that at 25°C the voltage across R_2 was 410 mV and note that at this temperature the voltage necessary to turn Q_4 on is about 600 mV, we see that the temperature increase required to turn Q_4 on is given by

$$\Delta T = \frac{600 - 410}{(0.62 + 2.3) \text{ mV/°C}} = \frac{190 \text{ mV}}{2.92 \text{ mV/°C}} = 65°C \qquad (3.145)$$

so we should expect that Q_4 will be turned on at about 90°C.

Actually, what happens is that as the temperature increases, the current through Q_4 starts to increase. This diverts more and more base drive away from Q_2, reducing the current through Q_1 and thus lessening the chip power dissipation. The rate of increase of chip temperature correspondingly decreases and the chip temperature asymptotically approaches an equilibrium value of about 90°C. Thus the thermal stabilization circuit involves an electrothermal feedback loop that senses the chip temperature and controls the chip power dissipation to set the chip temperature to the desired value.

Transistors Q_8 and Q_9 together with diode D_2 are a startup circuit to provide the initial bias to the temperature-stabilization circuit. Once diode D_1 has been biased to breakdown (V_Z), the voltage across Q_8 will be zero and Q_8 will turn off, which will disconnect the startup circuit from the rest of the thermal stabilization circuit. After this happens, the current through D_1 is provided by the multiple-collector PNP transistor Q_5. The current through Q_5 is controlled by D_1, Q_6, Q_7, R_1, and R_2 and thus is almost completely independent of the supply voltage. We see, therefore, that this circuit is "self-biasing," so changes in the supply voltage will have virtually no effect on the control of the chip temperature and thus on the reference voltage.

We have seen that the initial turn-on current of the thermal stabilization circuit is 140 mA at 25°C. As the temperature approaches the stabilization temperature of 90°C, this current decreases. Since the temperature rise from 25° to 90°C is 65°C, and the thermal impedance is 220°C/W, the steady-state power dissipation of the chip when thermally stabilized is 65°C/(200 W/°C) = 300 mW. If a supply voltage of 15 V for the stabilization circuit of 15 V is used, the corresponding current is about 20 mA.

As a result of the small thermal mass and of the large thermal impedance of the LM199 series devices, the warm-up time is quite short. The warm-up time for the regulator output voltage to reach within 0.05% of its operating voltage that is nominally 6.95 V is only 3 s starting from 25°C and with a supply voltage of 30 V.

Voltage reference circuit. The voltage reference circuit is shown in Figure 3.42b and uses a temperature-compensated zener diode as the basic voltage reference. The zener diode is D_3, and its positive temperature coefficient is compensated by the negative temperature coefficient of the base-to-emitter drop of Q_{13}. The reference voltage consisting of the voltage drop of the zener diode of 6.3 V and the base-to-emitter drop of Q_{13} of 0.65 V appears across the external terminals of the device as a total voltage of 6.95 V typical (6.8 minimum, 7.1 maximum).

Most of the device current, however, does not go through D_3 and Q_{13} but rather through Q_{11}. Any change in the output voltage produces a corresponding change in the V_{BE} of Q_{13}. Transistor Q_{14} acts as a current source active load for Q_{13}, and thus changes in the collector current of Q_{13} produce an equal change in the base current of Q_{12}. This base current change is multiplied by the current gain of Q_{12} and drives the base of Q_{11}, where it is again multiplied by the current gain. Thus any changes in the output voltage result principally in changes in the current through Q_{11}. As a result of the high gain of Q_{13} and the current gains of Q_{12} and Q_{11}, the rate of change of output current with respect to output voltage, dI_O/dV_O, which is the dynamic output conductance, is relatively large. The large value of the dynamic output conductance, g_o, thus means that the device will have a small dynamic output resistance, r_o. For the LM199 series of devices, this is specified as 0.5 Ω typical, 1.0 Ω maximum, at a current level of 1 mA.

The temperature coefficient of the terminal voltage is specified as 0.3 ppm/°C typical, 1.0 ppm/°C maximum. This corresponds to 2 μV/°C typical, 7 μV/°C maximum. This compares very favorably with the temperature coefficients of the non-thermally stabilized voltage references, which generally are in the range from 10 to 100 μV/°C.

Applications of the thermally stabilized voltage reference. In Figure 3.43, a functional block diagram of the thermally stabilized voltage reference is shown. Diode D_{sub} is a consequence of the fact that the thermal-stabilization circuit and the voltage reference circuit share the same IC chip. For proper operation of the device, this diode should never be forward biased in order to keep the two circuits electrically

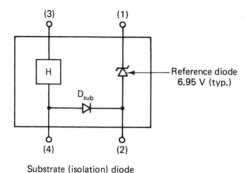

H

D_{sub}

Reference diode
6.95 V (typ.)

(4) (2)

Substrate (isolation) diode

Figure 3.43 Functional diagram of thermally stabilized voltage reference.

isolated. In addition, the breakdown voltage of this diode is 40 V (min.), so that no more than 40-V reverse bias should be applied across this diode.

In Figure 3.44, a buffered voltage reference is shown using the thermally stabilized voltage reference and an operational amplifier to provide isolation of the voltage reference device from the load. The operational amplifier circuit also provides for step-up of the reference voltage to the desired value, as given by $V_O = V_{REF}(1 + R_2/R_1)$. To preserve the low-temperature-coefficient characteristics of the voltage reference, the operational amplifier should be chosen to have a very low offset-voltage temperature coefficient, $TC_{V(OS)}$. This should preferably be down in the range of about 1 μV/°C or less.

Since the output voltage is a function of the R_2/R_1 resistor ratio, great care should be taken in the selection and placement of these two resistors. The resistors should preferably have low temperature coefficients of resistance (TCR), and even more important, very well matched temperature coefficients so that the resistor values will track each other with temperature changes such that the resistance ratio remains constant. The placement of these resistors in the circuit with respect to considerations of heat flow and temperature differentials is also very important. In general, the

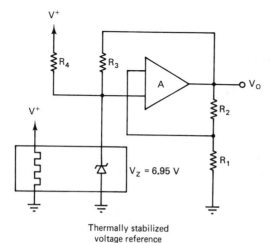

Thermally stabilized
voltage reference

Figure 3.44 Buffered voltage reference using thermally stabilized voltage reference.

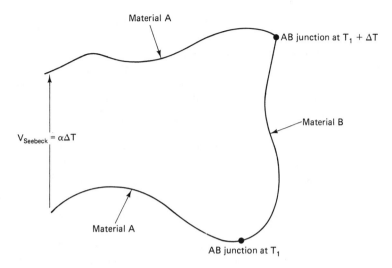

Figure 3.45 Seebeck effect: a thermocouple.

placement of the resistors should be such that the temperature difference between the two resistors is minimized.

Another consideration that can be of importance when very low temperature coefficients are of interest is that of thermocouple effects, that is, the thermoelectric voltage that is generated as the result of electrical contact between two dissimilar materials (the Seebeck effect). The Seebeck coefficient is the thermoelectric voltage produced per degree Celsius temperature difference for a junction between two given materials, as illustrated in Figure 3.45. In this figure, α is the Seebeck coefficient (μV/°C), so the thermoelectric voltage that results from a temperature difference of ΔT between the two junctions between metals A and B will be $V_{\text{Seebeck}} = \alpha \, \Delta T$.

The Seebeck coefficient between the Kovar pins of the IC and the copper leads or the printed-circuit-board copper metallization pattern is about 30 μV/°C. Therefore, as shown in Figure 3.46, a temperature differential of as little as 0.1°C between the pin connections of the voltage reference can lead to a thermoelectric voltage with a temperature coefficient of about 3 μV/°C, which can represent a serious degradation in the voltage reference performance. Therefore, very careful layout of the circuit

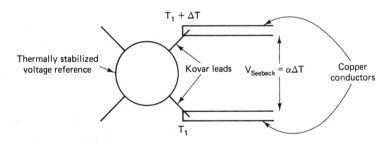

Figure 3.46 Generation of Seebeck voltage due to a temperature differential at pins of thermally stabilized voltage reference.

is important, with great attention paid to minimizing thermal gradients and temperature differentials.

To consider a simple design example for the buffered voltage reference of Figure 3.44, we will assume a supply voltage of $+15$ V and that the desired scaled-up reference voltage output of the circuit is 10 V. The specified operating range of the LM199 series is 0.5 to 10 mA. Here we will choose a current of 1.0 mA to flow through the voltage reference circuit so that R_3 becomes $R_3 = (V_o - V_Z)/1.0$ mA $= (10 - 6.95)/1.0$ mA $= 3$ kΩ.

Resistor R_4 is used for the initial startup of the circuit. It is used to supply a small initial bias current to the voltage reference circuit to produce a sufficiently large voltage drop across the voltage reference such that the output voltage, V_O, will be large enough such that sufficient current will flow through R_3 to bring the reference circuit up to its full reference voltage. Resistor R_4 should be small enough such that the initial current flow will be large enough to initiate operation of this circuit, but on the other hand it should not be too small such that variations in the supply voltage V^+ will result in significant variations in the current through the reference circuit. Since a current of 10 to 100 μA should be large enough to initiate operation of the circuit, a suitable value for R_4 is in the range of about 80 to 800 kΩ.

For the design of a resistive voltage divider, the basic requirement is that $1 + R_2/R_1 = V_O/V_Z$. This therefore fixes the ratio of the two resistors. For the absolute values of R_1 and R_2, an important consideration is the effect of the operational amplifier input bias current and the temperature coefficient thereof. The values of R_1 and R_2 should be chosen to be small enough such that the maximum expected bias current will not produce a significant change in the output voltage due to the flow of the bias current through R_1 and R_2. Perhaps even more important, the values of the two resistors should be small enough such that the temperature coefficient of the bias current will not produce an unacceptable degradation of the overall temperature coefficient of the circuit.

A reasonable choice of operational amplifier for this circuit would be the OP-77 (Precision Monolithics, Inc.), or equivalent. This amplifier is characterized by a very low offset voltage [10 μV (typ.), 25 μV (max.)] and even more important, a very low offset-voltage temperature coefficient of 0.1 μV/°C (typ.) and 0.3 μV/°C (max.). For a 10-V output voltage, this temperature coefficient will correspond to 0.01 ppm/°C (typ.) and 0.03 ppm/°C (max.), so this operational amplifier will not produce any significant increase in the overall temperature coefficient.

The input bias current of this operational amplifier is 1.2 nA (typ.) and 2.0 nA (max.). The temperature coefficient of the bias current is about 8 pA/°C (typ.). The contribution to the temperature coefficient of the output voltage due to the bias current temperature coefficient is given by $dV_O/dT = R_2(dI_B/dT)$. We see therefore that as long as R_2 is not excessively large (less than 1 MΩ), this contribution to the overall temperature coefficient will be small.

A reasonable choice for $R_1 + R_2$ is 50 kΩ, as this will not draw off an excessive current from the amplifier output. Since $1 + R_2/R_1 = (R_1 + R_2)/R_1 = 10$ V/6.95 V, we have that $R_1 = 50$ k$\Omega \times 0.695 = 34.75$ kΩ. Therefore, $R_2 = 50$ k$\Omega - 34.75$ k$\Omega = 15.25$ kΩ.

Some other thermally stabilized voltage references are the LM199A/299A/399A series with a temperature coefficient of only 0.2 ppm/°C (typ.), 0.5 ppm/°C (max.), and the LM3999, with a temperature coefficient of 2 ppm/°C (typ.), 5 ppm/°C (max.). These voltage references have internal circuitry that is identical to that of the LM199 and have nominal reference voltages of 6.95 V.

It is of interest now to compare these thermally stabilized voltage references with the LM129/329 series of voltage references. These devices have a voltage reference circuit that is identical to that of the above-mentioned thermally stabilized references and differ from them only in that there is no thermal stabilization circuit.

As a result of the chip temperature not being stabilized, the temperature coefficient of the reference voltage will be substantially greater than that of the thermally stabilized voltage references. The temperature coefficient for the best of the LM129/329 series is that of the LM129A and 329A, which is 6 ppm/°C (typ.) and 10 ppm/°C (max.). For the LM329D, the temperature coefficient is 50 ppm/°C (typ.) and 100 ppm/°C (max.). The nominal reference voltage is 6.95 V for all these devices.

We see therefore that the temperature coefficient is much greater in the absence of thermal stabilization of the chip temperature. Indeed, the temperature coefficient is greater by a factor in the range of about 30 to 100. Since $dV_{REF}/dT = (dV_{REF}/dT_{chip})(dT_{chip}/dT_{ambient})$, we can conclude that the thermal stabilization circuit has a temperature stabilization coefficient in the range from 0.01 to 0.03. This means that a 1°C change in the ambient temperature will result in a chip temperature change of only 0.01 to 0.03°C.

3.3.7 Sine-wave Reference

The voltage references considered thus far are dc references. Voltage references that have a sinusoidal waveform are also available. An example of a sine-wave reference is the SWR200 (Thaler Corporation). This IC produces a sinusoidal waveform with an rms value of 7.071 V with a tolerance of ±0.05%, and a temperature coefficient of 2 ppm/°C over a temperature range of −55° to +125°C. The sinusoidal waveform is generated by a phase shift oscillator with the frequency controlled by two external capacitors. The frequency can be set to values between 400 Hz to 10 kHz. The output of the oscillator is supplied to a chopper amplifier that acts like a precision full-wave rectifier to produce an output voltage that is the absolute value of the oscillator waveform. The chopper amplifier output then is compared to a precision dc reference voltage in an integrating amplifier to generate a dc error voltage. This error voltage then goes to a JFET, which acts as a voltage-variable resistance to control the gain of the phase shift oscillator. The result is a sinusoidal waveform with very high accuracy and a total harmonic distortion of less than 0.5%.

PROBLEMS

For all problems, use the following parameters unless otherwise indicated or implied.

1. Assume all transistors of the same type to be identical.
2. For NPN transistors: β = 100 (typ.), 50 (min.). For PNP transistors: β = 50 (typ.), 25 (min.).

3. Early voltage V_A: 200 V for both NPN and PNP transistors.
4. Base-to-emitter voltage temperature coefficient, $TC_{V(BE)} = dV_{BE}/dT = -2.2$ mV/°C.
5. Collector-base breakdown voltage: 50 V (min.).
6. $V_{BE} = 0.7$ V.
7. V_T (thermal voltage) $= 25$ mV.
8. Resistor temperature coefficient, $TCR = (1/R)(dR/dT) = +2000$ ppm/°C.
9. Assume all resistors to have the same temperature coefficient and to be at the same temperature.
10. Supply voltages: $V^+ = +10$ V, $V^- = -10$ V.

Current Source Problems

3.1. (*Diode-biased current sink*)   Refer to Figure P3.1.

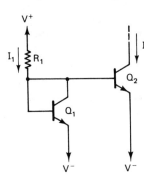

Figure P3.1

(a) Show that $I_2 = I_1/(1 + 2/\beta)$.
(b) For $I_1 = 1.0$ mA, find R_1.   (*Ans.:* 19.3 kΩ)
(c) Find I_2 (typ., max., and min.).   [*Ans.:* 0.96 (min.), 0.98 (typ.), 1.00 (max.) mA]
(d) Show that the temperature coefficient of I_2, $TC_{I_2} = (1/I_2)(dI_2/dT)$ will be given by $TC_{I_2} \simeq -TC_{R1} - (1/V_{supply})(dV_{BE}/dT)$, where $V_{supply} = V^+ - V^-$.
(e) Find TC_{I_2}.   (*Ans.:* -1890 ppm/°C or -0.189%/°C)
(f) Find the dynamic output resistance r_o and the dynamic output conductance g_o of the current sink.   (*Ans.:* $r_o = 200$ kΩ, $g_o = 5$ μS)
(g) Find the percent of change in the output current per 1-V change in the output voltage.   (*Ans.:* 0.5%/V)
(h) Find the voltage compliance range.   (*Ans.:* -9.8 to $+40$ V)
(i) If the offset voltage between Q_1 and Q_2 is ± 1.0 mV (max.), find the resulting difference produced between I_1 and I_2.   (*Ans.:* 0.041 mA or 4.1%).

3.2. (*Wilson current mirror current sink*)   Refer to Figure P3.2.
(a) Show that $I_2 = I_1[1 + (1/\beta_1) + (1/\beta_3) - (2/\beta_2)]$, neglecting the second-order terms in $1/\beta$ (the $1/\beta^2$ terms).
(b) If $\beta = 100$ and the maximum β mismatch between any two transistors is $\pm 5\%$, find the maximum mismatch between I_2 and I_1.   (*Ans.:* $\pm 0.105\%$)
(c) Find R_1 for $I_1 = 1.0$ mA.   (*Ans.:* 18.6 kΩ)
(d) Find the voltage compliance range.   (*Ans.:* -9.1 to $+41$ V)
(e) Show that $TC_{I_2} = -TC_{R1} - (2/V_{supply})(dV_{BE}/dT)$.
(f) Find TC_{I_2}.   (*Ans.:* -1780 ppm/°C or -0.18%/°C)
(g) Find r_o and g_o.   (*Ans.:* 20 MΩ, 50 nS)
(h) Find the percent of change in I_2 per 1-V change in V_{C2}.   (*Ans.:* 0.005%/V)

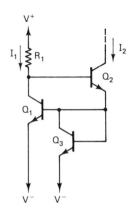

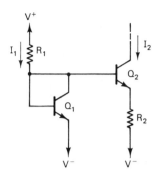

Figure P3.2   Figure P3.3

3.3. (*Current sink for low current levels*)   Refer to Figure P3.3.
   (a) Show that $I_2 = (V_T/R_2) \ln (I_1/I_2)$.
   (b) Show that $dI_2/I_2 = (dI_1/I_1)/[1 + \ln (I_1/I_2)]$.
   (c) Show that $dI_1/I_1 = dV_{supply}/V_{supply}$ and that as a result, $dI_2/I_2 = (dV_{supply}/V_{supply})/[1 + \ln (I_1/I_2)]$.
   (d) Find R_1 for $I_1 = 1.0$ mA.   (*Ans.*: 19.3 kΩ)
   (e) Find R_2 for $I_2 = 10$ μA.   (*Ans.*: 11.5 kΩ)
   (f) Find the percent of change in I_2 per 1-V change in the supply voltage.   (*Ans.*: 0.89%/V)
   (g) Find the voltage compliance range.   (*Ans.*: −9.7 to +41 V)
   (h) Find r_o and g_o and the percent of change in I_2 per 1-V change in V_{C_2}.   (*Ans.*: 108 MΩ, 9.26 nS, 0.0926%/V)
   (i) Show that TC_{I_2} is given approximately by

$$TC_{I_2} \simeq \frac{(1/T) - (TC_{R_2}) - \dfrac{(TC_{R_1}) + (1/V_{supply})dV_{BE}/dT}{\ln (I_1/I_2)}}{1 + 1/\ln (I_1/I_2)}$$

   (j) Find TC_{I_2}.   (*Ans.*: +758 ppm/°C or +0.0758%/°C)
3.4. (*Multiple current sinks*)   Refer to Figure P3.4.   *Given:* $R_B = 4$ kΩ, $I_A = 1.0$ mA, $I_1 = 1.0$ mA, $I_2 = 2.0$ mA, $I_3 = 10$ mA, and $I_4 = 0.1$ mA.
   (a) Find R_A, R_1, R_2, R_3, and R_4, assuming that the transistor emitter areas are scaled such that the *current densities are equal in* Q_A, Q_1, Q_2, Q_3, and Q_4.   (*Ans.*: 14.6 kΩ, 4 kΩ, 2 kΩ, 400, 40 kΩ)
   (b) If all the transistor emitter–base junction areas are *equal*, find R_1 through R_4.   (*Ans.*: 4 kΩ, 1.991 kΩ, 394, 40.575 kΩ)
   (c) Find the voltage compliance range.   (*Ans.*: −5.8 to +44 V)
   (d) Find the maximum difference that there will be between I_A and I_1 due to the effect of the base currents.   (*Ans.*: 0.564%)
   (e) Show that the g_o for all of the current sinks is given approximately by $g_o = I_C/(1.24 \times 10^4$ V).
   (f) Find the percent of change in the output current of each of the current sinks per 1-V change in output voltage.   (*Ans.*: 0.00807%/V)
   (g) What is the function of Q_B?

Chap. 3   Problems   **203**

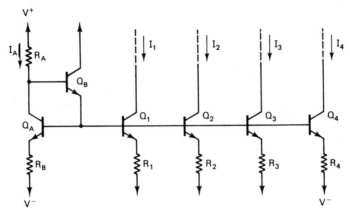

Figure P3.4

3.5. (*Current source using PNP transistor*)   Refer to Figure P3.5.
  (a) Show that $I_2 \simeq \alpha_{pnp}(I_o - I_1)$ and that $I_o = (1/R_1)(V^+ - V_{EB} - V_R)$.
  (b) For $V_R = +5.0$ V, $V_1 = 0$ V, $I_1 = 1.0$ mA, and $\alpha_{pnp} = 0.98$, find R_1 for $I_2 = 1.0$
  mA.   (*Ans.*: 2.13 kΩ)
  (c) Find the voltage compliance range.   (*Ans.*: −45 to +5.5 V)
  (d) **(1)** Find the transfer conductance dI_2/dV_R.   (*Ans.*: $-\alpha/R_1 = -460$ μS)
      (2) Find the transfer coefficient dI_2/dI_1.   (*Ans.*: $-\alpha_{pnp} = -0.98$)

3.6. (*Current source using PNP transistor*)   Refer to Figure P3.6.
  (a) Show that $I_2 \simeq \alpha_{pnp}(V^+ - V_1)/R_2$.
  (b) Given that $V_1 = +5.0$ V, find R_1 for $I_2 = 1.0$ mA.   (*Ans.*: 4.9 kΩ)
  (c) Find the minimum allowable value for I_1 if $\alpha_{pnp} = 0.95$ (min.).   (*Ans.*: 50 μA)
  (d) If the I_1 current sink is replaced by a resistor R_1, find R_1 for $I_1 = 1.0$ mA.   (*Ans.*:
  $R_1 = 14.3$ kΩ)
  (e) Find the voltage compliance range.   (*Ans.*: −45 to +4.8 V)
  (f) Find the transfer conductance dI_2/dV_1.   (*Ans.*: −0.20 mS)

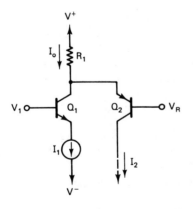

Figure P3.5 ·

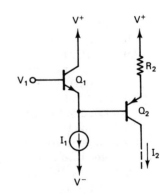

Figure P3.6

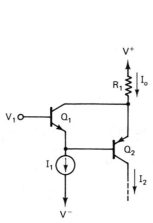

Figure P3.7

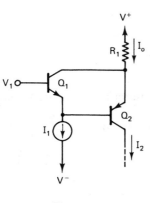

Figure P3.8

3.7. (*Current source using PNP transistor*) Refer to Figure P3.7.
  (a) Show that $I_2 = I_o - I_1$ and therefore is independent of the current gain of the PNP transistor.
  (b) Show that $I_o = (V^+ - V_1)/R_1$.
  (c) If $V_1 = +5.0$ V, $I_1 = 1.0$ mA, and $I_2 = 1.0$ mA, find R_1. (*Ans.*: 2.5 kΩ)
  (d) Find the transfer conductance dI_2/dV_1. (*Ans.*: $g = -1/R_1 = -0.40$ mS)
  (e) Minimum allowable value for I_1 if $\alpha_{pnp} = 0.95$ (min.). (*Ans.*: 50 μA)
  (f) Find the voltage compliance range. (*Ans.*: -45 to $+4.8$ V)
  (g) Find the dynamic output resistance, conductance, and the percentage of change in I_2 per 1-V change in V_{C_2}. (*Ans.* 6.7 MΩ, 146 nS, 0.015%/V or 148 ppm/V)
  (h) Find the temperature coefficient of I_2 if $TC_{I_1} = -500$ ppm/°C. (*Ans.*: -3500 ppm/°C)

3.8. (*Current sink biased current sink*) Refer to Figure P3.8.
  (a) Find R_1 for $I_1 = 1.0$ mA. (*Ans.*: 17.9 kΩ)
  (b) Find R_4 for $I_5 = 10$ μA. (*Ans.*: 11.5 kΩ)
  (c) Find the voltage compliance range. (*Ans.*: -8.4 to $+42$ V)
  (d) Find r_o and g_o. (*Ans.*: 2.02 GΩ, 0.495 nS)
  (e) Find the percent change in I_5 per 1-V change in V_{C_5}. (*Ans.*: 0.005%/V or 49.5 ppm/V)
  (f) If $C_{cb} + 2.0$ pF and $C_{cs} = 2.0$ pF, find the dynamic output impedance and admittance of this current sink at a frequency of 1.0 MHz. (*Ans.*: $y_o = +j25.1$ μS, $z_o = -j39.8$ kΩ)

3.9. (*Diode-compensated current source*) Refer to Figure P3.9.
  (a) Show that for a very large open-loop gain A_{OL} that $I_N = (V_{REF} - V^-)/R_N$, where $N = 1, 2, 3, \ldots$, and that the base-to-emitter voltage temperature dependence is compensated for by the diode-connected transistor in the feedback loop of the

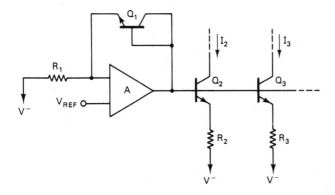

Figure P3.9

operational amplifier. Assume that transistor active (emitter) areas are scaled for equal current densities in all transistors.

(b) Show that the exact expression for I_N is given by

$$I_N = \frac{1}{R_N}\left(\frac{V_{\text{REF}}}{1 + 1/A_{OL}} - V^- - V_{BE}\frac{1}{A_{OL} + 1}\right)$$

(c) Find the voltage compliance range for $V_{\text{REF}} = -5.0$ V.   (*Ans.:* -4.8 to $+45$ V)

(d) Find the minimum allowable value for V_{REF}.   (*Ans.:* -10 V)

(e) If the resistors $R_2, R_3, \ldots$ are temperature-compensated resistors with a *TCR* of ± 50 ppm/°C (max.) and V_{REF} has a temperature coefficient of ± 25 ppm/°C (max.), find the temperature coefficient of I_N.   [*Ans.:* $TC_{I_N} = \pm 75$ ppm/°C (max.)]

3.10. (*Feedback-compensated current source*)   Refer to Figure P3.10.

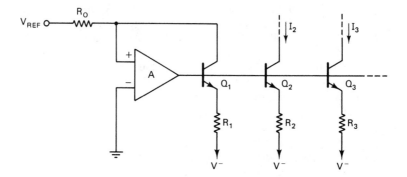

Figure P3.10

(a) Show that, if A_{OL} is very large, $I_N = (R_1/R_N)(V_{\text{REF}}/R_O)$, assuming that the transistor active areas are scaled such that the current densities are equal in all transistors.

(b) Show that the more exact expression for I_N is

$$I_N = \frac{R_1}{R_N}\left[\frac{V_{\text{REF}}}{R_O + R_1/A_{OL}} - \frac{V_{BE} + V^-}{A_{OL}(R_O + R_1/A_{OL})}\right]$$

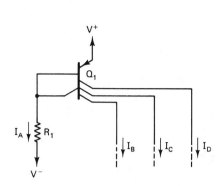

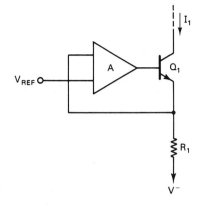

Figure P3.11   **Figure P3.12**

3.11. (*Multiple-collector lateral PNP current source*)   Refer to Figure P3.11.
  (a) If all collector active "lengths" are equal, show that

$$I_B = I_C = I_D = \frac{I_A}{1 + (4/\beta_{pnp})}$$

  (b) If $\beta_{pnp} = 25$ (min.), find the maximum difference between I_A and I_B.   (*Ans.:* 13.8%)
  (c) Find R_1 for $I_A = 1.0$ mA.   (*Ans.:* 19.3 kΩ)
  (d) Find the voltage compliance range.   (*Ans.:* -41 to $+9.8$ V)

3.12. (*Operational-amplifier feedback current sink*)   Refer to Figure P3.12.
  (a) Show that for large A_{OL}, $I_1 = (V_{REF} - V^-)/R_1$.
  (b) Show that a more exact expression for I_1 is given by

$$I_1 = \frac{1}{R_1}\left(\frac{V_{REF} - V^-}{1 + 1/A_{OL}} - \frac{V_{BE} + V^-}{A_{OL} + 1}\right)$$

  (c) If $A_{OL} = 100$ dB (min.) and $\beta = 25$ (min.), 50 (typ.), find the maximum difference or error between I_1 and $(V_{REF} - V^-)/R_1$.   [*Ans.:* 2% (typ.), 4% (max.)]
  (d) Find the voltage compliance range for $V_{REF} = -5.0$ V.   (*Ans.:* -4.8 to $+45$ V)

3.13. (*Darlington feedback current source*)   Refer to Figure P3.13.
  (a) If $A_{OL} = 10^5$ (min.), find the maximum difference or error between I_2 and $(V_{REF} - V^-)/R_1$ for $V_{REF} = 5.0$ V.   [*Ans.:* Difference due to finite A_{OL} is 4.3×10^{-4}% (max.) and difference due to β is 0.04% (max.), 0.01% (typ.), total difference will be 0.04% (max.)]
  (b) Find the voltage compliance range for $V_{REF} = -5.0$ V.   (*Ans.:* -4.1 to $+45$ V)

3.14. (*JFET feedback current source*)   Refer to Figure P3.14.
  (a) Show that

$$I_1 = \frac{1}{R_1}\left(\frac{V_{REF} - V^-}{1 + 1/A_{OL}} - \frac{V^- + V_{GS}}{A_{OL} + 1}\right)$$

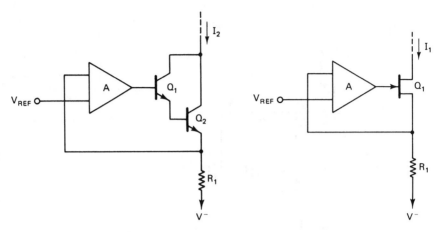

| Figure P3.13 | Figure P3.14 |

(b) For $V_{\text{REF}} = +5.0$ V and $V_{GS} = -2.0$ V, find the maximum difference or error between I_1 and $(V_{\text{REF}} - V^-)/R_1$. Use $A_{OL} = 100$ dB (min.).   [*Ans.*: $2 \times 10^{-4}\%$ (max.)]

(c) If $I_{DSS} = 10$ mA and $V_P = -6.0$ V, find V_{GS} when $I_1 = 2.0$ mA and find the maximum value that I_1 can have.   (*Ans.*: -3.32 V, 10 mA)

3.15 (*Resistor-biased temperature-compensated current sink*)   Refer to Figure P3.15.

(a) Show that

$$I_1 = \frac{1}{R_3} \left(\frac{V_{\text{supply}} R_2}{R_1 + R_2} - V_{BE} \right)$$

where $V_{\text{supply}} = V^+ - V^-$.

(b) Assuming that all resistors have identical *TCR* values and that they track in temperature, show that the temperature coefficient for I_1 is given by

$$TC_{I_1} = \frac{1}{I_1} \frac{dI_1}{dT} = -TCR - \frac{1}{V_{R_3}} \frac{dV_{BE}}{dT}$$

(c) Find V_{R_3} for a zero temperature coefficient for I_1.   (*Ans.*: 1.1 V)

(d) For $I_1 = 1.0$ mA and $I_{R_1} = I_{R_2} = 50 I_{B(\text{MAX})}$, find R_1, R_2, and R_3.   (*Ans.*: 18.2 kΩ, 1.8 kΩ, 1.1 kΩ)

3.16 (*Current source that is relatively independent of supply voltage*).   Refer to Figure P3.16.

(a) Show that I_2 is given by

$$I_2 = \frac{(V_{BE_1}/R_2) + (I_{R_1}/\beta_1)}{1 + (1/\beta_2) + (1/\beta_1\beta_2)} \simeq \frac{V_{BE_1}}{R_2}$$

(b) If $V_{BE} = 660$ mV at 1.0 mA for Q_1 and Q_2, find R_1 and R_2 for $I_1 = 250$ μA, $I_2 = 100$ μA, and $V^+ = 10$ V.   (*Ans.*: 35.1 kΩ, 6.25 kΩ)

(c) Given that $\beta = 50$ (min.) for Q_1 and Q_2, as β varies from its minimum value to infinity, what will be the resulting variation in I_2?   (*Ans.*: $\Delta I_2 = -2.9$ μA or -2.9%)

(d) Find the change in I_2 per 1-V change in the supply voltage V^+. Express the result in both absolute and percentage terms.   (*Ans.*: 0.455 μA/V or 0.455/V)

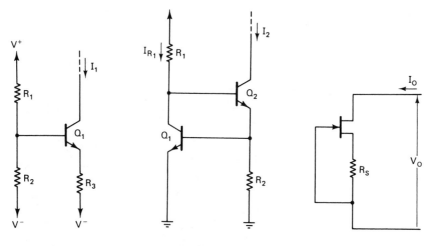

| Figure P3.15 | Figure P3.16 | Figure P3.17 |

(e) If $TCR = +2000$ ppm/°C and $TCV_{BE} = -2.2$ mV/°C, find TC_{I_2}.   (*Ans.*: -5520 ppm/°C or $-0.552\%/°C$)

3.17 (*JFET current regulator diode*)   Refer to Figure P3.17.   Given JFET with $I_{DDS} = 1.0$ mA and $V_P = -4.0$ V.   The dynamic drain-to-source resistance r_{ds} is 25 kΩ at $I_{DS} = 1.0$ mA.   Find R_s, r_{ds}, the dynamic output resistance r_o, and the percent change in I_O per 1-V change in V_O for the following values of I_O:

(a) 1.0 mA.   (*Ans.*: $R_S = 0$, $r_{ds} = 25$ kΩ, $r_o = 25$ kΩ, 4.0%/V)
(b) 0.50 mA.   (*Ans.*: 2.34 kΩ, 50 kΩ, 91.4 kΩ, 2.19%/V)
(c) 0.25 mA.   (*Ans.*: 8.0 kΩ, 100 kΩ, 300 kΩ, 1.33%/V)
(d) 0.10 mA.   (*Ans.*: 27.4 kΩ, 250 kΩ, 1.33 MΩ, 0.751%/V)
(e) 0.03 mA.   (*Ans.*: 110 kΩ, 833 kΩ, 8.79 MΩ, 0.379%/V)

3.18 Given Figure P3.3 with $I_1 = 1.0$ mA.   Use a computer program to determine I_2 for the following values of R_2: 300 Ω, 1000 Ω, 3000 Ω, 10 kΩ, and 30 kΩ.

Voltage Source Problems

3.19 Refer to Figure P3.19.
(a) Find R_1 and R_2 for a load voltage $V_L = +5.0$ V and $I_{R_1} \simeq I_{R_2} = 1.0$ mA.   (*Ans.*: $R_1 = 4.3$ kΩ, $R_2 = 5.0$ kΩ)
(b) Find the change in the load voltage, ΔV_L, for the output current going from a no-load value of 1.0 mA to a full-load value of 10 mA.   [*Ans.*: *load regulation* $= \Delta V_L = -0.265$ V (typ.), -0.473 V (max.)]
(c) Find the *dynamic output resistance* for a load current of 10 mA.   [*Ans.*: $r_o = 25.6$ Ω (typ.), 48.7 Ω (max.)].
(d) Find the *dynamic output resistance* at a load current of 1.0 mA.   [*Ans.*: $r_o = 48.1$ Ω (typ.), 71.2 Ω (max.)]
(e) Find the no-load to full-load *load regulation* if Q_1 is replaced by a Darlington configuration.   [*Ans.*: $\Delta V_L = -0.117$ V (typ.), -0.123 V (max.)]
(f) If the Darlington configuration of part (e) is as shown in Figure P3.19b with Q_{1A} biased by R_3 at a quiescent level of 1.0 mA, find R_3 and the *load regulation*.   [*Ans.*:

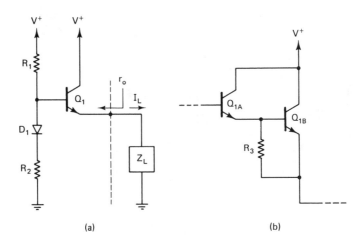

Figure P3.19

$R_3 = 700 \ \Omega$, load regulation $= \Delta V_L$ (with $I_{NL} = 2$ mA and $I_{FL} = 11$ mA): -0.062 V (typ.), -0.070 V (max.)]

(g) Find the *dynamic output resistance* r_o for the circuit of part (f) at a load current of 10 mA.   [*Ans.:* $r_o = 3.26 \ \Omega$ (typ.), $4.20 \ \Omega$ (max.)]

3.20 Refer to Figure P3.20.

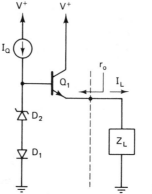

Figure P3.20

(a) If V_Z (breakdown voltage) of D_2 is 6.2 V, find V_L.   (*Ans.:* 6.2 V)

(b) If $I_Q = 1.0$ mA and $Z_z = 10 \ \Omega$, find the *load regulation* going from $I_{L(NL)} = 1.0$ mA to $I_{L(FL)} = 10$ mA.   [*Ans.:* $\Delta V_L = -0.061$ V (typ.), -0.064 V (max.)]

(c) Find the *dyamic output resistance* for $I_L = 10$ mA.   [*Ans.:* $r_o = 2.85 \ \Omega$ (typ.), $3.20 \ \Omega$ (max.)]

(d) If Q_1 is replaced by the Darlington configuration of Figure P3.19b with the quiescent current of Q_{1A} being set by R_3 at 10 mA, find R_3 and the *dynamic output resistance* at $I_L = 100$ mA.   [*Ans.:* $r_o = 0.306 \ \Omega$ (typ.), $0.334 \ \Omega$ (max.), $R_3 = 70 \ \Omega$]

(e) Find the *load regulation* for the case of part (d) with $I_{L(NL)} = 20$ mA and $I_{L(FL)} = 110$ mA.   [*Ans.:* $\Delta V_L = -0.060$ V (typ.), -0.063 V (max.)]

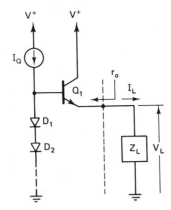

Figure P3.21

3.21 Refer to Figure P3.21.
  (a) If there are six diodes (D_1 to D_6) driven by I_Q, find V_L.   (*Ans.*: $V_L = 3.5$ V)
  (b) If $I_Q = 2.5$ mA, find the *load regulation* for $I_{L(NL)} = 1.0$ mA and $I_{L(FL)} = 10$ mA.   [*Ans.*: ($\Delta V_L = -63.0$ mV (typ.), -68.4 mV (max.)]
  (c) Find the *dynamic output resistance* for $I_L = 10$ mA.   [*Ans.*: $r_o = 3.1$ Ω (typ.), 3.7 Ω (max.)]

3.22 Refer to Figure P3.22. Given $I_Q = 12$ mA and $I_{R_1} \simeq I_{R_2} = 1.0$ mA.
  (a) Find R_1 and R_2 for $V_O = 2.0$ V.   (*Ans.*: $R_1 = 1300$ Ω, $R_2 = 700$ Ω)
  (b) Find the *current compliance range*.   (*Ans.*: $I_L = 0$ to almost $+11$ mA)
  (c) Find the *load regulation* for $I_{NL} = 0$ and $I_{FL} = 10$ mA.   (*Ans.*: $\Delta V_O = -171$ mV or -8.6%)
  (d) Show that the dynamic output conductance, $g_o = dI_O/dV_O$, is given by $g_o = 1/(R_1 + R_2) + R_2/(R_1 + R_2)g_m$.
  (e) Find g_o and r_o for $I_O = 10$ mA.   (*Ans.*: $g_o = 14.5$ mS, $r_o = 69.0$ Ω)
  (f) Find g_o and r_o for $I_O = 0$.   (*Ans.*: $g_o = 154.5$ mS, $r_o = 6.5$ Ω)
  (g) Sketch and dimension the I_O versus V_O graph for this circuit.

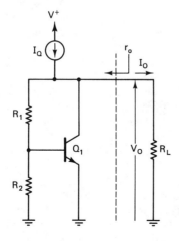

Figure P3.22

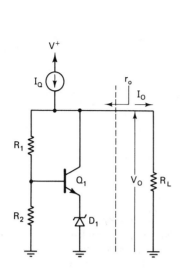

Figure P3.23

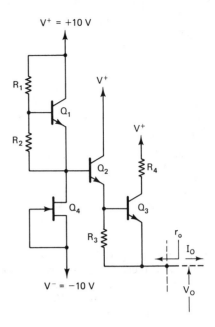

Figure P3.24

3.23 Refer to Figure P3.23. Given $I_Q = 12$ mA, $I_{R_1} = I_{R_2} = 1.0$ mA, $V_Z = 6.3$ V, and $Z_Z = 10$ Ω.

(a) Find R_1 and R_2 for $V_O = 10$ V.   (*Ans.*: $R_1 = 3$ kΩ, $R_2 = 7$ kΩ)

(b) Find the *current compliance range*.   (*Ans.*: $I_O = 0$ to almost 11 mA)

(c) Find the *load regulation* for $I_{NL} = 0$ and $I_{FL} = 10$ mA.   (*Ans.*: $\Delta V_O = -0.228$ V or -2.3%)

(d) Show that the dynamic output conductance is given by

$$g_o = \frac{1}{R_1 + R_2} + \frac{R_2}{R_1 + R_2}\frac{g_m}{1 + g_m Z_Z}$$

(e) Find g_o and r_o at $I_O = I_{FL} = 10$ mA.   (*Ans.*: $g_o = 20.1$ mS, $r_o = 49.75$ Ω)

(f) Find g_o and r_o at $I_O = I_{NL} = 0$.   (*Ans.*: $g_o = 57.1$ mS, $r_o = 17.5$ Ω)

3.24 Refer to Figure P3.24. Given $I_{DSS}(Q_4) = 6.0$ mA, $I_{R_1} \simeq I_{R_2} = 1.0$ mA, $I_2 = 5.0$ mA (quiescent current of Q_2).

(a) Find R_1 and R_2 for $V_O = 6.0$ V.   (*Ans.*: $R_1 = 1.9$ kΩ, $R_2 = 700$ Ω)

(b) Find R_3.   (*Ans.*: $R_3 = 140$ Ω)

(c) Find the *load regulation* for $I_{O(NL)} = 7.0$ mA and $I_{O(IFL)} = 105$ mA.   (*Ans.*: $\Delta V_o = -98$ mV or -1.63%)

(d) Find the *dynamic output resistance* at $I_O = 50$ mA.   [*Ans.*: $r_o = 0.611$ Ω (max.), 0.553 Ω (typ.)]

3.25 Refer to Figure P3.25. Given $V_{REF} = +5.00$ V, $I_{R_2} = 1.0$ mA.

(a) Find R_1 and R_2 for $V_O = +10.0$ V.   (*Ans.*: $R_1 = 5$ kΩ, $R_2 = 5$ kΩ)

(b) If $A_{OL}(0) = 100$ dB (min.) and the open-loop output resistance is 100 Ω, find the dynamic output resistance, r_o.   [*Ans.*: $r_o = 2.0$ mΩ (max.)]

(c) Find the *load regulation* for $I_{O(NL)} = 1.0$ mA and $I_{O(FL)} = 20$ mA.   [*Ans.*: $\Delta V_O = -38$ μV (max.) $= -3.8$ ppm $= -0.00038\%$]

(d) If $TC_{V(REF)} = 10$ ppm/°C, find TC_{V_O}.   (*Ans.*: 100 μV/°C or 10 ppm/°C)

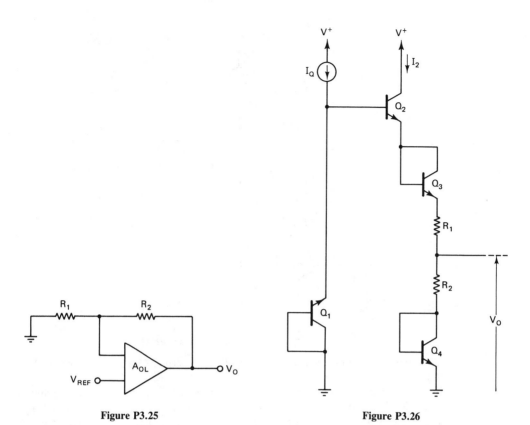

Figure P3.25   **Figure P3.26**

Temperature-Compensated Voltage Reference Problems

3.26 Refer to Figure P3.26.
  (a) Show that $V_O = V_Z[R_2/(R_1 + R_2)] - V_{BE}[2R_2 - R_1)/(R_1 + R_2)]$.
  (b) Show that for $TC_{V_O} = dV_O/dT = 0$ the requirement for the R_1/R_2 resistor ratio is $R_1/R_2 = 2 + [(dV_Z/dT)/(-dV_{BE}/dT)]$.
  (c) Given that $V_Z = 6.3\,\text{V}, dV_Z/dT = +3.0\,\text{mV/°C}, dV_{BE}/dT = -2.2\,\text{mV/°C}, \text{and } I_2 = 1.0$ mA, find R_1, R_2, and V_O for $TC_{V_O} = 0$.   *(Ans.: $R_1 = 3.2375\,\text{k}\Omega, R_2 = 962.5\,\Omega,$ $V_O = 1.66\,\text{V}$)*
  (d) If the temperature coefficient of the zener voltage dV_Z/dT or of the base-to-emitter forward voltage drop dV_{BE}/dT is off from the design center value by $\pm 10\%$, find the resulting value of TC_{V_O}.   *(Ans.: $TC_{V_O} = \pm 69\,\mu\text{V/°C or 41 ppm/°C}$)*
  (e) If the R_1/R_2 resistance ratio is off from the design center value by $\pm 1\%$, find the resulting value of TC_{V_O}. *(Hint: Let R_2 be 1% greater than the value given above for the $TC_{V_O} = 0$ condition and solve for TC_{V_O}. Check your answer by now letting R_1 be 1% greater than the value given above for the $TC_{V_O} = 0$ condition and solve for TC_{V_O}).* *(Ans.: $TC_{V_O} = \pm 17\,\mu\text{V/°C or} \pm 10.1\,\text{ppm/°C}$)*

3.27 Refer to Figure P3.27. Given $I_1 = 1.0\,\text{mA}, I_2 = 0.10\,\text{mA}$, and $V_{BE} = 0.7\,\text{V}$.
  (a) Find R_1, R_2, R_3, and V_O for the zero temperature coefficient of V_O.   *(Ans.: $R_1 = 1.866$ $\text{k}\Omega, R_2 = 11.66\,\text{k}\Omega, R_3 = 576\,\Omega, V_O = 2.566\,\text{V}$)*
  (b) If the *ratio tolerance* of the IC resistors is $\pm 2\%$, what will TC_{V_O} be (assuming that

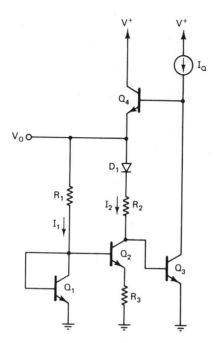

Figure P3.27

otherwise TC_{V_O} would be zero)?   (*Ans.: TC_{V_O} = 80 μV/°C or 31 ppm/°C or 0.0031%/°C*)

(c) If the *absolute value tolerance* of the IC resistors is ±5%, find TC_{V_O} (assuming that otherwise TC_{V_O} would be zero). [*Hint:* Note that $TC_{V_O} = dV_O/dT = (n + 1)(dV_{BE}/dT) + (R_2/R_3)(V_T/T) \ln (I_1/I_2)$ and that $I_1/I_2 = \exp (I_2R_3/V_T)$, so that $V_T \ln (I_1/I_2) = I_2R_3$ and $TC_{V_O} = (n + 1)(dV_{BE}/dT) + (R_2R_3)I_2R_3/T.$]   (*Ans.: TC_{V_O} = 201 μV/°C or 78 ppm/°C or 0.0078%/°C*)

(d) If the deviation of V_{BE} from the design center value is ±10 mV, find TC_{V_O} (assuming that otherwise TC_{V_O} would be zero).   (*Ans.: TC_{V_O} = 67 μV/°C or 26 ppm/°C or 0.0026%/°C*)

REFERENCES

BOYLESTAD, R., and L. NASHELSKY, *Electronic Devices and Circuit Theory*, Prentice-Hall, Englewood Cliffs, N.J., 1982.

GLASER, A. B., and G. E. SUBAK-SHARPE, *Integrated Circuit Engineering: Design Fabrication and Application*, Addison-Wesley, Reading, Mass., 1977.

GRAY, P. R. and R. G. MEYER, *Analylsis and Design of Analog Integrated Circuits*, 2nd ed., Wiley, New York, 1984.

GREBENE, A. B., *Analog Integrated Circuit Design*, Van Nostrand Reinhold, New York, 1972.

———, *Bipolar and MOS Analog Integrated Circuit Design*, Wiley, New York, 1984.

HAMILTON, D. J., and W. G. HOWARD, *Basic Integrated Circuit Engineering*, McGraw-Hill, New York, 1975.

IRVINE, R. G., *Operational Amplifiers—Characteristics and Applications*, 2nd ed., Prentice-Hall, Englewood Cliffs, N.J., 1987.

MANASSE, F. K., *Semiconductor Electronics Design*, Prentice-Hall, Englewood Cliffs, N.J., 1977.

4

Differential Amplifiers

The differential amplifier is a very important transistor amplifier stage configuration and is widely used in various types of analog ICs, such as operational amplifiers, voltage comparators, voltage regulators, video amplifiers, and balanced modulators and demodulators. Differential amplifiers are also the basis of the emitter-coupled digital logic (ECL) gates. The differential amplifier will be the first or input stage of operational amplifiers and other ICs and will therefore determine many of the important performance characteristics of the IC, such as the offset voltage V_{OS}, the input bias current I_B, the input offset current I_{OS}, the input impedance, and the common-mode rejection ratio (CMRR).

In this chapter the basic operation of differential amplifiers is investigated. Differential amplifiers using bipolar transistors are studied first, followed by an examination of differential amplifiers using JFETs and MOSFETs. This is followed by a discussion of active load circuits for differential amplifiers and the presentation of representative "current mirror" active load circuits. For some applications, the use of the Darlington compound transistor configuration for differential amplifiers is beneficial, and this will be the subject of a brief discussion in the latter part of this chapter.

4.1 ANALYSIS OF THE BIPOLAR TRANSISTOR DIFFERENTIAL AMPLIFIER

For the analysis of the differential amplifier, we will consider the basic differential-amplifier circuit shown in Figure 4.1. In Figure 4.2, the differential amplifier with load resistances is shown. For this analysis it will be assumed that both transistors of the differential-amplifier pair Q_1 and Q_2 are operated in the active mode and that the base currents will be small compared to the collector currents. The fundamental relationship to be used for this analysis will be the exponential relationship between the collector current and the base-to-emitter voltage of a transistor. For the collector current of Q_1, we can write

$$I_1 = I_{C_1} = I_{CO_1} \exp\left(\frac{V_{BE_1}}{V_T}\right) \tag{4.1}$$

where the preexponential constant in this relationship for Q_1 is I_{CO_1}, and V_{BE_1} is the base-to-emitter voltage of Q_1, so $V_{BE_1} = V_{B_1} - V_{E_1}$. In a similar fashion, we can write for Q_2 the equation

$$I_2 = I_{C_2} = I_{CO_2} \exp\left(\frac{V_{BE_2}}{V_T}\right)$$

If transistors Q_1 and Q_2 were exactly identical and operated with identical collector voltages, I_{CO_1} would be equal to I_{CO_2}. However, even two transistors right next to each other on an IC chip will not be exactly identical. For purposes of convenience, let us define the *offset voltage*, V_{OS}, by the following relationship:

$$\frac{I_{CO_2}}{I_{CO_1}} = \exp\left(\frac{V_{OS}}{V_T}\right)$$

so that
$$V_{OS} = V_T \ln \frac{I_{CO_2}}{I_{CO_1}} \tag{4.2}$$

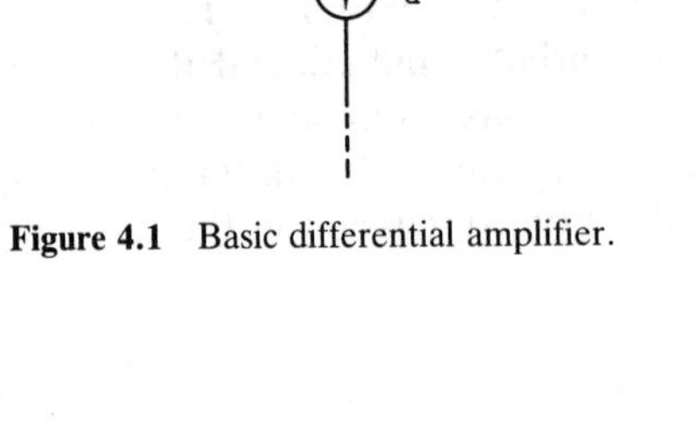

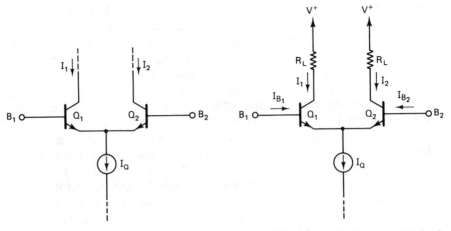

Figure 4.1 Basic differential amplifier.

Figure 4.2 Differential amplifier with load resistances.

Note that if the two transistors are exactly identical $V_{OS} = 0$. Using the definition of the offset voltage, we can rewrite the equation for I_2 as

$$I_2 = I_{CO_2} \exp\left(\frac{V_{BE_2}}{V_T}\right) = I_{CO_1} \exp\left(\frac{V_{BE_2} + V_{OS}}{V_T}\right)$$

Since $I_1 + I_2 = I_Q$, we have

$$I_Q = I_{CO_1}\left[\exp\left(\frac{V_{BE_1}}{V_T}\right) + \exp\left(\frac{V_{BE_2} + V_{OS}}{V_T}\right)\right] \tag{4.3}$$

so

$$I_{CO_1} = \frac{I_Q}{\exp\left(V_{BE_1}/V_T\right) + \exp\left[(V_{BE_2} + V_{OS})/V_T\right]} \tag{4.4}$$

If we now substitute this expression for I_{CO_1} into the expression for I_1, we obtain

$$I_1 = \frac{I_Q \exp\left(V_{BE_1}/V_T\right)}{\exp\left(V_{BE_1}/V_T\right) + \exp\left[(V_{BE_2} + V_{OS})/V_T\right]} \tag{4.5}$$

Dividing numerator and denominator by $\exp\left(V_{BE_1}/V_T\right)$ yields

$$I_1 = \frac{I_Q}{1 + \exp\left[(V_{BE_2} - V_{BE_1} + V_{OS})/V_T\right]} \tag{4.6}$$

If we now solve for I_2, we obtain

$$I_2 = I_{CO_2} \exp\left(\frac{V_{BE_2}}{V_T}\right) = I_{CO_1} \exp\left(\frac{V_{OS}}{V_T}\right) \exp\left(\frac{V_{BE_2}}{V_T}\right)$$

$$= \frac{I_Q \exp\left(V_{BE_2} + V_{OS}/V_T\right)}{\exp\left(V_{BE_1}/V_T\right) + \exp\left[(V_{BE_2} + V_{OS})/V_T\right]} \tag{4.7}$$

so

$$I_2 = \frac{I_Q}{1 + \exp\left[(V_{BE_1} - V_{BE_2} - V_{OS})/V_T\right]} \tag{4.8}$$

Since $V_{BE_1} = V_{B_1} - V_{E_1}$ and $V_{BE_2} = V_{B_2} - V_{E_2}$, and noting that $V_{E_1} = V_{E_2}$, we have that $V_{BE_1} - V_{BE_2} = V_{B_1} - V_{B_2}$.

Let us define the *differential input voltage* or *difference-mode input voltage* as $V_i = V_{B_1} - V_{B_2}$. In terms of the differential input voltage V_i, we can express I_1 and I_2 as

$$\boxed{I_1 = \frac{I_Q}{1 + \exp\left[(-V_i + V_{OS})/V_T\right]} = \frac{I_Q}{1 + \exp\left[-(V_i - V_{OS})/V_T\right]}} \tag{4.9}$$

and

$$\boxed{I_2 = \frac{I_Q}{1 + \exp\left[(V_i - V_{OS})/V_T\right]}} \tag{4.10}$$

In Figure 4.3, graphs of I_1 and I_2 versus $V_i - V_{OS}$ are presented. Note that, when $V_i = V_{OS}$, I_1 and I_2 are equal to each other and equal to $I_Q/2$. The value of V_{OS} for

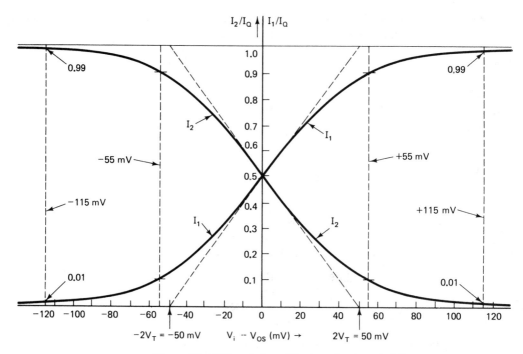

Figure 4.3 Differential-amplifier transfer characteristics.

IC transistors is normally on the order of 1 or 2 mV. Thus, when $V_i = V_{OS}$, the differential amplifier is balanced in that the current of the current source (or "sink") I_Q splits equally between the two transistors of the differential-amplifier pair.

Let us now determine the value of the differential input voltage that will cause 90% of the total current to flow through Q_1 and the remaining 10% through Q_2. For this condition, we have that $I_1 = 0.9I_Q$ and $I_2 = 0.1I_Q$, giving

$$0.1I_Q = \frac{I_Q}{1 + \exp\left[(V_i - V_{OS})/V_T\right]}$$

so

$$\exp\left(\frac{V_i - V_{OS}}{V_T}\right) = 9 \qquad (4.11)$$

and therefore

$$V_i - V_{OS} = V_T \ln 9 = 25 \text{ mV} \times 2.1972 \simeq \underline{55 \text{ mV}} \qquad (4.12)$$

For the opposite condition, that is, $I_1 = 0.1I_Q$ and $I_2 = 0.91I_Q$, we would have $V_i - V_{OS} = \underline{55 \text{ mV}}$. The total voltage change ΔV_1 required to shift the differential-amplifier current distribution from $I_1 = 0.9I_Q$ and $I_2 = 0.1I_Q$ to the opposite case of $I_1 = 0.1I_Q$ and $I_2 = 0.9I_Q$ is called the *transition voltage* and will therefore have a value of about $2 \times 55 \text{ mV} = \underline{110 \text{ mV}}$. We see that it does not require a very large voltage to produce a major shift in the current distribution of a differential amplifier.

Differential Amplifiers   Chap. 4

Referring to the equations for I_1 and I_2 and looking at the graphical presentation of the transfer characteristics of the differential amplifier, we see that as V_i increases in either direction more and more current flows through one transistor and less and less through the other. At no point, however, does all the current flow through one transistor, with the other one being cut off.

To consider another example for $I_1 = 0.99 I_Q$ and $I_2 = 0.01 I_Q$, we have

$$0.01 = \frac{1}{1 + \exp\left[(V_i - V_{OS})/V_T\right]} \tag{4.13}$$

so $V_i - V_{OS} = 115\ \text{mV}$. For $I_1 = 0.01 I_Q$ and $I_2 = 0.99 I_Q$, we would similarly have $V_i - V_{OS} = -115\ \text{mV}$. For an even more extreme case, if $I_1 = 0.999 I_Q$ and $I_2 = 0.001 I_Q$, the required differential input voltage is $V_i - V_{OS} = \underline{173\ \text{mV}}$.

If the biasing source for the differential amplifier that supplies the bias current I_Q is an ideal constant-current source, than I_Q is independent of the voltage across the current source and therefore is independent of the input voltages V_{B_1} and V_{B_2}. Looking at the equations for I_1 and I_2, we see that if indeed I_Q is a constant then I_1 and I_2 will be a function only of the differential input voltage, $V_i = V_{B_1} - V_{B_2}$, and will not respond to any common-mode component of the input voltage. Thus the amplifier is a true *differential*, or *difference amplifier*, responding only to the difference in the voltages applied to the two input terminals, B_1 and B_2, and not responding at all to any voltage that is common to the two inputs. The *differential*, or *difference-mode input voltage* has already been defined as $V_i = V_{B_1} = V_{B_2}$. Let us define the *common-mode input voltage* as the average of the two input voltages so that the *common-mode input voltage* $= V_{CM} = (V_{B_1} + V_{B_2})/2$. If $V_{B_1} = -V_{B_2}$, the common-mode component of the input voltage is zero and we have a pure difference-mode input voltage. If, on the other hand, $V_{B_1} = V_{B_2}$, the difference-mode component is zero and we now have a pure common-mode input voltage.

4.1.1 Transfer Conductances

Looking at the equations for I_1 and I_2 and the graph of I_1 and I_2 versus V_i, we see that the differential amplifier is truly a nonlinear device in terms of the relationship between the output currents, I_1 and I_2, and the input voltage, V_i. However, over a limited region of the transfer characteristic curves for I_1 versus V_i or I_2 versus V_i, we see that the relationship between the currents and the input voltage can be considered to be approximately linear. Looking at the transfer characteristics graph, we see that we have a characteristic curve that is approximately linear over a range of input voltages of from about $V_i - V_{OS} = -30\ \text{mV}$ to $V_i - V_{OS} = +30\ \text{mV}$, or a total span of about 60 mV. Thus, for small-signal ac conditions, the device can be considered to act as an approximately linear device as far as the relationships between the ac input voltage and the ac output currents are concerned.

In this section, expressions for the differential-amplifier transfer conductances will be obtained. *A transfer conductance* is the rate of change of a current at one port of a multiport device with respect to the voltage applied at another port. If we refer to Figure 4.4, which shows the differential amplifier in block diagram form, we

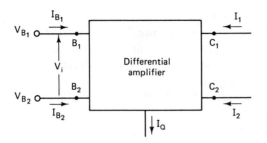

Figure 4.4 Differential-amplifier currents and voltages.

can define several transfer conductances as

$$g_{f11} = \frac{dI_1}{dV_{B_1}}, \qquad g_{f21} = \frac{dI_2}{dV_{B_1}}$$

$$g_{f12} = \frac{dI_1}{dV_{B_2}}, \qquad g_{f22} = \frac{dI_2}{dV_{B_2}} \tag{4.14}$$

We will see in the forthcoming analysis that all these transfer conductances for the differential amplifier are equal in magnitude, so $|g_{f11}| = |g_{f22}| = |g_{f12}| = |g_{f21}| = g_f$ and therefore differ only in terms of the algebraic sign.

The expression for I_1 has been given as

$$I_1 = \frac{I_Q}{1 + \exp\left[-(V_i - V_{OS})/V_T\right]} \tag{4.15}$$

where $V_i = V_{B_1} - V_{B_2}$, so

$$g_{f11} = \frac{dI_1}{dV_{B_1}} = \frac{dI_1}{dV_i}\frac{dV_i}{dV_{B_1}} = \frac{dI_1}{dV_i} \tag{4.16}$$

Performing the indicated operation yields

$$g_{f11} = \frac{dI_1}{dV_i} = \frac{I_1(1/V_T)\exp\left[-(V_i - V_{os})/V_T\right]}{1 + \exp\left[-(V_i - V_{os})/V_T\right]}$$

$$= \frac{I_1/V_T}{1 + \exp\left[(V_i - V_{os})/V_T\right]} \tag{4.17}$$

Noting that

$$I_2 = \frac{I_Q}{1 + \exp\left[(V_i - V_{os})/V_T\right]}$$

we can express g_{f11} as

$$g_{f11} = \frac{I_1 I_2}{I_Q V_T} \tag{4.18}$$

Since g_{f22} can be obtained by simply interchanging subscripts, we have

$$g_{f22} = \frac{I_2 I_1}{I_Q V_T} = g_{f11} \tag{4.19}$$

Since I_1 and I_2 will have the same algebraic sign as I_Q, $g_{f11} = g_{f22}$ will be algebraically positive, so $g_{f11} = g_{f22} = g_f$. For g_{f12}, we have

$$g_{f12} = \frac{dI_1}{dV_{B_2}} = \frac{dI_1}{dV_i} \frac{dV_i}{dV_{B_2}} \tag{4.20}$$

Since $V_i = V_{B_1} - V_{B_2}$,

$$\frac{dV_i}{dV_{B_2}} = -1$$

so that

$$g_{f12} = \frac{-dI_1}{dV_i} = \frac{-dI_1}{dV_{B_1}} = -g_{f11} = \frac{-I_1 I_2}{I_Q V_T} \tag{4.21}$$

Again upon interchanging of subscripts we obtain g_{f21} as

$$g_{f21} = \frac{-I_2 I_1}{I_Q V_T} = g_{f12} = -g_{f11} = -g_{f22} \tag{4.22}$$

We can therefore summarize the expressions for the transfer conductances of the differential amplifier very simply as

$$\boxed{g_f = g_{f11} = g_{f22} = -g_{f12} = -g_{f21} = \frac{I_1 I_2}{I_Q V_T}} \tag{4.23}$$

Since $I_1 + I_2 = I_Q$, we can also write g_f as

$$g_f = \frac{I_1(I_Q - I_1)}{I_Q V_T} = \frac{I_2(I_Q - I_2)}{I_Q V_T} \tag{4.24}$$

The value of g_f will be a maximum for $I_1 = I_2 = I_Q/2$ and be equal to

$$\boxed{g_{f(max)} = \frac{I_Q}{4V_T} = \frac{I_Q}{100 \text{ mV}}} \tag{4.25}$$

Figure 4.5 shows a graph of the transfer conductance as a function of the quiescent current through either Q_1 or Q_2. Note that there is a broad maximum where $I_1 = I_2 = I_Q/2$, but when either I_1 or I_2 becomes small compared to I_Q, the transfer conductance can become very small.

The transfer conductance can also be expressed in terms of the quiescent voltage between the two input terminals, $V_i = V_{B_1} - V_{B_2}$. Since

$$I_1 = \frac{I_Q}{1 + \exp\left[-(V_i - V_{os})/V_T\right]} \quad \text{and} \quad I_2 = \frac{I_Q}{1 + \exp\left[(V_i - V_{os})/V_T\right]} \tag{4.26}$$

we have that

$$g_f = \frac{I_1 I_2}{I_Q V_T} = \frac{I_Q}{V_T} \frac{1}{[1 + \exp(-x)][1 + \exp(+x)]}, \quad \text{where } x = \frac{V_i - V_{os}}{V_T} \tag{4.27}$$

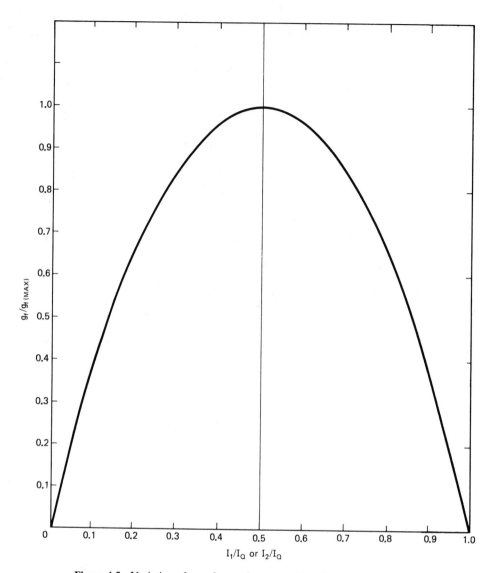

Figure 4.5 Variation of transfer conductance with quiescent current levels.

Since

$$[1 + \exp(-x)][1 + \exp(x)] = 2 + \exp(x) + \exp(-x)$$

$$= 2\left[1 + \frac{\exp(x) + \exp(-x)}{2}\right] \qquad (4.28)$$

$$= 2(1 + \cosh x)$$

we have

$$g_f = \frac{I_Q/V_T}{2[1 + \cosh(V_i - V_{os})/V_T]} \qquad (4.29)$$

Differential Amplifiers   Chap. 4

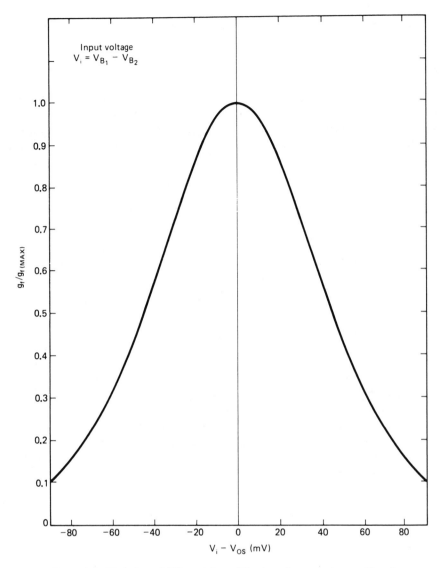

Figure 4.6 Variation of differential-amplifier transfer conductance with quiescent voltage.

Figure 4.6 is a graph of the variation of the transfer conductance g_f as a function of a dc bias voltage V_i applied between the differential amplifier input terminals. Note that g_f is a maximum when the bias voltage cancels out the offset voltage ($V_i - V_{OS} = 0$) so as to balance the circuit and produce the condition that $I_1 = I_2 = I_Q/2$. The transfer conductance decreases rapidly for values of $V_i - V_{OS}$ greater than 10 mV in magnitude. This variation in the transfer conductance produces a corresponding variation in the voltage gain of the differential-amplifier circuit and will be useful for *automatic gain control* (AGC) or *automatic volume control* (AVC) applications.

4.1.2 Output Voltage and Voltage Gain

Figure 4.4 is a block diagram of a differential amplifier. Figure 4.7 shows a simple ac small-signal equivalent circuit for the circuit of Figure 4.4. Load resistances of value R_L are placed in series with the collectors of Q_1 and Q_2 to produce ac output voltages from the ac component of the collector currents of Q_1 and Q_2. The ac output voltage at the collector of Q_1 is $v_{o1} = -g_f R_L v_i$, so the corresponding ac voltage gain is

$$A_{V1} = \frac{v_{o1}}{v_i} = -g_f R_L \qquad (4.30)$$

Similarly, the ac voltage at the collector of Q_2 is $v_{o2} = +g_f R_L v_i$, and the voltage gain with respect to this output voltage is

$$A_{V2} = \frac{v_{o2}}{v_i} = +g_f R_L \qquad (4.31)$$

Note that these two output voltages, and therefore the corresponding voltage gains, are equal in magnitude but opposite in algebraic sign. These two output voltages are called *single-ended* or *unbalanced outputs* since they are taken with one side of the output voltage at ground potential.

Figure 4.8 shows the case of a *double-ended* or *balanced output* wherein the output voltage is taken between the two collectors instead of between either collector and ground. The output voltage for this case is $v_o = v_{o1} - v_{o2} = -2g_f R_L v_i$, and the corresponding voltage gain is

$$A_V = \frac{v_o}{v_i} = -2g_f R_L \qquad (4.32)$$

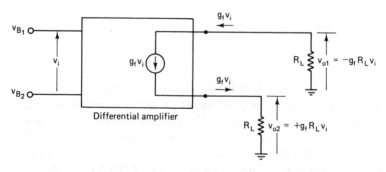

Figure 4.7 Differential amplifier with single-ended outputs.

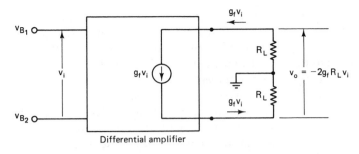

Figure 4.8 Differential amplifier with double-ended (balanced) output.

Note that the output voltage and the voltage gain for the double-ended case is exactly twice that obtained for the single-ended situation.

If this differential amplifier is to be used to drive a balanced load, such as another differential amplifier with input resistance R_i, the net load resistance driven by the differential amplifier between the two collectors is given by $R_{L(\text{NET})} = 2R_L \| R_i$ and the voltage gain is

$$A_V = -g_f R_{L(\text{NET})} = \frac{-g_f}{G_{L(\text{NET})}} = \frac{-g_f}{G_L/2 + G_i} \qquad (4.33)$$

In the more general case in which complex impedances are involved, we have

$$A_V = -g_f Z_{L(\text{NET})} = \frac{-g_f}{Y_{L(\text{NET})}} = \frac{-g_f}{Y_L/2 + Y_i} \qquad (4.34)$$

4.1.3 Common-mode Transfer Conductance

From the previous discussion of the differential-amplifier transfer characteristics, it appears that the ideal differential amplifier will be sensitive only to the difference-mode component of the input signal and will be completely unresponsive to the common-mode signal. That this will be the case can be verified by investigating the differential-amplifier response to a common-mode signal in which the same voltage is applied to both bases.

If the same voltage is applied to the bases of Q_1 and Q_2 of Figure 4.1, the voltage at the emitters will follow the voltage at the bases and will therefore appear across the current source I_Q. If the I_Q current source is indeed an ideal current source, then this voltage change across the current source will *not* result in any change of current. Since the total differential-amplifier current will remain constant at I_Q and the division of current between Q_1 and Q_2 is not affected by the common-mode input voltage, we see that the collector currents I_1 and I_2 will thus be unaffected by the common-mode input signal. The *common-mode transfer conductance*, which is defined as the rate of change of collector current with respect to a common-mode input voltage, will therefore be zero.

We will now consider the case of a nonideal current source in which a change in the voltage produces a change in the current. A real current source can be represented as the parallel combination of an ideal current source and a small dynamic conductance g_o. This conductance is the *dynamic output conductance* of the current source, and the change in the current of the current source ΔI_Q is equal to the change in the voltage across the current source times the output conductance g_o.

The application of a common-mode input voltage Δv_{CM} to the differential amplifier will produce a change in the voltage across the current source of approximately Δv_{CM} since the voltage at the two emitters will closely follow the common-mode base voltage. The change in the current of the current source is therefore $\Delta I_Q = \Delta v_{CM} \times g_o$. If the differential amplifier is balanced, this current change will be split equally between Q_1 and Q_2, and thus the change in the collector currents is $\Delta I_1 = \Delta I_2 = \Delta I_Q/2 = \Delta v_{CM}(g_o/2)$. The *common-mode dynamic transfer conductance* $g_{f(CM)}$ is therefore given by

$$ g_{f(CM)} = \frac{\Delta I_1}{\Delta v_{CM}} = \frac{\Delta I_2}{\Delta v_{CM}} = \frac{g_o}{2} \qquad (4.35) $$

For the more general situation in which the differential amplifier is not balanced such that the two collector currents are not equal, we have that

$$ I_1 = \frac{I_Q}{1 + \exp\left[-(V_i - V_{OS})/V_T\right]} \quad \text{and} \quad I_2 = \frac{I_Q}{1 + \exp\left(V_i - V_{OS}\right)/V_T} \qquad (4.36) $$

We can therefore express the common-mode transfer conductance as

$$ g_{f1(CM)} = \frac{dI_1}{dV_{CM}} = \frac{dI_1}{dI_Q}\frac{dI_Q}{dV_{CM}} = \frac{1}{1 + \exp\left[-(V_i - V_{OS})/V_T\right]} g_o = \frac{I_1}{I_Q} g_o \qquad (4.37) $$

and similarly,

$$ g_{f2(CM)} = \frac{dI_2}{dV_{CM}} = \frac{I_2}{I_Q} g_o \qquad (4.38) $$

The common-mode transfer conductance will be very much smaller than the difference-mode transfer conductance, usually by a factor of 10^4 or 10^5. It is this large ratio of the difference-mode transfer conductance g_f to the common-mode transfer conductance $g_{f(CM)}$ that is principally responsible for the very large common-mode rejection ratio (CMRR) values of operational amplifiers.

4.1.4 Input Conductance

The discussion in the previous sections has dealt with the differential-amplifier dynamic transfer conductances, which involve the rate of change of the *output* currents I_1 and I_2 with respect to the input voltages V_{B_1} and V_{B_2}. We will now consider the dynamic input conductance, which is the rate of change of the *input* currents I_{B_1} and I_{B_2} to the input voltages. If we refer back to Figure 4.2, we note that the input or base currents of the differential amplifier are related to the output or collector currents by the ac current gain of the transistor β or h_{fe}, as given by $dI_1/dI_{B_1} = \beta$ and $dI_2/dI_{B_2} = \beta$. This relationship will be used in the derivation of the expressions for the input conductance.

The *difference-mode input conductance* g_i will be defined as the rate of change of the input current with respect to the difference-mode input voltage and is given by $g_i = dI_{B_1}/dV_i = -dI_{B_2}/dV_i = dI_{B_1}/dI_1 \times dI_1/dV_i$. Since $dI_{B_1}/dI_1 = 1/\beta$ and dI_1/dV_i is the transfer conductance g_f, the equation for g_i can be written very simply as

$$g_i = \frac{g_f}{\beta} \tag{4.39}$$

If we note that $g_f = I_1 I_2 / I_Q V_T$, we see that g_i can be written as $g_i = I_1 I_2 / \beta I_Q V_T$. The *difference-mode input resistance* r_i is the reciprocal of the input conductance, so $r_i = 1/g_i = \beta I_Q V_T / I_1 I_2$. Since $I_Q = I_1 + I_2$, this can be rewritten as $r_i = \beta V_T(I_1 + I_2)/I_1 I_2 = \beta V_T(1/I_1 + 1/I_2) = V_T(\beta/I_1 + \beta/I_2)$. Since $I_1/I_{B_1} \simeq I_2/I_{B_2} \simeq \beta$, we can now express r_i very simply as

$$r_i = \frac{V_T}{I_{B_1}} + \frac{V_T}{I_{B_2}} \tag{4.40}$$

For the usual case of $I_1 \simeq I_2$ so that $I_{B_1} \simeq I_{B_2}$, this becomes

$$r_i = \frac{2V_T}{I_B} \tag{4.41}$$

where I_B $(\simeq I_{B_1} \simeq I_{B_2})$ is the input bias current of the differential amplifier. We see, therefore, that there is a very simple and direct relationship between the input bias (or base) current of the differential amplifier and the dynamic input resistance. For example, if the input bias current is 50 nA, the dynamic difference-mode input resistance will be given by $r_i = 2V_T/I_B = 50$ mV/50 nA $= 1$ MΩ. In Figure 4.9, an ac

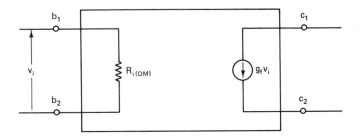

Figure 4.9 Differential-amplifier ac small-signal equivalent circuit, including the difference-mode input resistance.

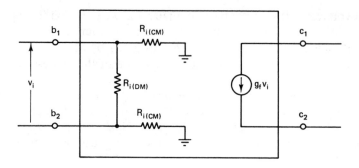

Figure 4.10 Differential-amplifier equivalent circuit, including the difference-mode and common-mode input resistances.

small-signal equivalent circuit for the differential amplifier is shown in which r_i is included.

The *common-mode input conductance* $g_{i(CM)}$ will be defined as the rate of change of the input current with respect to the common-mode input voltage. Thus we have that

$$g_{i_1(CM)} = \frac{dI_{B_1}}{dV_{CM}} = \frac{dI_{B_1}}{dI_1}\frac{dI_1}{dV_{CM}} = \frac{g_{f_1(CM)}}{\beta} \qquad (4.42)$$

and similarly

$$g_{i_2(CM)} = \frac{dI_{B_2}}{dV_{CM}} = \frac{dI_{B_1}}{dI_2}\frac{dI_2}{dV_{CM}} = \frac{g_{f_2(CM)}}{\beta} \qquad (4.43)$$

For the usual case in which $I_1 \simeq I_2 \simeq I_Q/2$, we have that $g_{f_1(CM)} = g_{f_2(CM)} = g_o/2\beta$ and thus

$$g_{i_1(CM)} = g_{i_2(CM)} = g_{i(CM)} = \frac{g_o}{2\beta} \qquad (4.44)$$

This is the dynamic conductance seen looking into each base terminal.

The *common-mode input resistance* $r_{i(CM)}$ is the reciprocal of the input conductance, so

$$r_{i(CM)} = \frac{1}{g_{i(CM)}} = \frac{2\beta}{g_o} = 2\beta r_o \qquad (4.45)$$

where g_o is the dynamic output conductance of the I_Q current source, and $r_o = 1/g_o$ is the dynamic output resistance of the current source. In Figure 4.10, the equivalent circuit for the differential amplifier is again presented, this time including both the difference-mode and common-mode input resistances. The common-mode input resistance will be a very large quantity, often in the gigaohm ($10^9 \ \Omega$) range.

Differential Amplifiers   Chap. 4

4.1.5 Balance of the Bipolar Differential Amplifier: Input Offset Voltage

Looking at the equation for the output currents, I_1 and I_2, of the differential amplifier, we note that when $V_i = 0$ the two currents will not be equal. To make the two currents equal, a small voltage equal to the offset voltage V_{OS} must be applied between the two input terminals such that $V_i = V_{OS}$.

The offset voltage has been defined by the equation

$$\exp\left(\frac{V_{OS}}{V_T}\right) = \frac{I_{CO_2}}{I_{CO_1}} \quad \text{so that} \quad V_{OS} = V_T \ln \frac{I_{CO_2}}{I_{CO_1}} \tag{4.46}$$

The principal factor in I_{CO} that is responsible for the variation of this quantity from one transistor to another is the effective base width, W_B'. Since I_{CO} is proportional to $1/W_B'$, we can express V_{OS} in terms of the base widths of transistors Q_1 and Q_2 as $V_{OS} = V_T \ln (W_{B_1}'/W_{B_2}')$. The base widths, although not equal, will normally be relatively close in value, generally within 10% of each other. Therefore, if we let $W_{B_1}' = W + \Delta W$ and $W_{B_2}' = W$, we have that

$$V_{OS} = V_T \ln \frac{W + \Delta W}{W} = V_T \ln \left(1 + \frac{\Delta W}{W}\right) \simeq V_T \frac{\Delta W}{W} \tag{4.47}$$

if $\Delta W/W \ll 1$, as is usually the case.

The current gain of a transistor β is inversely proportional to the base width W_B'. The primary cause of the unit-to-unit difference in the current gain is the difference in the base widths. Therefore, we can write $\beta_2/\beta_1 \simeq W_{B_1}'/W_{B_2}'$, and thus the fraction difference in the current gains $\Delta\beta/\beta$ is equal to the fractional difference in the base widths as given by $\Delta\beta/\beta \simeq \Delta W/W$. As a result, we see that there is some relationship between the offset voltage and the β mismatch.

To consider a representative example, let us assume that there is a 10% difference in the base widths of a pair of IC transistors. The corresponding offset voltage is $V_{OS} \simeq V_T \Delta W/W = 25 \text{ mV} \times 0.1 = 2.5 \text{ mV}$. In addition to this offset voltage, the difference in the current gains of these two transistors is expected to be about 10%.

Temperature coefficient of the offset voltage. The temperature coefficient of the offset voltage is defined as $TCV_{OS} = dV_{OS}/dT$. We have already seen that V_{OS} can be expressed in terms of the ratio of the effective base widths as $V_{OS} = V_T \ln (W_{B_1}'/W_{B_2}')$. Since the ratio of the base widths is relatively independent of temperature, and $V_T = kT/q$ so that $dV_T/dT = k/q = V_T/T$, we have that

$$\boxed{TCV_{OS} = \frac{dV_{OS}}{dT} = \frac{V_T}{T} \ln \frac{W_{B_1}'}{W_{B_2}'} \simeq \frac{V_{OS}}{T}} \tag{4.48}$$

For example, if $V_{OS} = 1.5 \text{ mV} = 1500 \text{ }\mu\text{V}$, the temperature coefficient is given by

$$TCV_{OS} = \frac{V_{OS}}{T} = \frac{1500 \text{ }\mu\text{V}}{300 \text{ K}} = 5 \text{ }\mu\text{V/K} = \underline{5 \text{ }\mu\text{V/}°\text{C}} \tag{4.49}$$

In percentage terms, we can say that the percent of change in the offset voltage per degree temperature change is given by

$$\frac{1}{V_{OS}}\frac{dV_{OS}}{dT} \times 100\% = \frac{1}{V_{OS}}\frac{V_{OS}}{T} \times 100\% = \frac{100\%}{300 \text{ K}}$$

$$= \underline{0.33\%/°C} \tag{4.50}$$

Note that this last expression for the percent of change in the offset voltage with temperature is independent of the magnitude of the offset voltage. Thus for a 10°C temperature rise the offset voltage will change by about 3.3%, and the offset voltage will change by about 10% for a 30°C temperature rise. This small but significant change in the offset voltage with temperature can be important, especially in applications requiring offset voltage compensation or nulling.

4.1.6 Effect of Collector Voltage on Differential-amplifier Balance

As a result of the base-width modulation effect (also known as the Early effect), a difference in the collector voltages of the two transistors that comprise the differential-amplifier pair can produce differences in the collector currents. Alternatively, the effect of the difference in the collector voltages can be expressed in terms of an input offset voltage.

The effect of the collector-to-base voltage, V_{CB}, of a transistor on the effective (electrical) base width of the transistor can be expressed in terms of the *base-width modulation factor* as $-(1/W)(dW/dV_{CB}) = 1/V_A$, where V_A is a *constant* having units of volts (see Appendix B). We note that $1/V_A$ is the fractional change in the base width, dW/W, per volt change in the collector-to-base voltage (V_{CB}).

As a result of a difference in the collector voltages of two transistors of amount $\Delta V_{CB} = V_{CB_1} - V_{CB_2}$, there is a difference in the base widths of the two transistors that is given approximately by $-(\Delta W/W) = \Delta V_{CB}/V_A$. When the offset voltage of a pair of transistors is considered, it can be expressed in terms of the effective base widths of the two transistors as

$$V_{OS} = V_T \ln \frac{W_2}{W_1} = V_T \ln \frac{W_1 + \Delta W}{W_1}$$

$$= V_T \ln \left(1 + \frac{\Delta W}{W_1}\right) \simeq V_T \frac{\Delta W}{W}, \qquad \text{for } \frac{\Delta W}{W} \ll 1 \tag{4.51}$$

The contribution of the difference in the collector-to-base voltages to the offset voltage is therefore given by

$$\Delta V_{OS} = V_T \left(\frac{-\Delta V_{CB}}{V_A}\right) = \frac{V_T(V_{CB_2} - V_{CB_1})}{V_A} \tag{4.52}$$

For an example, let us consider a pair of transistors for which the base-width modulation voltage, V_A, has a value of 250 V, which is a typical value for IC NPN transistors. The change in the offset voltage, ΔV_{OS}, produced by a difference in the collector-to-base voltages of the two transistors, is given by

$$\Delta V_{OS} = V_T \left(\frac{-\Delta V_{CB}}{V_A} \right) = \frac{25 \text{ mV}}{250 \text{ V}} (-\Delta V_{CB}) = (0.10 \text{ mV/V})(-\Delta V_{CB})$$

Therefore, the offset voltage changes by 0.10 mV for every 1-V change in ΔV_{CB}. A difference of 10 V in the V_{CB} values for the two transistors is therefore responsible for a shift of 1.0 mV in the offset voltage.

4.2 FIELD-EFFECT TRANSISTOR DIFFERENTIAL AMPLIFIERS

Differential amplifiers using field-effect transistors operate in a fashion that is basically similar to bipolar transistor differential amplifiers. The FET differential amplifiers offer the advantages of a very high input impedance ($\sim 10^9$ to 10^{12} Ω) and a very low input bias current ($\sim 10^{-9}$ to 10^{-12} A). A disadvantage of the FET differential amplifiers is the lower transfer conductance and therefore the lower voltage gain that is available. Another disadvantage is the somewhat higher offset voltage of a FET transistor pair compared to that of a pair of bipolar transistors.

4.2.1 JFET Differential Amplifiers

We will first consider the JFET differential amplifier shown in Figure 4.11. For simplicity, we will assume that both transistors have identical characteristics. For each transistor operating in the active region, we have the transfer relationship given by

$$I_{DS} = I_{DSS} \left(1 - \frac{V_{GS}}{V_P} \right)^2$$

so

$$1 - \frac{V_{GS}}{V_P} = \sqrt{\frac{I_{DS}}{I_{DSS}}} \tag{4.53}$$

and thus

$$V_{GS} = V_P \left(1 - \sqrt{\frac{I_{DS}}{I_{DSS}}} \right) \tag{4.54}$$

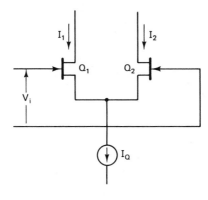

Figure 4.11 Junction field-effect transistor differential amplifier.

Under quiescent conditions the two JFET currents are equal and are given by $I_1 = I_{DS_1} = I_Q/2$ and $I_2 = I_{DS_2} = I_Q/2$. In response to a difference-mode input voltage V_i, the two drain currents change from the quiescent value by an amount ΔI, given by $I_1 = I_Q/2 + \Delta I$ and $I_2 = I_Q/2 - \Delta I$. The corresponding gate-to-source voltages of Q_1 and Q_2 are given by

$$V_{GS_1} = V_P \left(1 - \sqrt{\frac{I_Q/2 + \Delta I}{I_{DSS}}} \right) \quad \text{and} \quad V_{GS_2} = V_P \left(1 - \sqrt{\frac{I_Q/2 - \Delta I}{I_{DSS}}} \right) \qquad (4.55)$$

The relationship between ΔI and the difference-mode input voltage V_i is given by

$$V_i = V_{G_1} - V_{G_2} = V_{GS_1} - V_{GS_2} = V_P \left(\sqrt{\frac{I_Q/2 - \Delta I}{I_{DSS}}} - \sqrt{\frac{I_Q/2 + \Delta I}{I_{DSS}}} \right)$$

$$= V_P \sqrt{\frac{I_Q}{I_{DSS}}} \left(\sqrt{\frac{1}{2} - \frac{\Delta I}{I_Q}} - \sqrt{\frac{1}{2} + \frac{\Delta I}{I_Q}} \right) \qquad (4.56)$$

where I_Q must be limited to a value less than I_{DSS}. Figure 4.12 shows a normalized graph of the transfer characteristic for the JFET differential amplifier.

For small values of V_i and $\Delta I/I_Q$, a linear approximate relationship between V_i and ΔI can be obtained by using the approximation that $\sqrt{1 + x} \simeq 1 + (x/2)$ for $x \ll 1$. Applying this to the relationship between V_i and ΔI gives

$$V_i \simeq V_P \sqrt{\frac{I_Q}{I_{DSS}}} \frac{1}{\sqrt{2}} \left[\left(1 - \frac{\Delta I}{I_Q} \right) - \left(1 + \frac{\Delta I}{I_Q} \right) \right] = \frac{-V_P(2\Delta I)}{\sqrt{2 I_Q I_{DSS}}} \qquad (4.57)$$

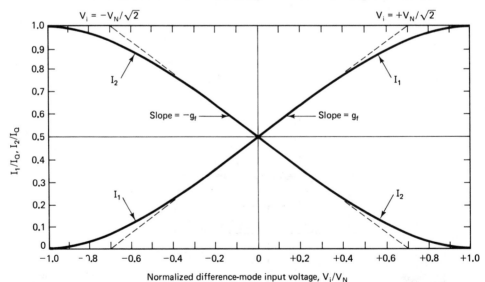

Normalization factor: $V_N = -V_P \sqrt{I_Q/I_{DSS}}$ for JFET, and $V_N = \sqrt{I_Q/K}$ for MOSFET

Figure 4.12 Field-effect transistor differential-amplifier transfer characteristic.

The dynamic transfer conductance of the differential amplifier, g_f, is therefore given by

$$g_f = \frac{\Delta I}{V_i} = \frac{\sqrt{2I_Q I_{DSS}}}{-2V_P} = \frac{\sqrt{I_{DSS} I_Q/2}}{-V_P} \tag{4.58}$$

4.2.2 MOSFET Differential Amplifier

A MOSFET differential amplifier is shown in Figure 4.13. We will again assume that the two transistors have identical characteristics. In the active region of operation the MOSFET transfer characteristic is given by $I_{DS} = K(V_{GS} - V_t)^2$, where V_t is the threshold voltage for channel formation (not the thermal voltage). Solving this equation for V_{GS} gives $V_{GS} = V_t + \sqrt{I_{DS}/K}$.

In response to a difference-mode input voltage V_i, the two drain currents become $I_1 = I_{DS_1} = I_Q/2 + \Delta I$ and $I_2 = I_{DS_2} = I_Q/2 - \Delta I$, so

$$V_{GS_1} = V_t + \sqrt{\frac{I_Q/2 + \Delta I}{K}} \quad \text{and} \quad V_{GS_2} = V_t + \sqrt{\frac{I_Q/2 - \Delta I}{K}} \tag{4.59}$$

The relationship between ΔI and the difference-mode input voltage V_i can thus be written as

$$V_i = V_{G_1} - V_{G_2} = V_{GS_1} - V_{GS_2} = \sqrt{\frac{I_Q/2 + \Delta I}{K}} - \sqrt{\frac{I_Q/2 - \Delta I}{K}}$$

$$= \sqrt{\frac{I_Q}{K}} \left(\sqrt{\frac{1}{2} + \frac{\Delta I}{I_Q}} - \sqrt{\frac{1}{2} - \frac{\Delta I}{I_Q}} \right) \tag{4.60}$$

We note that this relationship has the same mathematical form as that obtained for the JFET differential-amplifier transfer characteristic. The normalized transfer characteristic shown in Figure 4.12 is also applicable to the MOSFET differential amplifier.

We can again obtain the differential-amplifier transfer conductance g_f by using the approximation that $\sqrt{1 + x} \simeq 1 + (x/2)$ for $x \ll 1$, giving

$$V_i \simeq \sqrt{\frac{I_Q}{2K}} \left[\left(1 + \frac{\Delta I}{I_Q} \right) - \left(1 - \frac{\Delta I}{I_Q} \right) \right] = \sqrt{\frac{I_Q}{2K}} \frac{2\Delta I}{I_Q} = \sqrt{\frac{2}{I_Q K}} \Delta I \tag{4.61}$$

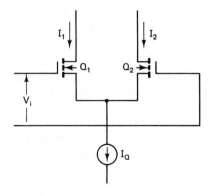

Figure 4.13 MOSFET differential amplifier.

The transfer conductance is therefore given by

$$g_f = \frac{\Delta I}{V_i} = \sqrt{\frac{KI_Q}{2}} = \frac{I_Q/2}{V_{GS} - V_t} \qquad (4.62)$$

If we look at the transfer characteristic graph of the FET differential amplifier, we note that over a large part of the normalized input voltage range an approximately linear transfer relationship is obtained. This approximately linear range extends from about $V_i = -0.5V_N$ to $V_i = +0.5V_N$, where V_N is the normalization factor as given by $V_N = -V_P\sqrt{I_Q/I_{DSS}}$ for JFET differential amplifier, and $V_N = \sqrt{I_Q/K} = (V_{GS} - V_t)\sqrt{2}$ for a MOSFET differential amplifier.

The common-mode transfer conductance for FET differential amplifiers is controlled by the output admittance of the I_Q current source, as is the case for the bipolar transistor differential amplifiers. The common-mode transfer conductance is given by the same equation as for the bipolar case, that is, $g_{f(CM)} = g_o/2$, where g_o is the output conductance of the current source that is used for the biasing of the differential amplifier.

4.2.3 FET Offset Voltage and Temperature Coefficient

Let us consider the case of a pair of IC JFETs, Q_1 and Q_2, both biased at $V_{GS} \cong 0$. The two drain currents will be designated as $I_{DS1} = I_{DSS}$ and $I_{DS2} = I_{DSS} + \Delta I_{DSS}$, respectively, where ΔI_{DSS} represents the small difference between the two zero-bias drain currents. To equalize the two drain currents, a small gate voltage differential or offset voltage, $\Delta V_{GS} = V_{OS}$, must be applied between the two gates. The offset voltage required to cancel the difference between the two drain currents ΔI_{DSS} is given by

$$\Delta I_{DSS} = g_{fso}\Delta V_{GS} = g_{fso}V_{OS} \qquad (4.63)$$

where $g_{fso} = 2I_{DSS}/(-V_P)$. Solving for the offset voltage gives

$$V_{OS} = -\frac{\Delta I_{DSS}}{2I_{DSS}}V_P \qquad (4.64)$$

If the IC processing variation is such that the standard drain currents of a pair of JFETs differ by 1%, and the pinch-off voltage is -4.0 V, the corresponding offset voltage is $V_{OS} = 0.01 \times 4 \, V/2 = 20$ mV. For the bipolar transistor case, the offset voltage is given by $V_{OS} = V_T(\Delta W_B/W_B)$, so a processing variation that results in a 1% difference in the base widths of the two transistors will produce an offset voltage of only 0.25 mV. We can therefore see that the offset voltage for JFETs will generally be considerably larger than for bipolar transistors. An analysis of the offset voltage for MOSFETs leads to results similar to those obtained for JFETs.

The temperature coefficient of the offset voltage of a JFET pair, TCV_{OS}, is principally due to the temperature coefficient of the pinch-off voltage V_P, which is mostly the result of the temperature dependence of the contact potential ϕ. For TCV_{OS}, we have

$$TCV_{OS} = \frac{dV_{OS}}{dT} = -\left(\frac{\Delta I_{DSS}}{2I_{DSS}}\right)\left(\frac{dV_P}{dT}\right) = V_{OS}\frac{1}{V_P}\frac{dV_P}{dT} = \frac{V_{OS}}{V_P}\frac{d\phi}{dT} \qquad (4.65)$$

The temperature coefficient of the contact potential, $d\phi/dT$, is generally about 1 mV/°C, so

$$TCV_{OS} \simeq \left(\frac{V_{OS}}{V_P}\right) 1 \text{ mV/°C} \qquad (4.66)$$

If, for example, $V_{OS} = 20$ mV and $V_P = 4.0$ mV, then $TCV_{OS} \simeq (20 \text{ mV/4V}) \times 1$ mV/°C $= 5$ μV/°C.

For a bipolar transistor pair, the temperature coefficient of the offset voltage has been given as $TCV_{OS} = V_{OS}/T$, if $V_{OS} = 1$ mV the corresponding temperature coefficient of the offset voltage is approximately 3 μV/°C. From this we see that, although the offset voltage for JFETs is often considerably larger than that for bipolar transistors, the temperature coefficient of the offset voltage for the two types of transistors can be quite comparable.

For a MOSFET transistor pair, an analysis similar to the foregoing for the JFET pair can be carried out and similar results will be obtained. An additional complication in the case of MOSFETs, however, is the drift of the threshold voltage due to ionic migration in the gate oxide. This can lead to both short- and long-term drifts in the offset voltage.

4.3 ACTIVE LOADS

The application of a difference-mode input voltage v_i to a differential amplifier will produce ac collector currents of $i_1 = i_2 = g_f v_i$. To transform these ac currents into an output voltage, a *load* must be used. The load can take the form of a *passive load*, which uses a pair of load resistors R_L connected between the collectors of the differential amplifier and the dc supply as shown in Figure 4.2, or an *active load*, which uses transistors for the current-to-voltage transformation.

The simplest type of load is the passive load using load resistors R_L. With this type of load, the ac output voltages at the two collectors are given by $v_{o_1} = -i_1 R_L = -g_f R_L v_i$ and $v_{o_2} = -i_2 R_L = g_f R_L v_i$, and the ac voltage gain for a single-ended output is $A_V = g_f R_L$. Thus, for a large voltage gain a large value of R_L is required. This can be seen more clearly by introducing the expression for g_f as $g_f = I_Q/4V_T$ so that $A_V = g_f R_L = I_Q R_L/4V_T = I_Q R_L/0.1$ V. Since very small values of I_Q are often used in differential amplifiers, such as down in the range of a few microamperes, we see that very large values of R_L will indeed be required (~ 1 MΩ) for substantial voltage gains. However, the use of these large values of load resistance carries with it some major disadvantages, especially for ICs.

1. For ICs the area taken up by large resistors is roughly proportional to the resistance value, so very large resistors take up an excessive amount of room on the IC chip.

2. Large IC resistors have large parasitic capacitances associated with them. The combination of a large resistance and a large parasitic capacitance value results in a very large RC time constant, which in turn severely limits the frequency response of the amplifier.

3. For the differential amplifier to operate properly, the transistors must remain in the active region and not go into saturation. This limits the maximum input voltage that can be applied to the bases of Q_1 and Q_2, for the base-to-collector junction must not be allowed to become forward biased by more than 0.5 V. A large value of load resistance produces a large dc voltage drop $(I_Q/2)R_L$, so the voltage at the collectors is $V_C = V^+ - (I_Q/2)R_L$ and thus is substantially less than the supply voltage V^+. This correspondingly reduces the input voltage range of the differential amplifier.

It is for these reasons that most IC differential amplifiers use *active loads*, which involve the use of transistors rather than resistors. A simple active-load configuration for a differential amplifier is the *current-mirror active load* shown in Figure 4.14. The active load is composed of transistors Q_3 and Q_4, with Q_3 being a diode-connected transistor. This combination of Q_3 and Q_4 is called a *current mirror*. We will assume Q_3 and Q_4 to be essentially identical transistors, and we note that the base-to-emitter voltages of these two transistors are the same. Since we have two identical transistors with identical base-to-emitter voltages, the collector currents of these two transistors

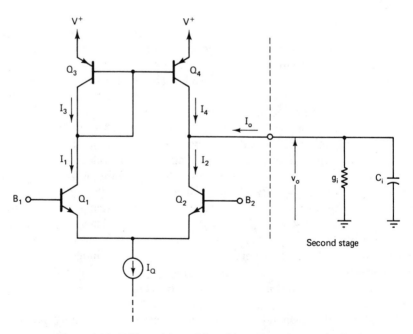

Figure 4.14 Differential amplifier with current-mirror active load.

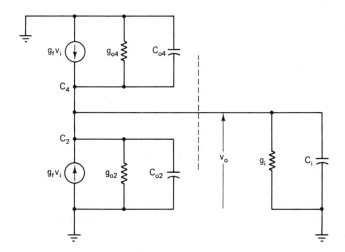

Figure 4.15 AC small-signal equivalent circuit at the C_2–C_4 node.

will be equal, so $I_3 = I_4$. Therefore, whatever current flows through Q_3 is "mirrored" by the same amount of current through Q_4.

We will first analyze this active-load circuit under quiescent conditions (no signal applied), and we will assume that the differential amplifier is balanced such that $I_1 = I_2$. Such $I_1 = I_2$ and $I_3 = I_4$, we have from Kirchhoff's current law that $I_O = I_{B_3} + I_{B_4} = (I_3 + I_4)/\beta_{PNP} = (I_1 + I_2)/\beta_{PNP} \cong I_Q/\beta_{PNP}$.

If an ac input voltage v_i is now applied to the differential amplifier, the various ac currents are given by $i_4 = i_3 = i_1 = g_f v_i$ and $i_2 = -g_f v_i$. Since $i_o = i_2 - i_4$, we have $i_o = -g_f v_i - g_f v_i = -2g_f v_i$. If the net resistance driven by this differential amplifier is R_L, the output voltage produced is $v_o = -i_o R_L = -2g_f R_L v_i$, so the ac voltage gain of this differential-amplifier stage is $A_V = v_o/v_i = 2g_f R_L$. Note that, even though the output of this stage is connected to only one side of the differential amplifier, both sides of the differential amplifier contribute to the output current and voltage by means of the action of the current-mirror active load.

For a more detailed analysis of the voltage gain of the differential amplifier with a current-mirror active load, we will make use of the ac small-signal equivalent-circuit diagram of part of the circuit, as shown in Figure 4.15. In this diagram g_{o2} and g_{o4} represent the dynamic output or collector conductances of Q_2 and Q_4, respectively, and C_{o2} and C_{o4} represent the output capacitances. The conductance g_i and the capacitance C_i represent the input conductance and capacitance of the following stage. The $g_f v_i$ current source associated with the C_4 node represents the ac current produced in Q_4, which is a reflection of the ac current through Q_3 and therefore of the ac current through Q_1.

Writing a node voltage equation (Kirchhoff's current law) for the C_2–C_4 node gives

$$2g_f v_i = v_o[(g_{o2} + g_{o4} + g_i) + j\omega(C_{o2} + C_{o4} + C_i)]$$

So

$$A_{V_1} = \frac{v_o}{v_i} = \frac{2g_f}{(g_{o2} + g_{o4} + g_i) + j\omega(C_{o2} + C_{o4} + C_i)} = \frac{2g_f}{g_{total} + j\omega C_{total}} \qquad (4.67)$$

where $g_{total} = g_{o2} + g_{o4} + g_i$ and $C_{total} = C_{o2} + C_{o4} + C_i$. The low-frequency gain is given by

$$A_{V_1}(0) = \frac{2g_f}{g_{total}} \tag{4.68}$$

The general expression for the gain of the differential-amplifier stage can be expressed as

$$A_{V_1}(f) = \frac{2g_f}{g_{total}[1 + j\omega(C_{total}/g_{total})]} = \frac{A_{V_1}(0)}{1 + j\omega\tau}$$

$$= \frac{A_{V_1}(0)}{1 + j(\omega/\omega_1)} = \frac{A_{V_1}(0)}{1 + j(f/f_1)} \tag{4.69}$$

where $\tau = C_{total}/g_{total}$ and $f_1 = 1/(2\pi\tau)$. We see that the gain will roll off with increasing frequency, with a breakpoint at f_1.

Let us now consider the circuit of Figure 4.16, in which part of the second gain stage comprised of transistors Q_6 and Q_7 is shown. We will again assume that all transistors of the *same type* are identical. Under quiescent conditions we see that, since $I_1 = I_2$ and $I_3 = I_4$, then $I_{B5} = I_{B6}$ and therefore $I_{E5} = I_{E6}$ since the transistor current gains are equal. Since $I_{E5} = I_{B3} + I_{B4}$ and $I_{E6} = I_{B7}$, we have that $I_{B7} = I_{B3} + I_{B4}$ and thus $I_{C7} = I_3 + I_4$, again because the transistor current gains are equal. Since $I_3 + I_4 \cong I_1 + I_2$, we therefore have that $I_{C7} \cong I_1 + I_2 \cong I_Q$. Thus the current source I_Q not only sets the quiescent current of the differential amplifier, but also sets the quiescent current of the Q_6–Q_7 second stage.

The ac low-frequency voltage gain of the first (differential amplifier) stage is

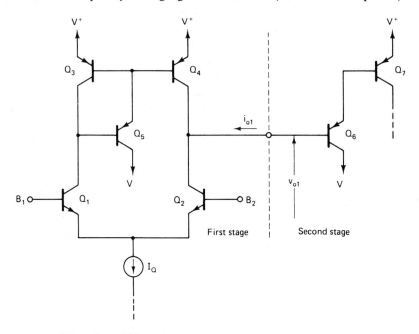

Figure 4.16 Differential amplifier with current-mirror active load.

Differential Amplifiers   Chap. 4

given by $A_{V_1}(0) = 2g_f/g_{\text{total}}$, where $g_{\text{total}} = g_{o2} + g_{o4} + g_{i6}$. For the transistor output conductances, we have the following equations:

$$g_{o2} = \frac{I_2}{V_{A(\text{NPN})}} = \frac{I_Q}{2V_{A(\text{NPN})}} \quad \text{and} \quad g_{o4} = \frac{I_4}{V_{A(\text{PNP})}} = \frac{I_Q}{2V_{A(\text{PNP})}} \quad (4.70)$$

where V_A is the transistor *Early voltage* or *base-width modulation coefficient* (see Appendix B).

The input conductance of the second stage g_{i6} as seen looking into the base of Q_6 is $g_{i6} = I_{B6}/2nV_T$, where n is a dimensionless factor typically of about 1.5. Since $I_{B6} = I_{C6}/\beta_6 = I_C/\beta_6\beta_7 = I_Q/\beta_6\beta_7$, this equation can be rewritten as $g_{i6} = I_Q/(\beta_6\beta_7 2nV_T)$. The factor of 2 multiplying V_T comes from the fact that the second stage is a compound transistor (Darlington) configuration with the input voltage being applied across two base-to-emitter junctions in series.

We will now consider a specific example, and we will choose the following representative parameter values: $V_{A(\text{NPN})} = V_{A(\text{PNP})} = 200$ V (min.) and $\beta = 50$ (min.) for all transistors, and we will choose $I_Q = 20$ μA as a reasonable value. With these values, we have

$$g_f = \frac{I_Q}{4V_T} = \frac{20\,\mu\text{A}}{100\,\text{mV}} = 200\,\mu\text{S}$$

$$g_{o2} = \frac{I_Q}{2V_{A(\text{NPN})}} = \frac{20\,\mu\text{A}}{400\,\text{V}} = 0.05\,\mu\text{A/V} = 0.05\,\mu\text{S}$$

$$= 50\,\text{nS (max.)}$$

$$g_{o4} = \frac{I_Q}{2V_{A(\text{PNP})}} = \frac{20\,\mu\text{A}}{400\,\text{V}} = 0.05\,\mu\text{A/V} = 0.05\,\mu\text{S} \quad (4.71)$$

$$= 50\,\text{nS (max.)}$$

$$g_{i6} = \frac{I_Q}{\beta_6\beta_7 2nV_T} = \frac{20\,\mu\text{A}}{(50 \times 50 \times 2 \times 1.5 \times 25\,\text{mV})}$$

$$= 107\,\text{nS (max.)}$$

Therefore, the total conductance g_{total} is $g_{\text{total}} = g_{o2} + g_{o4} + g_{i6} = 50$ nS + 50 nS + 107 nS = 207 nS (max.). The low-frequency voltage gain of the first (differential amplifier) stage is therefore

$$A_{V_1}(0) = \frac{2g_f}{g_{\text{total}}} = 2 \times 200\,\mu\text{S}/207\,\text{nS (max.)} = \underline{1932\,\text{(min.)}} \quad (4.72)$$

so we see that with an active load a very large voltage gain can indeed be obtained from just a single amplifier stage.

From this analysis we see that the dynamic conductance seen looking into the active load is only 50 nS (max.), which corresponds to a dynamic resistance of $r_{o4} = 20$ MΩ (min.), which permits the first stage to have a very large voltage gain. In spite of this very large dynamic resistance value, the dc voltage drop across the active load is only $2V_{BE} \simeq 1.2$ V. If a load resistance R_L of this large value of 20 MΩ were actually used in this circuit, the dc voltage drop that would result would be 10 μA × 20 MΩ = 200 V! Furthermore, this large resistor would take up an enormous amount of area on the IC chip, and the associated parasitic capacitance would be extremely large. The combination of the large resistance and the very large parasitic capacitance would lead to a very poor frequency response.

We note that the voltage drop across the active-load transistors Q_3 and Q_4 is $2V_{BE} \simeq 1.2$ V, so the voltage at the collectors of Q_1 and Q_2 is $V^+ - 2V_{BE} \simeq V^+ - 1.2$ V. The voltage drop across the base–emitter junction of a transistor varies logarithmically with current level, with a 10 : 1 current change resulting in an increase of only 60 mV in the V_{BE} drop. As a result, the voltage drop across the active load under actual operating conditions remains fairly constant at about 1.2 V.

The base voltage of Q_1 and Q_2 can go up to 0.5 V above the collector voltage before these transistors go into saturation. Therefore, the input voltage range of the differential amplifier extends up to $V^+ - 1.2 + 0.5 = V^+ - 0.7$ V and thus up to about 0.7 V away from the positive supply voltage.

The active load uses only two or three transistors and so takes up very little area on the IC chip. The parasitic capacitance of the active load, which is the output or collector capacitance of Q_4, is only about 3 to 10 pF and thus is relatively small. The active load can permit the amplifier stage to have a gain well in excess of 1000, but at the same time results in a dc voltage drop of no more than about 1.2 V. The active load therefore suffers none of the disadvantages of the passive load. In addition, it is of interest to note that all the conductances associated with the first stage are proportional to I_Q, so the voltage gain of this stage is independent of the quiescent current. As a result, very small values of I_Q can be used ($\lesssim 20$ μA) while still maintaining a large voltage gain. The small value of I_Q is desirable, for it will result in a small value for I_{BIAS} and a large value for the input resistance. For the example just considered, the input bias (or base) current is $I_B = 10$ μA/50 (min.) = 200 nA (max.), and the difference-mode input resistance is $r_i = 2V_T/I_B = 50$ mV/0.2 μA = 250 MΩ (min.). A very small value of I_Q is, however, undesirable because it results in a poorer frequency-domain and time-domain response for the amplifier. In many cases where a small value for I_{BIAS} is needed, the best approach is to use a JFET or MOSFET differential amplifier that is operated at a relatively high value of I_Q.

Finally, it should be noted that with this active load the voltage level at the collectors of the two transistors of the differential amplifier remains approximately constant and equal. This minimizes the contribution of the offset voltage that results from large differences in the collector voltages of a pair of transistors.

4.3.1 MOSFET Differential Amplifier with Current-Mirror Active Load

Let us now consider the MOSFET differential amplifier with a MOSFET current-mirror active load as shown in Figure 4.17. The low-frequency voltage gain of this stage is given by

$$A_{V_1}(0) = \frac{v_{o1}}{v_i} = \frac{2g_f}{g_{\text{total}}} = \frac{2g_f}{(g_{o2} + g_{o4})} \tag{4.73}$$

where $v_i = v_{g1} - v_{g2}$ is the difference-mode input voltage, g_f is the differential amplifier transfer conductance as given by $g_f = \sqrt{KI_Q/2} = (I_Q/2)(V_{GS} - V_t)$, and g_{o2} and g_{o4} are the dynamic output conductances of Q_2 and Q_4, respectively. These output conductances are given by

$$g_{o2} = g_{ds2} = \frac{I_{DS2}}{V_{A(\text{NMOS})}} = \frac{I_Q/2}{V_{A(\text{NMOS})}} \tag{4.74}$$

and

$$g_{o4} = g_{ds4} = \frac{I_{DS4}}{V_{A(\text{PMOS})}} = \frac{I_Q/2}{V_{A(\text{PMOS})}}$$

The transistor parameter $1/V_A$ is the channel length modulation coefficient as described in Appendix B. The factor of 2 in the equation for $A_{V_1}(0)$ comes from the current-doubling effect of the current mirror, just as in the case of the bipolar counterpart of this circuit.

After substitution of the above relationships for g_f, g_{o2}, and g_{o4}, the equation for A_{V_1} becomes

$$A_{V_1} = \frac{I_Q/(V_{GS} - V_t)}{(I_Q/2)[(1/V_{A(\text{NMOS})}) + (1/V_{A(\text{PMOS})})]}$$

$$= \frac{2[1/(V_{GS} - V_t)]}{(1/V_{A(\text{NMOS})}) + (1/V_{A(\text{PMOS})})} \tag{4.75}$$

As an example, if $V_{GS} - V_t = 1$ V and $V_{A(\text{NMOS})} = V_{A(\text{PMOS})} = 60$ V, we obtain a low-frequency voltage gain of $A_{V_1}(0) = 60$. This is to be compared to voltage gains into the range of 1000 that can be obtained for the bipolar transistor differential amplifiers.

We note that the voltage gain is directly proportional to $1/(V_{GS} - V_t)$. Since

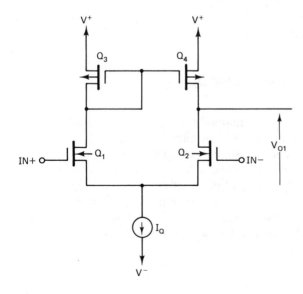

Figure 4.17 MOSFET differential amplifier with current-mirror active load.

$I_{DS} = K(V_{GS} - V_t)^2$, the voltage gain is proportional to $1/\sqrt{I_{DS}} = \sqrt{2/I_Q}$; so by operating the amplifier at a lower quiescent current level, the low-frequency voltage gain can be increased. This increase in the gain, however, is at the expense of a reduced bandwidth. If we consider the effect of a capacitive load on the performance of this circuit, the equation for A_{V_1} becomes $A_{V_1} = 2g_f/(g_{total} + j\omega C_L)$, where C_L is the net capacitance at the D_2–D_4 node and includes not only the load capacitance, but also the output capacitances of Q_2 and Q_4. The equation for A_{V_1} can now be rewritten as

$$A_{V_1} = \frac{2g_f}{g_{total}[1 + j(\omega C_L/g_{total})]} = \frac{A_{V_1}(0)}{1 + j(\omega/\omega_1)} = \frac{A_{V_1}(0)}{1 + j(f/f_1)} \tag{4.76}$$

where
$$\omega_1 = g_{total}/C_L = \left[\frac{I_Q}{2C_L}\right]\left[\frac{1}{V_{A(NMOS)}} + \frac{1}{V_{A(PMOS)}}\right]$$

and
$$f_1 = \omega_1/2\pi = 3\text{-dB bandwidth} \tag{4.77}$$

From this we see that the bandwidth is directly proportional to the quiescent current I_Q.

As an example, let us again take $V_{A(NMOS)} = V_{A(PMOS)}$, and we will use a value of $I_Q = 60$ μA and $C_L = 5$ pF. The 3-dB bandwidth of this differential-amplifier stage is $f_1 = 32$ kHz.

For frequencies well above f_1, the voltage gain becomes approximately $A_{V_1} = 2g_f/(j\omega C_L)$. The unity gain frequency f_u is therefore given by

$$f_u = 2g_f/2\pi C_L = \frac{I_Q/(V_{GS} - V_t)}{2\pi C_L}$$

If we again use $I_Q = 60$ μA, $C_L = 5$ pF, and $V_{GS} - V_t = 1$ V, the unity gain frequency is $f_u = 1.91$ MHz. Note that this same result can be obtained from $f_u = A_{V_1}(0)f_1 = 60 \times 32$ kHz $= 1.91$ MHz.

4.4 DIFFERENTIAL AMPLIFIERS USING COMPOUND TRANSISTORS

For many applications the use of a compound-transistor configuration is found to be advantageous for differential amplifiers. In particular, the use of the Darlington configuration differential amplifier, as shown in Figure 4.18, offers the possibility of a much higher input impedance and a much lower input bias current than would otherwise be the case. One drawback of the Darlington differential amplifier is the somewhat higher offset voltage, V_{OS}, since four transistors are now involved in the differential amplifier. On a statistical basis, the offset voltage is expected to be, on the average, about $\sqrt{2}$ times larger than for the equivalent, two-transistor, differential amplifier.

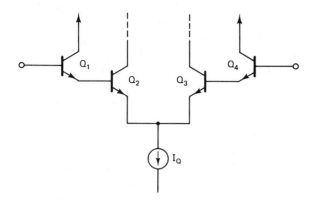

Figure 4.18 Darlington differential amplifier.

For the Darlington circuit of Figure 4.19, we have that

$$V_{BE} = V_{B_1E_1} + V_{B_2E_2} = V_T \ln\left(\frac{I_1}{I_{CO_1}}\right) + V_T \ln\left(\frac{I_2}{I_{CO_2}}\right)$$

$$= V_T \ln\left(\frac{I_1 I_2}{I_{CO_1} I_{CO_2}}\right) \tag{4.78}$$

Since $I_2 \cong \beta_2 I_1$, this can be rewritten as

$$V_{BE} = V_T \ln\left(\frac{I_2^2}{\beta_2 I_{CO_1} I_{CO_2}}\right)$$

Solving this for $I_C \cong I_2$ gives

$$I_C \cong I_2 \cong \sqrt{\beta_2 I_{CO_1} I_{CO_2}}\; e^{V_{BE}/2V_T}$$

$$= I'_{CO}\, e^{V_{BE}/2V_T}$$

So

$$I_C \propto e^{V_{BE}/2V_T} \tag{4.79}$$

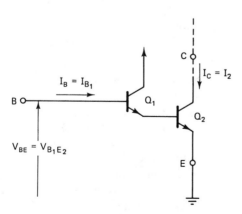

Figure 4.19 Darlington compound transistor configuration.

Note that the exponential relationship just given will be valid only if both transistors that comprise the Darlington circuit are operated in the active mode (EB junction "on" and CB junction "off").

If we now compare this exponential characteristic given for the Darlington circuit with that obtained for a single transistor, we see that the relationship for the Darlington case differs only by having the factor of $2V_T$ in the exponent rather than V_T as in the single-transistor case. As a result, if we examine the equations developed previously for the differential amplifier, we see that all of them can be applied to the case of the Darlington differential amplifier by simply replacing V_T by $2V_T$ in all the equations.

The most important difference in the two types of differential amplifiers has to do not with the change from V_T to $2V_T$, but rather with the extraordinarily large current gain of the Darlington configuration. For the Darlington circuit shown in Figure 4.18, the overall current gain, $\beta = I_2/I_B$, is equal to the product of the current gains of the two transistors that make up the Darlington circuit, as given by

$$\text{current gain} = \beta = \frac{I_2}{I_{B_1}} = \frac{I_2}{I_{B_2}}\frac{I_{B_2}}{I_{E_1}}\frac{I_{E_1}}{I_{B_1}} = \beta_2 \times 1 \times (\beta_1 + 1) \simeq \beta_2\beta_1 \qquad (4.80)$$

Since the current-gain values of the individual transistors are generally on the order of 100, the overall current gain, β, of the Darlington circuit is on the order of 10,000.

As a result of the very high current gain values available with the Darlington circuit, the input bias current, $I_{\text{bias}} = I_B$, is much lower than that of the corresponding two-transistor differential amplifier. Since the input resistance of the differential amplifier is inversely proportional to the bias (base) current, it is correspondingly much higher than for the two-transistor differential-amplifier case.

4.5 DIFFERENTIAL AMPLIFIER WITH INPUT VOLTAGE RANGE THAT INCLUDES GROUND POTENTIAL

One interesting possibility that is made available with the use of the Darlington differential-amplifier configuration is shown in Figure 4.20. This differential amplifier can be operated with the input base terminals (B_1 and B_4) at, or even slightly below, ground potential (down to about -0.5 V). As a result, this differential amplifier can be operated with just a single supply voltage.

In Figure 4.20 some of the circuit voltages have been indicated. For transistors Q_1 and Q_4 to be in the active mode, we see that the necessary condition is that V_{B_1} and V_{B_4} both be (algebraically) above -0.5 V so that the collector–base junctions of these two transistors remain turned "off." If $V_{B_1} = V_{B_4} \geq -0.5$ V, we have that $V_{B_2} = V_{B_3} \geq -0.5 + 0.6 = +0.1$ V. Since $V_{C_2} = V_{C_3} = +0.6$ V, the collector-to-base voltages of Q_2 and Q_3 are $V_{BC_2} = V_{BC_3} \geq -0.5$ V, so the collector–base junctions of Q_2 and Q_3 will also be "off." Thus, as long as the two input terminals B_1 and B_4 are no more than 0.5 V below ground potential, all the transistors in the differential-amplifier circuit will be in the active mode.

We have seen that this differential amplifier can be operated with the input voltage down to as much as 0.5 V below ground. Because of this, for ac signal amplitudes not exceeding 0.5 V, operation of this circuit with a single supply voltage

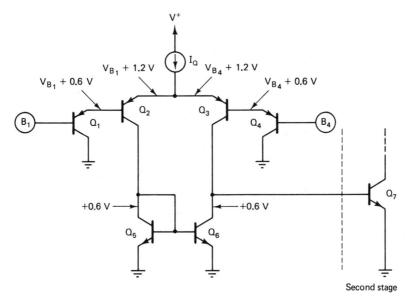

Figure 4.20 Differential amplifier with input voltage range that includes ground potential.

is possible. Many single-supply op amps and voltage comparators use differential amplifiers of this type.

4.6 AXIS OF SYMMETRY

The differential amplifier can be analyzed in many cases by splitting the circuit down the axis of symmetry into two equivalent amplifier circuits. Let us consider the circuit of Figure 4.21a. If the quiescent currents I_1 and I_2 are equal, the application of two equal-in-magnitude but opposite-in-algebraic-sign, small-signal ($\lesssim 25$ mV) ac voltages to the two bases ($+v_i/2$ to B_1 and $-v_i/2$ to B_2) will result in ac currents that are equal in magnitude but opposite in algebraic sign for Q_1 and Q_2. The net ac current flow through the current source output resistance, r_o, will be *zero*; as a result the ac voltage drop across the current source will be zero. The emitters of Q_1 and Q_2 are therefore at *ac ground potential*.

The circuit of Figure 4.21a can now be redrawn in the form of Figure 4.21b. Notice that one-half of the total base-to-base input voltage of v_i appears across each transistor. If v_i is the total (base-to-base) input voltage, the two ac collector currents are

$$i_1 = g_m \frac{v_i}{2} = \frac{(I_Q/2)\,(v_i/2)}{V_T} = \frac{I_Q}{4V_T}\,v_i = g_f v_i$$

and

$$i_2 = g_m \frac{-v_i}{2} = \frac{(I_Q/2)\,(-v_i/2)}{V_T} = -\frac{I_Q}{4V_T}\,v_i = -g_f v_i$$

(4.81)

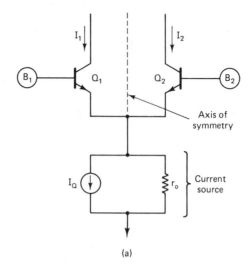

(a)

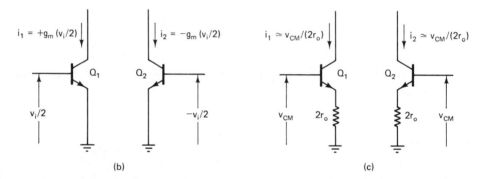

(b)   (c)

Figure 4.21 Analysis of differential amplifier using the axis of symmetry: (a) differential amplifier; (b) ac small-signal equivalent circuit for pure difference-mode input signal; (c) small-signal equivalent circuit for common-mode input signal.

where g_f is the dynamic transfer conductance of the differential amplifier as obtained by the general analysis presented earlier. Note that these results are entirely consistent with that obtained in that general analysis.

The case just considered is for a small-signal ac difference-mode input signal. Let us now consider what happens for the case of a common-mode input voltage, v_{CM}, applied to the two bases. For this case, the ac currents on both sides of the differential amplifier are exactly the same in terms of both magnitude and algebraic sign, so the ac current through current source output resistance is twice that of either transistor. The ac voltage across the current source is therefore $2i_c r_O$. The differential amplifier can now be split down the axis of symmetry into two equal halves, as shown in Figure 4.21c. Note that the ac voltage across r_O in the equivalent circuit of Figure 4.21c is the same as of the original differential amplifier.

The ac collector current, i_c, of either half-circuit is related to the common-mode input voltage by

$$i_c = \frac{g_m v_{CM}}{1 + g_m 2 r_O} \simeq \frac{v_{CM}}{2 r_O} = \frac{v_{CM} g_O}{2}, \qquad \text{since } g_m(2 r_O) \gg 1 \qquad (4.82)$$

If we now inspect the corresponding results obtained earlier in the general differential-amplifier analysis, we see that the same equation is obtained.

The general case of a combination of small-signal input voltages, v_1 and v_2, that are neither purely difference-mode nor common-mode signals can be handled by resolving the input voltage into difference-mode and common-mode components using the following relationships:

$$v_{DM} = v_1 - v_2 \quad \text{and} \quad v_{CM} = \frac{v_1 + v_2}{2} \qquad (4.83)$$

Note that the technique just described of splitting the differential amplifier down the axis of symmetry can be used *only* when the quiescent currents (I_1 and I_2) on the two sides of the amplifier are equal and the difference-mode component of the input voltage is a small signal (amplitude limited to about 25 mV). If these two conditions are not satisfied, the more general analysis must be used.

The foregoing discussion has been in terms of a bipolar transistor type of differential amplifier. The same axis of symmetry technique can, however, be applied equally well to both the JFET and MOSFET differential amplifiers. To use this technique for the FET differential amplifiers, the difference-mode input voltage amplitude should not exceed about $0.5\, V_N$, where $V_N = V_P \sqrt{I_Q / I_{DSS}}$ for a JFET differential amplifier, and $V_N = \sqrt{I_Q / K}$ for a MOSFET differential amplifier.

From the foregoing discussion, it is now easy to see the relationship between the dynamic forward transfer conductance g_f of a differential amplifier and the dynamic transfer conductance g_m or g_{fs} of the transistors that comprise the differential amplifier. For the case of the balanced differential amplifier that can be split into two equal halves down the axis of symmetry, the situation is the same as two single-ended transistors, each with an input voltage of one-half of the difference-mode input signal, $v_i/2$. The resulting collector current is given by $i_c = g_m v_i/2 = (g_m/2) v_i$ for the bipolar transistor case. Similarly, for an FET differential amplifier, the drain current is given by $i_{ds} = g_{fs} v_i/2 = (g_{fs}/2) v_i$. Since the transfer conductance of the differential amplifier is the ratio of the output current to the difference-mode input voltage, we see that the differential-amplifier transfer conductance is equal to one-half of the transfer conductance of the individual transistors that comprise the differential amplifier.

PROBLEMS

4.1. (*Differential amplifier*) Given: NPN differential amplifier with current sink biasing and passive load (Figure P4.1); $V_{\text{SUPPLY}} = \pm 15$ V; $I_Q = 200\ \mu$A; $I_{Q3} = 1.0$ mA; $h_{FE} = 200$ (min.); $V_{BE} = 650$ mV at 1.0 mA; $BV_{EBO} = 7.0$ V (min.); $V_{C1(Q)} = V_{C2(Q)} = +9.0$ V; assume all transistors identical, unless indicated otherwise; output conductance of current sink (Q_4) = 307 nS; $V_A = 250$ V.

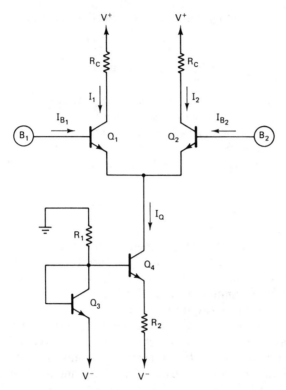

Figure P4.1

(a) Find R_1, R_2, and R_C.   (*Ans.:* 14.35 kΩ, 201 Ω, 60 kΩ)

(b) Find the difference-mode dynamic transfer conductance, g_f.   (*Ans.:* g_f = 2.0 mS)

(c) Find the difference-mode voltage gains: v_{c_1}/v_{b_1}, v_{c_2}/v_{b_1}, and $(v_{c_1} - v_{c_2})/v_{b_1}$, where v_{b_1} is the small-signal input voltage.   (*Ans.:* −120 V, +120 V, −240 V)

(d) Find the input bias current, I_{BIAS}.   [*Ans.:* I_{BIAS} = 500 nA (max.)]

(e) Find the difference-mode dynamic input resistance, R_i.   [*Ans.:* 100 kΩ (min.)]

(f) Find the common-mode input resistance, $R_{i(CM)}$.   [*Ans.:* 1.3 GΩ (min.)]

(g) Find the common-mode dynamic transfer conductance, $g_{f(CM)}$.   (*Ans.:* 153 nS)

(h) Find the common-mode voltage gain and CMRR.   (*Ans.:* $A_{V(CM)}$ = −0.0092; CMRR = 82.3 dB)

(i) Find the input voltage range.   (*Ans.:* (1) Either V_{B_1} or V_{B_2}, or both, must be greater than −14.2 V; (2) if $V_{B_1} = V_{B_2}$, then V_B must be less than +9.5 V, and if $V_{B_1} \neq V_{B_2}$, then both must be less than + 3.5 V)

(j) If $V_{OS} = 0$, $V_{B_1} = +10$ mV, and $V_{B_2} = +30$ mV, find I_1 and I_2.   (*Ans.:* 62 μA, 138 μA)

(k) Find g_f and the difference-mode voltage gain (single-ended output) for the conditions of part (j).   (*Ans.:* 1.71 mS, 103)

(l) If $V_{OS} = 2.0$ mV, find $V_{C1(Q)}$ and $V_{C2(Q)}$ ($V_i = V_{B_1} - V_{B_2} = 0$).   (*Ans.:* +9.24 V, +8.76 V)

(m) If $V_{OS} = 2.0$ mV, find $TC_{V(OS)}$.   (*Ans.:* 6.7 μV/°C)

(n) If this differential amplifier is to drive another differential amplifier that has a difference-mode input resistance of 100 kΩ, find the voltage gain (balanced output), $A_V = (v_{c_1} - v_{c_2})/v_{b_1}$.   (*Ans.:* −109 V)

(o) Find V_{C_1} and V_{C_2} for the conditions of $I_{B_1} = 0$ and $V_{B_2} = 0$.   (*Ans.: +15 V, +3.0 V*)

(p) If $V_{B_1} = 0$ and $V_{B_2} = -50$ mV, find I_1 and I_2.   (*Ans.: 176 µA, 24 µA*)

(q) Find the output conductance of the current sink (Q_4).   (*Ans.: 307 nS*)

4.2. (*JFET differential amplifier*) Given a JFET differential amplifier as shown in Figure P4.2 with $I_Q = 0.5$ mA. The differential amplifier transistors Q_1 and Q_2 and the current source active load transistors Q_3 and Q_4 have the following parameters: $I_{DSS} = 1.0$ mA, $V_P = -4.0$ V, and $r_{ds} = 200$ kΩ.

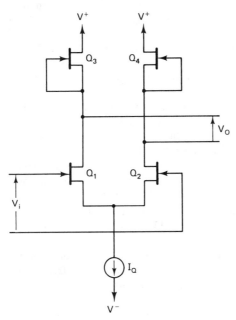

Figure P4.2

(a) Find the dynamic transfer conductance of the differential amplifier, g_f.   (*Ans.: g_f = 0.125 mS*)

(b) Find the ac small-signal voltage gain for a balanced output, v_o/v_i.   (*Ans.: A_V = 25*)

(c) What difference-mode input voltage is required to shift 90% of the total differential-amplifier current I_Q to one side of the differential amplifier?   (*Ans.: ±1.79 V*)

(d) Repeat part (c) for 99% of the total current.   (*Ans.: ±2.57 V*)

4.3. (*MOSFET differential amplifier*) Given: A differential amplifier uses MOSFETs that have a threshold voltage of +3.0 V and $I_{DS} = 1.0$ mA at $V_{GS} = +7.0$ V. The total differential-amplifier current is $I_Q = 1.0$ mA.

(a) Find the dynamic transfer conductance of the differential amplifier, g_f.   (*Ans.: 0.177 mS*)

(b) Find the input voltage required to shift 90% of the current to one side of the differential amplifier.   (*Ans.: ±2.52 V*)

(c) Find the input voltage required to shift 99% of the current to one side.   (*Ans.: ±3.58 V*)

4.4. (*Differential amplifier with current-mirror active load*) Given: Differential amplifier with current-mirror active load circuit of Figure 4.16.

(a) Show that a general expression for the ac small-signal voltage gain, $A_v = v_{o_1}/v_i$, is given by

$$A_v = \frac{\beta_6\beta_7}{(1/n) + \beta_6\beta_7 \, V_T[(1/V_{AN}) + (1/V_{AP})]}$$

(b) If $\beta_6 = \beta_7 = 70$ (min.), $n = 1.5$, and $V_{AN} = V_{AP} = 150$ V (min.), find A_V. [*Ans.:* 2130 (min.)]

(c) Repeat part (b) for $\beta_6 = \beta_7 = 50$ (min.). [*Ans.:* 1667 (min.)]

(d) Repeat part (b) for $\beta_6 = \beta_7 = 30$ (min.). [*Ans.:* 931 (min.)]

(e) If $I_Q = 40$ µA and $V^+ = 12$ V, find I_{BIAS} for $\beta_1 = \beta_2 = 50$ (min.). [*Ans.:* 400 nA (max.)]

(f) Find the difference-mode input resistance for the above conditions. [*Ans.:* 125 kΩ (min.)]

(g) Find the upper limit of the input voltage range for this differential amplifier. (*Ans.:* +11.1 V)

(h) For parts (b) to (d), find the maximum voltage gain that is obtained as the transistor current gain β goes to infinity. (*Ans.:* 3000)

(i) If transistors Q_1 and Q_2 are each replaced by a Darlington transistor pair, find the voltage gain if $\beta = 50$ (min.) for all transistors, and find the voltage gain that is approached as β goes to infinity. [*Ans.:* 1071 (min.), 1500 (max.)]

(j) Find I_{BIAS} and the difference-mode input resistance for the case of part (i) with $I_Q = 40$ µA. [*Ans.:* 8 nA (max.), 12.5 MΩ (min.)]

4.5. (*MOSFET differential amplifier with current-mirror active load*) Given: MOSFET differential amplifier circuit of Figure 4.17. For the NMOS transistors Q_1 and Q_2, the threshold voltage V_t is +1.0 V, and at $I_{DS} = 1.0$ mA the gate-to-source voltage V_{GS} is +3.0 V. The net capacitance at the Q_2–Q_4 node, including the load capacitance, is 5 pF. Assume $V_{A(NMOS)} = V_{A(PMOS)} = 50$ V. Find the low-frequency voltage gain $A_{V1}(0)$, the 3-dB bandwidth, and the unity gain frequency f_u for the following values of quiescent current I_Q:

(a) $I_Q = 2.0$ mA   (*Ans.:* 25, 1.27 MHz, 31.8 MHz)
(b) $I_Q = 200$ µA   (*Ans.:* 79, 127 kHz, 10.0 MHz)
(c) $I_Q = 60$ µA   (*Ans.:* 144, 38 kHz, 5.5 MHz)
(d) $I_Q = 20$ µA   (*Ans.:* 250, 12.7 kHz, 3.18 MHz)

REFERENCES

BOYLSTAD, R., and L. NASHELSKY, *Electronic Devices and Circuit Theory*, Prentice-Hall, Englewood Cliffs, N.J., 1982.

FITCHEN, F. C., *Electronic Integrated Circuits and Systems*, Van Nostrand Reinhold, New York, 1970.

GIACOLETTO, L. J., *Differential Amplifiers*, Wiley, New York, 1970.

GLASER, A. B., and G. E. SUBAK-SHARPE, *Integrated Circuit Engineering*, Addison-Wesley, Reading, Mass., 1977.

GRAEME, J. G., G. E. TOBEY, and L. P. HUELSMAN, *Operational Amplifiers—Design and Applications*, McGraw-Hill, New York, 1971.

GRAY, P. R., and R. G. MEYER, *Analysis and Design of Analog Integrated Circuits*, 2nd ed., Wiley, New York, 1984.

GRINICH, V. H., and H. G. JACKSON, *Introduction to Integrated Circuits*, McGraw-Hill, New York, 1975.

HAMILTON, D. J., and W. G. HOWARD, *Basic Integrated Circuit Engineering*, McGraw-Hill, New York, 1975.

LENK, J. D., *Manual for M.O.S. Users*, Prentice-Hall, Englewood Cliffs, N.J., 1975.

MIDDLEBROOK, R. D., *Differential Amplifiers*, Wiley, New York, 1963.

MILLMAN, J., *Microelectronics*, McGraw-Hill, New York, 1979.

MOTOROLA, INC., *Analysis and Design of Integrated Circuits*, McGraw-Hill, New York, 1967.

ROBERGE, J. K., *Operational Amplifiers*, Wiley, New York, 1975.

WAIT, J. V., *Introduction to Operational Amplifiers: Theory and Applications*, McGraw-Hill, New York, 1975.

5

Operational-amplifier Characteristics and Applications

5.1 INTEGRATED CIRCUITS

An integrated circuit (IC) is an electronic device in which there is more than one circuit component in the same package. Most ICs contain many transistors together with diodes, resistors, and capacitors. Integrated circuits may contain tens, hundreds, or even many thousands of transistors. A *monolithic IC* is one in which all the components are contained on a single-crystal chip of silicon. This chip typically measures from 1 mm × 1 mm × 0.25 mm thick for the smaller ICs to 5 mm × 5 mm × 0.3 mm thick for the larger size ICs.

In the case of a *hybrid IC* there is more than one chip in the package. The chips may be monolithic ICs, discrete transistors, or diode chips, or some combination thereof. There can also be chip capacitors. The chips are usually mounted on an insulating ceramic substrate, usually alumina (Al_2O_3), and are interconnected by a thin-film or thick-film conductor pattern than has been deposited on the ceramic substrate. There can also be thin-film or thick-film resistor patterns deposited on the ceramic substrate. Thin-film patterns are deposited by vacuum evaporation techniques and are usually about 1 μm in thickness, whereas thick-film patterns are applied by a silk-screening method and are usually 10 to 30 μm in thickness.

Integrated circuits can also be classified as to function, the two principal categories being *digital* and *analog* (or *linear*) ICs. A digital IC is one in which all the transistors operate in a switching mode being either off (cutoff mode) or on (saturation mode) to represent the high and low (or 1 and 0) digital logic levels. The transistors during the switching transient pass very rapidly through the active region. Virtually all digital ICs are of the monolithic type and contain just transistors and resistors, with both bipolar and MOSFET transistors being used. Some MOSFET digital ICs contain as many as several hundred thousand transistors, all on one silicon chip.

Analog or *linear* ICs operate on signal voltages and currents in analog form, and the transistors operate mostly in the active (or linear) mode of operation. There are many different types of analog ICs, such as operational amplifiers, audio power amplifiers, voltage regulators, voltage references, video amplifiers, radio-frequency amplifiers, voltage comparators, modulators and demodulators, logarithmic converters, multipliers, function generators, voltage-controlled oscillators, phase-locked loops, digital-to-analog and anaolog-to-digital converters, and other devices. The majority of analog ICs are of the monolithic construction, although there are many hybrid ICs of importance. In this chapter, operational amplifiers are studied, and in later chapters other types of analog ICs are investigated.

5.2 INTRODUCTION TO OPERATIONAL AMPLIFIERS

An operational amplifier (op amp) is an integrated circuit that produces an output voltage v_o that is an amplified replica of the difference between two input voltages, v_1 and v_2. An op amp can be ideally characterized by the input–output transfer function given by $v_o = A_{OL}(v_1 - v_2)$, where A_{OL} is the *open-loop gain* of the op amp. Most op amps are of the monolithic type, although there are literally hundreds of different types of op amps, and op amps are available from dozens of different manufacturers.

The term operational amplifier comes from one of the earliest uses of these devices, dating back to the early and middle 1960s in analog computers. Op amps were used in conjunction with other circuit components, principally resistors and capacitors, to perform various mathematical operations, such as addition, subtraction, multiplication, integration, and differentiation—hence the name "operational amplifier." The applications of op amps over the years since then has vastly expanded from that, however, and op amps are used to perform a multitude of tasks, as will be seen later.

A basic circuit representation of an op amp is shown in Figure 5.1a. The triangular-shaped "box" represents the op amp, which itself is a multistage amplifier containing generally 10 to 100 transistors. The output voltage v_o is related to the

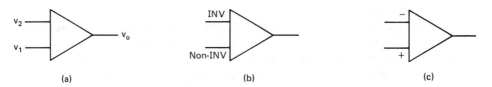

Figure 5.1 Operational amplifier symbols: (a) basic operational amplifier symbol; (b) symbol with input polarities indicated explicitly; (c) symbol with input polarities indicated explicitly.

two input voltages by the basic relationship $v_o = A_{OL}(v_1 - v_2)$. Note that the op amp is in theory sensitive only to the difference between the two input voltages, which is called the *difference-mode* input voltage, $v_i = v_1 - v_2$. The op amp is ideally completely insensitive to any voltage component that is common to the inputs, which is called the *common-mode* input voltage, and is given by $v_{i(CM)} = (v_1 + v_2)/2$.

The open-loop gain of the op amp is a positive dimensionless constant and is usually very large, often in the range 10^5 to 10^6 at low frequencies (0 to 30 Hz). If voltage v_1 is applied alone and $v_2 = 0$, we have that $v_o = A_{OL}v_1$, so that v_o will be an amplified *noninverted* replica of the input voltage v_1. If, on the other hand, v_2 is now applied alone with $v_1 = 0$, then $v_o = -A_{OL}v_2$ and the output voltage v_o will now be an amplified and *inverted* replica of the input voltage. Hence we can call the lower (v_1) input terminal the *noninverting* or *positive-gain input terminal,* and the upper input terminal (v_2) can be called the *inverting* or *negative-gain terminal.* This is the usual convention and will be assumed to be the case unless otherwise indicated. The two input terminals may be designated explicitly as shown in Figure 5.1b and c. If there is no explicit indication with respect to the inverting and noninverting input terminals, the usual convention will be assumed to hold.

The op amp is a multistage electronic amplifier containing many transistors. As is the case with any type of electronic amplifier or system, there must be a dc power supply voltage applied to bias the various transistors properly. This power supply connection is shown explicitly in Figure 5.2a. Usually, however, the power supply connections are omitted from the diagram in order to simplify the diagram, although these connections are always assumed to be present. Most op amps are operated with a *dual* or *split* power supply, as shown in Figure 5.2a. The two supply voltages, V^+ and V^-, are usually of the same magnitude (but of opposite algebraic sign). The supply voltages can usually range from as little as ± 5 V up to a maximum of ± 18 V, with ± 10 to ± 15 V being the most commonly used range of values for the supply voltages. Some op amps can operate with supply voltages as small as ± 3 V, and other high-voltage op amps can operate with supply voltages substantially greater than ± 18 V.

The use of a dual or split supply offers the advantage of allowing the input and output voltages to swing both above and below ground potential for a *bipolar* output voltage swing. Some op amps can be operated with a single supply, as shown in Figure 5.2b. This has the advantage of a simpler power supply configuration, but the output voltage can now swing only in the positive direction and cannot go below ground potential (0 V), so only a *unipolar* output voltage swing can be obtained. Furthermore, the allowable input voltage range is usually limited such that the input voltage cannot be allowed to drop more than about 0.5 V below ground for proper operation of the circuit.

In any case, the output voltage of an op amp cannot swing beyond either supply voltage. Usually, the maximum attainable output voltage swing will be about 1 V less than the supply voltage in either direction, although under heavily loaded conditions the output voltage swing may be reduced even further. In the case of single-supply operation, the upper limit remains as stated above, but the lower limit of the output voltage will be ground potential (0 V). For op amps with a CMOS output stage, it is possible, under very lightly loaded conditions (such as when driving FET

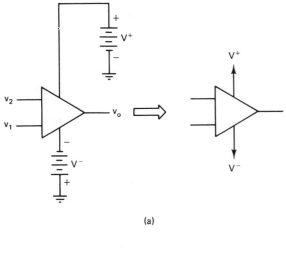

(a)

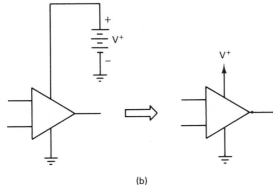

(b)

Figure 5.2 (a) Split-supply operation of operational amplifier; (b) single-supply operation.

input devices), for the output voltage to swing to within just a few millivolts of both supply voltages. This is often referred to as a *rail-to-rail* output voltage swing.

The ideal input–output transfer characteristic of an op amp is shown in Figure 5.3. Note the linear or amplifying region wherein $v_o = A_{OL}v_i$, which is bounded on either extreme by the saturation regions, in which the output voltage is limited by the supply voltage and is no longer responsive to changes in the input voltage. Since the value of A_{OL} is so very large, especially at the lower frequencies, where values of 10^5 to 10^7 are possible, the width of the linear or amplifying region is very small and is given by $\Delta v_i \simeq (V^+ - V^- - 2\,\text{V})/A_{OL}$. Thus, typically, we will have a supply voltage of around ± 10 V, so $\Delta v_i \simeq 20\ \text{V}/A_{OL} \simeq 20$ to $200\ \mu\text{V}$. Therefore, if the output voltage is to be an amplified replica of the input voltage, the amplitude of the input voltage must be kept very small, generally less than 1 mV. If this is not the case, the op amp will enter the saturated region during part of the output-voltage swing and the output-voltage waveform will be clipped, usually resulting in severe signal distortion. It is because of this limitation, as well as a number of other factors to be discussed later, that the op amp is usually operated in a *feedback* or *closed-loop* configuration in which a fraction of the output voltage is fed back to the inverting

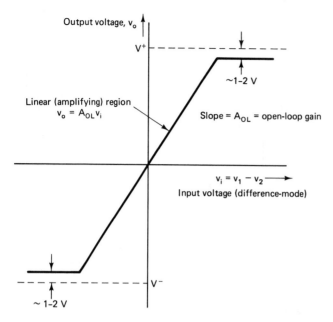

Figure 5.3 Operational amplifier input–output transfer characteristic.

input terminal as shown in Figure 5.4. This produces a condition of *negative feedback* and offers a number of very substantial advantages.

Under these conditions the op amp is operating as a closed-loop system and the fraction of the output voltage that is fed back to the inverting input terminal is called the *feedback factor F* and for the circuit of Figure 5.4 is given by $F = v_{fb}/v_o = Z_1/(Z_1 + Z_2)$ as obtained from the simple voltage-division relationship. Note that although the basic relationship that $v_o = A_{OL}v_i$ is still valid the difference-mode input voltage v_i is *no longer* given by $v_i = v_1 - v_2$, but is now given by $v_i = v_1 - v_2' - v_{fb} = v_1 - v_2' - Fv_o$, so $v_o = A_{OL}v_i = A_{OL}(v_1 - v_2' - Fv_o)$, and thus $v_o(1 + FA_{OL}) = A_{OL}(v_1 - v_2')$. Solving for v_o gives

$$v_o = \frac{A_{OL}}{1 + FA_{OL}}(v_1 - v_2') = A_{CL}(v_1 - v_2') \tag{5.1}$$

where A_{CL} is the *closed-loop gain*. Note that since the signal is no longer applied directly to the inverting input terminal, but rather by means of the Z_1-Z_2 voltage divider, the voltage v_2' is related to the signal voltage v_2 by $v_2' = v_2[Z_2/(Z_1 + Z_2)]$, as again obtained from the simple voltage-division relationship.

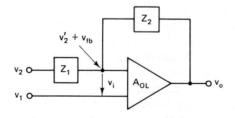

Figure 5.4 Closed-loop (negative feedback) operational-amplifier system.

For this case of negative feedback, we see that the closed-loop gain $A_{CL} = A_{OL}/(1 + FA_{OL})$ will be smaller than the open-loop gain, A_{OL}. The quantity FA_{OL} is called the *loop gain*. For the case of the large loop gain such that $FA_{OL} \gg 1$, we have that $A_{CL} \simeq A_{OL}/FA_{OL} = 1/F$, so the closed-loop gain is relatively independent of the open-loop gain A_{OL} and dependent principally on the parameters of the feedback network. For the circuit of Figure 5.4, we have that $F = Z_1/(Z_1 + Z_2)$, so for large loop gains we have that $A_{CL} = 1/F = (Z_1 + Z_2)/Z_1 = 1 + Z_2/Z_1$, and thus A_{CL} is controlled principally by the ratio of the two impedances, Z_1 and Z_2. The gain that the signal applied to the noninverting input terminal, v_1, obtains is therefore $1 + (Z_2/Z_1)$. The net gain with respect to the other input signal v_2, however, is negative and is also modified by the Z_1–Z_2 voltage divider to become

$$\frac{Z_2}{Z_1 + Z_2}\left[-\left(1 + \frac{Z_2}{Z_1}\right)\right] = -\frac{Z_2}{Z_1}$$

In the sections to follow, we consider the analysis of closed-loop op-amp systems using the *node voltage circuit equations*, which represent basically applications of Kirchhoff's current law. This will first be applied for the case in which the open-loop gain is assumed to be very large ($A_{OL} \to \infty$), which will be called the *infinite open-loop gain case*. Although this case will never be obtained in practice, the results so obtained will be applicable as an excellent approximation for most actual situations. Then this node voltage method will be applied for the case in which the open-loop gain is assumed to be finite. After this, the effects of some of the nonideal op-amp characteristics on the circuit performance of op amps are considered.

5.3 ANALYSIS OF CLOSED-LOOP OPERATIONAL-AMPLIFIER CIRCUITS USING THE NODE VOLTAGE EQUATIONS

For the op-amp analysis to follow we will use the node voltage equations, which are basically statements of Kirchhoff's current law. We will assume that the op amp is ideal in that there will be no currents flowing into or out of the two input terminals. We will use the circuit of Figure 5.5, and for the first analysis we will let the open-loop gain of the op amp go to infinity (the "infinite-gain approximation"), and then an analysis will be provided for the case of finite open-loop gain.

Ideal operational amplifier with infinite gain. If we let the open-loop voltage gain of the op amp go to infinity, then for a finite output voltage v_o the input voltage v_i will go to zero as given by $v_i = v_o/A_{OL} \to 0$ as $A_{OL} \to \infty$. Therefore, for the node x and node y voltage we have that $v_x = v_y = v_1$. Writing nodal equations for the currents through the admittances Y_1 and Y_2 gives $(v_2 - v_1)Y_1 = (v_1 - v_o)Y_2$, so $v_o Y_2 = -v_2 Y_1 + v_1(Y_1 + Y_2)$. Solving for v_o gives

$$\boxed{v_o = v_1\left(1 + \frac{Y_1}{Y_2}\right) - v_2\frac{Y_1}{Y_2}} \tag{5.2}$$

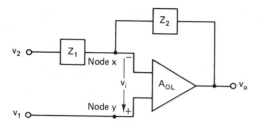

Figure 5.5 Circuit for node voltage analysis of operational amplifier.

and in terms of impedances this will be

$$v_o = v_1\left(1 + \frac{Z_2}{Z_1}\right) - v_2\frac{Z_2}{Z_1}$$ (5.3)

Ideal operational amplifier but with finite open-loop gain. For the case of a finite value of A_{OL}, we have that $v_i = v_o/A_{OL}$, so $v_x = v_1 - v_o/A_{OL}$. Writing the node voltage equations for this case gives $(v_2 - v_1 + v_o/A_{OL})Y_1 = (v_1 - v_o/A_{OL} - v_o)Y_2$, so upon collecting terms in v_o we obtain $v_o(Y_2 + Y_2/A_{OL} + Y_1/A_{OL}) = -v_2Y_1 + v_1(Y_2 + Y_1)$. Solving this for the output voltage v_o gives

$$
\begin{aligned}
v_o &= \frac{v_1(Y_2 + Y_1) - v_2Y_1}{Y_2 + (1/A_{OL})(Y_2 + Y_1)} = \frac{v_1(1 + Y_1/Y_2) - v_2(Y_1/Y_2)}{1 + (1/A_{OL})(1 + Y_1/Y_2)} \\
&= \frac{v_1(1 + Z_2/Z_1) - v_2(Z_2/Z_1)}{1 + (1/A_{OL})(1 + Z_2/Z_1)}
\end{aligned}
$$ (5.4)

Let us consider the simple op-amp system of Figure 5.6. For the output voltage v_o, we have that $v_o = A_{CL}v_1 = (1 + R_2/R_1)v_1$. In Figure 5.7, both the open-loop and closed-loop input–output transfer characteristics are shown. Note that since A_{CL} can be made to be much smaller than A_{OL} the input voltage range for linear operation in the closed-loop case can be made to be very much larger than that for the open-loop case.

The open-loop gain of op amps usually exhibits a very large unit-to-unit variation among devices of the same type, with the manufacturers specifying only minimum and sometimes maximum values for A_{OL}. The A_{OL} unit-to-unit range can often extend over a span of 3 : 1 or even 10 : 1. The open-loop gain also exhibits a strong frequency dependence and can vary from values in the 10^6 range at low frequencies (0 to 10 Hz) all the way down to values below unity at frequencies in excess of a few

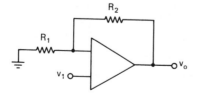

Figure 5.6 Noninverting operational-amplifier circuit.

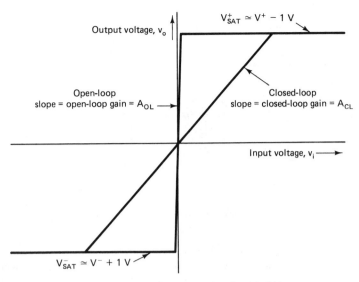

Figure 5.7 Input–output transfer characteristics.

megahertz. Furthermore, the open-loop gain is sensitive to supply voltage variations and temperature effects.

In the negative-feedback closed-loop situation, we see that the closed-loop gain will become relatively indepedent of A_{OL} and dependent principally on the parameters of the feedback loop. In the specific case under consideration, we have that $A_{CL} = 1 + R_2/R_1$. Since the resistance ratio can be very accurately set and is relatively independent of temperature, voltage level, and frequency, we see that the use of negative feedback can produce a closed-loop gain that is both *accurately set* and *stable*.

In the case of Figure 5.6, we have a noninverting amplifier with a gain of $A_{CL} = 1 + R_2/R_1$. An interesting special case of this is the circuit of Figure 5.8, for which the gain is +1. This op-amp circuit is called a *voltage follower* since the output voltage accurately follows the voltage. It might at first be thought that this circuit would be of little or no practical value since the gain is unity. It turns out, however, that this is a very useful and very widely used circuit. This is a result of the very large input impedance Z_i and very small output impedance Z_o of this op-amp configuration. This will allow relatively low impedance loads to be driven by relatively high impedance sources without any substantial signal loss, as illustrated by the example shown in Figure 5.9. In the situation of Figure 5.9a, where a 100-kΩ source is driving a load impedance of only 100 Ω, the ratio of output voltage to signal voltage

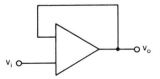

Figure 5.8 Voltage follower.

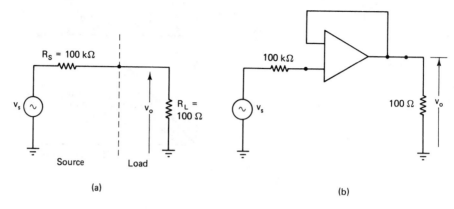

Figure 5.9 Application of voltage follower.

is given by $v_o/v_s = 100\ \Omega/(100\ \Omega + 100\ k\Omega) \approx 0.001$, so we see that a very severe attenuation of the signal occurs. In Figure 5.9b, a voltage follower is interposed between source and load. As long as $Z_i \gg 100\ k\Omega$ and $Z_O \ll 100\ \Omega$, which are conditions that are readily obtainable in practice, we have that $v_o \approx v_s$, and there will be very little signal attenuation.

In Figure 5.10a, another simple op-amp circuit is shown. For this circuit, $v_o = A_{CL}v_2 = -(R_2/R_1)v_2$, so this is an *inverting amplifier* with a voltage gain of $-R_2/R_1$. Again we note that the gain is controlled by the resistance ratio and is relatively independent of the op-amp open-loop gain. We also note that the gain magnitudes for the two cases just considered are not equal, the gain for the noninverting case being $1 + R_2/R_1$ and the gain for the inverting case being $-R_2/R_1$.

Let us now consider the circuit of Figure 5.10b. For this circuit, the gain accorded voltage v_2 is $-R_2/R_1$, as before. We note, however, that the signal voltage v_1 is attenuated by the R_1-R_2 voltage divider so that the voltage that appears at the noninverting input terminal of the op amp is $v_1R_2/(R_1 + R_2)$. The output voltage resulting from v_1 is therefore given by $v_o = v_1R_2/(R_1 + R_2)(1 + R_2/R_1) = v_1(R_2/R_1)$.

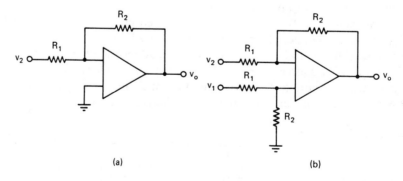

Figure 5.10 (a) Inverting amplifier; (b) difference amplifier.

The net output voltage resulting from the simultaneous application of both v_1 and v_2 will produce an output voltage given by

$$v_o = \frac{R_2}{R_1} v_1 - \frac{R_2}{R_1} v_2 = \frac{R_2}{R_1} (v_1 - v_2) \qquad (5.5)$$

so we see that the voltage gain accorded both v_1 and v_2 will now be equal in magnitude (but still opposite in algebraic sign). This amplifier configuration can be described as a *difference amplifier* since the output voltage is a function only of the *difference-mode* component of the input signals and is insensitive to any *common-mode* signal component.

5.3.1 Superposition Theorem and Virtual Ground

The superposition theorem and the concept of virtual ground can both be very useful for the analysis of operational-amplifier circuits. Let us consider as an example the circuit of Figure 5.11. For simplicity, we will assume the case of an ideal, infinite-gain operational amplifier.

Making use of the superposition theorem, we will determine the output voltage resulting from each of the input voltages acting alone, the others being set equal to zero, and then obtain the algebraic summation of these separate contributions to the

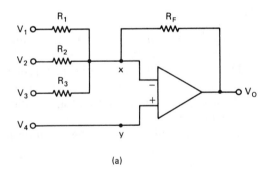

(a)

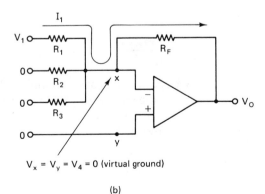

$V_x = V_y = V_4 = 0$ (virtual ground)

(b)

Figure 5.11 Operational-amplifier circuit analysis example: (a) operational-amplifier circuit with four input voltages; (b) circuit with V_1 active and symbol with input polarities indicated explicitly, $V_2 = V_3 = V_4 = 0$.

net output voltage. With V_1 acting alone, we have the situation shown in Figure 5.11b. Since $V_X - V_Y = V_o/A_{OL}$ goes to zero as A_{OL} goes to infinity, and $V_Y = V_4 = 0$, then $V_X = 0$ and the inverting input terminal is at ground potential. Since the inverting input terminal is at ground potential, but there is no direct connection between this terminal and the circuit ground, this will be called a *virtual ground*. We have that $I_1 = (V_1 - V_X)/R_1 = V_1/R_1$ since $V_X = 0$.

For the case of an ideal operational amplifier, no current flows into or out of the input terminals, so the current through R_F is equal to I_1. We therefore have that $V_O = V_X - I_1 R_F = -I_1 R_F$, since again $V_X = 0$. Therefore, the output voltage due to V_1 acting alone is $V_O = -I_1 R_F = -V_1(R_F/R_1)$. Similarly, with V_2 acting alone we have that $V_O = -V_2(R_F/R_2)$, and for V_3 we have $V_O = -V_3(R_F/R_3)$.

With V_4 active and $V_1 = V_2 = V_3 = 0$, we see that R_1, R_2, and R_3 are all connected in parallel between the amplifier inverting input terminal and ground. As a result, the closed-loop gain with respect to V_4 can be written as $V_O/V_4 = 1 + R_F/(R_1\|R_2\|R_3)$. The net output voltage resulting from the combined action of all four input voltages can now be expressed as

$$V_O = -V_1\frac{R_F}{R_1} - V_2\frac{R_F}{R_2} - V_3\frac{R_F}{R_3} + V_4\left(1 + \frac{R_F}{R_1\|R_2\|R_3}\right) \qquad (5.6)$$

5.4 GAIN ERROR AND GAIN STABILITY

Looking at the equation for the output voltage for the case of the op amp with finite open-loop gain, it is apparent that the output voltage and therefore the closed-loop gain are functions of the open-loop gain. It should also be apparent that, as the open-loop gain gets very large (with respect to $1 + Z_2/Z_1$), the closed-loop gain becomes less and less dependent on the open-loop gain and approaches the value obtained for the infinite gain approximation.

Considering, for example, the case in which $v_2 = 0$, the equation for v_o is

$$v_o = v_1\frac{1 + Z_2/Z_1}{1 + (1/A_{OL})(1 + Z_2/Z_1)}$$

so the closed-loop gain is

$$A_{CL} = \frac{v_o}{v_1} = \frac{1 + Z_2/Z_1}{1 + (1/A_{OL})(1 + Z_2/Z_1)}$$

As the open-loop gain A_{OL} becomes very large and approaches infinity, the value of the closed-loop gain A_{CL} will approach a limiting value to be denoted by $A_{CL}(\infty)$. The value of $A_{CL}(\infty)$ for the case under consideration is $A_{CL}(\infty) = 1 + Z_2/Z_1$. Using this relationship, we can now rewrite the expression for A_{CL} as

$$A_{CL} = \frac{A_{CL}(\infty)}{1 + A_{CL}(\infty)/A_{OL}}$$

From this equation, we note that, for small values of A_{OL} such that $A_{OL} << A_{CL}(\infty)$, we have that $A_{CL} \simeq A_{OL}$. When $A_{OL} = A_{CL}(\infty)$, then $A_{CL} = \frac{1}{2}A_{CL}(\infty)$. For large values of open-loop gain such that $A_{OL} \gg A_{CL}(\infty)$, which is the usual case of interest, A_{CL} approaches $A_{CL}(\infty)$ and the equation for A_{CL} can be written in approximate form as

$$A_{CL} = \frac{A_{CL}(\infty)}{1 + A_{CL}(\infty)/A_{OL}} \simeq A_{CL}(\infty)\left[1 - \frac{A_{CL}(\infty)}{A_{OL}}\right] \simeq A_{CL}(\infty)(1 - \epsilon) \qquad (5.7)$$

where ϵ is the *gain error*. The gain error is defined as the *fractional change* in the closed-loop gain as the open-loop gain varies from some specified finite value up to infinity. The gain error can thus be expressed as

$$\text{gain error} = \epsilon = \frac{A_{CL}(\infty) - A_{CL}}{A_{CL}(\infty)} \qquad (5.8)$$

From the equation given above for A_{CL}, it can readily be verified that $\epsilon \simeq A_{CL}(\infty)/A_{OL}$. We see that large values of A_{OL} with respect to the closed-loop gain will lead to small values for the gain error.

Example. To consider an example, let us take the case of an op amp with $Z_2 = 99$ kΩ, $Z_1 = 1.0$ kΩ, and an open-loop gain of $A_{OL} = 10^5$ (min.). The gain error for this case is approximately given by

$$\epsilon \simeq \frac{A_{CL}(\infty)}{A_{OL}} = 100/10^5 = 0.001 \text{ or } 0.10\%$$

We can therefore say that the maximum variation in the closed-loop gain as the open-loop gain varies from a minimum value of 10^5 up to infinity is only 0.10%. Thus the closed-loop gain is stabilized against changes in the open-loop gain, and a large change in the open-loop gain produces only a very small variation in the closed-loop gain.

To look at this question of gain stability from a slightly different viewpoint, let us start off again with the expression for A_{CL} as given by

$$A_{CL} = \frac{A_{CL}(\infty)}{1 + A_{CL}(\infty)/A_{OL}} \qquad (5.9)$$

and take the derivative of A_{CL} with respect to A_{OL}. Doing this gives

$$\frac{dA_{CL}}{dA_{OL}} = \frac{-A_{CL}(\infty)}{[1 + A_{CL}(\infty)/A_{OL}]^2} \frac{-A_{CL}(\infty)}{A_{OL}^2} = \frac{A_{CL}^2}{A_{OL}^2} \qquad (5.10)$$

This can be rewritten in the form

$$\boxed{\frac{dA_{CL}}{A_{CL}} = \frac{A_{CL}}{A_{OL}} \frac{dA_{OL}}{A_{OL}}} \qquad (5.11)$$

From this we see that the *fractional* (or percentage) change in the closed-loop gain, dA_{CL}/A_{CL}, that results from a fractional change in the open-loop gain, dA_{OL}/A_{OL}, is equal to the fractional (or percentage) change in the open-loop gain multiplied by the factor A_{CL}/A_{OL}.

Going back now to the example just considered for which $A_{CL} = 100$ and

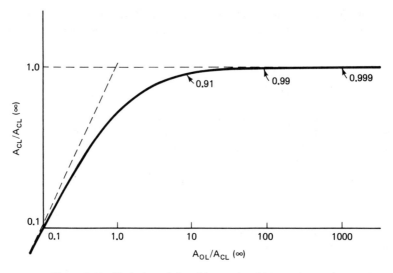

Figure 5.12 Variation of closed-loop gain with open-loop gain.

$A_{OL} = 10^5$, we have that $dA_{CL}/A_{CL} = (100/10^5)(dA_{OL}/A_{OL}) = 0.001(dA_{OL}/A_{OL})$. Thus, if the open-loop gain changes by 10%, the corresponding change in the closed-loop gain will be only 0.01%.

From the previous discussion we have seen that the gain error can be reduced and the gain stability improved by making the open-loop gain large compared to the closed-loop gain. Indeed, the factor by which the gain error or the gain stability is improved is the same factor by which the gain is sacrificed in going from open-loop conditions to closed-loop conditions.

Figure 5.12 is a doubly normalized graph of the closed-loop gain as a function of the open-loop gain. Note that as A_{OL} increases A_{CL} becomes increasingly *less* dependent on the open-loop gain. When $A_{OL} = 10A_{CL}(\infty)$ the closed-loop gain is $0.91A_{CL}(\infty)$. For $A_{OL} = 100A_{CL}(\infty)$, the closed-loop gain is $0.99A_{CL}(\infty)$, and for $A_{OL} = 1000A_{CL}(\infty)$, the closed-loop gain is $0.999A_{CL}(\infty)$. We see therefore that as A_{OL} increases the closed-loop gain asymptotically approaches $A_{CL}(\infty)$, and that the rate of change of A_{CL} with respect to A_{OL} decreases rapidly. At the other extreme, as A_{OL} becomes small compared to $A_{CL}(\infty)$, we see that the closed-loop gain asymptotically approaches A_{OL}.

5.5 FREQUENCY RESPONSE

An op amp is a multistage electronic amplifier, and the variation of the open-loop gain with frequency can generally be expressed in the form

$$A_{OL}(f) = \frac{A_{OL}(0)}{(1 + jf/f_1)(1 + jf/f_2)(1 + jf/f_3) \cdots} \tag{5.12}$$

where $A_{OL}(0)$ is the zero-frequency value of the open-loop gain and the breakpoint frequencies are in the sequence $f_1 < f_2 < f_3 < \cdots$. For most op amps, the first breakpoint frequency, f_1, is very small (~10 Hz) compared to f_2 (~1 to 3 MHz) and the other breakpoint frequencies. The frequency range that is of greatest interest is usually that for which $f^2 \gg f_1^2$ and $f^2 \ll f_2^2, f_3^2$, and so on. For this frequency range, the expression for A_{OL} can be written as $A_{OL} \simeq A_{OL}(0)/j(f/f_1) = A_{OL}(0)f_1/jf$. The frequency condition for this approximation is generally satisfied for frequencies such that f is at least one-half of a decade ($\sqrt{10} : 1$ or approximately $3 : 1$) away from f_1 and f_2 such that $3f_1 \lesssim f \lesssim f_2/3$.

The *unity-gain frequency* f_u is defined as the frequency at which the open-loop gain as given in the approximate expression above has dropped to unity. Therefore, we have that at $f = f_u$, $A_{OL}(0)f_1/f_u = 1$, so

$$\boxed{f_u = A_{OL}(0)f_1} \qquad (5.13)$$

The open-loop gain can therefore be expressed in terms of the unity-gain frequency as

$$\boxed{A_{OL} \simeq \frac{f_u}{jf}} \qquad (5.14)$$

Figure 5.13 shows the open-loop gain versus frequency characteristics (Bode plot). The frequency response of the op amp is flat down to zero frequency (dc).

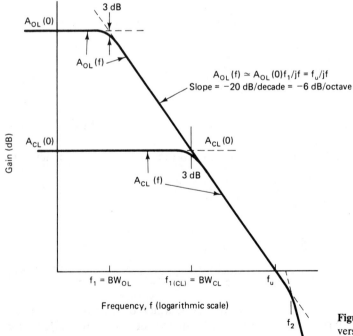

Figure 5.13 Operational-amplifier gain versus frequency characteristics (Bode plot).

At $f = f_1$ (the first breakpoint frequency), the open-loop gain is 3 dB down from the zero-frequency value of $A_{OL}(0)$, so the open-loop 3-dB bandwidth is $BW_{OL} = f_1$.

The closed-loop gain for the simple noninverting amplifier case considered before is given by

$$A_{CL} = \frac{1 + R_2/R_1}{1 + (1/A_{OL})(1 + R_2/R_1)} \tag{5.15}$$

Assuming that the zero-frequency open-loop gain $A_{OL}(0)$ is large enough, which is virtually always the case, the zero-frequency value of the closed-loop gain $A_{CL}(0)$ is given by $A_{CL}(0) = 1 + R_2/R_1$. As a result, the closed-loop gain as a function of frequency can now be expressed as

$$A_{CL}(f) = \frac{A_{CL}(0)}{1 + A_{CL}(0)/A_{OL}} \tag{5.16}$$

If we now introduce the approximate expression for $A_{OL} = f_u/jf$, we obtain

$$A_{CL}(f) = \frac{A_{CL}(0)}{1 + jfA_{CL}(0)/f_u} \tag{5.17}$$

We see that when $f = 0$ the closed-loop gain is indeed $A_{CL}(0)$, and as f increases, $A_{CL}(f)$ decreases monotonically, as expected.

When $fA_{CL}(0)/f_u = 1$, the expression for $A_{CL}(f)$ becomes

$$A_{CL}(f) = \frac{A_{CL}(0)}{1 + j1} = \frac{A_{CL}(0)}{\sqrt{2}\ \angle +45°} \tag{5.18}$$

Thus A_{CL} will be down by a factor of $\sqrt{2}$ or 3 dB from the zero-frequency value at the frequency given by

$$\boxed{f = \frac{f_u}{A_{CL}(0)}} \tag{5.19}$$

so this corresponds to the *half-power* or *3-dB frequency* for the closed-loop gain and will be denoted as $f_{1(CL)}$.

The expression above for the variation of closed-loop gain with frequency can be rewritten as

$$A_{CL}(f) = \frac{A_{CL}(0)}{1 + jA_{CL}(0)/|A_{OL}|} \tag{5.20}$$

The *magnitude* of the closed-loop gain is

$$A_{CL}(f) = \frac{A_{CL}(0)}{\sqrt{1 + [A_{CL}(0)/A_{OL}]^2}} \tag{5.21}$$

From this last equation we see that, at the frequency at which the open-loop gain has decreased to the point where it is equal to $A_{CL}(0)$, the closed-loop gain is down by a factor of $\sqrt{2}$ or 3 dB below the zero-frequency value of $A_{CL}(0)$. This frequency therefore corresponds to the closed-loop half-power or 3-dB frequency. Looking at the Bode plot of Figure 5.13, we see that the closed-loop 3-dB frequency $f_{1(CL)}$ will thus occur at the point of intersection of the open-loop gain curve, $A_{OL}(f)$, with the horizontal line corresponding to $A_{CL}(0)$.

The 3-dB bandwidth of a system is the range of frequencies over which the gain remains within 3 dB of the maximum value. Since op amps are direct-coupled or dc amplifiers with no coupling capacitors between stages and no bypass capacitors, the open-loop frequency response remains flat down to zero frequency (dc). The range of frequencies over which the gain stays within 3 dB of the maximum value therefore extends from 0 up to the 3 dB (or half-power) frequency. Therefore, the 3-dB bandwidth is equal to the 3-dB frequency f_1 and thus $BW_{OL} = f_1$, where BW_{OL} is the *open-loop 3-dB bandwidth*.

Since the open-loop gain remains flat down to zero frequency, the closed-loop gain is similarly flat down to zero frequency. Therefore, the *closed-loop 3-dB bandwidth*, BW_{CL}, is equal to the closed-loop 3-dB frequency, $f_{1(CL)}$, so $BW_{CL} = f_{1(CL)}$. Since $BW_{CL} = f_{1(CL)} = f_u/A_{CL}(0)$, the closed-loop gain–bandwidth product is equal to the unity-gain frequency as given by

$$\boxed{A_{CL}(0) \times BW_{CL} = f_u} \tag{5.22}$$

For f_u, we have also had the relationship that

$$\boxed{f_u = A_{OL}(0)f_1 = A_{OL}(0) \times BW_{OL}} \tag{5.23}$$

so we see that the open-loop gain–bandwidth product is also equal to the unity-gain frequency. For either gain–bandwidth product case given above, the gain is the zero-frequency (dc) gain and the bandwidth is the 3-dB bandwidth. Thus we see that for both the closed-loop and open-loop situations the gain–bandwidth product is a constant and equal to the unity-gain frequency.

From the gain–bandwidth product relationship, we see that by sacrificing gain in going from the open-loop situation to the closed-loop case there will be a corresponding improvement in the bandwidth. Indeed, by whatever factor the gain is sacrificed by using negative feedback, the bandwidth will be improved by the same factor.

If we take the ratio of the closed-loop bandwidth to the open-loop bandwidth, we obtain

$$\frac{BW_{CL}}{BW_{OL}} = \frac{A_{OL}(0)}{A_{CL}(0)} \tag{5.24}$$

We thus see that the factor by which the bandwidth is improved in going from open-loop conditions to closed-loop conditions is exactly the factor by which the gain is sacrificed in going from the open-loop gain to the closed-loop gain.

To consider some simple examples of bandwidth improvement, let us assume a representative value for f_u of 1.0 MHz. For a closed-loop gain of $A_{CL}(0) = 1000$ (or 60 dB), the closed-loop bandwidth is 1.0 MHz/1000 = 1.0 kHz. For $A_{CL}(0) = 100$, the bandwidth is 10 kHz, and for $A_{CL}(0) = 10$ it increases to 100 kHz. Finally, for the unity-gain voltage follower case, the bandwidth is equal to f_u or 1.0 MHz. In contrast to these bandwidth values, if the open-loop gain for $A_{OL}(0)$ is 10^5 (or 100 dB), the open-loop bandwidth is only 10 Hz!

For the bandwidth relationships above, it is to be understood that f_u is the *extrapolated unity-gain frequency* as obtained from the extrapolation of the $A_{OL} = f_u/jf$ response line down to the 0-dB (unity-gain) level. In most cases, this will correspond to the actual frequency at which the gain is unity. However, in the cases in which f_2 is less than f_u, the value of f_u can be substantially greater than the value of frequency at which the gain has dropped to unity.

For the bandwidth relationships given above to be valid, it is necessary that the second op-amp breakpoint frequency f_2 be substantially above the closed-loop bandwidth BW_{CL}. If this is not the case, the closed-loop bandwidth can be considerably less than that predicted by the equations above.

For frequencies that are substantially above $f_{1(CL)} = BW_{CL}$, such that the open-loop gain has dropped considerably below $A_{CL}(0)$, we see from the equation for $A_{CL}(f)$ that the $A_{CL}(f)$ curve will asymptotically approach the A_{OL} curve as shown in the Bode plot. The closed-loop gain starts off at $A_{CL}(0)$ at low frequencies, is down 3 dB at $f_{1(CL)} = BW_{CL} = f_u/A_{CL}(0)$, and then will asymptotically approach the A_{OL} curve, dropping at a rate of 20 dB/decade.

5.6 TIME-DOMAIN RESPONSE

To evaluate the frequency response of a system, a sinusoidal excitation or signal is applied to the system and the response or output of the system is measured as a function of the frequency of the signal source. A graph of the input-to-output gain or transfer ratio is then usually drawn. For the *time-domain* or *transient* response, a step-function excitation is applied and the graph or display of the output response as a function of time is obtained. The time-domain response of most systems is usually most conveniently described by the *rise time*, which is the time required for the output response to rise from a specified lower value to some other specified upper value. The most commonly used lower and upper values are 10% and 90% of the maximum or ultimate response level, respectively. Unless otherwise indicated or implied, the rise time, t_{rise}, will be understood to be the 10% to 90% rise time. This is illustrated in Figure 5.14 for the case of a simple system that is characterized by a single time constant.

Although a step-function excitation and the corresponding response are the simplest to evaluate mathematically, from a practical standpoint the most convenient type of excitation is often a repetitive rectangular pulse waveform of suitable pulse duration and pulse period. This type of signal can be obtained from a pulse generator or square-wave generator, and the input and output waveforms can be viewed on an oscilloscope.

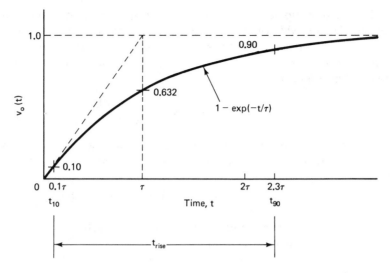

Figure 5.14 Time-domain response characteristics (normalized).

For a system that has a gain versus frequency response characterized by an equation of the form

$$A(f) = \frac{A(0)}{1 + jf/f_1} \tag{5.25}$$

where f_1 is the half-power frequency or 3-dB bandwidth of the system, the corresponding response of the system to a unit-step-function excitation is given by $v_o(t) = A(0)[1 - \exp(-t/\tau)]$, where $v_o(t)$ is the system response as a function of time and τ is the system time constant. The time constant of the system is related to the 3-dB frequency by $\tau = 1/\omega_1 = 1/(2\pi f_1)$.

For the 10% to 90% rise time, we have that at the 10% response point $0.1 = 1 - \exp(-t_{10}/\tau)$, so $t_{10} = 0.105\tau$. Similarly, at the 90% response point $0.9 = 1 - \exp(-t_{90}/\tau)$, so $t_{90} = 2.303\tau$. We therefore have that $t_{\text{rise}} = t_{90} - t_{10} = 2.2\tau$. Since $\tau = 1/(2\pi f_1)$, we can express the rise time in terms of the bandwidth as $t_{\text{rise}} = 2.2\tau = 2.2/(2\pi f_1) = 0.35/f_1 = 0.35/BW$. Thus the rise time can be obtained from the 3-dB bandwidth, and vice versa. This relationship between rise time and bandwidth can most conveniently be described in terms of the rise-time bandwidth product as given by

$$\boxed{t_{\text{rise}} \times BW = 0.35} \tag{5.26}$$

The closed-loop frequency response characteristics for most op-amp systems can adequately be described by an equation of the type given above with a single breakpoint frequency f_1, which is the closed-loop bandwidth of the system. As a result, for these cases the simple rise time–bandwidth product relationship given above can be used. For example, if the closed-loop bandwidth is 10 kHz, the corresponding rise time is $t_{\text{rise}} = 0.35/10 \text{ kHz} = 35 \text{ μs}$.

The relationship between the frequency response equation and the time-domain response equation will now be derived. If a system has a frequency response given by $A(\omega) = A(0)/(1 + j\omega/\omega_1)$, the corresponding expression in terms of the Laplace transform variable s is $A(s) = A(0)/(1 + s/\omega_1)$. The response of the system to a unit-step-function excitation $v_i(s) = 1/s$ is

$$v_o(s) = A(s)v_i(s) = \frac{A(0)}{s(1 + s/\omega_1)} \tag{5.27}$$

To obtain the response in the time domain, we will now take the inverse Laplace transform of the foregoing expression for $v_o(s)$ to obtain $v_o(t)$ as given by

$$v_o(t) = A(0)[1 - \exp(\omega_1 t)] = A(0)\left[1 - \exp\left(\frac{-t}{\tau}\right)\right] \tag{5.28}$$

where $\tau = 1/\omega_1 = 1/2\pi f_1$), thus demonstrating the relationship between the time constant and the 3–dB frequency f_1.

5.7 OFFSET VOLTAGE

For the ideal op amp, we have that $v_o = A_{OL}v_i$, so $v_o = 0$ when $v_i = 0$. In any real op amp, however, there are various component mismatches and other circuit imbalances within the op amp so that with zero input voltage ($v_i = 0$) the output voltage will not be zero. To produce a condition of zero output voltage, a small input voltage must be applied, this input voltage being equal to the *input offset voltage*, V_{OS}. The transfer function for the real op amp must now be expressed as $v_o = A_{OL}(v_i - V_{OS})$.

The offset voltage V_{OS} is a small dc voltage generally in the range of about 1 mV, although some op amps may have maximum offset voltage specifications as large as 5 or 10 mV. Other *precision op amps* may have maximum V_{OS} specifications as low as in the range from 10 to 100 μV. If we consider a large number of op amps of the same type, the distribution of offset voltage values will be a symmetrical Gaussian or bell-shaped curve with a mean value of zero and truncated at the maximum V_{OS} values specified by the manufacturer. An individual op amp is thus just as likely to have a V_{OS} of one polarity as it is to have one of the opposite polarity. The specification sheet for the op amp will list a guaranteed maximum value for the offset voltage under certain specified conditions, and sometimes a typical value will also be listed.

For the purposes of analysis, the effect of the offset voltage for a real op amp

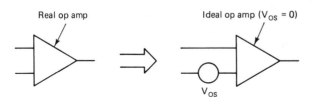

Real op amp   Ideal op amp ($V_{OS} = 0$)

V_{OS}

Figure 5.15 Offset voltage representation.

can be considered by replacing the real op amp by an ideal op amp ($V_{OS} = 0$) and by placing a dc voltage source of strength V_{OS} in series with one of the input terminals of the op amp, as shown in Figure 5.15. It is usually most convenient to insert the V_{OS} source in series with the noninverting input terminal.

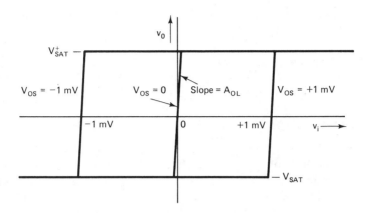

Figure 5.16 Open-loop transfer characteristics showing effect of V_{OS}.

Figure 5.16 gives the open-loop input–output transfer characteristics for three different cases of V_{OS}. The magnitude of V_{OS} is usually such that under open-loop conditions the output voltage of the op amp will be saturated in either the high (V_{SAT}^{+}) or low (V_{SAT}^{-}) states, even when no input voltage is applied. Figure 5.17 gives the transfer characteristics under closed-loop conditions. As a result of the greatly expanded input voltage range, the output voltage can be kept well away from saturation. Thus again we see the advantage of closed-loop operation of the op amp.

To illustrate the effect of the offset voltage on the output voltage for the closed-loop op-amp system, let us consider the simple circuit of Figure 5.18. To determine the effect of V_{OS}, we will now consider the op amp to be ideal and insert a voltage source of strength V_{OS} in series with the noninverting input terminal. The resulting output voltage is $v_o = V_{OS}A_{CL} = V_{OS}(1 + R_2/R_1)$. In general, we can say that the

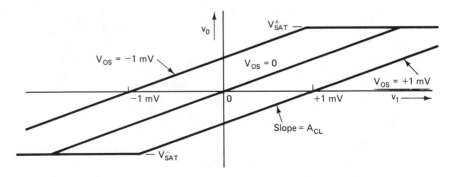

Figure 5.17 Closed-loop transfer characteristics showing effect of V_{OS}.

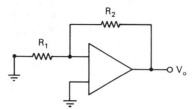

Figure 5.18 Circuit for the calculation of V_O due to V_{OS}.

output voltage component that results from the input offset voltage is equal to the offset voltage V_{OS} multiplied by the closed-loop gain $A_{CL}(0)$, taken with respect to the noninverting input terminal.

5.7.1 Offset Voltage Compensation

For many applications, especially those in which the input signal level is large compared to the offset voltage V_{OS}, the effect of V_{OS} presents no serious problem. Also, precision op amps are available with maximum input offset voltage values down in the range from 10 to 100 μV. Furthermore, since V_{OS} is a dc voltage, in many cases the effect of V_{OS} can be removed by the use of capacitive or other coupling such that only the ac component of the output voltage is obtained. Nevertheless, there are situations where it is desirable to compensate for or cancel out the offset voltage. We will now examine some methods for offset voltage compensation.

Figure 5.19 shows a simple means of V_{OS} compensation. By the use of potentiometer R_3 and the resistive voltage divider R_4 and R_5, a suitably small and adjustable voltage of either polarity can be applied to the noninverting input terminal of the op amp to cancel out V_{OS}. With the circuit as shown and with supply voltages of ±15 V, the range of cancellation voltages will extend from −15 mV to +15 mV, which should prove to be a sufficient voltage range to cover almost all V_{OS} situations. Indeed, it may be desirable to reduce this voltage range by making R_5 smaller, such as 50 Ω or even 20 Ω, so as to make the offset voltage compensation adjustment more precise. Figure 5.19b shows a similar compensation circuit, but this time the compensation voltage is applied to the inverting input terminal.

Some op amps provide specific offset voltage compensation (or *nulling*) terminals, as shown in Figure 5.19c. A potentiometer, generally in the range of 20 to 100 kΩ, is connected between the op-amp V_{OS} compensation (or V_{OS} null) terminals. The wiper arm of the potentiometer is connected to the negative supply, V^-.

For any of the compensation techniques described, the compensation is usually carried out with no signal applied to the input terminals and a sensitive dc voltmeter connected to the output. The potentiometer is then adjusted to the point where the output voltage becomes zero (that is, is "nulled out").

5.7.2 Offset Voltage Temperature Coefficient

The input offset voltage is a function of device temperature. The rate of change of the offset voltage with temperature is denoted by

$$\text{temperature coefficient of } V_{OS} = TC_{V_{OS}} = \frac{dV_{OS}}{dT} \qquad (5.29)$$

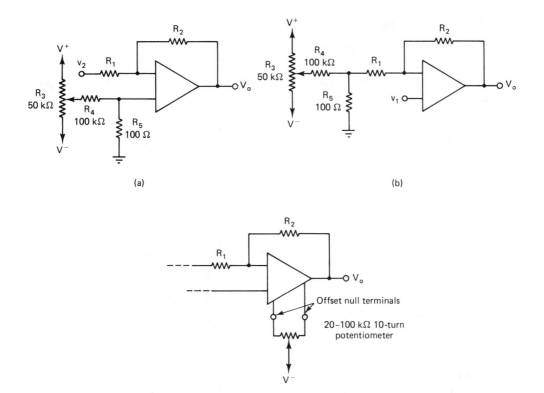

Figure 5.19 (a) and (b) Circuits for offset voltage compensation; (c) operational amplifier with offset null terminals.

For op amps with bipolar transistor input stages, the $TC_{V_{OS}}$ is related to the offset voltage itself by the following approximate relationship: $TC_{V_{OS}} \simeq V_{OS}/T$, where T is the absolute temperature ($\sim$300 K in most cases). Thus, for $V_{OS} = 1$ mV, a temperature coefficient of about $TC_{V_{OS}} = 1\,\text{mV}/300\,\text{K} = 1000\,\mu\text{V}/300\,\text{K} = 3\,\mu\text{V/K} = 3$ μV/°C is expected. Thus a temperature change of 10°C will result in a drift of the input offset voltage of about 30 μV.

As a result of the variation of the input offset voltage with temperature, the V_{OS} compensation techniques described earlier are completely effective at only one temperature. As the temperature of the device drifts in either direction from the temperature at which the compensation was initially made, the effect of the offset voltage will again be felt. Nevertheless, the net effect of the offset voltage will generally be substantially reduced by the compensation techniques. For example, if the offset voltage is 1 mV and is nulled out by a compensation circuit at a given temperature, a subsequent temperature drift of 10°C may result in an effective uncompensated offset voltage of around 30 μV. This is, however, some 30 times smaller than the initial V_{OS}, so the offset voltage compensation is indeed worthwhile.

5.8 INPUT BIAS CURRENT

For any operational amplifier to operate properly, a current must be allowed to flow into (or out of) the two input terminals. This current is the *input bias current* and is shown in Figure 5.20. The two currents I_{B_1} and I_{B_2} in Figure 5.20 will not be exactly equal to each other, but will generally be within about 10% of each other. The *input bias current* I_{BIAS} or I_B by strict definition is the average of the two input currents as given by

$$I_B = \frac{I_{B_1} + I_{B_2}}{2} \tag{5.30}$$

The difference between these two currents is called the *input offset current* I_{OS}, given by

$$I_{OS} = I_{B_1} - I_{B_2} \tag{5.31}$$

The algebraic sign of the offset current is usually of no importance, with equal probabilities for either algebraic sign.

For some op amps, I_B is of positive polarity (flowing into the op amp), whereas for other op amps it is of negative polarity (flowing out of the op amp). For op amps that use bipolar transistor input stages, the input bias current can range from 10 μA down to as little as a few nanoamperes. For op amps with field-effect transistor input stages, the input bias current will be very small, even down to the range of a few picoamperes for some op amps. On the manufacturer's specification sheet for a particular op amp, the usual specification for the bias current is the guaranteed maximum value for I_B under certain specified operating conditions. Sometimes a typical value for I_B is also indicated. Similarly, for the input offset current I_{OS} a guaranteed maximum value is specified, with a typical value sometimes being given as well.

To determine the effect of the input bias current on the output voltage of an op amp, let us consider the basic circuit of Figure 5.21. We will evaluate V_o using the superposition theorem, in that we will consider the effect on V_o of each current, I_{B_1} and I_{B_2}, acting separately and then take the algebraic summation of the two results to obtain the net output voltage V_o. The current I_{B_1} in flowing through R_3 produces a voltage drop $-I_{B_1}R_3$ that appears at the noninverting input terminal of the op amp.

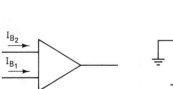

Figure 5.20 Input bias current.

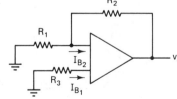

Figure 5.21 Circuit for the calculation of the effects of the input bias current.

This voltage is multiplied by the closed-loop gain of $1 + R_2/R_1$ to appear at the output as a voltage $V_o = (-I_{B_1}R_3)(1 + R_2/R_1)$. The component of output voltage that results from I_{B_2} acting alone (I_{B_1} is now considered to be zero) can be determined by first noting that the voltage at the noninverting input terminal is now zero, and hence the inverting input terminal is also essentially at ground potential (a "virtual ground"). The voltage across R_1 is therefore essentially zero, so there will be no current through R_1. All of current I_{B_2} is therefore flowing through R_2, producing a voltage drop $I_{B_2}R_2$. Since the voltage at the inverting input terminal is essentially zero and the voltage drop across R_2 is $I_{B_2}R_2$, the output voltage due to I_{B_2} is $V_o = I_{B_2}R_2$. The *net output voltage* due to both I_{B_1} and I_{B_2} is therefore

$$V_o = I_{B_2}R_2 - I_{B_1}R_3\left(1 + \frac{R_2}{R_1}\right) \qquad (5.32)$$

Since I_{B_1} and I_{B_2} are generally close in value, we see that if the coefficients of the I_{B_1} and I_{B_2} terms are equal there will be near cancellation of the effects of the bias current. For this to happen, we require that $R_2 = R_3(1 + R_2/R_1)$ so that $R_1R_2/(R_1 + R_2) = R_3$. Since $R_1R_2/(R_1 + R_2)$ is the parallel combination of R_1 and R_2, we see that this requirement means that R_3 should be equal to the parallel combination of R_1 and R_2. Under these conditions the output voltage due to the bias currents is given by

$$V_o = R_2(I_{B_2} - I_{B_1}) = -R_2I_{OS} \qquad (5.33)$$

Since the offset current is usually a small fraction of the bias current, this can represent a substantial improvement, so it is very often desirable to make R_3 equal to the parallel combination of R_1 and R_2.

The bias current usually poses a problem only for very large values of R_1, R_2, and R_3. If this is the case, the situation can most easily be remedied by using an op amp with a very low value of input bias current, such as various types of FET-input op amps. Such op amps are available with I_B values down as low as 10 pA. It is also possible to introduce various compensation circuits, including some similar to the offset voltage compensation circuits.

One basic problem with any bias current compensation circuit is that the bias current is a function of temperature. The rate of change of bias current with temperature is the *temperature coefficient of the bias current* as given by $TC_{I_B} = dI_B/dT$. Similarly, the offset current also exhibits a temperature coefficient as expressed by $TC_{I_{os}} = dI_{OS}/dT$.

5.9 COMMON-MODE GAIN

An op amp is ideally sensitive only to the difference-mode signal v_i applied to its input terminals and is completely unresponsive to any common-mode input voltage. In the ideal op-amp case, we have that $v_o = A_{OL}v_i$, where v_i is the difference-mode input voltage, $v_i = v_x - v_y$, as shown in Figure 5.22. The open-loop gain is the

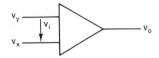

Figure 5.22 Difference-mode input voltage.

same as the difference-mode gain A_{DM}, since both are taken with respect to the difference-mode signal. Any real op amp, however, will exhibit some small response to the common-mode component of the input voltages as given by $v_{cm} = (v_x + v_y)/2$. The output voltage resulting from this common-mode input voltage is given by $v_o = A_{CM}v_{cm}$, where A_{CM} is the common-mode voltage gain. The common-mode gain, however, is very much smaller than the difference-mode gain. The ratio of the difference-mode gain, A_{DM} or A_{OL}, to the common-mode gain, A_{CM}, is the *common-mode rejection ratio* (CMRR) and is usually expressed in decibels (dB). CMRR values in the range of 80 dB (10^4) to 120 dB (10^6) are typical for op amps.

To evaluate the effect of the common-mode gain in a closed-loop op-amp situation, let us consider the circuit of Figure 5.23. For the infinite open-loop gain (A_{OL}) case, we have seen that this is a simple difference amplifier with $v_o = (R_2/R_1)(v_1 - v_2)$, so this circuit is sensitive only to the difference-mode component of the input signal and is completely insensitive to any common-mode component. We will now analyze this circuit for the case in which both the open-loop difference-mode gain A_{DM} (or A_{OL}) and the common-mode gain A_{CM} are finite (nonzero) in value.

From the circuit of Figure 5.23, we have the following equations:

$$(1) \quad v_o = A_{DM}(v_x - v_y) + \frac{A_{CM}(v_x + v_y)}{2} \tag{5.34}$$

$$(2) \quad v_x = \frac{R_2 v_1}{(R_1 + R_2)}$$

$$(3) \quad v_y = \frac{v_2 R_2}{R_1 + R_2} + \frac{v_o R_1}{R_1 + R_2} \tag{5.35}$$

Upon substituting equations (2) and (3) into equation (1), we obtain

$$v_o = A_{DM}\left[\frac{(v_1 - v_2)R_2}{R_1 + R_2} - \frac{v_o R_1}{R_1 + R_2}\right]$$

$$+ A_{CM}\left[\frac{v_1 + v_2}{2}\frac{R_2}{R_1 + R_2} + \frac{v_o R_1}{2(R_1 + R_2)}\right] \tag{5.36}$$

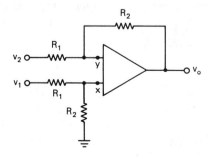

Figure 5.23 Difference amplifier circuit for the analysis of the effect of the common-mode gain.

Collecting terms in v_o on the left side gives

$$v_o\left[1 + \frac{A_{DM}R_1}{R_1 + R_2} - \frac{A_{CM}R_1}{2(R_1 + R_2)}\right]$$

$$= A_{DM}(v_1 - v_2)\frac{R_2}{R_1 + R_2} + A_{CM}\frac{v_1 + v_2}{2}\frac{R_2}{R_1 + R_2} \qquad (5.37)$$

Multiplying both sides by the factor $(R_1 + R_2)/R_1$ gives

$$v_o\left(\frac{R_1 + R_2}{R_1} + A_{DM} + \frac{A_{CM}}{2}\right) = A_{DM}(v_1 - v_2)\frac{R_2}{R_1} + A_{CM}\frac{v_1 + v_2}{2}\frac{R_2}{R_1} \qquad (5.38)$$

Solving for v_o now yields

$$v_o = \frac{(R_2/R_1)[A_{DM}(v_1 - v_2) + A_{CM}(v_1 + v_2)/2]}{A_{DM} + (R_1 + R_2)/R_1 - A_{CM}/2} \qquad (5.39)$$

Now dividing through numerator and denominator by A_{DM} gives

$$v_o = \frac{(R_2/R_1)[(v_1 - v_2) + (A_{CM}/A_{DM})(v_1 + v_2)/2]}{1 + [(R_1 + R_2)/A_{DM}R_1] - A_{CM}/2A_{DM}} \qquad (5.40)$$

Note that if A_{OL} (that is, A_{DM}) goes to infinity (the infinite-gain case) this equation for v_o reduces to simply $v_o = (R_2/R_1)(v_1 - v_2)$, as obtained previously. For the usual case of a very large open-loop gain such that $A_{OL} = A_{DM} \gg (R_1 + R_2)/R_1$ and $A_{DM} \gg A_{CM}$, the equation for v_o can be expressed approximately as

$$v_o = \frac{R_2}{R_1}\left[(v_1 - v_2) + \frac{(A_{CM}/A_{DM})(v_1 + v_2)}{2}\right] \qquad (5.41)$$

We see from this result that the common-mode component of the input voltage has some influence on the output, but the effect is reduced by the factor of A_{DM}/A_{CM} (the common-mode rejection ratio) with respect to the difference-mode component. For most cases, the CMRR is such that the effect of the common-mode signal component is very small indeed. For example, for a CMRR of 100 dB, the gain accorded the common-mode signal component is 100,000 times smaller than the gain received by the difference-mode signal. Finally, it should be noted that in many cases the major contribution to the common-mode gain of a difference amplifier will be mismatches in the R_1–R_2 resistor ratio.

5.10 INPUT IMPEDANCE

The ideal op amp has an infinite input impedance Z_i such that the op amp can be driven by a signal source of any impedance level without any signal loss due to a high value of the source impedance. Any real op amp has a finite input impedance Z_i, which in some cases can seriously affect the operation of the circuit.

For the analysis of the effect of Z_i, the equivalent-circuit representation of Figure 5.24a will be used. The input impedance Z_i can be represented as a combination of

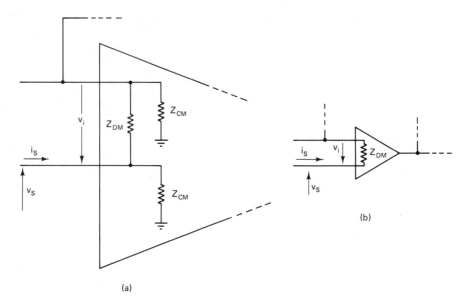

(a)

(b)

Figure 5.24 (a) Equivalent circuit representation of input impedance; (b) evaluation of the effect of Z_{DM}.

a *difference-mode input impedance* Z_{DM}, and a *common-mode input impedance* Z_{CM}, both of which are dynamic or ac impedances.

We will evaluate the effect of the difference-mode impedance Z_{DM} under closed-loop conditions with reference to the circuit of Figure 5.24b. The ac current i_s flowing into the noninverting input terminal is given by

$$i_s = \frac{v_i}{Z_{DM}} = \frac{v_o/A_{OL}}{Z_{DM}} = \frac{v_s A_{CL}/A_{OL}}{Z_{DM}}$$

Solving this for the *closed-loop input impedance* $Z_{i(CL)}$, we have

$$Z_{i(CL)} = \frac{v_s}{i_s} = \frac{A_{OL}}{A_{CL}} Z_{DM} \tag{5.42}$$

Thus the closed-loop input impedance resulting from Z_{DM} is increased by a factor of (A_{OL}/A_{CL}) over the open-loop value. This can represent a very substantial improvement. We note again in this case as well that the factor of improvement is exactly that factor by which the gain is sacrificed in going from open-loop conditions to closed-loop conditions.

The open-loop value of Z_{DM} for op amps is generally in the range from 100 kΩ up to as large as many gigaohms (10^9 Ω) for the case of FET-input op amps. These large Z_{DM} values when coupled with the very large A_{OL}/A_{CL} ratios available result in extremely large values for $Z_{i(CL)}$ such that the difference-mode input impedance very rarely produces any significant loading effect on the signal source and can therefore be neglected from further consideration.

We will now consider the effect of the common-mode input impedance Z_{CM}, and we will make use of Figure 5.24a. For large values of the open-loop gain, the voltage v_i across Z_{DM} is very small, such that the current through Z_{DM} is negligible. Therefore, the signal current i_s flows principally through Z_{CM} and is given by $i_s = v_s/Z_{CM}$. The closed-loop input impedance is therefore given by $Z_{i(CL)} = v_s/i_s = Z_{CM}$.

The value of Z_{CM} for op amps is generally in the range of several megohms up to 100 GΩ (10^{11} Ω) for FET op amps in parallel with a capacitance on the order of 3 to 10 pF.

The net closed-loop input impedance considering the effects of both Z_{DM} and Z_{CM} is the *parallel combination* of Z_{CM} and $(A_{OL}/A_{CL})Z_{DM}$. Although Z_{CM} is much larger than Z_{DM}, the A_{OL}/A_{CL} factor usually makes the $(A_{OL}/A_{CL})Z_{DM}$ impedance very much greater than Z_{CM}, so the net closed-loop input impedance will indeed be approximately equal to just Z_{CM} alone as expressed by $Z_{i(CL)} \simeq Z_{CM}$.

To consider a very simple example of the effect of the op-amp input impedance on the circuit performance, let us refer to the voltage-follower circuit of Figure 5.25. Let us assume that the op-amp open-loop gain is very large and that the common-mode input impedance Z_{CM} is 1 GΩ, in parallel with 10 pF. The source resistance is chosen to be 50 MΩ for this example, so at low frequencies the 50-MΩ source resistance and the 1000-MΩ closed-loop input impedance of the op amp constitute a simple resistive voltage divider with a voltage-division ratio of $1000/(1000 + 50) = 1000/1050 = 0.952$. We therefore have that the gain of this system at low frequencies is $v_o/v_s = 0.952$.

At higher frequencies, the op-amp input impedance will look primarily capacitative, so at 10 kHz, for example, we have that

$$Z_{i(CL)} = Z_i \simeq \frac{1}{j\omega C_i} = \frac{1}{j(2\pi \times 10 \text{ kHz}) \times 10 \text{ pF}} = -j1.59 \text{ M}\Omega \qquad (5.43)$$

The system gain at 10 kHz is therefore given by

$$\frac{v_o}{v_s} = \frac{-j1.59 \text{ M}\Omega}{50 \text{ M}\Omega - j1.59 \text{ M}\Omega} = -j\left(\frac{1.59}{50}\right) = -j0.0318 = 0.0318 \; \underline{/-90°} \qquad (5.44)$$

This represents a very severe signal attenuation resulting from the combination of the op-amp input capacitance C_i and the high value of source resistance R_S. Indeed, R_S and C_i can be considered to form a simple RC low-pass filter with a time constant of $\tau = (50 \parallel 1000) \text{ M}\Omega \times 10 \text{ pF} \simeq 500$ μs and a breakpoint frequency or 3-dB bandwidth of only $BW \simeq 1/(2\pi500 \text{ μs}) = 318$ Hz.

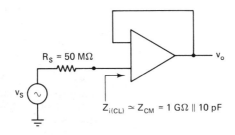

Figure 5.25 Input impedance example.

5.11 OUTPUT IMPEDANCE

The ideal op amp acts as a voltage source of strength $A_{OL}v_i$ with a zero output impedance. For this case, the output voltage v_o is not a function of the load impedance Z_L that is driven by the op amp. Any real op amp has a finite *output impedance* Z_O such that the output voltage and therefore the closed-loop gain is a function of the load impedance.

To analyze the effect of the finite output impedance Z_O on the circuit perform-ance of an op amp, the equivalent circuit representation of Figure 5.26 will be used. The op-amp output impedance Z_o and the load impedance Z_L constitute a voltage divider that results in a voltage-division ratio of $Z_L/(Z_o + Z_L)$, so as a result the output voltage is given by $v_o = A_{OL}v_i \times Z_L/(Z_L + Z_O) = A'_{OL}v_i$, where $A'_{OL} = A_{OL}Z_L/(Z_L + Z_O)$. The output voltage can now be expressed in terms of the signal voltage v_s by

$$v_o = \frac{1 + R_2/R_1}{1 + (1 + R_2/R_1)/A'_{OL}} \, v_s \qquad (5.45)$$

If we say that $1 + R_2/R_1 = A_{CL}$, then we have that

$$v_o = \frac{A_{CL}v_s}{1 + A_{CL}/A'_{OL}} = \frac{A_{CL}v_s}{1 + [A_{CL}(Z_L + Z_O)/A_{OL}Z_L]}$$

$$= \frac{A_{CL}Z_Lv_s}{Z_L + (A_{CL}/A_{OL})(Z_L + Z_O)} \qquad (5.46)$$

This can be rewritten as

$$v_o = A_{CL}v_s \frac{Z_L}{Z_L + (A_{CL}/A_{OL})Z_L + (A_{CL}/A_{OL})Z_O} \qquad (5.47)$$

Noting that $A_{CL}/A_{OL} \ll 1$ this can be written approximately as

$$v_o \simeq A_{CL}v_s \frac{Z_L}{Z_L + (A_{CL}/A_{OL})Z_O} = A_{CL}v_s \frac{Z_L}{Z_L + Z_{O(CL)}} \qquad (5.48)$$

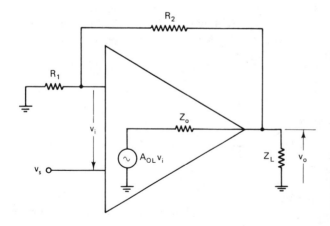

Figure 5.26 Equivalent circuit for the evaluation of the closed-loop output impedance.

where
$$Z_{O(CL)} = \frac{A_{CL}}{A_{OL}} Z_O \qquad (5.49)$$

is the *closed-loop output impedance*. From this last result we see that under closed-loop conditions the effective voltage-division ratio for v_o due to the voltage-divider action of Z_L and Z_O is $Z_L/(Z_L + Z_{O(CL)})$, so the effective output impedance is $Z_{O(CL)}$.

The open-loop output impedance of op amps is generally on the order of 10 to 100 Ω. Since A_{CL} is usually very much smaller than A_{OL}, we see that the closed-loop output impedance can be made to be very small indeed, with values down in the milliohm range being readily obtainable. We must again take note of the fact that the factor of improvement, A_{CL}/A_{OL}, in the output impedance is exactly the same factor by which the gain is sacrificed in going from open-loop to closed-loop conditions.

5.12 POWER DISSIPATION AND CURRENT LIMIT

For any electronic device there is a *maximum power dissipation*, $P_{d(MAX)}$ limitation. If the power dissipated in the device goes beyond the $P_{d(MAX)}$ value, the resulting temperature rise in the device may be such that the device temperature will exceed the maximum allowable temperature rating $T_{J(MAX)}$ and the device will be subject to permanent irreversible damage.

For op amps packaged in the plastic dual-in-line package (the DIP package), the $P_{d(MAX)}$ rating is about 500 to 750 mW. For other packages, the $P_{d(MAX)}$ rating can be as high as around 1 W. For some ICs with heat sinks attached to provide more efficient heat flow away from the device by either conduction or convection, maximum power dissipation ratings of up to 10 W are possible.

Most op amps have some internal circuitry to limit the output current to a safe maximum value such that the power dissipated in the device will not exceed the maximum power dissipation rating, even under a worst-case situation. Let us consider the op-amp circuit of Figure 5.27a. The worst-case situation from the standpoint of op-amp power dissipation occurs when the supply voltage is at its maximum rated value and the load resistance R_L is zero. In this case the full supply voltage of V^+ (or V^-) appears across the output stage of the op amp. The power dissipation of the op amp is $P_d = V^+ I_O + P_{d(Q)}$, where $P_{d(Q)}$ is the *quiescent power dissipation* of the op amp. This is the power dissipation of the op amp when it is not supplying any current to the load. Normally, this quiescent power dissipation $P_{d(Q)}$ is very small, often less than 1 mW, so it is not an important factor in the determination of the maximum allowable output current.

The maximum supply voltage rating of most op amps is around ± 18 V, so we will choose this value. If the $P_{d(MAX)}$ rating of the op amp is a representative value of 500 mW, the output current of the op amp should be limited to a value of

$$I_{o(MAX)} = I_{CL} = \frac{P_{d(MAX)}}{V^+} = \frac{500 \text{ mW}}{18 \text{ V}} = 28 \text{ mA} \qquad (5.50)$$

A reasonable choice for the current limit I_{CL} is therefore in the range from 20 to 25

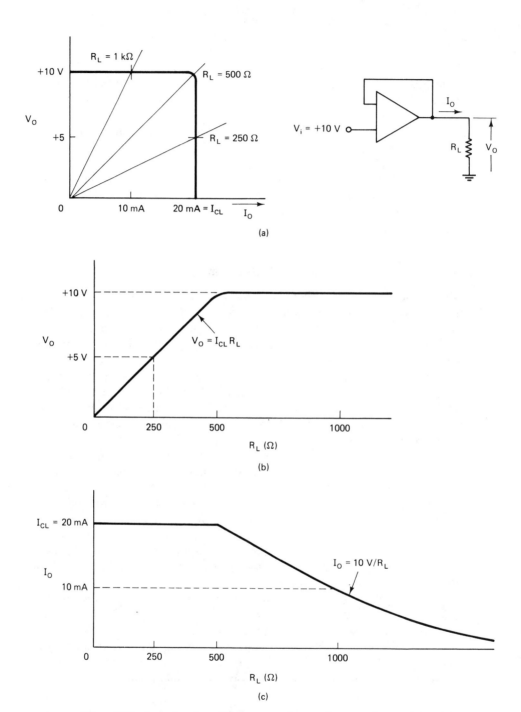

Figure 5.27 Operational-amplifier output voltage and current characteristics: (a) V_O versus I_O output characteristic; (b) V_O versus R_L; (c) I_O versus R_L.

Operational-amplifier Characteristics and Applications   Chap. 5

mA, which allows for a suitable safety factor, and indeed this is the typical range of current limit values for most op amps.

In Figure 5.27, the output voltage and current characteristics are present for an op amp with internal current-limiting circuitry. A current-limit value of 20 mA is chosen, and the output voltage is assumed to be $+10$ V when the op amp is not operating in the current-limited region. In Figure 5.27a, the V_o versus I_o output characteristic is shown. Note that for load resistance values greater than 500 Ω the output current is less than 20 mA, and the op amp will not be operating in the current-limited region. The output voltage is now essentially independent of R_L. For R_L less than 500 Ω, however, the op amp is driven into the current-limited mode of operation and the output voltage decreases linearly with R_L, whereas the output current remains relatively constant at 20 mA. This behavior is also evident from inspection of Figure 5.27b and c.

Most op amps are designed to be operated with split power supplies and will have internal current limiting in both directions so that the maximum current that the op amp can source is limited to I_{CL}^+, and the maximum current that the op amp is allowed to sink is I_{CL}^-. Usually, the two current are approximately equal and in the range from 20 to 25 mA.

5.13 EFFECTS OF FEEDBACK ON DISTORTION

The closed-loop gain of an amplifier is given by $A_{CL} = A_{OL}/(1 + FA_{OL})$ in terms of the open-loop gain A_{OL} and the feedback factor F. From this we see that for a large loop gain such that $FA_{OL} \gg 1$ the closed-loop gain is given approximately by $A_{CL} \simeq 1/F$. Since the feedback factor F is usually determined by a resistive voltage divider, it will be a constant and virtually independent of the input signal level, whereas the open-loop gain will reflect the nonlinearities of the various stages of the amplifier, especially that of the output stage, where the signal level is the largest. Therefore, since $A_{CL} \simeq 1/F$, we see that the effect of feedback is to "linearize" the amplifier and thus to cause the output to be a more exact replica of the input signal. In other words, the amplifier distortion is greatly reduced.

The fractional variation in the closed-loop gain due to a fractional change in the open-loop gain can be obtained from the foregoing expression for A_{CL} and expressed in the form

$$\frac{dA_{CL}}{A_{CL}} = \frac{dA_{OL}}{A_{OL}} \frac{1}{1 + FA_{OL}} = \frac{dA_{OL}}{A_{OL}} \frac{A_{CL}}{A_{OL}} \tag{5.51}$$

Thus any instantaneous variations in the open-loop gain resulting from the nonlinearity of the open-loop v_o versus v_i transfer characteristics result in a considerably smaller fractional change in the closed-loop gain. We see that, in effect, the nonlinearity or distortion of the amplifier has been reduced by the factor A_{CL}/A_{OL}. There is, however, a corresponding loss in the gain, but this can easily be made up by increasing the gain of the stages preceding this amplifier. Since this increased gain occurs where the signal level is considerably lower than in the output stage, the problem of distortion is considerably reduced.

For a representative example, let us consider an amplifier with $A_{OL} = 1000$ and $A_{CL} = 50$. The distortion of the output signal under open-loop conditions with an output voltage of 10 V peak-to-peak will be assumed to be 10%. Under closed-loop conditions, the distortion is reduced to only 10% × (50/1000) = 0.5%, with the same output voltage swing. The input signal level of $v_s = v_o/A_{CL} = 10$ V/50 = 0.2 V (peak-to-peak) is sufficiently small such that there is no significant distortion introduced by any preamplifier stages that may be required.

To look at this from a slightly different approach, let us write the expression for the v_o versus v_i transfer characteristics under open-loop conditions in the form of $v_o = a_1 v_i + a_2 v_i^2 + a_3 v_i^3 + \cdots$. The a_1 term represents the linear response of the amplifier, and all the other terms represent the amplifier nonlinearity. The open-loop voltage gain is therefore given by

$$A_{OL} = \frac{v_o}{v_i} = a_1 + a_2 v_i + a_3 v_i^2 + \cdots \qquad (5.52)$$

We see from this that the open-loop gain is a function of the signal level, so we indeed have a nonlinear response characteristic and the distortion that results therefrom. For the closed-loop gain, we have the expression

$$A_{CL} = \frac{A_{OL}}{1 + FA_{OL}} = \frac{1/F}{1 + (1/FA_{OL})} \simeq \frac{1}{F}\left(1 - \frac{1}{FA_{OL}}\right)$$

since FA_{OL} is large compared to unity. If we insert the foregoing expression for A_{OL} into this last equation, we obtain

$$\begin{aligned} A_{CL} &\simeq \frac{1}{F}\left(1 - \frac{1/F}{a_1 + a_2 v_i + a_3 v_i^2 + \cdots}\right) \\ &\simeq \frac{1}{F}\left\{1 - \frac{1/F}{a_1[1 + (a_2/a_1)\, v_i + (a_3/a_1)v_i^2 + \cdots}\right\} \end{aligned} \qquad (5.53)$$

Since the a_2/a_1, a_3/a_1, and so on, terms are small compared to unity, we can rewrite this last expression as

$$A_{CL} \simeq \frac{1}{F}\left[1 - \frac{1}{a_1 F}\left(1 - \frac{a_2}{a_1} v_i - \frac{a_3}{a_1} v_i^2 - \cdots\right)\right] \qquad (5.54)$$

Since $A_{OL} \simeq a_1$ and $A_{CL} \simeq 1/F$, we can write an equation for the output voltage v_o as

$$v_o \simeq A_{CL} v_i \left[1 - \frac{A_{CL}}{A_{OL}}\left(1 - \frac{a_2}{a_1} v_i - \frac{a_3}{a_1} v_i^2 - \cdots\right)\right] \qquad (5.55)$$

From this last expression we see that the nonlinearities of the v_o versus v_i transfer characteristic have been reduced by the A_{CL}/A_{OL} factor, and if $A_{CL} \ll A_{OL}$, a very linear response characteristic can indeed be obtained.

5.14 POWER SUPPLY REJECTION RATIO

For the ideal op amp, the output voltage v_o is independent of the dc supply voltage. For any real op amp, however, the supply voltage influences the output voltage due to both the variation of the open-loop gain A_{OL} with the supply voltage and a feed-through of supply voltage fluctuations into the signal path of the circuit. The open-loop gain of an op amp increases slowly with supply voltage, a typical variation being an increase in A_{OL} by a 2 : 1 or 3 : 1 ratio for a supply voltage variation from a minimum value of ± 5 V up to a maximum rated value of ± 18 V. Although this is a significant variation, it must be remembered that the closed-loop gain is relatively independent of the open-loop gain, so the resulting effect on the closed-loop gain, and therefore on the output voltage, is usually very small.

A principal cause of supply voltage fluctuations is the ac ripple voltage from a dc power supply that uses a rectifier to convert an ac line voltage to dc followed by a filter circuit. The most important ripple component is usually at 120 Hz. This ac ripple voltage can produce a small contribution to the output voltage. This effect of the power supply ripple voltage or other voltage fluctuations can be referred back to the input of the op amp and can be described in terms of an effective change in the offset voltage.

The *power supply rejection ratio* (PSRR) is defined as the ratio of change in the input offset voltage resulting from a change in the supply voltage to the change in the supply voltage and is usually expressed in decibels (dB). The PSRR is usually very small, typically in the range of -80 dB (10^{-4}) to -100 dB (10^{-5}). For example, with a PSRR of -100 dB, a 1-V ac ripple on the power supply voltage is equivalent to an input voltage of only 0.01 mV.

5.15 NOISE

In any communications system, unwanted and extraneous signals that interfere with the desired signal will set a lower limit to the strength of the desired signal that can reliably and accurately be received and understood. The interfering signals that are of a *random* nature and are not emitted by any other communications system are called *noise*. It is the random, unpredictable nature of noise that makes it so difficult to deal with and eliminate. Interfering signals that have a reasonably fixed or pre-dictable nature, such as having a constant frequency or band of frequencies, can be reduced by special signal-processing techniques, such as the use of band-pass and band-stop filters. Although there are techniques for noise reduction, it can never be entirely eliminated. Furthermore, the signal processing that is required for noise reduction generally involves a trade-off of some other aspect of overall system per-formance. A good example of this trade-off is the reduction in the system bandwidth in order to reduce the overall noise level. The reduced system bandwidth will cor-respondingly reduce the information transmission rate of which the system is capable.

In an electronic communications system, noise is generated internally both within the transmitter and, much more importantly, in the receiver as a result of a number

of random processes. These processes include the random thermal motion of electrons and holes and the random nature of the electron–hole generation and recombination processes in semiconductor devices. In addition, noise generated by external
sources may enter the receiver together with the desired signal. Some examples of
external noise sources are lightning discharges, electrical machinery, and the ignition
systems of gasoline engines.

The most critical component of an electronic communications system from the
standpoint of the ability of the system to detect weak signals is the first or input stage
of the receiver. It is here that the signal is the weakest and therefore the most
susceptible to interference by noise, either internally generated in the first stage or
coming into the first stage together with the signal.

The ratio of the signal voltage (rms) to the noise voltage (rms) is called the
signal-to-noise ratio (SNR). At the receiver input, there is a certain SNR depending
on the signal strength and the external noise level. The internally generated noise
in the first stage adds to the noise already present, so there is generally a significant
degradation of the SNR in the first stage. It is of course true that the signal is
amplified by the first stage, but the noise is also amplified by the same factor; so,
due to the additional noise that is generated in the first stage, there will actually be
some decrease or loss in the SNR.

Following the first stage, there will be further amplification of both the signal
and the noise by the other stages of the receiver. In each stage, additional noise is
added to the signal due to the internally generated noise of that stage. As a result,
the SNR continues to decrease as the signal proceeds through the various stages of
the amplifier. Nevertheless, it is the first stage that is generally the most important
by far in determining the SNR of the receiver. If the gain of the first stage is denoted
by A_1, that of the second stage by A_2, and so on, the overall amplifier gain will be
$A_T = A_1 A_2 A_3 \cdots$. If the incoming signal level is S_i and the noise level is N_i, the
SNR at the receiver input is $(SNR)_i = S_i/N_i$. The noise added by the first stage is
N_1, so the SNR following the first stage is $A_1 S_i / A_1 (N_i + N_1) = S_i/(N_i + N_1)$.

At the output of the second stage, the SNR is given by

$$(SNR)_2 = \frac{A_1 A_2 S_i}{A_1 A_2 (N_i + N_1) + A_2 N_2} = \frac{S_i}{(N_i + N_1) + (N_2/A_1)} \tag{5.56}$$

where N_2 is the noise added by the second stage. Similarly, at the output of the
third stage we have

$$(SNR)_3 = \frac{A_1 A_2 A_3 S_i}{A_1 A_2 A_3 (N_i + N_1) + A_2 A_3 N_2 + A_3 N_3}$$

$$= \frac{S_i}{(N_i + N_1) + (N_2/A_1) + (N_3/A_1 A_2)} \tag{5.57}$$

We see from this that for reasonably large gains per stage the SNR of the first stage
will largely determine the overall SNR of the receiver.

The *sensitivity* of a system is a measure of the ability of the system to detect or
recover weak signals. It is usually described in terms of the lowest input signal level

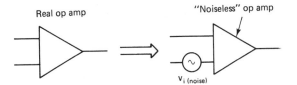

Figure 5.28 Equivalent input noise voltage.

that will result in an acceptable SNR. In many cases the minimum SNR needed to detect the presence of a signal is equal to unity, for which case the sensitivity is the signal level that is equal to the noise present at the first amplifier stage.

In the ideal situation, the amplified signal that is produced by the receiver is an exact replica of the original signal produced by the transmitter. This is never the case in practice, due to the various nonlinearities in the system that result in distortion and the presence of noise.

In general, to produce an acceptable signal level, an SNR of at least around 10 (20 dB) is necessary, and for some applications SNR values as high as 30 or 40 (30 to 36 dB) are required. For example, to produce a good-quality television picture free of any apparent noise, an SNR of about 30 (36 dB) is required. If the SNR drops below 10 (20 dB), the amount of noise (or "snow") in the picture becomes quite noticeable. For SNR values less than about 3 (10 dB), the noise will be very objectionable, and if the SNR drops below about 1 or 2 (0 to 6 dB), it will be very difficult to discern the picture at all.

There are two basic sources of noise in an electronic amplifier system, *thermal noise* and *shot noise*. Thermal noise (or Johnson noise) is due to the random thermal motion of charge carriers (electrons and holes) in the various circuit components of the amplifier, principally the resistors and transistors. Shot noise is the result of random fluctuations in the flow of current through electronic devices. Both thermal noise and shot noise are random functions of time with a zero average value, and both have a Gaussian probability distribution function.

In any electronic device or component, there are minute random voltage and current fluctuations known as noise. An op amp is no exception to this. Electrical noise generated within the op amp, especially that produced by the transistors of the first or input stage of the op amp, will appear together with the desired signal as part of the output voltage. The noise generated within an op amp can most easily be described and its effect evaluated by replacing the real op amp by an ideal "noiseless" op amp with an *equivalent input noise voltage* source, $v_{i(noise)}$, in series with one of the input terminals as shown in Figure 5.28. It is usually most convenient to place the equivalent noise source in series with the noninverting input terminal.

The noise voltage is a completely random function of time with a zero average or dc value. The noise voltage can, however, be specified in terms of its rms (root mean square) value over a given frequency range or bandwidth. The noise voltage will have a flat spectral distribution, except at low frequencies ($\lesssim 100$ Hz), as shown in Figure 5.29. The rms voltage over a given bandwidth range is proportional to the *square root* of the bandwidth.

The equivalent input noise voltage is usually expressed in terms of its *spectral density*, which is the rms noise voltage per unit bandwidth and is often given in such

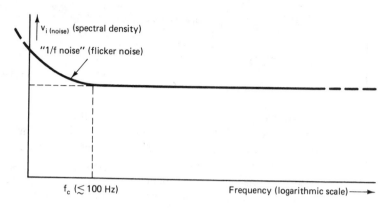

Figure 5.29 Noise voltage spectral distribution.

units as $nV/\sqrt{Hz}$. A typical value for the equivalent input noise voltage of an op amp is $20 \ nV/\sqrt{Hz}$. The effective bandwidth to be used for noise calculations (the noise bandwidth) is related to the 3-dB bandwidth of the system under consideration by $(BW)_{noise} = (\pi/2)(BW)_{3dB}$ for systems characterized by a single breakpoint frequency (a 20-dB/decade roll-off). For example, an op amp with an equivalent input noise voltage (spectral density) of $20 \ nV/\sqrt{Hz}$ and a 10-kHz 3-dB bandwidth has a total input noise voltage of $20 \ nV/\sqrt{Hz} \times \sqrt{(\pi/2) \times 10 \ kHz} = 2.5 \ \mu V$ rms. Therefore, the input signal level that is needed to produce a *signal-to-noise ratio* of 10 : 1 is 25 μV. A signal level of 2.5 μV results in an SNR of unity, which results in a barely detectable, extremely noisy signal. If the SNR drops below unity, the signal becomes buried in the noise and in most cases is not recoverable. The input noise voltage therefore determines the *sensitivity* of the system for detecting low-level signals.

One other op-amp noise source of interest is due to the input bias current I_B. The input bias current is essentially a dc current, but it will exhibit minute random fluctuations due to the particulate nature of electrical current flow. These minute fluctuations constitute a noise current called the *shot noise*, i_{sn}, which like the noise voltage has a zero average or dc value, but can be described in terms of its rms value. It will have a reasonably flat spectral distribution, except at low frequencies ($\lesssim 100$ Hz), and the rms value will be proportional to the square root of the bandwidth. The shot noise current is usually described in terms of its spectral density, which is the shot noise current per unit bandwidth, often specified in units of $pA/\sqrt{Hz}$ or sometimes $fA/\sqrt{Hz}$ (1 fA $= 10^{-15}$ A). The shot noise current (spectral density) is related to the bias current by the simple relationship

$$i_{sn} \text{ (rms, spectral density)} = \sqrt{2eI_B}$$

where e is the electronic charge (1.6×10^{-19} C). For example, if $I_B = 100$ nA, the shot noise current (rms, spectral density) is $i_{sn} = 0.18 \ pA/\sqrt{Hz}$.

One other noise source of importance in op-amp systems is the noise voltage developed in resistors, known as the *thermal noise voltage* because it is proportional

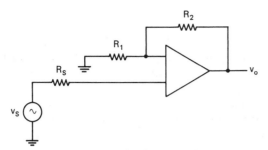

Figure 5.30 Circuit for noise calculation example.

to the square root of the temperature. This is a random voltage of zero average or dc value and can be described by its rms value over a specified bandwidth. The rms thermal noise voltage is proportional to the square root of the bandwidth and is usually specified in terms of the spectral density, which is the thermal noise voltage per unit bandwidth. The rms thermal noise voltage spectral density is given by $v_{th} = \sqrt{4kTR}$, where k is Boltzmann's constant (1.38×10^{-23}/K), T is the absolute temperature (K), and R is the resistance value. At a temperature of 293 K (20°C), this equation becomes $v_{th} = \sqrt{1.6 \times 10^{-20} R}$. For $R = 1$ MΩ, this gives $v_{th} = 127$ nV/$\sqrt{\text{Hz}}$. The various (uncorrelated) noise sources in a system can be combined by taking the square root of the sum of the squares of the various noise voltages, wherever they appear in series.

To consider some simple examples of noise calculations in an op-amp system, we will use the circuit of Figure 5.30. We will consider an op amp with $v_{i(\text{noise})} = 20$ nV/$\sqrt{\text{Hz}}$ and $I_B = 100$ nA such that $i_{sn} = 0.18$ pA/$\sqrt{\text{Hz}}$. We will assume a closed-loop bandwidth of 10 kHz and also, for simplicity, that R_1 and R_2 are sufficiently small ($\leq$10 kΩ) such that they will not contribute significantly to the overall noise.

When R_S is small ($\leq$10 kΩ), the dominant noise source is the equivalent input noise voltage of the op amp of 20 nV/$\sqrt{\text{Hz}}$ that results in a net rms noise voltage of 2.5 μV for the 10-kHz 3-dB bandwidth.

If $R_S = 250$ kΩ, it produces a thermal noise voltage of 63 nV/$\sqrt{\text{Hz}}$, so it now becomes the dominant noise source. The noise voltage resulting from the flow of the shot noise current through R_S is 0.18 pA/$\sqrt{\text{Hz}}$ × 250 kΩ = 45 nV/$\sqrt{\text{Hz}}$. The net noise voltage, considering the effects of all three noise sources, is now 80 nV/$\sqrt{\text{Hz}}$, or 10 μV for the 10-kHz bandwidth.

When $R_S = 1$ MΩ, the thermal noise voltage is now up to $v_{th} = 127$ nV/$\sqrt{\text{Hz}}$, and the flow of the shot noise current through R_S produces a voltage drop if $i_{sn}R_S = 0.18$ pA/$\sqrt{\text{Hz}}$ × 1 mΩ = 180 nV/$\sqrt{\text{Hz}}$; so the shot noise current is now the most important noise source. The net noise voltage due to all three sources is 221 nV/$\sqrt{\text{Hz}}$, or 28 μV for the 10-kHz bandwidth.

We see from this example that when the resistance value is very large a low value for the shot noise current can be a very desirable feature. This, in turn, means that the input bias current should be very small. This requirement can be satisfied most readily by using a FET-input op amp for which I_B values down as low as 10 pA can be obtained.

5.16 DEFINITION OF TERMS

Now that some of the basic characteristics of op amps have been discussed, a list of definitions of various terms relating to op amps is presented. Some of these characteristics have already been discussed, and some will be considered in later sections.

1. *Amplifier-to-amplifier coupling*: In some cases there is more than one op amp on the same IC chip, such as the case of the 747 device, which consists of two complete 741-type op amps on the same IC chip. Another example is a quad op amp such as the 124 type, in which there are four op amps on one chip.

   The amplifier-to-amplifier coupling is a measure of the degree of interaction between the op amps on a chip. It is usually expressed as the ratio of the change in the input offset voltage of one op amp as a result of the change in the output voltage of another op amp on the same chip. This ratio is usually expressed in decibels (dB), with the frequency or frequency range at which the measurement is made being specified.

2. *Bandwidth, 3-dB*: This is the frequency at which the voltage gain of an op amp has dropped by 3 dB from its zero-frequency value. Numerically, this corresponds to the voltage gain being 0.7071 ($1/\sqrt{2}$) times the zero-frequency value. The *open-loop* bandwidth is the frequency at which the open-loop gain of the op amp has decreased by 3 dB from the zero-frequency value of the open-loop gain, $A_{OL}(0)$. For internally compensated op amps, this bandwidth is typically just on the order of 10 Hz. The *closed-loop* bandwidth, BW_{CL}, is the frequency at which the closed-loop gain has dropped 3 dB from the zero-frequency value, $A_{CL}(0)$. The closed-loop bandwidth is generally much higher than the open-loop bandwidth, BW_{OL}.

3. *Common-mode input impedance, $Z_{i(CM)}$*: This is the ratio of the change in the common-mode input voltage to the change in the current at either input terminal.

4. *Common-mode rejection ratio (CMRR)*: This is the ratio of the difference-mode voltage gain to the common-mode voltage gain, so CMRR $= A_{DM}/A_{CM}$. This ratio is usually expressed in decibels, so CMRR$_{dB} = 20 \log_{10}(A_{DM}/A_C)$.

5. *Common-mode voltage gain, A_{CM}*: This is the ratio of the change in the output voltage to the common-mode input voltage of the op amp. The common-mode input voltage is the average of the two input voltages of the op amp.

$$V_{DM} = V_A - V_B$$
$$V_{CM} = (V_A + V_B)/2$$
$$A_{DM} = A_{OL} = \Delta V_O/\Delta V_{DM}$$
$$A_{CM} = \Delta V_O/\Delta V_{CM}$$

6. *Difference-mode voltage gain, A_{DM}*: This is the ratio of the change in the output voltage to the change in the voltage between the two input terminals of the op amp. The difference-mode voltage gain is essentially identical to the op-amp open-loop voltage gain, A_{OL}.

7. *Equivalent input noise current*: The input current (the bias current) of an op

amp is a dc current, but will not be absolutely constant. There will be very minute fluctuations of a random nature in the input currents. These minute random fluctuations can be considered to be a noise current that is superimposed on the dc value of the input current and can be represented by a current source external to the op amp. The input noise current is commonly expressed in terms of a spectral density, typically in units of $pA/\sqrt{Hz}$.

8. *Equivalent input noise voltage*: The various circuit components within an op amp generate electrical noise. This noise is amplified together with the desired signal in the op amp and appears at the output. The effect of the internally generated noise in an op amp can be expressed in terms of an equivalent noise voltage applied across the input terminals of the op amp. The *equivalent input noise voltage*, $v_{i(noise)}$, is a random voltage of zero dc value, but it will have some rms value. The noise voltage is generally proportional to the square root of the amplifier (closed-loop) bandwidth, so the frequency range involved must be included as part of the specification of $v_{i(noise)}$. Very often the spectral density of the noise voltage is specified. This is the noise voltage expressed on a per unit bandwidth (1 Hz) basis, with the usual units being expressed as $nV/\sqrt{Hz}$.

9. *Full-power bandwidth (FPBW)*: The full-power bandwidth of an op amp is the frequency at which the closed-loop voltage gain has decreased by 3 dB below the zero-frequency value of the closed-loop gain under large-signal conditions such that the amplifier response is slew rate limited.

10. *Harmonic distortion*: As a result of nonlinearities in the op amp, the output voltage waveform will not be an exact replica of the input voltage. A quantitative index of the amount of distortion produced by an op amp under specified conditions can be obtained under conditions of a sinusoidal input signal by obtaining the ratios of the amplitudes of the various harmonic terms in the output voltage to the fundamental component of the output voltage. Such ratios are usually expressed on a percentage basis, such as percent second harmonic distortion, and so on. An index of the total amount of distortion can be obtained from the following equation:

$$\% \text{ THD} = \frac{\sqrt{(V_2^2 + V_3^2 + V_4^2 + \cdots)} \times 100\%}{V_1}$$

where % THD is the percent *total harmonic distortion*, and V_1, V_2, V_3, and so on, are the amplitudes of the fundamental (V_1), second harmonic (V_2), third harmonic (V_3), and so on, of the output voltage of the op amp.

11. *Input bias current, I_B or I_{BIAS}*: For an op amp to operate properly, there must be some quiescent (dc) current flowing into (or out of) the two input terminals. The average of the two input currents is called the input bias current, I_B or I_{BIAS}. For bipolar input op amps, I_B is generally in the range of 1 µA to 1 nA. For JFET op amps, it is on the order of 1 to 10 pA, and for MOSFET input op amps it can be even smaller.

12. *Input bias current temperature coefficient, TCI_B*: This is the rate of change of the input bias current with temperature as given by $TCI_B = dI_B/dT$.

13. *Input impedance, Z_i*: The *difference-mode input impedance* is the ratio of the change in the input voltage to the change of the input current on either input with the other input terminal grounded (ac). This is the ac (dynamic) input impedance looking into either input terminal with the other input terminal grounded.

14. *Input offset current, I_{OS} or I_{OFFSET}*: This is a dc current that is the difference between the two input currents. The value of I_{OS} is generally about 10% to 20% of I_B.

15. *Input offset voltage, V_{OS} or V_{OFFSET}*: This is the voltage that must be applied between the two input terminals of the op amp to obtain zero output voltage. A primary source of V_{OS} is the mismatch between the transistors making up the input stage difference amplifier. For op amps with a bipolar transistor (NPN or PNP) input stage, V_{OS} is generally on the order of 1 mV, although for some op amps it can be down as low as about 10 to 100 μV. For FET input stage op amps, the V_{OS} values are generally somewhat larger, typically around 5 mV.

16. *Input offset voltage temperature coefficient, $TC_{V_{OS}}$*: This is the rate of change of V_{OS} with temperature. For bipolar transistor input stage op amps, $TC_{V_{OS}}$ is given approximately by $TCV_{OS} \simeq V_{OS}/T$, so if $V_{OS} = 1.0$ mV $= 1000$ μV, $TC_{V_{OS}} \simeq 3$ μV/°C.

17. *Input voltage range*: This is the range of voltages applied to the input terminals of the op amp for which the device will operate within specifications. This is sometimes expressed in terms of the *common-mode input voltage range* and the *difference-mode voltage range*. The common-mode voltage range is the range of input voltages that can be applied to either or both input terminals and still have the op amp operate properly. The difference-mode voltage range is the maximum voltage that can be applied between the two input terminals and still allow for proper operation of the op amp.

18. *Large-signal voltage gain*: This is the ratio of the output voltage swing to the change in the voltage applied between the two input terminals required to change the output voltage from zero to some specified value. This is a dimensionless quantity, is sometimes expressed in "units" of V/mV, and is essentially the open-loop gain of the op amp, $A_{OL}(0)$.

19. *Long-term stability*: This is the drift of the input offset voltage as a function of time and is typically expressed in units of μV/month.

20. *Operating temperature range*: This is the range of temperatures over which the op amp will operate within the stated specifications. Many manufacturers specify and test op amps for the following three temperature ranges:

Commercial:	0° to +70°C
Industrial:	−25° to +85°C
Military:	−55° to +125°C

21. *Output impedance, Z_O*: This is the ratio of the change in the output voltage to the change in the output current.

22. *Output short-circuit current*: This is the maximum current that can be supplied by an op amp to a load. For most op amps, this maximum current is set by an internal current-limiting circuit so as to protect the device against excessive power dissipation. Generally, there are two somewhat different values of maximum output current for an op amp, one value corresponding to the device supplying current (sourcing current) to a load, and the other corresponding to current coming into the op amp (sinking current) from the load.

23. *Output voltage swing*: This is the peak output voltage swing, referred to zero, that can be obtained without clipping. It is generally limited by the supply voltage such that the output voltage cannot get to less than about 1 or 2 V of either the positive or the negative supply voltage.

24. *Power supply rejection ratio (PSRR)*: This is the ratio of the change in the input offset voltage to the change in the power supply voltage producing this change. The PSRR is usually expressed in decibels (dB).

25. *Settling time*: This is the time interval between the application of a step-function input voltage and the time when the output voltage has settled to a value that is within a specified range (error band) of the final output voltage.

26. *Slew rate*: This is the maximum rate of change of the output voltage with time when a large-amplitude step-function voltage is applied to the input terminals. The slew rate of an op amp is approximately inversely proportional to the size of the compensation capacitor, C_{COMP}, that is used for feedback stability. The slew rate is usually expressed in units of volts per microsecond (V/µs).

27. *Supply current*: This is the quiescent current required to operate the op amp supplied by the power supply. This quiescent current is measured under conditions of no load (no output current) and the output voltage at (or near) zero.

28. *Thermal feedback coefficient*: The temperature of an IC chip is a function of the IC power dissipation, and in turn the input offset voltage is a function of the chip temperature. The ratio of the change in the input offset voltage to the change in the power dissipation is called the thermal feedback coefficient. Typical units for this quantity are nV/mW or µV/mW.

29. *Transient response*: This term refers to the closed-loop step-function response of the amplifier under specified small-signal conditions. The transient response is usually expressed in terms of the 10% to 90% rise time.

30. *Unity-gain frequency, f_u*: This is the frequency at which the open-loop voltage gain, A_{OL}, of the op amp has decreased to unity. The value of f_u is typically in the range of about 1 to 10 MHz.

Some of the preceding definitions dealing with large-signal frequency-domain and time-domain response characteristics, such as the slewing rate and the full-power bandwidth (FPBW), as well as some other definitions, will be considered in more detail later when the internal circuitry of op amps is investigated.

5.17 COMPARISON OF IDEAL AND ACTUAL OPERATIONAL-AMPLIFIER CHARACTERISTICS

The basic operation of operational amplifiers has now been discussed and some of the nonideal characteristics of real devices have been investigated. The difference between the behavior of ideal and real operational amplifiers will now be summarized.

The ideal operational amplifier can be represented as a voltage-controlled voltage source as shown by the equivalent circuit model of Figure 5.31a. The input–output transfer relationship is simply $v_o = A_{OL}v_i$, where $v_i = v_1 - v_2$ is the difference-mode input voltage. The open-loop gain A_{OL} is a positive constant. The input impedance is infinite and the output impedance is zero. The operational amplifier therefore does not have any loading effect on a signal source, so $v_i = v_s$, and the load impedance Z_L does not affect the output voltage so that $v_o = A_{OL}v_i$. Thus the gain is $v_o/v_s = A_{OL}$ and is completely independent of the source and load impedances. The open-loop gain A_{OL} is considered to be a constant, independent of frequency, temperature, supply voltage, and so on, and there is no offset voltage, bias current, or internally generated noise voltages or currents; so the output voltage will be an exact, but amplified, replica of the signal voltage.

The nonideal or real operational amplifier can be represented as in Figure 5.31b. In this case, the voltage gain from v_s to v_o is affected by the source and load imped-

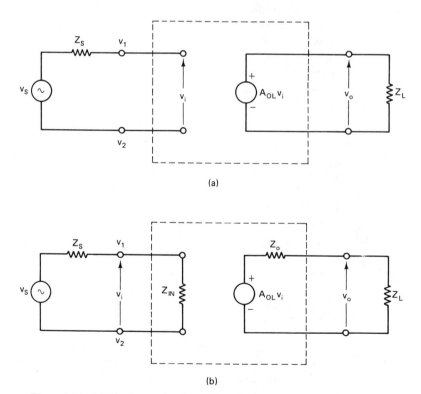

(a)

(b)

Figure 5.31 (a) Ideal operational amplifier; (b) nonideal operational amplifier.

ances. Furthermore, the open-loop gain is not a constant, but will vary with frequency, temperature, supply voltage, and so on. As a result of this, the output voltage is not an exact replica of the signal voltage. Some nonideal characteristics of a real operational amplifier are as follows:

1. Input offset voltage (V_{OS})
2. Input bias current (I_B) and offset current (I_{OS})
3. Input impedance (difference mode and common mode)
4. Limited input voltage range
5. Output impedance
6. Limited output voltage swing
7. Limited bandwidth and a finite response time
8. Limited slewing rate (SR) and full-power bandwidth (FPBW)
9. Finite common-mode rejection ratio (CMRR)
10. Finite power supply rejection ratio (PSRR)
11. Internally generated noise voltages and currents
12. Temperature dependence of parameters, including A_{OL}, V_{OS}, I_B, and others

5.18 OPERATIONAL-AMPLIFIER STABILITY

Let us consider the closed-loop op-amp system shown in Figure 5.32 together with the following definitions:

$$A_{OL} = \text{open-loop gain}$$

$$F = \text{feedback factor} = \text{fraction of the output voltage that is fed back to the input}$$

$$FA_{OL} = \text{loop gain} = \text{signal gain around the feedback loop}$$

$$A_{CL} = \text{closed-loop gain}$$

For the system of Figure 5.32 the relationship between the output voltage v_o and the input or signal voltage v_s can be written as $v_o = A_{OL}v_s - FA_{OL}v_o$, so $v_o(1 + FA_{OL}) = A_{OL}v_s$, and thus the closed-loop gain is given by

$$A_{CL} = \frac{v_o}{v_s} = \frac{A_{OL}}{1 + FA_{OL}} = \frac{\text{open-loop gain}}{1 + \text{loop gain}} \qquad (5.58)$$

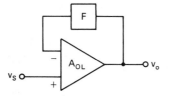

Figure 5.32 Operational-amplifier closed-loop configuration.

If we divide both the numerator and the denominator by the loop gain FA_{OL}, we obtain

$$A_{CL} = \frac{1/F}{1 + (1/FA_{OL})} \tag{5.59}$$

If the loop gain is large compared to unity ($FA_{OL} \gg 1$), the closed-loop gain is approximately given by $A_{CL} \simeq 1/F$. We see that the closed-loop gain becomes approximately equal to the reciprocal of the feedback factor F and thus is dependent essentially only on the parameters of the feedback network and is relatively independent of the amplifier open-loop gain.

At the other extreme, we note that as the loop gain becomes small compared to unity ($FA_{OL} \ll 1$) the closed-loop gain approaches the open-loop gain. We also note that as long as the loop gain is algebraically positive the closed-loop gain is less than the open-loop gain. For large values of loop gain, the closed-loop gain is much smaller than the open-loop gain.

The equation for the closed-loop gain for an op amp with negative feedback is given by $A_{CL} = A_{OL}/(1 + FA_{OL})$, and we see that the closed-loop gain is less than the open-loop gain. The op amp is a multistage electronic amplifier, and as the frequency increases the input-to-output phase shift ϕ increases monotonically with frequency. At some frequency, to be called $f_{180°}$, the phase shift will have reached 180°. At this frequency of $f_{180°}$ what was the inverting input at low frequencies becomes, in effect, the noninverting input terminal, and vice versa for the other input terminal. Thus what was *negative feedback* at low frequencies becomes actually *positive feedback* at $f = f_{180°}$. At $f = f_{180°}$ the loop gain becomes algebraically negative, and the closed-loop gain expression can be written as

$$A_{CL} = \frac{A_{OL}}{1 - |FA_{OL}|}$$

As long as the magnitude of the loop gain is less than unity ($|FA_{OL}| < 1$), the system is *stable* with a finite closed-loop gain. However, if the magnitude of the loop gain at $f = f_{180°}$ is equal to or greater than unity ($FA_{OL} \geq 1$ at $f = f_{180°}$), the closed-loop gain goes to infinity and the system is said to be *unstable* (or *oscillatory*).

If the system is unstable with an infinite closed-loop gain, the system can produce an output voltage with zero signal input. The voltage fed back from the output to the input via the feedback network F is sufficient to drive the system, so the system is supplying its own input voltage by means of the positive feedback loop that exists at $f = f_{180°}$. The output voltage waveform is generally a sine wave or a square wave with a frequency of $f = f_{180°}$. The op amp is now operating as an *oscillator* rather than an *amplifier* and is relatively unresponsive to any signal input.

The low-frequency value of the closed-loop gain is related to the feedback factor F by $A_{CL}(0) \simeq 1/F$. Therefore, the stability condition that $|FA_{OL}| < 1$ at $f = f_{180°}$ can be expressed as $|A_{OL}|/A_{CL}(0) < 1$ at $f = f_{180°}$, or as follows:

$$\boxed{\text{For stability:} \quad A_{CL}(0) > |A_{OL}| \qquad \text{at } f = f_{180°}} \tag{5.60}$$

Since the open-loop gain decreases *monotonically* with increasing frequency, if the open-loop gain becomes equal to $A_{CL}(0)$ at a frequency that is less than $f_{180°}$, then at $f = f_{180°}$ the open-loop gain will surely be less than $A_{CL}(0)$. Therefore, the stability condition can be reexpressed as

$$\text{For stability:} \quad |A_{OL}| = A_{CL}(0) \quad \text{at } f < f_{180°} \qquad (5.61)$$

That is, *the frequency at which A_{OL} becomes equal to $A_{CL}(0)$ should be less than $f_{180°}$*.

The foregoing stability conditions hold true only for the situations in which there is no significant phase shift contributed by the feedback network F. For the usual case of a feedback network containing only resistive elements, this requirement will be satisfied.

We will now consider another derivation for the closed-loop gain equation that will give us further insight into the question of stability. Let us consider again the system of Figure 5.32. An input signal v_s produces an output voltage $v_o = A_{OL}v_s$. A fraction F of this output voltage is fed back to the input, and after passing through the op amp with gain $-A_{OL}$ it appears at the output as $(A_{OL}v_s)(-FA_{OL})$. A fraction F of this output voltage component is then fed back to the input and, after passing through the op amp with gain $-A_{OL}$, appears at the output as $(A_{OL}v_s)(-FA_{OL})$ $(-FA_{OL}) = (A_{OL}v_s)(-FA_{OL})^2$. This process keeps repeating indefinitely and a *net* output voltage is produced as given by

$$v_o = (A_{OL}v_s) + (A_{OL}v_s)(-FA_{OL}) + (A_{OL}v_s)(-FA_{OL})^2 + \cdots$$
$$= (A_{OL}v_s)[1 + (-FA_{OL}) + (-FA_{OL})^2 + (-FA_{OL})^3 + \cdots] \qquad (5.62)$$
$$= (A_{OL}v_s)(1 + x + x^2 + x^3 + \cdots)$$

where $x = -FA_{OL}$.

As long as $x < 1$, this infinite power series is convergent and can be written in closed form as

$$v_o = (A_{OL}v_s) \frac{1}{1 - x} = \frac{A_{OL}}{1 + FA_{OL}} v_s$$

so

$$A_{CL} = \frac{A_{OL}}{1 + FA_{OL}} \qquad (5.63)$$

and A_{CL} is finite; thus the system is stable. This last expression for A_{CL} can be recognized as being identical to the one obtained previously. If $x \geq 1\underline{/0°}$, the series is not convergent, so the closed loop will go to infinity. Thus, for stability, we must have that $x = -FA_{OL} < 1\underline{/0°}$. Since F is a positive real quantity and A_{OL} becomes negative ($\underline{/A_{OL}} = 180°$) at $f = f_{180°}$, this stability condition can be written as

$$\text{For stability:} \quad F|A_{OL}| < 1 \quad \text{at } f = f_{180°} \qquad (5.64)$$

Since $A_{CL}(0) = 1/F$, it can also be expressed as $A_{CL}(0) > |A_{OL}|$ at $f = f_{180°}$. This is the same stability condition as that obtained in the previous analysis.

5.18.1 Determination of $f_{180°}$

From the preceding discussions, it is apparent that the value of $f_{180°}$ and the open-loop gain at $f = f_{180°}$ are key factors in the stability of op-amp and other feedback systems. We will now develop an approximate analytical expression for $f_{180°}$ and find the corresponding value of the open-loop gain. We will consider the open-loop frequency response of the op amp to be of the form $A_{OL}(f) = A_{OL}(0)/(1 + jf/f_1)(1 + jf/f_2)(1 + jf/f_3) \ldots$, where the breakpoints are in the sequence $f_1 < f_2 < f_3 < \cdots$, and f_1 is considerably smaller than all the rest, so $f_1^2 \ll f_2^2$.

At $f = f_1$, the phase angle of A_{OL} is approximately $-45°$. For frequencies in the range given by $f_1^2 < f^2 < f_2^2$, the phase angle is approximately $-90°$. The phase angle is in the neighborhood of $-135°$ at $f = f_2$, although it may be somewhat larger if f_3 is close to f_2. For the remainder of this discussion, let us assume that f_4 is well beyond f_3. If this is the case, the phase angle of A_{OL} will reach $-180°$ somewhere between f_2 and f_3. We will now determine the frequency at which the phase angle will be $180°$ ($f_{180°}$).

The phase angle of A_{OL} is given by

$$- \angle A_{OL} = \angle 1 + jf/f_1 + \angle 1 + jf/f_2 + \angle 1 + jf/f_3 + \cdots$$
$$= \tan^{-1} \frac{f}{f_1} + \tan^{-1} \frac{f}{f_2} + \tan^{-1} \frac{f}{f_3} + \cdots \qquad (5.65)$$

For frequencies such that $f^2 > f_1^2$, $\tan^{-1}(f/f_1) \simeq 90°$. Furthermore, assuming that $f_4^2 > f_3^2$, the phase angle contributions of f_4 and all the other higher breakpoint frequencies are small. Therefore, for $- \angle A_{OL} = 180°$, we have

$$180° \simeq 90° + \tan^{-1} \frac{f}{f_2} + \tan^{-1} \frac{f}{f_3} \qquad (5.66a)$$

so
$$\tan^{-1} \frac{f}{f_2} + \tan^{-1} \frac{f}{f_3} \simeq 90° \qquad (5.66b)$$

Looking at the triangle shown in Figure 5.33, we see that, since $\tan \theta = x$, $\tan(90° - \theta) = 1/x$; so if we take the inverse tangent functions, we obtain $\theta = \tan^{-1} x$ and $90° - \theta = \tan^{-1}(1/x)$. If we now add these two inverse tangent functions together, we obtain $\tan^{-1} x + \tan^{-1}(1/x) = 90°$. Therefore, for $\tan^{-1}(f/f_2) + \tan^{-1}(f/f_3) = 90°$, we must have that if $f/f_2 = x$; then $f/f_3 = 1/x$ and thus $f/f_2 = f_3/f$, giving $f^2 = f_2 f_3$. Thus $f \simeq \sqrt{f_2 f_3}$ as the frequency at which the total phase angle of the open-loop gain is equal to $180°$.

It has been determined that the frequency at which the phase angle of the open-loop gain is $180°$ is given approximately by $f = \sqrt{f_2 f_3}$, subject to the conditions that f_2 and f_3 be well separated from both f_1 and f_4. The condition with respect to f_1 is usually very well satisfied. The condition with respect to f_4 and the rest of the higher breakpoint frequencies is sometimes not well satisfied. For those cases, then, $f_{180°}$ occurs at a frequency somewhat lower than $\sqrt{f_2 f_3}$, but usually not less than f_2. We note that $f_{180°} = \sqrt{f_2 f_3}$ means that $f_{180°}$ is the *geometric mean* of f_2 and f_3, and on a logarithmic frequency scale $f_{180°}$, in terms of actual distance, is exactly *halfway* between f_2 and f_3.

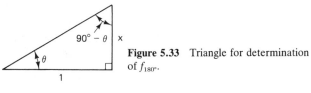

Figure 5.33 Triangle for determination of $f_{180°}$.

We will now determine the value of the open-loop gain at $f = f_{180°} = \sqrt{f_2 f_3}$. If we substitute $f = \sqrt{f_2 f_3}$ into the expression for A_{OL}, we obtain

$$A_{OL}(f = f_{180°}) \simeq \frac{A_{OL}(0)}{(1 + j\sqrt{f_2 f_3}/f_1)(1 + j\sqrt{f_2 f_3}/f_2)(1 + j\sqrt{f_2 f_3}/f_3) \cdots} \tag{5.67}$$

If we now drop the higher breakpoint frequency terms (f_4, f_5, etc.) as being small, and take the product of the second and third terms in the denominator, we obtain

$$\begin{aligned}
A_{OL}(f_{180°}) &\simeq \frac{A_{OL}(0)}{(j\sqrt{f_2 f_3}/f_1)(1 - 1 + j\sqrt{f_3/f_2} + j\sqrt{f_2/f_3})} \\
&\simeq \frac{A_{OL}(0)}{-(\sqrt{f_2 f_3}/f_1)(\sqrt{f_3/f_2} + \sqrt{f_2/f_3})} \\
&\simeq \frac{A_{OL}(0) \angle 180°}{(f_2 + f_3)/f_1} = \frac{A_{OL}(0)f_1 \angle 180°}{f_2 + f_3}
\end{aligned} \tag{5.68}$$

If we use the definition $A_{OL}(0)f_1 = f_u =$ (extrapolated) unity-gain frequency, we can write the expression for $A_{OL}(f_{180°})$ as

$$A_{OL}(f_{180°}) \simeq \frac{f_u}{f_2 + f_3} \angle 180° \tag{5.69}$$

Note that the phase angle of the open-loop gain does indeed turn out to be 180°, as expected. We see that at $f = f_{180°}$ the magnitude of the open-loop gain is smaller than the zero-frequency value by a factor $f_1/(f_2 + f_3)$. In terms of the unity-gain frequency, the magnitude of the open-loop gain at $f = f_{180°}$ can be expressed very simply as $A_{OL}(f_{180°}) \simeq f_u/(f_2 + f_3)$.

The stability condition can now be expressed as

$$\text{For stability:} \quad \text{at } f = f_{180°}, \; |FA_{OL}| < 1 \tag{5.70}$$

and since $F = 1/A_{CL}(0)$, this becomes $A_{CL}(0) > A_{CL}(f_{180°})$. Since $A_{OL}(f_{180°}) = A_{OL}(0)f_1/(f_2 + f_3) = f_u/(f_2 + f_3)$, this gives

$$\text{For stability:} \quad A_{CL}(0) > \frac{A_{OL}(0)f_1}{f_2 + f_3}$$

$$\text{or in terms of } f_u, \quad A_{CL}(0) > \frac{f_u}{f_2 + f_3} \tag{5.71}$$

5.18.2 Phase Margin

For many practical amplifier designs, it is desirable to follow for some safety margin in the design for stable operation. The term *phase margin* will be defined as follows:

$$\text{phase margin} = 180° - \underline{/FA_{OL}(f)}$$

where f is the frequency at which the magnitude of the loop gain is unity. A value of phase margin that is frequently chosen is 45°. With this choice of phase margin, the angle of the loop gain at the frequency at which $|FA_{OL}| = 1$ is 135° and is thus 45° away from the critical phase angle of 180°.

If the breakpoint frequencies are well separated such that $f_1^2 \ll f_2^2 \ll f_3^2$, then the frequency at which the phase angle of the amplifier open-loop gain is 135° is approximately f_2. At this frequency, we have a contribution of 90° from the $(1 + jf/f_1)$ term and 45° from the $(1 + jf/f_2)$ term, so the total phase angle is 135° (assuming that the phase-angle contribution from the other terms is small).

At $f = f_2$, the value of the open-loop gain is given by

$$A_{OL}(f_2) = \frac{A_{OL}(0)}{(1 + jf_2/f_1)(1 + jf_2/f_2)} = \frac{A_{OL}(0)}{j(f_2/f_1)(1 + j1)} = \frac{A_{OL}(0)}{(f_2/f_1)\sqrt{2}\ \underline{/135°}}$$

$$= \frac{A_{OL}(0)f_1}{(\sqrt{2}\,f_2)\underline{/-135°}} \tag{5.72}$$

Since $f_u = A_{OL}(0)f_1$, this last equation can be rewritten as

$$\boxed{A_{OL}(f_2) = \frac{f_u}{\sqrt{2}\,f_2}} \tag{5.73}$$

For a phase margin of at least 45°, we must satisfy the condition $|FA_{OL}| \leq 1$ at $f_{135°}$. Since $f = 1/A_{CL}(0)$, this can be rewritten as $A_{CL}(0) \geq |A_{OL}(f_{135°})|$. If we now use the expression obtained above for $A_{OL}(f_{135°})$, we have

$$\boxed{A_{CL}(0) \geq \frac{A_{OL}f_1}{\sqrt{2}\,f_2} = \frac{f_u}{\sqrt{2}\,f_2}, \qquad \text{for a phase margin} \geq 45°} \tag{5.74}$$

The preceding expressions for $A_{OL}(f_{135°})$ and the corresponding phase margin condition were based on the assumption that the second breakpoint frequency (f_2) was well separated from its closest neighbors (f_1 and f_3). The condition with respect to f_1 is usually very well satisfied, but there can be cases in which f_3 may be close to f_2. If this is the case, then $f_{135°}$ can be less than f_2. Let us consider, as an example, the *worst-case* condition, in which $f_3 = f_2$. For this case, at $f = f_{135°}$ we must have that $\tan^{-1}(f/f_1) + \tan^{-1}(f/f_2) + \tan^{-1}(f/f_3) = 135°$. Since $f_2 \gg f_1$, $\tan^{-1}(f/f_1) \approx 90°$, and noting that for this case $f_2 = f_3$ we now have $2\tan^{-1}(f/f_2) = 45°$; so $\tan^{-1}(f/f_2) = 22.5°$, giving $f_{135°} = f_2 \tan(22.5°) = 0.4142f_2$. Therefore, we see that in this worst-case condition $f_{135°}$ is a little less than one-half of f_2.

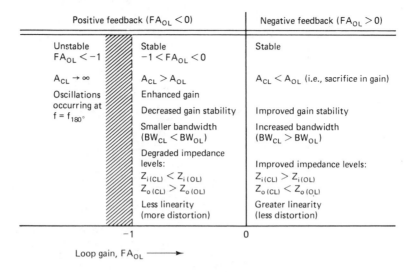

-1 0

Loop gain, FA_{OL} ⟶

Figure 5.34 Characteristics of closed-loop (feedback) systems.

The value of the open-loop gain under these conditions (at $f = f_{135°}$) is given by

$$A_{OL}(f_{135°}) = \frac{A_{OL}(0)}{(j0.4142f_2/f_1)(1 + j0.4142f_2/f_2)^2} = \frac{A_{OL}(0)f_1}{0.4853f_2 \; \angle 135°}$$

$$= \frac{A_{OL}(0)f_1}{(0.4853f_2)\angle -135°} = \frac{f_u}{(0.4853f_2)\angle -135°} \qquad (5.75)$$

As a result, for a 45° phase margin, the condition on the closed-loop gain is $A_{CL}(0) \geq f_u/0.4853f_2$ for a phase margin = 45°.

In Figure 5.34, the important characteristics of negative and positive feedback are summarized. Figure 5.35 shows a representative open-loop op-amp Bode plot. The critical frequencies of $f_{180°}$ and $f_{135°}$ are indicated as well as the open-loop gains at these frequencies. If the closed-loop gain $A_{CL}(0)$ falls below the $A_{CL}(f_{180°})$ line, the op amp is operating in the unstable region. For stability, it is necessary that $A_{CL}(0) > A_{OL}(f_{180°})$. If $A_{CL}(0) > A_{CL}(f_{135°})$, not only is stability ensured, but the phase margin is greater than 45°. If the unity-gain frequency f_u is less than f_2, the op amp is stable with a 45° phase margin for all closed-loop gain situations with resistive feedback elements, even down to the worst-case condition of unity feedback ($F = 1$), which corresponds to a closed-loop gain of unity. This situation in which $f_u < f_2$ will generally be the case for internally compensated op amps.

5.18.3 Operational-amplifier Compensation

The open-loop gain of op amps at low frequencies $[A_{OL}(0)]$ is usually extremely large ($\gtrsim 100,000$). As a result, for almost all op-amp circuits it will be the case that $A_{CL}(0) \ll A_{OL}(0)$. If we now take a look at the stability conditions, we see that it

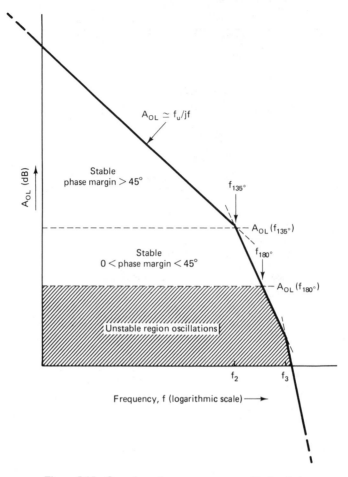

Figure 5.35 Open-loop frequency response (Bode plot).

is necessary that the first breakpoint frequency (f_1) be much smaller than the second breakpoint frequency (f_2). As the closed-loop gain decreases, so must the value of f_1. For example, if $A_{OL}(0) = 100,000$ (100 dB), $f_2 = 1.0$ MHz, and $f_3 = 4.0$ MHz, for stable operation with a closed-loop gain, $A_{CL}(0) = 100$, the value of f_1 must be no greater than 5.0 kHz. If stable operation with closed-loop gain values as small as 1.0 is desired (a unity-gain voltage follower), the value of f_1 must be restricted to a maximum of 50 Hz. If it is desired to operate the voltage follower with a 45° phase margin, then f_1 must be no greater than 14 Hz.

To achieve such low values for the first breakpoint frequency, it is necessary to either use an *internally compensated* op amp or else modify the frequency response of the op amp by adding sufficient capacitance across the *compensation terminals* of the op amp. In either case, a roll-off network is produced that will introduce a factor of the form $1/(1 + jf/f_1)$ into the open-loop gain versus frequency characteristics of the op amp. The roll-off network can be considered to be basically similar to a simple

RC low-pass network that has a transfer characteristic expressed by

$$\frac{v_o}{v_i} = \frac{1/j\omega C}{R + (1/j\omega C)} = \frac{1}{1 + j\omega RC}$$

$$= \frac{1}{1 + j(f/f_1)}, \quad \text{where } f_1 = \frac{1}{2\pi RC} \tag{5.76}$$

Internally compensated operational amplifiers. In internally compensated op amps, the value of the first breakpoint frequency f_1 is usually chosen such that the op amp will be stable, with a phase margin of at least 45° for all closed-loop conditions that involve feedback networks containing only resistive elements. In particular, for the worst-case condition of the unity-gain voltage follower for which $A_{CL}(0) = 1$, a phase margin of at least 45° is obtained. As a result, for any other closed-loop configuration (with resistive elements in the feedback loop), the phase margin is greater than 45°.

For the internally compensated op amps, the required value of f_1 is usually in the range from 3 to 30 Hz, often around 10 Hz. Since this is so much lower than the other breakpoint frequencies, the open-loop 3–dB bandwidth is essentially determined by just f_1, and therefore we have that $BW_{OL} \simeq f_1$.

The compensation in the internally compensated op amps is usually achieved by the addition of a small (30 to 50 pF) capacitor to the silicon chip during the manufacturing process. This capacitor added to the appropriate points in the circuit shifts the breakpoint frequency f_1 all the way down to about 10 Hz.

Operational amplifiers that are not internally compensated. Internally compensated op amps have the advantage that they are simple to use since they will be stable (with a 45° phase margin) under any feedback condition that involves just resistive elements in the feedback network. They do have the disadvantage that they result in a lower closed-loop bandwidth for all closed-loop gains above unity than would otherwise be the case. This is because the value of f_1 (and hence the closed-loop bandwidth) is chosen on the basis of producing a 45° phase margin under unity-gain conditions. This value of f_1 therefore does not correspond to the value needed for other closed-loop gain conditions.

For an op amp that is not internally compensated, the compensation is usually achieved by adding a small capacitor (usually in the range from 5 to 50 pF) to specified compensation terminals. The net result, in terms of the op-amp circuitry, is the same as for the internally compensated op amp except that now the compensation capacitor is an external element and can thus be chosen for optimum circuit performance for any given closed-loop gain configuration. For a unity-gain voltage follower, there is essentially no difference between the bandwidth obtained with the op amp that is not internally compensated compared to the device that is internally compensated, and thus the internally compensated op amp is the one to choose. For closed-loop gains greater than unity, it may be possible to obtain closed-loop bandwidths that are substantially higher by using a noninternally compensated op amp together with the correct choice of compensation capacitor. In some cases the closed-loop bandwidth obtained with the noninternally compensated op amp can be as much as 5 or even 10 times higher than the values that would be obtained by using an internally compensated op amp.

Decompensated operational amplifiers. A decompensated operational amplifier is one in which the degree of compensation is less than that required for the fully compensated op amp. For the decompensated op amp, the gain is stable with a 45° phase margin down to some specified closed-loop gain value that is greater than unity gain, typically in the range of 2 to 10. Decompensated op amps offer the advantage of a wider bandwidth than the fully compensated amplifiers and do not require any external compensation network.

5.18.4 Gain Peaking

At the critical frequency of $f = f_{180°}$, what was negative feedback at low frequencies has turned around 180° to become positive feedback. If the loop gain FA_{OL} is equal to or greater than unity in magnitude at this critical frequency, the op amp will be able to supply its own input and the circuit will break into oscillations. We have already examined the conditions necessary for amplifier stability. However, even if the circuit is indeed stable, the positive-feedback condition that exists in the region of $f_{180°}$ can cause the closed-loop gain in this frequency range to be higher than the low-frequency value of the closed-loop gain $A_{CL}(0)$. If this is the case, a condition of *gain peaking*, as shown in Figure 5.36, is said to exist. In this section we examine gain peaking by obtaining an expression for the quantity $A_{CL}(f_{180°})/A_{CL}(0)$. If this quantity is greater than unity, there will be gain peaking. From the expression for $A_{CL}(f_{180°})/A_{CL}(0)$, the required conditions for no gain peaking will be obtained.

We have that $A_{CL} = A_{OL}/(1 + FA_{OL})$. At $f = 0$ this becomes $A_{CL}(0) = A_{OL}(0)/$

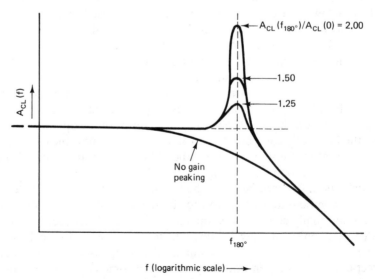

Figure 5.36 Gain peaking due to inadequate phase margin.

$[1 + FA_{OL}(0)] \simeq 1/F$, since $FA_{OL}(0) \gg 1$, so $F \simeq 1/A_{CL}(0)$. We can therefore express A_{CL} as $A_{CL} = A_{OL}/[1 + A_{OL}/A_{CL}(0)]$, and in normalized fashion as

$$\frac{A_{CL}}{A_{CL}(0)} = \frac{A_{OL}/A_{CL}(0)}{1 + A_{OL}/A_{CL}(0)} = \frac{A_{OL}}{A_{CL}(0) + A_{OL}} \tag{5.77}$$

At $f = f_{180°}$, we have that

$$A_{OL}(f_{180°}) = \frac{f_u}{f_2 + f_3} \angle 180° = \frac{-f_u}{f_2 + f_3} \tag{5.78}$$

so for the normalized closed-loop gain we have

$$\frac{A_{CL}(f_{180°})}{A_{CL}(0)} = \frac{-f_u/(f_2 + f_3)}{A_{CL}(0) - f_u/(f_2 + f_3)} = \frac{-1}{[A_{CL}(0)(f_2 + f_3)/f_u] - 1} \tag{5.79}$$

For stability, the closed-loop gain at $f = f_{180°}$ must remain finite, so we require that $A_{CL}(0) > f_u/(f_2 + f_3)$. This is the same stability requirement that was given earlier.

In order not to have any gain peaking, we must require that $A_{CL}(f_{180°})/A_{CL}(0) \leq 1$. For this to be the case, then $A_{CL}(0)(f_2 + f_3)/f_u - 1 \geq 1$, so that $A_{CL}(0)(f_2 + f_3)/f_u \geq 2$, giving the requirement that $A_{CL}(0) \geq 2f_u/(f_2 + f_3)$. Note that the lower limit on the closed-loop gain, $A_{CL}(0)$, obtained here for no gain peaking is exactly *twice* the minimum allowable $A_{CL}(0)$ value for stability. In terms of the f_u values for a given value of $A_{CL}(0)$, the f_u value for no gain peaking is one-half of the maximum allowable f_u value for stability.

Examples of gain peaking. Let us consider the worst-case situation of $A_{CL}(0) = 1$ (a unity-gain voltage-follower circuit). Let $f_2 = 1.0$ MHz and $f_3 = 2.0$ MHz so that $f_2 + f_3 = 3.0$ MHz. For stability, the requirement therefore is that $f_u < 3.0$ MHz.

Case 1: $f_u = 2.5$ MHz. For this case, we have that $A_{CL}(f_{180°}) = 1/[(3 \text{ MHz}/2.5 \text{ MHz}) - 1] = 5.0$. Thus we have a 400% gain peaking factor!

Case 2: $f_u = 2.0$ MHz. For this case, $A_{CL}(f_{180°}) = 1/[(3 \text{ MHz}/2.0 \text{ MHz}) - 1] = 2.0$, so we now have 100% gain peaking.

Case 3: $f_u = 1.75$ MHz. For this case, $A_{CL}(f_{180°}) = 1/[(3 \text{ MHz}/1.75 \text{ MHz}) - 1] = 1.40$, so there is now 40% gain peaking.

Case 4: $f_u = (f_2 + f_3)/2 = 1.5$ MHz. For this case, $A_{CL}(f_{180°}) = 1/[(3 \text{ MHz}/1.5 \text{ MHz}) - 1] = 1.00$, so there is now no gain peaking. This case represents the borderline situation between the conditions of gain peaking and no gain peaking.

Case 5: $f_u = 1.0$ MHz. Now we have that $A_{CL}(f_{180°}) = 1/[(3 \text{ MHz}/1 \text{ MHz}) - 1] = 0.50$, so not only is there no gain peaking, but the gain at $f = f_{180°}$ is down substantially below $A_{CL}(0)$.

The condition for a 45° phase margin has been given as $A_{CL}(0) \gtrsim f_u/\sqrt{2} f_2$, for the case of $f_3^2 \gg f_2^2$, and in the worst-case situation of $f_3 = f_2$ the requirement becomes $A_{CL}(0) = f_u/(0.485f_2)$. In terms of f_u, we have $f_u \leq \sqrt{2} \, f_2 A_{CL}(0)$ for the case in which f_3 is well separated from f_2, and $f_u \leq 0.485 f_2 A_{CL}(0)$ for the worst-case situation of $f_3 = f_2$. Therefore, for the example above, we see that for a 45° phase margin an f_u value of about 1.0 MHz maximum is required. We note that with this value of f_u there is no gain peaking, and, in general, there is no gain peaking if the phase margin is at least 45°.

We have examined the gain peaking situation at the critical frequency of $f = f_{180°}$. From that analysis we see that to prevent gain peaking the requirement is that $A_{CL}(0)(f_2 + f_3)/f_u \geq 2$, so $f_u \leq A_{CL}(0)(f_2 + f_3)/2$. For the worst-case situation of $f_2 = f_3$, this becomes $f_u \leq A_{CL}(0)f_2$. At the other extreme, however, when f_3 is remote from f_2, there may still be gain peaking even if the conditions above are satisfied, although the gain peaking will not occur at $f_{180°}$.

To examine this situation, let us assume that

$$A_{OL} = \frac{A_{OL}(0)}{(jf/f_1)(1 + jf/f_2)} = \frac{f_u}{jf(1 + jf/f_2)} \tag{5.80}$$

Since

$$A_{CL} = \frac{A_{OL}}{1 + FA_{OL}} = \frac{1/F}{1 + (1/FA_{OL})} = \frac{A_{CL}(0)}{1 + A_{CL}(0)/A_{OL}} \tag{5.81a}$$

we have that

$$\frac{A_{CL}}{A_{CL}(0)} = \frac{1}{1 + A_{CL}(0)/A_{OL}} = \frac{1}{1 + [A_{CL}(0)/f_u]jf[1 + j(f/f_2)]} \tag{5.81b}$$

$$= \frac{1}{1 + [A_{CL}(0)/f_u][jf - (f^2/f_2)]}$$

$$= \frac{1}{1 - (f^2/f_2 f_u)A_{CL}(0) + jf A_{CL}(0)/f_u} \tag{5.81c}$$

When $f = \sqrt{f_2 f_u/A_{CL}(0)}$, the normalized closed-loop gain is given by

$$\frac{A_{CL}}{A_{CL}(0)} = \frac{1}{j\sqrt{[f_u f_2/A_{CL}(0)]}\,[A_{CL}(0)/f_u]}$$

$$= \frac{1}{j\sqrt{(f_2/f_u)}A_{CL}(0)} = -j\sqrt{\frac{f_u}{f_2 A_{CL}(0)}} \tag{5.82}$$

Therefore, for no gain peaking we require that $f_u < f_2 A_{CL}(0)$. When this condition is compared to the previous condition pertaining to gain peaking at $f_{180°}$, we see that in general if $f_u < f_2 A_{CL}(0)$ there will be no gain peaking.

Figure 5.36 shows the closed-loop frequency response characteristics of an op amp with various amounts of gain peaking. Figure 5.37 shows the op amp response to a rectangular pulse input voltage. Gain peaking in the frequency-domain response has as its counterpart *overshoot* in the time-domain response. In addition to the overshoot, there is often a damped oscillatory type of transient response known as *ringing*.

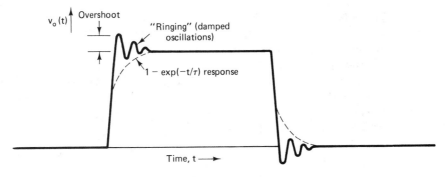

Figure 5.37 Time-domain response showing overshoot and ringing.

5.18.5 Effect of Load Capacitance and Input Capacitance on Stability

Internally compensated operational amplifiers are normally stable with a phase margin of at least 45° for all closed-loop situations involving just resistive elements in the feedback loop. A large load capacitance C_L can, however, have a destabilizing effect on the op amp. The combination of the load capacitance and the open-loop output resistance R_o of the op amp results in a low-pass or *lag* network as shown in Figure 5.38. This lag network has a transfer function given by

$$T(f) = \frac{1}{1 + j\omega\tau_L} = \frac{1}{1 + jf/f_L} \tag{5.83}$$

where $\tau_L = (R_o\|R_L)C_L$ and $f_L = 1/(2\pi\tau_L)$. This adds a new breakpoint frequency to the loop gain, which can now be written as

$$\text{loop gain} = FA_{OL} = \frac{FA_{OL}(0)}{(1 + jf/f_1)(1 + jf/f_2)\cdots(1 + jf/f_L)} \tag{5.84}$$

As long as f_L is substantially larger than f_2, it will have no serious effect on the operation and stability of the circuit. However, if C_L is such that f_L drops below f_2, the sequence of frequencies in the stability equations must be reordered, with f_L

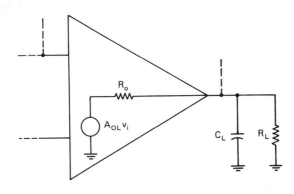

Figure 5.38 Effect of operational-amplifier capacitance on amplifier stability.

replacing f_2, f_2 replacing f_3, and so on. The condition for a 45° phase margin in this case becomes $A_{CL}(0) > f_u/\sqrt{2}\,f_L$. Thus, as C_L increases and f_L correspondingly decreases, this condition may no longer be satisfied. Indeed, with a large enough value of C_L the op amp may become unstable and break into oscillations.

As an example of the effect of load capacitance, let us consider an op amp with $f_u = 1.0$ MHz and an open-loop output resistance of $R_o = 100\ \Omega$. Taking the case of a unity-gain voltage-follower circuit, as f_L drops below f_2 the requirement for stability with a 45° phase margin becomes

$$A_{CL}(0) = 1.0 > \frac{f_u}{\sqrt{2}\,f_L} = \frac{1.0\ \text{MHz}}{\sqrt{2}\,f_L} = \frac{707\ \text{kHz}}{f_L} \tag{5.85}$$

Therefore, $f_L = 1/[2\pi(R_o\|R_L)C_L]$ should be greater than 707 kHz. Taking the worst-case condition of $R_L \gg R_o$, the corresponding requirement on the load capacitance is that $C_L < 2.25$ nF.

The combination of large-value resistors in the feedback circuit and the input capacitance C_i of the op amp can also act to destabilize the op amp. The combination of the feedback resistors R_1 and R_2 and the input capacitance C_i results in a low-pass or lag network, as shown in Figure 5.39. The transfer function of this lag network is given by

$$T(f) = \frac{R_1/(R_1 + R_2)}{1 + j\omega\tau_i} = \frac{R_1/(R_1 + R_2)}{1 + j(f/f_i)} \tag{5.86}$$

where the time constant τ_i is given by $\tau_i = (R_1\|R_2)C_i$ and $f_i = 1/(2\pi\tau_i)$.

This transfer function is part of the loop gain and thus adds a new breakpoint frequency to the loop gain, which can now be written as

$$\text{loop gain} = FA_{OL} = \frac{R_1/(R_1 + R_2)A_{OL}(0)}{(1 + jf/f_1)(1 + jf/f_2) \cdots (1 + jf/f_i)} \tag{5.87}$$

The situation is now similar to that encountered with respect to the effect of the load capacitance. As long as f_i is substantially above f_2, it will have no significant effect on the op-amp stability. If f_i drops below f_2, however, it must now be considered, and the sequence of breakpoint frequencies in the stability equations must be reordered, with f_i replacing f_2, f_2 replacing f_3, and so on.

As an example of the effect of the input capacitance on stability, let us take an op amp with $f_u = 1.0$ MHz and $C_i = 5.0$ pF. For the case of a voltage-follower

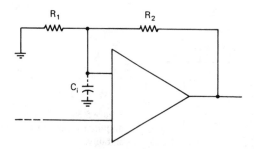

Figure 5.39 Effect of input capacitance on amplifier stability.

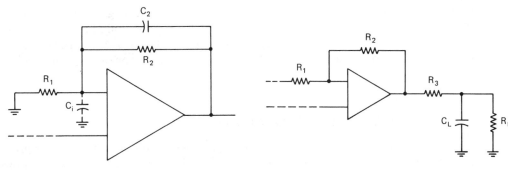

Figure 5.40 Compensation for input capacitance.

Figure 5.41 Use of resistor to isolate load capacitance.

circuit with $A_{CL}(0) = 1$, the condition for stability with a 45° phase margin becomes $A_{CL}(0) = 1 > f_u/\sqrt{2}f_i = 1.0 \text{ MHz}/\sqrt{2}f_i = 707 \text{ kHz}/f_i$. The corresponding requirement in terms of R_1 and R_2 is that $f_i = 1/[2\pi(R_1\|R_2)C_i] > 707 \text{ kHz}$, so that $(R_1\|R_2) < 45$ kΩ.

The effect of C_i can be compensated for by using a small feedback capacitor C_2 in parallel with the R_2 feedback resistor, as shown in Figure 5.40. The feedback factor for this circuit can now be written as $F = Z_1/(Z_1 + Z_2) = 1/[1 + (Z_2/Z_1)]$, where $Z_1 = R_1/(1 + j\omega\tau_1)$ and $Z_2 = R_2/(1 + j\omega\tau_2)$ with $\tau_1 = R_1C_i$ and $\tau_2 = R_2C_2$. Since the impedance ratio is $Z_2/Z_1 = (R_2/R_1) \times (1 + j\omega\tau_1)/(1 + j\omega\tau_2)$, we see that, if the two time constants are equal such that $\tau_1 = \tau_2$, the impedance ratio and therefore the feedback factor becomes independent of frequency, and the effect of C_i is therefore canceled out. The basic requirement on C_2 can therefore be written as $R_1C_i = R_2C_2$.

The same basic technique described can be used in some cases to compensate for the effects of the load capacitance. If $R_2 \gg R_1$, the required condition for this compensation is that $\tau_2 = \tau_L$, where $\tau_2 = R_2C_2$ and $\tau_L = R_oC_L$.

Another technique that can be used to minimize the destabilization effects of large load capacitances is to use a small resistor R_3 between the op amp and the load, as shown in Figure 5.41. This provides some degree of isolation between the op amp and C_L, and, generally, resistance values on the order of 100 Ω will be sufficient. The use of this isolating resistor can ensure stability of the op amp, but it carries with it the disadvantages of reducing the output voltage, increasing the output impedance of the circuit, and reducing the overall bandwidth.

5.18.6 Power Supply Decoupling

In a typical situation where the power supply of an electronic system is used to supply power to several op amps or other types of amplifiers or other devices, there can be an interaction between the various amplifiers in the system due to the nonzero output impedance of the power supply. This includes the effects of the resistance and inductance of the leads connecting the power supply to the various amplifiers and other devices that comprise the system, including the power supply ground return leads. This parasitic coupling produces feedback loops that can easily result in insta-bility in the system. These effects can be minimized by the strategic use of bypass (decoupling) capacitors and by the judicious routing of the signal ground connections. The use of separate lines for the signal and the power supply ground returns can also prove beneficial. The use of bypass capacitors in the 0.01- to 0.1-μF range connected between every IC power supply pin (V^+ and V^-) and ground is an often recommended practice.

5.19 OPERATIONAL-AMPLIFIER APPLICATIONS

A number of representative operational-amplifier applications are shown in the problems at the end of this chapter. Some brief remarks concerning these applications are given below. Also, a brief discussion of some of the circuits relating to the evaluation of op-amp characteristics is included. *Note:* The figures referenced in this section can be found at the end of this chapter in the Problem section.

1. *Difference amplifier* (see Figure P5.1): In this circuit the output voltage is proportional to the difference of the two input voltages. In the ideal situation any common-mode component of the two input voltages is totally rejected. In the actual situation, because of the finite common-mode gain of the operational amplifier and the mismatches in the resistance ratios, there is some small common-mode response. Notice also that the input resistance with respect to the V_A and V_B inputs is not the same, being R_A for the V_A input and $R_A + R_F$ for the V_B input.

2. *Summing amplifier* (see Figure P5.2): A number of input voltages can be arithmetically combined with various weighting factors.

3. *Current-to-voltage converter* or *transimpedance amplifier* (see Figure P5.3): This circuit produces an output voltage that is directly proportional to the input current. Note that the voltage drop at the input V_i is $V_i = -V_o/A_{OL} = I_i R_F/A_{OL}$, so the effective input resistance looking into this circuit is $R_i = V_i/I_i = R_F/A_{OL}$. Therefore, even if R_F is relatively large, the effective input resistance R_i is very small, due to the very large open-loop gain of the amplifier. This very small input resistance is of great advantage since it will not produce a significant effect on the circuit in which the current is to be measured. Note also that the output voltage is substantially independent of the load that is driven by this circuit.

4. *Voltage-to-current converter* or *voltage-controlled current source* (see Figure P5.4): The current through the load resistance is independent of the value of the load resistance, but is directly proportional to the input voltage, so this circuit will act as a voltage-controlled constant current source. Note, however, that neither end of R_L can be grounded.

5. *Low-pass active filter* or *integrator* (see Figure P5.5): This circuit is a simple single-pole low-pass active filter with a 3-dB bandwidth given by $BW = 1/(2\pi R_F C_F)$. At low frequencies the gain asymptotically approaches the zero-frequency gain value of $A_{CL}(0) = -R_F/R_A$. At high frequencies the gain will asymptotically approach a response given by $A_{CL}(\omega) = 1/j\omega R_A C_F$ and thus will drop off at a rate of -20 dB/decade or -6 dB/octave.

Looking at this in the time domain, this circuit can be used as an integrator. In the limiting case in which R_F goes to infinity, we have $V_o = -(1/C_F) \int_{-\infty}^{t} i\, dt = -(1/C_F) \int_{-\infty}^{t} (V_S/R_A)\, dt = -(1/R_A C_F) \int_{-\infty}^{t} V_S\, dt$, so this circuit will indeed act as an ideal integrator. Since the dc gain under these conditions is A_{OL}, the output voltage is very strongly affected by the op-amp V_{OS} and is either $A_{OL}V_{OS}$ or more likely is saturated near the positive or negative supply voltages. To prevent this from happening, a feedback resistor R_F of suitable size is usually included in the circuit, although it makes the circuit a less than ideal integrator.

6. *High-pass active filter* or *differentiator* (see Figure P5.6): This circuit is a simple single-pole high-pass active filter with a breakpoint frequency of $f_{bp} = 1/(2\pi R_A C_A)$. The gain will asymptotically approach $-R_F/R_A$ at high frequencies and will asymptotically go to zero at low frequencies.

   Looking at the operation of this circuit in the time domain and letting R_A go to zero, we have that $i_s = C_A(dv_s/dt)$, so $v_o = -i_s R_F = -R_F C_A (dv_s/dt)$. We see that this circuit will act as an ideal differentiator. With $R_A = 0$, however, the gain at high frequencies is relatively large. This large high-frequency gain can cause problems due to circuit noise and other factors, so a resistor R_A of some suitable size is often included in the circuit, although this results in a less than ideal differentiator characteristic.

7. *Precision detector* or *rectifier* (see Figure P5.7): The input voltage V_i that is required to turn the diode on is $V_i = V_D/A_{OL} \simeq 0.5 \text{ V}/A_{OL}$, so the effect of the diode forward voltage drop is essentially nullified. As a result, this circuit can detect (or rectify) very low-level signals.

8. *Precision full-wave rectifier* (see Figure P5.8): Note again the effect of the op amp in essentially nullifying the forward voltage drop of the diode. The first and third op amps are voltage followers for source and load isolation, respectively.

9. *Precision peak detector* (see Figure P5.9): Again note the action of the op amp in reducing the effect of the diode forward voltage drop. The second op amp is used for load isolation such that the load resistance will not affect the discharge rate of the capacitor.

10. *Logarithmic converter* (see Figure P5.10): This circuit uses nonlinear elements (transistors) in the feedback loop to obtain a nonlinear (logarithmic) transfer characteristic. The gain of the difference amplifier (A_3) is set so as to obtain a logarithmic conversion scale factor of 1.0 V/decade.

11. *Exponential amplifier* or *antilogarithmic converter* (see Figure P5.11): This circuit produces an output voltage that is an exponential function of the input voltage. Note that the overall transfer characteristic of this circuit represents the inverse function of the feedback device.

12. *Current integrator* or *charge amplifier* (see Figure P5.12): This is another integrator circuit in which the output voltage is directly proportional to the net flow of charge Q into the circuit. Note that the voltage drop across the input terminals is very small. Note also that the V_{OS} of the op amp can cause problems so that it may be desirable to connect a feedback resistor R_F across C_F, although this would make this circuit a less than ideal current integrator.

13. *Schmitt trigger* (see Figure P5.13): This circuit uses positive feedback and will operate in the switching mode with the output voltage being in either of two states, V_O^+ and V_O^-. For most op amps the high state V_O^+ is near the positive supply V^+ and the low output state V_O^- is close to the negative supply voltage V^-. The voltage applied to the noninverting input terminal is a function of the fixed reference voltage V_{REF} and the output voltage V_O. As a result, the input threshold voltage for switching is a function of the output voltage, and this results in the hysteresis loop of the input–output transfer characteristic.

14. *Positive voltage regulator* (see Figure P5.14): In this circuit the output voltage is limited to values less than V_Z. The voltage follower is used to keep load resistance or current variations from affecting the output voltage (to give good load regulation).

15. *Positive voltage regulator with V_O greater than V_Z* (see Figure P5.15): In this circuit the output voltage can be greater than V_Z, but cannot be any less than V_Z. Note that the current through the zener diode is given by $I_Z = I_{R3} = V_O R_2 / (R_1 + R_2)$, so it is essentially independent of the supply voltage. As a result, the output voltage is relatively insensitive to changes in the supply voltage (good line regulation) as long as the supply voltage is above the minimum value required for the operation of this circuit.

16. *High current voltage regulator with current limiting* (see Figure P5.16): The maximum current output of this circuit is not limited by the maximum current capability of the op amp, but is increased by the net current gain of the Q_1–Q_2 Darlington circuit. Transistor Q_3 and resistor R_{CL} are used for current limiting to limit I_L to a safe value to prevent excessive power dissipation in Q_2.

17. *Constant-current sink* (see Figure P5.17): The output current I_O is approximately independent of the output voltage $V_O = V_{C2}$ as long as both Q_1 and Q_2 remain in the active region. This requires that $V_{CE2} > 0.9 \text{ V} + V_{REF}$ and less than the collector-to-emitter breakdown voltage. Note that $I_O + I_{B1} - I_{BIAS} = I_{R1} = (V_{REF} + V_{OS})/R_1$, so for precision operation down to low current levels a small V_{OS} and I_{BIAS} is desirable. The high net current gain of the Darlington pair makes I_{B1} very small compared to I_O, which is also a desirable condition.

18. *Precision constant-current source for low current levels* (see Figure P5.18): In this circuit, $I_O + I_G - I_{BIAS} = (V_{REF} + V_{OS})/R_1$, where I_G is the gate current of the JFET. Again we see that for precision operation down to very low current levels the op-amp offset voltage and bias current should be very small. The use of a JFET is of advantage in this circuit due to the very small gate current I_G, which is usually less than 1 nA.

19. *Operational amplifier with electronic gain control* (see Figure P5.19): The drain-to-source resistance of a JFET when $V_{DS} \lesssim V_P/3$ is a function of the gate-to-source voltage V_{GS} and is given approximately by $r_{ds} = r_{ds(ON)}/(1 - \sqrt{V_{GS}/V_p})$, where $r_{ds(ON)}$ is the drain-to-source resistance when $V_{GS} = 0$. This voltage-variable resistance characteristic of the JFET can be used for the electronic gain control of an op-amp circuit. Note, however, that the closed-loop gain is a nonlinear function of the gain control voltage, $-V_{BIAS}$.

20. *Operational amplifier with electronic gain control* (see Figure P5.20): Again in this circuit the voltage-variable resistance characteristics of JFETs are used to control the closed-loop gain of an op-amp circuit. Note that in this circuit, however, the gain is a linear function of the control voltage.

21. *Tracking voltage regulator* (see Figure P5.21): This circuit produces two voltages of equal magnitude proportional to the input voltage V_{REF}, but of opposite algebraic sign. The two transistors are used just to increase the output current capability of the circuit.

22. *Precision phase splitter with high input impedance and low output impedance* (see Figure P5.22): This circuit produces two output voltages of equal magnitude, but of opposite algebraic sign.

23. *Instrumentation amplifier with high input impedance and low output impedance* (see Figure P5.23): This is basically a difference amplifier, but with a very high input impedance for both inputs.

24. *Instrumentation amplifier with high input impedance and low output impedance* (see Figure P5.24): This again is a difference amplifier with a very high input impedance for both input channels and a very low output impedance.

25. *Exponential converter* or *antilogarithmic amplifier* (see Figure P5.25): This circuit produces an output voltage that increases at an exponential rate with increasing input voltage. Note that the input voltage V_i can be of either positive or negative polarity, but the output voltage is of positive polarity in either case.

26. *Circuit for raising a variable to a power with logarithmic techniques* (see Figure P5.26): This circuit uses a combination of a logarithmic converter and an anti-logarithmic converter to produce the input–output relationship $V_O = (V_i)^{R_F/R_L}$. Note that the exponent R_F/R_L is not restricted to integer values and it can be less than unity or greater than unity as required.

27. *Amplifier with exponential gain control* (see Figure P5.27): The exponential gain control characteristics of this circuit make it possible to obtain a very large variation in the closed-loop gain with only a relatively small variation in the gain control (or AGC) voltage.

28. *Function generator* (see Figure P5.28): Note that the functional relation obtained from this circuit is the inverse of the functional relationship of the device in the feedback loop.

29. *Dump-and-integrate circuit* (see Figure P5.29): In the dump-and-integrate circuit of Figure P5.29, transistor Q_1 is turned on prior to the integration time to discharge capacitor C_1 and set the output voltage to zero. Then the transistor is turned off. The output voltage is then equal to the integral of the input voltage as given by

$$V_O = -\int_0^t \frac{1}{R_1 C_1} V_i \, dt$$

30. *Precision voltage-controlled limiting (clipping or bounding) circuit* (see Figure P5.30): This is a noninverting unity-gain amplifier circuit that clips the output voltage at V_{REF} such that $V_O = V_S$ for $V_S < V_{REF}$ and $V_O = V_{REF}$ for $V_S > V_{REF}$, where V_{REF} can be of either polarity. The forward voltage drop of the diode

is, in effect, divided by the amplifier open-loop gain such that it has very little effect on the operation of this circuit. If diode D_1 is reversed, the transfer relationship is $V_O = V_S$ for $V_S > V_{REF}$ and $V_O = V_{REF}$ for $V_S < V_{REF}$.

31. *Precision voltage-controlled clamping circuit* (see Figure P5.31): This circuit produces an output voltage that has the same ac variation as the input voltage, but the dc level is shifted such that the output voltage will not drop below V_{REF}. The voltage V_{REF} can be of either polarity. If diode D_1 is reversed, the output voltage again has the same ac variation as the input, but the shift in the dc level is such that the output voltage will not rise above V_{REF}.

The diode forward voltage drop is, in effect, divided by the open-loop gain of A_2 such that it has a negligible effect on the performance of the circuit in most cases. This circuit can be used to restore the dc level of a signal that has lost its dc level due to passing through a coupling capacitor. When used in this application, it is often called a *dc restorer*.

32. *Voltage-controlled gain polarity switching circuit* (see Figure P5.32): This circuit is an inverting amplifier with a closed-loop gain given by $A_{CL} = -(1 + R_2/R_1)$ when transistor Q_1 is turned on ($V_{control} = 0$) and a noninverting amplifier with $A_{CL} = 1 + R_2/R_1$ when Q_1 is turned off ($V_{control} < V_P$). The magnitudes of the two voltage gains are approximately equal subject to the condition that $r_{ds} \ll R_4$.

33. *Voltage-controlled impedance multiplier* (see Figure P5.33): The feedback impedance Z_F of this circuit is transformed to appear as an input impedance of value $Z_i = Z_F/(1 + A)$, and correspondingly the input admittance is given by $Y_i = (1 + A)Y_F$. If the feedback impedance is a capacitor of value C_F, the input capacitance of this circuit is $C_i = C_F(1 + A)$ so that this circuit can act as a capacitance multiplier. If the gain of the amplifier A is varied by the gain control voltage, the input capacitance is a voltage-variable capacitance.

34. *Inductance simulator* (see Figure P5.34): The input impedance of this circuit is proportional to the reciprocal of the feedback impedance Z_F and is given by $Z_i = R_2R_3/Z_F$. If Z_F is a capacitor of value C_F, the input impedance is given by $Z_i = j\omega R_2R_3C_F = j\omega L_{eq}$, so the input impedance appears as an inductance of value $L_{eq} = R_2R_3C_F$.

35. *Analog signal multiplexing circuit* (see Figure P5.35): For this circuit, ϕ_1 through ϕ_N are nonoverlapping active low clock pulses such that transistors Q_1 through Q_N are "on" except when the clock pulses applied to one of the transistors is active, which will turn that transistor "off." This then permits the signal transmission from the selected input to the output. The resistance R_1 should be chosen such that $R_1 \gg r_{ds(ON)}$.

One variation of this circuit is to use a series-shunt switch arrangement with additional FETs in series with R_1 and driven with active high-clock pulses such that when a shunt switch transistor is "on" the corresponding series switch transistor is "off," and vice versa.

36. *Symmetrical bipolar limiter* (see Figure P5.36): This circuit provides symmetrical limiting of the output voltage. This is done without excessive current drain from the op-amp output due to the limiting diodes being in the feedback loop rather than between the output terminal and ground.

37. *Constant-amplitude phase shifter* (see Figure P5.37): This circuit has a closed-loop voltage gain that has a magnitude of unity, independent of frequency. The phase shift, however, changes with frequency, ranging from 0° at dc to a value approaching −180° at high frequencies. This circuit can be used as a time-delay circuit, with the time delay T_d being given by $T_d = \phi/\omega$, where ϕ is the phase shift.

38. *Bridge amplifier* (see Figure P5.38): This circuit is useful for thermometry. If a piezoresistive element is used in the bridge, this circuit can be used as a pressure or strain transducer.

39. *Active RC bandpass amplifier* (see Figure P5.39): This circuit can be used as a bandpass amplifier to amplify a narrow band of frequencies (the passband) and to reject frequencies that are outside the passband. This circuit is especially advantageous for low-frequency applications since only *RC* elements and no inductors are required.

40. *Square-wave oscillator* (see Figure P5.40): This simple circuit produces a square wave with a peak-to-peak amplitude that is almost equal to the net supply voltage $(V^+ - V^-)$ and with a frequency determined by the R_1C_1 time constant. Note the combined use of positive and negative feedback in this circuit.

41. *Precision triangle wave generator* (see Figure P5.41): This circuit produces a triangular waveform that has very straight sides. Op amp A_1 is a Schmitt trigger, A_2 is an integrator, and A_3 serves as an inverting amplifier.

42. *Sample-and-hold circuit* (see Figure P5.42): This is a simple sample-and-hold circuit that can sample the input voltage over a short time interval and then hold that sampled value over an extended period of time. Note the use of the two voltage followers for source and load isolation.

43. *Howland current source* (see Figure P5.43): This is a voltage-controlled constant-current source that produces an output current of $I_L = (V_1 - V_2)/R_1$. The output current can be of either polarity, and one side of the load resistance that is driven by this current source can be grounded.

44. *Circuit to produce an output voltage that increases linearly with temperature* (see Figure P5.44): This is an interesting circuit that produces an output voltage that is directly proportional to the absolute temperature.

45. *Voltage regulator* (see Figure P5.45): This is a simple voltage regulator circuit for $V_O > V_{\text{REF}}$. By adding a transistor or a Darlington transistor pair on the output side of the op amp, the current output capability of this circuit can be greatly increased.

46. *Voltage regulator with current foldback* (see Figure P5.46): This somewhat more complex voltage regulator circuit has a current foldback current-limiting characteristic to increase the output current available to drive the load, yet still provide full protection to the output transistors.

47. *Analog multiplier* (see Figure P5.47): With a combination of three op amps and

a set of four matched transistors, a circuit is obtained that can be used as a multiplier, a divider, or as a square-rooting circuit. Note that this circuit can be connected as a four-quadrant multiplier in which input voltages of either polarity can be accepted. This circuit is also available as a monolithic IC.

48. *Phase shift oscillator* (see Figure P5.48): This is a feedback oscillator. For oscillations to occur, the net gain around the feedback loop must be greater than unity at the frequency at which the net phase shift around the feedback loop is zero. The frequency of oscillation is given by $f_{osc} = 1/(2\pi\sqrt{3}R_1C_1)$, and the gain condition is satisfied if $R_3/R_2 > 8$. If the net gain around the feedback loop is adjusted to a value that is just slightly above unity, the output voltage will be a relatively undistorted sinusoidal waveform.

This circuit also works if voltage-followers A_2 and A_3 are omitted. The frequency of oscillation is then given by $f_{osc} = 1/(2\pi\sqrt{6}\ R_1C_1)$, and the gain condition will require that $R_3/R_2 > 29$.

49. *Wien bridge oscillator* (see Figure P5.49): This circuit has both a positive and negative feedback loop. For oscillations to occur, the net feedback must be positive with a phase angle of zero. This gives the required condition that $(R_3/R_4) > (R_1/R_2) + (C_1/C_2)$, and the corresponding frequency of oscillation is $f_{osc} = 1/(2\pi\sqrt{R_1R_2C_1C_2})$. If the R_3/R_4 ratio is adjusted to a value that is just slightly above that required for oscillations, then a relatively undistorted sinusoidal waveform will be obtained.

50. *Optoelectronic sensor* (see Figure P5.50): This simple light-sensing circuit consists of a photodiode and a current-to-voltage converter op-amp circuit. The photocurrent produced by the photodiode is a linear function of the light intensity, so the resulting output voltage of this circuit is also a linear function of the light intensity. The photodiode bias voltage could be set to zero, but a reverse bias voltage across the photodiode will have the advantage of reducing the junction capacitance and thereby decreasing the response time of this circuit.

51. *Circuit for evaluating the power supply rejection ratio (PSRR)* (see Figure P5.54): The coupling capacitor C_C is used to block the output voltage component that is the result of the dc input offset voltage of the op amp. As a result, only the ac output voltage that results from the power supply ac ripple is measured.

52. *Circuit for obtaining the open-loop gain* (see Figure P5.55): Note that, although it is the open-loop gain of the op amp that is to be measured, this circuit is nevertheless operated as a closed-loop system. This is to prevent the output voltage from being driven into saturation near either supply voltage due to the effect of the input offset voltage V_{OS}.

53. *Evaluation of the equivalent input noise voltage* (see Figure P5.56): This op amp is operated with a high closed-loop gain in order to greatly amplify the input noise voltage to make it easier to measure.

54. *Common-mode input impedance* (see Figure P5.57): This simple voltage-follower circuit illustrates the effect of the common-mode input resistance and the input capacitance on the overall circuit performance and can be used for the measurement of these two parameters.

55. *Common-mode gain* (see Figure P5.59): If the op amp has a nonzero common-mode gain or if the resistance ratios are not exactly equal, this circuit will not act as an ideal difference amplifier, but there will be some response to a common-mode input voltage.

PROBLEMS

See pages 310 to 317 for a brief discussion of these circuits. For the following problems assume ideal, infinite-gain operational amplifiers, unless otherwise indicated or implied.

Operational Amplifier Applications

5.1. (*Difference amplifier*) Referring to Figure P5.1, show that $V_o = (R_F/R_A)(V_B - V_A)$.

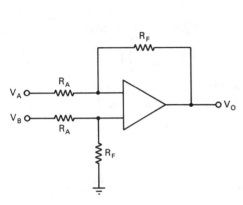

Figure P5.1

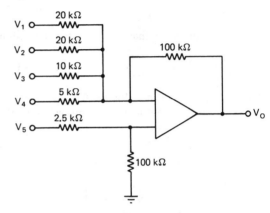

Figure P5.2

5.2. (*Summing amplifier*) Referring to Figure P5.2, show that $V_o = -5V_1 - 5V_2 - 10V_3 - 20V_4 + 40V_5$.

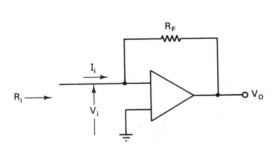

Figure P5.3

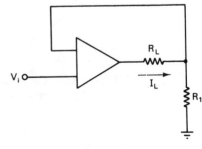

Figure P5.4

5.3. (*Current-to-voltage converter* or *transimpedance amplifier*) Refer to Figure P5.3.
(a) Show that

$$V_O = \frac{-I_iR_F}{1 + (1/A_{OL})} \simeq -I_iR_F, \qquad \text{for } A_{OL} \gg 1$$

(b) Show that the input resistance is given by

$$R_i = \frac{R_F}{1 + A_{OL}} \simeq \frac{R_F}{A_{OL}}, \qquad \text{for } A_{OL} \gg 1$$

5.4 (*Voltage-to-current converter* or *voltage-controlled current source*)   Referring to Figure P5.4, show that

$$I_L = \frac{V_i}{R_1[1 + (1/A_{OL})(1 + R_L/R_1)]}$$

$$\simeq \frac{V_i}{R_1}, \quad \text{if } A_{OL} \gg 1 + \frac{R_L}{R_1}$$

5.5. (*Low-pass active filter* or *integrator*)   Refer to Figure P5.5.

    (a) Show that for sinusoidal excitation

$$A_{CL} = \frac{V_o}{V_S} = -\frac{R_F}{R_A}\frac{1}{1 + j\omega R_F C_F}$$

    (b) Show that if V_S is a step-function input of amplitude V_S, V_O is given by

$$V_O = -V_S \frac{R_F}{R_A}\left[1 - \exp\left(-\frac{t}{R_F C_F}\right)\right]$$

and that, for $t \lesssim 0.1\, R_F C_F$, $V_O \simeq -V_S(t/R_A C_F)$, so therefore for $t \lesssim 0.1 R_F C_F$ the output voltage is approximately the integral of the input voltage.

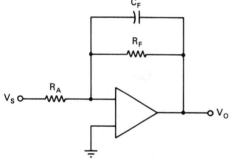

Figure P5.5

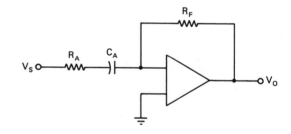

Figure P5.6

5.6. (*High-pass active filter* or *differentiator*)   Refer to Figure P5.6.

    (a) Show that for sinusoidal excitation

$$A_{CL} = \frac{-j\omega C_A R_F}{1 + j\omega R_A C_A}$$

    (b) Show that if V_S is a step-function input the output voltage is given by

$$V_O = -V_S \frac{R_F}{R_A}\exp\left(-\frac{t}{R_A C_A}\right)$$

and therefore, for $t \gtrsim 0.1 R_A C_A$, $V_O \simeq -V_S(R_F/R_A)$.

5.7. (*Precision detector* or *rectifier*)   Referring to Figure P5.7, show that $V_O = V_i$ for $V_i > 0$ (actually $> V_D/A_{OL}$, where V_D is the forward voltage drop of the diode), and $V_o = 0$ for $V_i < 0$.

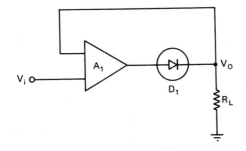

Figure P5.7

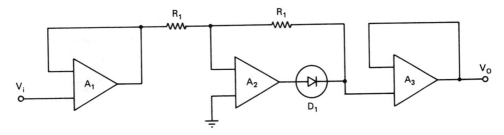

Figure P5.8

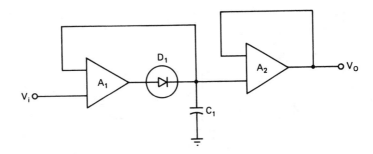

Figure P5.9

5.8. (*Precision full-wave rectifier*)   Referring to Figure P5.8, show that $V_O = |V_i|$.

5.9. (*Precision peak detector*)   Referring to Figure P5.9, show that V_o is the positive peak value of V_i.   What is the function of the second op amp?

5.10. (*Logarithmic converter*)   Referring to Figure P5.10, show that (at room temperature)

$$V_O = (1.0 \text{ V}) \log_{10} \frac{V_2 R_1}{V_1 R_2}$$

Assume that Q_1 and Q_2 are matched transistors.

5.11. (*Exponential amplifier* or *antilogarithmic converter*)   Referring to Figure P5.11, show that $V_O = V_R \, 10^{-(v_i R_2 / K R_1)}$, where K is a constant having units of volts.

5.12. (*Current integrator* or *charge amplifier*)   Referring to Figure P5.12, show that

$$V_O = -\frac{1}{C} \int_{-\infty}^{t} I_i dt = -\frac{Q_i}{C}$$

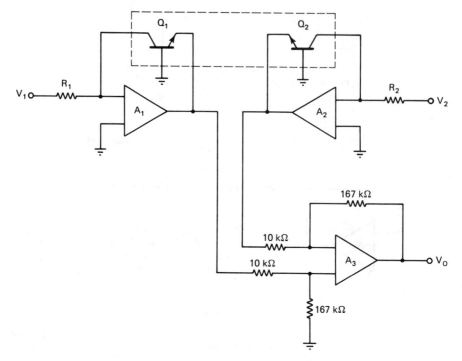

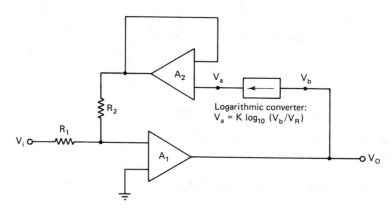

Figure P5.10

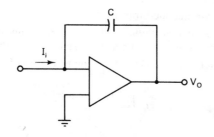

Logarithmic converter:
$V_a = K \log_{10} (V_b/V_R)$

Figure P5.11

Figure P5.12

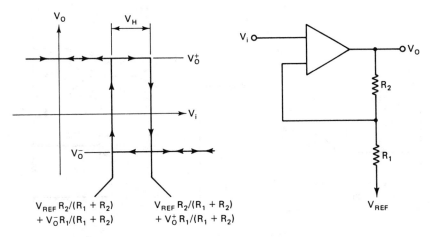

Figure P5.13

5.13. (*Schmitt trigger*)   Referring to Figure P5.13, show that the transfer characteristics given are obtained for the circuit shown where V_o^+ and V_o^- are the saturation output voltage levels of the amplifier.

5.14. (*Positive voltage regulator*)   Referring to Figure P5.14, show that $V_O = V_Z R_2/(R_1 + R_2)$, where V_Z is the breakdown voltage of the zener diode.

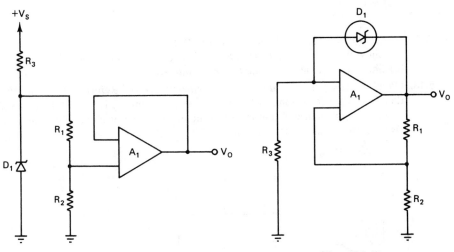

Figure P5.14   **Figure P5.15**

5.15. (*Positive voltage regulator with V_o greater than V_Z*)   Referring to Figure P5.15, show that $V_O = V_Z(1 + R_2/R_1)$.

5.16. (*High-current voltage regulator with current limiting*)   Refer to Figure P5.16.
   (a) Show that $V_L = V_{REF} (1 + R_1/R_2)$.
   (b) Show that $V_{L(MAX)} = V^+ - 0.9$ V.
   (c) Show that $I_{L(MAX)} \simeq 600$ mV/R_{CL}.
   (d) If $V^+ = 20$ V (max.) and $P_{d(MAX)}$ for Q_2 is 50 W, find the appropriate value for $I_{L(MAX)}$.   (*Ans.*: 2.5 A)

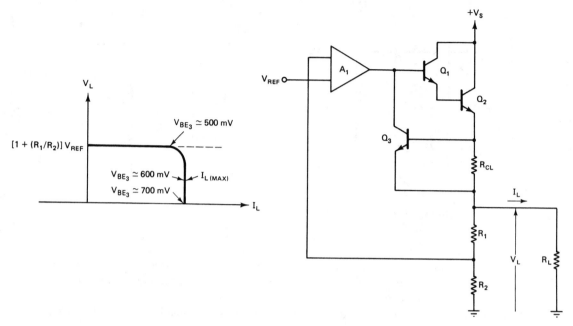

Figure P5.16

5.17 (*Constant-current sink*)   Refer to Figure P5.17.
  (a) Show that $I_o = V_{REF}/R_1$ as long as V_C is greater than $+0.9$ V $+ V_{REF}$.
  (b) Show an equivalent *constant-current source*.

5.18. (*Precision constant-current sink for low current levels*)   Referring to Figure P5.18, show
  that $I_o = V_{REF}/R_1$ as long as $V_D > V_{REF} + V_p$, where V_p is the JFET pinch-off voltage.
  Show an equivalent constant-current source.   Why is this circuit characterized as a *precision* current sink for low current levels?

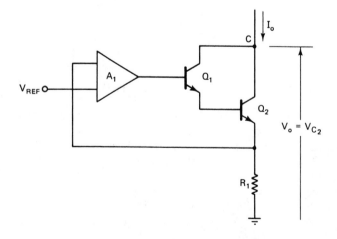

Figure P5.17

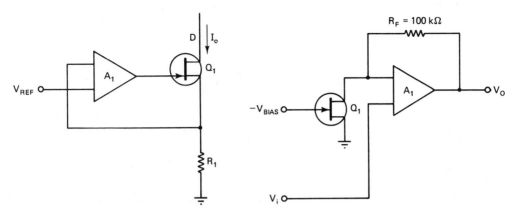

Figure P5.18   Figure P5.19

5.19. (*Op amp with electronic gain control*)   Refer to Figure P5.19. Given $r_{ds(ON)} = 1000\ \Omega$ for Q_1.

(a) Show that for small V_i (V_i small compared to the JFET pinch-off voltage)

$$A_{CL} = \frac{V_O}{V_i} = 1 + \left(\frac{R_F}{r_{ds(ON)}}\right)\left(1 - \sqrt{\frac{V_{BIAS}}{V_P}}\right), \qquad \text{for } V_{BIAS} < V_P$$

(b) Find $A_{CL(MAX)}$ and $A_{CL(MIN)}$.   (*Ans.*: 101, 1)

5.20. (*Op amp with electronic gain control*)   Refer to Figure P5.20.

(a) If both V_R and V_S are small compared to the JFET pinch-off voltage, show that the drain-to-source resistance of both FETs is the same and is equal to $r_{ds} = (V_R/V_C)R_1$ (for $V_C > 0$).

(b) Show that under the conditions above the voltage gain, V_O/V_S, is given by $V_O/V_S = 1 + (R_2/R_1)(V_C/V_R)$.

(c) If $V_{\text{pinch-off}} = 10$ V (min.) and $r_{ds(ON)} = 100\ \Omega$ (max.), find R_2 for a gain range of 1 to 500 (min.).   [*Ans.*: $R_2 = 50$ kΩ (min.)].

(d) If $V_R = +2.0$ V and $V_C = 10$ V (max.), find the required value of R_1 such that r_{ds} can be driven down to its minimum value of $r_{ds(ON)}$.   (*Ans.*: $R_1 = 500\ \Omega$)

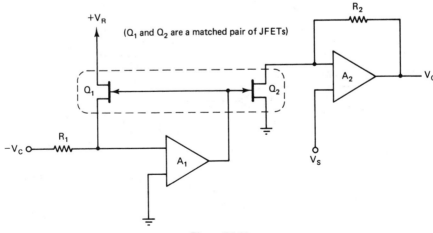

Figure P5.20

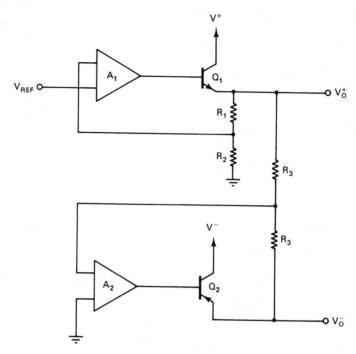

Figure P5.21

5.21. (*Tracking voltage regulator*) Referring to Figure P5.21, show that $V_O^+ = V_{REF}(1 + R_1/R_2)$ and that $V_O^- = -V_O^+$, so this voltage regulator will produce two voltages equal in magnitude but opposite in algebraic sign. (Note that this will be true even if V_S^+ and V_S^- are not equal in magnitude.)

5.22. (*Precision phase splitter with high input impedance and low output impedance*) Referring to Figure P5.22, show that $|V_O|/V_i = 1 + R_2/R_1$.

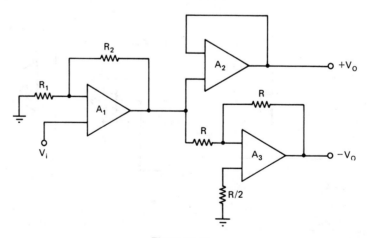

Figure P5.22

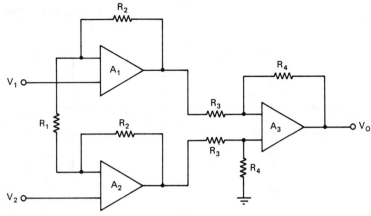

Figure P5.23

5.23. (*Instrumentation amplifier with high input impedance and low output imped-ance*)   Referring to Figure P5.23, show that $V_O = (R_4/R_3)(1 + 2R_2/R_1)(V_2 - V_1)$.

5.24. (*Instrumentation amplifier with high input impedance and low output imped-ance*)   Referring to Figure P5.24, show that $V_O = (R_2/R_1)(V_2 - V_1)$.

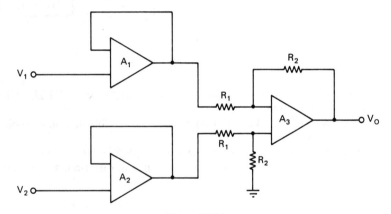

Figure P5.24

5.25. (*Exponential converter* or *antilogarithmic converter*)   Refer to Figure P5.25.
   (a) Show that $V_O = R_3 I_R \exp\left[-V_i R_2/V_T(R_1 + R_2)\right]$.
   (b) If $I_R = 10\ \mu\text{A}$, $R_3 = 100\ \text{k}\Omega$, $R_1 = 160\ \text{k}\Omega$, and $R_2 = 10\ \text{k}\Omega$, show that $V_O = (1.0\ \text{V})10^{-(V_i/1.0\ \text{V})}$, so V_O changes by a factor of 10 for every 1.0-V change in V_i and is 1.0 V when $V_i = 0$.

5.26. (*Circuit for raising a variable to a power with logarithmic techniques*)   Refer to Figure P5.26.   Given:

For the logarithmic converter: $V_O = (1.0\ \text{V}) \log_{10}(V_i/1.0\ \text{V})$.

For the antilogarithmic converter: $V_O = (1.0\ \text{V})10^{-(V_i/1.0\ \text{V})}$.

Show that $V_O = (V_i)^{R_F/R_L}$.

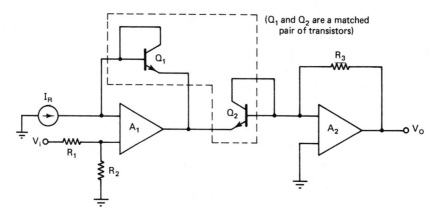

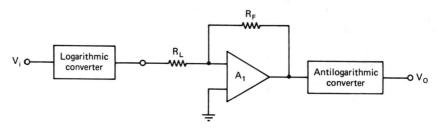

Figure P5.25

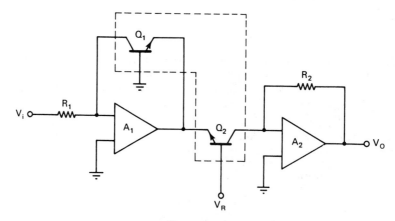

Figure P5.26

5.27. (*Amplifier with exponential gain control*)   Refer to Figure P5.27.   Given that Q_1 and Q_2 are matched transistors.

   (a) Show that $V_O = V_i(R_2/R_1) \exp(V_R/V_T)$, so the voltage gain is exponentially dependent on V_R.

   (b) Find the range in voltage gain available with $R_1 = R_2 = 10\ \text{M}\Omega$ for V_R ranging from 0 up to 200 mV.   (*Ans.:* Voltage gain will vary from 1.0 to 2980.)

Figure P5.27

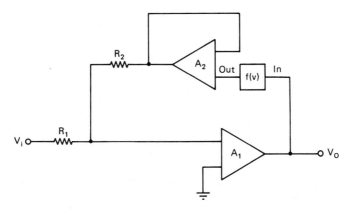

Figure P5.28

5.28. (*Function generator*)   Referring to Figure P5.28, show that, if $f(v) = av^n$, V_O is given by $V_o = (-R_2V_i/aR_1)^{1/n}$.

5.29. (*Dump-and-integrate circuit*)   Refer to Figure P5.29.   In the dump-and-integrate circuit, transistor Q_1 is turned on prior to the integration time to discharge capacitor C_1 and set the output voltage to zero.   Then the transistor is turned off.

  (a) Show that

$$V_o = -\int_0^t \frac{1}{R_1C_1} V_i \, dt$$

  (b) If V_i is $+10$ V and $R_1 = 1$ kΩ, find C_1 such that the output voltage at the end of a 1.0-ms integration period is -10 V.   (*Ans.*: 1.0 μF)

  (c) Describe the effect that the finite open-loop gain, the input offset voltage, and the input bias current have on the accuracy of this integrator circuit.

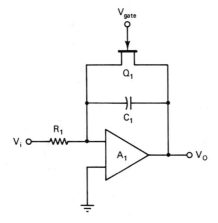

Figure P5.29

5.30. [*Precision voltage-controlled limiting (clipping* or *bounding) circuit*]   Refer to Figure P5.30.

  (a) Draw the V_O versus V_S transfer characteristic and show that (1) $V_O = V_S$ for $V_S \leq V_{REF}$ and (2) $V_O = V_{REF}$ for $V_S \geq V_{REF}$.

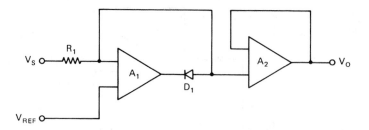

Figure P5.30

(b) Repeat part (a) for the case in which diode D_1 is turned around.

(c) Draw a bilateral limiter circuit with clipping levels at V_{REF1} and V_{REF2}. (*Hint:* Consider using two cascaded clipping circuits.)

(d) Why is this circuit called a *precision* limiting or clipping circuit? What effect will the forward voltage drop of diode D_1 have on the limiting characteristic?

5.31. (*Precision voltage-controlled clamping circuit*)   Refer to Figure P5.31.

(a) Show that the output voltage V_O will have the same ac variation as the input voltage V_S, but that the dc level will be shifted such that the output voltage will not drop below V_{REF}.

(b) If $V_S(t) = 10V \sin(\omega t)$ and $V_{REF} = +5.0$ V, find $V_O(t)$.   [*Ans.:* $V_O(t) = 10V \sin (\omega t) + 15$ V]

(c) Repeat part (b) for $V_{REF} = 0$.   [*Ans.:* $V_O(t) = 10V \sin(\omega t) + 10$ V]

(d) Repeat part (b) for $V_{REF} = -5.0$ V.   [*Ans.:* $V_O(t) = 10V \sin(\omega t) + 5$ V]

(e) If diode D_1 is turned around (reversed polarity), show that the output voltage V_O will again have the same ac variation as the input voltage, but that now the dc level will be shifted such that the output voltage will not rise above V_{REF}.

(f) Repeat part (b), but with the diode turned around.   [*Ans.:* $V_O(t) = 10$ V sin $(\omega t) - 5$ V]

(g) Why is this called a *precision* clamping circuit? What effect does the forward voltage drop of the diode have on the performance of this circuit? Can the clamping level voltage V_{REF} be of either positive or negative polarity?

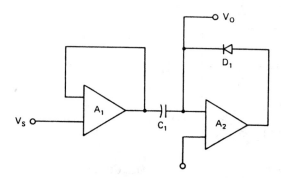

Figure P5.31

5.32. (*Voltage-controlled gain polarity switching circuit*)   Refer to Figure P5.32.

(a) Assuming that the drain-to-source resistance of Q_1, $r_{ds(ON)}$, is very small compared to R_4 when Q_1 is turned on ($V_{control} = 0$), show that the closed-loop gain of this circuit is given by $A_{CL} = V_o/V_s = -(1 + R_2/R_1)$ *when Q_1 is turned on*, and $A_{CL} = V_o/V_s = 1 + R_2/R_1$ *when Q_1 is turned off*.

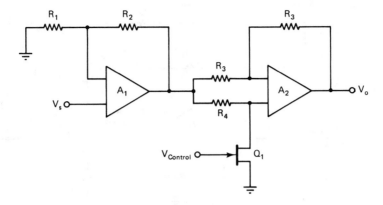

Figure P5.32

(b) If $r_{ds(ON)} = 100 \ \Omega$ (max) for Q_1, find R_4 such that the absolute values of the closed-loop gains given above differ by no more than 1%.   (*Ans.*: $R_4 = 20 \ \text{k}\Omega$)

(c) Discuss the relative advantages and disadvantages, if any, of replacing Q_1 by a bipolar transistor.

5.33. (*Voltage-controlled impedance multiplier*)   Refer to Figure P5.33.

(a) Show that the input impedance of this circuit is given by

$$Z_i = \frac{v_i}{i_i} = \frac{Z_F}{1 + A}$$

and that $Y_i = (1 + A)Y_F$.

(b) If the feedback impedance is a capacitor of value C_F, show that the input capacitance of this circuit is $C_i = C_F(1 + A)$.

(c) How is the impedance transformation produced by this circuit related to the Miller effect?

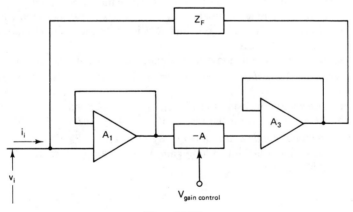

Figure P5.33

5.34. (*Inductance simulator*)   Refer to Figure P5.34.

(a) Show that the input impedance of this circuit is given by

$$Z_i = \frac{v_i}{i_i} = \frac{R_2 R_3}{Z_F} = R_2 R_3 Y_F$$

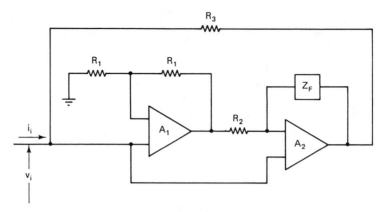

Figure P5.34

so the input impedance is proportional to the reciprocal of the feedback impedance Z_F.

(b) If Z_F is a capacitor of value C_F, show that the input impedance is given by $Z_i = j\omega R_2 R_3 C_F = j\omega L_{eq}$, so the input impedance of this circuit appears as an inductance of value $L_{eq} = R_2 R_3 C_F$.

(c) For the circuit of this problem, the closed-loop voltage gain of A_1 is 2. Repeat part (a) for the more general case in which the closed-loop gain of A_1 is K. [*Ans.*: $Z_i = R_2 R_3 / (K - 1) Z_F$]

(d) The input impedance of this circuit is to appear to be an inductance of 1.0 mH. If $R_1 = R_2 = R_3 = 1000 \ \Omega$, find the value of C_F that is required. (*Ans.*: 1.0 nF)

5.35. (*Analog signal multiplexing circuit*) Refer to Figure P5.35.

(a) Given that the JFET pinch-off voltage is -5 V and the $r_{ds(ON)}$ is much less than R_1 and that ϕ_1 through ϕ_N are nonoverlapping active low (0 to -10 V) clock pulses, show that during the time when ϕ_i is active the output voltage is given by $V_O = -(R_3/R_2)V_{Si}$, where ϕ_i represents any of the clock pulses between 1 and N.

(b) If $r_{ds(ON)} = 100 \ \Omega$ (max), find R_1 such that the channel crosstalk will not exceed 1%. (*Ans.*: $R_1 \gtrsim 10 \text{ k}\Omega$)

(c) Describe the function of the voltage followers, A_1 through A_N.

5.36. (*Symmetrical bipolar limiter*) Refer to Figure P5.36. Given that $V_Z = 9.4$ V for D_1 and D_2.

(a) Draw and dimension the V_O versus V_i transfer characteristics.

(b) Repeat part (a) for the case of $V_Z = 9.4$ V for D_1 and 4.4 V for D_2.

5.37. (*Constant-amplitude phase shifter*) Refer to Figure P5.37.

(a) Assuming a high-open-loop gain, ideal op amp, show that the closed-loop voltage gain is given by

$$A_{CL} = \frac{v_o}{v_i} = 1.0 \ \angle -2 \tan^{-1}(\omega C_2/G_2)$$

$$= 1.0 \ \angle -2 \tan^{-1}(\omega R_2 C_2)$$

(b) Find A_{CL} for $R_1 = 2.0 \text{ k}\Omega$, $R_2 = 1.0 \text{ k}\Omega$, $C_2 = 10$ nF, and $f = 10$ kHz. (*Ans.*: $A_{CL} = 1.0 \ \angle -64.3°$)

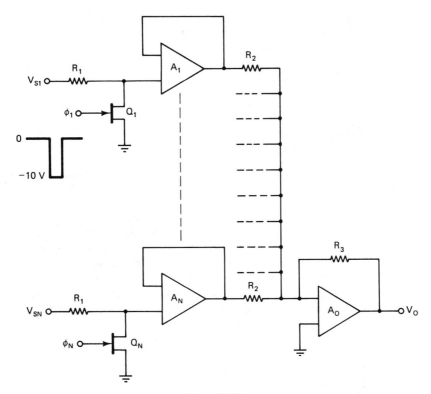

Figure P5.35

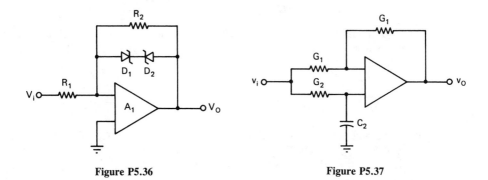

Figure P5.36   **Figure P5.37**

5.38. (*Bridge amplifier*)   Refer to Figure P5.38.
   (a) Show that

$$V_O = \left(1 + \frac{G_1}{G_2}\right) \frac{-\Delta G}{2G_3 + \Delta G} V^+$$

   (*Hint:* Use nodal analysis.)
   (b) We will now consider an application of the bridge amplifier.   The bridge amplifier

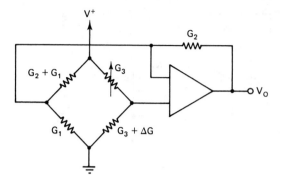

Figure P5.38

will be used to measure small changes in temperature. Given that $R_1 = 1.0$ kΩ, $R_2 = 9.0$ kΩ; $G_3 + \Delta G$ is a resistor with a temperature coefficient of 0.20%/°C; G_3 is a resistor similar to $G_3 + \Delta G$, but kept at a constant temperature and adjusted (with the aid of a trimming potentiometer) so that $V_O = 0$ at $T = 25°C$; and $V^+ = 10$ V. Find V_O as a function of temperature. [*Ans.*: $V_O = (0.10 \text{ V/°C})(T - 25°C)$]

5.39. (*Active RC bandpass amplifier*) Refer to Figure P5.39.

(a) Show that the closed-loop gain is given by

$$A_{CL} = \frac{V_O}{V_i} = \frac{-y_1 y_3}{(y_1 + y_2 + y_3 + y_4) y_5 + y_3 y_4}$$

(b) If $y_1 = G_1$, $y_2 = G_2$, $y_3 = j\omega C_3$, $y_4 = j\omega C_4$, and $y_5 = G_5$, obtain an equation for A_{CL}. [*Ans.*:

$$A_{CL} = \frac{-j\omega C_3 G_1}{G_5(G_1 + G_2) - \omega^2 C_3 C_4 + j\omega G_5(C_3 + C_4)}]$$

(c) Find the frequency at which the gain is a maximum and the value of the gain at that frequency. [*Ans.*: At $\omega_o = \sqrt{G_5(G_1 + G_2)/(C_3 C_4)}$ the gain will have a maximum value of

$$A_{CL(MAX)} = -\left(\frac{C_3}{C_3 + C_4}\right)\frac{G_1}{G_5} = -\left(\frac{C_3}{C_3 + C_4}\right)\frac{R_5}{R_1}]$$

(d) Given that the 3-dB bandwidth (in *rad/s*) is $BW = (C_3 + C_4)/(C_3 C_4 R_5)$, find R_1, R_2, and R_5 if $f_o = 100$ Hz, $BW = 10$ Hz, $A_{CL(MAX)} = -10$, and $C_3 = C_4 = 1.0$ μF. (*Ans.*: $R_1 = 1.59$ kΩ, $R_2 = 83.8$ Ω, $R_5 = 31.8$ kΩ)

5.40. (*Square-wave oscillator*) Refer to Figure P5.40. Given that the positive and negative saturation voltages of the op amp are equal.

(a) Show that the frequency of oscillation is given by

$$f_{osc} = \frac{1}{2R_1 C_1 \ln 3}$$

(b) Draw the waveform of V_O and V_{C1}.

5.41. (*Precision linear triangle wave generator*) Refer to Figure P5.41. Given that the output voltage of A_1 will saturate at V_H in the high state and V_L in the low state, and A_1 can be either a voltage comparator or an operational amplifier.

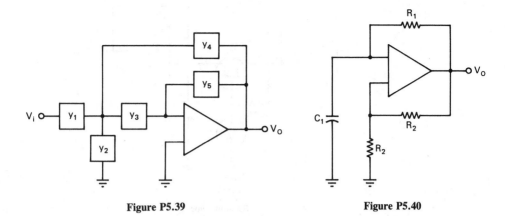

Figure P5.39　　　　　　　　　　　**Figure P5.40**

(a) Show that the output voltage V_O is a linear triangular waveform with a peak-to-peak voltage of $V_{O(P-P)} = (V_H - V_L)R_5/(R_4 + R_5)$.

(b) Show that the period T of the triangular waveform is given by

$$T = (R_1 C_1)(R_2/R_3) \left[\frac{V_{O(P-P)}}{V_H} + \frac{V_{O(P-P)}}{-V_L} \right]$$

$$= (R_1 C_1) \left[\frac{R_2}{R_3} \right] \left[\frac{R_5}{R_4 + R_5} \right] \left[\frac{V_H}{-V_L} + \frac{-V_L}{V_H} \right]$$

(c) If $V_{REF} = 0$, $V_H = +10$ V, $V_L = -10$ V, $R_4 = R_5$, $R_2 = R_3$, $R_1 = 10$ kΩ, and $C_1 = 10$ nF, (1) draw and dimension the output voltage waveform, and (2) find the frequency of oscillation.　　　(*Ans.*: $f_{osc} = 10$ kHz)

(d) Describe the effect that voltage V_{REF} will have on the output voltage.

(e) Why is this called a *precision linear triangular wave generator*? What factors will limit the maximum frequency of oscillation and the linearity of the waveform?

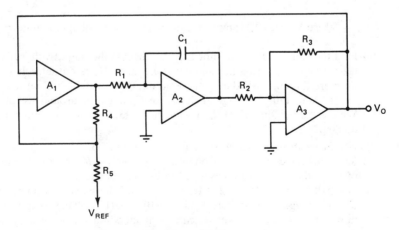

Figure P5.41

5.42. (*Sample-and-hold circuit*)　Refer to Figure P5.42.　Given for the JFET: $r_{ds(ON)} = 300$ Ω (max.), $V_p = -3.0$ V (max.), and $I_{DS(OFF)} = 500$ pA (max.)　Given for the op amp:

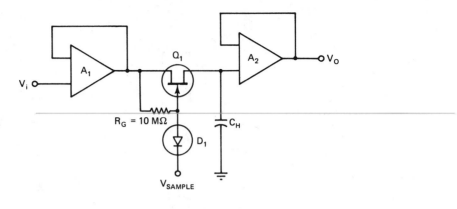

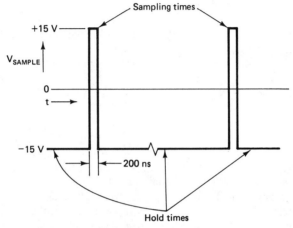

Figure P5.42

$I_{\text{BIAS}} = 500$ pA (max.), sampling rate $= 100$ kHz, sampling time $= 200$ ns, and $V_{\text{supply}} = \pm 15$ V.

   (a) Find the value of the capacitor C_H such that **(1)** the output voltage, V_O, reaches at least 98% of V_i during the *sampling time*, and **(2)** the output voltage decreases by no more than 0.1 mV during the hold interval.   (*Ans.:* C must be between the limits 100 to 170 pF.   A good choice for C_H might be around 130 pF.)
   (b) What is the function of **(1)** R_G, **(2)** the diode, **(3)** the first op amp, and **(4)** the second op amp?

5.43. (*Howland current source circuit*)   Refer to Figure P5.43.
   (a) Show that the current I_L is given by $I_L = (V_1 - V_2)/R_1$ and is therefore independent of R_L so that this circuit acts as a constant-current source.
   (b) Given $R_1 = 1$ kΩ, $R_2 = 250$ Ω, $V_2 = 0$, and the maximum operational amplifier output voltage swing is from -10 to $+10$ V.   The voltage compliance range is the range of load voltages over which this circuit will act as a constant-current source. Find the voltage compliance range.   (*Ans.:* -8.0 to $+8.0$ V)
   (c) If the maximum input voltage range for V_1 is -10 to $+10$ V, find the maximum range of the output current I_L.   Use the conditions of part (b).   (*Ans.:* -10 to $+10$ mA)

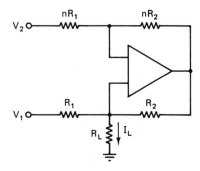

Figure P5.43

5.44. (*Circuit to produce an output voltage that increases linearly with temperature*) Referring to Figure P5.44, show that V_O is given by $V_O = [(R_3/R_2)(k/q)(\ln 2)]\,T = (1.0\ \text{mV/°C})T$, so V_O will increase linearly with temperature with a coefficient of $1.0\ \text{mV/K} = 1\ \text{mV/°C}$ and T is the absolute temperature (K).

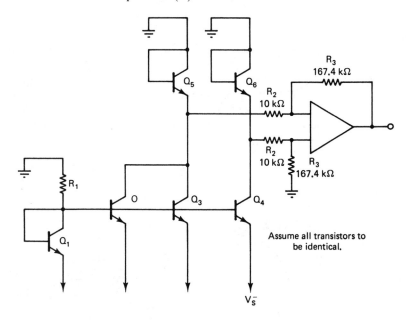

Figure P5.44

5.45. (*Voltage regulator*) Refer to Figure P5.45.

(a) If $R_1 + R_2 = 10\ \text{k}\Omega$, find R_1 and R_2 for $V_o = +10.0\ \text{V}$.   (*Ans.*: $R_1 = 8.2\ \text{k}\Omega$, $R_2 = 1.8\ \text{k}\Omega$)

(b) If $A_{OL}(0) = 10{,}000$, $f_u = 1.0\ \text{MHz}$, and $R_{o(OL)} = 50\ \Omega$, find $Z_{o(CL)}$ at (1) dc and (2) 10 kHz.   (*Ans.*: (1) $28 + j0$ mΩ, (2) $0 + j2.8\ \Omega$)

(c) If the reference voltage has a temperature coefficient given by $TC_{V(\text{REF})} = 10\ \mu\text{V/°C}$, find the temperature coefficient of the regulated output voltage.   (*Ans.*: $TC_{Vo} = 55.6\ \mu\text{V/°C}$)

(d) Find the *load regulation* (change in V_o) as the output current I_o goes from a no-load value of $I_{o(\text{NL})} = 0$ to a full-load value of $I_{o(\text{FL})} = 10\ \text{mA}$. Express the result in terms of ΔV_o and the percentage of change in V_o.   (*Ans.*: $-280\ \mu\text{V}$ or -0.0028%)

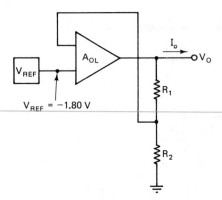

$V_{REF} = -1.80 \text{ V}$

Figure P5.45

5.46. (*Voltage regulator with current foldback*)   Refer to Figure P5.46.

(a) Show that current limiting will occur when

$$\frac{I_O R_{CL} R_2}{R_1 + R_2} - \frac{V_O R_1}{R_1 + R_2} \simeq 0.6 \text{ V}$$

(b) Given that $P_{D(MAX)}$ of Q_2 is 5.0 W, $V_{IN} = +15$ V.   Find the value of $I_{O(MAX)}$ for $V_O = +10$ V, and for $V_O = 0$.   (*Ans.:* 1.0 A, 0.33 A)

(c) Given that $R_1 + R_2 = 1.0$ kΩ, find R_1, R_2, and R_{CL} such that the current limiting corresponding to the conditions above will be obtained.   (*Ans.:* $R_1 = 120 \ \Omega$, $R_2 = 880 \ \Omega$, $R_{CL} = 2.05 \ \Omega$)

(d) Draw the V_O versus I_O characteristics of the voltage regulator corresponding to the conditions above.

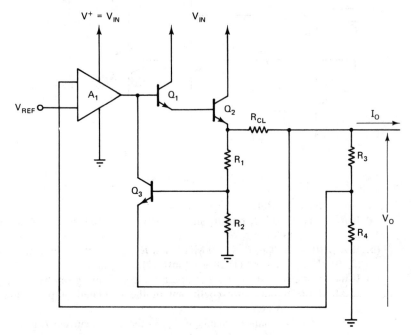

Figure P5.46

(e) If $R_3 + R_4 = 2.0$ kΩ and $V_{REF} = 1.80$ V, find R_3 and R_4 for a regulated output voltage of 10 V. (*Ans.*: $R_3 = 1640\ \Omega$, $R_4 = 360\ \Omega$)

5.47. (*Analog multiplier*)
  (a) For the circuit shown in Figure P5.47a, assume all transistors to have identical characteristics and all op amps to be ideal with very high open-loop gains. Show that the following relationship will be obtained, with all currents being positive:

$$I_1 I_2 = I_3 I_4$$

This circuit is available in monolithic IC form as the RC4200 (Raytheon).
  (b) (*Four-quadrant multiplier*) For the circuit shown in Figure P5.47b, show that the following relationship is obtained.

$$V_o = \frac{V_X V_Y}{V_R} \frac{R_2 R_o}{R_1^2}$$

If $R_1 = R_2$, $R_o = 20$ kΩ, and $V_R = +10$ V, find V_O. (*Ans.*: $V_O = V_X V_Y/10$ V)
  If all four currents (I_1 through I_4) must be positive and limited to a maximum value of 1.0 mA for the accurate operation of this circuit, what will be the corresponding restrictions on V_X and V_Y? (*Ans.*: -10 V $< V_X < +10$ V, -10 V $< V_Y < +10$ V)
  Why is this circuit called a four-quadrant multiplier?
  (c) (*One-quadrant analog divider*) For the circuit shown in Figure P5.47c, with V_X, V_Z, and V_R being of positive polarity, show that the following relationship is obtained:

$$V_O = \frac{V_X}{V_Z} V_R \frac{R_4 R_0}{R_1 R_2}$$

Why is this circuit called a one-quadrant divider?
  (d) (*Square-rooting circuit*) To produce a square-rooting circuit, V_Z in the divider circuit (Figure P5.47c) is connected to V_O such that $V_Z = V_O$. Show that V_O will now become proportional to the square root of V_X as given by

$$V_O = \sqrt{\frac{V_X V_R R_0 R_4}{R_1 R_2}}$$

If all resistors are equal and $V_R = +10$ V, find V_O. (*Ans.*: $V_O = \sqrt{10 V_X}$)

5.48. (*Phase-shift oscillator*) Refer to Figure P5.48. For oscillators to occur, the net gain around the feedback loop must be greater than unity and the net phase shift around the feedback loop must be zero.
  (a) Show that the frequency of oscillation is given by

$$f_{osc} = \frac{1}{2\pi\sqrt{3}\ R_1 C_1}$$

and the condition for oscillation to occur is that $R_3/R_2 > 8$.
  (b) If $R_1 = R_2 = 10$ kΩ and $C_1 = 1.0$ nF, find the frequency of oscillation and the value of R_3 required for oscillations to occur. (*Ans.*: 9.19 kHz, 80 kΩ)
  (c) If the net voltage gain around the feedback loop is adjusted to a value just slightly greater than unity, the output voltage will be a relatively undistorted sinusoidal waveform. Design a circuit that will automatically do this.

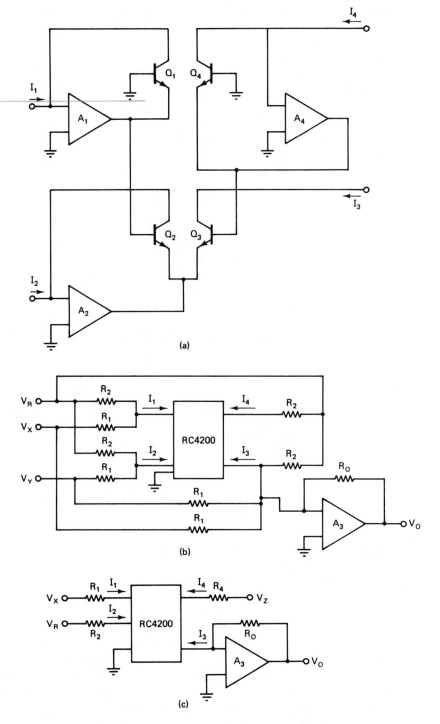

(a)

(b)

(c)

Figure P5.47

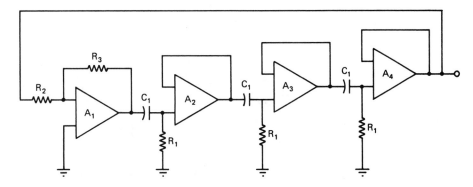

Figure P5.48

5.49. (*Wien bridge oscillator*) Refer to Figure P5.49. For this circuit to oscillate, the *net* feedback must be positive and have a phase angle of zero.

(a) Show that the frequency of oscillation is given by

$$f_{osc} = \frac{1}{2\pi\sqrt{R_1 R_2 C_1 C_2}}$$

and the condition for oscillation is

$$\frac{R_3}{R_4} > \frac{R_1}{R_2} + \frac{C_2}{C_1}$$

(b) If $R_1 = R_2 = R_4 = 10$ kΩ and $C_1 = C_2 = 1.0$ nF, find the frequency of oscillation and the value of R_3 required for oscillations to occur.   (*Ans.*: 15.9 kHz, 20 kΩ)

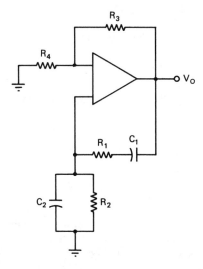

Figure P5.49

5.50. (*Optoelectronic sensor*) Given that the photodiode in Figure P5.50 has an active area of 10 mm^2 and a current responsivity (ratio of output current to incident optical power) of 0.5 A/W. Find the output voltage V_o for an incident optical power density of 100 nW/cm^2.   (*Ans.*: 5 mV)

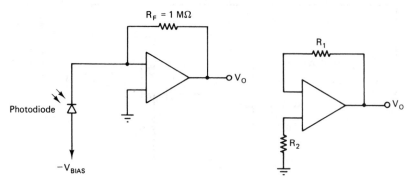

Figure P5.50   **Figure P5.51**

5.51. (*Circuit for obtaining V_{OS}, I_B, and I_{OS}*)   Refer to Figure P5.51.   Given:

$$V_O = +2.2 \text{ mV when } R_1 = R_2 = 0$$

$$V_O = +20 \text{ mV when } R_1 = R_2 = 100 \text{ M}\Omega$$

$$V_O = -120 \text{ mV when } R_1 = 0 \text{ and } R_2 = 100 \text{ m}\Omega$$

Find:
   (a) V_{OS}.   (*Ans.*: 2.2 mV)
   (b) I_{BIAS}.   (*Ans.*: 1.2 nA)
   (c) I_{OS}.   (*Ans.*: 178 pA)

5.52. (*Effect of supply voltage on output voltage swing*)   Find the output voltage v_o for the circuits of Figure P5.52.
   (a) (*Ans.*: v_o will be near $+11$ or -11 V)
   (b) (*Ans.*: v_o will be near $+11$ V if I_B is positive and near -11 V if I_B is negative)
   (c) (*Ans.*: v_o will be approximately a square wave with an amplitude of ± 11 V)

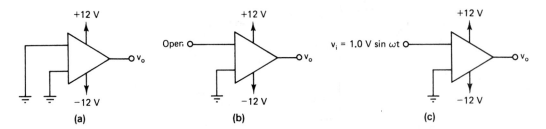

Figure P5.52

5.53. (*Effect of current limiting on output voltage*)   Refer to Figure P5.53.   Given: $I_{CL} = \pm 20$ mA.   Find v_o.   [*Ans.*: v_o will be a 10-V amplitude sine wave clipped at the ± 2.0-V level and will therefore look approximately like a square wave of ± 2.0-V amplitude]

5.54. (*Circuit for evaluating PSRR*)   Refer to Figure P5.54.   Given: $V_O = 2.0$ mV (rms) at 1 kHz.   Find PSRR (in dB).   (*Ans.*: 94 dB)

5.55. (*Circuit for obtaining open-loop gain*)   Refer to Figure P5.55.   Given $V_O = 5.0$ V (ac) and $V_1 = 20$ mV (ac).   Find the open-loop gain of the op amp.   (*Ans.*: $A_{OL} = 250,000$ or 108 dB)

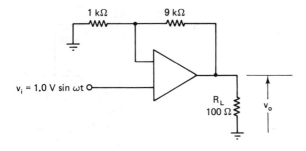

Figure P5.53

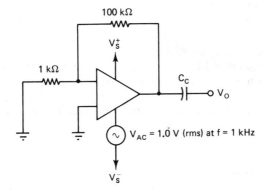

Figure P5.54

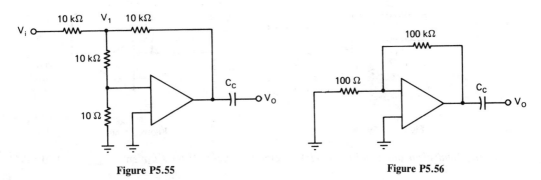

Figure P5.55   Figure P5.56

5.56. (*Evaluation of equivalent input noise voltage*)   Refer to Figure P5.56.   Given: $V_{O(\text{noise})} = 2.5$ mV (rms).   Bandwidth (3 dB) = 10 kHz.
   (a) Find the amplifier equivalent input noise voltage.   (*Ans.*: 2.5 μV)
   (b) Find the amplifier equivalent input noise voltage spectral density.   [*Ans.*: 20 nV/ (Hz)$^{1/2}$]

5.57. (*Common-mode input impedance*)   Refer to Figure P5.57.   Given: $f_u = 1.0$ MHz, $A_{OL}(0) = 106$ dB (min.), $z_{i(\text{CM})} = [1.0 \text{ G}\Omega \parallel 4.0 \text{ pF}]$.   Find:
   (a) $A_{CL}(0)$.   (*Ans.*: 0.999)
   (b) 3-dB bandwidth.   (*Ans.*: 40 kHz)
   (c) 0.5-dB bandwidth.   (*Ans.*: 14 kHz)

5.58. (*Closed-loop gain versus frequency*)   Refer to Figure P5.58.   Given $f_u = 1.0$ MHz, $A_{OL}(0) = 106$ dB (min.).   Find.

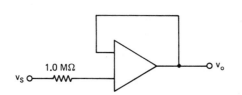

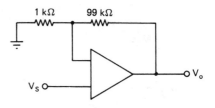

Figure P5.57	Figure P5.58

   (a) Open-loop 3-dB bandwidth.   [*Ans.*: 5.0 Hz (max.)]
   (b) A_{CL} at 1.0 kHz.   (*Ans.*: 99.5 $\angle -5.7°$)
   (c) A_{CL} at 10 kHz.   (*Ans.*: 71 $\angle -45°$)
   (d) A_{CL} at 100 kHz.   (*Ans.*: 9.95 $\angle -84.3°$)

5.59. (*Common-mode gain*)   Refer to Figure P5.59.
   (a) Find v_o if CMRR = 100 dB at 1 kHz.   (*Ans.*: v_o = 100 μV at 1 kHz)
   (b) Find v_o resulting from a ±1% mismatch between the two R_2 resistors.   (*Ans.*: ±9 mV)
   (c) Find v_o resulting from a ±1% mismatch between the two R_1 resistors.   (*Ans.*: ±9 mV)

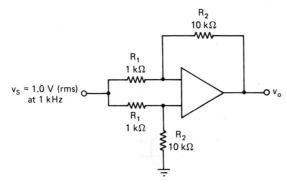

Figure P5.59

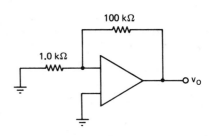

Figure P5.60

5.60. (*Equivalent input noise voltage*)   Refer to Figure P5.60.   Given: $v_{o(\text{noise})}$ = 0.20 mV (rms)
   (a) Find the equivalent input noise voltage, $v_{i(\text{noise})}$.   [*Ans.*: 2.0 μV (rms)]
   (b) If the amplifier has a unity-gain frequency of 1.0 MHz, find the equivalent input noise voltage spectral density.   [*Ans.*: 16 nV/(Hz)$^{1/2}$] (*Hint:* Remember the $\pi/2$ conversion factor to go from 3-dB bandwidth to equivalent noise bandwidth.)

5.61. (*Amplifier-to-amplifier coupling*)   Referring to Figure P5.61, find the amplifier-to-amplifier coupling (in dB).   (*Ans.*: −94 dB)

5.62. (*Power dissipation and temperature rise*)   Refer to Figure P5.62. Given an op amp with $T_{J(\text{MAX})}$ = 125°C, thermal impedance = Θ = 100°C/W, I_{CL} = ±50 mA, $TC_{V_{OS}}$ = 5 μV/°C, and quiescent supply current = $I_{S(Q)}$ = 1.0 mA. Find:
   (a) Maximum power dissipation, $P_{D(\text{MAX})}$ (for T_{ambient} = 25°C).   [*Ans.*: $P_{D(\text{MAX})}$ = 1.0 W]
   (b) Quiescent power dissipation, $P_{D(Q)}$.   [*Ans.*: $P_{D(Q)}$ = 36 mW]

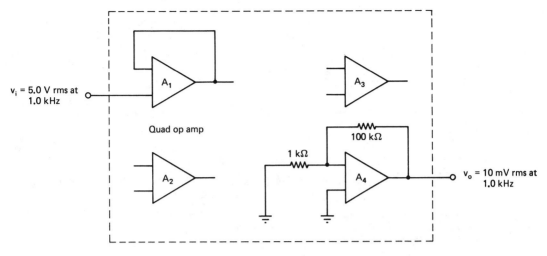

Figure P5.61

(c) Maximum allowable current limit for full protection of op amp if maximum rated supply voltage is ± 18 V.   (*Ans.*: $I_{CL} = \pm 55$ mA)
(d) Power delivered to load, P_O.   (*Ans.*: $P_O = 400$ mW)
(e) Op-amp power dissipation.   (*Ans.*: $P_D = 356$ mW)
(f) Temperature rise of op-amp chip above ambient temperature, ΔT_J.   (*Ans.*: $\Delta T_J = 35.6°C$)
(g) Change in the offset voltage result from the temperature rise above.   (*Ans.*: $\Delta V_{OS} = 178$ μV)

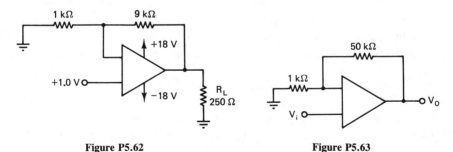

Figure P5.62   **Figure P5.63**

5.63. (*Bandwidth and rise-time calculations*)   Refer to Figure P5.63.   Given: $f_u = 1.0$ MHz and $f_2 > f_u$.   Find:
   (a) 3-dB bandwidth.   (*Ans.*: 19.6 kHz)
   (b) Rise time.   (*Ans.*: 17.9 μs)

5.64. (*Compensation capacitance calculations*)   Refer to Figure P5.64.   Given that when $C_{COMP} = 40$ pF, $A_{OL} = 0$ dB at $f = f_2 = 1.0$ MHz.   Assume that $f_u \propto 1/C_{COMP}$.
   (a) Find the 3-dB bandwidth and the rise time (for small-signal conditions).   (*Ans.*: 354 kHz, 1.0 μs)
   (b) Find the smallest value of C_{COMP} that can be used to maintain a phase margin of at least 45°.   (*Ans.*: 10 pF)

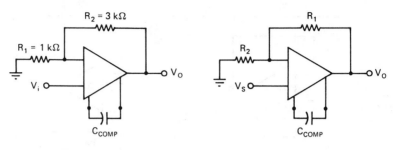

Figure P5.64   **Figure P5.65**

(c) Find the corresponding bandwidth and rise time when this value of C_{COMP} is used.   (*Ans.*: $BW = 1.4$ MHz, $t_{\text{rise}} = 248$ ns)

5.65. (*Compensation capacitance calculations*)   Refer to Figure P5.65.   Given: $f_u = 1.0$ MHz when $C_{\text{COMP}} = 30$ pF, and $f_u = 21$ MHz when $C_{\text{COMP}} = 0$ (extrapolated values) and $f_2 = 1.0$ MHz.   Assume that $f_u \propto 1/(C_{\text{COMP}} + C_{\text{parasitic}})$.   Find:

  (a) f_u for $C_{\text{COMP}} = 3.0$ pF.   (*Ans.*: 7.0 MHz)
  (b) The 3-dB bandwidth if $R_1 = 39$ kΩ, $R_2 = 1.0$ kΩ, and $C_{\text{COMP}} = 30$ pF.   (*Ans.*: 25 kHz)
  (c) Repeat part (b) for $C_{\text{COMP}} = 3.0$ pF.   (*Ans.*: 175 kHz)
  (d) If $R_1 = 30$ kΩ and $R_2 = 10$ kΩ, find C_{COMP} for a 45° phase margin.   What will be the corresponding bandwidth?   [*Ans.*: $C_{\text{COMP}} = 4.07$ pF (min.), $BW \simeq 1.0$ MHz]

REFERENCES

BARNA, A., *Operational Amplifiers*, Wiley, New York, 1971.

———, and D. I. PORAT, *Operational Amplifiers*, 2nd ed., Wiley, New York, 1989.

CARR, J. J., *Linear IC/Op Amp Handbook*, TAB Books, Blue Ridge Summit, Pa., 1983.

CONNELLY, J. A., *Analog Integrated Circuits*, Wiley, New York, 1975.

FITCHEN, F., *Electronic Integrated Circuits and Systems*, Van Nostrand Reinhold, New York, 1970.

FRANCO, S., *Design with Operational Amplifiers and Analog Integrated Circuits*, McGraw-Hill, New York, 1988.

FREDERIKSON, T. M., *Intuitive Operational Amplifiers*, McGraw-Hill, New York, 1988.

GLASER, A. B., and G. E. SUBAK-SHARPE, *Integrated Circuit Engineering*, Addison-Wesley, Reading, Mass., 1977.

GRAEME, J. G., *Applications of Operational Amplifiers*, McGraw-Hill, New York, 1973.

GRAY, P. R., and R. G. MEYER, *Analysis and Design of Analog Integrated Circuits*, Wiley, New York, 1984.

GRINICH, V. H., and H. G. JACKSON, *Introduction to Integrated Circuits*, McGraw-Hill, New York, 1975.

HAMILTON, D. J., and W. G. HOWARD, *Basic Integrated Circuit Engineering*, McGraw-Hill, New York, 1975.

HNATEK, E. R., *A User's Handbook of Integrated Circuits*, Wiley, New York, 1973.

HUGHES, F. W., *Op Amp Handbook*, Prentice-Hall, Englewood Cliffs, N.J., 1981.

IRVINE, R. G., *Operational Amplifiers—Characteristics and Applications*, 2nd ed., Prentice-Hall, Englewood Cliffs, N.J., 1987.

JACOB, J. M., *Applications and Design with Analog Integrated Circuits*, Prentice-Hall, Englewood Cliffs, N.J., 1982.

JUNG, W. G., *IC Op-Amp Cookbook*, Howard W. Sams, Indianapolis, Ind., 1986.

KENNEDY, E. J., *Operational Amplifier Circuits*, Holt, Rinehart, and Winston, New York, 1988.

LENK, J. D., *Handbook of Integrated Circuits*, Prentice-Hall, Englewood Cliffs, N.J., 1978.

———, *Manual for Integrated Circuit Users*, Prentice-Hall, Englewood Cliffs, N.J., 1973.

———, *Manual for Operational Amplifier Users*, Prentice-Hall, Englewood Cliffs, N.J., 1976.

MANASSE, F. K., *Semiconductor Electronics Design*, Prentice-Hall, Englewood Cliffs, N.J., 1977.

MCMENAMIN, J. M., *Linear Integrated Circuits: Operation and Applications*, Prentice-Hall, Englewood Cliffs, N.J., 1985.

MILLMAN, J., *Microelectronics*, McGraw-Hill, New York, 1979.

RIPS, E. M., *Discrete and Integrated Electronics*, Prentice-Hall, Englewood Cliffs, N.J., 1986.

ROBERGE, J. K., *Operational Amplifiers*, Wiley, New York, 1975.

SEDRA, A. S., and K. C. SMITH, *Microelectronic Circuits*, Holt, Rinehart and Winston, New York, 1982.

SEIPPEL, R. G., *Operational Amplifiers*, Prentice-Hall, Englewood Cliffs, N.J., 1983.

UNITED TECHNICAL PUBLICATIONS, *Modern Applications of Integrated Circuits*, TAB Books, Blue Ridge Summit, Pa., 1974.

WAIT, J. V., *Introduction to Operational Amplifier Theory and Applications*, McGraw-Hill, New York, 1975.

WONG, Y. J., and W. E. OTT, *Function Circuits*, McGraw-Hill, New York, 1976.

6

IC Active Filters and Switched-capacitor Circuits

6.1 ACTIVE FILTERS

An important category of op-amp applications is that of active filters. In the following discussion, various low-pass, high-pass, bandpass, and band-reject active filters will be presented together with the appropriate design equations.

The ideal frequency selective filter is a device or system that has an input-to-output transfer characteristic that is constant over the specified passband and provides zero signal transmission in the stop bands. In Figure 6.1, the transfer characteristics of some ideal and actual low-pass, high-pass, bandpass, and band-reject filters are shown.

Passive filters use only passive elements, resistors, capacitors, and inductors. An active filter uses one or more active devices, usually an operational amplifier, in the filter circuit. Some advantages of active filters over their passive counterparts are:

1. *Gain:* With active filters, transfer functions with maximum values greater than unity are possible.
2. *Minimal loading effects:* The transfer characteristics of an active filter can be substantially independent of the load driven by the filter or of the source that drives the filter.

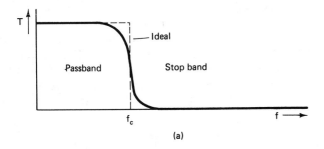

(a)

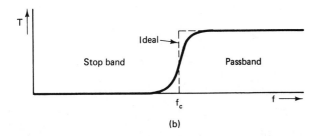

(b)

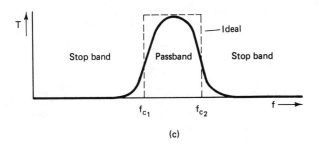

(c)

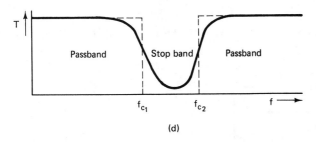

(d)

Figure 6.1 Frequency selective filter characteristics: (a) low-pass filter; (b) high-pass filter; (c) bandpass filter; (d) band-reject filter.

3. *Inductorless filters:* With active filters, only resistive and capacitative elements are required; no inductors are needed. This can be an especially useful feature for operation at relatively low frequencies (<10 kHz) where otherwise large inductors would be required.

6.1.1 Analysis of General Two-pole Active Filter Circuit

Let us consider the circuit of Figure 6.2. If we assume an ideal infinite-gain op amp, the output voltage V_o is equal to the voltage at node B, so $V_o = V_B$. The node voltage equation for node A is $V_{in}Y_1 + V_oY_3 + V_oY_2 = V_A(Y_1 + Y_2 + Y_3)$ and for nobe B we have that $V_AY_2 = V_o(Y_2 + Y_4)$. Solving this last node voltage equation for V_A gives $V_A = V_o(Y_2 + Y_4)/Y_2$. Substituting this back into the equation for node A gives

$$V_{in}Y_1 + V_o(Y_2 + Y_2) = \frac{V_o(Y_2 + Y_4)(Y_1 + Y_2 + Y_3)}{Y_2} \tag{6.1}$$

Multiplying this equation through by Y_2 yields

$$V_{in}Y_1Y_2 + V_o(Y_2^2 + Y_2Y_3)$$
$$= V_o(Y_1Y_2 + Y_2^2 + Y_2Y_3 + Y_1Y_4 + Y_2Y_4 + Y_3Y_4) \tag{6.2}$$

and upon cancellation of like terms this reduces to

$$V_{in}Y_1Y_2 = V_o(Y_1Y_2 + Y_1Y_4 + Y_3Y_4) \tag{6.3}$$

Solving for the filter transfer function $T = V_o/V_{in}$ gives

$$
\begin{aligned}
T = \frac{V_o}{V_{in}} &= \frac{Y_1Y_2}{Y_1Y_2 + Y_1Y_4 + Y_2Y_4 + Y_3Y_4} \\
&= \frac{Y_1Y_2}{Y_1Y_2 + Y_4(Y_1 + Y_2 + Y_3)}
\end{aligned} \tag{6.4}
$$

Two-pole low-pass active filter. We will now consider the special case of a two-pole low-pass active filter for which $Y_1 = G_1$, $Y_2 = G_2$, $Y_3 = sC_3$, and $Y_4 = sC_4$, as shown in Figure 6.3. The transfer function for this case is

$$T = \frac{G_1G_2}{G_1G_2 + sC_4(G_1 + G_2 + sC_3)} \tag{6.5}$$

Note that at zero frequency where $s = j\omega = 0$, we have that $T = 1$, and as the

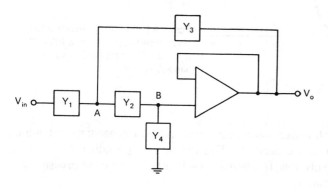

Figure 6.2 General two-pole active filter circuit.

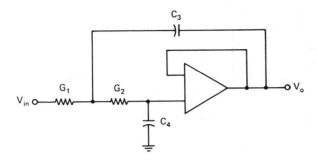

Figure 6.3 Two-pole active low-pass filter.

frequency ($s = j\omega$) goes to infinity, T will go to zero, so this will indeed act as a low-pass filter.

For simplicity, let us now further specify that $G_1 = G_2 = G = 1/R$ so that the transfer function can now be written as

$$T = \frac{G^2}{G^2 + sC_4(2G + sC_3)} = \frac{1}{1 + (sC_4/G)(2 + sC_2/G)}$$

$$= \frac{1}{1 + sRC_4(2 + sRC_3)} = \frac{1}{1 + s\tau_4(2 + s\tau_3)}$$

(6.6)

where $\tau_4 = RC_4$ and $\tau_3 = RC_3$. To examine the response in the frequency domain, we let $s = j\omega$ and obtain

$$T = \frac{1}{1 + j\omega\tau_4(2 + j\omega\tau_3)} = \frac{1}{1 - \omega^2\tau_3\tau_4 + j(2\omega\tau_4)}$$

(6.7)

The square of the magnitude of the filter transfer function is

$$|T|^2 = \frac{1}{1 - 2\omega^2\tau_3\tau_4 + \omega^4\tau_3^2\tau_4^2 + 4\omega^2\tau_4^2}$$

$$= \frac{1}{1 + \omega^2(4\tau_4^2 - 2\tau_3\tau_4) + \omega^4\tau_3^2\tau_4^2}$$

(6.8)

If we want a *maximally flat* filter frequency response that falls off monotonically with frequency, we first set $d|T|^2/d\omega = 0$ and then solve for the condition of a zero slope ocurring only at $\omega = 0$. Doing this gives $2\omega(4\tau_4^2 - 2\tau_3\tau_4) + 4\omega^3(\tau_3^2\tau_4^2) = 0$, so $2\tau_4^2 - \tau_3\tau_4 + \omega^2\tau_3^2\tau_4^2 = 0$. For the zero slope to occur only at $\omega = 0$, we must therefore require that $2\tau_4 = \tau_3$ and thus

$$\boxed{C_3 = 2C_4}$$

(6.9)

Under these conditions, the square of the transfer function magnitude becomes

$$|T|^2 = \frac{1}{1 + 4\omega^4\tau_4^4} = \frac{1}{1 + 4(\omega\tau_4)^4}$$

(6.10)

Sec. 6.1   Active Filters

349

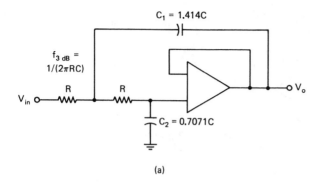

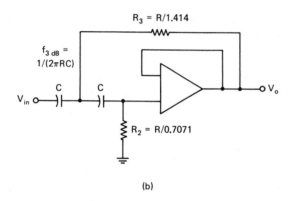

Figure 6.4 (a) Two-pole Butterworth low-pass active filter; (b) two-pole Butterworth high-pass active filter.

The 3-dB point will occur when $|T|^2 = \frac{1}{2}$, so $4(\omega\tau_4)^4 = 1$ and thus $\omega\tau_4 = 1/\sqrt{2}$, and the *3-dB bandwidth* or *cutoff frequency* is given by

$$\boxed{\omega_{3\mathrm{dB}} = \frac{1}{\sqrt{2}\,\tau_4} = \frac{0.7071}{RC_4}} \tag{6.11}$$

This type of filter with a maximally flat transfer characteristic in the passband is called a Butterworth filter, and in Figure 6.4a this two-pole Butterworth active filter circuit is shown. Figure 6.5 is a Bode plot of the frequency response of this filter. Note that the high-frequency response drops off as $1/\omega^2$ or at a 40-dB/decade (12-dB/octave) rate. This is to be compared to the case of a simple single-pole RC low-pass network for which the high-frequency response in the stop band drops off as $1/\omega$ or at a 20-dB/decade (6-dB/octave) rate. A corresponding two-pole high-pass Butterworth active filter is shown in Figure 6.4b. The 3-dB frequency for both the low-pass and high-pass filters is given by $f_{3\mathrm{dB}} = 1/(2\pi RC)$, with the passband being in the region $f < f_{3\mathrm{dB}}$ for the low-pass case and $f > f_{3\mathrm{dB}}$ in the high-pass case.

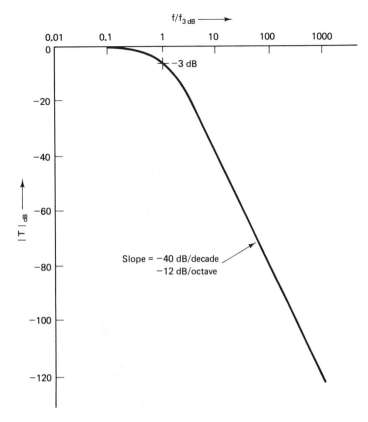

Figure 6.5 Frequency response to two-pole Butterworth filters. (From L. G. Cowles, *A Source Book of Modern Transistor Circuits*, Prentice-Hall, Englewood Cliffs, N.J., 1976.)

6.1.2 Higher-order Butterworth Active Filters

The filter order is the number of poles. An N-pole active low-pass filter will have a high-frequency asymptotic response that falls off at a rate of $N \times 20$ dB/decade or $N \times 6$ dB/octave, as shown in Figure 6.6a. Correspondingly, an N-pole high-pass filter will have a low-frequency stop-band response that increases at a rate of $N \times 20$ dB/decade (or $N \times 6$ dB/octave), as shown in Figure 6.6b.

In Figure 6.7 a number of low-pass and high-pass Butterworth active filter circuits are shown, ranging from the simple case of a single-pole ($N = 1$) filter to the more complex five-pole ($N = 5$) filter. For every case the 3-dB frequency is given by $f_{3dB} = 1/(2\pi RC)$.

Active filter circuits with an even greater number of poles are possible. In general, an N-pole active filter where N is odd will have one three-pole section and $(N - 3)/2$ two-pole sections. If N is even, the filter will have $N/2$ two-pole sections.

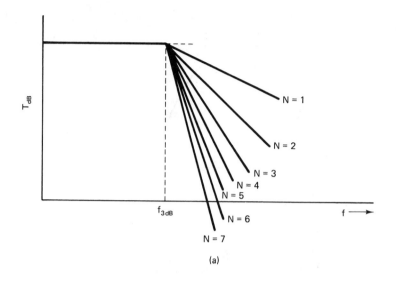

(a)

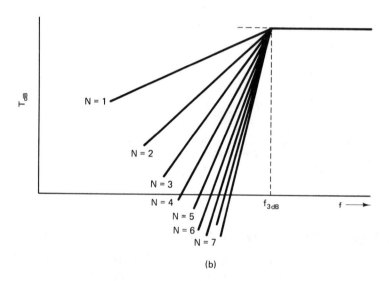

(b)

Figure 6.6 Frequency response characteristics of *N*-pole filters: (a) low-pass filters; (b) high-pass filters.

In Table 6.1, capacitance and resistance values are given for Butterworth active filters with *N*-values ranging from 2 to 10. For a two-pole section, the circuit configuration is as shown in Figure 6.4 and for a three-pole section the circuit arrangement is as shown in Figure 6.7c for the low-pass case and Figure 6.7d for a high-pass filter section. For all cases, the 3-dB frequency is given by $f_{3dB} = 1/(2\pi RC)$. Note that a higher-order Butterworth filter is not obtained by simply cascading a number of lower-order filters.

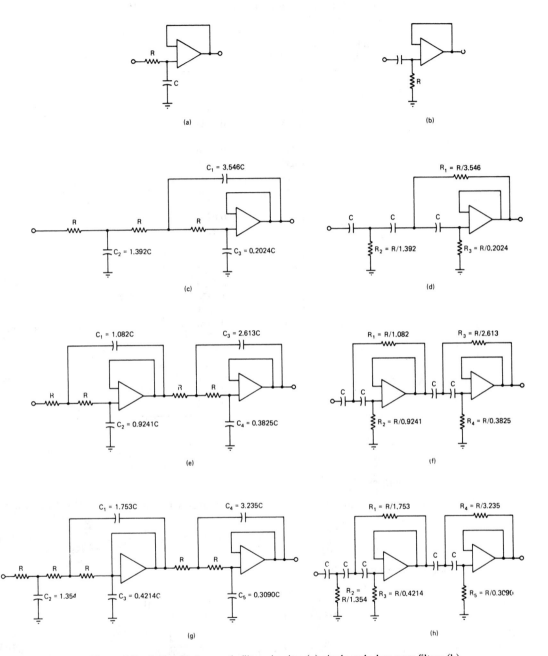

Figure 6.7 Active Butterworth filter circuits: (a) single-pole low-pass filter; (b) single-pole high-pass filter; (c) three-pole low-pass filter; (d) three-pole high-pass filter; (e) four-pole low-pass filter; (f) four-pole high-pass filter; (g) five-pole low-pass filter; (h) five-pole high-pass filter.

TABLE 6.1 COMPONENT VALUES FOR BUTTERWORTH ACTIVE FILTERS

Number of poles	C_1/C or R/R_1	C_2/C or R/R_2	C_3/C or R/R_3
2	1.414	0.7071	
3	3.546	1.392	0.2024
4	1.082	0.9241	
	2.613	0.3825	
5	1.753	1.354	0.4214
	3.235	0.3090	
6	1.035	0.9660	
	1.414	0.7071	
	3.863	0.2588	
7	1.531	1.336	0.4885
	1.604	0.6235	
	4.493	0.2225	
8	1.020	0.9809	
	1.202	0.8313	
	1.800	0.5557	
	5.125	0.1950	
9	1.455	1.327	0.5170
	1.305	0.7661	
	2.000	0.5000	
	5.758	0.1736	
10	1.012	0.9874	
	1.122	0.8908	
	1.414	0.7071	
	2.202	0.4540	
	6.390	0.1563	

6.1.3 Bandpass Active Filters

Let us now consider the operational amplifier circuit of Figure 6.8. We have that

$$v_{o2} = -\frac{v_o(1/sC)}{R_2} = -\frac{v_o}{sCR_2} \quad \text{and} \quad v_{o3} = -v_{o2} = \frac{v_o}{sCR_2} \tag{6.12}$$

At node (x) the node voltage equation is

$$\frac{v_{in}}{R_4} + \frac{v_o}{R_1} + sCv_o + \frac{v_o}{sCR_2R_3} = 0 \tag{6.13}$$

so

$$\frac{v_{in}}{R_4} = -v_o\left(\frac{1}{R_1} + sC + \frac{1}{sCR_2R_3}\right) \tag{6.14}$$

The voltage gain of this circuit is therefore

$$A_V = \frac{V_o}{V_{in}} = \frac{-1/R_4}{(1/R_1) + sC + 1/(SCR_2R_3)} \tag{6.15}$$

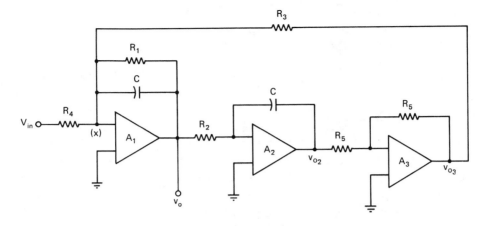

Figure 6.8 Bandpass amplifier.

In the frequency domain, $s = j\omega$ and this equation becomes

$$A_V(\omega) = \frac{-1/R_4}{(1/R_1) + j\omega C + 1/j\omega CR_2R_3} \tag{6.16}$$

Let us now consider the *RLC* parallel resonant circuit of Figure 6.9. For this circuit we have that

$$A_V(\omega) = \frac{V_o}{V_{in}} = -\frac{G_f}{Y} = \frac{-G_f}{(1/R_1) + j\omega C + 1/j\omega L} \tag{6.17}$$

If we let $G_f = 1/R_4$ and $L = CR_2R_3$, this circuit is equivalent to the bandpass amplifier circuit considered above. The maximum gain or gain at resonance is

$$\boxed{A_{V\text{MAX}} = A_V(\omega_o) = -G_f R_1 = -\frac{R_1}{R_4}} \tag{6.18}$$

The frequency at which the maximum gain occurs or the *resonant frequency* is given by

$$\omega_o^2 = \frac{1}{LC} = \frac{1}{C^2 R_2 R_3} \tag{6.19}$$

so

$$\boxed{\omega_o = \frac{1}{C\sqrt{R_2 R_3}}} \tag{6.20}$$

The Q-factor at resonance (ω_o) is given by

$$Q_o = \frac{R_1}{\omega_o L} = \omega_o CR_1 = \frac{CR_1}{C\sqrt{R_2 R_3}} = \frac{R_1}{\sqrt{R_2 R_3}} \tag{6.21}$$

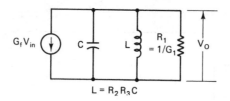

Figure 6.9 Equivalent representation of bandpass amplifier.

The 3-dB bandwidth is

$$BW = \frac{f_o}{Q_o} = \frac{\omega_o}{2\pi Q_o} = \frac{\omega_o}{2\pi \omega_o C R_1} = \frac{1}{2\pi C R_1} \tag{6.22}$$

Design example for bandpass amplifier. For a design example of this bandpass active filter, let us assume the following requirements: $f_o = 1000$ Hz, $BW = 10$ Hz, and $A_{V(MAX)} = -50$. Note that since there are five components (R_1, R_2, R_3, R_4, and C) to be determined and three design parameters, two of the component values can be chosen arbitrarily. Let us choose $C = 1.0$ μF and we will also set $R_2 = R_3$.

For the 3-dB bandwidth, we have that $BW = 1/(2\pi R_1 C)$, so solving for R_1 gives $R_1 = \underline{15.916 \text{ k}\Omega}$. Since $A_{V(MAX)} = A_V(\omega_o) = -R_1/R_4 = -50$, we obtain that $R_4 = \underline{318 \ \Omega}$. For the resonant radian frequency, we have that $\omega_o = 1/C\sqrt{R_2 R_3} = 1/CR_2$ since $R_2 = R_3$. Solving this for R_2 gives $R_2 = R_3 = \underline{159.2 \ \Omega}$. We see that reasonable component values are obtained. If a capacitance value of $C = 0.1$ μF had been chosen, all the resistance values above will be increased by a factor of 10, which will still represent acceptable values.

Bandpass amplifier using just one operational amplifier. For a second example of a bandpass active filter, let us consider the general operational amplifier circuit of Figure 6.10. If we write the node voltage equations for this circuit, we obtain the result given by

$$A_{CL} = \frac{V_o}{V_{in}} = \frac{-Y_1 Y_2}{Y_2 Y_3 + Y_5(Y_1 + Y_2 + Y_3 + Y_4)} \tag{6.23}$$

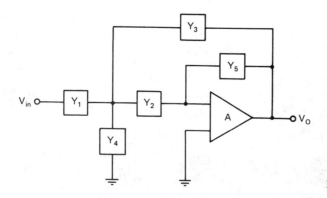

Figure 6.10 Active filter circuit.

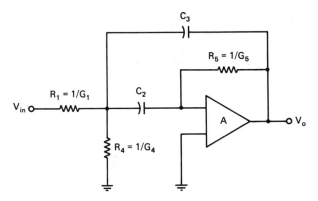

Figure 6.11 Bandpass active filter.

For this circuit to be a bandpass active filter (Figure 6.11), we let $Y_1 = G_1$, $Y_2 = sC_2$, $Y_3 = sC_3$, $Y_4 = G_4$, and $Y_5 = G_5$. With these circuit parameters, the equation for the gain becomes

$$A_{CL} = \frac{-sG_1C_2}{s^2C_2C_3 + G_5(G_1 + sC_2 + sC_3 + G_4)} \tag{6.24}$$

After dividing through by sC_2, this equation can be rewritten as

$$A_{CL} = \frac{-G_1}{sC_3 + (G_1 + G_4)G_5/sC_2 + G_5(C_2 + C_3)/C_2} \tag{6.25}$$

We will now compare the gain equation obtained for this bandpass active filter to that obtained for the parallel resonant RLC circuit of Figure 6.12. For that circuit we have that

$$A_V = \frac{V_o}{V_{in}} = -\frac{G_1}{Y} = \frac{-G_1}{sC_3 + 1/sL + G} \tag{6.26}$$

The two circuits are equivalent if we let

$$L = \frac{C_2}{G_5(G_1 + G_4)} \quad \text{and} \quad G = \frac{G_5(C_2 + C_3)}{C_2} \tag{6.27}$$

The maximum gain or gain at resonance is given by

$$A_{CL(MAX)} = A_{CL}(\omega_o) = -\frac{G_1}{G} = -\frac{(G_1/G_5)C_2}{C_2 + C_3} = -\frac{(R_5/R_1)C_2}{C_2 + C_3} \tag{6.28}$$

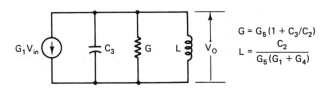

$$G = G_5(1 + C_3/C_2)$$

$$L = \frac{C_2}{G_5(G_1 + G_4)}$$

Figure 6.12 Equivalent representation of bandpass active filter.

The resonant frequency is given by

$$\omega_o^2 = \frac{1}{LC_3} = \frac{G_5(G_1 + G_4)}{C_2C_3} \tag{6.29}$$

The Q-factor at resonance is

$$Q_o = \frac{R}{\omega_o L} = R\omega_o C_3 = \frac{\omega_o C_3}{G} = \frac{\omega_o C_3 C_2}{(C_2 + C_3)G_5} \tag{6.30}$$

The 3-dB bandwidth is given by

$$BW = \frac{f_o}{Q_o} = \frac{\omega_o}{2\pi Q_o} = \frac{\omega_o}{2\pi R\omega_o C_3} = \frac{1}{2\pi RC_3} = \frac{G}{2\pi C_3} = \frac{G_5(C_2 + C_3)}{2\pi C_2 C_3} \tag{6.31}$$

Note that we have five component values (G_1, C_2, C_3, G_4, and G_5) and three design parameters ($A_{CL(MAX)}$, f_o, and BW). If, for convenience, we set $C_2 = C_3 = C$, the equations above become

$$A_{CL(MAX)} = A_{CL}(\omega_o) = -\frac{R_5}{2R_1} \tag{6.32}$$

$$\omega_o = \frac{\sqrt{G_5(G_1 + G_4)}}{C} \tag{6.33}$$

$$BW = \frac{G_5}{\pi C} = \frac{1}{\pi R_5 C} \tag{6.34}$$

Note that there are now three design parameters and four circuit component values, so one additional circuit parameter can still be specified arbitrarily.

6.1.4 Band-reject Active Filters

A good example of a band-reject (or band-stop) active filter is the "boot-strapped twin-T active filter" shown in Figure 6.13. This filter produces the frequency response characteristics shown in Figure 6.14. The notch frequency f_o is given by

$$f_o = \frac{1}{2\pi RC} \tag{6.35}$$

and the Q-factor is $Q_o = 1/4(1 - k)$, where $k = R_2/(R_1 + R_2)$. The 3-dB band-

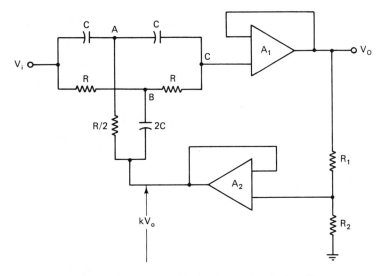

Figure 6.13 Boot-strapped twin-T active band-reject filter.

width is therefore given by

$$BW = \frac{f_o}{Q_o} = 4f_o(1 - k) \tag{6.36}$$

As k approaches unity the bandwidth of this filter circuit will become very small, although it will never go to zero because of resistor and capacitor mismatches. For the case of $k = 1$, the second op amp (A_2) and resistors R_1 and R_2 can be eliminated, and a direct feedback connection can be made from the output of A_1 back to the twin-T network.

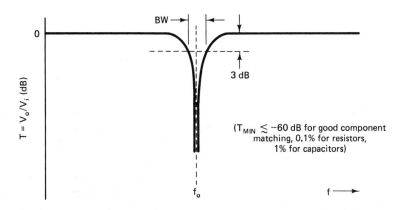

Figure 6.14 Frequency response characteristics of twin-T active band-reject (notch) filter.

For the analysis of this active filter circuit, we will first write the node voltage equations for nodes A, B, and C, which gives

$$\text{Node } A: \quad sCV_i + sCV_o + 2kGV_o = 2(sC + G)V_A \tag{6.37a}$$

$$\text{Node } B: \quad GV_i + GV_o + 2ksCV_o = 2(G + sC)V_B \tag{6.37b}$$

$$\text{Node } C: \quad sCV_A + GV_B = (G + sC)V_o \tag{6.37c}$$

From these three node voltage equations, the transfer function for this active filter can be obtained as

$$T = \frac{V_o}{V_i} = \frac{G^2 + s^2C^2}{G^2 + s^2C^2 + 4(1 - k)sCG} = \frac{s^2 + (G/C)^2}{s^2 + (G/C)^2 + 4(1 - k)s(G/C)} \tag{6.38}$$

For the frequency-domain response, we let $s = j\omega$ and we also define ω_o as $\omega_o = G/C = 1/RC$. The transfer function can now be written as

$$T(\omega) = \frac{\omega^2 - \omega_o^2}{\omega^2 - \omega_o^2 - 4(1 - k)j\omega\omega_o} \tag{6.39}$$

From this last equation we see that at $\omega = \omega_o$ the transfer function will be zero. As ω goes to zero, the transfer function will approach unity, and as ω goes to infinity, the transfer function will also asymptotically approach unity. In actual practice, the high-frequency response will, of course, be limited by the frequency response of the operational amplifier.

At the 3-dB points, we have that $T = 1/\sqrt{2}$, so $\pm 4(1 - k)\omega\omega_o = \omega^2 - \omega_o^2$ and thus $\omega^2 \pm 4(1 - k)\omega\omega_o - \omega_o^2 = 0$. This quadratic equation can be rewritten as $(\omega/\omega_o)^2 \pm 4(1 - k)(\omega/\omega_o) - 1 = 0$ and in terms of the frequency (Hz) this becomes $(f/f_o)^2 \pm 4(1 - k)(f/f_o) - 1 = 0$. Solving this equation for the upper and lower 3-dB frequencies, we obtain

$$f_u = f_o[\sqrt{1 + 4(1 - k)^2} + 2(1 - k)]$$

and

$$f_L = f_o[\sqrt{1 + 4(1 - k)^2} - 2(1 - k)] \tag{6.40}$$

The 3-dB bandwidth is

$$\boxed{BW = f_u - f_L = 4(1 - k)f_o} \tag{6.41}$$

and the corresponding Q-factor is $Q_o = f_o/BW = 1/[4(1 - k)]$.

We see that as k approaches unity the Q-factor becomes very large and the bandwidth approaches zero. In practice, the Q-factor and the minimum bandwidth are limited by mismatches between the resistors and capacitors and also by the frequency and phase shift response of the operational amplifiers.

One major applications area of the band-reject or *notch* filter is the elimination of the power-line frequency interference (60 Hz) in systems. The very narrow bandwidth and the very large attenuation at the notch frequency are useful for this type of application.

Filter response types. In addition to the basic filter types, such as low pass, high pass, and bandpass, there are a variety of filter response types. These include the Butterworth, Chebyshev, Bessel, and elliptic. The Butterworth filter is also known as a maximally flat filter in that the response is monotonic with a passband that is very flat with no ripple. The Chebyshev filter exhibits a sharper roll-off near the cutoff frequency, but does exhibit some ripple in the passband, typically 0.1 to 0.5 dB. The Bessel filter is characterized by a phase shift that is a linear function of frequency. Since time delay Δt is related to phase shift θ by $\Delta t = d\theta/d\omega$, a linear variation of phase shift with frequency will mean that all frequency components in the signal will experience the same time delay, so there will be no distortion of the signal. The Bessel filter does, however, have a less rapid roll-off near the cutoff frequency than does the Butterworth filter. The elliptic filter has the steepest roll-off near the cutoff frequency, but there is some ripple in both the passband and the stopband, together with a very nonlinear phase versus frequency characteristic. For all the basic filter types, the maximum slope of the roll-off in the stopband is 20 dB/decade for every pole in the filter response characteristic. Thus for a four-pole filter the maximum slope in the stopband will be -80 dB/decade for low-pass filters and $+80$ dB/decade for high-pass filters.

6.1.5 Active Filter ICs

Active filter hybrid ICs that contain all the active devices and most of the passive components needed for various types of filters are available. An example of an active filter IC is the UAF11/UAF21 (Burr-Brown). This hybrid IC contains three op amps, two 1-nF capacitors, and five 100-kΩ thin-film resistors; with the addition of four external resistors, it becomes a two-pole active filter with bandpass, low-pass, and high-pass outputs. By summing the low-pass and high-pass outputs with an external op amp, a band-reject filter can be obtained. The UAF11 can be operated over the frequency range from 0.001 Hz to 20 kHz, and the UAF21 can go up to 200 kHz; both ICs offer Q-values of from 0.5 up to 500. The UAF41 (Burr-Brown) is similar to the UAF11 and UAF21, except that an additional op amp is included so that the band-reject function can be implemented without any external op amp being needed.

6.2 SWITCHED-CAPACITOR FILTERS

Active *RC* filters using ICs have the advantages of not requiring inductors and of offering easy implementation of various high-performance low-pass, high-pass, band-pass, and band-stop filters. The resistor and capacitor values needed for these filters are generally much too large for fabrication on a monolithic IC chip. Large-value resistors ($\gtrsim 10$ kΩ) take up an excessive amount of chip area and there are also problems with respect to the large parasitic capacitance, the absolute-value tolerance, and the temperature coefficient. The upper limit for capacitors on a monolithic IC chip is generally around 100 pF because of the area requirements. There are also problems with respect to the absolute-value tolerance and the temperature coefficient.

Therefore, conventional active filters are either in the form of a hybrid IC using resistor and capacitor chips, or else they use the combination of monolithic IC operational amplifiers together with discrete resistors and capacitors.

The switched-capacitor filter offers an attractive alternative to the conventional *RC* active filter. The large resistor values required are easily simulated by the combination of small-value capacitors (~1 to 10 pF) and MOS switching transistors. The equivalent resistance values can be obtained such that the filter capacitance values will be small enough to be easily incorporated on a monolithic IC chip. As a result, the entire active filter circuit can be in the form of a monolithic integrated circuit.

6.2.1 Analysis of Switched-capacitor Circuit

Let us consider the switched-capacitor circuits of Figure 6.15. A two-phase clock will provide the complementary but nonoverlapping ϕ_1 and ϕ_2 clock pulses, and when a clock pulse is high the corresponding transistor will be turned on. The clock frequency will be assumed to be large compared to the signal frequency.

We will first consider the circuit shown in Figure 6.15a. When ϕ_1 is high and ϕ_2 is low, capacitor C_1 will charge up to a voltage of $v_1 - v_2$. In doing this, the charge supplied to the capacitor and therefore passing through the circuit is $Q_1 = C_1(v_1 - v_2)$. When ϕ_1 goes low and ϕ_2 goes high, C_1 will discharge back to zero. The current flow through capacitor i_1 is equal to the rate at which charge is transferred through the circuit via C_1 and is therefore given by

$$i_1 = \frac{Q_1}{T_c} = \frac{C_1(v_1 - v_2)}{T_c} = f_c C_1(v_1 - v_2) = \frac{v_1 - v_2}{R_{eq}} \qquad (6.42)$$

where T_c is the clock period, $f_c = 1/T_c$ is the clock frequency, and R_{eq} is the equivalent resistance as given by $R_{eq} = T_c/C_1 = 1/f_c C_1$.

From this last result we see that a large resistance can be simulated by the use of a small IC capacitor and a suitable clock frequency. For example, an equivalent resistance of 1.0 MΩ can be obtained with a 10-pF MOS capacitor and a clock frequency of 100 kHz.

Another, closely related switched capacitor circuit is shown in Figure 6.15b. When ϕ_1 is high, capacitor C_1 charges up to v_1, corresponding to a charge transfer of $Q_1 = C_1 v_1$. When ϕ_1 goes low and ϕ_2 goes high, the capacitor charges up in the opposite direction to v_2. The charge transfer from the output of the circuit back into C_1 is $Q_1' = C_1(v_2 - v_1)$. The current flow i_1 is the time average of this charge transfer and is given by

$$i_1 = -\frac{Q_1'}{T_c} = -\frac{C_1(v_1 - v_2)}{T_c} = -f_c C_1(v_1 - v_2) = -\frac{v_1 - v_2}{R_{eq}} \qquad (6.43)$$

The two circuits just considered used a series switched capacitor. In Figure 6.15c, a shunt switched-capacitor circuit is shown. This circuit can be considered to

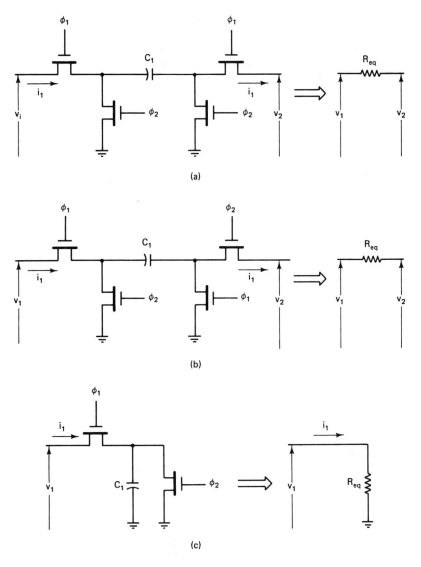

Figure 6.15 Switched-capacitor circuits: (a) noninverting switched-capacitor circuit; (b) inverting switched-capacitor circuit; (c) shunt-capacitor circuit.

be a special case of the circuit of Figure 6.15a, in which $v_2 = 0$. For this circuit, capacitor C_1 charges up to $Q_1 = C_1 v_1$ when ϕ_1 is high. Then when ϕ_1 goes low and ϕ_2 goes high, C_1 discharges to zero through the ϕ_2 transistor. Current i_1 is therefore given by

$$i_1 = \frac{Q_1}{T_c} = \frac{C_1 v_1}{T_c} = f_c C_1 v_1 = \frac{v_1}{R_{eq}}, \qquad \text{where } R_{eq} = \frac{1}{f_c C_1} \qquad (6.44)$$

6.2.2 Examples of Switched-capacitor Filters

Various types of low-pass, high-pass, bandpass, and band-stop active filter circuits can be implemented using switched capacitors to replace the resistors so that an *all-capacitor* filter circuit can be obtained. The various time constants that govern the response characteristics of the filter circuits are of the form given by

$$\tau = R_{eq}C_2 = \frac{C_2/C_1}{f_c} \tag{6.45}$$

We see that the filter time constants can easily and accurately be controlled by the clock frequency, so various types of programmable and tracking filters can be obtained. The filter characteristics can be electronically varied over a wide range by control of the clock frequency.

For small-area MOS capacitors of around 1 pF or less, a ratio accuracy in the range from 1% to 2% can be expected, and for larger-area MOS capacitors of around 15 pF or greater, the ratio accuracy improves to about 0.1%. As a result of this

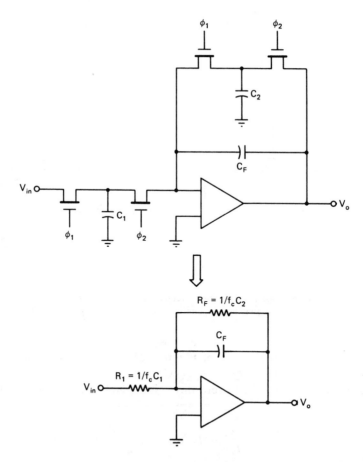

Figure 6.16 Switched-capacitor low-pass filter.

capacitance-ratio accuracy that is obtainable and the close control of the clock frequency, very precise control of the filter characteristics can be obtained.

A simple example of a switched-capacitor active filter is the low-pass filter of Figure 6.16. For this filter, the transfer function is given by

$$T(f) = \frac{V_o}{V_{in}} = \frac{R_F/R_1}{1 + j\omega R_F C_F} = \frac{R_F/R_1}{1 + j(f/f_L)} \tag{6.46}$$

where $R_F = 1/f_c C_2$, $R_1 = 1/f_c C_1$, and $f_L = 1/(2\pi R_F C_F) = f_c C_2/(2\pi C_F)$. The transfer function can thus be rewritten as

$$T(f) = \frac{C_1}{C_2} \frac{1}{1 + j(f/f_L)} \tag{6.47}$$

We see that the filter characteristics will be a function only of the capacitance ratio and the clock frequency.

6.2.3 Clock Frequency

The switched-capacitor circuit is a sampled data system in that the analog input signal does not pass through the system as a continuous signal, but rather as a series of pulses at the sampling frequency. As in the case of any sampled data system, the sampling frequency must be at least twice the highest analog signal frequency in order that the analog signal can be reconstructed at the output (the Nyquist theorem). If this condition is not satisfied, a condition known as *aliasing* will be present and the reconstructed analog signal will be distorted. To prevent this condition of aliasing, the analog signal must be band limited, and in many cases a low-pass filter known as an *antialiasing filter* must be used at the input. In many cases the ratio of sampling frequency to analog signal bandwidth is set to a value considerably greater than the minimum value of 2 in order to ease the filtering requirements. Indeed, in the case of many switched-capacitor filter ICs, a frequency ratio of the clock frequency to the filter corner frequency or bandpass center frequency in the range of 50 : 1 to as high as 100 : 1 is used, with the sampling frequency being one-half of the clock frequency. On the output side of the system, a low-pass filter is needed to convert the current pulses back into a continuous analog signal. In some cases this low-pass filter is already an inherent part of the system, so no additional filtering is needed.

The maximum sampling frequency of a switched-capacitor circuit is limited by the time required to charge the MOS capacitors via the MOSFET switches. To charge a capacitor to 99% of an input voltage step requires a charging time of $T_c = 4.6\tau$, where the $R-C$ time constant $\tau = RC$ is the product of the switch "on" resistance and the MOS capacitance. To charge the capacitance up to the 99.9% level requires a charging time of $T_c = 6.9\tau$. For a switched-capacitor circuit, the charging resistance R will be the MOSFET "on" resistance and is generally on the order of 1000 Ω. The capacitance is the MOS capacitance. The minimum value is generally around 1 pF based on tolerance considerations due to the dimensional accuracy limitations of the photolithographic process. The maximum capacitance should generally not be more than about 30 pF so as to limit the area taken up by the capacitor on the chip. As a result, the charging times will range from about 10 ns up to around 200 ns. Since the charging time is one-half of the total sampling period, the corresponding value for the maximum allowable sampling frequency ranges from about 2 MHz up to 50 MHz. Thus the switched-capacitor circuit is capable of operation at analog signal frequencies of several hundred kilohertz. In practice, the actual upper-frequency capability of the switched-capacitor circuit may be limited by the op amps in the circuit. The maximum analog signal frequency of many switched-capacitor filters is usually about 10 or 20 kHz.

6.2.4 Switched-capacitor Filter ICs

Switched-capacitor filter ICs are self-contained monolithic active filters that require no external resistors or capacitors. The filter capacitors are small-value MOS capacitors on the IC chip. The large-value resistors required are simulated by the switched-capacitor technique using small MOS capacitors and MOS transistor switches, all contained on the IC chip. Bandpass, band-reject, low-pass, and high-pass filters are available. In addition to not requiring any external components, another major advantage of the switched-capacitor filters is that the center frequency of the bandpass and band-reject filters and the 3-dB frequency of the high-pass and low-pass filters are directly proportional to the clocking frequency supplied to the device.

Some examples of monolithic switched-capacitor low-pass filter integrated circuits are the RF5609A Elliptic Low-pass Filter and the RF5613A Linear Phase Low-pass Filter (EG&G Reticon). These filters have corner frequencies that are adjustable over the range from 10 Hz to 25 kHz by variation of the clock frequency. The ratio of the clock frequency to the corner frequency is 100 (typ.) for the RF5609 and 128 for the RF5613. The filter input impedance is 10 MΩ (min.) in parallel with 50 pF (max.), and the output impedance is 10 Ω (typ.), 250 Ω (max.). The transfer ratio in the passband (insertion loss) is 0 dB (± 0.4 dB), so it is very easy to interface this filter with other systems.

The RF5611A is a monolithic high-pass swutched-capacitor filter. It is a five-pole Chebyshev filter with a 30-dB per decade roll-off at low frequencies (that is, the stopband) and less than 0.6-dB ripple in the passband. The corner frequency can be varied from 10 Hz to 5 kHz, and the ratio of the clock frequency to the corner frequency is 500 (typ.).

The RF5612A (EG&G Reticon) is a monolithic switched-capacitor notch filter with a frequency response characteristic as shown in Figure 6.17. This is a four-pole notch filter with an attenuation at the notch frequency (notch rejection) of 45 dB (min.). The notch frequency can be varied over the range from 10 Hz to 2.7 kHz, and the ratio of the clock frequency to the corner frequency is 1000 (typ.).

The three filters described all have similar input and output impedances and insertion losses, so they can be easily incorporated into various systems. The analog signal sampling frequency is one-half the clock frequency for all cases, so the analog signal frequencies should be limited to a maximum of one-fourth of the clock frequency to prevent aliasing.

Other examples of switched-capacitor ICs are the MF4 and MF6 (National Semiconductor). The MF4 is a four-pole Butterworth low-pass active filter in an eight-pin mini-DIP package. The MF6 is similar to the MF4 except that it is a six-pole Butterworth low-pass filter and comes in a 14-pin DIP package. The ratio of

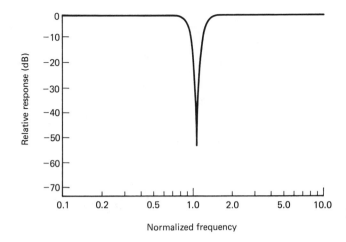

Figure 6.17 Frequency response of RF5612 notch filter (EG&G Reticon).

the clock frequency to the filter cutoff frequency is 50 : 1 for the MF4-50 and MF6-50 and 100 : 1 for the MF4-100 and MF6-100. A clock input can be supplied or an external resistor and capacitor can be used to set the frequency of an internal clock. The cutoff frequency can be varied over the range from 0.1 Hz to 20 kHz, and the gain in the passband is unity so that additional MF4 or MF6 sections can be easily cascaded for higher-order filtering.

Some other low-pass switched-capacitor filters, their order and type, and the maximum corner frequency f_o are as follows: XR-1003 (EXAR) fourth-order Bessel with f_o = 15 kHz; XR-1005 (EXAR) fourth-order Chebyshev with f_o = 15 kHz; LTC1062 (Linear Technology) fifth-order maximally flat with f_o = 20 kHz; MC145414 (Motorola) fifth-order elliptic with f_o = 10 kHz; LMF60 (National Semiconductor) sixth-order Butterworth with f_o = 30 kHz; RF6069A (EG&G Reticon) seventh-order elliptic with f_o = 25 kHz; LTC1064-1 (Linear Technology) eighth-order elliptic with f_o = 50 kHz; LTC1062-2 (Linear Technology) eighth-order Butterworth with f_o = 140 kHz.

Examples of some other high-pass switched-capacitor filters are the MAX268 (Maxim) fourth-order universal with f_o = 140 kHz, the MF8 (National Semiconductor) fourth-order universal with f_o = 20 kHz, and the TSG8751 (SGS-Thomson) fourth-order elliptic with f_o = 12 kHz.

6.2.5 Switched-capacitor Bandpass Filters

Switched-capacitor bandpass filters are also available, as for example the RF5614A/RF5615A/RF5616A (EG&G Reticon) series of monolithic ICs. These ICs come in an eight-pin mini-DIP package and are six-pole Chebyshev bandpass filters. The bandpass characteristics are shown in Figure 6.18. The center frequency F_o can be varied over a very wide range, from 1 Hz up to 20 kHz, by variation of the clock frequency. The ratio of the clock frequency to the center frequency is 54.5 (typ.).

The -3-dB and -40-dB frequencies of these three active filter ICs are given in Table 6.2. Note that one-third of an octave is a frequency ratio of $(2)^{1/3} = 1.26$.

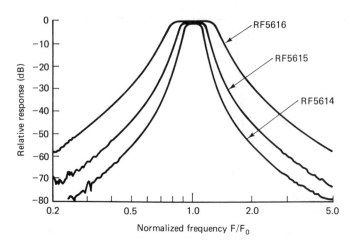

Figure 6.18 Normalized frequency response for RF5614, RF5615, and RF5616 (EG&G Reticon).

The ratio of the upper -3-dB frequency to the lower -3-dB frequency for the RF5614A is 1.25, so we see that this corresponds closely to one-third of an octave. For the RF5615A, which is a one-half octave filter, the ratio of the upper to lower -3-dB frequencies is 1.45. This corresponds closely to one-half of an octave, which is a frequency ratio of $(2)^{1/2} = 1.414$. The RF5616A is a full-octave bandpass filter with a -3-dB frequency ratio of 1.94. We see that this is close to a full octave.

For these bandpass filters, the clock frequency is approximately $54F_o$, and the analog signal sampling rate is one-half the clock frequency, or $27F_o$. Therefore, to prevent any signal aliasing, the maximum analog signal frequency should be limited to one-half the sampling rate, or $13.5F_o$.

These filters have nominal unity gain (± 0.5 dB) at the center frequency. They can accommodate input signals up to 4 V rms and have a dynamic range of 86 dB.

Some other examples of monolithic switched-capacitor bandpass filters are the RM5604A/RM5605A/RM5606A series (EG&G Reticon). These are all six-pole Chebyshev bandpass filters. The RM5604A contains three stagger-tuned one-third octave filters. The center frequencies of the three filters are in the ratio of $0.8 : 1.0 : 1.25$, and the frequency response curves are as shown in Figure 6.19. The three filter sections can be connected in cascade to form a stagger-tuned filter having a relatively

TABLE 6.2 SWITCHED-CAPACITOR BANDPASS FILTER CHARACTERISTICS

IC	Type	-3-dB Frequencies Lower	Upper	-40-dB Frequencies Lower	Upper
RF5614A	1/3 octave	$0.89F_o$	$1.11F_o$	$0.64F_o$	$1.52F_o$
RF5615A	1/2 octave	$0.82F_o$	$1.19F_o$	$0.51F_o$	$1.96F_o$
RF5616A	Full octave	$0.72F_o$	$1.40F_o$	$0.31F_o$	$3.29F_o$

IC Active Filters and Switched-capacitor Circuits   Chap. 6

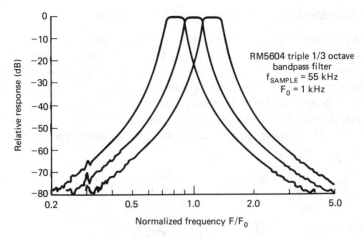

Figure 6.19 Frequency response of the RM5604 (EG&G Reticon).

wide and flat passband covering a full octave, but with a rapid attenuation outside the passband. The RM5605A contains two stagger-tuned one-half octave bandpass filters that when cascaded produce a passband covering a full octave, as shown in Figure 6.20. The RM5606A is a single full-octave bandpass filter. For all these devices, the center frequency of the passband can be varied over the range from 10 Hz to 10 kHz, and the ratio of the clock frequency to the passband center frequency is 109 (typ.). The insertion loss is 0 dB (± 0.5 dB), and the signal distortion is less than 0.1% over a 76-dB dynamic range for a 10-V peak-to-peak input signal.

One interesting application area for switched-capacitor bandpass filter ICs is spectrum analysis, including harmonic analysis and noise analysis. By simply sweeping the clock frequency at a suitable rate over a suitable range, the bandpass filter can be used for the measurement of the amount of signal present in a given frequency range.

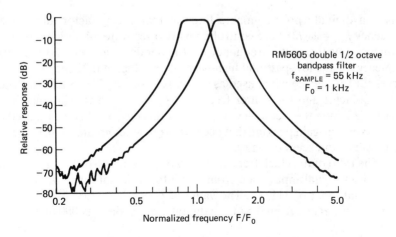

Figure 6.20 Frequency response of the RM5605 (EG&G Reticon).

6.2.6 Switched-capacitor Universal Active Filters

With a switched-capacitor universal active filter, any type of classical filter configuration can be implemented, including low-pass, bandpass, high-pass, all-pass, and notch filters. Since these are switched-capacitor types of filters, the characteristic frequencies can easily be controlled by variation of the clock frequency, so a high degree of design flexibility is available with these devices.

An example of a switched-capacitor universal active filter is the RU5620A (EG&G Reticon). The transfer functions for the various filter types are as follows:

Low pass: $\qquad T(s) = \dfrac{\omega_o^2}{s^2 + (\omega_o/Q)s + \omega_o^2}$

High pass: $\qquad T(s) = \dfrac{s^2}{s^2 + (\omega_o/Q)s + \omega_o^2}$

Bandpass: $\qquad T(s) = \dfrac{-(\omega_o/Q)s}{s^2 + (\omega_o/Q)s + \omega_o^2}$

Notch: $\qquad T(s) = \dfrac{s^2 + \omega_o^2}{s^2 + (\omega_o/Q)s + \omega_o^2}$

All pass: $\qquad T(s) = \dfrac{s^2 - (\omega_o/Q)s + \omega_o^2}{s^2 + (\omega_o/Q)s + \omega_o^2}$

Low pass elliptic: $\quad T(s) = \dfrac{(\omega_o/\omega_z)s^2 + \omega_o^2}{s^2 + (\omega_o/Q)s + \omega_o^2}$

High pass elliptic: $\quad T(s) = \dfrac{s^2 + (\omega_z/\omega_o)\omega^2}{s^2 + (\omega_o/Q)s + \omega_o^2}$

This is a digitally programmable filter in that the Q-factor and the characteristic frequency f_o ($= \omega_o/2\pi$) are controlled by two separate 5-bit digital input codes. The frequency f_o is the center frequency for the bandpass and notch filters and the corner frequency for the low-pass and high-pass filters. Figure 6.21 shows frequency response curves for this device when used as a bandpass filter. The 3-dB bandwidth is related to the Q-factor and the center frequency f_o by the usual relationship, $BW = f_o/Q$. Figure 6.22 gives the frequency response of this device when configured as a low-pass filter. Note the gain peaking that occurs for Q-values greater than 0.707 and especially for Q values in excess of unity.

The ratio of the clock frequency f_c to the characteristic frequency f_o can be varied by the 5-bit digital input code from 200.0 for a digital input of 00000 down to 50.0 for a digital input of 11111. The value of Q can be varied by a second 5-bit digital input over the range from 0.57 for a digital input code of 00000 up to 85.0 for a digital

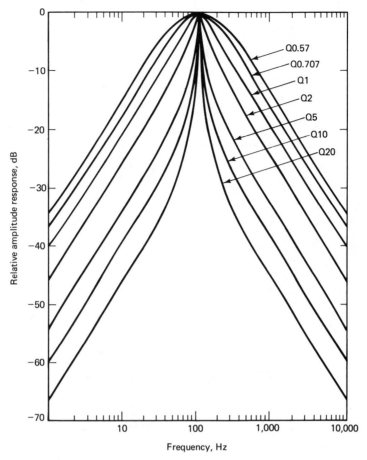

Figure 6.21 Typical bandpass amplitude response for the RU5620A (EG&G Reticon).

input of 11111. The allowable range of clock frequencies is from 500 Hz up to 1250 kHz, and the range of f_o is from 10 Hz up to 20 kHz.

The RU5621A and the RF5622A (EG&G Reticon) are resistor programmable switched-capacitor universal active filters. The RF5621A is a four-pole filter and can be used to implement the same filter types as the RU5620A that was discussed in the preceding section. The RU5622A is a six-pole filter and can also implement all the filter types except for the all-pass filter. The ratio of the clock frequency to the characteristic frequency f_o and the value of Q can be set by external resistor ratios. The characteristic (center or corner) frequency can be varied over a range of from 1 Hz to 30 kHz, and the value of Q can be adjusted from 0.5 to 500.

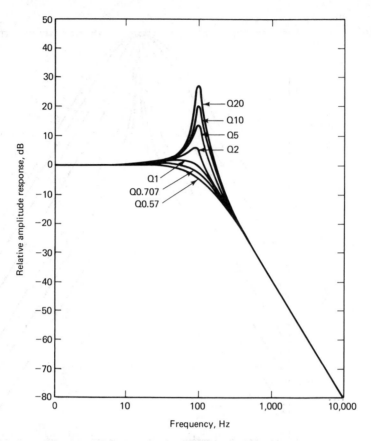

Figure 6.22 Typical low-pass amplitude response for the RU5620A (EG&G Reticon).

PROBLEMS

For all these problems, assume ideal infinite gain op amps unless otherwise indicated.

6.1. Design a single-pole low-pass filter with a gain of 20 in the passband and a cutoff (3 dB) frequency of 10 kHz.

6.2. Design a single-pole low-pass filter with a gain of -20 in the passband and a cutoff (3 dB) frequency of 10 kHz.

6.3. Design a single-pole high-pass filter with a gain of 10 in the passband and a cutoff (3 dB) frequency of 20 kHz.

6.4. It is required that a filter have a cutoff frequency of 10 kHz and that at 20 kHz the gain be down by at least 30 dB. Find the minimum number of poles required for the filter. (*Ans.:* five poles)

6.5. Find the component values needed for a three-pole Butterworth active low-pass filter with a cutoff frequency of 10 kHz. Use $R = 10$ kΩ. (*Ans.:* $C_1 = 5.64$ nF, $C_2 = 2.215$ nF, $C_3 = 322$ pF)

6.6. For the bandpass active filter of Figure 6.8, $f_o = 5$ kHz, $BW = 20$ Hz, and $A_{V(MAX)} = -25$. Let $C = 0.1$ μF, $R_5 = 10$ kΩ, and $R_2 = R_3$. Find R_1 through R_4. (*Ans.:* $R_1 = 79.6$ kΩ, $R_2 = R_3 = 318$ Ω, $R_4 = 3.18$ kΩ)

6.7. For the bandpass active filter of Figure 6.11, $f_o = 500$ Hz, $BW = 20$ Hz, and $A_{V(MAX)} = -25$. Let $C_2 = C_3 = 0.1$ μF. Find the component values needed. (*Ans.*: $R_5 = 159$ kΩ, $R_1 = 3.18$ kΩ, $R_4 = 65$ Ω)

6.8. For the boot-strapped twin-T band-reject filter of Figure 6.13, the center frequency is 120 Hz and the bandwidth is to be 1.0 Hz. If $C = 1.0$ μF and $R_2 = 10$ kΩ, find R and R_1. (*Ans.*: $R = 1.33$ kΩ, $R_1 = 21$ Ω)

6.9. (*Switched-capacitor low-pass filter*) For the switched-capacitor low-pass filter of Figure 6.16, the gain in the passband is 10 and the cutoff frequency is 10 kHz. The clock frequency is to be 10 times the cutoff frequency and $C_F = 50$ pF. Find R_F, C_1, and C_2. (*Ans.*: $R_F = 318$ kΩ, $C_1 = 314$ pF, $C_2 = 31.4$ pF)

REFERENCES

ALLEN, P. E., and E. SANCHEZ-SINENCIO, *Switched Capacitor Circuits*, Van Nostrand Reinhold, New York, 1984.

BOWRON, P., *Active Filters for Communications and Instrumentation*, McGraw-Hill, New York, 1979.

FRANCO, S., *Design with Operational Amplifiers and Analog Integrated Circuits*, McGraw-Hill, New York, 1988.

GARRETT, P. H., *Analog I/O Design*, Prentice-Hall, Englewood Cliffs, N.J., 1981.

GREBENE, A. B., *Bipolar and MOS Analog Integrated Circuit Design*, Wiley, New York, 1984.

HAYKIN, S. S., *Synthesis of RC Active Filters*, McGraw-Hill, New York, 1969.

HILBURN, J. L., and D. E. JOHNSON, *Manual of Active Filter Design*, McGraw-Hill, New York, 1973.

HUELSMAN, L. P., and P. E. ALLEN, *Introduction to the Theory and Design of Active Filters*, McGraw-Hill, New York, 1980.

JOHNSON, D. E., and J. L. HILBURN, *Rapid Practical Designs of Active Filters*, Wiley, New York, 1975.

———, and V. JAYAKUMAR, *Operational Amplifier Circuits—Design and Applications*, Prentice-Hall, Englewood Cliffs, N.J., 1982.

KENNEDY, E. J., *Operational Amplifier Circuits*, Holt, Rinehart. and Winston, New York, 1988.

LAM, H. Y.-F., *Analog and Digital Filters*, Prentice-Hall, Englewood Cliffs, N.J., 1979.

McMENAMIN, J. M., *Linear Integrated Circuits: Operation and Applications*, Prentice-Hall, Englewood Cliffs, N.J., 1985.

NATARAJAN, S., *Theory and Design of Linear Active Networks*, Macmillan, New York, 1987.

RUTKOWSKI, G. B., *Integrated Circuit Operational Amplifiers*, 2d ed., Prentice Hall, Englewood Cliffs, N.J., 1984.

TEDESCHI, F. P., *The Active Filter Handbook*, TAB Books, Blue Ridge Summit, Pa., 1979.

TEMES, G. C., *Integrated Analog Filters*, IEEE Press, New York, 1987.

TSIVIDIS, Y. P., and P. ANTOGNETTI, *Design of MOS VLSI Circuits for Telecommunications*, Chapter 9, Switched-Capacitor Filter Synthesis, by A. S. SEDRA, and Chapter 10, Performance Limitations in Switched-Capacitor Filters, by P. R. GRAY and R. CASTELLO, Prentice-Hall, Englewood Cliffs, N.J., 1985.

VAN VALKENBERG, *Analog Filter Design*, Holt, Rinehart, and Winston, New York, 1982.

WILLIAMS, A. B., *Active Filter Design*, Artech House, Dedham, Mass., 1975.

7

Operational-amplifier Circuit Design

7.1 OPERATIONAL-AMPLIFIER CIRCUIT ANALYSIS

We will now start to examine the *internal* circuitry of op amps. Much of this circuitry will be closely related to that of other analog ICs. A basic block diagram representative of most op amps is shown in Figure 7.1. There are two gain stages followed by an emitter-follower output stage. The voltage gain of each of the two gain stages is generally in the range from 300 to 1000, and the emitter-follower output stage has a gain of around unity, so the overall op-amp low-frequency open-loop voltage gain A_{CL} (0) is on the order of 10^5 to 10^6 (100 to 120 dB).

The emitter-follower output stage provides impedance transformation so that relatively low values of load resistance can be driven. This stage is usually of a *push–pull* configuration such that the op amp can *source* ($i_o > 0$) as well as *sink* ($i_o < 0$) output currents.

As is the case with all ICs, all the stages are direct coupled, there being no coupling or bypass capacitors in the circuit. Much of the design of the circuit relies on the close matching of transistors and resistors that are possible with ICs. This close matching comes about because the components are very close to each other on the same small silicon chip, and they have undergone identical manufacturing processing. In addition, since capacitors and large-value (above 50 kΩ) resistors occupy large areas on IC chips and also exhibit large undesirable parasitic effects, these components are usually avoided and the emphasis in the circuit design is on maximizing the number of transistors (and diodes) as compared to resistors.

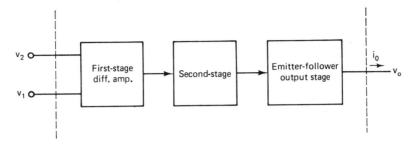

Figure 7.1 Basic block diagram of an operational amplifier.

For feedback stability, it is desirable to minimize the total open-loop phase shift in the amplifier. This will result in a larger value for $f_{180°}$ and thus allow for a greater value for the unity-gain frequency f_u, which will mean a larger gain–bandwidth product. Since each gain stage contributes to the total amplifier phase shift, it is therefore desirable to minimize the total number of stages while maintaining a large overall open-loop voltage gain. The earliest IC operational amplifiers had three gain stages, but almost all present-day op amps have just two gain stages; so for a very large open-loop gain (10^5 to 10^6) the voltage gain of each stage must be very large.

The principal exception to the general rule of two gain stages is op amps that use FETs. The FETs can be used just in the input stage or throughout the entire amplifier circuit, and these op amps offer the advantages of a very high input impedance and a very low input bias current. However, as a result of the much lower transfer conductance and therefore voltage gain available with FETs as compared to bipolar transistors, some FET op amps need three gain stages.

7.1.1 Operational-amplifier Open-loop Frequency Response

Now that the basic configuration of the op amp has been examined, the open-loop frequency response will be investigated. For this purpose, the op-amp representation of Figure 7.2 will be used, in which the second gain stage and the emitter-follower output stage are represented as amplifiers with gains of $-A_2$ and $+1$, respectively. A feedback or compensation capacitor C_c is shown connected across the second stage, although it could just as well be connected across both the second and output stages with essentially the same effect.

By inspection of this circuit, we have that

$$i_c \simeq i_{o_1} = 2g_f v_i \tag{7.1}$$

and
$$i_c = (v_o - v_{i_2})j\omega C_c = \left(v_o + \frac{v_o}{A_2}\right)j\omega C_c \tag{7.2}$$

$$= v_o\left(1 + \frac{1}{A_2}\right)j\omega C_c$$

so
$$v_o = \frac{i_c}{j\omega C_c(1 + 1/A_2)} = \frac{2g_f v_i}{j\omega C_c(1 + 1/A_2)} \tag{7.3}$$

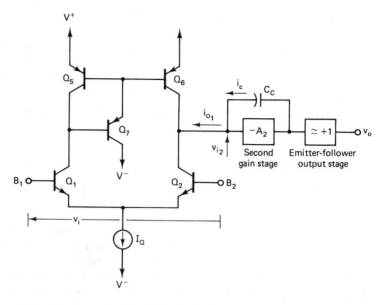

Figure 7.2 Operational-amplifier circuit for analysis of frequency response.

If $A_2 \gg 1$, as will usually be the case, the preceding equation can be written as

$$v_o \simeq \frac{2g_f v_i}{j\omega C_c} \tag{7.4}$$

so the equation for the op-amp open-loop gain can be written very simply as

$$A_{OL} = \frac{v_o}{v_i} \simeq \frac{2g_f}{j\omega C_c} \tag{7.5}$$

At the unity-gain frequency, the magnitude of the open-loop gain is unity, so $2g_f/\omega_u C_c = 1$ and thus

$$\omega_u = \frac{2g_f}{C_c} \tag{7.6}$$

where ω_u is the *unity-gain radian frequency*. The corresponding value of f_u is

$$f_u = \frac{\omega_u}{2\pi} = \frac{2g_f}{2\pi C_c} \tag{7.7}$$

The capacitance C_c is principally the capacitance that is purposely added to the circuit to establish the required value for f_u, either as an on-chip capacitor in the case of internally compensated op amps or as an external capacitor. There will also be some parasitic circuit capacitance on the order of a few picofarads such that C_c should be considered to be given by $C_c = C_{comp} + C_p$, where C_{comp} is the compensation capacitance added to the circuit and C_p is the parasitic capacitance ($\sim$1 to 3 pF) already present in the circuit.

The equation for A_{OL} given above is of course not valid at very low frequencies (below f_1) at which the gain approaches a limiting value of $A_{OL}(0)$. It is also not valid at frequencies above the op-amp second breakpoint frequency f_2, at which point the gain of the second stage $(-A_2)$ is no longer large compared to unity. In a later section where an example of an op-amp circuit is discussed in detail, a more extensive analysis of the frequency response will be carried out and the effect of the second breakpoint frequency f_2 will be examined.

As an example of the relationship between f_u and C_c, let us consider an op amp for which $I_Q = 20$ μA and specify that $f_u = 1.25$ MHz. For this value of I_Q, the differential-amplifier transfer conductance g_f is $g_f = I_Q/4V_T = 20$ μA/100 mV = 200 μS. The required value for C_c is given by $f_u = 2g_f/2\pi C_c = 400$ μS/$(2\pi C_c)$, so solving for C_c gives $C_c = 400$ μS/$(2\pi \times 1.25$ MHz$) = 51$ pF. Thus only a relatively small capacitance value is required, and this capacitance value is small enough such that it can be incorporated on the IC chip without occupying an excessive amount of area. Note that if a Darlington differential-amplifier configuration were used the equation for g_f above would have to be modified to become $g_f = I_Q/4(2V_T) = I_Q/8V_T$.

7.1.2 Pole-zero Compensation

In the compensation technique just discussed, a compensation capacitor is added across the second stage. This capacitance plus the small parasitic capacitance already present shifts the lowest breakpoint frequency (or pole) down to a lower value to achieve the required stability. This reduction in the lowest pole of the op amp produces a corresponding reduction in the unity-gain frequency f_u, and therefore results in a reduced bandwidth for the system.

In some op amps a resistor R_C is added in series with the compensation capacitor C_C. The impedance of the feedback network becomes

$$Z_C = \frac{1}{j\omega C_C} + R_C = \frac{1 + j\omega R_C C_C}{j\omega C_C} = \frac{1 + j(f/f_z)}{j\omega C_C}$$

where $f_z = 1/(2\pi R_C C_C)$. Thus the feedback network now introduces a zero at frequency f_z. The open-loop gain expression can now be written as

$$A_{OL}(f) \cong \frac{A_{OL}(0)\,[1 + j(f/f_z)]}{j(f/f_1)\,[1 + j(f/f_2)][1 + j(f/f_3)]\cdots}$$

If $f_z = f_2$, then the zero at f_z will cancel the pole at f_2, so, in effect, the second breakpoint frequency of the op amp becomes f_3. Since f_3 is greater than f_2 and often substantially greater, the result is that f_u can be larger. This will lead to an improved frequency- and time-domain response.

7.1.3 Slewing Rate

The *slewing rate* (SR) of an op amp is the maximum rate at which the output voltage can change with time, $(dv_o/dt)_{MAX}$. The output voltage of the op amp can be related to the current through C_c as

$$i_o = C_c \frac{d(v_o - v_{i2})}{dt} \simeq C_c \frac{dv_o}{dt} \tag{7.8}$$

if the gain of the second stage is large such that $v_{i2} = v_o/A_2 \ll v_o$, as will usually be the case. Solving for dv_o/dt, we now have that

$$\frac{dv_o}{dt} \simeq \frac{i_c}{C_c} \simeq \frac{i_{o1}}{C_c} \tag{7.9}$$

The output current of the first stage is limited by the I_Q current source to a maximum limit of $i_{o1(MAX)} = +I_Q$ in one direction and $-I_Q$ in the other direction. As a result, the rate of change of output voltage is limited to a maximum value given by

$$\boxed{\text{slewing rate } (SR) = \left(\frac{dv_o}{dt}\right)_{MAX} = \frac{i_{o1(MAX)}}{C_c} = \pm \frac{I_Q}{C_c}} \tag{7.10}$$

There is a direct relationship between the slewing rate and the unity-gain frequency for the bipolar transistor case since $f_u = 2g_f/(2\pi C_c)$ and $g_f = I_Q/4V_T$. We have that

$$f_u = \frac{2g_f}{2\pi C_c} = \frac{2(I_Q/4V_T)}{2\pi C_c} = \frac{I_Q/C_c}{4\pi V_T} = \frac{SR}{4\pi V_T} \tag{7.11}$$

For the case of a Darlington differential amplifier, $g_f = I_Q/8V_T$, so the corresponding relationship between f_u and the slewing rate becomes $f_u = SR/8\pi V_T$.

To consider an example of a slewing-rate calculation, let us choose $I_Q = 20\ \mu A$ and $f_u = 1.25\ MHz$, as given above. The value of C_c needed is 51 pF, and the slewing rate is

$$SR = \frac{I_Q}{C_c} = 20\ \mu A/51\ pF = 3.9 \times 10^5\ V/s = \underline{0.39\ V/\mu s} \tag{7.12}$$

Most op amps have slewing rates in the range from 0.5 to 3 V/μs, and there are op amps available with slewing rates as high as 7000 V/μs. In particular, op amps with FET input stages generally have very high slewing rates.

An op amp will become slewing-rate limited when the rate of change of the output voltage dv_o/dt as predicted by the small-signal response equations exceeds the slewing rate. For a step-function type of input voltage, we have that $v_o(t) = V_{O(MAX)}$ $[1 - \exp(-t/\tau)]$, so

$$\left(\frac{dv_o}{dt}\right)_{MAX} = \left(\frac{dv_o}{dt}\right)_{t=0} = \frac{V_{O(MAX)}}{\tau} \tag{7.13a}$$

Thus if

$$\boxed{\frac{V_{O(MAX)}}{\tau} > SR} \tag{7.13b}$$

the op amp will be *slewing-rate limited* and the output voltage will increase linearly with time at the slewing rate. An example of this situation is shown in Figure 7.3.

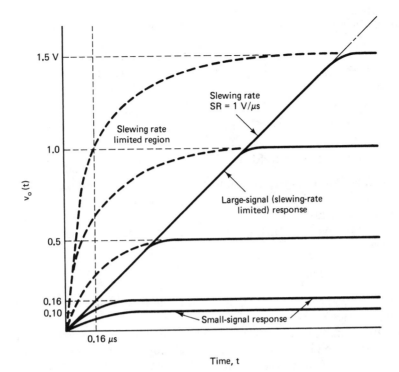

Figure 7.3 Step-function response characteristics.

The op-amp closed-loop bandwidth is related to the closed-loop time constant by $\tau = 1/(2\pi BW_{CL})$ so that $BW_{CL} = 1/(2\pi\tau)$. For the example of Figure 7.3, an op-amp slewing rate of 1 V/μs is chosen and a closed-loop bandwidth of $BW_{CL} = 1.0$ MHz is assumed, so the corresponding time constant is $\tau = 1/(2\pi BW_{CL}) = 0.16$ μs. Thus, for $V_{O(MAX)} > SR \times \tau = 1$ V/μs $\times$ 0.16 μs = 0.16 V, the op amp will be operating under slewing-rate-limited conditions.

In Figure 7.4 the large-signal or slewing-rate-limited response of an op amp to a rectangular pulse input voltage is shown. Note the trapezoidal-shaped response

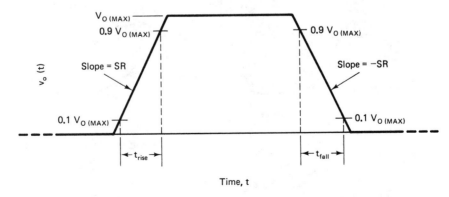

Figure 7.4 Slewing-rate-limited response to rectangular pulse voltage input.

characteristics. The time for the full-scale response (0 to $V_{O(MAX)}$) is approximately $V_{O(MAX)}/SR$, so the 10% to 90% rise time is approximately $t_{rise} = 0.8\ V_{O(MAX)}/SR$.

The small-signal rise time for the example above is $t_{rise} = 2.2\tau = 0.35/BW_{CL} = 0.35\ \mu s$. However, under large-signal slewing-rate-limited conditions the rise time will approach $t_{rise} = 0.8\ V_{O(MAX)}/SR$. For a $V_{O(MAX)}$ of 10 V, the rise time is $t_{rise} = 0.8 \times 10\ \text{V}/1\ \text{V}/\mu s = 8\ \mu s$, which is much longer than the small-signal rise time.

7.1.4 Full-power Bandwidth

Let us now consider the application of a sinusoidal input voltage to an op amp. Under small-signal conditions such that the output is not slewing rate limited, the output voltage can be expressed as $v_o(t) = V_{O(MAX)} \sin \omega t$. The maximum rate of change of output voltage with time occurs at the zero crossings and is

$$\left(\frac{dv_o}{dt}\right)_{MAX} = \omega V_{O(MAX)}\ (\cos \omega t)_{MAX} = \pm\ \omega V_{O(MAX)} \tag{7.14}$$

If $\omega V_{O(MAX)} > SR$, the op amp is slewing rate limited and the maximum slope of the output voltage is limited to the slewing rate.

In Figure 7.5, some output voltage waveforms are shown for various values of $V_{O(MAX)}$. When $\omega V_{O(MAX)} > SR$, the op-amp output becomes slewing rate limited and the output voltage waveform starts to become distorted from a sinusoidal shape to a waveform approaching a triangular wave, with sides having slopes of $\pm SR$. The peak value of the output voltage is limited to a value of $SR \times T/4 = SR/4f$, where $T = 1/f$ is the period of the waveform.

We see that the peak output voltage swing is limited to a value of $\pm SR/4f$ under large-signal slewing-rate-limited conditions, and thus as the frequency increases the peak output voltage swing decreases rapidly. Figure 7.6 shows the variation of the peak output voltage swing as a function of frequency. The *full-power bandwidth* (FPBW) is the frequency at which the op-amp output becomes slewing rate limited

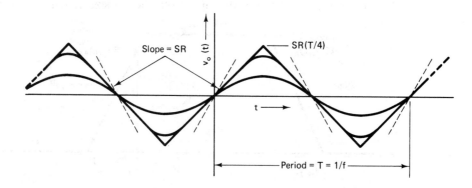

Figure 7.5 Full-power bandwidth.

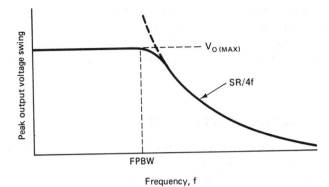

Figure 7.6 Peak output voltage swing versus frequency.

and is obtainable from the condition that $\omega V_{O(MAX)} = SR$, so $\omega = SR/V_{O(MAX)}$ and thus

$$\boxed{\text{FPBW} = \frac{\omega}{2\pi} = \frac{SR}{2\pi V_{O(MAX)}}} \tag{7.15}$$

The full-power bandwidth can be considerably smaller than the bandwidth under small-signal non-slewing-rate-limiting conditions. For example, if $BW_{CL} = 1.0\,\text{MHz}$, $SR = 1.0\,\text{V/}\mu\text{s}$, and $V_{O(MAX)} = 10\,\text{V}$, we have that

$$\text{FPBW} = \frac{SR}{2\pi V_{O(MAX)}} = \frac{1.0\,\text{V/}\mu\text{s}}{2\pi 10\,\text{V}} = \underline{16\,\text{kHz}} \tag{7.16}$$

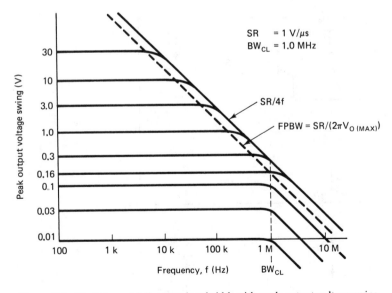

Figure 7.7 Variation of full-power bandwidth with peak output voltage swing.

This value for the full-power bandwidth is to be compared with the small-signal bandwidth of 1.0 MHz, so we see that the FPBW can be much smaller than the small-signal bandwidth. Figure 7.7 shows the variation of the full-power bandwidth as a function of the maximum output-voltage swing $V_{O(MAX)}$, with both variables being on logarithmic scales.

7.2 OPERATIONAL-AMPLIFIER CIRCUIT ANALYSIS EXAMPLE

Consider the operational-amplifier circuit shown in Figure 7.8. The circuit consists of two gain stages and a push–pull Darlington emitter-follower output stage. The first gain stage consists of the differential-amplifier transistor pair Q_1 and Q_2, which is biased by the current source comprised of Q_{12} and Q_{13} and drives the current-mirror active load comprised of Q_3 and Q_4 and the input transistor to the next stage, Q_6.

The second stage consists of the Q_6–Q_7 Darlington compound-transistor configuration, in which Q_6 is an emitter follower (common collector) and Q_7 is in the common-emitter configuration. This stage has as an active load current source transistor Q_{14} (which in turn is biased by Q_{12} and R_1) and drives the output stage transistors.

The output stage is a complementary push–pull emitter-follower configuration consisting of a Darlington NPN emitter-follower compound transistor Q_8 and Q_9, for

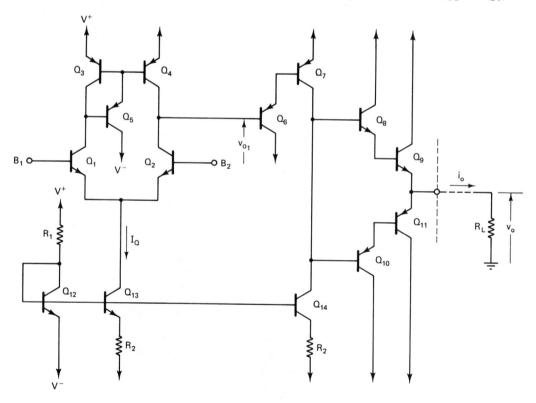

Figure 7.8 Operational-amplifier circuit example.

sourcing currents into the load, and hence to produce the positive-going output voltage swing, and a Darlington PNP emitter-follower compound transistor, Q_{10} and Q_{11}, for sinking currents from the load, and hence to produce the negative-going output voltage swing.

If we assume all transistors of the same type to be identical, then under quiescent conditions ($v_i = v_{B_1} - v_{B_2} = 0$) we have that $I_{C_1} = I_{C_2}$ and $I_{C_3} = I_{C_4}$; so by applying Kirchhoff's current law to the $C_1 - C_3 - B_5$ node and the $C_2 - C_4 - B_6$ node, we obtain that $I_{B_5} = I_{B_6}$. Since the transistor current gain values are equal, we have that $I_{E_6} = I_{E_5}$, so $I_{B_7} = I_{B_3} + I_{B_4}$. Again noting the equality of the transistor current gains, we have that $I_{C_7} = I_{C_3} + I_{C_4}$. Since the base currents are small compared to the emitter and collector currents, we can say that $I_{C_3} + I_{C_4} = I_{C_1} + I_{C_2} = I_Q$, so $I_{C_7} = I_Q$. Since $I_{C_{14}} = I_{C_{13}} = I_Q$, by applying Kirchhoff's current law to the $C_7 - C_{14} - B_8 - B_{10}$ node, we have that $I_{B_8} + I_{B_{10}} = 0$. We now note that Q_8 and Q_{10} cannot both be "on" at the same time, so we must have that $I_{B_8} = 0$ and $I_{B_{10}} = 0$ under these (quiescent) conditions. As a result of these two base currents being zero, we have that the two emitter currents I_{E_9} and $I_{E_{11}}$ are zero, and thus $I_o = 0$. Therefore, $V_o = 0$ under quiescent conditions ($V_i = 0$). Thus with *zero input* ($V_i = 0$) we get *zero output* ($V_o = 0$) for this amplifier, which is precisely the condition that is desired.

7.2.1 Voltage-gain Analysis

The voltage gain of the first (differential amplifier) stage is given by

$$A_{v1} = \frac{v_{o1}}{v_i} = \frac{2g_f}{g_{\text{total}}} \tag{7.17}$$

where g_{total} is the total conductance from the $C_2 - C_4 - B_6$ node to ground and is given by $g_{\text{total}} = g_{i6} + g_{o2} + g_{o4}$. The input conductance of Q_6 is given by

$$g_{i6} = \frac{I_{B_6}}{2nV_T} = \frac{I_Q}{2n\beta_6\beta_7 V_T} \tag{7.18}$$

The output conductances of Q_4 and Q_2 are given by

$$g_{o4} = g_{ce4} = \frac{I_{C_4}}{V_{AP}} = \frac{I_Q}{2V_{AP}} \quad \text{and} \quad g_{o2} = g_{e2} = \frac{I_{C_2}}{V_{AN}} = \frac{I_Q}{2V_{AN}} \tag{7.19}$$

where V_{AP} and V_{AN} are the Early voltages of the PNP and NPN transistors, respectively (see Appendix B). The voltage gain of the first stage is thus given by

$$A_{V1} = \frac{2g_f}{g_{\text{total}}} = \frac{2(I_Q/4V_T)}{(I_Q/2n\beta_6\beta_7 V_T) + (I_Q/2V_{AP}) + (I_Q/2V_{AN})} \tag{7.20a}$$

$$= \frac{1/V_T}{(1/n\beta_6\beta_7 V_T) + (1/V_{AP}) + (1/V_{AN})} \tag{7.20b}$$

$$= \frac{\beta_6\beta_7}{1/n + \beta_6\beta_7 V_T[(1/V_{AP}) + (1/V_{AN})]} \tag{7.20c}$$

The voltage gain of the second stage is given by

$$A_{V_2} = \frac{v_o}{v_{i_6}} = \frac{v_o}{v_{o_1}} = -g_{f_{6,7}} R_{L_7} \tag{7.21}$$

The effective load driven by the Q_6–Q_7 compound transistor configuration is the transformed value of the load resistance, R_L, which is given by $R_{L_7} = \beta_{8,10}\beta_{9,11}R_L$. We are assuming here that the emitter-follower output stage has approximately unity gain, so for the gain of the second stage we have that

$$A_{V_2} = \left(\frac{I_Q}{2V_T}\right)\beta_{8,10}\beta_{9,11}R_L \tag{7.22}$$

Thus the overall gain of the amplifier circuit is given by

$$A_{OL}(0) = A_{V(\text{TOTAL})} = A_{V_1}A_{V_2} = \frac{-\beta_6\beta_7\beta_{8,10}\beta_{9,11}I_Q R_L/(2V_T)}{1/n + \beta_6\beta_7 V_T[(1/V_{AP}) + (1/V_{AN})]} \tag{7.23}$$

To now consider a numerical example, let us assume $\beta = 50$ (min.), $n = 1.5$ for all transistors, and $V_{AN} = V_{AP} = 200$ V. The β product thus becomes $\beta_6\beta_7\beta_{8,10}\beta_{9,11} = (50)^4 = 6.25 \times 10^6$ (min.), and we also have that $\beta_6\beta_7 V_T = 62.5$ V. Therefore, the total voltage gain is

$$A_{V(\text{TOTAL})} = \frac{6.25 \times 10^6 \text{ (min.)}(I_Q R_L/50 \text{ mV})}{1/n + 62.5[(1/200) + (1/200)]}$$

$$= \frac{6.25 \times 10^6 \text{ (min.)}(I_Q R_L/50 \text{ mV})}{1.292} \tag{7.24}$$

If $R_L = 1.0$ kΩ and $I_Q = 20$ μA are chosen as typical values for this example, we have that $I_Q R_L/50$ mV $= 20$ mV/50 mV $= 0.4$, so

$$A_{V(\text{TOTAL})} = 6.25 \times 10^6 \text{ (min.)} \times \frac{0.4}{1.292} = \underline{1.935 \times 10^6 \text{ (min.)}} \tag{7.25}$$

Note that this is the low-frequency gain $A_{OL}(0)$ and that the gain will fall off at higher frequencies.

7.2.2 Operational-amplifier Circuit with a Darlington Input Stage

In many cases it may be advantageous to use a Darlington differential-amplifier configuration as shown in Figure 7.9 to replace the differential amplifier that comprises just a pair of transistors.

If a Darlington differential amplifier is used in place of the simple two-transistor differential amplifier, the basic operation of the circuit will be the same as before. The two most important changes will be that the effective current gain (β) of the Darlington compound-transistor configuration will be the product of the individual transistor current gains ($\beta = \beta_A\beta_B$) of the two transistors that comprise the compound configuration. This will give rise to a very large current gain value and thus corre-

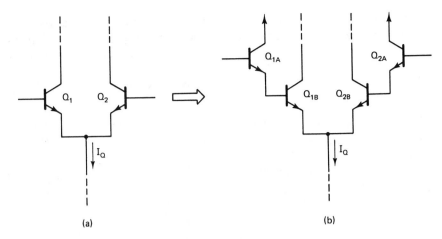

Figure 7.9 Darlington differential amplifier: (a) simple differential amplifier; (b) Darlington differential amplifier.

spondingly to a much reduced input bias current for the amplifier and a much larger input resistance.

In the equations for the differential amplifier, V_T must be replaced by $2V_T$ wherever it appears, so the dynamic forward transfer conductance of the differential amplifier becomes $g_f = I_Q/V_T = I_Q/200 \, \text{mV}$. Since the transfer conductance is reduced to one-half of its previous value, the voltage gain will also be reduced to one-half of its previous value. In addition to these changes, there will also be some small changes in the input voltage range.

Note on the input resistance of Q_6. If we take into account the fact that the dependence of the base current of a transistor in the active mode on the base-to-emitter voltage is given by $I_B \propto \exp{(V_{BE}/nV_T)}$, where n is a dimensionless constant with values between 1 and 2 and with a typical value of about 1.5 or 1.6, we note that the input conductance is given by $g_i = g_{be} = I_B/nV_T$. For the compound Darlington common-emitter configuration (such as Q_6 and Q_7), the input conductance is correspondingly $g_{i_6} = I_B/2nV_T$.

7.2.3 Effect of the Transistor Current Gain (β) on the Voltage Gain

From inspection of the equation for the voltage gain, we may note that the voltage gain is indeed highly dependent on the transistor β values. Indeed, if all the transistor β values turn out to be twice the minimum value, the total voltage gain will be 16 times larger than the voltage gain obtained when the β values are all at the minimum values. Thus we may expect very great unit-to-unit variations in the open-loop gain of operational amplifiers. Operational amplifiers, are, however, usually operated in a closed-loop (negative feedback) configuration such that the closed-loop gain is determined primarily by the feedback factor and is relatively independent of the open-loop gain, as long as the loop gain is much greater than unity. Thus, as long as the open-loop gain of the amplifier is sufficiently large to ensure that the loop gain will be much greater than unity, the closed-loop gain will be relatively independent of the open-loop gain and thus relatively independent of the transistor current gain values.

7.2.4 Voltage Gain of the Darlington Emitter-follower Output Stage

We will now consider the voltage gain of the Darlington emitter-follower push–pull output stage consisting of the NPN compound transistor Q_8 and Q_9 and the PNP compound transistor Q_{10} and Q_{11}.

The voltage gain of a simple (single transistor) emitter-follower stage, from base (input) to emitter (output), is given by

$$A_V = \frac{g_m R_L}{1 + g_m R_L} = \frac{R_L}{(1/g_m) + R_L} = \frac{R_L}{(V_T/I) + R_L} \qquad (7.26)$$

where R_L is the net load driven by the emitter-follower transistor, and I is the current through the transistor. For a Darlington (compound) emitter-follower configuration, we simply replace V_T by $2V_T$ in the expressions above to obtain

$$A_{V(\text{Darlington EF})} = \frac{R_L}{(2V_T/I) + R_L} = \frac{1}{1 + (2V_T/IR_L)} \qquad (7.27)$$

Since the output voltage $V_o = IR_L$, this can be expressed as

$$A_{V(\text{DEF})} = \frac{1}{1 + (2V_T/V_o)} \qquad (7.28)$$

From this last expression we see that, as long as the output voltage satisfies the condition $V_o \gg 2V_T = 50$ mV, the output-stage voltage gain is approximately (but always somewhat less than) unity.

We must note, however, that whenever the output voltage drops below $2V_T = 50$ mV, the output-stage voltage gain decreases to values that are substantially less than unity. This reduced value of gain produces a type of distortion known as *crossover distortion*. This is shown in Figure 7.10.

The circuit of Figure 7.11, showing the addition of transistor Q_{15} and resistors R_3 and R_4, is one that can be used to minimize crossover distortion. With this circuit there will be a small quiescent current flowing through the output-stage transistors even when $I_o = 0$, this being the result of the voltage drop produced by Q_{15} in conjunction with R_3 and R_4. If we call this quiescent current (with $I_o = 0$) $I_{Q_{EF}}$, then the *minimum* voltage gain of the emitter-follower output stage is given by

$$A_{V(\text{DEFmin})} = \frac{R_L}{R_L + (2V_T/I_{\min})} = \frac{R_L}{R_L + (2V_T/I_{Q_{EF}})} \qquad (7.29)$$

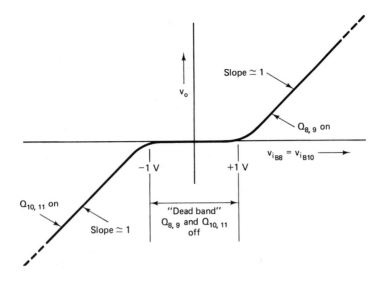

Figure 7.10 Emitter-follower output stage characteristics showing crossover distortion.

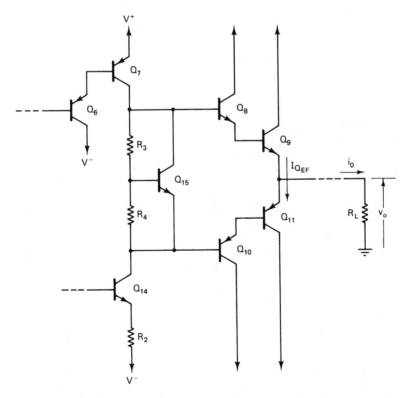

Figure 7.11 Circuit for minimizing crossover distortion.

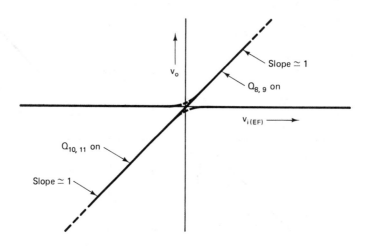

Figure 7.12 Minimization of crossover distortion by the biasing of the output stage at the threshold of conduction.

For example, if $I_{Q_{EF}} = 1.0$ mA, then $2V_T/I_{Q_{EF}} = 50$ mV/1.0 mA $= 50$ Ω, so

$$A_{V(\text{DEF}_{\min})} = \frac{R_L}{R_L + 50\ \Omega} \tag{7.30}$$

So for $R_L = 1.0$ kΩ, we have

$$A_{V(\text{DEF}_{\min})} = \frac{1000}{1050} \simeq 0.95 \tag{7.31}$$

This virtually eliminates the crossover distortion, as shown by Figure 7.12.

We should also note that op amps are usually operated in a closed-loop configuration and that the closed-loop gain is usually relatively independent of the open-loop gain. The variation of closed-loop gain with respect to open-loop gain has been given by $dA_{CL}/A_{CL} = (A_{CL}/A_{OL})(dA_{OL}/A_{OL})$. Therefore, as long as $A_{CL} \ll A_{OL}$, a variation in A_{OL} will have relatively little effect on A_{CL}. Looking at it from a slightly different viewpoint, we have that $A_{CL} = A_{OL}/(1 + FA_{OL})$, so as long as the *loop gain*, FA_{OL}, is large compared to unity, we have that $A_{CL} \simeq 1/F$, and the closed-loop gain is relatively independent of the open-loop gain. Thus, as long as the open-loop is maintained above some suitable minimum value, we have that $A_{CL} \simeq 1/F$, so the nonlinear distortion in the open-loop gain characteristics will have relatively little impact on the closed-loop characteristics.

7.2.5 Current Limiting

In Figure 7.13, the output stage is modified to incorporate circuitry for bilateral output current limiting using Q_{16} and Q_{17} together with the two current-limiting resistors R_{CL_1} and R_{CL_2}. This circuit limits the maximum current that the op amp can either

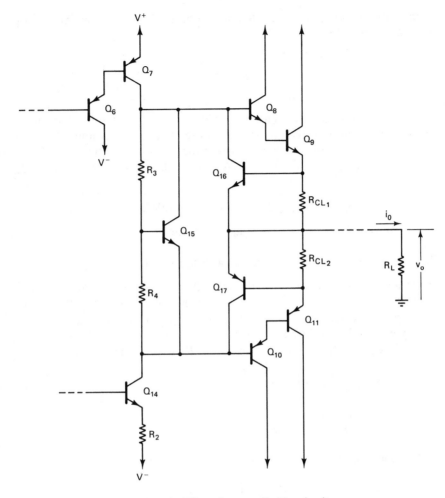

Figure 7.13 Bilateral current-limiting circuit.

source or sink to a safe maximum value, even under the worst-case condition of a
short circuit across the output ($v_o = 0$).

As long as the output current is such that the voltage drop across either current-
limiting resistor is less than 0.5 V, both Q_{16} and Q_{17} will be *off* and will therefore
have no effect on the operation of the circuit. However, as the output current
increases such that the voltage drop across either R_{CL_1} or R_{CL_2} rises above 0.5 V,
then Q_{16} or Q_{17}, respectively, will start to turn on. As either current limit transistor
starts to go into conduction, it starts to divert base drive away from the Darlington
emitter follower, Q_8–Q_9 or Q_{10}–Q_{11}. As the output current increases even more,
the increased voltage drop across R_{CL} causes the current diverted through Q_{16} or Q_{17}

to increase at an exponential rate and ultimately limits the maximum output current to a value given by

$$I_{O(\text{MAX})} = I_{CL} = \frac{V_{BE(\text{ON})}}{R_{CL}} \simeq \frac{0.6 \text{ to } 0.7 \text{ V}}{R_{CL}} \qquad (7.32)$$

This will be *bilateral* current limiting with Q_{16} and R_{CL_1} acting to limit the maximum current that the op amp can source to a safe maximum value, and Q_{17} and R_{CL_2} acting to limit the maximum current that the op amp can sink. If, for example, a value of $I_{O(\text{MAX})} = I_{CL} = \pm 20$ mA is needed, the suitable choice for R_{CL} is given by

$$R_{CL} = V_{BE(\text{ON})}/I_{CL} \simeq 600 \text{ to } 700 \text{ mV}/20 \text{ mA} \simeq \underline{30 \text{ to } 35 \text{ } \Omega}$$

so we see that R_{CL} is a relatively small resistor.

7.2.6 Effect of Supply Voltage on Voltage Gain

If we look at the equation for the overall gain of the op amp, we do not see any explicit dependence of the voltage gain on the supply voltage. We must note, however, that the current I_Q, upon which the gain is directly dependent, is related to the supply voltage via an interrelationship with I_{12} (the current through Q_{12}).

Since $I_Q = I_{13} = I_{14} = I_{12} \exp(-I_Q R_2/V_T)$, we have that

$$\begin{aligned} \frac{dI_Q}{dV_S} &= \frac{dI_{12}}{dV_s} \exp\left(-\frac{I_Q R_2}{V_T}\right) + \left[I_{12} \exp\left(-\frac{I_Q R_2}{V_T}\right)\right]\left[-\frac{R_2}{V_T}\frac{dI_Q}{dV_S}\right] \\ &= \frac{dI_{12}}{dV_S}\frac{I_Q}{I_{12}} + (I_Q)\left(-\frac{R_2}{V_T}\frac{dI_Q}{dV_S}\right) \end{aligned} \qquad (7.33)$$

so by collecting terms in I_Q we have

$$\frac{1}{I_Q}\frac{dI_Q}{dV_S}\left(1 + \frac{I_Q R_2}{V_T}\right) = \frac{1}{I_{12}}\frac{dI_{12}}{dV_S} \qquad (7.34)$$

Since $I_{12} = (2V_S - V_{BE})/R_1 \simeq 2V_S/R_1$, we obtain

$$\frac{1}{I_{12}}\frac{dI_{12}}{dV_S} = \frac{R_1}{2V_S}\frac{2}{R_1} = \frac{1}{V_S} \qquad (7.35)$$

so we now have

$$\frac{1}{I_Q}\frac{dI_Q}{dV_S} = \frac{1/V_S}{1 + (I_Q R_2/V_T)} \qquad (7.36)$$

Since $\exp(I_Q R_2/V_T) = I_{12}/I_Q$, this can be rewritten as

$$\frac{dI_Q}{I_Q} = \frac{dV_S}{V_S}\frac{1}{1 + \ln(I_{12}/I_Q)} \qquad (7.37)$$

Since the voltage gain is directly proportional to I_Q, we have that the fractional change in the voltage gain is equal to the fractional change in I_Q, so $dI_Q/I_Q = dA_V/A_V$ and we finally obtain

$$\frac{dA_V}{A_V} = \frac{dI_Q}{I_Q} = \frac{dV_S}{V_S}\frac{1}{1 + \ln(I_{12}/I_Q)} \tag{7.38}$$

For example, if $I_{12} = 1.0$ mA $= 1000$ μA and $I_Q = 25$ μA, then

$$\ln\frac{I_{12}}{I_Q} = \ln\frac{1000\ \mu A}{25\ \mu A} = \ln 40 = 3.69 \tag{7.39}$$

so

$$\frac{dA_V}{A_V} = \frac{dI_Q}{I_Q} = \frac{dV_S}{V_S}\left(\frac{1}{4.69}\right) = 0.213\left(\frac{dV_S}{V_S}\right) \tag{7.40}$$

Thus if $V_S = \pm 15$ V, a 1.0-V change in V_S will produce a fractional change in the open-loop voltage gain given by

$$\frac{dA_V}{A_V} = 0.213\left(\frac{1}{15}\right) = 0.0142 \tag{7.41}$$

which corresponds to a 1.42% change, which is quite small. This is the change in the *open-loop* gain. The change in the closed-loop gain will be much smaller, usually by a very large factor, remembering that $dA_{CL}/A_{CL} = (dA_{OL}/A_{OL})(A_{CL}/A_{OL})$.

Minimum voltage to operate op amp. The minimum supply voltage required to operate this op amp is governed by the basic requirement that all the transistors in the circuit operate in the active mode. Therefore, we must require that the collector-to-emitter voltage be greater than the saturation value, $V_{CE(SAT)}$, and that the base-to-emitter voltage drop be the value needed to operate the transistor in the active region. If we choose $V_{BE} = 700$ mV and $V_{CE(SAT)} = 200$ mV as conservative values to use in the determination of the minimum required operating voltage, we have the following requirements to meet:

$$V_S^+ - V_S^- = V_{BE_{3,4}} + V_{BE_5} + V_{CE_{1,2}} + V_{CE_{13}} + V_{R_2}$$

$$= (700 + 700 + 200 + 200 + 92)\ \text{mV} = 1892\ \text{mV} \simeq 2.0\ \text{V} \tag{7.42}$$

as a necessary condition, and

$$V_S^+ - V_S^- = V_{CE_7} + V_{BE_8} + V_{BE_9} + V_{EB_{11}} + V_{EB_{10}} + V_{CE_{14}} + V_{R_2}$$

$$= (200 + 700 + 700 + 700 + 700 + 200 + 92)\ \text{mV} \tag{7.43}$$

$$= 3292\ \text{mV} \simeq \underline{3.3\ \text{V}}$$

as a necessary *and sufficient* condition. Thus it appears that this op amp can be operated with a supply voltage, $V_S^+ - V_S^-$, as low as about 3.3 V.

Voltage gain variation. We will now investigate the variation in the open-loop gain of the op amp as the supply voltage varies from a minimum of 3.3 V (total supply voltage) to a maximum of 30 V ($V_S = \pm 15$ V), assuming that at $V_S = \pm 15$

V we have $I_{12} = 1.0$ mA and $I_Q = 25$ μA, corresponding to $R_1 = 29.3$ kΩ and $R_2 = 3.69$ kΩ. At $V_S = \pm 1.65$ V, we have that

$$I_{12} = \frac{3.3 - 0.7}{29.3 \text{ k}\Omega} = 0.08874 \text{ mA} = 88.74 \text{ μA} \tag{7.44}$$

By using an iterative calculation, we obtain that I_Q will now be 13.0 μA. Thus I_Q has dropped by a factor of $13/25 = \underline{0.52}$, and we would expect the open-loop voltage gain to be reduced correspondingly by the same factor. Thus, in going from a total supply voltage of 3.3 V up to a total supply voltage of 30 V, the open-loop voltage gain increases by almost a factor of 2. The corresponding change in the closed-loop gain is, however, generally very much less.

Since the unity-gain frequency f_u is given by $\omega_u = 2g_f/C_f$ and since g_f is directly proportional to I_Q, we should see a similar variation in the unity-gain frequency. Thus, if $f_u = 1.0$ MHz when we have a total supply voltage of 30 V across the amplifier, the value of f_u drops to about 520 kHz when we have a total supply voltage of 3.3 V across the amplifier. Since the slewing rate is given by $SR = I_Q/C_F$, we should expect to see a similar variation in the slewing rate.

7.2.7 Common-mode Rejection Ratio

The common-mode dynamic transfer conductances of a differential amplifier are given by $g_{f1(CM)} = (I_1/I_Q)g_o$ and $g_{f2(CM)} = (I_2/I_Q)g_o$, where g_o is the output conductance of the current source that is used to bias the differential amplifier. For the op-amp circuit under consideration, $g_o = g_{o13}$. Since $i_3 \simeq i_1$ and $i_4 = i_3$, we have that $i_o = i_4 - i_2 \simeq i_1 - i_2$. Therefore, in response to a common-mode input voltage, we have that $i_1 = g_{f1(CM)}v_{iCM}$ and $i_2 = g_{f2(CM)}v_{iCM}$, so

$$i_o \simeq i_1 - i_2 = [g_{f1(CM)} - g_{f2(CM)}]v_{iCM} = \frac{I_1 - I_2}{I_Q} g_{o13} v_{iCM} \tag{7.45}$$

The common-mode output of the differential-amplifier stage is therefore given by

$$v_{o1(CM)} = \frac{i_{oCM}}{g_{\text{total}}} = \frac{I_1 - I_2}{I_Q} \frac{g_{o13}}{g_{\text{total}}} v_{iCM} \tag{7.46}$$

and the common-mode gain of the differential-amplifier stage is thus

$$A_{V1(CM)} = \frac{v_{o1(CM)}}{v_{iCM}} = \frac{I_1 - I_2}{I_Q} \frac{g_{o13}}{g_{\text{total}}} \tag{7.47}$$

If we now note that the gain with respect to the common-mode signal is the same as that accorded to the difference-mode signal beyond the first stage, we can say that the common-mode rejection ratio is determined by the ratio of the difference-mode gain to the common-mode gain in the first stage. We therefore have $A_{VDM}/A_{VCM} = A_{V1(DM)}/A_{V1(CM)}$. Since $A_{V1(DM)} = 2g_f/g_{\text{total}}$, we now have

$$\text{CMRR} = \frac{A_{V1(DM)}}{A_{V1(CM)}} = \frac{2g_f/g_{\text{total}}}{[(I_1 - I_2)/I_Q](g_{o13}/g_{\text{total}})} = \frac{2g_f}{[(I_1 - I_2)/I_Q]g_{o13}} \tag{7.48}$$

If we let V_i' represent the sum of the difference-mode input voltage and the input offset voltage, we have $V_i' = V_i + V_{OS}$. The resulting collector currents are $I_1 = I_Q + \Delta I$ and $I_2 = I_Q - \Delta I$, where $\Delta I \simeq g_f V_i'$. Therefore, $I_1 - I_2 \simeq 2g_f V_i'$. The common-mode rejection ratio can now be written as

$$\text{CMRR} \simeq \frac{2g_f}{(2g_f V_i'/I_Q)g_{o13}} = \frac{I_Q}{g_{o13}V_i'} \tag{7.49}$$

The output conductance of the current sink transistor Q_{13} is given by

$$g_{o13} = \frac{I_Q}{V_A[1 + \ln(I_{12}/I_Q)]} \tag{7.50}$$

where V_A is the Early voltage. Substituting this into the CMRR expression gives

$$\boxed{\text{CMRR} = \frac{I_Q}{g_{o13}V_i'} = \frac{V_A}{V_i'}\left(1 + \ln\frac{I_{12}}{I_Q}\right)} \tag{7.51}$$

To consider a representative example, let us take $V_A = 200$ V, $I_{12} = 1.0$ mA, and $I_Q = 25$ μA. For V_i' we will use a unit value of 1.0 mV. For these values we obtain CMRR $\simeq (200 \text{ V}/1.0 \text{ mV})(4.69) = 938{,}000$, corresponding to 119 dB. The actual value of the CMRR may be substantially lower than this as a result of other imbalances in the circuit and as a result of capacitive feedthrough, especially at higher frequencies.

7.2.8 Power Supply Rejection Ratio

The power supply rejection ratio (PSRR) is the ratio of the change in the input offset voltage to the change in the power supply voltage that produces this change and is usually expressed in decibels.

If we inspect the operational-amplifier circuit under consideration, we see that a change in the supply voltage can produce a change in the offset voltage by several means. First, a change in the supply voltage V_S will change the current through Q_{12} and thereby change the quiescent current I_Q, which in turn will produce a change in the input offset voltage. This will turn out to be by far the most important contribution to the PSRR.

A second but much smaller contribution to the PSRR is the change in I_Q due to the change in the voltage across the current source comprised of Q_{13} and R_2. A third, even smaller contribution to the PSRR is the change in the current through Q_7 resulting from the change in the voltage across Q_7.

We will now investigate the first and most important contribution to the PSRR, the change in I_{12} due to the change in the supply voltage, which in turn results in a change in I_Q. We obtained earlier that

$$\frac{dI_Q}{I_Q} = \frac{dV_S}{V_S}\frac{1}{1 + \ln(I_{12}/I_Q)} \tag{7.52}$$

for the type of current source under consideration here.

The output current of the differential amplifier stage I_O is given by $I_O = I_4 - I_2 \simeq I_1 - I_2 = 2g_f V_i'$, where $V_i' = V_i + V_{OS}$. Since $g_f = I_Q/4V_T$, this can be rewritten as $I_O = (I_Q/2V_T)V_i'$. The required change in V_i' to cancel out a change in I_O produced by a change in I_Q is given by $dV_i'/V_i' = -dI_Q/I_Q$. We therefore have that

$$\frac{dV_i'}{V_i'} = -\frac{dI_Q}{I_Q} = -\frac{dV_S}{V_S} \frac{1}{1 + \ln (I_{12}/I_Q)} \tag{7.53}$$

The power supply rejection ratio $dV_{OS}/dV_S = dV_i'/dV_S$ is therefore given by

$$\boxed{PSRR = \left| \frac{dV_{OS}}{dV_S} \right| = \frac{V_i'}{V_S} \frac{1}{1 + \ln (I_{12}/I_Q)}} \tag{7.54a}$$

To consider a representative example, let us take $I_{12} = 1.0$ mA, $I_Q = 25$ μA, $V_S = \pm 15$ V, and for V_i' we will use a unit value of 1.0 mV. With these values we obtain

$$PSRR = \frac{1.0 \text{ mV}}{15 \text{ V}} \frac{1}{1 + \ln 40} = \frac{1}{7 \times 10^4} \tag{7.54b}$$

which corresponds to -97 dB.

The change in the voltage across Q_{13} also produces a change in I_Q and thus contributes to the overall PSRR. To evaluate this, we note that the change in I_Q resulting from the change in the voltage across Q_{13} and R_2 is given by

$$dI_Q = g_{o13} \, dV_S = \frac{I_Q}{V_A[1 + \ln (I_{12}/I_Q)]} \, dV_S \tag{7.55a}$$

The change in V_i' required to compensate for the change in I_Q is, as before, given by $dV_i'/V_i' = dI_Q/I_Q$, so we obtain

$$\frac{dV_i'}{V_i'} = -\frac{dI_Q}{I_Q} = -\frac{dV_S}{V_A} \frac{1}{1 + \ln (I_{12}/I_Q)} \tag{7.55b}$$

Therefore, this contribution to the PSRR is given by

$$PSRR = \left| \frac{dV_i'}{dV_S} \right| = \frac{V_i'}{V_A} \frac{1}{1 + \ln (I_{12}/I_Q)} \tag{7.56}$$

If we now use the same values as used in the previous example together with $V_A = 200$ V, we obtain

$$PSRR = \frac{1.0 \text{ mV}}{200 \text{ V}} \frac{1}{1 + \ln 40} = \frac{1}{9.4 \times 10^5} \tag{7.57}$$

We therefore see that this contribution to the overall PSRR is very small, and the dominant contribution is still due to the change in I_{12} resulting from the change in the supply voltage.

We will now consider the contribution of the change in the current through Q_7 to the PSRR. The change in the current through Q_7, ΔI_7, that results from a change in the supply voltage, ΔV_S, is given by

$$\Delta I_7 = g_{o7} \, \Delta V_S = g_{ce7} \, \Delta V_S = \frac{I_7}{V_A} \Delta V_S = \frac{I_Q}{V_A} \Delta V_S \tag{7.58}$$

where V_A is the Early voltage of Q_7. The corresponding change in I_{B_6} that is required to cancel out the effect of the change in the current through Q_7 produced by the change in the supply voltage is given by

$$\Delta I_{B_6} = \frac{\Delta I_7}{\beta_6 \beta_7} \tag{7.59}$$

This change in I_{B_6} will in turn require a change in the input offset voltage, given by the following relationship:

$$\Delta I_{B_6} = 2 g_f \, \Delta V_{OS} = \frac{I_Q}{2 V_T} \Delta V_{OS} \tag{7.60}$$

As a result, we now obtain

$$\frac{I_Q}{2 V_T} \Delta V_{OS} = \frac{\Delta I_7}{\beta_6 \beta_7} = \frac{(I_Q/V_A) \, \Delta V_S}{\beta_6 \beta_7} \tag{7.61}$$

Solving for the ratio $\Delta V_S / \Delta V_{OS}$, we obtain

$$\frac{1}{\text{PSRR}} = \frac{\Delta V_S}{\Delta V_{OS}} = \frac{V_A (\beta_6 \beta_7)}{2 V_T} = \frac{V_A (\beta_6 \beta_7)}{50 \text{ mV}} \tag{7.62}$$

For example, if we again choose $V_A = 200$ V and let $\beta_6 = 50$ (min.) and $\beta_7 = 50$ (min.), we obtain

$$\frac{1}{\text{PSRR}} = \frac{200 \text{ V}}{50 \text{ mV}} (50 \times 50)(\text{min.})$$

$$= 4 \times 10^3 \times 2.5 \times 10^3 = 1 \times 10^7 \tag{7.63}$$

We therefore see that this contribution to the net value of the PSRR is very small indeed.

Finally, if we were to investigate the effect of the change in the voltage across Q_{14} and R_2 on the input offset voltage, we have essentially the same result as that given with respect to Q_7.

In summary, it indeed does appear that the major contributor to the PSRR is the result of the change in I_{12}, and therefore of I_{13} ($= I_Q$) due to a change in the supply voltage.

7.2.9 Frequency Response Analysis

The op-amp equivalent circuit shown in Figure 7.14 will be used for the analysis of the frequency response of the circuit of Figure 7.8. The capacitance C_{O_1} represents the summation of all the capacitances between the C_2–C_4–B_6 node and ground.

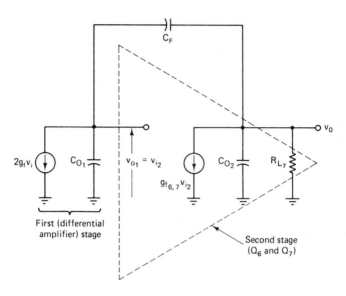

Figure 7.14 Operational amplifier equivalent circuit for the analysis of frequency response.

Capacitance C_{O_2} is the sum of all the capacitances between the C_7–C_{14}–B_8–B_{10} node and ground.

If we write a node voltage equation for the input node of the second stage (the C_2–C_4–B_6 node), we obtain

$$(V_o - V_{i_2})(j\omega C_F) = 2g_f v_i + V_{i_2}(j\omega C_{O_1}) \qquad (7.64)$$

Since $V_o = -A_{V_2}V_{i_2}$, we have that $V_{i_2} = -V_o/A_{V_2}$. If we now collect all terms of which V_O is a factor on the left side, we obtain

$$V_O\left[j\omega C_F\left(1 + \frac{1}{A_{V_2}}\right) + \frac{j\omega C_{O_1}}{A_{V_2}}\right] = 2g_f V_i \qquad (7.65)$$

Therefore, the open-loop gain, $A_{OL} = V_O/V_i$, is given by

$$A_{OL} = \frac{V_O}{V_i} = \frac{2g_f}{j\omega C_F(1 + 1/A_{V_2}) + j\omega C_{O_1}/A_{V_2}}$$

$$= \frac{2g_f/j\omega C_F}{1 + (1/A_{V_2}) + (C_{O_1}/C_F)/A_{V_2}} \qquad (7.66)$$

Since the (extrapolated) unity-gain frequency is given by the equation $2g_f/\omega_u C_F = 1$, we have that $\omega_u = 2g_f/C_F$, so we can write the voltage-gain expressions as

$$A_{OL} = \frac{\omega_u/j\omega}{1 + [(1 + C_{O_1}/C_F)/A_{V_2}]} = \frac{f_u/jf}{1 + [(1 + C_{O_1}/C_F)/A_{V_2}]} \qquad (7.67)$$

From this equation we see that as long as the voltage gain of the second stage, A_{V_2}, is large compared to unity, the frequency response of the second stage will not have an important effect on the overall frequency response. However, when this is no longer the case, the frequency response characteristics of the second gain stage must be taken into consideration. Since the voltage gain of the second gain stage at low frequencies is very large [$A_{V_2}(0) \gg 1$], typically on the order of 1000, we see that the frequency range wherein $A_{V_2} \lesssim 1 + (C_{O_1}/C_F)$ obtains will be considerably above the 3-dB (or breakpoint) frequency of the second stage. Indeed, in this frequency region we can express the voltage gain of the second stage as

$$A_{V_2} = -g_{f_{6,7}}Z_{L_7} = -\frac{g_{f_{6,7}}}{Y_{L_7}} = \frac{-g_{f_{6,7}}}{G_{L_7} + j\omega C_{O_2}} \qquad (7.68)$$

$$\simeq -\frac{g_{f_{6,7}}}{j\omega C_{O_2}}$$

where $g_{f_{6,7}}$ is the dynamic transfer conductance of the Q_6–Q_7 compound transistor configuration was given by $g_{f_{6,7}} = I_7/2V_T = I_Q/2V_T$, since the quiescent current of Q_7 is I_Q. If we define f_2 as the unity-gain frequency of the second stage, we have that $g_{f_{6,7}}/\omega_2 C_{O_2} = 1$, so $\omega_2 \simeq g_{f_{6,7}}/C_{O_2}$, and we can therefore express the voltage gain of the second stage (in the frequency range of interest) as

$$A_{V_2} \simeq \frac{\omega_2}{j\omega} = \frac{f_2}{jf} \qquad (7.69)$$

The overall gain of the op amp can now be expressed as

$$A_{OL} \simeq \frac{f_u/jf}{1 + [(1 + C_{O_1}/C_F)/(f_2/jf)]} = \frac{f_u/jf}{1 + [jf(1 + C_{O_1}/C_F)/f_2]} \qquad (7.70)$$

Let us define f_2' as $f_2' = f_2/(1 + C_{O_1}/C_F)$ so that we can now write

$$A_{OL} = \frac{f_u/jf}{1 + j(f/f_2')} \qquad (7.71)$$

For a 45° phase margin, we require that $|A_{OL}| \leq 1$ when $f = f_{135°}$. By inspection of the last equation for A_{OL}, we see that at $f = f_2'$ the phase angle of A_{OL} is $-135°$; therefore, $f_{135°} = f_2'$. The value of A_{OL} at $f = f_{135°} = f_2'$ is given by

$$|A_{OL}(f = f_2')| = \frac{f_u/f_2'}{|1 + j(1)|} = \frac{f_u}{\sqrt{2}f_2'} \qquad (7.72)$$

Thus, to satisfy the requirement that $|A_{OL}| \leq 1$ at $f = f_{135°}$, we have that $f_u/(\sqrt{2}f_2') \leq 1$, which gives $f_u \leq \sqrt{2}f_2'$ or, correspondingly, $\omega_u \leq \sqrt{2}\omega_2'$.

We have already stated that $\omega_u = 2g_f/C_F$, and we now note that $g_f = $ dynamic transfer conductance of the first (differential amplifier) stage $= I_Q/4V_T$, so that $\omega_u = (I_Q/2V_T)/C_F$. For ω_2' we have

$$\omega_2' = \frac{\omega_2}{1 + C_{O_1}/C_F} = \frac{g_{f_{6,7}}/C_{O_2}}{1 + C_{O_1}/C_F} \qquad (7.73)$$

where $g_{f6,7} = I_7/2V_T = I_Q/2V_T$. If we now substitute these relationships into the equation relating f_u to f_2', we obtain

$$\frac{I_Q/2V_T}{C_F} \leq \frac{\sqrt{2}I_Q/2V_T}{C_{O_2}(1 + C_{O_1}/C_F)} \tag{7.74}$$

After canceling out common factors, I_Q and $2V_T$, this becomes

$$\frac{1}{C_F} \leq \frac{\sqrt{2}}{C_{O_2}(1 + C_{O_1}/C_F)} \tag{7.75}$$

so the required feedback capacitance is given by

$$C_F \geq \frac{C_{O_2}}{\sqrt{2}}\left(1 + \frac{C_{O_1}}{C_F}\right) \tag{7.76}$$

The capacitance C_{O_2} is the summation of all the capacitances between the C_7–C_{14}–B_8–B_{10} node and ground. Since C_F is connected to this node, it will be part of C_{O_2}. We can thus express C_{O_2} as $C_{O_2} = C_{O_2}' + C_F$, where C_{O_2}' is the summation of all the capacitances except C_F. The inequality for C_F can now be written as

$$C_F \geq \frac{(C_{O_2}' + C_F)(1 + C_{O_1}/C_F)}{\sqrt{2}} = \frac{C_{O_2}' + C_{O_1} + C_{O_2}'C_{O_1}/C_F + C_F}{\sqrt{2}} \tag{7.77}$$

By multiplying through this last equation by $2C_F$ and collecting terms, we obtain a quadratic equation in C_F as given by

$$C_F^2(\sqrt{2} - 1) - (C_{O_1} + C_{O_2}')C_F - C_{O_1}C_{O_2}' = 0 \tag{7.78}$$

So C_F is given by

$$C_F = \frac{C_{O_1} + C_{O_2}'}{2(\sqrt{2} - 1)}\left(1 + \sqrt{1 + \frac{4C_{O_1}C_{O_2}'(\sqrt{2} - 1)}{(C_{O_1} + C_{O_2}')^2}}\right) \tag{7.79}$$

For C_{O_1} we have

$$C_{O_1} = C_{CB_2} + C_{CS_2} + C_{CB_4} + C_{CB_6} + C_{BE_6} \tag{7.80}$$

where C_{CS} is the collector-to-substrate capacitance. For C_{O_2}' we have

$$C_{O_2}' = C_{CB_7} + C_{CB_{14}} + C_{CS_{14}} + C_{CB_8} + C_{CS_8} + C_{CB_{10}} \tag{7.81}$$

Note that the base-to-emitter capacitances of Q_8 and Q_{10} are not involved since the voltage gain of the emitter-follower stage can be considered to be close to unity.

For a typical example, let us assume that for all transistors involved $C_{CB} = 1.0$ pF (typ.), 1.5 pF (max.); $C_{CS} = 1.0$ pF (typ.), 1.5 pF (max.); and $C_{BE} = 10$ pF (typ.), 15 pF (max.). With these capacitance values, we obtain $C_{O_1} = 14$ pF (typ.), 21 pF (max.), and $C_{O_2}' = 6$ pF (typ.), 9 pF (max.). Solving for C_F using the quadratic formula given above, we obtain $C_F = 52.2$ pF (typ.), 78.3 pF (max.). The corresponding unity-gain frequencies are given by $f_u = 2g_f/2\pi C_F$. If $I_Q = 25$ μA, then $g_f = I_Q/4V_T = 250$ μs. For a conservative design, we would use $C_F = C_{F(max)} = 78.3$ pF, which gives us a unity-gain frequency of $f_u = 1.02$ MHz.

7.2.10 Equivalent Input Noise Voltage

Due to the finite charge on the electron, electrical current flow will not be absolutely constant, but will always exhibit minute random fluctuations called *shot noise*. The mean-squared shot noise current per unit bandwidth is given by the equation $\overline{i_{noise}^2} = 2qI$, where q is the electronic charge of 1.6×10^{-19} and I is the average current.

As a result of the high voltage gain of the first (differential amplifier) stage, virtually all the noise at the output of the amplifier has its origin in the first stage. The mean-squared output noise current of the first stage is contributed to by Q_1 and Q_2 of the differential amplifier and Q_3 and Q_4 of the active load. The noise currents of these four transistors are uncorrelated, so the mean-squared noise currents are directly additive. The mean-squared output noise current of the first stage is given by $\overline{i_{o(noise)}^2} = \overline{i_1^2} + \overline{i_2^2} + \overline{i_3^2} + \overline{i_4^2}$. For $\overline{i_1^2}$ we have that $\overline{i_1^2} = 2qI_{C_1} = 2q(I_Q/2) = qI_Q$, and we will obtain similar expressions for the other three noise currents. The output noise current is therefore given by $\overline{i_{o(noise)}^2} = 4qI_Q$ (A²/Hz).

The equivalent input noise voltage $v_{i(noise)}$ is a noise voltage that when applied to the input of the amplifier produces a noise current equal to $i_{o(noise)}$. These two quantities are related by the dynamic transfer conductance g_f as given by $\overline{i_{o(noise)}^2} = (2g_f)^2 \overline{v_{i(noise)}^2}$, where $g_f = I_Q/4V_T$ and the factor of 2 comes from the current doubling action of the current-mirror active load. Solving this equation for the *mean-squared equivalent input noise voltage* gives

$$\overline{v_{i(noise)}^2} = \frac{\overline{i_{o(noise)}^2}}{(2g_f)^2} = \frac{4qI_Q}{(I_Q/2V_T)^2} = 16qV_T\frac{V_T}{I_Q} \tag{7.82}$$

Since $V_T = kT/q$, $qV_T = kT$, so

$$\boxed{\overline{v_{i(noise)}^2} = 16kT\frac{V_T}{I_Q}} \tag{7.83}$$

At $T = 290$ K $= 17°C$, $4kT = 1.60 \times 10^{-20}$ J, so $v_{i(noise)}$ can be written as

$$\overline{v_{i(noise)}^2} = 6.4 \times 10^{-20}\frac{V_T}{I_Q} \quad \text{V²/Hz} \tag{7.84}$$

In addition to the noise produced by the random fluctuations of the transistor currents, called the *shot noise*, there is also a significant amount of noise due to the *thermal noise voltage* of the transistor base spreading resistances, $r_{bb'}$. For the two transistors of the differential amplifier, the mean-squared thermal noise voltage due to the two base resistances will be $\overline{v_{th}^2} = 4kT(2r_{bb'})$. The *total* mean-squared equivalent input noise voltage of the amplifier is now given by

$$\boxed{\overline{v_{i(noise)}^2} = 16kT\frac{V_T}{I_Q} + 4kT(2r_{bb'}) = 8kT\left[2\left(\frac{V_T}{I_Q}\right) + r_{bb'}\right]} \tag{7.85}$$

If we use $I_Q = 25$ μA and $r_{bb'} = 150$ Ω as representative values, we obtain an equivalent mean-squared input noise voltage of

$$v_{i(\text{noise})}^2 = 3.20 \times 10^{-20}[2(25 \text{ mV}/25 \text{ μA}) + 150] \text{ Ω}$$

$$= 3.20 \times 10^{-20} (2000 + 150) \text{ Ω} \qquad (7.86)$$

$$= \underline{6.88 \times 10^{-17} \text{ V}^2/\text{Hz}}$$

The rms noise voltage is

$$v_{i(\text{noise})} \text{ (rms, spectral density)} = \underline{8.295 \text{ nV}/\sqrt{\text{Hz}}} \qquad (7.87)$$

The equivalent noise bandwidth is equal to the 3-dB bandwidth multiplied by $\pi/2$ for systems characterized by a single breakpoint frequency. Therefore, if for the example above the system 3-dB bandwidth is 10 kHz, the equivalent noise bandwidth is 15.71 kHz. The total rms noise voltage over this system bandwidth is given by

$$v_{i(\text{noise})} \text{ (rms)} = 8.295 \text{ nV}/\sqrt{\text{Hz}} \times \sqrt{15.71 \text{ kHz}} = \underline{1.04 \text{ μV}} \qquad (7.88)$$

If the input stage is a Darlington configuration, the transfer conductance g_f will be one-half of the value used in the preceding case. In addition, due to the use of four transistors in the differential amplifier input stage compared to two transistors in the preceding case, the thermal noise voltage will now result from a total resistance of $4r_{bb'}$. As a result, the mean-squared equivalent input noise voltage of a system using a Darlington differential-amplifier input stage is given by

$$\overline{v_{i(\text{noise})}^2} = 16kT\left(\frac{2V_T}{I_Q} + r_{bb'}\right) \qquad (7.89)$$

If we use the same values as in the preceding example ($I_Q = 25$ μA and $r_{bb'} = 150$ Ω), we now obtain

$$\overline{v_{i(\text{noise})}^2} = 6.40 \times 10^{-20}(2150 \text{ Ω}) = 1.377 \times 10^{-16} \text{ V}^2/\text{Hz} \qquad (7.90)$$

The rms noise voltage is

$$v_{i(\text{noise})} \text{ (rms)} = \underline{11.7 \text{ nV}/\sqrt{\text{Hz}}}$$

Over a 10-kHz 3-dB system bandwidth, the total noise voltage is $\underline{1.47 \text{ μV}}$.

7.2.11 Input Noise Current

The input bias current of an amplifier exhibits minute random fluctuations (shot noise) just as is the case for the collector currents. The rms input noise current (spectral density) is given by $i_{i(\text{noise})} \text{ (rms)} = \sqrt{2qI_B}$, where I_B is the input bias (or base) current.

If $I_Q = 25$ μA and $\beta = 50$ (min.), the input noise current for the two-transistor differential amplifier is

$$i_{i(\text{noise})} = \sqrt{2qI_B} = \sqrt{\frac{2q(I_Q/2)}{\beta}} = 0.283 \text{ pA}/\sqrt{\text{Hz}} \qquad (7.91)$$

For a Darlington differential amplifier with the same I_Q and β values, the input noise current is only 0.040 pA/$\sqrt{\text{Hz}}$ and is therefore much smaller.

Thus, although the input noise voltage of the Darlington circuit is $\sqrt{2}$ times as large as the two-transistor differential amplifier, the input noise current is much smaller. For situations involving very large values of source resistance, the Darlington circuit may be the preferable choice from the standpoint of noise and the signal-to-noise ratio. For these situations, the use of JFET or MOSFET operational amplifier should also be considered as a possible choice.

7.3 PRECISION OPERATIONAL AMPLIFIERS

Several basic techniques are used to obtain operational amplifiers with input offset voltages that are down in the microvolt range. One technique is to improve the matching of transistors in the input stage of the operational amplifier by means of the design and layout of these transistors. Another approach is to compensate for transistor mismatches by the adjustment or trimming of on-chip resistors during the manufacturing process. The third basic approach is the self-nulling or "chopper-stabilized" operational amplifier. These three basic approaches will now be discussed.

The matching between transistors on an IC chip can be improved by increasing the transistor area and the base width. The larger dimensions reduce the statistical effects of the IC wafer-processing variations. Another technique that is very effective in reducing the effects of processing variations is to use a four-transistor differential-amplifier configuration as shown in Figure 7.15a. The transistor layout is shown in Figure 7.15b, and it is to be noted that diametrically opposite transistors are connected in parallel. As a result of this geometry, the effects of process variations can be greatly reduced.

The larger transistor geometries and the parallel connection of transistors require that larger quiescent currents be used for the biasing of these transistors. The larger base widths lead to lower β values, which, in combination with the larger quiescent currents, lead to substantially larger base currents, and therefore a large input bias current, I_{BIAS}.

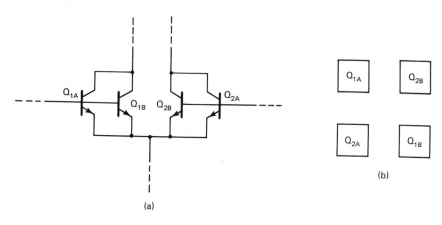

Figure 7.15 Four-transistor differential-amplifier configuration to minimize effects of processing variations: (a) four-transistor differential amplifier; (b) transistor layout.

Some operational amplifiers, such as the OP-7 and the OP-77 (Precision Monolithic, Inc.), have a base current cancellation circuit to mitigate this problem. A simplified diagram of the cancellation circuit is shown in Figure 7.16. This circuit uses a cascode differential amplifier comprised of transistors $Q1$ through $Q4$. For this analysis, we will assume that all transistors of the same type are identical. This being the case, we note that the base currents of $Q1$ and $Q3$ are equal. If we now turn our attention to the $Q5$–$Q7$ current mirror, we note that $I_{B3} = I_{C7} + I_{B7} + I_{B5}$. Since $I_{C5} = I_{C7}$, we have that $I_{B5} = I_{B7} = I_{C5}/\beta$, and thus $I_{B3} = I_{C5} + I_{C5}(2/\beta)$ $= I_{C5}[1 + (2/\beta)]$. Solving for I_{C5} gives $I_{C5} = I_{B3}/[1 + (2/\beta)] \cong I_{B3}[1 - (2/\beta)]$ since $\beta \gg 1$. Noting that $I_{B3} = I_{B1}$ and that $I_{\text{BIAS 1}} = I_{B1} - I_{C5}$, we obtain $I_{\text{BIAS 1}} \cong$ $I_{B1} - I_{B1}[1 - (2/\beta)] = I_{B1}(2/\beta)$. For the other side of the differential amplifier with transistors $Q2$, $Q4$, $Q6$, and $Q8$, a similar result can be obtained. From this last result, we see that a very considerable reduction in the input bias current can indeed be achieved. The last result predicts a bias current reduction by a factor of around 50 to 100. In practice, the reduction, while still very substantial, will be less than this due to mismatches between transistors.

The trimming or adjustment of on-chip resistors can be used in addition to the

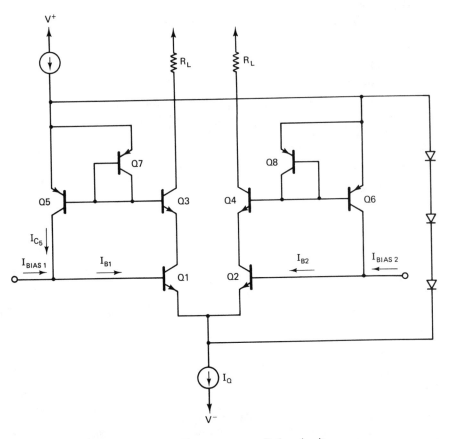

Figure 7.16 Bias current cancellation circuit.

Operational-amplifier Circuit Design   Chap. 7

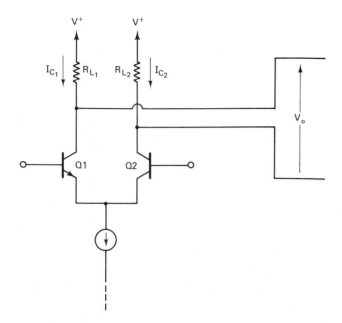

Figure 7.17 Trimming of differential amplifier load resistance to compensate for offset voltage.

better matching of transistors to achieve reduced values for V_{OS}. Let us consider the simple case shown in Figure 7.17. The net output voltage is given by $V_O = V_{C1} - V_{C2} = (V^+ - I_{C1}R_{L1}) - (V^+ - I_{C2}R_{L2}) = I_{C2}R_{L2} - I_{C1}R_{L1}$. To make $V_O = 0$ when $V_1 = 0$, we therefore must require that $R_{L2}/R_{L1} = I_{C1}/I_{C2}$. The collector current ratio of the $Q1-Q2$ pair is given by $I_{C1}/I_{C2} = \exp(V_{OS}/V_T)$, so if $R_{L2}/R_{L1} = \exp(V_{OS}/V_T)$, the effect of the offset voltage will be compensated for. In practice, the compensation will never be exact, especially due to the temperature coefficient of V_{OS}.

The load resistances can be trimmed or adjusted on the chip by laser trimming in the case of thin-film resistors, as was shown in Figure 2.52, in which a resistor is trimmed to increase its resistance value. This is done in an automated closed-loop process during which time the operational amplifier output voltage is monitored in order to null out the output voltage.

A second method of on-chip resistor adjustment can be used with diffused resistors as well as with thin-film resistors. In this case the resistor is actually made up of a string of series-connected smaller resistors, each of which is shunted by a small-area zener diode. In the actual circuit operation, these zener diodes are reverse biased, but at a voltage level far below the breakdown voltage, so they have no effect on the circuit. For the resistance adjustment, a short-duration, high-intensity pulse of current is passed through one or more selected zener diodes in the reverse-bias direction. This current pulse produces very intense localized heating in the region of the reverse-biased PN junction. This causes the alloying of the adjacent metallized contacts into the silicon and thereby results in a permanent short-circuit condition. The resistor segment that is shunted by that zener diode is therefore shorted out, and the net resistance of the series string is reduced by the amount of that shorted-out segment.

An example of a precision operational amplifier that uses resistor trimming at the wafer stage and bias current cancellation is the OP-77 (Precision Monolithics, Inc.). This operational amplifier has a V_{OS} of 10 μV (typ.) and 25 μV (max.) in the OP-77A grade, and in the OP-77B grade the values are 20 μV (typ.) and 60 μV (max.). The temperature coefficient of V_{OS} is 0.1 μV/°C (typ.) and 0.3 μV/°C (max.) in the OP-77A grade. The input bias current (OP-77A) is specified as 0.2 nA (min.), 1.2 (typ.), 2.0 nA (max.). Note that the input bias current can have values that are negative as well as positive due to the nature of the bias current cancellation circuit. This operational amplifier also features a very large open-loop voltage gain, which enhances the precision operation of this operational amplifier in various closed-loop situations and produces a more linear closed-loop transfer characteristic. For the OP-77A the open-loop gain A_{OL} (0) is specified as 6000 V/mV = 6×10^6 (typ.) and 2000 V/mV = 2×10^6 (min.).

7.3.1 Self-nulling Precision Operational Amplifiers

A very important category of precision operational amplifiers that offers input offset voltages down in the range of 10 μV or less is the self-nulling or chopper-stabilized operational amplifier. To understand how the self-nulling or auto-nulling process works, let us first consider the simple circuit of Figure 7.18a. We will assume that the operational amplifier has a transfer characteristic given by $V_O = A_{OL}(V_i - V_{OS})$. Under dc steady-state conditions, the voltage across capacitor C_1 will become equal to the output voltage V_O, so $V_{C1} = V_i = -V_O$, and thus $-V_i = A_{OL}(V_i - V_{OS})$. Therefore, $V_i(A_{OL} + 1) = A_{OL}V_{OS}$ and thus $V_C = V_i = V_{OS}[A_{OL}/(A_{OL} + 1] \cong V_{OS}$. We therefore see that capacitor C_1 will charge up to a voltage level that is very close to V_{OS}.

Now let us consider the circuit of Figure 7.18b, where capacitor C_1 has been disconnected from the inverting input terminal of the operational amplifier and is now in series with the signal. We now have that $V_O = A_{OL}(V_i - V_{OS}) = A_{OL}[(V_1 - V_2) + V_{C1} - V_{OS}]$. Since $V_{C1} \cong V_{OS}$, this reduces to simply $V_O = A_{OL}(V_1 - V_2)$, and thus V_{OS} is effectively cancelled out. During the time that the circuit is connected in this way, capacitor C_1 discharges due to the input bias current of the operational amplifier as well as the internal leakage current of the capacitor. In view of this capacitor discharge current and also the temperature and time drift of V_{OS}, for this technique to be really effective the nulling process of Figure 7.18 cannot be done just once, but must be repeated often, typically at a rate of 100 Hz. This self-nulling of the operational amplifier will, of course, produce an interruption of the signal flow. To get around this problem, two basic techniques are used, both using a combination of two operational amplifiers. One technique uses the commutating autozero (CAZ) operational amplifier, and the other technique uses the chopper-stabilized operational amplifier, using a *main operational amplifier* for the continuous signal flow and a *nulling operational amplifier* that performs the offset nulling both on itself and the main amplifier.

The commutating autozero (CAZ) amplifier will be considered first, and reference is made to Figure 7.19a, in which amplifier 1 has its offset voltage nulled by means of the voltage produced across C_1 while at the same time the signal is being fed into amplifier 2 and the output voltage taken from this amplifier. Then all the

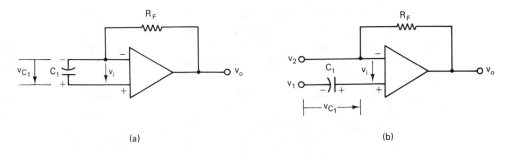

Figure 7.18   Operational amplifier input offset voltage self-nulling.

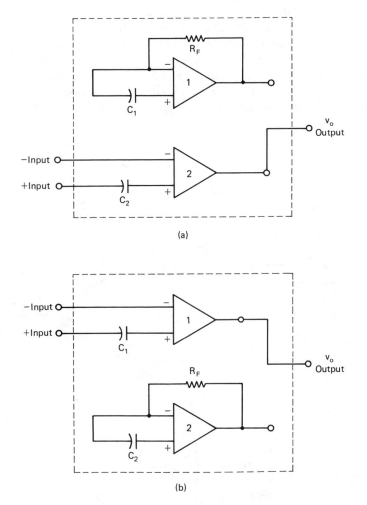

(a)

(b)

Figure 7.19   Basic diagram of a commutating autozero operational-amplifier system.

connections are switched to produce the situation of Figure 7.19b, where now the offset voltage of amplifier 2 is being nulled out and the signal transmission is via amplifier 1. The situation switches back and forth between these two modes of operation on every half-cycle of the commutation clock, which typically operates in the range of 100 to 300 Hz.

Examples of CAZ operational amplifiers are the ICL7600 and ICL7601 (Intersil), which feature an input offset voltage specification of 2 μV (typ.) and an offset voltage temperature coefficient of only 0.005 μV/°C (typ.) and 0.2 μV/°C (max.). The long-term drift is specified as 0.2 μV/year (typ.). These devices use CMOS analog switches in a parallel NMOS–PMOS configuration to minimize the switch-on resistance variation with voltage level. Figure 7.20 is a simplified diagram showing the analog switch connections to each of the two internal operational amplifiers that make up this CAZ amplifier. The voltages applied to the analog switches are complementary, but nonoverlapping square waves of phases ϕ and $\bar{\phi}$ of the commutation oscillator output. When the voltage applied to the NMOS side of the CMOS switch is high and the complementary voltage applied to the PMOS side is therefore low, both transistors that make up the switch are on. Therefore, when ϕ is high, switches $S1$, $S3$, and $S5$ are on, and switches $S2$ and $S4$ are off. In this mode, input signal V_1, via switch $S1$, will be coupled in series with C_1 to the noninverting input terminal of the operational amplifier, and input signal V_2 will be coupled via switch $S3$ to the

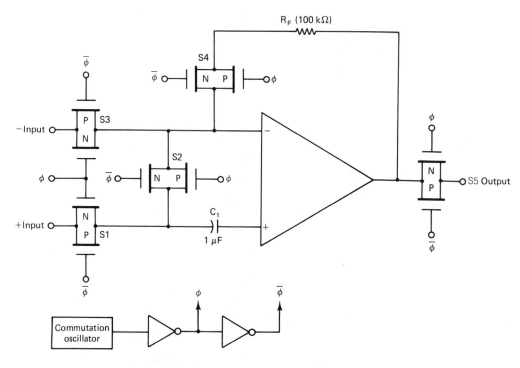

Figure 7.20 Simplified diagram showing the analog switch connections to each of the two internal amplifiers (ICL7600 Intersil).

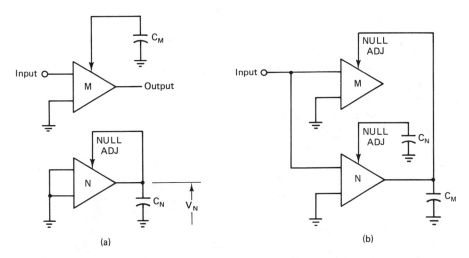

Figure 7.21 Simplified operation of the chopper-stabilized operational amplifier: (a) step 1, nulling of the nulling amplifier; (b) step 2, nulling of the main amplifier.

inverting input terminal. The output voltage of this operational amplifier is connected to the amplifier output terminal via switch $S5$. When ϕ goes low, switches $S1$, $S3$, and $S5$ are off and switches $S2$ and $S4$ turn on. This causes both input signals, V_1 and V_2, and the output V_O to be disconnected from this operational amplifier and be connected to the other operational amplifier. At the same time, the feedback resistor R_F is connected to the inverting input terminal and capacitor C_1 is connected between the two input terminals. The other operational amplifier operates in a complementary fashion.

The ICL7600 operates at a nominal commutation frequency of 160 Hz, with 2560 Hz also being available by means of pin selection. As is the case with any chopper-stabilized type of operational amplifier, the operation of the analog switches produces small commutation transients, principally at the switching frequency and harmonics thereof. These transients contribute to the total noise of the system. The switching transient noise contribution decreases with increasing commutation frequency, although at the same time the input offset voltage increases. With charge storage capacitors of $C_1 = C_2 = 1$ μF, the optimum commutation frequency for most applications is around 160 Hz.

The ICL 7605/7606 (Intersil) is another CAZ operational amplifier and offers a V_{OS} of ± 2 μV (typ.) and ± 5 μV (max.) with a temperature coefficient of 0.01 μV/°C (typ.) and 0.2 μV/°C (max.). The long-term stability is quoted as being 0.5 μV/year. This operational amplifier is intended for low-frequency applications, generally from dc to 10 Hz.

Another type of self-nulling or chopper-stabilized operational amplifer uses a main amplifier for the continuity of the signal transmission and a separate nulling amplifier to carry out the offset nulling activity. Before we look at the complete diagram of this type of operational amplifier, let us first investigate the circuit shown in Figure 7.21. Figure 7.21a shows two amplifiers, the main amplifier M, and the nulling amplifier N. The signal transmission is through the main amplifier, and in step 1 of the process the input terminals of the nulling amplifier are shorted together and the output voltage of the nulling amplifier is applied across the capacitor C_N. This capacitor is also connected to the nulling adjustment (NULL ADJUST) terminal of the nulling amplifier. It will be assumed that the change in the voltage applied to this terminal will produce an equal change in V_{OS} to be called ΔV_{OS}. The output voltage is therefore given by $V_O = A_{OL}(V_{OS} - \Delta V_{OS})$, where ΔV_{OS} is the offset nulling voltage supplied by capacitor C_N to the offset null terminal. Since $\Delta V_{OS} = V_O$, this gives $V_O(1 + A_{OL}) = A_{OL}V_{OS}$, so $\Delta V_{OS} = V_O = V_{OS}A_{OL}/(1 + A_{OL})$. The net result on the amplifier transfer characteristics is that the effective input offset voltage is $V'_{OS} = V_{OS} - \Delta V_{OS} = V_{OS} - V_{OS}A_{OL}/(1 + A_{OL}) = V_{OS}/1 + A_{OL})$. Since A_{OL} is the dc value of the open-loop gain in the preceding analysis and is therefore very large, we see that the offset voltage is very effectively nulled out. Now that the offset voltage of the nulling amplifier has been nulled out, we can go on to to step 2 of the process. In step 2, the input of the nulling amplifier is connected across the input terminals of the main amplifier, and the output of the nulling amplifier is disconnected from capacitor C_N and now connected to capacitor C_M. The charge stored on capacitor C_N acts to keep the nulling amplifier offset voltage nulled out during this step. The nulling amplifier senses the voltage across the input terminals of the main amplifier and charges up capacitor C_M. Since the main amplifier also has a high open-loop gain, the voltage across the input terminals is principally the input offset voltage of that amplifier. Note that since the offset voltage of the nulling amplifier has already been nulled out it will therefore not be added to the offset voltage of the main amplifier during this process. The nulling amplifier senses the offset voltage only of the main amplifier, amplifies it, and applies the result to capacitor C_M, which is also connected to the nulling offset terminal (NULL ADJUST) of the main amplifier. The action is the same as described before for the nulling amplifier. The offset voltage of the main amplifier is nulled out and remains so as a result of the charge stored on capacitor C_M. The process then continues to alternate between that of step 1 for the offset voltage nulling of the nulling amplifier and step 2 for the offset voltage nulling of the main amplifier.

Now let us turn our attention to Figure 7.22, which shows the main amplifier (M), the nulling amplifier (N), and the analog switches, $S1$ through $S4$. These switches are driven by a system clock that produces complementary clock phases ϕ and $\overline{\phi}$. When clock phase ϕ is high, switches $S1$ and $S3$ are on, and switches $S2$ and $S4$ are off. During this clock phase, the input signal is disconnected from the nulling amplifier, the input terminals of the nulling amplifier are shorted together, and the output voltage of the nulling amplifier is connected to capacitor C_N. This activates the self-nulling process for the nulling amplifier. Then on the next clock phase ϕ is high, switches $S1$ and $S3$ are off, and switches $S2$ and $S4$ are on. The nulling amplifier is now connected across the input terminals of the main amplifier, and the output of the nulling amplifier goes to capacitor C_M, which is connected to the nulling offset (NULL ADJ) terminal of the main amplifier. This activates the nulling of the main amplifier. This process then continues, switching back and forth on alternate clock phases.

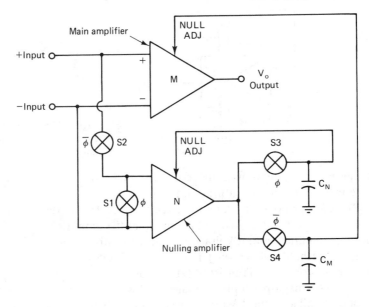

Figure 7.22 Simplified circuit of the chopper-stabilized operational amplifier.

Output clamping. As long as the main amplifier operates in the linear region, the difference-mode voltage that appears across the main amplifier input terminals is equal to V_O/A_{OL} and is very small. If, however, the amplifier output saturates due to an output voltage swing getting too close to either supply voltage, this is no longer the case and a substantial voltage can then be present. In spite of the presence of a negative feedback loop, the voltage at the inverting input terminal will no longer closely follow the voltage at the noninverting input terminal, and a large difference-mode input voltage will result. The large voltage difference will be misinterpreted by the offset nulling circuit as an extraordinarily large offset voltage, and the voltage that is produced across capacitor C_M by the nulling amplifier will correspondingly increase to a large value. Capacitors C_M and C_N are quite large, generally around 0.1 μF. Therefore, after the overload condition that produced the saturation of the output voltage of the main amplifier is gone, it will take a long time (about 100 ms to 1 s) for the offset correction capacitor C_M to discharge back to its normal voltage level. Thus it will take a correspondingly long time for the system to recover from this overload condition.

An output voltage limiting or clamping circuit is often incorporated into the output circuit of the main amplifier of a chopper-stabilized operational amplifier to keep this overload condition from happening. An example of an output clamping circuit is shown in Figure 7.23. This circuit uses MOSFET transistors $Q1$ and $Q2$ as analog switches to keep the output from saturating near the supply voltage. They do this by turning on when the output voltage reaches a value close to the saturation level and then providing a very low resistance feedback path from the output to the

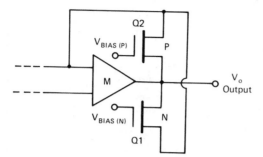

Figure 7.23 Output clamping circuit for the chopper-stabilized operational amplifier.

inverting input. This drastically reduces the closed-loop gain such that the output voltage will not reach the saturation level.

To consider how this clamping circuit works, let us consider a specific example and refer to Figure 7.24. We will assume that the amplifier supply voltages are ± 12 V and the amplifier output voltage will start to saturate at the ± 10-V level. Let us assume a threshold voltage of $+1$ V for the NMOS transistor $Q1$ and -1 V for the PMOS transistor $Q2$. Transistor $Q1$ is biased by voltage $V_{\text{BIAS(N)}}$, which is a few volts above the negative supply voltage, and we will assume a value of -8 V for this example. Similarly, the PMOS transistor $Q2$ is biased by a voltage $V_{\text{BIAS(P)}}$ that is a few volts below the positive supply voltage, and we will assume a value of $+8$ V. The sources of both transistors are connected to the output terminal of the amplifier. These transistors are both normally off. When the output voltage V_O increases in the negative direction such that the source of $Q1$ drops below -9 V, the gate-to-source voltage of $Q1$ increases above $+1$ V and therefore above the threshold value. Thus transistor $Q1$ turns on. This provides a low-resistance feedback path from the output to the input and drastically reduces the closed-loop gain and thereby prevents the output voltages from reaching the saturation level of -10 V.

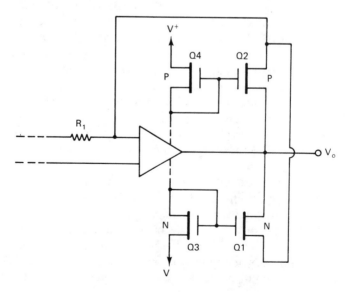

Figure 7.24 Biasing arrangement for the output clamping circuit.

PROBLEMS

7.1. (*Two-stage amplifier with active loads and complementary push-pull emitter-follower output stage*)   Refer to Figure P7.1.   Given: $V_{\text{SUPPLY}} = \pm 12$ V, $I_4 = 500$ μA, $I_Q = 10$ μA, $\beta_{\text{NPN}} = 100$ (min.), $\beta_{\text{PNP}} = 50$ (min.), and assume that $V_{BE} \approx 0.6$ V wherever applicable.

   (a) Find R_3 and R_4.   (*Ans.:* $R_3 = 9.8$ kΩ, $R_4 = 22.8$ kΩ)

   (b) What value should R_{11} have in order that $V_o = 0$ when $V_i = 0$?   (*Ans.:* 9.8 kΩ)

   (c) Find I_{BIAS}.   [*Ans.:* 50 nA (max.)]

   (d) Find the input resistance, R_i.   [*Ans.:* 1.0 MΩ (min.)]

   (e) Find the input voltage range (common mode).   (*Ans.:* +11.3 to −11.1 V)

   (f) If the emitter–base junction breakdown voltage, BV_{EBO}, is 7.0 V, what is the maximum allowable difference-mode voltage?   (*Ans.:* ±7.6 V)

   (g) Which input (B_1 or B_2) is the inverting input?   Explain.

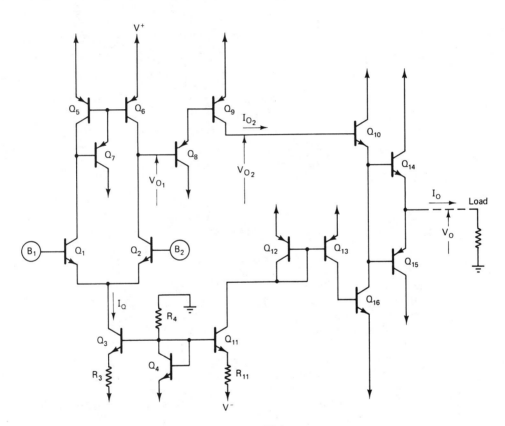

Figure P7.1

7.2. (*Operational-amplifier circuit, LM124 type*)   Refer to Figure P7.2.   Given: $V_s = +10$ V, $\beta_{\text{NPN}} = 100$ (min.), and $\beta_{\text{PNP}} = 50$ (min.).

   (a) Find R_6, R_5, R_{12}, R_{13}, and R_{10}.   (*Ans.:* $R_6 = 18.8$ kΩ, $R_5 = 18.4$ kΩ, $R_{12} = 30$ kΩ, $R_{13} = 402$ Ω, $R_{10} = 30$ kΩ)

   (b) Find R_{SC} for a short-circuit current limit of 20 mA.   (*Ans.:* $R_{SC} = 25$ Ω)

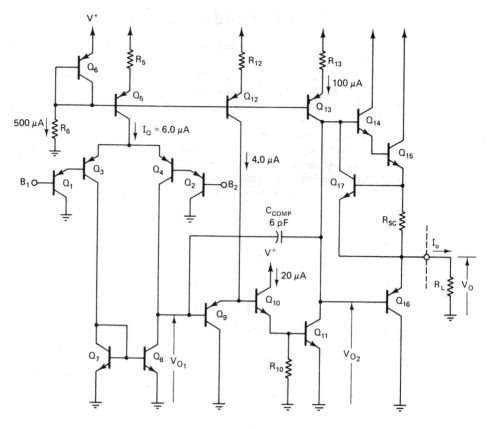

Figure P7.2

(c) Find I_{BIAS}.   [*Ans.*: I_{BIAS} = 1.2 nA (max.)]
(d) Find the input resistance, R_1.   [(*Ans.*: 83 MΩ (min.)]
(e) Find the unity-gain frequency, f_u.   (*Ans.*: f_u = 1.6 MHz)
(f) Find the slew rate.   (*Ans.*: SR = 1.0 V/μs)
(g) Find A_{OL} at f = 10 kHz.   (*Ans.*: $-j160$)
(h) Find the input (common mode) voltage range.   (*Ans.*: -0.5 to $+8.5$ V)
(i) Find the maximum output-voltage swing.   (*Ans.*: $+8.6$ to 0 V)
(j) Find the quiescent power dissipation.   (*Ans.*: 6.3 mW)
(k) For a closed-loop gain of 30, find the 3-dB bandwidth and the rise time.   (*Ans.*: 53 kHz, 6.6 μs)
(l) For an output-voltage peak swing of 3.0 V, at what frequency will the op amp become slew-rate limited?   (*Ans.*: 53 kHz)

7.3. (*Operational-amplifier circuit*)   Refer to Figure P7.3.   Given: operational-amplifier circuit with β = 50 (min.) for all transistors, V_S^+ = 15 V, V_S^- = -15 V, I_{12} = 1.0 mA, I_Q = 25 μA, n = 1.5 (factor in exponent for the base current), and V_{ANPN} = V_{APNP} = 250 V (the early voltage).
(a) Find R_1 and R_2.   (*Ans.*: R_1 = 29.3 kΩ, R_2 = 3.69 kΩ)
(b) Find g_f (differential amplifier).   (*Ans.*: g_f = 250 μS)

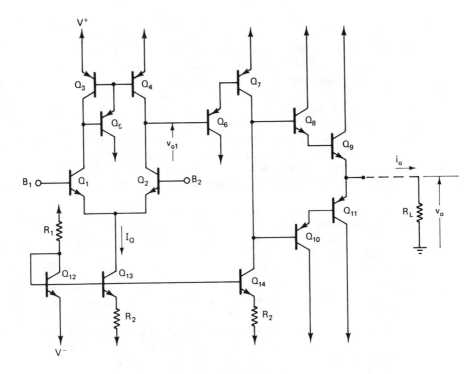

Figure P7.3

(c) Find the input-voltage range (common mode) and the maximum output-voltage swing. (*Ans.*: +14.1 to −14.0 V, +13.4 to −13.3 V)

(d) Find the input bias current and the input resistance (difference mode). (*Ans.*: 250 nA (max.), 200 kΩ (min.)]

(e) Find the quiescent current drain and the power dissipation (with $V_o = 0$). (*Ans.*: 1.05 mA, $P_{d(Q)} = 31.5$ mW)

(f) Find the input conductance of Q_6, the output conductance of Q_4 and Q_2, and the total conductance driven by the first stage. [*Ans.*: $g_{i_6} = 133$ nS (max.); $g_{o_4} = 50$ nS, $g_{o_2} = 50$ nS; $g_{total} = 233$ nS (max.)]

(g) Find the voltage gain of the first stage, v_{o1}/v_i. [*Ans.*: $A_{v1} = v_{o1}/v_i = 2146$ (min.)]

(h) If $R_L = 1.0$ kΩ, find the voltage gain of the second stage [assume hat emitter-follower gain (Q_8 and Q_9 and Q_{10} and Q_{11}) is approximately unity]. [*Ans.*: $A_{V2} = v_o/v_{o1} = 1250$ (min.)]

(i) Find the open-loop gain of the op amp, $A_{OL}(0)$. ($R_L = 1.0$ kΩ) Express the result numerically and in decibels. [*Ans.*: $A_{OL}(0) = 2.68 \times 10^6$ (min.), 128.6 dB (min.)]

(j) Show that the voltage gain of the emitter-follower output stage ($Q_{8,9}$ and $Q_{10,11}$) will be close to unity for output voltage swings of about 1 V or greater.

(k) Show that B_1 is the inverting input and B_2 is the noninverting input of the op amp.

(l) If a Darlington differenital amplifier as shown in Figure 7.9 is used, describe how the answers above would change (qualitatively and quantitatively).

(m) If the output stage is modified by the addition of R_3, R_4, and Q_{15} as shown in Figure 7.11, find R_3 and R_4 if $I_9 = I_{11} = I_Q$ (typ.) under quiescent conditions and $I_{R3} =$

$I_{R4} = I_Q/2$. Assume that $V_{BE} = 600$ mV at $I_C = I_Q$ for the NPN and PNP transistors and $\beta = 100$ (typ.).   (*Ans.*: $R_3 = 127$ kΩ, $R_4 = 46$ kΩ)

(n) If the output stage is further modified by the addition of R_{CL}, Q_{16}, and Q_{17} as shown in Figure 7.13, find R_{CL} for a current limit of 25 mA.   (*Ans.*: $R_{CL} = 24$ Ω)

(o) If the maximum supply voltage is ± 18 V, find the corresponding maximum power dissipation if $I_{CL} = 25$ mA.   (*Ans.*: $P_{d(MAX)} = 495$ mW)

(p) Find the required value for the compensation capacitor (to be connected between B_6 and C_7) for a unity gain frequency of 1.0 MHz.   (*Ans.*: $C_{COMP} = 80$ pF)

(q) Find the slewing rate and the full-power bandwidth (FPBW) for the above for a peak output swing of ± 10 V.   (*Ans.*: 0.3125 V/μs, 4.97 kHz)

(r) Find the output conductance of the current source that biases the differential amplifier (Q_{13}), and the corresponding output resistance.   (*Ans.*: $g_{o13} = 22$ nS, $r_{o13} = 44.1$ MΩ)

(s) Find the common-mode transfer conductance, g_{fCM}.   (*Ans.*: $g_{fCM} = 11$ nS

(t) Find the common-mode gain and the common-mode rejection ratio assuming that the offset voltage of the Q_1–Q_2 pair is 1.0 mV ($R_L = 1.0$ kΩ).   [*Ans.*: $A_{VCM} = 2.3$ (min.), CMRR = 121 dB]

(u) The proper operation of the op-amp circuit is dependent on the matching of circuit components that is inherent in devices on IC chips. The components are made very close together on the same IC chip and have undergone exactly the same processing so that a very high degree of matching should be expected. If there were *exact* matching, the input offset voltage of this op amp would be zero. Find the contribution to the input offset voltage (V_{OS}) of the following component mismatches:

   (1) Q_1 and Q_2: a 1-mV offset voltage.   (*Ans.*: 1 mV)
   (2) Q_3 and Q_4: a 1-m V offset voltage.   (*Ans.*: 1 mV)
   (3) Q_5 and Q_6: a 10% β mismatch.   (*Ans.*: 2 μV)
   (4) Q_7 and $Q_{3,4}$: a 10% β mismatch.   (*Ans.*: 2 μV)
   (5) A mismatch between the emitter resistors (R_2) of Q_{13} and Q_{14} sufficient to cause the current of Q_{14} to differ from that of Q_{13} by 10%.   (*Ans.*: 2 μV)
   (6) What is the corresponding resistor mismatch (R_2)?   (*Ans.*: 14%)

(v) In the voltage-follower configuration, the input impedance seen looking into the amplifier will be essentially the common-mode input impedance because the voltage that is present at the two input terminals will be essentially the same. Find the common-mode input impedance at zero frequency.   [*Ans.*: $Z_{iCM}(0) = R_{iCM} = 4.55$ GΩ (min.)]

(w) If $C_{CB} = 1.0$ pF and $C_{CS} = 1.0$ pF, such that $C_i = 2.0$ pF, find Z_{iCM} at 1.0 kHz.   (*Ans.*: $Z_{iCM} = -j80$ MΩ) Find the breakpoint frequency for Z_{iCM}.   (*Ans.*: $f = 17.5$ Hz)

(x) Find the equivalent input noise voltage (spectral density). Use $r_{bb'} = 200$ Ω. Find the total rms noise over a 20-kHz 3-dB system bandwidth and the input signal level needed to produce a 10 : 1 signal-to-noise ratio.   (*Ans.*: 8.3905 nV/$\sqrt{\text{Hz}}$, 1.49 μV, 14.9 μV)

(y) This op amp is driven from a signal source that has a 1000-Ω source resistance. Using the noise voltage obtained in part (x) together with the additional noise due to the 1000-Ω source resistance, find the input signal level required to produce a 10 : 1 signal-to-noise ratio with a 20-kHz 3-dB bandwidth.   (*Ans.*: 16.5 μV) What would be the result in the case of a 10-kΩ source resistance?   (*Ans.*: 27.4 μV)

(z) The *noise figure* (NF) of an amplifier is the factor (expressed in decibels) by which the amplifier decreases the signal-to-noise ratio of the system below what would be obtained if the amplifier were noiseless. Find the noise figure for the two preceding problems.   (*Ans.*: 7.4 dB, 1.6 dB)

7.4. An op amp has $A_{OL}(0) = 100$ dB, $f_1 = 12$ Hz, $f_2 = 2.5$ MHz, and $f_3 = 5.0$ MHz.
  (a) Find the unity-gain frequency f_u.   (*Ans.:* 1.2 MHz)
  (b) If the breakpoint frequencies given above are for $C_{COMP} = 40$ pF, find the corresponding values for a compensation capacitance of 200 pF.   (*Ans.:* 2.4 Hz, 2.5 MHz, 5.0 MHz)

7.5. An op amp has a slewing rate of 2.0 V/μs. Find the full-power bandwidth (FPBW) for a peak output voltage swing of ± 10 V.   (*Ans.:* 31.8 kHz)

7.6. An op amp has a slewing rate of 2.0 V/μs. Find the rise time for an output voltage of 10-V amplitude resulting from a rectangular pulse input, if the op amp is slewing rate limited.   (*Ans.:* 4.0 μs)

7.7. An op amp has a slewing rate of 1.0 V/μs when $C_{COMP} = 25$ pF. Find the slewing rate when C_{COMP} is increased to 100 pF.   (*Ans.:* 0.25 V/μs)

7.8. MOSFET-input op amps use an internal pair of gate protection diodes, connected back to back in series, across the input terminals of the circuit. What is the function of these diodes? Do they have any effect on the characteristics of the op amp? Explain.

7.9. Given: For the op-amp circuit for this problem (Figure P7.9), $R_1 \parallel R_2 = R_S$.
  (a) Show that the signal-to-noise ratio (SNR) at the amplifier output is given by

$$(\text{SNR})^2 = \left[\frac{v_{o(\text{signal})}}{v_{o(\text{noise})}} \right]^2 = \frac{v_s^2}{[(v_{i(\text{noise})})^2 + 2(i_{i(\text{noise})}R_S)^2 + 8kTR_S](\pi/2)BW}$$

  where $v_{i(\text{noise})}$ and $i_{i(\text{noise})}$ are the equivalent input noise voltage and current of the op amp, respectively.
  (b) If $v_{i(\text{noise})} = 20$ nV/$\sqrt{\text{Hz}}$, $i_{i(\text{noise})} = 0.1$ pA/$\sqrt{\text{Hz}}$, and $R_S = 100$ Ω, find v_s for a 10 : 1 SNR. The 3-dB bandwidth is 10 kHz.   (*Ans.:* 25.2 μV)
  (c) Repeat for $R_S = 1000$ Ω.   (*Ans.:* 26.0 μV)
  (d) Repeat for $R_S = 10$ kΩ.   (*Ans.:* 33.7 μV)
  (e) Repeat for $R_S = 30$ kΩ.   (*Ans.:* 46.5 μV)
  (f) Repeat for $R_S = 100$ kΩ.   (*Ans.:* 77.2 μV)

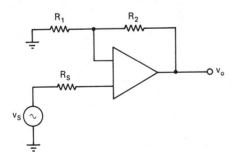

Figure P7.9

7.10. (Pole-zero compensation) An op amp has $f_2 = 1.0$ MHz and $f_3 = 2.5$ MHz. The transfer conductance of the input differential amplifier stage is 200 μS. Find C_C and R_C necessary for pole-zero compensation and a 45° phase margin at unity gain.   (*Ans.:* $C_C = 9.0$ pF, $R_C = 17.7$ kΩ)

REFERENCES

BARNA, A., and D. I. PORAT, *Operational Amplifiers,* 2nd ed., Wiley, New York, 1989.

CONNELLY, J. A., *Analog Integrated Circuits,* Wiley, New York, 1975.

COWLES, L. G., *Sourcebook of Modern Transistor Circuits,* Chapter 7, Prentice-Hall, Englewood Cliffs, N.J., 1976.

FITCHEN, F. C., *Electronic Integrated Circuits and Systems,* Van Nostrand Reinhold, New York, 1970.

FRANCO, S., *Design with Operational Amplifiers and Analog Integrated Circuits,* McGraw-Hill, New York, 1988.

GLASER, A. B., and G. E. SUBAK-SHARPE, *Integrated Circuit Engineering,* Addison-Wesley, Reading, Mass., 1977.

GRAEME, J. G., G. E. TOBEY, and L. P. HUELSMAN, *Operational Amplifiers—Design and Applications,* McGraw-Hill, New York, 1971.

GRAY, P. R., and R. G. MEYER, *Analysis and Design of Analog Integrated Circuits,* Wiley, New York, 1984.

GREBENE, A. B., *Analog Integrated Circuit Design,* Van Nostrand Reinhold, New York, 1972.

HAMILTON, D. J., and W. G. HOWARD, *Basic Integrated Circuit Engineering,* McGraw-Hill, New York, 1975.

IRVINE, R. G., *Operational Amplifiers—Characteristics and Applications,* 2nd ed., Prentice-Hall, Englewood Cliffs, N.J., 1987.

JOHNSON, D. E., and V. JAYAKUMAR, *Operational Amplifier Circuits—Design and Applications,* Prentice-Hall, Englewood Cliffs, N.J., 1982.

NATARAJAN, S., *Theory and Design of Linear Active Networks,* Macmillan, New York, 1987.

ROBERGE, J. K., *Operational Amplifiers,* Wiley, New York, 1975.

RUTKOWSKI, G. B., *Integrated Circuit Operational Amplifiers,* 2nd ed., Prentice-Hall, Englewood Cliffs, N.J., 1984.

SEDRA, A. S., and K. C. SMITH, *Microelectronic Circuits,* Holt, Rinehart and Winston, New York, 1982.

SEVIN, L. J., JR., *Field-Effect Transistors,* McGraw-Hill, New York, 1965.

WALLMARK, J. T., and H. JOHNSON, *Field-Effect Transistors,* Prentice-Hall, Englewood Cliffs, N.J., 1966.

WOJSLAW, C. F., and E. A. MOUSTAKAS, *Operational Amplifiers,* Wiley, New York, 1986.

8

Field-effect Transistor Operational Amplifiers

8.1 INTRODUCTION

The op-amp circuits discussed thus far have been bipolar op amps in that they contain only bipolar transistors. Op amps using field-effect transistors (FETs) in the input stage can offer some very significant advantages over bipolar op amps, especially in such areas as input impedance, input bias and offset currents, and slewing rate.

Bipolar transistors have much higher transfer conductances than those of FETs. The transfer conductance of bipolar devices is often some 30 to 100 times larger than that of FETs operating at comparable current levels. As a result the voltage gain for bipolar transistors will be substantially higher than that obtained with FETs.

The input resistance seen looking into the base of a bipolar transistor is limited to only moderate values, generally in the kilohm range, by the fact that the base–emitter junction is *forward biased*. In the case of a *junction field-effect transistor* (JFET) the input terminal is the *gate*, which is one side of the *reverse-biased* gate–channel junction. As a result, extremely high values of input resistance can be obtained, generally well up into the gigaohm ($10^9 \ \Omega$) range. The input current of the device is the *gate current* I_G and is the reverse leakage current of the gate–channel junction. This current will usually be well below 1 nA (10^{-9} A) and is often around 10 pA. In contrast to this, the input or base current of a bipolar transistor is related to the output or collector current by $I_B = I_C/\beta$ and will generally be somewhere in the microampere range.

In the case of the *metal oxide–silicon FET* (MOSFET), the gate electrode is separated from the rest of the device by a thin insulating layer of silicon dioxide (SiO$_2$). The input resistance is much larger than in the case of the JFET, with values well up into the teraohm (10^{12} Ω) range usually being obtained. Correspondingly, the gate current I_G is extremely small, often below 1 pA. Thus a differential amplifier input stage using FETs will have a very high input impedance and a very low input bias current.

A JFET differential amplifier is shown in Figure 8.1a, and a MOSFET differential amplifier is presented in Figure 8.1b. The devices shown are N-channel FETs, although P-channel devices can also be used. Various types of loads can be used, including active loads with bipolar or with FET transistors. In the case of bipolar differential amplifiers the input bias current and the input resistance are both directly dependent on the quiescent current I_Q as given by

$$I_B = \frac{I_C}{\beta} = \frac{I_Q}{2\beta} \quad \text{and} \quad R_i = \frac{2V_T}{I_B} = \frac{2\beta V_T}{I_C} = \frac{4\beta V_T}{I_Q} \tag{8.1}$$

For field-effect transistors the gate current is not only very small, but it is also not directly dependent on the differential-amplifier quiescent current level, so relatively large values of I_Q can be used. The input resistance is very large and is also not directly affected by I_Q. The large values of I_Q possible in the case of FET differential amplifiers can make available very large slewing rates (~50 to 75 V/μs) and also relatively large values for the unity-gain frequency (~20 MHz).

We will now consider some examples of JFET and MOSFET op amps. Some of these op amps use the FETs only in the input stage, but others use FETs throughout the entire amplifier circuit.

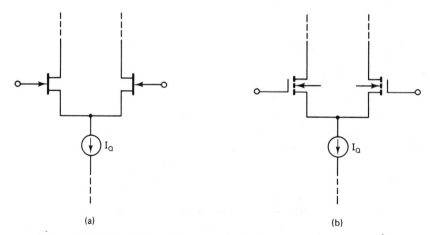

(a) (b)

Figure 8.1 Differential amplifier using field-effect transistors: (a) JFET differential amplifier; (b) MOSFET differential amplifier.

8.2 JFET Operational Amplifiers

8.2.1 Hybrid FET Operational Amplifiers: The LH0022/42/52

A representative example of a FET input hybrid op amp is the LH0022/42/52 (National Semiconductor) series. This op amp uses a pair of JFET transistors only for the input-stage differential amplifier, and the rest of the circuit uses bipolar transistors. A simplified diagram of this op amp is shown in Figure 8.2.

The input-stage differential amplifier consists of Q_1 through Q_4, in which Q_1 and Q_2 are N-channel JFETs operating in a *common-drain* or *source-follower* configuration. These JFETs drive Q_3 and Q_4, which operate in the *common-base* configuration. Thus the differential amplifier uses a common-drain/common-base compound JFET/bipolar transistor configuration, which provides the very high input resistance and very low input current characteristic of JFETs, while producing the high-voltage gain that is obtainable from bipolar transistors. This compound differential amplifier drives an active load consisting of Q_5 through Q_7, operating as a current mirror.

The rest of this op amp is of conventional design, with the second gain stage composed of Q_{16} and Q_{17} in a Darlington common-emitter configuration with a 30-

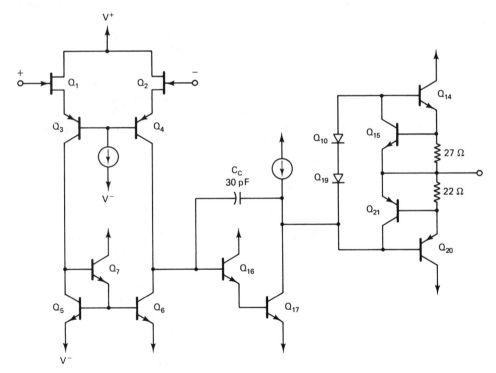

Figure 8.2 Hybrid JFET-input operational amplifier: LH0022/42/52 series (National Semiconductor).

TABLE 8.1 LH0042C[a]

Parameter	Min.	Typ.	Max.	Units
Offset voltage, V_{os}		6.0	20	mV
Temperature coefficient of V_{os}, TCV_{os}		10		$\mu V/°C$
Input bias current, I_B		15	50	pA
Input offset current, I_{os}		2	10	pA
Difference-mode input resistance		10^{12}		Ω
Common-mode input resistance		10^{12}		Ω
Input capacitance, C_i		4.0		pF
Open-loop voltage gain, $A_{OL}(0)$	25	100		V/mV
($R_L = 1$ kΩ)				
CMRR	70	80		dB
PSRR	70	80		dB
Unity-gain frequency, f_u		1.0		MHz
Slewing rate, SR	1.0	3.0		V/μs

[a]$T = 25°C$, $V^+ = +15$, $V^- = -15$.

pF feedback or compensation capacitor. The output stage is a complementary push–pull emitter follower using Q_{14} and Q_{20}. The voltage drop produced across Q_{10} and Q_{19} produces a *prebias* across Q_{14} and Q_{20} so as to bias these transistors slightly into conduction under quiescent conditions, thereby minimizing crossover distortion. Transistors Q_{15} and Q_{21} together with the associated 27- and 20-Ω resistors are used for current limiting.

Some of the important specifications of the LH0042C are listed in Table 8.1. This device is the lowest-cost unit of these series and is specified for operation over the industrial temperature range from $-25°$ to $+85°C$. We take note in particular of the high value of input resistance of 10^{12} Ω (typ.) and the very small values of bias and offset currents of 15 pA and 2 pA (typ.), respectively, that are direct results of the use of the JFET input transistors.

8.2.2 Another Hybrid Operational Amplifier: The LH0062

A second example of a hybrid FET op amp is the LH0062 (National Semiconductor). A simplified schematic diagram of this device is shown in Figure 8.3. The input differential-amplifier stage consists of a pair of N-channel JFETs Q_1 and Q_2 biased by a current source transistor Q_3 and driving a passive load consisting of resistors R_1 and R_2 (20 kΩ). The quiescent current of the differential amplifier is set at 600 mV/600 Ω = 1.0 mA by the combination of D_1(1.2 V), Q_3, and R_5 (600 Ω), and this current is relatively independent of the supply voltage.

The second gain stage is a bipolar differential amplifier using transistors Q_4 and Q_5 with the 6-kΩ emitter resistors, R_3 and R_4, setting the quiescent current level through each side at approximately a level given by (0.5 mA $\times$ 20 kΩ $-$ 0.6 V)/6 kΩ = 1.6 mA. These emitter resistors also produce negative feedback within the second stage, which acts to increase the input impedance and to extend the frequency

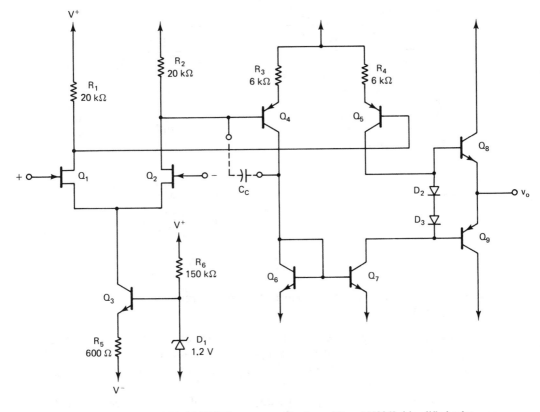

Figure 8.3 Hybrid JFET-input operational amplifier: LH0062 (simplified schematic diagram) (National Semiconductor).

response of this stage. The second stage drives a current-mirror active load comprised of Q_6 and Q_7.

The output stage is a complementary push–pull emitter-follower circuit using Q_8 and Q_9. The voltage drop that is produced across diodes D_2 and D_3 is used to prebias the output stage transistors so as to minimize crossover distortion.

This device is specified for operation over the industrial temperature range from $-25°$ to $+85°C$. Some of the important parameters of this device are listed in Table 8.2 for a supply voltage of ± 15 V and at $T = 25°C$.

This op amp has the very high input impedance and very low input bias current that are characteristic of FET-input op amps. We note also that this op amp has a large value of unity-gain frequency f_u [15 MHz (typ.)] and a very large slewing rate [50 (min.), 75 (typ.) V/μs]. This very large slewing rate is made possible by the large value of quiescent current ($I_Q = 1$ mA) of the JFET input stage. We note again that a large I_Q can be used for JFET differential amplifiers without seriously affecting the input (gate) current or the input resistance. This is not the case with bipolar differential amplifiers, for which I_Q must be kept small in order to have a low input bias current and a high input resistance.

TABLE 8.2 LH0062C

Parameter	Min.	Typ.	Max.	Units
Offset voltage, V_{OS}		10	15	mV
Temperature coefficient of V_{OS}, TCV_{OS}		10	35	$\mu V/°C$
Input bias current, I_B		10	65	pA
Input offset current, I_{OS}		1	5	pA
Difference-mode input resistance		10^{12}		Ω
Common-mode input resistance		10^{12}		Ω
Input capacitance, C_i		4		pF
Open-loop voltage gain, $A_{OL}(0)$	25	160		V/mV
(R_L = 2 kΩ)				
CMRR	70	90		dB
PSRR	70	90		dB
Unity-gain frequency, f_u		15		MHz
Slewing rate, SR	50	75		V/μs

8.2.3 Monolithic FET-input Operational Amplifiers: The LF155/156/157

As a third example of a JFET op amp, we will consider the LF155/156/157 (National Semiconductor) series of devices. A simplified schematic diagram of this IC is shown in Figure 8.4. This is a monolithic IC with all transistors, both JFET and bipolar, on the same chip.

The input differential-amplifier stage consists of a pair of P-channel JFETs Q_1 and Q_2 biased by a current source I_Q. This differential amplifier drives a current source active load using Q_3 and Q_4, which are diode-connected JFETS ($V_{GS} = 0$) operating as current regulator diodes.

The second stage is a bipolar differential amplifier with current sink biasing and a current source active load. This stage drives a Darlington emitter-follower output stage with an internal current-limiting circuit. Transistor Q_{10}, together with Q_{11}, D_1, D_2, and the I_{Q4} current source, is used for sinking currents from the load.

An on-chip feedback capacitor across the second stage provides frequency compensation. This capacitor is 10 pF for the LF155 and LF156 device series and provides for stabilization down to the unity-gain case with a unity-gain frequency of 2.5 MHz for the LF155 and 5 MHz for the LF156. The corresponding slewing rates are 5 V/μs for the LF155 and 12 V/μs for the LF156. The LF157 is a decompensated op amp with a feedback capacitor of only 2 pF. As a result of this, the op amp is stable (with a 45° phase margin) only for closed-loop gains of 5 or greater. However, the unity-gain frequency (or gain–bandwidth product) is now up to 20 MHz (typ.) and the slewing rate is 50 V/μs (typ.).

The LF355/356/357 device series (National Semiconductor) is the same as the LF155/156/157 series except that the 300-series devices are specified over the commercial temperature range. Some of the important characteristics of the LF355/356/357 op amps are tabulated in Table 8.3 for a supply voltage of ± 15 V and at $T = 25°C$.

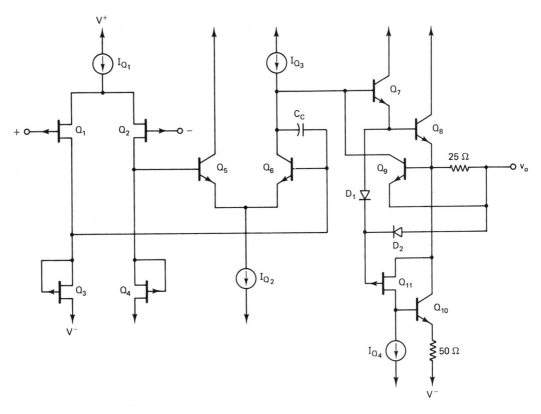

Figure 8.4 Simplified schematic diagram of a monolithic JFET-input operational amplifier: LF155/156/157 series (National Semiconductor).

TABLE 8.3 LF355/356/357

Parameter	Min.	Typ.	Max.	Units
Input offset voltage, V_{OS}		3	10	mV
Temperature coefficient of V_{OS}, TCV_{OS}		5		μV/°C
Input bias current, I_B		30	200	pA
Input offset current, I_{OS}		3	50	pA
Input resistance		10^{12}		Ω
Input capacitance, C_i		3		pF
Open-loop voltage gain, $A_{OL}(0)$ ($R_L = 2$ kΩ)	25	200		V/mV
CMRR	80	100		dB
PSRR	80	100		dB
Unity-gain frequency, f_u (gain–bandwidth product)				
LF355		2.5		MHz
LF356		5		MHz
LF357		20		MHz
Slewing rate				
LF355		5		V/μs
LF356		12		V/μs
LF357 ($A_V = 5$)		50		V/μs

8.24 Quad FET-input Operational Amplifiers: The LF347

The LF347 (National Semiconductor) is an example of a monolithic JFET-input quad operational amplifier. There are four electrically independent op amps on the same silicon IC chip. The four op amps share only the positive and negative supply terminals. This device is available in a 14-pin DIP package. A simplified schematic diagram of this device is shown in Figure 8.5.

The input stage is a P-channel JFET differential amplifier that drives a bipolar transistor current mirror active load. The second stage is a common-emitter transistor with a current source active load. A 10-pF compensation capacitor C_c is placed as a feedback capacitor across the second stage to ensure stability. The output stage is a complementary push–pull emitter-follower configuration. The voltage drop across the two diodes is used to bias the output stage for the minimization of crossover distortion.

The LF347 is specified over the commercial temperature range from 0° to 70°C. Listed in Table 8.4 are some of the important characteristics of this device when operated with a supply voltage of ± 15 V and at a temperature of 25°C.

We again note the very low input bias current, very large input resistance, and high slewing rate that are characteristic of FET-input op amps. The amplifier-to-

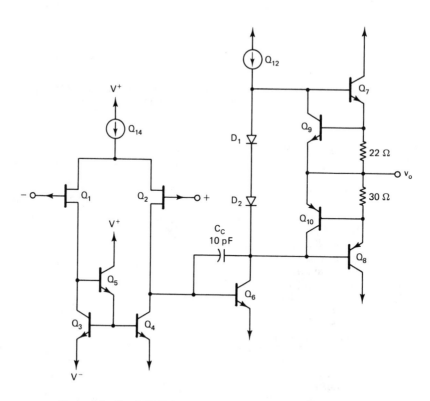

Figure 8.5 Quad JFET-input operational amplifier: LF347 (National Semiconductor).

TABLE 8.4 LF347

Parameter	Min.	Typ.	Max.	Units
Input offset voltage, V_{OS}		5	10	mV
Temperature coefficient of V_{OS}, TCV_{OS}		10		μV/°C
Input bias current, I_B		50	200	pA
Input offset current, I_{OS}		25	100	pA
Input resistance		10^{12}		Ω
Open-loop voltage gain, $A_{OL}(0)$	25	100		V/mV
$\quad (R_L = 2$ k$\Omega)$				
CMRR	70	100		dB
PSRR	70	100		dB
Gain–bandwidth product, f_u		4		MHz
Slewing rate, SR		13		V/μs
Amplifier-to-amplifier coupling		-120		dB
Equivalent input noise voltage		16		nV/$\sqrt{\text{Hz}}$
$\quad (f = 1$ kHz$)$				
Input noise current ($f = 1$ kHz)		0.01		pA/$\sqrt{\text{Hz}}$

amplifier coupling coefficient of -120 dB indicates a very high degree of electrical isolation between the four op amps that share the same IC chip.

The LF353 is a monolithic dual op amp with circuitry and characteristics similar to the LF347 quad op amp, and the LF351 is a single op amp that is also similar to the LF347.

8.2.5 Monolithic Bipolar-JFET Construction

The monolithic FET-input op amps just considered have both bipolar transistors and JFETs on the same IC chip. In Figure 8.6 a cross-sectional view is presented showing an NPN transistor and a P-channel JFET of the type used for these ICs. The NPN transistor is of the standard IC construction.

The JFET uses the same processing steps as the bipolar transistor, including the P-type and N-type diffusions, except that there is a boron ion implantation to produce the thin channel region of the JFET. This implantation is a relatively low dosage, low energy implantation to produce a very thin channel region that will have a low pinch-off voltage V_p and at the same time produce an acceptably high value for I_{DSS}. The uniformity of the implantation also results in well-matched JFETs.

A good example of what can be achieved with well-matched ion-implanted JFETs in the input stage is the OPA111 (Burr-Brown) with a TCV_{OS} of only 1 μV/°C and an input bias current of only 1 pA. This op amp has a dielectrically isolated JFET differential amplifier input stage that is biased by a 800-μA current source. Each side of the differential amplifier drives a 4-kΩ resistive load. These load resistances are laser trimmed to null out the input offset voltage, resulting in a maximum offset voltage of only 250 μV. Another important feature of this op-amp design is its low noise characteristics. The equivalent input noise voltage is only 6 nV/$\sqrt{\text{Hz}}$, and the low input bias current results in an input noise current of only 1.6 fA/$\sqrt{\text{Hz}}$ (0.0016 pA/$\sqrt{\text{Hz}}$). Other features of this op amp are an open-loop gain of 125 dB, a unity gain frequency of 2 MHz, and a slewing rate of 2 V/μs.

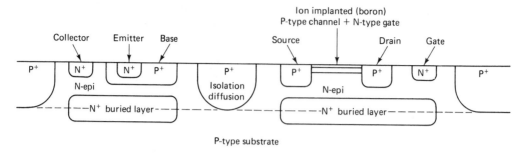

Figure 8.6 Monolithic bipolar-JFET construction: cross-sectional view.

8.3 MOSFET OPERATIONAL AMPLIFIERS

8.3.1 The CA3130

An interesting example of a monolithic MOSFET operational amplifier is the CA3130 (RCA), which contains both MOSFETs and bipolar transistors on the same chip. A simplified schematic diagram of this device is shown in Figure 8.7a and a block diagram is given in Figure 8.7b.

The input stage is a differential amplifier consisting of a pair of P-channel enhancement-mode MOSFETs, Q_6 and Q_7. This differential amplifier is biased by a 200-μA MOSFET current source, and it drives a current-mirror active load using Q_9 and Q_{10}. Resistors R_5 and R_6 (1 kΩ) in the active load circuit are used together with an externally connected potentiometer (across pins 1 and 5) for offset voltage nulling. The quiescent voltage drop across these resistors is 100 μA × 1 kΩ = 100 mV, so a ±100-mV offset adjustment range is available. Diodes D_5 and D_8 are connected between the input terminals to protect the thin gate oxide of the input MOSFETs against excessive voltage spikes and static electricity discharge, which could cause breakdown of the oxide layer and result in irreversible damage to the transistors. The voltage gain of this first stage is only about 5, this low value being the result of the relatively low transfer conductance of the MOSFETs.

The second stage consists of a bipolar transistor connected as a common-emitter amplifier, with a 200-μA MOSFET current source serving as the active load. As a result of the high dynamic impedance provided by the current source active load and the very high input impedance seen looking into the MOSFET output stage, the voltage gain of the second stage is very large (~6000).

The output stage is a *complementary-symmetry* pair of MOSFETs (CMOS) with Q_8 being the P-channel device (PMOS) and Q_{12} being the N-channel (NMOS) transistor. As the voltage at the collector of Q_{11} goes up above the quiescent level, the NMOS transistor Q_{12} turns on and the PMOS device Q_8 turns off. This drives the output voltage down toward the negative supply. Conversely, as the voltage at the collector of Q_{11} goes down, the PMOS transistor Q_8 is turned on and the NMOS transistor Q_{12} is turned off. This pulls the output voltage up toward the positive supply. The drain-to-source resistance of the CMOS output transistors is about 300 Ω when the transistor is turned on, so for large-value load resistances ($\geq$1 MΩ) the

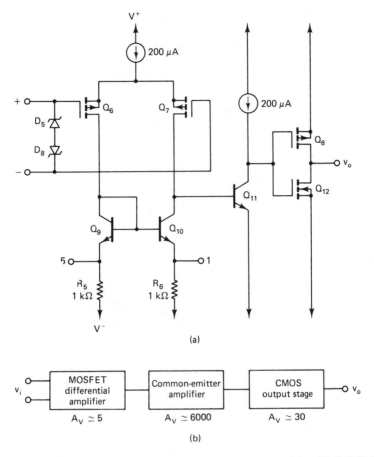

(a)

MOSFET differential amplifier	Common-emitter amplifier	CMOS output stage
$A_V \simeq 5$	$A_V \simeq 6000$	$A_V \simeq 30$

(b)

Figure 8.7 MOSFET-input, CMOS-output operational amplifier, CA3130 (RCA Corporation): (a) simplified schematic diagram; (b) block diagram.

output voltage can be driven to within a few millivolts of either supply voltage. As the load resistance drops down into the kilohm range, the output voltage swing, of course, gets smaller. The voltage gain of this CMOS output stage is approximately 30.

Some of the important specifications for the CA3130 are listed in Table 8.5 for operating conditions of a ± 15-V supply and at a temperature of 25°C.

Figure 8.8 is a diagram of the current source circuitry of the CA3130. Transistor Q_1 is a diode-connected MOSFET, which together with Q_2 constitutes a current mirror. The current through Q_1 is set at 100 μA by the combination of the voltage drops across the zener diode Z_1, diodes D_1 through D_4, and the 40-kΩ resistor R_1. Transistors Q_2, Q_3, Q_4, and Q_5 are scaled to have twice the channel width of Q_1, so the current through these transistors is twice that of Q_1, or 200 μA. Transistor pairs Q_2–Q_4 and Q_3–Q_5 are cascode-connected transistors, which results in a lower output conductance for the two current sources. This leads to a better common-mode rejec-

TABLE 8.5 CA3130

Parameter	Min.	Typ.	Max.	Units
Input offset voltage, V_{OS}		8	15	mV
Input bias current, I_B		5	50	pA
Input offset current, I_{OS}		0.5	30	pA
Input resistance		1.5		TΩ
Input capacitance, C_i		4.3		pF
Open-loop voltage gain, $A_{OL}(0)$	50	320		V/mV
($R_L = 2$ kΩ)				
CMRR	70	90		dB
PSRR	70	90		dB
Unity-gain frequency, f_u		15		MHz
Slewing rate, SR		10		V/μs

tion ratio for the differential amplifier stage and a higher voltage gain for the second stage.

For bipolar transistors, the base voltage must not be allowed to go beyond the collector voltage such that the collector–base voltage becomes forward biased by more than 0.5 V and the transistor goes into saturation. This restriction can severely

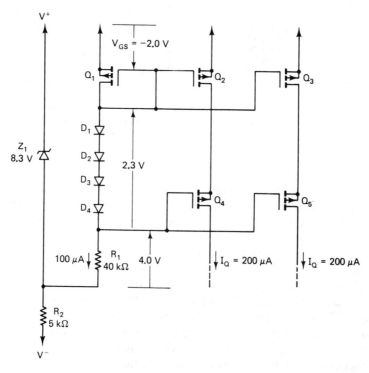

Figure 8.8 Cascode MOSFET current source circuit for the CA3130 (RCA Corporation).

limit the input voltage range of bipolar-input op amps such that operation with a split power supply is required. Unlike the case of the bipolar transistor, the gate of a MOSFET can go well beyond the drain voltage. As a result of this, the input voltage range of the CA3130 extends down to 0.5 V below the negative supply voltage. This fact makes possible the operation of this device with a single positive supply voltage.

8.3.2 The CA3140

A second example of a MOSFET-input op amp is the CA3140 (RCA). A simplified schematic diagram of this device is given in Figure 8.9. The input stage is a P-channel MOSFET differential amplifier, Q_9 and Q_{10}, biased by a 200-μA current source and driving a current-mirror active load comprised of Q_{11} and Q_{12}. Resistors R_4 and R_5 (500 Ω) are used together with an externally connected potentiometer ($\sim$10 kΩ) between pins 1 and 5 for offset-voltage nulling. The voltage drop across R_4 and R_5 of 100 μA $\times$ 500 Ω = 50 mV provides for a $\pm$50-mV offset adjustment range. Diodes D_3 and D_4 are used for the protection of the MOSFET gate oxides. The voltage gain of this first stage is approximately 10; this relatively low voltage gain again is the result of the low transfer conductance of the MOSFET transistors.

The second stage uses transistor Q_{13} in a common-emitter configuration with a 200-μA current source active load. A 12-pF feedback capacitor across this stage is

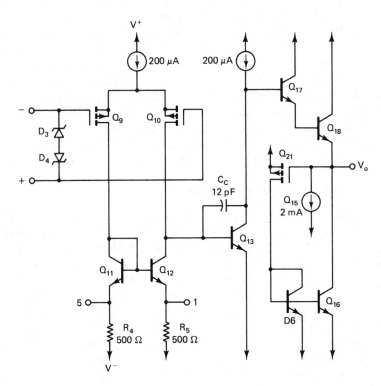

Figure 8.9 MOSFET-input operational amplifier: CA3140 (RCA Corporation).

TABLE 8.6 CA3140

Parameter	Min.	Typ.	Max.	Units
Input offset voltage, V_{OS}		8	15	mV
Input bias current, I_B		10	50	pA
Input offset current, I_{OS}		0.5	30	pA
Input resistance		1.5		TΩ
Open-loop voltage gain, $A_{OL}(0)$	20	100		V/mV
($R_L = 2$ kΩ)				
CMRR	70	90		dB
PSRR	76	80		dB
Unity-gain frequency, f_u		4.5		MHz
Slewing rate, SR		9		V/μs

used for feedback stabilization. As a result of the high dynamic impedance presented by the active load and the high input impedance of the Darlington emitter-follower output stage, the voltage gain of the second stage will be about 10,000. Since the voltage gain of the output stage is around unity, the overall amplifier open-loop gain, $A_{OL}(0)$, is about 100,000, or 100 dB.

The output stage uses a Darlington NPN emitter follower Q_{17}–Q_{18} for sourcing output currents, thus driving the output voltage up in the positive direction above the quiescent level. For sinking load currents, transistor Q_{16} is used. As the current delivered by Q_{18} drops below the 2-mA current level of the Q_{15} current sink, the difference between these two currents will be the load current sunk by the op amp. This causes the output voltage to drop below the quiescent level, which causes the MOSFET Q_{21} to turn on. As a result of Q_{21} turning on, current is supplied to the diode-connected transistor D_6, which together with Q_{16} constitutes a current mirror. Thus the current flowing through Q_{21} and thence through D_6 produces a similar current flow through Q_{16}, which can now sink the major part of the output current.

Some of the important specifications for the CA3140 are listed in Table 8.6 for the case of operation with a supply voltage of ±15 V and at $T = 25°C$.

8.3.3 Quad CMOS Operational Amplifier

An example of an all-MOSFET quad operational amplifier is the MC14573 (Motorola). A simplified circuit diagram is shown in Figure 8.10. Transistors Q_1 and Q_2 constitute a PMOS differential amplifier, with NMOS transistors Q_3 and Q_4 serving as a current-mirror active load. The differential amplifier is biased by the Q_5–Q_6 current mirror with a quiescent current I_Q given approximately by $I_Q \simeq I_{SET} = (V^+ - V^- - V_{GSs})/R_{SET} \simeq (V^+ - V^- - 1.0 \ V)/R_{SET}$, where R_{SET} is an external programming resistor. The second and output stage uses Q_7 in a common-source configuration, and Q_8 serves as a current source active load.

The op amp features an open-loop gain, $A_{OL}(0)$, of 90 dB, an amplifier-to-amplifier coupling (or channel separation) of −100 dB, an input bias current of 1 nA (max.), and an offset current of 200 pA (max.). When operated with $I_{SET} = 40$ μA, the slewing rate is 2.5 V/μs.

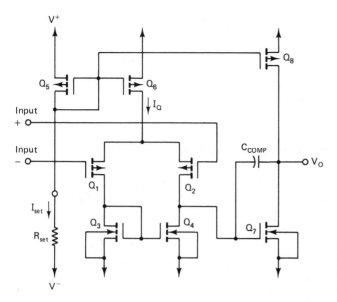

Figure 8.10 CMOS operational amplifier: MC14573 (Motorola), simplified circuit diagram. (Courtesy Motorola Semiconductor Products, Inc.)

We will now consider a simple ac analysis of this operational amplifier circuit. The low-frequency voltage gain of the first stage is given by $A_{V_1}(0) = 2g_f/g_{\text{total}} = 2g_f/(g_{o2} + g_{o4})$. The transfer conductance of the differential amplifier is given by $g_f = \sqrt{KI_Q/2} = (I_Q/2)/(V_{GS} - V_t)$, and the output conductances of Q_2 and Q_4 are $g_{o2} = g_{ds2} = (I_Q/2)/V_{A(\text{PMOS})}$ and $g_{o4} = g_{ds4} = (I_Q/2)/V_{A(\text{NMOS})}$, respectively. The voltage gain can therefore be written as

$$A_{V_1}(0) = \frac{I_Q/(V_{GS} - V_t)}{(I_Q/2)\left[\dfrac{1}{V_{A(\text{NMOS})}} + \dfrac{1}{V_{A(\text{PMOS})}}\right]} = \frac{2/(V_{GS} - V_t)}{\dfrac{1}{V_{A(\text{NMOS})}} + \dfrac{1}{V_{A(\text{PMOS})}}} \qquad (8.2)$$

For the MC14573 at $I_Q = I_{\text{SET}} = 40$ μA, the value of V_{GS} is approximately 1 V. If we assume $V_t = 0.5$ V and $V_{A(\text{NMOS})} = V_{A(\text{PMOS})} = 75$ V as reasonable values, we obtain a voltage gain of $A_{V_1}(0) = 212$ for the first stage.

The voltage gain for the second stage is given by

$$A_{V_2}(0) = g_{fs7}/(g_{\text{total}})_7 = g_{fs7}/(g_{o7} + g_{o8} + g_L) \qquad (8.3)$$

where g_{fs7} = transfer conductance of $Q_7 = 2K(V_{GS} - V_t) = 2I_{DS7}/(V_{GS} - V_t) = 2I_Q/(V_{GS} - V_t)$. The conductance g_{o7} is the output conductance of Q_7 as given by $g_{o7} = g_{ds7} = I_{DS7}/V_{A(\text{NMOS})} = I_Q/V_{A(\text{NMOS})}$, and g_{o8} is the output conductance of the current source active load transistor Q_8 as given by $g_{o8} = g_{ds8} = I_{DS8}/V_{A(\text{PMOS})} = I_Q/V_{A(\text{PMOS})}$. The conductance g_L represents conductance of the load driven by this operational amplifier. For the rest of the analysis, we will assume $g_L = 0$. We now obtain for the voltage gain of the second stage the expression

$$A_{V_2}(0) = \frac{2I_Q/(V_{GS} - V_t)}{\dfrac{I_Q}{V_{A(\text{NMOS})}} + \dfrac{I_Q}{V_{A(\text{PMOS})}}} = \frac{2/(V_{GS} - V_t)}{\dfrac{1}{V_{A(\text{NMOS})}} + \dfrac{1}{V_{A(\text{PMOS})}}} \qquad (8.4)$$

If we use the same parameters as used for the first stage, we now obtain a voltage gain of $A_{V_2}(0) = 150$.

The overall amplifier open-loop voltage gain $A_{OL}(0)$ will be $A_{OL}(0) = A_{V_1}(0) \times A_{V_2}(0) = 212 \times 150 = 31,820$ which corresponds to 90 dB. This is close to the manufacturer's specification of 90 dB (typ.). Thus we see that although low voltage gains are generally associated with MOSFET amplifiers, the use of active loads can provide for an acceptably large open-loop gain for this two-stage operational-amplifier circuit.

The unity-gain frequency for this operational-amplifier circuit can be obtained from the equation $f_u = 2g_f/(2\pi C_C)$, where g_f is the transfer conductance of the differential amplifier and C_C is the compensation or feedback capacitor placed across the second stage. If we substitute in the equation for g_f, we obtain $f_u = I_Q/[2\pi C_C(V_{GS} - V_t)]$. If we assume $I_Q = I_{SET} = 50$ μA and use a value of 16 pF for C_C, we obtain a unity-gain frequency of $f_u = 0.80$ MHz. The corresponding slewing rate is given by slewing rate $= SR = I_Q/C_C = 40$ μA/16 pF $= 2.5$ V/μs, which is in agreement with the manufacturer's specification.

8.3.4 Low-voltage and Micropower CMOS Operational Amplifiers

In the case of operational amplifiers with bipolar transistor input and output stages, the input voltage range can generally extend to no closer than 1 V from either supply. The output voltage swing similarly can generally reach no closer than 1 V from either supply. This limitation usually arises from the need to allow for a base-to-emitter voltage drop of about 0.7 V and a collector-to-emitter voltage of 0.2 V for all the transistors operating in the active region. One exception to the general rule just cited are the single-supply operational amplifiers for which, when operated with a single positive supply voltage, the input voltage range can include ground, and the output voltage can go down to ground. For the low-voltage operation of operational amplifiers, especially with supply voltages of 5 V or less, this limitation on the input-voltage range and the output-voltage swing can be quite serious. For example, consider the case of a bipolar operational amplifier operated with a supply voltage of ±2.5 V. We will assume that the input and output voltages cannot approach any closer than 1.5 V away from the supply. For this case, the maximum output voltage swing is ±1 V or 2 V peak-to-peak. The maximum input voltage swing that can be accommodated by this operational amplifier is similarly ±1 V, or 2 V peak-to-peak.

CMOS operational amplifiers are available that do not have the same restriction on the input voltage range and the output voltage swing as is the case with bipolar operational amplifiers. Indeed, these CMOS operational amplifiers offer *rail-to-rail* input and output voltage ranges. This means that the input voltage range extends from the positive supply rail all the way down to the negative supply rail. Similarly, the output voltage can swing from very close to the positive supply rail all the way down to very close to the negative supply rail. This rail-to-rail output-voltage swing is, however, the case only for lightly loaded conditions. When the operational amplifier is driving a heavier load (a lower load resistance), the output-voltage swing is substantially less than rail-to-rail.

The low-voltage operation of the CMOS operational amplifiers, as well as the ability to operate these devices at quiescent current levels down in the microampere range, results in the possibility of micropower operational amplifiers with quiescent power dissipation levels down in the microwatt range. Some examples of low-voltage micropower operational amplifiers are the ALD1701/2701/4701 series and also the ALD1706 (Advanced Linear Devices). These operational amplifiers have a rail-to-rail input-voltage range and a rail-to-rail output-voltage swing capability. For all these operational amplifiers, supply voltage range is from ± 1.0 V (min.) to ± 6.0 V (max.) for dual-supply operation and 2.0 V (min.) to 12.0 V (max.) for single-supply operation. The ALD1701 is a single operational amplifier, the ALD2701 is a dual version of the ALD1701, and the ALD4701 is a quad operational amplifier version of the ALD1701. At a supply voltage of ± 2.5 V, the quiescent supply current for the ALD1701 is 120 μA (min.) and 250 μA (max.) and is twice that for the dual version (ALD2701) and four times that for the quad version (ALD4701). Thus, the quiescent power dissipation is only 600 μW (typ.) and 1250 μW (max.) for the ALD1701. With the ± 2.5-V supply and a load resistance of 100 kΩ, the output-voltage swing is ± 2.48 V (typ.), ± 2.40 V (min.) for the ALD1701 series.

The ALD1706 is described as an "Ultra micropower rail-to-rail CMOS operational amplifier." The supply voltage range is the same as for the ALD1701 series, but quiescent supply current is specified as only 20 μA (typ.) and 40 μA (max.) when operated with a ± 2.5-V supply. The corresponding quiescent power dissipation is just 100 μW (typ.) and 200 μW (max.). The FET input stage and low-voltage operation of this operational amplifier results in an input bias current of only 1.0 pA (typ.) and an input resistance of 10^{12} Ω.

A major trade-off in the design of the low-voltage, low-current operational amplifiers is, of course, speed and bandwidth. The ALD1701 series has a gain–bandwidth product of 700 kHz and a slewing rate of 0.7 V/μs. The ALD1706, with a quiescent current of 20 μA as compared to the 120 μA for the ALD1701, has a gain–bandwidth product of only 400 kHz and a slewing rate of 0.17 V/μs. The preceding values are for a ± 2.5-V supply. With a ± 1-V supply, the gain–bandwidth product for the ALD1706 drops to just 300 kHz.

8.3.5 TLC1078 CMOS Operational Amplifier

The TLC1078 (Texas Instruments) is a good example of an all-MOSFET operational amplifier. A simplified schematic of this device is shown in Figure 8.11. The input stage is a common-source PMOS differential amplifier comprised of transistors $Q1$ and $Q5$ and biased by the PMOS current-source transistor $Q3$. This stage drives a current-mirror active load comprised of NMOS transistors $Q2$ and $Q4$. The second stage uses NMOS transistor $Q7$ in a common-source configuration, with $Q6$ acting as a current-source active load. The output stage uses a pair of NMOS transistors that operates in a push–pull fashion. Transistor $Q8$ acts as a source follower to source current out into the load, and $Q9$ acts as a common-source amplifier to sink current from the load. Transistor $Q8$ is driven directly by $Q7$, whereas $Q9$ is driven by the differential amplifier. As a result, the voltages applied to the gates of the push–pull output transistors, $Q8$ and $Q9$, will differ by 180° due to the phase inversion produced

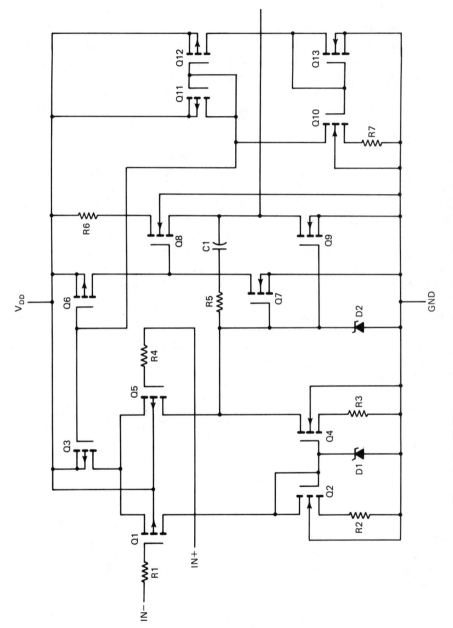

Figure 8.11 CMOS operational amplifier: TLC1078 (Texas Instruments).

by the common-source amplifier stage of $Q7$. Note also that the voltage gain of the signal in going from the operational amplifier input to the source of $Q8$ is approximately equal to the voltage gain from the input to the drain of $Q9$. In the first case, the signal coming from the differential amplifier goes through a common-source amplifier ($Q7$) and then is applied to the gate of the source-follower output transistor $Q8$. In the second case, the signal goes from the differential amplifier to the common-source output transistor $Q9$.

The R_5–C_1 circuit is a frequency-compensation network. This is a pole-zero type of circuit with the C_1 acting to reduce the lowest pole or breakpoint frequency of the system to a value low enough to ensure stability and the combination of R_5 and C_1 producing a zero that can cancel out the second pole of the amplifier open-loop frequency response.

Biasing circuit. Transistors $Q10$ through $Q13$ constitute the dc biasing circuit for this operational amplifier, as shown separately in Figure 8.12. We will assume all transistors of the same type to be identical, except for the channel widths of $Q10$ and $Q13$. Since $V_{GS11} = V_{GS12} = V_{GS3} = V_{GS6}$, we have that $I_{11} = I_{12} = I_3 = I_6 = I_Q$. Therefore, $I_{13} = I_{12} = I_{10} = I_Q$. For $Q10$ and $Q13$, we have that $V_{GS10} = V_{GS13} - I_Q R_7$. The drain currents of $Q10$ and $Q13$ are given by $I_{10} = K_{10}(V_{GS10} - V_T)^2$ and $I_{13} = K_{13}(V_{GS13} - V_T)^2$. Since $V_{GS10} = V_{GS13} - I_Q R_7$ and $I_{10} = I_{13} = I_Q$, we have that $K_{10}(V_{GS13} - I_Q R_7 - V_T)^2 = K_{13}(V_{GS13} - V_T)^2$, so that $V_{GS13} - I_Q R_7 - V_T = (K_{13}/K_{10})^{1/2}(V_{GS13} - V_T)$. Solving this for $I_Q R_7$ gives $I_Q R_7 = (V_{GS13} - V_T)[1 - (K_{13}/K_{10})^{1/2}]$. The ratio of the MOSFET constants is equal to the ratio of the channel widths of $Q13$ and $Q10$ as given by $K_{13}/K_{10} = W_{13}/W_{10}$. Thus we obtain that $I_Q R_7 = (V_{GS13} - V_T)[1 - K_{13}/K_{10})^{1/2}]$ and thus $I_Q = [(V_{GS13} - V_T)/R_7][1 - (K_{13}/K_{10})^{1/2}]$. Note that I_Q, which sets the operational amplifier biasing level, is independent of the supply voltage.

To consider an example, let us assume that we need $I_Q = 10$ μA, and we will use a channel width ratio of $W_{13}/W_{10} = 1/4$. We will also assume that the NMOS transistors (except $Q10$) have $V_{GS} - V_T = 1$ V at a drain current of 10 μA. Solving for the required value of R_7 gives $R_7 = [(V_{GS13} - V_T)/I_Q][1 - (K_{13}/K_{10})^{1/2}] = [1$ V/10 μA$][1 - (1/4)^{1/2}] = 1$ V $\times$ 0.5/10 μA $= 50$ kΩ.

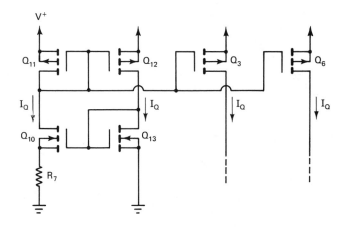

Figure 8.12 Detail of biasing circuit of the TLC1078 CMOS operational amplifier.

Single-supply operation. The TLC1078 can be operated from a single positive supply voltage, and therefore the input-voltage range should include ground potential. For the proper operation of this circuit, all the transistors of the differential amplifier circuit should operate in the active region. To investigate this, let us assume as a reasonable example that the gate-to-source voltage of $Q2$ and of $Q4$ is $+1$ V, and the threshold voltage of PMOS transistors $Q1$ and $Q5$ is -1 V. For the operation of a MOSFET in the active region, the channel should be pinched off at the drain end. Therefore, the gate-to-drain voltage of $Q1$ and $Q5$ should be greater than -1 V. Therefore, the gate voltage of $Q1$ and $Q5$ should be greater than -1 V $+ V_{G2,4} = -1$ V $+ 1$ V $= 0$ V. Therefore, we see that the input voltage range includes ground potential, and thus single-supply operation of this operational amplifier is possible. The actual input-voltage range for this device is specified as from -0.3 V up to 0.8 V below the positive supply voltage. The 0.8-V difference between the input voltage and the positive supply is necessary for transistor $Q1$ and $Q5$ to be turned on.

Output-voltage swing. The output stage of this operational amplifier consists of NMOS transistors $Q8$ and $Q9$ for sourcing and sinking load currents, respectively. This allows for a large output voltage swing, especially under lightly loaded conditions. A graph of the low-level output voltage versus the current sunk by the operational amplifier is shown in Figure 8.13. We note that the output resistance is about 100 Ω, so when the operational amplifier is sinking 1 mA from a load, the low-level output voltage is approximately 0.1 V and drops to zero as the output current goes to zero.

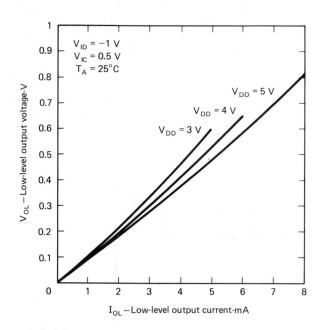

Figure 8.13 Low-level output voltage versus low-level output current (Texas Instruments).

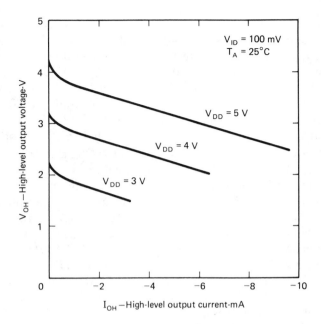

Figure 8.14 High-level output voltage versus high-level output current (Texas Instruments).

In Figure 8.14 the variation of the high-level output voltage with the operational-amplifier source current is shown. When sourcing currents out to a load, the high-level output voltage is about 0.8 V below the positive supply voltage under very lightly loaded conditions ($I_O < 0.1$ mA). From the circuit diagram, we can see why this is so. For a large difference-mode input voltage, transistor $Q7$ is turned off. This causes the gate of the output transistor $Q8$ to rise to a level very close to V^+. For $Q8$ to be turned on in order to source currents out to the load, the gate-to-source voltage of $Q8$ must be greater than the threshold voltage. If $V_T \cong +0.8$ V, we see that the maximum value of the output voltage is about 0.8 V below V^+.

The small values of threshold voltages of both the NMOS and PMOS transistors are obtained by using polycrystalline silicon for the gate electrodes. These small threshold voltages allow for a large common-mode input-voltage range, a large output-voltage swing, and also operation at low supply voltages. Indeed, the TLC1078C can be operated with single-supply voltages ranging from a minimum of 1.4 V up to a maximum of 16 V. The input bias current is specified as a very low 0.6 pA (typ.) with a $+5$-V supply and 0.7 pA (typ.) with a 10-V supply.

8.3.6 Additional Examples of CMOS Operational Amplifiers

The ICL76XX series from Intersil is low-power internally compensated CMOS op amps that can operate with supply voltages ranging from only ± 1 V up to ± 8 V. The quiescent supply current level is pin-programmable to allow quiescent current selections of 1 mA, 100 μA, and 10 μA. A quiescent power consumption of a mere

20 μW is possible by operation at a supply voltage of 2 V and supply current of 10 μA. The output-voltage swing under lightly loaded conditions is essentially rail-to-rail. The gain–bandwidth product and slewing rates for the various quiescent current (I_Q) selections are 44 kHz and 0.016 V/μs for I_Q = 10 μA, 480 kHz and 0.16 V/μs for I_Q = 100 μA, and 1.4 MHz and 1.6 V/μs for I_Q = 1 mA. The input bias current is 1 pA (typ.) and 50 pA (max.), and the input resistance is 10^{12} Ω (typ.). Devices in the ICL76XX series include the ICL7611/ICL7612 single op amp, ICL7621 dual op amp, ICL7631 triple op amp, and ICL7641/ICL7642 quad op amp.

The TLC25XX and TLC27XX series of CMOS op amps from Texas Instruments offers single (X = 1), dual (X = 2), and quad (X = 4) op amps with a wide choice of quiescent supply currents. The low bias current versions, the TLC25LX and TLC27LX, operate with a supply current of just 10 μA per op amp in the package and have a unity gain frequency of 100 kHz and slewing rate of 0.04 V/μs. The medium-bias op amps are the TLC25MX and TLC27MX, with a supply current of 150 μA per op amp and a unity gain frequency of 700 kHz and slewing rate of 0.6 V/μs. The high-bias op amps are the TLC25X and TLC27X operating at 1 mA per op amp and with a unity gain frequency of 2.3 MHz and slewing rate of 4.5 V/μs.

The OP-80 from Precision Monolithics, Inc., is a CMOS op amp that features an extremely low input bias current of only 60 fA (max.) at 25°C and 20 pA (max.) at 125°C. The input offset voltage is also relatively low for a CMOS op amp, being 1 mV (max.) for the OP-80E.

PROBLEMS

8.1. (*JFET-input operational amplifier*) Given: An op amp has an input stage that is a JFET differential amplifier with a current-mirror active load. The JFETs have I_{DSS} = 400 μA and V_P = −4.0 V. The differential amplifier is biased by a current sink of I_Q = 200 μA.

  (a) Find the compensation capacitance needed for a unity-gain frequency of f_u = 2.5 MHz.   (*Ans.*: 6.4 pF)
  (b) Find the slewing rate (*SR*).   (*Ans.*: 31.4 V/μs)
  (c) Find the full-power bandwidth (FPBW) for a peak output voltage swing of ±10 V.   (*Ans.*: 500 kHz)
  (d) If the JFETs have a channel-length modulation coefficient of $1/V_A$ = 1/(50 V), find the maximum possible gain that can be obtained from this JFET input stage (with a bipolar current-mirror active load). Compare this result with what is generally available with a bipolar transistor input stage.   (*Ans.*: 50)
  (e) Operational amplifiers with JFET input stages often have much higher slewing rates than those of all-bipolar operational amplifiers. Explain.
  (f) If the input bias current of the JFET input operational amplifier is 1.0 nA (max.) at 25°C and doubles for every 11°C temperature rise, what will I_{BIAS} be at 100°C?   [*Ans.*: 113 nA (max.)]
  (g) Repeat part (f) for 125°C.   [*Ans.*: 545 nA (max.)]
  (h) The input bias current of a JFET-input op amp is very small at 25°C, but increases at an exponential rate with increasing temperature. In contrast, the input bias current of an all-bipolar amplifier is much higher at 25°C, and actually decreases very slowly with increasing temperature. Explain.

8.2. (*CMOS operational amplifier*) Given the CMOS operational amplifier circuit of Figure 8.10 with a supply voltage of ± 15 V and $I_{SET} = 1$ mA. Assume that the threshold voltages are $+2$ V for the NMOS transistors and -3 V for the PMOS transistors. For the NMOS transistors, $V_{GS} = +4$ V at $I_{DS} = 1$ mA, and for the PMOS transistors, $V_{GS} = -6$ V at $I_{DS} = -1$ mA. Assume a channel-length modulation voltage of $V_A = 100$ V for all transistors.

   (a) Find R_{SET}. (*Ans.*: $R_{SET} = 24$ kΩ)

   (b) Find the g_f of the differential amplifier. (*Ans.*: $g_f = 235$ μS)

   (c) Find the net load conductance driven by the differential amplifier. (*Ans.*: $g_{total} = 10$ μS)

   (d) Find the voltage gain of the first (differential amplifier) stage. (*Ans.*: $A_{V1} = 47.1$)

   (e) Find the transfer conductance and the voltage gain of the second stage ($Q7$) under no-load conditions. (*Ans.*: $g_{f2} = 1000$ μS, $A_{V2} = 50$)

   (f) Find the total operational amplifier open-loop dc gain under no-load conditions. [*Ans.*: $A_{OL}(0) = 2357$]

8.3. (*CMOS op amp output resistance*) Op amps with CMOS output stages offer the advantage of permitting almost rail-to-rail output-voltage swings under lightly loaded conditions. However, under other load conditions the relatively high drain-to-source resistance of MOSFETs can severely restrict the maximum output-voltage swing. Furthermore, loads with a substantial capacitive component can lead to bandwidth and slewing-rate limitations. An op amp operates with a ± 12-V supply and has a CMOS output stage with $r_{ds(ON)} = 1000$ Ω for both the NMOS and PMOS transistors.

   (a) Find the minimum value of load resistance for a peak-to-peak output voltage swing of 20 V. (*Ans.*: $R_L = 5000$ Ω)

   (b) If the load capacitance is 1 nF, find the maximum slewing rate and full-power bandwidth that can be obtained. (*Ans.*: $SR = 12$ V/μs, FPBW = 191 kHz)

8.4. (*Effect of temperature on input bias current*) FET op amps have very low input bias currents at room temperature, but I_B increases exponentially with temperature, approximately doubling for every 10°C rise in temperature. If a FET op amp has $I_B = 1$ pA at 25°C, find I_B at **(a)** 75°C, **(b)** 100°C, and **(c)** 125°C. (*Ans.*: 32 pA, 181 pA, 1.02 nA)

8.5. (*MOSFET current source*) A MOSFET op amp uses the biasing circuit of Figure 8.12. Given that $V_T = 1.0$ V, $I_{DS} = 5$ μA at $V_{GS} = 1.5$ V for all MOSFETs except $Q10$, and a channel-width ratio of $W_{10}/W_{13} = 4$, find R_7 for $I_Q = 5$ mA. (*Ans.*: 200 kΩ)

8.6. (*MOSFET threshold voltage*) A MOSFET has $I_{DS} = 100$ μA at $V_{GS} = 2.0$ V, and $I_{DS} = 400$ μA at $V_{GS} = 3.0$ V. Find **(a)** the threshold voltage, and **(b)** I_{DS} at $V_{GS} = 4.0$ V. (*Ans.*: $V_T = 1.0$ V, $I_{DS} = 900$ μA)

8.7. (*MOSFET drain-to-source resistance*) A MOSFET is operating in the saturated region and has $I_{DS} = 100$ μA at $V_{DS} = 5.0$ V and $I_{DS} = 105$ μA at $V_{DS} = 10$ V. Find **(a)** the channel-length modulation coefficient, and **(b)** the dynamic drain-to-source resistance at $I_{DS} = 100$ μA. (*Ans.*: 1/100 V, $r_{ds} = 1$ MΩ)

REFERENCES

ALLEN, P. E., and D. H. HOLBERG, *CMOS Analog Design*, Holt, Rinehart and Winston, New York, 1987.

GREBENE, A. B., *Bipolar and MOS Analog Integrated Circuit Design*, Wiley, New York, 1984.

HASKARD, M. R., and I. C. MAY, *Analog VLSI Design*, Prentice-Hall, Englewood Cliffs, N.J., 1988.

ONG, D. G., *Modern MOS Technology*, McGraw-Hill, New York, 1984.

OXNER, E. S., *FET Technology and Applications*, Marcel Dekker, New York, 1989.

TSIVIDIS, Y. P., and P. ANTOGNETTI, *Design of MOS VLSI Circuits for Telecommunications*, Chapter 3, CMOS Operational Amplifiers, by B. J. HOSTICKA, Prentice-Hall, Englewood Cliffs, N.J., 1985.

_____, and _____, *Design of MOS VLSI Circuits for Telecommunications*, Chapter 5, NMOS Operational Amplifiers, by B. D. SENDEROWICZ, Prentice-Hall, Englewood Cliffs, N.J., 1985.

WALSH, M. J. (ed.), *Choosing and Using CMOS*, Chapter 7, Analogue Techniques, by JOHN PENNOCK, McGraw-Hill, New York, 1985.

9

Current-feedback, Norton, and Transconductance Operational Amplifiers

In this chapter, three important types of op amps will be considered, the current-feedback (CFB) operational amplifier, the Norton or *current differencing* operational amplifier, and the operational transconductance amplifier (OTA). The op amps considered up to this point have been of the conventional voltage-controlled voltage source or VCVS type as represented by the equivalent circuit model of Figure 9.1a. For the ideal VCVS op amp, the transfer function is given by $v_o = A_{OL}v_i$, where $v_i = v_1 - v_2$ is the difference-mode input voltage.

The operational transconductance amplifier (OTA) acts as a voltage-controlled current source or VCCS and can be represented by the equivalent circuit model of Figure 9.1b. The OTA has a transfer relationship given by $i_o = g_f v_i$, where again $v_i = v_1 - v_2$ is the difference-mode input voltage.

The Norton or *current differencing* operational amplifier is basically a current-controlled voltage source or CCVS with a transfer relationship given by $v_o = r_f(i_1 - i_2)$, where r_f is the dynamic transfer resistance of the device. The ideal Norton op amp can be represented by the equivalent circuit model of Figure 9.1c.

The current-feedback (CFB) op amp has a transfer relationship given by $v_o = Z_T i_{INV}$, where i_{INV} is the net current at the inverting input node as shown in the equivalent circuit model of Figure 9.2.

The three op-amp types to be discussed in this chapter offer many performance and applications opportunities beyond those of the conventional VCVS op amps.

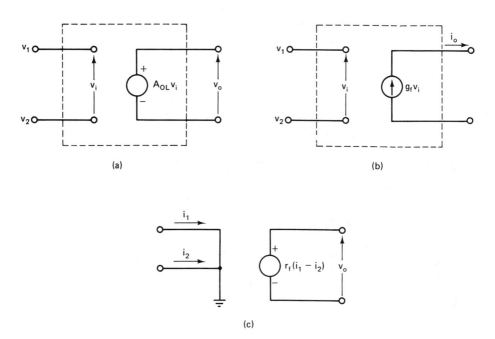

Figure 9.1 Ideal operational-amplifier equivalent circuit models: (a) voltage-controlled voltage source (VCVS) model: (b) voltage-controlled current source (VCCS) model: (c) current-controlled voltage source (CCVS) model.

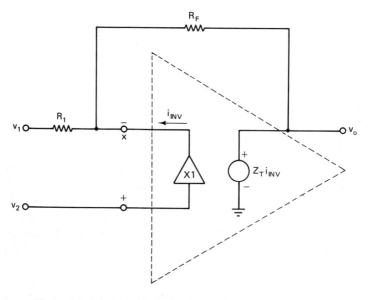

Figure 9.2 Ideal current-feedback operational-amplifier model.

Operational Amplifiers   Chap. 9

The CFB op amp, in particular, offers very wide bandwidth and slewing-rate performance. The Norton op amps can readily be employed in low-voltage single-supply applications. In the case of the OTA, the control of the gain by an externally supplied biasing current offers many application possibilities in such areas as automatic gain control, modulators, and multipliers.

9.1 CURRENT-FEEDBACK OPERATIONAL AMPLIFIERS

The current-feedback operational amplifier (CFB op amp) has an internal circuit design that differs considerably from the conventional voltage-controlled voltage source type of operational amplifier. Nevertheless, the external circuitry and applications are similar to that for the VCVS operational amplifier. The CFB operational amplifier offers as one of its principal features a very large closed-loop bandwidth, generally up in the 50- to 200-MHz range. Furthermore, the bandwidth is relatively independent of the closed-loop gain. Related to the large bandwidth is a very fast time-domain response, with rise and fall times generally down in the 2- to 10-ns range, and with settling times on the order of 10 ns for a 0.1% band. The slewing rates are up in the 500- to 5000-V/μs range, and the full-power bandwidth values are up in the 50- to 100-MHz range.

Figure 9.2 shows a small-signal model of the ideal CFB operational amplifier, together with the feedback resistor R_F and resistor R_1. Internally, there is a unity voltage gain buffer between the noninverting input terminal, where voltage v_2 is applied, and the inverting input terminal (node X). This unity-gain buffer causes the voltage at the inverting input terminal to follow the voltage applied to the noninverting input terminal, so that $V_X = v_2$. For the ideal CFB operational amplifier, the input impedance at the noninverting input terminal is infinite, and the impedance level at the inverting input terminal will be assumed to be zero. On the output side of this CFB operational amplifier model, there is a current-controlled voltage source of strength $v_o = z_T i_{INV}$, where z_T is the forward dynamic transfer impedance of the device and i_{INV} is the net current flowing out of the inverting input terminal.

For an initial look at the analysis of the CFB operational amplifier, let us consider the circuit of Figure 9.3. We note that the voltage at the two input nodes is equal, so in this case we have that node X will be a virtual ground. We therefore have that $i_{FB} + i_S + i_{INV} = 0$, where i_{FB} is the feedback current and i_S is the current due to the application of the signal voltage v_S. The output voltage v_o is related to i_{INV} by $v_o = i_{INV} z_T$. Generally, for CFB operational amplifiers the transfer impedance is very large, especially at low frequencies. Let us consider the ideal case in which z_T goes to infinity. For a finite value of v_o, this means that the input current i_{INV} will go to zero and we will have $i_{INV} = 0$, and therefore $i_S + i_{FB} = 0$. Thus, the action of this feedback loop is to adjust the value of v_o so as to make the input current i_{INV} equal to zero and thus to make the sum of i_S and i_{FB} equal to zero. This is analogous to the action of the feedback loop of the VCVS type of operational amplifier in which the action of the feedback loop is to adjust the output voltage to such a value that the difference-mode input voltage is reduced to zero, and therefore the voltages at the two input nodes become equal. Since $i_S + i_{FB} = 0$, $i_S = v_S/R_1$, and $i_{FB} = v_o/R_F$, we

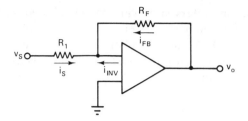

Figure 9.3 Inverting current-feedback operational-amplifier circuit.

now have that $v_S/R_1 = -v_o/R_F$, so the closed-loop voltage gain is given by $A_{CL} = v_o/v_S = -R_F/R_1$. Note that this expression for the closed-loop voltage gain is the same as that obtained for the conventional VCVS type of operational amplifier.

We can now see why this is called a CFB operational amplifier. We note that the output voltage is a function of the net current going out of the inverting input node (i_{INV}). This inverting input terminal is a low-impedance current-summing node where the feedback current from the output is summed with the current due to the input signal. The action of the feedback loop is to make the sum of these two currents equal to zero. In the case of a VCVS operational amplifier, the output voltage is proportional to the difference-mode voltage that appears between the two input terminals. The inverting input terminal is a high-impedance voltage-summing node where the feedback voltage is summed with the voltages that result from the application of the input signals.

Now let us consider a more general analysis of the CFB operational amplifier, this time considering the case of input signals being applied to both input terminals and also considering the effect of a finite value for the transfer impedance z_T. We refer again to the circuit of Figure 9.2. Writing a node voltage equation for node X gives

$$v_1 G_1 + v_o G_F + i_{INV} = v_X(G_1 + G_F) \tag{9.1}$$

We note that due to the unity-gain buffer we have that $v_x = v_2$, and we also note that $v_o = z_T i_{INV}$, so $i_{INV} = v_o/z_T = y_T v_o$. Therefore, the equation for node X can be rewritten as

$$v_1 G_1 + v_o G_F + y_T v_o = v_2(G_1 + G_F) \tag{9.2}$$

and thus for the output voltage we obtain

$$v_o = \frac{v_2(G_1 + G_F) - v_1 G_1}{G_F + y_T} \tag{9.3}$$

This can be rewritten as

$$v_o = \frac{v_2(G_1 + G_F) - v_1 G_1}{G_F(1 + y_T/G_F)} \tag{9.4}$$

$$= \frac{v_2(1 + G_1/G_F) - v_1(G_1/G_F)}{1 + y_T/G_F} \tag{9.5}$$

If we now rewrite this expression in terms of resistances, we obtain

$$v_o = \frac{v_2(1 + R_F/R_1) - v_1(R_F/R_1)}{1 + R_F/z_T} \tag{9.6}$$

Since the feedback resistor R_F is generally down in the range of less than 2000 Ω, and the transfer impedance at low frequencies is generally up in the range of 300 kΩ to 3 MΩ, we have that at the lower frequencies $z_T \gg R_F$, so the equation for v_o reduces to

$$v_o = v_2\left(1 + \frac{R_F}{R_1}\right) - v_1\frac{R_F}{R_1}$$

This result is exactly the same as that obtained for the conventional VCVS type of operational amplifier. Note that the error in the gain calculation that results from not taking the R_F/z_T term in the denominator into account is *independent* of the closed-loop gain, which is controlled by the R_F/R_1 ratio. At the lower frequencies where $z_T \cong z_T(0) = r_T$, where r_T is the dynamic transfer resistance, the gain error ε is given by $\varepsilon = R_F/r_T$. To consider a representative example, let $R_F = 1000$ Ω and $r_T = 1$ MΩ. For this case the gain error is $= 1000$ $\Omega/1$ M$\Omega = 10^{-3} = 0.1\%$. Thus we see that in this representative example the gain error is very small and is independent of the closed-loop gain as long as the feedback resistor R_F is not changed.

9.1.1 CFB Operational Amplifier Frequency Response

For the analysis of the frequency response of the CFB operational amplifier, we will use the following expression or the variation of the transfer impedance with frequency:

$$z_T = \frac{z_T(0)}{1 + j(f/f_1)} = \frac{r_T}{1 + j(f/f_1)} \tag{9.7}$$

This variation of the transfer impedance with frequency will be derived later when the internal circuitry of the CFB operational amplifier is analyzed. For simplicity, we will consider the case of just one input signal, so we will let $v_1 = 0$ and will therefore have that

$$v_o = \frac{v_2(1 + R_F/R_1)}{1 + (R_F/z_T)} \tag{9.8}$$

and thus the closed-loop gain is

$$A_{CL} = \frac{1 + (R_F/R_1)}{1 + (R_F/z_T)} \tag{9.9}$$

At low frequencies where $z_T \cong z_T(0) = r_T \gg R_F$, we have that $A_{CL}(0) \cong 1 + (R_F/R_1)$, and therefore the expression for the closed-loop gain can be expressed as

$$A_{CL} = \frac{A_{CL}(0)}{1 + R_F/z_T} \tag{9.10}$$

Introducing the expression for z_T, the gain equation can be written as

$$\frac{A_{CL}}{A_{CL}(0)} = \frac{1}{1 + (R_F/r_T)(1 + jf/f_1)} = \frac{1}{1 + (R_F/r_T) + j(R_F/r_T)(f/f_1)} \qquad (9.11)$$

Since $R_F/r_T \ll 1$, the closed-loop gain expression can be rewritten as

$$\frac{A_{CL}}{A_{CL}(0)} \cong \frac{1}{1 + (R_F/r_T)(jf/f_1)} = \frac{1}{1 + j(f/f_{1CL})} \qquad (9.12)$$

where f_{1CL} is the closed-loop 3-dB bandwidth, BW_{CL}, as is given by $BW_{CL} = f_{1CL} = (r_T/R_F)f_1$. Since $r_T/R_F \gg 1$, we see that the closed-loop bandwidth is much greater than f_1.

Since $A_{CL}(0) = 1 + R_F/R_1$ and $BW_{CL} = (r_T/R_F)f_1$, we see that the closed-loop bandwidth is *independent* of the closed-loop gain, and a large closed-loop gain and a large closed-loop bandwidth can simultaneously be achieved. To put it another way, the CFB operational amplifier offers a very large gain–bandwidth (GBW) product.

In the analysis of the CFB operational amplifier circuit, we will obtain the result that $f_1 = 1/(2\pi r_T C_T)$, so $r_T f_1 = 1/(2\pi C_T)$; therefore, $BW_{CL} = (r_T/R_F)f_1 = 1/(2\pi R_F C_T)$. For a representative example, let us assume $A_{CL}(0) = 10$, $R_F = 1000 \ \Omega$, and use a value of $C_T = 1.6$ pF. We obtain a closed-loop bandwidth of $BW_{CL} = 1/(2\pi \times 1 \ \text{k}\Omega \times 1.6 \ \text{pF}) = 100$ MHz. For a closed-loop gain of 10, the required value of R_1 is 111 Ω. The gain–bandwidth product is a very impressive 1000 MHz.

From the preceding discussion, it might seem that by making the feedback resistor ever smaller extraordinarily large bandwidths could be achieved. There are, however, some basic limitations to the ultimate bandwidth that can be achieved. One limitation is the existence of other higher-order breakpoint frequencies (poles) in the transfer function of the operational amplifier. Another limitation is the effect of the nonzero input resistance at the inverting input terminal; this effect will be analyzed next. Another consideration is the fact that as R_F and therefore R_1 is made smaller, the feedback network will act as more of a load on the amplifier output, drawing more current as R_F is decreased. This leads to increased power consumption, heat dissipation, and a decrease in the amount of current available to drive the load.

9.1.2 Analysis of CFB Internal Circuit

We will now investigate the internal circuitry of a CFB operational amplifier. A representative example is shown in Figure 9.4. For this initial discussion, we will assume that all transistors of the same type are identical. The unity-gain buffer between the noninverting ($+$INPUT) and inverting input ($-$INPUT) terminals is comprised of transistors $Q1$ through $Q4$, all connected as emitter followers.

Writing Kirchoff's voltage law around the path $Q1$, $Q3$, $Q4$, and $Q2$, we obtain

$$V_{EB1} - V_{BE3} - V_{EB4} + V_{BE2} = 0 \qquad (9.13)$$

and thus

$$V_T \ln \frac{I_Q}{I_{COP}} - V_T \ln \frac{I_3}{I_{CON}} - V_T \ln \frac{I_4}{I_{COP}} + V_T \ln \frac{I_Q}{I_{CON}} = 0 \qquad (9.14)$$

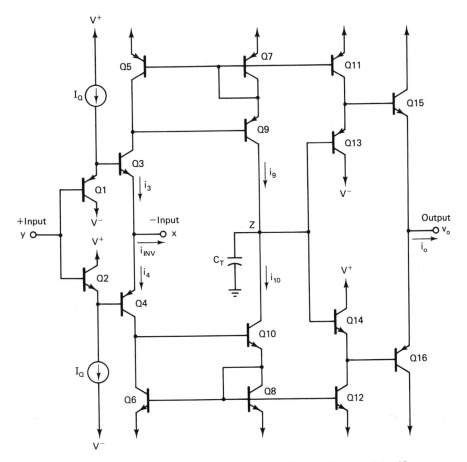

Figure 9.4 Simplified diagram of a current-feedback operational amplifier (Comlinear Corp.).

After canceling out V_T and combining the logarithmic terms, we obtain

$$\ln \frac{I_Q^2}{I_3 I_4} = 0$$

and thus

$$I_3 I_4 = I_Q^2 \tag{9.15}$$

If we assume that under quiescent conditions $I_{INV} = 0$ then $I_3 = I_4 = I_Q$. The voltage between the inverting and noninverting input terminals is given by

$$V_x - V_y = V_{OS} = V_{EB1} - V_{BE3} = V_T \ln \frac{I_Q}{I_{COP}} - V_T \ln \frac{I_3}{I_{CON}} \tag{9.16}$$

$$= V_T \ln \frac{I_{CON}}{I_{COP}}, \quad \text{since } I_3 = I_Q$$

Thus there is just a small offset voltage between the two input terminals, generally no more than a few millivolts. Aside from this small dc offset voltage, the voltage at node X will closely follow the voltage at the $+$INPUT (node Y). Furthermore, as a result of the cascaded emitter-follower arrangement, the impedance looking into the $+$INPUT terminal is very large, generally greater than $100\,\text{k}\Omega$, and the impedance looking into the $-$INPUT terminal is very small, generally less than $30\,\Omega$.

The PNP transistor configuration consisting of $Q5$, $Q7$, and $Q9$ and the NPN configuration using transistors $Q6$, $Q8$, and $Q10$ are two Wilson current mirrors. The action of these two current mirrors is to make $i_9 = i_3$ and $i_{10} = i_4$. Since $i_3 - i_4 = i_{INV}$, we have that $i_9 - i_{10} = i_{INV}$. Thus the action of the two current mirrors is to make the net current coming into node Z equal to i_{INV}.

Let us now turn our attention to node Z and the output stage of this operational amplifier. The output stage uses transistors $Q11$ through $Q16$ in the form of a complementary Darlington push–pull emitter-follower configuration. Transistor $Q15$ is used for sourcing currents out to the load, and $Q16$ is used for sinking load currents. These two emitter-follower output transistors are driven from node Z via the emitter-follower transistors $Q13$ and $Q14$, respectively. Transistors $Q11$ and $Q12$ serve as current-source active loads for $Q13$ and $Q14$, respectively. The Darlington emitter-follower configuration of the output stage produces a very large impedance transformation between the load driven by this operational amplifier and node Z, such that the impedance level at node Z is very large. As we will see, this contributes to having a large transfer impedance z_T for this circuit.

Let us look at quiecent conditions for this amplifier, for which we have that the voltage at the $+$INPUT terminal (node Y) is zero. For simplicity, let us assume that all transistor junction drops are equal. Under quiescent conditions with the voltage

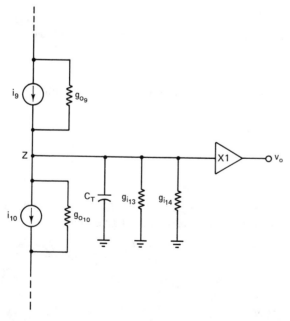

Figure 9.5 AC small-signal model for node Z.

Operational Amplifiers   Chap. 9

applied to the $+$INPUT terminal equal to zero, the voltage at node X will also be zero and therefore $i_{\text{INV}} = 0$. Therefore, $i_3 = i_4$, and as a result of the action of the Wilson current mirrors, $i_9 = i_{10}$. The net current coming into node Z is therefore zero, so the voltage at node Z is zero, and therefore the output voltage is zero. The cascaded PNP–NPN arrangement for the output stage results in the voltage at the output v_o being equal to the voltage at node Z, so that $v_o = 0$. Therefore, with zero input voltage we obtain a zero output voltage under these ideal conditions of perfectly matched transistors. For a real operational amplifier, there will of course be transistor mismatches and, as a result, a small offset voltage.

Figure 9.5 shows an ac small-signal model for node Z. In this model g_{o9} and g_{o10} represent the dynamic output conductances of $Q9$ and $Q10$, respectively. Conductances g_{i13} and g_{i14} represent the dynamic input conductances looking into the bases of $Q13$ and $Q14$, respectively. Capacitance C_T is the net capacitance between node Z and ground. The net current coming into node Z is $i_9 - i_{10} = i_3 - i_4 = i_{\text{INV}}$. If we write the node voltage equation for node Z, we obtain

$$i_{\text{INV}} = v_Z(g_{o9} + g_{o10} + g_{i13} + g_{i14} + j\omega C_T) \tag{9.17}$$
$$= v_Z(g_T + j\omega C_T) = v_Z y_T$$

Solving for v_Z and noting that, since the voltage gain of the emitter-follower output stage is close to unity, $v_o = v_z$, we obtain

$$v_o = i_{\text{INV}}/y_T = i_{\text{INV}} z_T \tag{9.18}$$

where z_T is the dynamic transfer impedance of the device as given by

$$z_T = \frac{1}{y_T} = \frac{1}{g_T + j\omega C_T} = \frac{1}{g_T(1 + j\omega C_T/g_T)} = \frac{r_T}{1 + j\omega r_T C_T} \tag{9.19}$$
$$= \frac{r_T}{1 + j(\omega/\omega_1)} = \frac{r_T}{1 + j(f/f_1)}$$

where $r_T = z_T(0)$ is the low-frequency value of z_T, $\omega_1 = 1/r_T C_T$, and $f_1 = 1/(2\pi r_T C_T)$ is the breakpoint or -3-dB frequency of the transfer impedance.

The output conductance of the Wilson current mirrors is given approximately by $g_o = I_C/\beta V_A = I_Q/\beta V_A$, where V_A is the Early voltage. The net input conductance that is seen looking into the bases of $Q13$ and $Q14$ is approximately equal to the transformed value of the load resistance as given by

$$g_{i13} + g_{i14} = G_L' = \frac{1}{R_L'} = \frac{1}{\beta_{(13,14)}\beta_{(15,16)}R_L} \tag{9.20}$$

Capacitance C_T is the net capacitance at node Z and is given by

$$C_T = C_{CB13} + C_{CB14} + C_{CB9} + C_{CB10}$$

Example. Let us now consider a representative example. We will assume that $I_Q = 2.0$ mA, $R_L = 100\ \Omega$, and for all transistors $\beta = 100$ (typ.) and $V_A = 100$ V. A collector–base junction capacitance of 0.4 pF will be assumed for all transistors. For the output conductances, we have $g_{o9} + g_{o10} = 2 \times 2\ \text{mA}/(100 \times 100\ \text{V}) = 4.0$

mA/(10,000 V) $= 4 \times 10^{-7}$ S $= 0.4$ μS. The net input conductance of the output stage is approximately the transformed value of the load resistance as given by $g_{i_{13}} + g_{i_{14}} = 1/R'_L = 1/(100 \times 100 \times 100 \, \Omega) = 1 \times 10^{-6}$ S $= 1.0 \, μ$S. The total conductance at node Z is therefore $g_T = 1.4 \, μ$S, and the transfer resistance is $r_T = 1/g_T = 0.714$ MΩ $= 714$ kΩ. This is the result for a load resistance of only 100 Ω, which is a relatively low value. As R_L increases and goes to infinity, the transfer resistance will increase and approach a limiting value of $r_T(\text{max.}) = 1/(g_{o_9} + g_{o_{10}}) = 1/(0.4 \, \text{mS}) = 2.5$ MΩ. From this example, we see that transfer resistances in the range of from several hundred kilohms up to several megohms can generally be expected.

When all the collector–base junction capacitances at node Z are added up, the result is that $C_T = 1.6$ pF. The transfer impedance will therefore have a breakpoint frequency at $f_1 = 1/(2\pi R_T C_T) = 1/(2\pi \times 714 \, \text{kΩ} \times 1.6 \, \text{pF}) = 139$ kHz. While this does not seem like a very large frequency, it should be remembered that this is not the closed-loop bandwidth, and indeed the closed-loop bandwidth can be very much larger than this, as was seen in the analysis of the frequency response of the CFB operational amplifier.

9.1.3 Effect of Input Resistance on Bandwidth

In the analysis of the CFB frequency response it appeared that the closed-loop bandwidth was completely independent of the closed-loop gain. In actuality, there will be some dependence of bandwidth on the gain. This is mainly due to the nonzero value of the input resistance r_{IN} at the inverting input terminal. To analyze the effect of this resistance, let us consider the equivalent circuit model of the CFB operational amplifier as shown in Figure 9.6. This is the same model as considered previously, except for the addition of r_{IN} at the inverting input terminal. At node X we have again the equation

$$v_1 G_1 + v_o G_F + i_{INV} = v_X(G_1 + G_F) \tag{9.21}$$

but this time

$$v_x = v_2 - i_{INV} r_{IN} = v_2 - y_T v_o r_{IN}$$

so

$$v_1 G_1 + v_o G_F + y_T v_o = v_2(G_1 + G_F) - y_T v_o r_{IN}(G_1 + G_F)$$

and thus

$$v_o[G_F + y_T + y_T r_{IN}(G_1 + G_F)] = v_2(G_1 + G_F) - v_1 G_1$$

and therefore

$$\begin{aligned}
v_o &= \frac{v_2(G_1 + G_F) - v_1 G_1}{G_F + y_T[1 + r_{IN}(G_1 + G_F)]} \\
&= \frac{v_2(1 + G_1/G_F) - v_1(G_1/G_F)}{1 + (y_T/G_F)[1 + r_{IN}(G_1 + G_F)]} \\
&= \frac{v_2(1 + R_F/R_1) - v_1(R_F/R_1)}{1 + (R_F/z_T)[1 + (r_{IN}/R_1) + (r_{IN}/R_F)]}
\end{aligned} \tag{9.22}$$

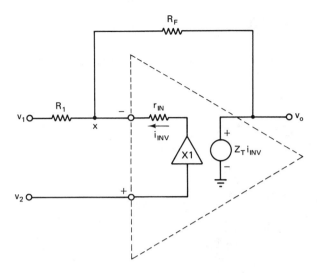

Figure 9.6 Effect of input resistance on voltage gain and bandwidth.

If $r_{IN} \ll [R_1 \parallel R_F]$, this will reduce to the previous result. From the circuit of Figure 9.4, we see that $r_{IN} = (1/2)(V_T/I_Q)$ so r_{IN} is generally down in the range of just 3 to 30 Ω. Thus r_{IN} is usually very small compared to $[R_1 \parallel R_F]$.

To examine the effect of r_{IN} on the closed-loop bandwidth, we will again for simplicity consider the case of $v_1 = 0$. From the result given above, we can now express the closed-loop gain as

$$A_{CL} = \frac{v_o}{v_2} = \frac{1 + (R_F/R_1)}{1 + (R_F/z_T)[1 + r_{IN}(G_1 + G_F)]} \tag{9.23}$$

Since $A_{CL}(0) = 1 + (R_F/R_1)$, this can be rewritten as

$$\frac{A_{CL}}{A_{CL}(0)} = \frac{1}{1 + (R_F/z_T)[1 + r_{IN}(G_1 + G_F)]}$$

$$= \frac{1}{1 + (R_F/r_T)[1 + j(f/f_1)][1 + r_{IN}(G_1 + G_F)]} \tag{9.24}$$

$$\cong \frac{1}{1 + (R_F/r_T)(j(f/f_1)[1 + r_{IN}(G_1 + G_F)]}$$

$$\cong \frac{1}{1 + j(f/f_{1CL})}$$

where the closed-loop bandwidth is now given by

$$BW_{CL} = f_{1CL} = \frac{f_1(r_T/R_F)}{1 + r_{IN}(G_1 + G_F)}$$

$$= \frac{f_1(r_T/R_F)}{1 + r_{IN}[(1/R_1) + 1/R_F)]} \tag{9.25}$$

$$= \frac{f_1(r_T/R_F)}{1 + (r_{IN}/R_F)[1 + (R_F/R_1)]}$$

TABLE 9.1

$A_{CL}(0)$	BW_{CL} (MHz)
1	96.6
3	88.6
5	86.5
10	76.5
30	52.4
50	39.8

Since $f_1 = 1/(2\pi r_T C_T)$, we have that $f_1 r_T/R_F = 1/(2\pi R_F C_T)$ and thus

$$
\begin{aligned}
BW_{CL} &= \frac{1}{2\pi R_F C_T} \times \frac{1}{1 + (r_{IN}/R_F)[1 + (R_F/R_1)]} \\[2mm]
&= \frac{1}{2\pi R_F C_T} \times \frac{1}{1 + (r_{IN}/R_F)A_{CL}(0)}
\end{aligned}
\tag{9.26}
$$

The first term on the right side of this equation is the same result as obtained previously in the case of $r_{IN} = 0$. We see from this last equation that there is some dependence of the closed-loop bandwidth on the closed-loop gain.

We will now consider a representative example to show the dependence of the bandwidth on the closed-loop gain. We will use the values given previously for R_F (1 kΩ) and C_T (1.6 pF), and assume a value for r_{IN} of 30 Ω. For this we obtain $BW_{CL} = 100\,\text{MHz}/[1 + (30/1000)A_{CL}(0)] = 100\,\text{MHz}/[1 + 0.03A_{CL}(0)]$. Bandwidth values obtained for various values of closed-loop gain are shown in Table 9.1, where we see that there is some decrease of closed-loop bandwidth with increasing closed-loop gain, especially for gains above 10. However, even at a gain of 30 the bandwidth is still no less than half the unity-gain bandwidth, and at a gain of 50 the bandwidth is about 40% of the unity-gain value.

9.1.4 Comparison of CFB and VCVS Operational Amplifier Frequency Response

We will now compare the frequency-response characteristics of the CFB and VCVS operational amplifiers to gain further insight into the factors controlling the frequency-domain and time-domain response. A fundamental factor to consider in all this is that the frequency response of any amplifier is directly related to the rate at which the capacitances in the circuit can be charged and discharged. This in turn is a function of the currents available to charge and discharge capacitances at the various nodes of the circuit.

We will first go through a simplified analysis of the CFB operational amplifier. We will use the model shown in Figure 9.7. This diagram shows a CFB operational amplifier connected as an inverting amplifier with a closed-loop gain of $-R_F/R_1$. First we will consider the case of $r_{IN} = 0$ such that the inverting input node of the operational amplifier is a zero impedance node at ground potential. As a result, $i_1 = v_1/R_1$ and $i_{FB} = v_o/R_F$. At this node we have that $i_1 + i_{FB} + i_{INV} = 0$, so $i_{INV} = -i_{FB} - i_1$.

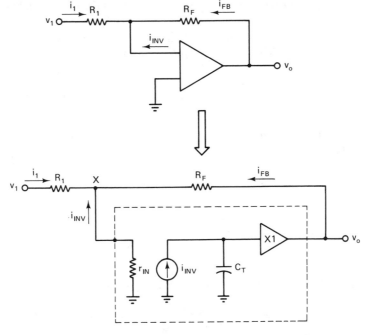

Figure 9.7 Simplified current-feedback operational-amplifier model for analysis of frequency response.

The output voltage is equal to the voltage developed across C_T, and we have that $dv_o/dt = i_{INV}/C_T$. Therefore, $dv_o/dt = -(i_{FB} + i_1)/C_T = -[v_o/R_FC_T)] - [v_1/(R_1C_T)]$, so $dv_o/dt + v_o/(R_FC_T)] = -v_1/(R_1C_T)$. Taking the Laplace transform of this gives $V_o(s)[s + 1/(R_FC_T)] = -V_1(s)/(R_1C_T)$, and thus

$$A_{CL}(s) = \frac{V_O(s)}{V_1(s)} = \frac{-1}{R_1C_T} \times \frac{1}{s + 1/(R_FC_T)} = \frac{-R_F/R_1}{1 + sR_FC_T} \qquad (9.27)$$

For the frequency-domain response, we set $s = j\omega$ and obtain

$$A_{CL}(\omega) = \frac{-R_F/R_1}{1 + j\omega R_FC_T} = \frac{-R_F/R_1}{1 + j(\omega/\omega_B)} = \frac{-R_F/R_1}{1 + j(f/f_B)} \qquad (9.28)$$

where f_B is the closed-loop bandwidth (BW_{CL}) as given by $BW_{CL} = f_B = 1/(2\pi R_FC_T)$, and the zero frequency closed-loop gain is given by $A_{CL}(0) = -R_F/R_1$. We see that it is the same result as obtained previously and we note again that the bandwidth is independent of the closed-loop gain. That result is due to the fact that the inverting input node is a zero-impedance node, so the feedback current i_{FB} is a function only of R_F and is independent of R_1. Note that the frequency-domain response (and time-domain response) is directly related to the rate at which the voltage across capacitance C_T can be changed and thus is related to the current available to charge (and discharge) C_T.

Voltage-controlled voltage source operational amplifier. We will now consider the situation of the VCVS operational amplifier as shown in Figure 9.8. For this device the inverting input node will be assumed to be an infinite-impedance node with zero current. The difference-mode input voltage v_i is given by

$$v_i = \frac{v_o R_1}{R_1 + R_F} + \frac{v_1 R_F}{R_1 + R_F} = F v_o + (1 - F) v_1 \tag{9.29}$$

where $F = \text{feedback factor} = R_1/(R_1 + R_F)$. The rate of change of the output voltage with time is given by

$$\frac{dv_o}{dt} = \frac{-g_f v_i}{C_T} = \frac{-g_f[F v_o + (1 - F) v_1]}{C_T} \tag{9.30}$$

and thus

$$\frac{dv_o}{dt} + \frac{g_f F v_o}{C_T} = \frac{-g_f(1 - F) v_1}{C_T}$$

Taking the Laplace transform of both sides yields

$$V_O(s)[s + g_f F/C_T] = -(g_f/C_T)(1 - F)V_1(s) \tag{9.31}$$

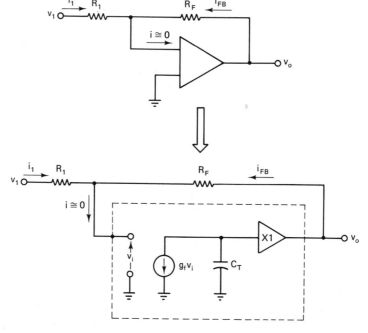

Figure 9.8 Simplified VCVS operational-amplifier model for analysis of frequency response.

and thus

$$A(s) = \frac{V_O(s)}{V_1(s)} = \frac{-(g_f/C_T)(1 - F)}{s + (g_f F/C_T)} = \frac{-(1 - F)/F}{1 + sC_T/(Fg_f)} \qquad (9.32)$$

We note that $(1 - F)/F = R_F/R_1$. For the frequency-domain response, we again set $s = j\omega$ and obtain

$$A_{CL}(\omega) = \frac{-R_F/R_1}{1 + j(\omega C_T/Fg_f)} = \frac{-R_F/R_1}{1 + j(\omega/\omega_B)} = \frac{-R_F/R_1}{1 + j(f/f_B)} \qquad (9.33)$$

where f_B is the closed-loop bandwidth as given by $BW_{CL} = f_B = Fg_f/(2\pi C_T)$. Since $F = R_1/(R_1 + R_F) = 1/[1 + (R_F/R_1)]$, this can be written as $BW_{CL} = g_f/\{2\pi C_T[1 + (R_F/R_1)]\}$. Since $A_{CL}(0) = -R_F/R_1$, we see that the closed-loop bandwidth of the VCVS operational amplifier is a function of the closed-loop gain and is approximately inversely proportional to that gain. We note that for the VCVS operational amplifier the feedback voltage from the output Fv_o that provides the current to charge capacitance C_T is proportional to the feedback factor $F = R_1/(R_1 + R_F)$, which is a function of both R_F and R_1. As the closed-loop gain is increased, the feedback factor is decreased, and as a result the current available to charge C_T also decreases, which results in a lower bandwidth.

9.1.5 CFB Operational Amplifier with Nonzero Resistance at the Inverting Input

Now let us return to the circuit of Figure 9.7 to reexamine the effect of a nonzero resistance at the inverting input terminal (node X). The node voltage equation at node X can be written as $v_1 G_1 + v_o G_F + i_{INV} = v_x(G_1 + G_F)$, and we note that $v_x = -i_{INV}r_{IN}$, so $v_1 G_1 + v_o G_F + i_{INV} + i_{INV}r_{IN}(G_1 + G_F) = 0$, and thus $i_{INV}(1 + r_{IN}G_1 + r_{IN}G_F) = -(v_1 G_1 + v_o G_F)$. Let $K = r_{IN}G_1 + r_{IN}G_F$ so that $i_{INV} = -(v_1 G_1 + v_o G_F)/(1 + K)$. We see that the current available to charge C_T has been reduced by a factor of $1 + K$. Since $dv_o/dt = -i_{INV}/C_T = (v_1 G_1 + v_o G_F)/[C_T(1 + K)]$ and, comparing this result with the previous one, we see that the bandwidth is reduced by the $1 + K$ factor to give $BW_{CL} = 1/[2\pi R_F C_T(1 + K)]$. Since $K = (r_{IN}G_1) + (r_{IN}G_F) = r_{IN}/[R_1 \parallel R_F] = r_{IN}/R_P$, where $R_P = [R_1 \parallel R_F]$ is the parallel combination of R_1 and R_F, we can write $BW_{CL} = 1/\{2\pi R_F C_T[1 + (r_{IN}/R_P)]\}$. From these results we see that the current available to charge C_T is now a function of both R_1 and R_F, so as a result the closed-loop bandwidth is now a function of the closed-loop gain. Note, in particular, that as R_1 decreases, which corresponds to increasing $A_{CL}(0)$, the component of i_{INV} that is due to v_o decreases as a result of the current division between r_{INV} and R_1.

9.1.6 CFB Operational Amplifier Slewing Rate

For the conventional VCVS type of operational amplifier, the slewing rate SR is given by the equation $SR = I_Q/C_{COMP}$ and is thus controlled by I_Q, the maximum current that the differential amplifier input stage can deliver to the compensation capacitor, and C_{COMP}, the size of the compensation capacitor. The maximum current that the differential amplifier can deliver to the compensation capacitor is usually equal to the strength of the current source that is used to bias the differential amplifier. This current source is often down in the range of 10 μA, and C_{COMP} is generally in the 10- to 50-pF range, so slewing rates in the range of 0.2 to 1 V/μs are often encountered. Some higher-speed VCVS-type operational amplifiers, however, have slewing rates that are up in the 500-V/μs range. The CFB operational amplifiers can, however, offer slewing rates that are even much higher than this.

CFB operational amplifiers do not require a large compensation capacitor. The capacitance that needs to be charged up is C_T at node Z in Figure 9.8 and is generally down in the 1- to 5-pF range. In addition, the current available to charge this capacitance is not limited to the strength of a constant-current source. For example, consider the case of the simplified CFB operational amplifier circuit model of Figure 9.9. Since $v_x = v_s$, the value of I_{INV} immediately after the application of the step is given by $I_{INV}(t = 0) = v_x/[R_1 \parallel R_F] = v_s(R_1 + R_F)/R_1 R_F = (v_s/R_F)[1 + (R_F/R_1)]$. Since $A_{CL}(0) = 1 + (R_F/R_1)$, this can be rewritten as $I_{INV}(t = 0) = v_s A_{CL}(0)/R_F$. The initial rate of change of the output voltage with time is therefore given by $SR = dv_o/dt \big|_{MAX} = I_{INV}(0)/C_T = v_s A_{CL}(0)/(R_F C_T)$. The final or steady-state value of the output voltage is given by v_o (final) $= v_s A_{CL}(0)$, so the slewing rate can now be expressed as $SR = v_o$ (final)/$R_F C_T$. Thus the slewing rate is not really limited since it will continue to go up as the signal level increases.

The small-signal closed-loop bandwidth is given approximately by $BW_{CL} = 1/(2\pi R_F C_T)$, so the slewing rate can be expressed as $SR = v_o$ (final) $\times (2\pi BW_{CL}) = v_o$ (final)ω_{CL}. Thus the slewing rate is given by the size of the output voltage step times the closed-loop radian bandwidth. Thus, for example, if $BW_{CL} = 100$ MHz and the output voltage step is 10 V, the slewing rate is given by $SR = 10\ V \times 2\pi \times 100$ MHz $= 6.3 \times 10^9$ V/s $= 6.3$ V/ns or 6300 V/μs. We see that this is a very high slewing rate and is about one or two orders of magnitude above the values obtained for the conventional VCVS operational amplifiers. Indeed, it may be stated that the CFB operational amplifiers are really not subject to slew-rate limiting at all, since the maximum rate of change of output voltage with time is not fixed by any internal current sources, but rather is a function of the output voltage.

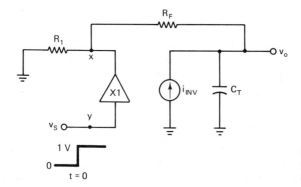

Figure 9.9 Circuit for the analysis of the slewing rate of a current-feedback operational amplifier.

9.1.7 Rise Time and Settling Time

From the analysis given previously for the frequency-response characteristics of the CFB operational amplifier, the following Laplace transform expression was obtained:

$$A_{CL}(s) = \frac{V_O(s)}{V_1(s)} = \frac{-1}{R_1 C_T} \times \frac{1}{s + 1/(R_F C_T)} = \frac{-R_F/R_1}{1 + sR_F C_T} \qquad (9.34)$$

The time-domain response can be obtained from this by taking the inverse transform. For a voltage step function input of amplitude V_s, we have that $V_1(s) = V_s/s$, so that $V_o(s) = A_{CL}(s)V_1(s) = A_{CL}(s)V_s/s$, and thus $V_o(s) = -(R_F/R_1)V_s/[s(1 + sR_F C_T)] = A_{CL}(0)V_s/[s(1 + sR_F C_T)]$. Taking the inverse transform of this to obtain the output voltage as a function of time yields $v_o(t) = A_{CL}(0)V_s\{1 - \exp[-t/(R_F C_T)]\}$. This is the familiar "single time constant" type of time-domain response. The 10% to 90% rise time is given by $t_{\mathrm{RISE}} = 2.2(R_F C_T)$. We might also note that the initial rate of change of output voltage with time is given by $dv_o/dt\,|_{t=0} = A_{CL}(0)V_s/(R_F C_T)$, and the expression for the slewing rate has been given as $SR = v_o\,(\text{final})/R_F C_T$. For the case under consideration, $v_o\,(\text{final}) = A_{CL}(0)V_s$, so $SR = A_{CL}(0)V_s/R_F C_T$, and thus there is really no difference between the small-signal response characteristics and the large-signal or slewing-rate-limited response. This again shows that, in contrast to VCVS operational amplifiers, CFB operational amplifiers do not really become slewing rate limited.

The settling time of an operational amplifier is the time required for the output voltage to settle within a specified percentage range or band that is centered about the final or steady-state value of the output voltage in response to a step function or rectangular pulse input. We have seen that the CFB operational amplifier exhibits an exponential time-domain response as given by $v_o(t) = A_{CL}(0)V_s\{1 - \exp[-t/(R_F C_T)]\} = A_{CL}(0)V_s[1 - \exp(-t/\tau)]$, where the time constant τ is given by $\tau = R_F C_T = 1/(2\pi BW_{CL})$.

Let us now consider an example of the calculation of the 0.1% settling time t_s for which we obtain $1 - \exp(-t_s/\tau) = 0.999$. Therefore, $\exp(t_s/\tau) = 1000$ and $t_s = \tau \ln(1000) = 6.9\tau$, so $t_s = 6.9/(2\pi BW_{CL}) \cong 1/BW_{CL}$. If $BW_{CL} = 100$ MHz, we obtain a 0.1% settling time of $t_s(0.1\%) \cong 1/BW_{CL} = 1/(100\text{ MHz}) = 10^{-8}s = 10$ ns. The actual settling time is usually somewhat longer than this due to various parasitic capacitances and other effects. Actual 0.1% settling times are generally in the 10- to 20-ns range for high-speed CFB operational amplifiers.

9.1.8 CFB Output Resistance

The analysis of the CFB operational amplifier carried out thus far has not indicated any dependence of the closed-loop gain on the load resistance. That will now be investigated. The transfer resistance r_T has been given as $r_T = 1/g_T = (1/g_{o9} + g_{o10} + g_{i13} + g_{i14})$. Since $g_{i13} + g_{i14}$ is due principally to the transformed value of the load resistance R_L, we will express the sum of these two input conductances as $g_{i13} + g_{i14} = G_L' = 1/R_L'$, so r_T becomes $r_T = 1/(g_{o9} + g_{o10} + G_L')$.

The output voltage is given by

$$V_O = \frac{(1 + R_F/R_1)V_2 - (R_F/R_1)V_1}{1 + R_F/r_T} = \frac{N}{D} \qquad (9.35)$$

For the denominator D, we have that $D = 1 + R_F[(1/r_{o_9}) + (1/r_{o_{10}}) + (1/R_L')]$. Since $r_{o_9} \gg R_F$ and $r_{o_{10}} \gg R_F$, the denominator can be written as $D \cong 1 + R_F/R_L'$, and therefore V_O can be expressed as

$$V_O = \frac{N}{D} = \frac{N}{1 + (R_F/R_L')} = N\frac{R_L'}{R_L' + R_F} \qquad (9.36)$$

Since $R_L' = \beta_{\text{NET}}R_L$, where $\beta_{\text{NET}} = \beta_{13,14}\beta_{15,16}$, this can be rewritten as

$$V_O = N\frac{\beta_{\text{NET}}R_L}{\beta_{\text{NET}}R_L + R_F} = N\frac{R_L}{R_L + (R_F/\beta_{\text{NET}})}$$

$$= N\frac{R_L}{R_L + r_O} \qquad (9.37)$$

where r_O is the effective closed-loop output resistance. Thus V_O can be expressed as

$$V_O = [(1 + R_F/R_1)V_2 - (R_F/R_1)V_1][R_L/(R_L + r_O)] \qquad (9.38)$$

From this last expression we see that, as long as the load resistance R_L is large compared to r_O, the load resistance will not have a significant effect on the voltage gain. If, for example, $R_F = 1000\ \Omega$ and $\beta = 100$ (min.) for all transistors, we obtain $r_O = 1000\ \Omega/[10,000\ \text{(min.)}] = 0.1\ \Omega$ (max.), and thus for any practical value of load resistance R_L the transfer function of the system is relatively independent of R_L.

9.1.10 Commercially Available CFB Operational Amplifiers

Table 9.2 lists some important characteristics of a variety of commercially available CFB operational amplifiers. The specifications are typical unless otherwise indicated. From this table we see the very impressive performance characteristics of the CFB operational amplifiers. It may be noted that the full-power bandwidth is up in the same range as the small-signal bandwidth, which is another indication of the fast slewing performance of these devices. Although the bandwidth values are given for a specified voltage gain, the bandwidth will not change very much with voltage gain. An example of this is the CLC 220A. For this device, typical values of the ac performance characteristics are given in Table 9.3. We see indeed that the bandwidth and the other ac parameters do not change much with the voltage gain.

Some of these CFB OAs are limited in terms of the maximum supply voltage. While many CFB OAs can operate with supply voltages up in the ±15- or ±20-V range, some are limited to ±5 or ±6 V. The low supply voltage rating is due to the low collector-to-emitter breakdown voltages of the high-speed, narrow base width transistors. An example of this is the AD9611 (Analog Devices), which has a maximum supply voltage rating of ±6 V and offers a small-signal bandwidth of 280 MHz at a gain of -5 and a full-power bandwidth of 210 MHz, but with just a 3-V peak-to-peak output voltage swing.

TABLE 9.2 COMMERCIALLY AVAILABLE CFB OPERATIONAL AMPLIFIERS

Mfr./Device	Bandwidth at A_v (MHz)		SR (V/ns)	Rise Time (ns) (step size)	Settling time (ns) (band)	FPBW (MHz) (V P-P)	Feedback Resistor (Ω)
Analog Devices							
AD 846	46	−1	0.45	10	110 (0.01%)	6.8 (20 V)	1000, external
AD 9610	100	−10	3.5	3.5 (5 V)	30 (0.02%)		1500, internal
AD 9611	280	−5	1.9	1.4 (3 V)	16 (0.05%)	210 (3 V)	1000, internal
Apex MicroTech							
WA-01	110	+30	4	3 (20 V)	20 (0.1%)	140 (20 V)	1500, internal
Comlinear							
CLC 103A	150	+20	6	4 (20 V)	10 (0.4%)	80 (20 V)	1500, internal
CLC 200A	95	+20	4	4 (20 V)	23 (0.1%)	25 (20 V)	2000, internal
CLC 201A	95	+20	4	4 (20 V)	18 (0.1%)	95 (10 V)	2000, internal
CLC 203A	170	+20	6	4 (20 V)	15 (0.2%)	62 (20 V)	1500, internal
CLC 205A	170	+20	2.4	4.8 (10 V)	22 (0.1%)	105 (10 V)	2000, internal
CLC 206A	180	+20	3.4	7 (20 V)	19 (0.1%)	70 (20 V)	2000, internal
CLC 220A	190	+20	7	1.9 (2 V)	10 (0.1%)		1500, internal
CLC 221A	170	+20	6.5	2.1 (2 V)	15 (0.1%)	125 (10 V)	1500, internal
CLC 231A	130	+5	3	2.5 (2 V)	12 (0.1%)	95 (10 V)	250, external
CLC 300A	85	+20	3	7 (20 V)	20 (0.8%)	40 (20 V)	1500, internal
CLC 400	150	+2	0.43	10 (5 V)	10 (0.1%)		250, external
CLC 401	150	+20	0.8	5 (5 V)	15 (0.1%)		1500, external
CLC 404	185	+6	2	2.6 (5 V)	10 (0.2%)		500, external
CLC 500	150	+2	0.8	5 (5 V)	10 (0.1%)		250, external
CLC 501	75	+32	1.2	5.5 (5 V)	12 (0.05%)		1500, external
Elantec							
EL 2020	30	+10	0.5	25 (1 V)	5.5 (1%)	8 (10 V)	680, external
EL 2022	165	+2	1.9	4.3 (10 V)			250, external
PMI	8 (min.) 5 (min.)	1 to 50 100	0.2				

TABLE 9.3

Parameter	Closed-loop gain					
	+4	+20	+50	-4	-20	-50
Bandwidth (MHz)	250	190	120	200	190	120
Rise time (ns) (2-V step)	1.6	1.9	2.3	1.6	1.9	2.3
Slewing rate (V/ns)	7	7	7	7	7	7
Settling time (0.1%)	10	8	10	8	8	10

It should be noted that devices listed with internal feedback resistors (R_F) have pins available at both ends so that external resistors can be added in series or in parallel with the internal R_F, so the net value of the feedback resistance can be made either smaller or larger than the internal R_F.

9.2 NORTON OPERATIONAL AMPLIFIERS

The operational amplifiers described thus far use a differential-amplifier input stage to produce an output voltage that is proportional to the difference-mode input voltage. In the Norton operational amplifier, a current-mirror circuit is used at the input to produce an output voltage for the amplifier that is proportional to the difference between the two input currents. This characteristic leads to many interesting applications, and the circuit configuration of the Norton amplifier is such that it can easily be biased from a single dc supply. In addition, a very large input voltage range can be obtained by the use of large-value resistors placed in series with the two amplifier input terminals.

The circuit symbol of the Norton operational amplifier is shown in Figure 9.10. Figure 9.11 shows the circuit diagram of the LM3900 (National Semiconductor) Norton amplifier. The LM3900 is actually a quad operational amplifier, and it can be operated with a single supply voltage as low as 4 V and as high as 36 V. It is internally compensated by means of capacitor C_1 to produce a unity-gain frequency of 2.5 MHz.

In the LM3900 circuit, transistors Q_1 and Q_2 constitute a simple current-mirror circuit, and if we neglect the base currents, we have that $I_{C_2} = I^+$. The base current of Q_3 is therefore be equal to the difference of the two input currents as given by $I_{B_3} = I^- - I_{C_2} = I^- - I^+$.

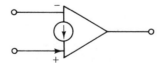

Figure 9.10 Norton amplifier circuit.

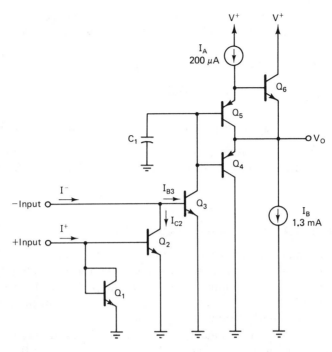

Figure 9.11 Norton amplifier (LM3900) circuit diagram (National Semiconductor).

This amplifier has a single gain stage in which Q_3 is used as a common-emitter amplifier. Transistors Q_5 and Q_6 are a compound PNP–NPN pair of emitter followers for the impedance transformation of the load resistance that is driven by this amplifier. The large impedance transformation ratio that is produced by this compound emitter-follower stage results in a very large voltage gain for the Q_3 amplifier stage.

Current sink I_B (1.3 mA) is used to sink output currents, so this circuit can both source as well as sink currents from the load. Under large-signal conditions, transistor Q_4 will be turned on to provide an additional current sinking capability.

For a simple voltage gain analysis of this circuit, we will assume a current gain of $\beta = 50$ (min.) for the NPN transistors. We will also assume that the voltage gain of the Q_5–Q_6 emitter-follower output stage is close to unity. The voltage gain provided by the Q_3 common-emitter gain stage is $A_V = -g_{m3} R_L'$, where R_L' is the transformed value of the load resistance R_L as given by $R_L' = \beta_5\beta_6 R_L$. The transfer conductance of Q_3 is given by

$$g_{m3} = \frac{I_{C3}}{V_T} = \frac{I_{B5}}{V_T} \simeq \frac{I_{E5}}{\beta_5 V_T} = \frac{200\ \mu A}{\beta_5 V_T} \tag{9.39}$$

Thus the voltage gain is given by

$$A_V = -g_{m3}R_L' = -\frac{200\ \mu A}{\beta_5 V_T} \beta_5\beta_6 R_L = \frac{200\ \mu A(\beta_6 R_L)}{25\ mV} \tag{9.40}$$

$$= -8\ mS \times \beta_6 R_L$$

For a load resistance of $R_L = 10 \text{ k}\Omega$, this gives a voltage gain of $A_V = -8 \text{ mS} \times 50 \,(\text{min.}) \times 10 \text{ k}\Omega = 4000 \,(\text{min.}) = 70 \text{ dB} \,(\text{min.})$, which is close to the manufacturer's specification.

In terms of currents, we have that $I_{B_3} = I^- - I^+$ and

$$\Delta I_O \cong -I_{B_3} \times \beta_3 \beta_5 \beta_6$$

so that

$$\Delta I_O \cong \beta_3 \beta_5 \beta_6 \, (I^+ - I^-) \sim 1 \times 10^5 \, (I^+ - I^-)$$

Thus the current gain is very large, and under closed-loop conditions we have that $I^+ \cong I^-$. This is analogous to the condition that $V^+ \cong V^-$ for the VCVS type of op amp.

9.2.1 Norton Amplifier Closed-loop Analysis

For the general closed-loop gain analysis of the Norton operational amplifier, we will use the circuit of Figure 9.12 and reference will also be made to Figure 9.11. The voltage levels at the two input nodes of the amplifier are both one base-to-emitter voltage drop ($V_{BE} \simeq 0.6 \text{ V}$) above ground. As a result of the very high gain of the circuit, the action of the feedback loop is such that the two input currents are essentially equal, so $I^+ = I^-$. This is analogous to the case of the ordinary operational amplifier, where the action of the feedback loop causes the voltages at the two input terminals to be essentially equal.

In general, the current through resistor R_N is given by $I_N = (V_N - V_{BE})/R_N$, where $N = 1, 2, 3, \ldots$. To simplify the writing of the equations, we will define V_N' as $V_N' = V_N - V_{BE}$. At node 1 we have

$$V_1' G_1 + V_3' G_3 + V_5' G_5 = I^+ \qquad (9.41)$$

and at node 2

$$V_o' G_F + V_2' G_2 + V_4' G_4 + V_6' G_6 = I^- \qquad (9.42)$$

Since $I^+ = I^-$, we obtain

$$V_o' G_F = V_1' G_1 + V_3' G_3 + V_5' G_5 - (V_2' G_2 + V_4' G_4 + V_6' G_6) \qquad (9.43)$$

so that

$$V_o' = \frac{V_1' G_1}{G_F} + \frac{V_3' G_3}{G_F} + \frac{V_5' G_5}{G_F} - \frac{V_2' G_2}{G_F} - \frac{V_4' G_4}{G_F} - \frac{V_6' G_6}{G_F} \qquad (9.44)$$

In terms of resistances, this equation for V_o' can be rewritten as

$$V_o' = \frac{V_1' R_F}{R_1} + \frac{V_3' R_F}{R_3} + \frac{V_5' R_F}{R_5} - \frac{V_2' R_F}{R_2} - \frac{V_4' R_F}{R_4} - \frac{V_6' R_F}{R_6} \qquad (9.45)$$

The input voltage for any input signal that is active ranges down to a minimum value of $V_{BE} \simeq 0.6 \text{ V}$. For the proper operation of the LM3900 Norton amplifier, the input currents I^+ and I^- should not go above 6 mA. As a result, the maximum input voltage will be limited solely by the requirement that neither input current

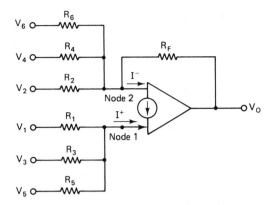

Figure 9.12 Circuit for the closed-loop analysis of the Norton operational amplifier.

exceed 6 mA. Therefore, with the choice of suitable large resistance values, very large input-voltage levels can be accommodated by this circuit.

For a simple example of a Norton amplifier application, let us consider the inverting ac amplifier shown in Figure 9.13. In this circuit the dc or quiescent output voltage level will be set at

$$\frac{V_{O(\text{DC})} - V_{BE}}{R_F} = \frac{V^+ - V_{BE}}{2R_F} \tag{9.46}$$

and thus

$$V_O = \frac{V^+}{2} + \frac{V_{BE}}{2} \cong \frac{V^+}{2}$$

This setting of the quiescent output voltage to a point approximately midway between the positive supply and ground allows for the maximum possible symmetrical output voltage swing. The ac voltage gain is $A_V = v_o/v_s = -R_F/R_1$. Values for R_F in the range from 100 kΩ to 10 MΩ are generally satisfactory.

Another quad single-supply Norton operational amplifier is the MC3401 (Motorola). Its circuitry is similar to that of the LM3900 considered earlier. This amplifier has an open-loop voltage gain of 60 dB (min.) [66 dB (typ.)], and the 3-pF on-chip compensation capacitor results in a unity-gain frequency of 5.0 MHz. It can be operated from a single supply voltage over the range from 5 to 18 V. Other quad Norton op amps are the MC3301 (Motorola) and the LM2900, LM3301, and LM3401 (National Semiconductor).

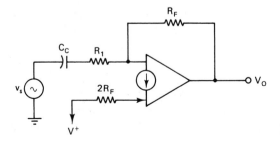

Figure 9.13 Inverting ac amplifier.

9.3 OPERATIONAL TRANSCONDUCTANCE AMPLIFIERS

The operational transconductance operational amplifier or OTA has a circuit design that is very much different from the conventional voltage-controlled voltage source or VCVS operational amplifier. The OTA is basically a voltage-controlled current source or VCCS. With reference to Figure 9.14, the transfer characteristic of the ideal OTA is given by $i_o = g_T(v_1 - v_2) = g_T v_i$, where i_o is the output current, g_T is the *dynamic forward transfer conductance* of the OTA, and $v_i = v_1 - v_2$ is the difference-mode input voltage of the device. The basic output quantity of the OTA is the current i_o, but it can easily be converted into an output voltage by means of a load resistance, such as is shown in Figure 9.14. For this circuit, $v_o = i_o R_L = g_T R_L v_i$, so the voltage gain is $A_v = v_o/v_i = g_T R_L$. Note that the OTA can be used in an open-loop configuration as shown in Figure 9.14 and can also be used in various closed-loop configurations that are similar to the VCVS operational amplifier. The transfer conductance of the OTA is controlled by an externally supplied *amplifier bias current*, I_{ABC}, and is given by $g_T = I_{ABC}/(2V_T) = I_{ABC}/50 \text{ mV} = (20/V) I_{ABC}$. This control of the OTA transfer conductance leads to many interesting applications for this device.

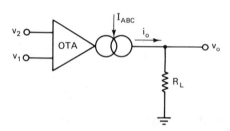

Figure 9.14 Basic operational transconductance amplifier circuit.

9.3.1 OTA Circuit Analysis

We will now investigate the internal circuitry of the OTA. Figure 9.15 is a simplified circuit diagram of an OTA. It consists of a differential amplifier using transistors $Q1$ and $Q2$ and four current mirrors comprised of $Q3-Q4$, $Q5-Q6$, $Q7-Q8$, and $Q9-Q10$. For this discussion, we will assume that all transistors of the same type are identical and we will neglect the base currents.

First considering the $Q3-Q4$ current mirror, we have that $i_4 = i_3 = I_{ABC}$. For the $Q1-Q2$ differential amplifier, we have that under small-signal conditions $i_1 = (i_4/2) + g_F v_i$ and $i_2 = (i_4/2) - g_F v_i$, where g_F is the dynamic forward transfer conductance of the differential amplifier, and v_i is the difference-mode input voltage as given by $v_i = v_{B1} - v_{B1}$. The transfer conductance g_F is given by $g_F = i_4/(4V_T)$, and since $i_4 = I_{ABC}$, this becomes $g_F = I_{ABC}/(4V_T)$. Since $i_8 = i_1$ and $i_{10} = i_6 = i_2$, the net output current i_o is given by $i_o = i_8 - i_{10} = i_1 - i_2 = 2g_F v_i = (I_{ABC}/2V_T)v_i = g_T v_i$, where $g_T = 2g_F = I_{ABC}/(2V_T) = I_{ABC}/50 \text{ mV} = (20/V)I_{ABC}$. Thus the OTA acts as a voltage-controlled current source.

The foregoing analysis is based on small-signal conditions with respect to the characteristics of the differential amplifier, such that the difference-mode input voltage

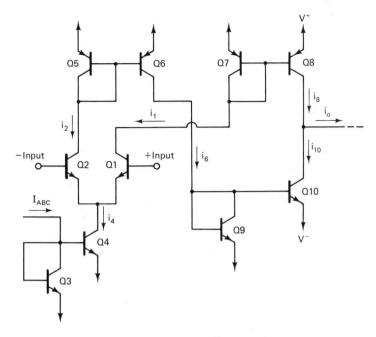

Figure 9.15 Simplified diagram of an operational transconductance amplifier.

is less than about 25 mV peak (or 50 mV peak-to-peak). To get around this small-signal limitation, OTAs are available with a modified differential-amplifier circuit that has a linearization circuit to greatly extend the dynamic input-voltage range for linear operation. This will be investigated later.

9.3.2 OTA Output Conductance

We have seen that the OTA can be modeled as a voltage-controlled current source. An important characteristic of any real approximation to the ideal constant current source is the dynamic output conductance, g_o. This is shown in the OTA model of Figure 9.16. From this circuit we see that the output voltage is given by $v_o = [g_T/(g_o + G_L)]v_i$, and thus the voltage gain is $A_v = g_T/(g_o + G_L)$, where $G_L = 1/R_L$ is the load conductance.

For the OTA circuit under consideration, the dynamic output conductance is the sum of the output or collector conductances of transistors $Q8$ and $Q10$. This is given by $g_o = g_{o8} + g_{o10} = (i_8/V_{A8}) + (i_{10}/V_{A10})$. Under quiescent conditions, $i_8 = i_{10} = i_4/2 = I_{ABC}/2$, so $g_o = (I_{ABC}/2)[(1/V_{A8}) + (1/V_{A10})]$. The maximum possible voltage gain attainable from an OTA is achieved under open-loop conditions and when R_L goes to infinity (and hence G_L goes to zero). Under these conditions, we have that $A_{v(\text{MAX})} = g_T/g_o$. If we insert into this last equation the expressions for g_T and g_o, the amplifier bias current I_{ABC} cancels out, and we obtain for A_v the equation $A_{v(\text{MAX})} = (1/V_T)/[(1/V_{A8}) + (1/V_{A10})]$. To consider a representative example, we will use a value of $V_A = 100$ V (MIN) for the Early voltages for both the

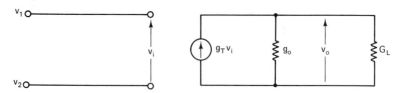

Figure 9.16 Operational transconductance amplifier equivalent circuit.

NPN and PNP transistors. With this choice, we have $A_{v(MAX)} = (V_A/2V_T) = [50\text{ V}$ (MIN)]$/25\text{ mV} = 2000$ (MIN). Although this is a moderately large gain value, for many applications a much larger value for the open-loop gain is required. This can readily be accomplished by replacing the simple current-mirror configuration in the circuit of Figure 9.15 by Wilson current mirrors. This is shown in Figure 9.17. This circuit operates in the same basic fashion as that considered earlier. The Wilson current mirror, however, provides for a much closer approximation to the ideal current

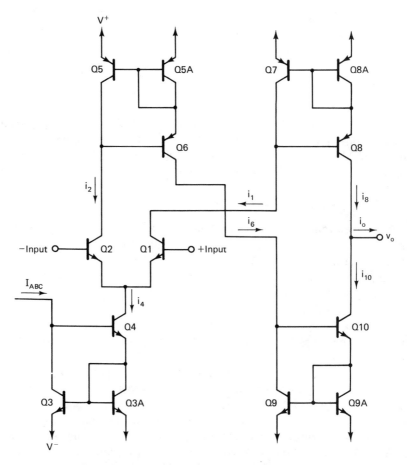

Figure 9.17 Operational transconductance amplifier with Wilson current mirrors.

source. The output conductance of the Wilson current mirror is shown in Chapter 3 to be given approximately by $g_o = I_c/(\beta V_A)$. We therefore see that the output conductance of the Wilson current mirror is approximately β times less than that of the simple two-transistor current mirror. Applying this to the case at hand, and still assuming that $V_{A8} = V_{A10} = V_A$, we have that $A_{v(MAX)} = g_T/g_o = [I_{ABC}/(2V_T)]/[2(I_{ABC}/2)/(\beta V_A)] = \beta V_A/(2V_T)$. If we assume $\beta = 50$ (MIN) and $V_A = 100$ V (MIN) for all transistors as before, we obtain a voltage gain of $A_{v(MAX)} = 50$ (MIN) $\times$ 100 V (MIN)/50 mV = 100,000 or 100 dB. We therefore see that open-loop gain values can be attained for the OTA that are comparable to the VCVS operational amplifier. To achieve these high gain values, a large load resistance is required. Some OTAs, such as the VA713 and VA2713 (VTC, Inc.) and the LM 13600 and LM 13700 have on-chip Darlington buffers that can be used with the OTA so that large open-loop gains are possible, even with moderate-size load resistances.

9.3.3 Biasing the OTA

The OTA is a programmable operational amplifier in the sense that the quiescent currents of the devices can be set or programmed by the external control of the amplifier bias current I_{ABC}. The total quiescent supply current for the OTA configuration of Figure 9.17 is given by $I_{SUPPLY} = 3I_{ABC}$, so for micropower applications a very low value of I_{ABC} can be chosen, even down to 1 μA. There is, however, a severe penalty to be paid for this micropower operation in the area of speed and bandwidth.

A simple circuit to set the OTA quiescent current level is shown in Figure 9.18, where we have that $I_{ABC} = (V^+ - V^- - 2V_{BE})/R_{ABC}$. For the case of a ± 5-V supply and a total quiescent supply current of 3 μA, the required value of $I_{ABC} = 1$ μA, and thus $R_{ABC} = (10$ V $- 1.2$ V)/1 μA $= 8.8$ MΩ. The quiescent power consumption of the OTA is only 30 μW. The major trade-off in this micropower operation is the loss of bandwidth and speed, since both are directly proportional to I_{ABC}. If, for example, $C_{NET} = 10$ pF and $I_{ABC} = 1$ μA, then the gain–bandwidth product or f_U will be 318 kHz and the slewing rate will be only 0.1 V/μs. In addition to the sluggish response of the OTA, there is also a penalty in terms of the open-loop gain. Since $g_T = I_{ABC}/2V_T = (20/V)I_{ABC}$, and $A_{VOL}(0) = g_T R_{NET} = (20/V)I_{ABC}R_{NET}$, there is a loss in the open-loop gains as I_{ABC} decreases, unless the value of the load resistance is correspondingly increased. This increase in the load

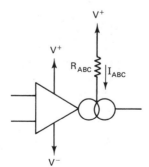

Figure 9.18 Circuit to set operational transconductance amplifier quiescent current level.

I_{ABC}	f_U (MHz)	SR (V/μs)
1 μA	0.318	0.1
10 μA	3.18	1.0
100 μA	31.8	10
1.0 mA	318	100

resistance, however, will not help to increase the gain–bandwidth product. Table 9.4 gives values of the gain–bandwidth product (f_U) and the slewing rate for various values of I_{ABC} for the case of C_{NET} = 10 pF. From this table we see that, although the device is quite slow in the micropower range, when the value of I_{ABC} is increased up to the milliampere range, very large bandwidths and slewing rates are attainable. With increasing values of I_{ABC}, the values of f_U and SR at first increase linearly with current up to about 1 mA, but beyond that start to level off due to other capacitances in the circuit and other effects.

For the closed-loop circuit shown in Figure 9.19a, the feedback resistance R_F has to be considered as part of the net load resistance R_{NET}, unless a buffer is used as shown in Figure 9.19b. For Figure 9.19a, the net load resistance driven by the OTA is $R_{NET} \cong [R_L \| R_F]$, so $A_{VOL}(0) = g_T R_{NET}$ and $A_{CL}(0) \cong 1 + (R_F/R_1)$ if $A_{VOL}(0) > A_{CL}(0)$. For the case of Figure 9.19b, where a buffer is used between the OTA output and the load, the open-loop gain is still given by $A_{VOL}(0) = g_T R_{NET}$, where R_{NET} is now $R_{NET} = [R_2 \| R_{IN} \text{(buffer)}]$; and since the buffer input resistance, R_{IN} (buffer), can be much greater than the load resistance R_L, a large open-loop can still be maintained, even down to relatively low values of I_{ABC}. Some OTAs have available on-chip Darlington emitter-follower buffers of the type shown in Figure 9.19c. For this type of buffer, we have that R_{IN} (buffer) $\cong \beta_1 \beta_2 R_L$, so impedance transformation ratios on the order of 1000 to 30,000 are generally possible; as a result, a large open-loop gain can be obtained, even at relatively low quiescent current levels. Note also that the buffers act to isolate the load capacitance from the OTA output node, so C_{NET} is reduced, which leads to a large gain–bandwidth product and a higher slewing rate.

If we look back at the OTA circuit diagram shown in Figure 9.15, we see that, aside from the input terminals, there is only one high-impedance node in the circuit, the output node at the junction of the collectors of $Q8$ and $Q10$. The other internal circuit nodes are low-impedance nodes since there is a forward-biased base–emitter junction between these nodes and ac ground. As a result, it is principally the net resistance and capacitance at the output node that determines the open-loop frequency response of the system.

To analyze the open-loop frequency response of the OTA, let us look at the ac small-signal model of Figure 9.20. In this diagram, g_o is the output conductance of

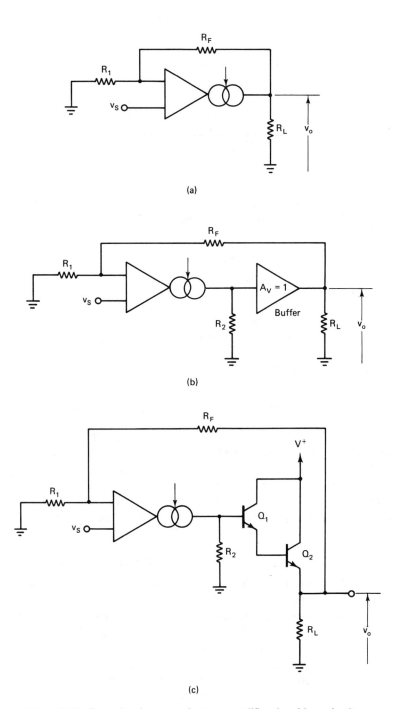

(a)

(b)

(c)

Figure 9.19 Operational transconductance amplifier closed-loop circuits.

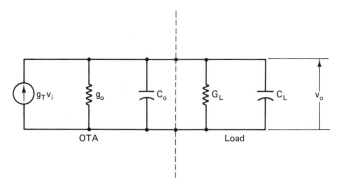

Figure 9.20 AC small-signal equivalent circuit for the analysis of operational transconductance amplifier frequency response.

the OTA circuit and is the sum of the output or collector conductances of $Q8$ and $Q10$, and C_o is the net output capacitance seen looking back into the OTA circuit and is given by $C_o = C_{CB_8} + C_{CB_{10}}$. The load driven by the OTA is represented by its input conductance G_L and capacitance C_L. If we let $g_{NET} = g_o + G_L$, $C_{NET} = C_o + C_L$, and $y_{NET} = g_{NET} + j\omega C_N$, we obtain $A_{VOL} = v_o/v_i = g_T/y_{NET} = g_T/(g_{NET} + j\omega C_{NET}) = (g_T/g_{NET})/[1 + j\omega(C_{NET}/g_{NET})]$. The zero-frequency value of the gain is $A_{VOL}(0) = (g_T/g_{NET})$. If we let $\omega_1 = g_{NET}/C_{NET}$ and $f_1 = g_{NET}/(2\pi C_{NET})$, then the equation for the open-loop gain can be written as

$$A_{VOL}(f) = \frac{A_{VOL}(0)}{1 + j(\omega/\omega_1)} = \frac{A_{VOL}(0)}{1 + j(f/f_1)} \tag{9.47}$$

The open-loop bandwidth is therefore given by $BW_{OL} = f_1 = g_{NET}/(2\pi C_{NET})$. Since it is usually the case that $G_L \gg g_{NET}$, we have that $BW_{OL} \cong G_L/(2\pi C_{NET}) = 1/(2\pi R_L C_{NET})$. For the case of $f \gg f_1 = BW_{OL}$, we have that $A_{VOL} \cong A_{VOL}(0)/(jf/f_1) = A_{VOL}(0)f_1/jf$. From this we see that the unity-gain frequency f_U, or gain–bandwidth product (GBW), is given by $f_U = A_{VOL}(0)f_1 = A_{VOL}(0)BW_{OL}$. If we make appropriate substitutions for the open-loop gain and bandwidth, we obtain

$$f_U = \text{GBW} = \frac{g_T}{g_{NET}} \times \frac{g_{NET}}{2\pi C_{NET}} = \frac{g_T}{2\pi C_{NET}} \tag{9.48}$$

Note that these results are similar to those obtained for the conventional VCVS type of operational amplifier.

The closed-loop gain–bandwidth product is equal to the open-loop gain–bandwidth product, so $A_{CL}(0)BW_{CL} = f_U$, so $BW_{CL} = g_T/[2\pi C_{NET}A_{CL}(0)]$. Since $g_T = (20/V)I_{ABC}$, this can be expressed as $BW_{CL} = (20/V)I_{ABC}/[2\pi C_{NET}A_{CL}(0)]$; so we see that the closed-loop bandwidth can be controlled directly by the amplifier bias current I_{ABC}. This can lead to some interesting applications, such as an active low-pass filter with the cutoff frequency controlled by I_{ABC}, which in turn can be controlled by an externally supplied control voltage.

When operating an operational amplifier under closed-loop conditions, the dif-ference-mode input voltage that appears across the terminals of the input-stage differential amplifier is equal to the output voltage divided by the open-loop gain and thus is generally very small, usually less than a few millivolts. As a result, the differential amplifier operates under small-signal conditions as a linear device. The OTA can be operated under closed-loop conditions, for which case the small-signal conditions will hold true. The OTA, however, can also be operated under open-loop conditions with a load resistance value such that the open-loop gain is a relatively low value. As a result, the difference-mode input voltage may no longer satisfy the small-signal conditions, especially if it exceeds a 25-mV peak value. This will lead to a nonlinear transfer characteristic and the signal distortion that results.

It is possible to improve the linearity of a differential amplifier. One simple way of doing this is to insert resistors in series with the emitters of the two transistors making up the differential amplifier, as shown in Figure 9.21a. These emitter resistors produce a negative feedback effect (*emitter degeneration*) and can produce a very substantial improvement in the linearity, but at the expense of a reduced transfer conductance. A similar technique can be used with FET differential amplifiers, as shown in Figure 9.21b. Another technique to improve linearity is to use a *diode linearization circuit*, an example of which is shown in Figure 9.22. To understand how this works, let us first review the basic transfer relationship for the differential amplifier as given by $I_1 = I_Q/\{1 + \exp[(V_{B2} - V_{B1})/V_T]\}$ and $I_2 = I_Q/\{1 + \exp[(V_{B1} - V_{B2})/V_T]\}$. We note the nonlinear relationship between the difference-mode input voltage, $V_i = V_{B1} - V_{B2}$, and the two collector currents. The diode linearization circuit is basically a logarithmic compression circuit that compensates for the exponential expansion in the differential-amplifier transfer equations.

To analyze the linearization circuit of Figure 9.22, we will assume that all transistors are identical. We note that $V_{B1} - V_{B2} = V_{BE3} - V_{BE4}$. Since $I_3 + I_4 = 2I_D$ and $I_3 = I_D + I_S$, we have that $I_4 = I_D - I_S$. We also note that $V_{BE3} = V_T \ln (I_3/I_{CO})$ and $V_{BE4} = V_T \ln (I_4/I_{CO})$, so $V_{B1} - V_{B2} = V_{BE3} - V_{BE4} = V_T \ln (I_3/I_4) = V_T \ln [(I_D + I_S)/(I_D - I_S)]$. If we now substitute this result for $V_{BE1} - V_{BE2}$ back

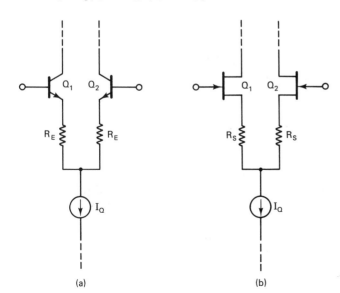

Figure 9.21 Differential-amplifier linearization.

(a)   (b)

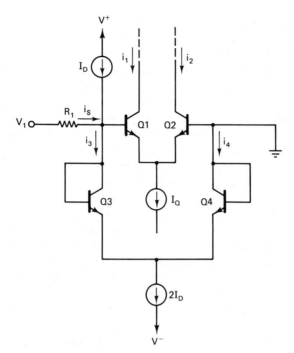

Figure 9.22 Differential-amplifier linearization circuit.

into the expressions for I_1 and I_2, we obtain

$$I_2 = \frac{I_Q}{1 + \exp{(V_{B1} - V_{B2})/V_T}} = \frac{I_Q}{1 + \exp{\{\ln{[(I_D + I_S)/(I_D - I_S)]}\}}}$$

$$= \frac{I_Q}{1 + (I_D + I_S)/(I_D - I_S)} = \frac{(I_D - I_S)I_Q}{2I_D} \qquad (9.49)$$

and similarly for I_1 the result is

$$I_1 = \frac{I_Q}{1 + \exp{(V_{B2} - V_{B1})/V_T}} = \frac{I_Q}{1 + \exp{\{\ln{[(I_D - I_S)/(I_D + I_S)]}\}}}$$

$$= \frac{I_Q}{1 + (I_D - I_S)/(I_D + I_S)} = \frac{(I_D + I_S)I_Q}{2I_D} \qquad (9.50)$$

Since $I_Q = I_{ABC}$ and the OTA output current is given by $I_O = I_1 - I_2$, we obtain $I_O = I_1 - I_2 = I_S I_{ABC}/I_D$. Thus the output current of the OTA is a linear function of the signal current I_S.

The impedance that is seen looking into this circuit is the sum of the dynamic resistances of the two linearization diode-connected transistors $Q3$ and $Q4$. This is given by $r_{IN} = 2V_T/I_D$, subject to the condition that $I_S \ll I_D$, although the condition that $I_S < I_D/3$ usually proves to be sufficient. Recommended values for I_D are usually in the 1- to 5-mA range. The current I_S is given by $I_S = V_S/(R_1 + r_{IN})$, and if R_1 satisfies the condition that $R_1 \gg r_{IN}$, then $I_S \cong V_S/R_1$. If we consider a value of 2 mA for I_D, the input resistance is $r_{IN} = 50$ mV/2 mA $= 25\ \Omega$; so for R_1 values greater than about 250 Ω this should produce very good linearity.

Since $I_O = I_S I_{ABC}/I_D$ and noting that $I_S = V_S/(R_1 + r_{IN})$ and $V_O = I_O R_L$, we can therefore obtain that $A_V = V_O/V_S = (I_{ABC}/I_D)[R_L/(R_1 + r_{IN})]$. If $R_1 >> r_{IN}$, as is the usual case, then this reduces to simply $A_V = (I_{ABC}/I_D)(R_L/R_1)$. This linear dependence of the voltage gain on I_{ABC} and I_D can be used for many interesting applications, including automatic gain control, multipliers, and amplitude-modulation circuits.

Commercially available OTAs include the LM 13600 and LM 13700 from National Semiconductor, the CA3060 and CA3280 from RCA, the NE5517 from Signetics and the VA703, VA713, VA2703, and VA2713 from VTC, Inc.

The LM 13600 is a dual OTA with linearizing diodes, includes two Darlington buffers, and comes in a 16-pin DIP package. The OTA circuit configurations use Wilson current mirrors and are similar to that of Figure 9.17. They can operate over a supply voltage range of from ± 2 to ± 22 V. The transfer conductance g_T can be varied over a range of 6 decades with excellent linearity by varying I_{ABC} from 2 nA up to the rated maximum of 2 mA. The output resistance is inversely proportional to I_{ABC} and has a value of 20 MΩ at $I_{ABC} = 100$ μA. The input and output capacitances are relatively independent of I_{ABC} and both have values of approximately 4.5 pF at a supply voltage of ± 15 V.

The CA3060 contains three OTAs in one IC 16-pin DIP package. It can operate with amplifier bias currents from 1 μA up to a maximum of 2 mA. With a supply voltage of ± 15 V, the output resistance is 200 MΩ (typ.) at $I_{ABC} = 1$ nA and is inversely proportional to I_{ABC}. The output capacitance is 4.5 pF (typ.), and the open-loop bandwidth is 20 kHz at $I_{ABC} = 1$ μA, 45 kHz at 10 μA, and 110 kHz at 100 μA. The CA3280 is a dual OTA in a 16-pin DIP package and has linearizing diodes. It can operate with amplifier bias currents from 1 μA up to a maximum of 10 mA. At $I_{ABC} = 100$ μA, the output resistance is 63 MΩ (typ.), the output capacitance is 7.5 pF (typ.), and the open-loop voltage gain is 100 dB under no-load conditions. With $I_{ABC} = 1$ mA and a load resistance of 100 Ω, the open-loop bandwidth is 9 MHz (typ.) and at this bias current level a slewing rate of 125 V/μs is obtained.

The VA703 is a high-speed OTA with linearizing diodes and comes in 8-pin DIP and metal can packages. It has a circuit configuration similar to that of Figure 9.17 and uses Wilson current mirrors. With a load resistance of 200 Ω and a load capacitance of 10 pF (max.), the open-loop bandwidth values are 10 MHz at $I_{ABC} = 20$ μA, 40 MHz at 100 μA, and 85 MHz at 1 mA and levels off at 90 MHz for currents above 2 mA. The VA713 is similar to the VA703, except that it includes a push–pull Darlington buffer. The VA2703 and VA2713 are dual versions of the VA703 and VA713, respectively.

9.4 DUAL, QUAD, AND SINGLE-SUPPLY OPERATIONAL AMPLIFIERS

There are many types of op amps available with various particular features, such as low offset voltage, low input bias current, low dc power supply current, wide bandwidth, high slewing rate, and low noise. The FET-input op amps discussed in previous sections offered some of these features.

Most op amps require both a positive and a negative dc supply voltage for proper operation. There are some op amps, however, that can be operated from a single positive dc supply voltage. Furthermore, the single dc supply voltage can be as small as +5 V. This can be a great convenience in many applications. Another category of op amps are the dual and the quad op amps. The dual op amp has two electrically independent op amps in one IC package, and the quad op amp offers the convenience of four independent op amps in the same package. In most cases, the op amps share the same silicon chip. Quad op amps are available in a 14-pin DIP package. The input and output terminals of the four op amps are separate, but they do share common positive dc supply and ground pins. Since the input, output, positive supply, and ground connections use up all the available pins, such features as offset nulling pins and pins for an external compensation capacitor are not present.

Since the four op amps generally share a common silicon chip, there will inevitably be some amount of electrical coupling between the op amps. This *amplifier-to-amplifier coupling*, or channel separation, is very small and generally down in the range of about −120 dB.

An example of an IC that is a quad op amp and can also be used with just a single positive dc supply voltage is the LM124/224/324 series (National Semiconductor). A schematic of this device is shown in Figure P6.2. This device can be operated from either single or dual supply voltages from as low as +3 V (or +1.5 and −1.5 V) to as high as +30 V (or +15 and −15 V). It is an internally compensated op amp with a unity-gain frequency of 1 MHz, a slewing rate of 0.5 V/μs, and a dc open-loop gain of 100 dB. The coupling between the four op amps on the chip is very small, the amplifier-to-amplifier coupling being down to about −120 dB.

Another interesting example of a quad op amp is the TL074/075 series (Texas Instruments), which comes in a 14-pin DIP package. Other devices in this same series are the TL072, which is a dual op amp in a mini-DIP 8-pin package, and the TL070 and TL071, also in 8-pin mini-DIP packages. These are not single-supply op amps, but do offer the special feature of a JFET input stage. The JFET input stage uses a pair of P-channel JFETs that are biased by a current source and drive a current-mirror active load. The second stage is a Darlington common-emitter configuration and has a current-source active load. The output stage is a complementary push–pull emitter follower with two diode-connected transistors used for the class AB biasing of the output stage to minimize the crossover distortion.

The JFET input stage results in a low input bias current of 30 pA and an offset current of 5 pA. The unity-gain frequency is 3 MHz and the slewing rate is 13 V/μs. This device is also characterized as a low-noise op amp. This is due principally to the low input bias current, which results in an input noise current of only 0.01 pA/$\sqrt{\text{Hz}}$. The equivalent input noise voltage is 18 nV/$\sqrt{\text{Hz}}$. The amplifier-to-amplifier coupling is rated at −120 dB for both the dual and the quad op amps.

An example of an all-bipolar transistor, single-supply quad op amp that offers a wide bandwidth and high slewing rate capability is the MC34074 (Motorola). This op amp has a unity-gain bandwidth of 4.5 MHz and a slewing rate of 13 V/μs. It can operate with a single positive dc supply voltage from as low as +3 V up to a maximum of +44 V and comes in a 14-pin DIP package.

9.5 HIGH-PERFORMANCE OPERATIONAL AMPLIFIERS

Many op amps exhibit various high-performance characteristics, such as very low input offset voltage (precision op amps), ultra-low input bias current, very wide bandwidth, very high slewing rate, very low quiescent current drain (micropower op amps), high voltage, and high output current. Examples of a few of these op amps will now be presented.

9.5.1 Precision Operational Amplifiers

A precision op amp is one that has a very low input offset voltage, generally less than 1 mV. Along with a low offset voltage, the temperature coefficient of the offset voltage is very small, generally less than 5 μV/°C. A good example of a precision op amp is the OP-77 (Precision Monolithics, Inc.), which has an offset voltage of only 10 μV typical and 25 μV maximum. The temperature coefficient of the offset voltage is only 0.1 μV/°C typical and 0.3 μV/°C maximum, and the long-term drift of the offset voltage is specified as 0.2 μV/month typical. This is a bipolar transistor op amp that uses laser-trimmed load resistances in the input differential amplifier stage to minimize the offset voltage. Two offset null pins are available for nulling out the offset voltage. A 20-kΩ potentiometer can be connected between the offset null pins with the wiper arm connected to the positive supply for the offset voltage cancellation. Also available from Precision Monolithics, Inc., are the OP-07, OP-27, OP-37, and OP-97 with similar input offset voltage characteristics.

Other examples of precision op amps are the LT1008C (Linear Technology, Inc.), 3510 (Burr-Brown), LM108A/208A/308A series (National Semiconductor), LM11C (National Semiconductor), AD547 (Analog Devices), AD611KH (Analog Devices), and LT1013 (Texas Instruments).

9.5.2 Ultra-low Input Current Operational Amplifiers

Operational amplifiers with FET input stages generally have input bias currents down in the range from 10 to 100 pA. An op amp with an input current below 10 pA can be considered to have an ultra-low input bias current. A good example of such an op amp is the OPA104 series (Burr-Brown) with maximum input bias currents less than 1 pA at 25°C. The OPA104CM has an input bias current of a mere 75 fA (0.075 pA) maximum at 25°C. Associated with the ultra-low input bias currents is a very high input impedance. The difference-mode input impedance is 10^{14} Ω in parallel with a capacitance of 0.5 pF, and the common-mode input impedance is specified as 10^{15} Ω in parallel with a capacitance of 1.0 pF. This op amp also has a relatively low input offset voltage of 200 μV typical and 500 μV maximum, with a temperature coefficient of 5 μV/°C typical and 10 μV/°C maximum.

Examples of other op amps with input bias currents of 1 pA or less, together with the input bias current specification (max.), are as follows: AD515L (Analog Devices), 75 fA; 3523L (Burr-Brown), 100 fA; AD515K (Analog Devices), 150 fA; 3523K (Burr-Brown), 250 fA; AD515J (Analog Devices), 300 fA; 3523J (Burr-Brown), 500 fA; AD545K/L/M (Analog devices), 1 pA; 3522L (Burr-Brown), 1 pA; and CA3420A (RCA), 1 pA.

9.5.3 Wide Bandwidth and Fast Slewing Operational Amplifiers

Most op amps have unity-gain frequencies (or gain–bandwidth products) in the range from 1 to 3 MHz with slewing rates in the range from 0.5 to 5 V/μs. There are many op amps that can be characterized as very wide bandwidth and fast slewing op amps. The CFB op amps are good examples of this. Another good example is the HA2539 (Harris Semiconductor) with a gain–bandwidth product of 600 MHz and a slewing rate of 600 V/μs. The NE5539 (Signetics) is a very wide bandwidth monolithic bipolar operational amplifier with a gain–bandwidth product of 1200 MHz at a gain of 7 and a slewing rate of 600 V/μs. The full-power bandwidth is 20 MHz with a peak-to-peak output of 3 V into a 150-Ω load. A very wide bandwidth hybrid op amp is the 3554 (Burr-Brown) with a gain–bandwidth product of 1700 MHz (at a gain of 1000) and a slewing rate of 1000 V/μs. The small-signal bandwidth is 22.5 MHz at a gain of 10, 7.25 MHz at a gain of 100, and 1.7 MHz at a gain of 1000. The full-power bandwidth is 19 MHz when driving a 100-Ω load with a 20-V peak-to-peak output voltage. Another wide bandwidth hybrid op amp is the LH0032 (National Semiconductor). This JFET-input op amp has a unity-gain frequency of 70 MHz and a slewing rate of 500 V/μs along with an input impedance of 10^{12} Ω and an input bias current of just 10 pA.

In Table 9.2 there is a listing of CFB op amps that offer very wide bandwidths and very high slewing rates. Some other wide bandwidth op amps together with their typical unity-gain bandwidths and slewing rates are given in Table 9.5.

TABLE 9.5

Op Amp	Manufacturer	Bandwidth (MHz)	Slewing Rate (V/μs)
3551J	Burr-Brown	50	250
AD380J	Analog Devices	40	330
OP37G	Precision Monolithics, Inc.	40	17
LF357	National Semiconductor	20	50
LM318	National Semiconductor	15	70

9.5.4 High-voltage Operational Amplifiers

Most op amps have maximum total supply voltage ratings of no more than 36 V and, correspondingly, a peak-to-peak output voltage swing that is limited to about 32 V. There are high-voltage op amps available, however, with supply voltage ratings much greater than this. An example of a high-voltage op amp is the 3584 (Burr-Brown) with a total supply voltage rating of 300 V and a maximum peak-to-peak output voltage swing of 290 V. This JFET input op amp offers a unity-gain bandwidth of 7 MHz, a slewing rate of 150 V/μs, and a full-power bandwidth of 135 kHz.

Other examples of high-voltage op amps together with the maximum supply voltages are as follows: PA85/PA88 (Apex Microtechnology), ±225 V; PA08V (Apex

Microtechnology), ± 175 V; PA08A/PA83A/PA83A/PA82J (Apex Microtechnology), ± 150 V; 3582/3583/3584J (Burr-Brown), ± 150 V; 3581J (Burr-Brown), ± 75 V; HA-2645 (Harris), ± 50 V; 3580J (Burr-Brown), ± 35 V.

9.5.5 Low-noise Operational Amplifiers

For systems with very low level input signals, the noise characteristics of the op amps are of great importance. Some examples of low-noise op amps, together with the typical equivalent input noise voltage and input noise current, are the following devices from Precision Monolithics, Inc.: OP-07 with 10 nV/$\sqrt{Hz}$ and 14 pA/$\sqrt{Hz}$, OP-27 and OP-37 both with 3 nV/$\sqrt{Hz}$ and 1 pA/$\sqrt{Hz}$, and the OP-77 with 10 nV/$\sqrt{Hz}$ and 0.2 pA/$\sqrt{Hz}$. Other low-noise op amps are the OPA101 and OPA102 from Burr-Brown with 10 nV/$\sqrt{Hz}$ and 2 fA/$\sqrt{Hz}$. The extremely low noise current of the OPA101 and OPA102 is the result of the low input bias current (10 pA max.) due to the JFET input stage. It should be noted that the noise current and noise voltage spectral density values increase at lower frequencies, especially below 1000 Hz.

PROBLEMS

9.1. (*CFB operational amplifier*) Given the CFB operational amplifier of Figure 9.3. Assume that all transistors of the same type are identical, $\beta = 100$ (MIN), and $V_A = 100$ V for all transistors. Assume $I_Q = 1.0$ mA and $R_F = 1000\ \Omega$.

(a) Find the dynamic transfer resistance r_T under no-load conditions.   [*Ans.:* $r_T = 5$ MΩ (MIN)]

(b) Find the dynamic transfer resistance r_T for $R_L = 100\ \Omega$.   [*Ans.:* $r_T = 833$ kΩ (min)]

(c) Find R_1 for $A_{CL}(0) = +5$.   (*Ans.:* $R_1 = 250\ \Omega$)

(d) Find the gain error for $R_F = 1000\ \Omega$ and $R_L = 100\ \Omega$.   [*Ans.:* 0.12% (max)]

(e) If $C_T = 3.0$ pF and $r_{IN} = 0$, find the closed-loop bandwidth.   (*Ans.:* $BW_{CL} = 53$ MHz)

(f) Find r_{IN}.   (*Ans.:* $r_{IN} = 12.6\ \Omega$)

(g) Find the closed-loop bandwidth for the following closed-loop gains: $A_{CL}(0) = 1, 5,$ 10, and 50.   (*Ans.:* 52.4, 49.9, 47.1, 32.55 MHz)

(h) Find the closed-loop output resistance.   [*Ans.:* $r_o = 0.1\ \Omega$ (MAX)]

(i) Find the change in the low-frequency voltage gain as R_L varies from 100 to infinity.   (*Ans.:* +0.1%)

(j) Find the slewing rate for an input voltage step of 1 V and a closed-loop gain of +10. Assume $C_T = 3.0$ pF.   (*Ans.:* 3.33 V/ns)

(k) Find the rise time.   (*Ans.:* 6.6 ns)

(l) Assuming the simple exponential type of response characteristics, find the 0.4% settling time.   (*Ans.:* 16.6 ns)

(m) Find the maximum, no-load output-voltage swing.   (*Ans.:* from V$^+$ − 0.9 V to V$^-$ + 0.9 V)

(n) Find the dynamic input resistance at the noninverting input terminal for a closed-loop gain of +5.   [*Ans.:* 2 MΩ (MIN)]

(o) Repeat part (n) for a gain of +50.   [*Ans.:* 200 kΩ (MIN)]

9.2. (*OTA amplifier with AGC*) For the circuit shown in Figure P9.2:

(a) Show that the voltage gain $A_v = v_o/v_s$ is given by $A_v = (20/V)(V^+ - V_c)(R_L/R_2)$.

(b) What is the purpose of resistor R_1?

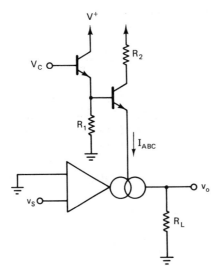

Figure P9.2 Operational
transconductance amplifier with AGC.

9.3. (*OTA amplifier with linear AGC/multiplier/amplitude modulator*)
  (a) For the circuit shown in Figure P9.3, show that the output voltage is given by $v_o = (20/V)(R_L/R_1)V_S V_C$.
  (b) Can this circuit be used as a four-quadrant multiplier? Explain.

9.4. (*Frequency-to-voltage converter/tachometer*)
  (a) For the circuit shown in Figure P9.4, A_1 is a voltage comparator with a high-state output voltage of V_H and a low state output voltage of V_L. Assuming a long time constant for the $R_2 C_2$ low-pass filter such that $R_2 C_2 > 1/f_{IN}$, show that the output voltage v_o is given by $v_o = (V_H - V_L - 0.7\ \text{V})\ R_2 C_1 f_{IN} = K_{FV} f_{IN}$.
  (b) If the frequency-to-voltage transfer coefficient, K_{FV}, is to be 1 V/kHz, $C_2 = 10\ \mu\text{F}$, $f_{IN}(\text{MIN}) = 100\ \text{Hz}$, and $R_2 C_2 = 50/f_{IN}$, find R_2 and C_1. (*Ans.:* $R_2 = 50\ \text{k}\Omega$, $C_1 = 1.04\ \text{nF}$)

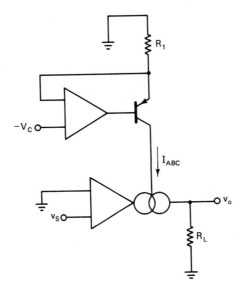

Figure P9.3 Operational
transconductance amplifier
with linear gain control.

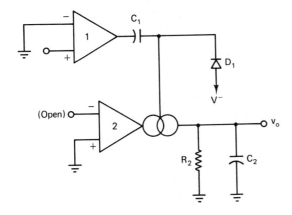

Figure P9.4 Frequency-to-voltage converter/tachometer.

9.5. (*OTA sample-and-hold circuit*) For the circuit of Figure P9.5, the input signal is a 5-V step. The sampling time t_S is 5 μs and the hold time t_H is 1 ms. The total leakage current from the hold capacitor C_H is 1 nA (MAX), and the droop during the hold time should not exceed 10 mV. Find C_H. (*Ans.:* 100 pF < C_H < 500 pF)

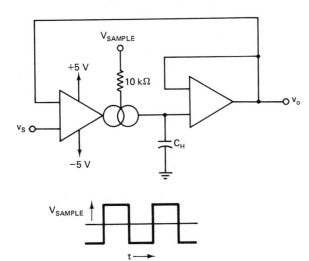

Figure P9.5 Operational transconductance amplifier sample-and-hold circuit.

9.6. [*Current-controlled current source (current amplifier) using an OTA*] Refer to the circuit of Figure P9.6.
    (a) Show that the current transfer ratio of this circuit is given by $A_i = i_o/i_{IN} = -R_F I_{ABC}/(2V_T) \cong -(20/V)R_F I_{ABC}$.

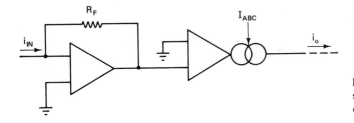

Figure P9.6 Current-controlled current source (current amplifier) using an operational transconductance amplifier.

(b) If $I_{ABC} = 1.0$ mA, find R_F for a current gain of -20.   (*Ans.:* $R_F = 1000\ \Omega$)

(c) Show a circuit that will produce a current gain of $+20$.

REFERENCES

FRANCO, S., *Design with Operational Amplifiers and Analog Integrated Circuits*, McGraw-Hill, New York, 1988.

IRVINE, R. G., *Operational Amplifiers—Characteristics and Applications*, 2nd ed., Prentice-Hall, Englewood Cliffs, N.J., 1987.

MORLEY, M. S., *The Linear IC Handbook*, TAB Books, Blue Ridge Summit, Pa., 1986.

10

Voltage Comparators

A comparator is an integrated circuit as shown in Figure 10.1 that is used to compare two input voltages V_1 and V_2 and to produce an output voltage $V_o = V_H$ if $V_1 > V_2$ and $V_o = V_L$ if $V_1 < V_2$, where V_H and V_L are two fixed output-voltage levels. In Figure 10.2 the V_o versus $V_i = V_1 - V_2$ transfer characteristics of a comparator are shown. Depending on the particular type of comparator and the supply voltage levels, V_H and V_L may be of opposite polarity with V_H positive and V_L negative, or they both may be positive, or they both may be negative.

The open-loop gain of comparators is usually very large, although generally less than that of operational amplifiers. Low-frequency voltage gains in the range from 3000 to 100,000 are typically the case for comparators. As a result, the required change of input voltage to produce a transition from one output state to the opposite state is very small, generally in the range of only 0.1 to 3 mV. As a consequence of this very small transition voltage range, the output of the comparator is virtually always saturated at either V_L or V_H.

A comparator can be thought of as being a simple one-bit analog-to-digital (A/D) converter, producing a digital 1 output ($V_o = V_H$) if the analog input voltage V_1 is above the reference level of $V_{\text{REF}} = V_2$. A digital 0 ($V_o = V_L$) results if the input voltage level falls below the reference level.

In many respects, comparators are very similar to op amps, and indeed op amps can be used as comparators. A comparator is, however, designed specifically to

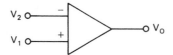

Figure 10.1 Voltage comparator.

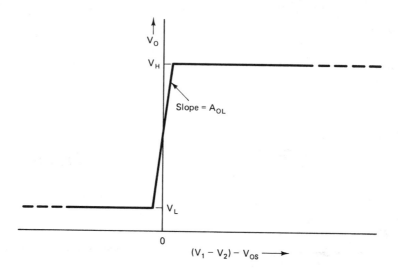

Figure 10.2 Comparator transfer characteristic.

operate under open-loop conditions, basically as a switching device. An op amp, on the other hand, is almost always used in a closed-loop configuration and is usually operated as a linear amplifier.

Being designed to be used in a closed-loop configuration, the frequency response characteristics of an op amp are generally tailored to ensure an adequate measure of stability against an oscillatory type of response. This results in a sacrifice being made in the bandwidth, rise time, and slewing rate of the device. In contrast to this, since a comparator operates as an open-loop device, no sacrifices have to be made in the frequency-response characteristics, so a very fast response time can be obtained.

An op amp is designed to produce a zero output voltage when the difference-mode input signal is nominally zero. A comparator, in contrast, operates between two fixed output-voltage levels, so the output voltage for zero input voltage is generally either V_H or V_L, depending on the polarity of the input offset voltage.

The output voltage of an op amp will saturate at levels that are generally about 1 or 2 V away from the positive and negative power supply voltage levels. The comparator output is often designed to provide some degree of flexibility in fixing the high- and low-state output voltage levels and for ease in interfacing with digital logic circuits.

There are many applications of comparators. These include pulse generators, square-wave and triangular-wave generators, pulse-width modulators, level detectors, zero-crossing detectors, pulse regenerators, line receivers, limit comparators, voltage-controlled oscillators, A/D converters, and time-delay generators.

10.1 COMPARATOR CHARACTERISTICS

The most important characteristic of a comparator is generally the response time or propagation delay time. This is the time between the input-voltage transition and some specified point on the output-voltage transition, as shown in Figure 10.3. A commonly used point is the 50% point, although sometimes the 90% point is specified. For the response time measurement, the noninverting input terminal is usually grounded ($V_{\text{REF}} = 0$). The input voltage is applied to the inverting input terminal and is such that a rapid transition is made from some level such as $+100$ mV or -100 mV to a voltage level of opposite polarity, the magnitude of which is called the input overdrive voltage. Input overdrive voltages of 2, 5, 10, 20, and 100 mV are commonly used.

Response times of voltage comparators generally range from about 1 μs down to as fast as about 10 ns. The open-loop voltage gain of comparators is generally in the range from 3000 to 100,000, so the input-voltage swing necessary to produce the output-voltage transition between the two saturated states is in the range of about 0.1 to 3 mV.

The offset voltage V_{OFFSET} or V_{OS} of a comparator is another important specification since it results in a shift in the transition point for the input voltage from V_{REF} to $V_{\text{REF}} + V_{\text{OFFSET}}$. The offset voltage for comparators is typically in the range from 1 to 10 mV.

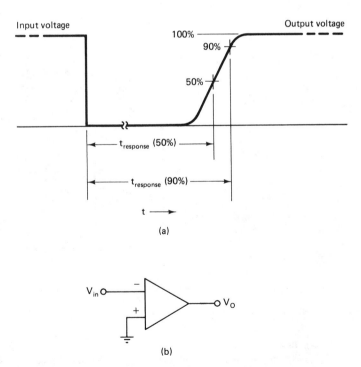

Figure 10.3 Voltage comparator response time: (a) input and output voltage waveforms; (b) test circuit.

10.2 COMPARATORS WITH POSITIVE FEEDBACK

Since comparators are not used as linear amplifiers, but rather as switching-mode devices, positive feedback can be used advantageously to increase the gain of the circuit and to produce some degree of hysteresis in the V_O versus V_i transfer characteristics. The increased gain reduces the input-voltage swing necessary to produce the output-voltage transition to negligibly small values.

A comparator circuit with positive feedback is shown in Figure 10.4a, and the transfer characteristics are presented in Figure 10.4b. The input-voltage level at which the low-to-high output transition takes place is

$$V_{i(L-H)} = \frac{V_{\text{REF}}R_1 + V_L R_2}{R_1 + R_2} \qquad (10.1)$$

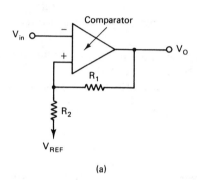

(a)

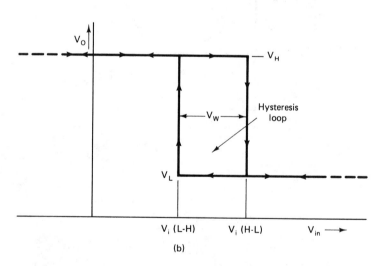

(b)

Figure 10.4 Voltage comparator with positive feedback: (a) circuit diagram; (b) transfer characteristics; (c) voltage comparator switching characteristics without hysteresis (no positive feedback); (d) voltage comparator switching characteristics with hysteresis (positive feedback present).

and the input-voltage level at which the high-to-low transition occurs is

$$V_{i(H-L)} = \frac{V_{REF} R_1 + V_H R_2}{R_1 + R_2} \tag{10.2}$$

The width of the hysteresis loop V_W is

$$V_W = \frac{(V_H - V_L)R_2}{R_1 + R_2} \tag{10.3}$$

For many applications the presence of this hysteresis loop is desirable to produce a more definitive switching action of the comparator with less *contact bounce* or *chattering*, as illustrated in Figure 10.4c. Note that in Figure 10.4d, which represents the case when hysteresis is present, if the width of the hysteresis loop V_W is made larger than the peak-to-peak fluctuation in the input voltage V_i, a definitive switching action occurs without the intermittent switch chattering that would otherwise be present.

The use of positive feedback to produce a hysteresis loop and to increase the closed-loop gain is especially useful for input voltages that change very slowly near the reference voltage level. If the comparator is part of some larger feedback control system, there could also be a problem with an oscillatory type of response if hysteresis were not used.

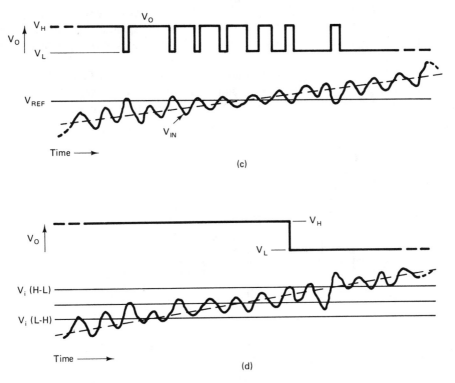

Figure 10.4 (continued)

For a very simple example of this problem, consider the case of a comparator that is part of a thermostatic control system. The output of the comparator is used to control the heating element or system, and the input voltage is proportional to the temperature. Starting with temperatures below the thermostat set point, the comparator input voltage will be less than the reference voltage and the comparator output will be in the high state, turning the heating system on. The temperature rises, producing a corresponding increase in the comparator input voltage. In the absence of hysteresis, as soon as the input voltage exceeds the reference voltage, the comparator output will be switched to the low state, thereby turning the heating system off. The resulting decrease in the temperature will very soon cause the heater to be turned back on again. This will therefore result in a rapid off–on cycling of the system, continuing indefinitely at a rapid rate as the temperature fluctuates very slightly above and below the set point. This rapid off–on switching is generally very undesirable. It can be very greatly reduced, however, by the introduction of a hysteresis loop of suitable width.

10.3 COMPARATOR CIRCUITRY

The internal circuitry of comparators is in many respects very similar to that of op amps, especially with respect to the differential-amplifier input stages. The major differences occur in the output stages.

For an example of a fairly simple comparator circuit, we will examine the LM139/239/339 series (National Semiconductor). These are quad comparators, which means that there are four electrically independent comparators in one package (DIP or flat package).

A circuit diagram of this comparator is presented in Figure 10.5. It may be noted that the input stage is very similar to that of the LM124/224/324 operational amplifiers (see Figure P7.2). Like these op amps, this comparator can be operated

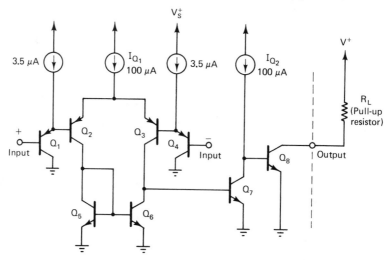

Figure 10.5 Voltage comparator circuit: LM139 type (National Semiconductor).

from a single positive supply voltage, although operation from a dual supply is also possible. This device can operate from supply voltages as low as 2.3 V up to as much as 36 V (or ± 18 V if operated from a dual supply).

10.3.1 Circuit Description: LM139

The differential-amplifier input stage consists of Q_1, Q_2, Q_3, and Q_4 connected as a Darlington (CC–CE) differential amplifier. The quiescent current of Q_2 and Q_3 is 50 μA each, and that of Q_1 and Q_4 is 3.5 μA + (50 μA/β_{PNP}). Note that the 3.5-μA current sources ensure that the quiescent currents of Q_1 and Q_4 are always at least 3.5 μA. As a result, the current gain (β) of these two transistors, as well as the dynamic transconductance (g_m), is always maintained at an acceptably high value.

Transistors Q_5 and Q_6 constitute a current-mirror active load for the input stage. The second gain stage is comprised of Q_7 in a common-emitter configuration with a current source (100 μA) active load. The third gain stage, which is also the output stage, consists of Q_8, which is also connected in a common-emitter configuration. Note that Q_8 has an uncommitted collector. In the actual operation of this comparator, there will generally be a load resistor (also called a *pull-up* resistor) connected from the output (the collector of Q_8) to some positive supply voltage, which can be the comparator supply voltage, V_S^+, although this is not necessarily always the case.

This comparator is usually operated open loop. Since the open-loop gain is very large, operation in this mode usually means that the second- and third-stage transistors will be driven back and forth between cutoff and saturation, passing very quickly through the active region.

If the input voltage is such that the $-$input is *higher* in voltage than the $+$input, this will cause the base current of Q_7 to be cut off, thus reducing the collector current of Q_7 to zero. This will cause all of the current I_{Q2} (100 μA) to become the base drive of Q_8. The output transistor Q_8 will thus be turned on and be in either the active region or the saturation region.

If the collector of Q_8 is connected to a positive supply voltage V^+ via a pull-up resistor, R_L, the saturation current is

$$I_{C(SAT)} = \frac{V^+ - V_{CE(SAT)}}{R_L} \simeq \frac{V^+}{R_L} \qquad (10.4)$$

If $\beta \times 100$ μA $\geqslant I_{C(SAT)}$, Q_8 is in the saturation region with $I_C = I_{C(SAT)}$. This is the usual case. If, however, $\beta \times 100$ μA $< I_{C(SAT)}$, then Q_8 is in the active region with $I_C = \beta \times 100$ μA. If Q_8 is driven into saturation, the output voltage is given by $V_O = V_{CE8} = V_{CE(SAT)}$. Therefore, the output voltage in the low state is close to ground potential. If, however, Q_8 is not in saturation, the output voltage is given by $V_O = V_{CE8} = V^+ - I_C R_L = V^+ - (\beta \times 100$ μA$)R_L$ and thus is dependent on the current gain value of Q_8.

If the input voltage is such that the $-$input is *lower* in voltage than the $+$input, the current through Q_2, and therefore that through Q_5 and Q_6, is lower than the current through Q_3. The difference between I_{C3} and I_{C6} ($\simeq I_{C5} \simeq I_{C2}$) will become the base current of Q_7. If the base current of Q_7 is such that $\beta_7 I_{B7} > 100$ μA, Q_7 is in saturation. If Q_7 is in saturation, Q_8 is cut off and the output voltage becomes $V_O = V_{C_8} \simeq V^+$. The high-state output voltage is essentially equal to V^+.

To switch the output voltage from the low state to the high state, the required change in the base current of Q_7 is thus from $I_{B_7} = 0$, corresponding to Q_7 cutoff and Q_8 in saturation, to $I_{B_7} = 100 \ \mu\text{A}/\beta_7$, corresponding to Q_7 in saturation and Q_8 in cutoff. The change in I_{B_7} in terms of the change in the input voltage, ΔV_i, is given by

$$\Delta I_{B_7} = \Delta I_{C_3} - \Delta I_{C_6} \simeq \Delta I_{C_3} - \Delta I_{C_5} \simeq \Delta I_{C_3} - \Delta I_{C_2} = 2g_f \, \Delta V_i \quad (10.5)$$

where g_f is the dynamic transfer conductance of the differential amplifier, which in this case is given approximately by $g_f = I_Q/4V_T$. Note that the voltage gain of the Q_1 and Q_4 emitter-follower stages is close to unity due to the 3.5-μA current sources that provide for the additional quiescent current for these transistors.

We can therefore now write

$$\Delta I_{B_7} = \frac{100 \ \mu\text{A}}{\beta_7} = 2g_f \, \Delta V_i = \frac{I_Q}{2V_T} \, \Delta V_i \quad (10.6)$$

so that

$$\Delta V_i = \frac{100 \ \mu\text{A}/\beta_7}{I_Q/2V_T} = \frac{100 \ \mu\text{A}/\beta_7}{100 \ \mu\text{A}/50 \ \text{mV}} = \frac{50 \ \text{mV}}{\beta_7} \quad (10.7)$$

This is the required change in the input voltage to switch the output from the low state ($V_O = V_{CE(SAT)} \simeq 0.1$ V) to the high state ($V_O \simeq V^+$). If, for example, $\beta = 50$ (min.), 100 (typ.), we have that $\Delta V_i = 50$ mV/$\beta_7 = 1.0$ mV (max.), 0.5 mV (typ.).

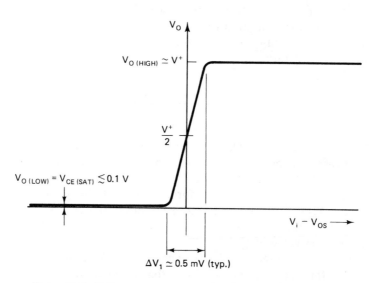

Figure 10.6 Voltage comparator transfer characteristic, V_o versus V_i.

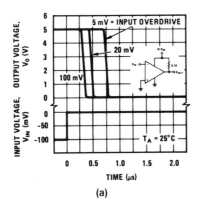
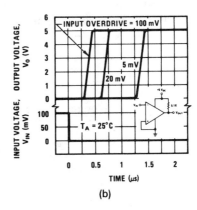

Figure 10.7 Response times for various input overdrives for the LM139 voltage comparator: (a) negative transition; (b) positive transition (National Semiconductor).

The V_O versus V_i transfer characteristic for this comparator is shown in Figure 10.6. The offset voltage, V_{OS}, is defined as the voltage required on the input to produce an output voltage that is midway between the low and high states, or approximately $V^+/2$.

The condition on the load resistance required to cause Q_8 to be in saturation for the low state is given by $\beta_8 \times 100\ \mu A \geqslant I_{C(SAT)} \simeq V^+/R_L$, so $R_L \gtrsim V^+/(\beta_8 \times 100\ \mu A)$. For example, if $\beta_8 = 50$ (min.), 100 (typ.) we have that $R_L \gtrsim V^+/\beta_8 \times 100\ \mu A = 1.0\ k\Omega$ (max.), 500 Ω (typ.), using $V^+ = 5.0$ V.

If instead of being connected to ground, the low side of the comparator circuit is connected to a negative supply voltage V^-, the low-state output voltage is shifted downward by V^-, to $V^- + V_{CE(SAT)} \simeq V^-$. As a result, the low-state voltage is approximately V^- and the high-state voltage is approximately V^+. If these two voltages are equal in magnitude, a symmetrical low-to-high transition is obtained; that is, the low-state and high-state voltages are equidistant from ground potential.

In Figure 10.7a, curves are shown of the response times for negative (high-to-low) output voltage transitions, and in Figure 10.7b curves are presented for positive (low-to-high) transitions. The supply voltage is $+5$ V and the load (pull-up) resistor is 5.1 kΩ. Curves are given for 5-, 20-, and 100-mV input-voltage overdrives.

The input-voltage transition for a negative output-voltage transition is from -100 mV to a level above ground ($V_{REF} = 0$) equal to the specified input-voltage overdrive. Thus, for a 20-mV overdrive the input-voltage swing is from -100 to $+20$ mV. For the positive-going output-voltage transition, the input-voltage swing is from $+100$ mV to a level below the reference voltage (0 V) equal to the specified input overdrive.

Some amount of input overdrive is necessary to ensure that there will be an output voltage transition, and it should be at least equal to the maximum expected value of the input offset voltage [5 mV (max.) for the LM139 comparators]. Increasing the input overdrive also reduces the response time, although increasing the overdrive above about 50 mV will result in very little improvement in the response time.

From the response time graphs for the LM139, we obtain the following set of response times (ns):

Input overdrive (mV)	Negative output transition	Positive transition
5	750	1375
20	500	700
100	300	375

Notice that the response can indeed be speeded up by increasing the input overdrive. It is also evident that the positive-going output transitions take appreciably longer than the negative-going transitions. This is the result of the fact that for the positive transition the output transistor Q_8 starts off in deep saturation and must then be driven out of saturation, across the active region, and then into cutoff. With a supply voltage of $+5$ V and a 5.1-kΩ pull-up resistor, the current through Q_8 is limited to about 1.0 mA when in saturation, whereas $\beta I_B = 150 \times 100~\mu\text{A} = 15$ mA, so we see that Q_8 is indeed driven into deep saturation. When Q_8 is in deep saturation, there is a considerable amount of stored charge in the transistor as a result of the large forward-bias voltage at the collector–base junction. The time required to remove this excess stored charge (the *storage time*) is responsible for the longer transition time.

For the negative-going output-voltage transition, it is Q_7 that must be driven out of saturation and into cutoff. However, the storage time for Q_7 is appreciably less than that of Q_8 due to the fact that the saturation current of Q_7 is limited to 100 μA as compared to 1.0 mA for Q_8. As a result of the storage times of Q_7 and Q_8, the LM139 is not a very fast comparator.

Figure 10.8 is a graph of the low-state output voltage as a function of the output sink current. We see that for currents less than 15 mA the output voltage is small (less than 1 V). In this range the transistor is in saturation, and we have that

$$V_{CE(\text{SAT})} = V_{CB} + V_{BE} + I_C r_{cc'} = -0.6~\text{V} + 0.7~\text{V} + I_C r_{cc'} = 0.1~\text{V} + I_C r_{cc'}$$

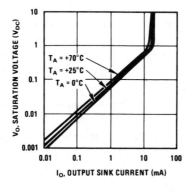

Figure 10.8 Output saturation voltage for the LM139 (National Semiconductor).

Voltage Comparators   Chap. 10

where $r_{cc'}$ is the collector bulk series resistance (*saturation resistance*). At $I_C = 10$ mA, $V_O = V_{CE(SAT)} = 0.8$ V, so $r_{cc'} = (0.8 - 0.1)$ V/10 mA $= 70\ \Omega$.

As the collector current ($I_C = I_0$) goes above 15 mA the output voltage increases very rapidly, for now the transistor is going out of the saturation region and into the active region. Since this transition occurs at $I_C = 15$ mA and the base current drive is $I_B = 100\ \mu$A, we can conclude that the current gain (β) of Q_8 is 150.

10.4 TRANSISTOR STORAGE TIME

We have seen that a major factor limiting the speed of a comparator is the transistor storage time. We will now look at the factors that determine the storage time and consider the techniques that can be used to reduce the storage time.

In the active mode of operation of a transistor, the emitter–base junction is biased "on" and the collector–base junction is biased "off." The emitter emits minority carriers (electrons for an NPN transistor) into the base, whereupon they flow across the base and are collected by the collector. In this mode of operation the collector only collects electrons and does not emit any electrons back into the base. This situation is illustrated in Figure 10.9a, and in Figure 10.9b the variation of the injected minority carrier density with distance is shown.

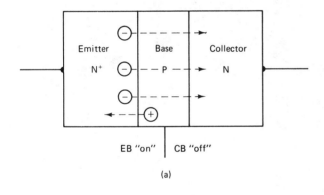

(a)

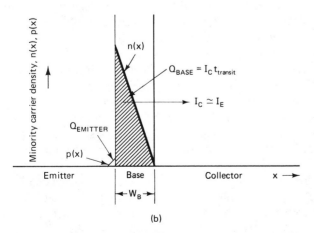

(b)

Figure 10.9 Transistor operation in the active mode.

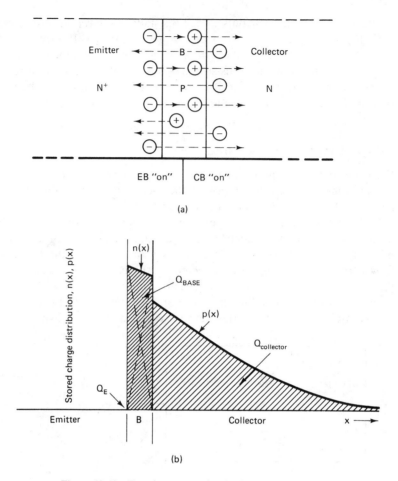

Figure 10.10 Transistor operation in the saturation mode.

The charge stored in the base region due to the electrons in transit across the base is given by $Q_{\text{base(active)}} = I_C t_{\text{transit}}$, where t_{transit} is the transit time of the electrons across the base region from the emitter to the collector. This time is very short, typically on the order of only 30 to 100 ps. There is also a very small amount of charge stored in the emitter region due to the injection of holes from the base into the emitter as a result of the forward-bias voltage across the emitter–base junction. This current injection is very small compared to the flow of electrons in the other direction due to the very large doping asymmetry of the N^+/P emitter–base junction. The very heavy doping of the emitter region also results in a very short lifetime for the injected holes, so as a net result the charge stored in the emitter region Q_E is very small.

When the transistor is driven into saturation, both junctions are turned "on," so the collector not only collects electrons, but also emits electrons back into the base. Most of these electrons then travel across the base region and are collected by the emitter, which collects most of the electrons emitted by the collector, but still also is emitting electrons back into the base region. This situation is illustrated in Figure 10.10.

When the transistor is driven deeply into saturation, both the emitter–base and the collector–base junctions are strongly forward biased, and there will be a high flow of electrons across the base region in both directions. As a result, the base region is flooded with electrons, and the stored charge in the base region is now very much greater than it was in the active mode of operation.

At the same time, there will also be a very substantial injection of holes from the base region into the collector. The large magnitude of this hole injection is the result of the collector region being much less heavily doped than the base region. As a result, when the collector–base junction is forward biased by more than about 0.5 V, there is a large flow of electrons from the collector into the base and a large flow of holes from the base into the collector. This large injection of holes into the collector region and the relatively long lifetime of holes in this region τ_p results in a large amount of stored charge.

The total stored charge in the transistor in the saturation mode is $Q_{\text{stored (sat)}} = Q_{\text{base}} + Q_{\text{emitter}} + Q_{\text{collector}}$. We have already noted that Q_{emitter} is very small. As a result of the very narrow base width (~ 0.3 to 1.0 μm) and the resulting very small transit time, we generally have that the major component of the total stored charge is $Q_{\text{collector}}$. This is the charge stored in the collector region due to the injection of holes from the base region. This stored charge is related to the base current approximately by $Q_{\text{stored (sat)}} \simeq Q_{\text{collector}} = I_{B(\text{sat})}\tau_p$, where τ_p is the lifetime of holes in the collector region. This lifetime is typically in the range from 1 to 10 μs, in contrast to the base transit time on the order of 30 to 100 ps. We therefore see that the stored charge in the transistor when operated in the saturation mode can be several orders of magnitude larger than when operated in the active mode.

When the transistor is now switched out of the saturation mode and across the active region into cutoff, the stored charge produces a continued flow of current through the transistor for the period of time required to remove this stored charge, as shown in Figure 10.11. This period of time of continued current flow is called the *storage time*, t_s. The stored charge is removed in two ways. One is by the reverse flow of base current $I_{B(R)}$, which directly removes the excess stored charge from the transistor. The other way is the recombination of the stored minority carriers with majority carriers. The rate at which the stored charge in the collector region recombines with electrons is $dQ/dt = Q_{\text{collector}/\tau p}$. The total rate at which the stored charge decreases with time is given by $dQ_{\text{stored}}/dt = -I_{B(R)} - Q_{\text{stored}}/\tau_p$, where $I_{B(R)}$ is the base current in the reverse (negative) direction and we have assumed that most of the stored charge is in the collector region.

If the base current during this switching transition is limited to a very small value, or is zero, the decay of the stored charge is given approximately by $dQ_{\text{stored}}/dt \simeq -Q_{\text{stored}}/\tau_p$, so the decay of the stored charge as a function of time is

$$Q_{\text{stored}}(t) = Q_{\text{stored}}(0) \exp\left(-\frac{t}{\tau_p}\right)$$

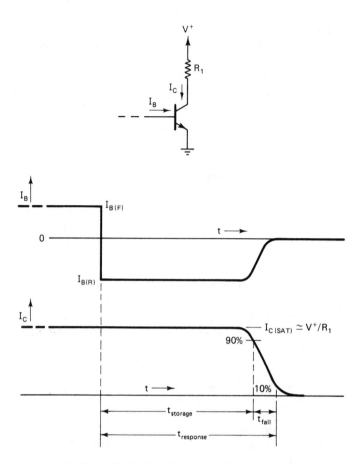

Figure 10.11 Transistor turn-off transition.

where $Q_{\text{stored}}(0)$ is the stored charge at the beginning of the switching transition. From this we see that it takes a time of about 3 or $4\tau_p$ to dissipate the stored charge. This results in a very long storage time, up to several microseconds.

If a substantial reverse base current is allowed to flow, the rate at which the stored charge is removed is principally controlled by the reverse-base current $I_{B(R)}$. For this case, we have that $dQ_{\text{stored}}/dt \simeq -I_{B(R)}$, so $Q_{\text{stored}}(t) \simeq Q_{\text{stored}}(0) - I_{B(R)}t$. The storage time is the time required to remove all the stored charge, so it is given by $t_{\text{storage}} \simeq Q_{\text{stored}}(0)/I_{B(R)}$. Since $Q_{\text{stored}}(0) \simeq Q_{\text{collector}}(0) = I_{B(\text{SAT})}\tau_p$, the storage time becomes $t_{\text{storage}} \simeq [I_{B(\text{SAT})}/I_{B(R)}]\tau_p$. Thus, by forcing a large reverse base current to flow out of the transistor, the storage time can be made to be very small. In many cases the storage time can be reduced to about 10 ns.

The fall-time portion of the total response time or switching time of the transistor is the time required to change the voltage across the emitter–base junction capacitance from $V_{BE(\text{sat})}$ to $V_{BE(\text{cutoff})}$. The emitter–base junction capacitance $C_{b'e}$ is generally in the range from 10 to 30 pF, and the voltage change necessary to drive the transistor from saturation to cutoff is in the range of about 0.2 to 0.4 V. The fall time can be obtained from the relationship

$$\Delta V_{BE} = \frac{\Delta Q_{b'e}}{C_{b'e}} = \frac{I_{B(R)}t_f}{C_{b'e}}$$

so that

$$t_f = \Delta V_{BE} \frac{C_{b'e}}{I_{B(R)}} = \frac{[V_{BE(\text{sat})} - V_{BE(\text{cutoff})}] C_{b'e}}{I_{B(R)}} \tag{10.8}$$

For example, if $V_{BE(SAT)} = 0.75$ V, $V_{BE(\text{cutoff})} = 0.55$ V, $C_{b'e} = 20$ pF, and $I_{B(R)} = 1.0$ mA, the fall time is $t_f = 0.2$ V $\times$ 20 pF/1.0 mA $= 4$ ns. We see that the reverse base current is important in determining the fall time.

10.5 TECHNIQUES TO REDUCE RESPONSE TIME

Several techniques can be used to speed up the transition of the transistor from saturation to cutoff, as follows:

1. Reduce the storage time by reducing the minority carrier lifetime
2. Decrease the storage and fall times by increasing the reverse base current
3. Reduce the charge stored in the base and collector regions by using a thinner and less heavily doped base region
4. Use a Schottky-barrier diode across the collector–base junction to prevent the collector–base junction from turning on

10.5.1 Minority Carrier Lifetime Reduction

The first technique of reducing the minority carrier lifetime involves adding a small amount of a *lifetime killer impurity* to the IC during the manufacturing process. This lifetime killer impurity acts to decrease the minority carrier lifetime and thereby reduces the stored charge, especially in the collector region. Gold is commonly used for this purpose, and it serves as a very effective recombination center in silicon in promoting or catalyzing the recombination of free electrons and holes. This increased recombination rate results in a reduced carrier lifetime. As a result, gold and other heavy metal impurities such as copper and iron are called *lifetime killers*.

With the addition of small amounts of gold to silicon (about 10^{14} to 10^{15} cm^{-3}), the minority carrier lifetime can be reduced from the undoped level of around 1 to 10 μs down to around 10 ns. This correspondingly results in a reduction of the storage time down to the range 10 ns.

The major disadvantage of the gold-doping process is that it is nonselective in that all the devices on the IC chip become gold-doped. The reduction in the minority carrier lifetime produced by the gold doping does have the beneficial effect of reducing the storage time, but there are some unwanted side effects as well. The reduction in the minority carrier lifetime reduces the current gain β of the transistors because of the increased recombination of minority carriers in transit across the base region.

For the narrow-base NPN transistors with base transit times in the range from 30 to 100 ps, the reduction of the minority carrier lifetime to around 10 ns will not prove to be too severe an effect. We still have that the minority carrier lifetime is considerably greater than the transit time, so most of the injected minority carriers will still make it safely across the base to the collector. However, the lateral and substrate PNP transistors will have base widths generally in the range 4 to 10 μm and therefore a much longer transit time, in the range from 10 to 50 ns. As a result, a reduction in the minority carrier lifetime to about 10 ns can produce a very substantial reduction in the transistor current gain.

Another major disadvantage of gold doping is the increase in the reverse leakage current of the diodes and transistors. One component of the leakage current of a reverse-biased PN junction is the flow of the thermally generated minority carriers across the junction. The generation rate of these minority carriers is inversely proportional to the minority carrier lifetime, so a decrease in the lifetime from about 10 μs to 10 ns can result in an increase in this leakage current component by a 1000 : 1 ratio.

10.5.2 Comparator Example: μA760

The transistor storage and fall times can be reduced substantially by increasing the reverse base current during the saturation to cutoff switching transition time. The storage time can also be minimized by limiting the amount of forward base drive that puts the transistor into saturation.

An example of a voltage comparator that uses these techniques together with the use of gold doping for minority carrier lifetime reduction is the μA760 (Fairchild). This device offers typical response times of 16 ns when driven with a 100-mV input step with a 5-mV overdrive. A simplified circuit diagram of this comparator is shown in Figure 10.12. The input stage is an NPN differential amplifier comprised of Q_1 and Q_2 with current sink biasing being provided by Q_3. With a typical supply voltage of ± 5.0 V, the quiescent current of this differential amplifier is set by Q_3 and Q_{16} to

$$I_Q = I_3 = I_{16} = \frac{V^- - V_{BE}}{R_{14} + R_{22}} = \frac{(5.0 - 0.7) \text{ V}}{4.75 \text{ k}\Omega} = 0.9 \text{ mA}$$

Resistors R_1 and R_2 (1 kΩ) provide a passive load for Q_1 and Q_2. The low value of these resistors helps to give this device a fast response time. A balanced output from the first stage is then fed to the differential amplifier second stage via emitter followers Q_4 and Q_6. These emitter-follower transistors, together with diodes D_1 through D_4 and the current sink transistors Q_5 and Q_7, provide dc level shifting. The current provided by Q_5 and Q_7 is 0.9 mA. Since the voltage drop across the 10-kΩ resistors (R_4 and R_6) plus the diode drop of D_2 and D_4 is sufficient to drive the diodes D_1 and D_3 into breakdown ($V_Z \simeq 6.5$ V), the quiescent voltage at the bases of Q_9 and Q_{10} is approximately

$$V_{B_9} = V_{B_{10}} = V^+ - I_1 R_1 - V_{BE_4} - V_{Z_1} + V_{D_2}$$

$$= 5.0 - (0.45 \text{ mA})(1 \text{ k}\Omega) - 0.7 - 6.5 - 0.7 = -1.95 \text{ V}$$

(10.9)

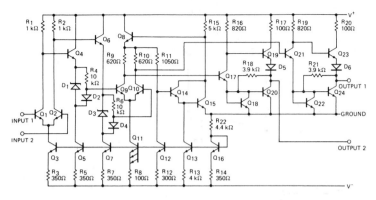

Figure 10.12 High-speed comparator circuit: μA760 (Fairchild Semiconductor).

The second stage, comprised of Q_9 and Q_{10}, is biased by Q_{11} acting as a current sink and providing a quiescent current level of $I_{11} = 0.9$ mA $\times$ (350/100) = 3.15 mA. The current through Q_{12} is $I_{12} = 0.9$ mA $\times$ (350/300) = 1.05 mA. The quiescent voltage at the emitter of Q_8 is therefore

$$V_{E8} = V_{BE15} + V_{BE14} + I_{12}R_{11} \simeq 0.6 + 0.7 + (1.05 \text{ mA})(1.05 \text{ k}\Omega) = +2.4 \text{ V}$$

The quiescent voltage level at the collectors of Q_9 and Q_{10} is then

$$V_{C9} = V_{C10} = V_{E8} - I_9R_9 \simeq 2.4 - (3.15 \text{ mA}/2)(620 \ \Omega) = 1.4 \text{ V}$$

The balanced output of the second stage drives two identical output circuits. These output circuits provide the complementary TTL-compatible output voltages. A detailed view of one of the two output circuits is shown in Figure 10.13. Note that the quiescent voltage level at the collectors of Q_9 and Q_{10} of $+1.4$ V as obtained above is at the proper value to interface with the output circuit.

If the input signal is such that the current through Q_9 decreases substantially below the quiescent level, the excess of the current through R_9 becomes the forward base drive of Q_{17} as indicated by $I_{B17(F)}$. This turns Q_{17} on and also turns Q_{18} on. Since Q_{18} and Q_{20} are connected as a current mirror, this also results in Q_{20} being turned on.

At the same time that Q_{20} is being turned on, Q_{19} is rapidly driven out of saturation by the sinking of the reverse base current of Q_{19}, $I_{B(R)19}$, by Q_{17}.

The base drive of Q_{17} is sufficient to drive it into saturation. As a result, the voltage at the C_{17}–B_{19} node is approximately

$$V_{B19} = V_{C17} = V_{BE18} + V_{CE(SAT)17} \simeq 0.9 \text{ V}$$

Since a voltage of at least $V_{BE19} + V_{D5} \simeq 1.2$ V is required at the base of Q_{19} in order to keep Q_{19} on, we see that Q_{19} will be turned off. The output is therefore driven down to the low state with Q_{20} sinking the load current.

Note that although Q_{20} may be driven into saturation, since Q_{18} and Q_{20} are connected in a current-mirror configuration, the base drive of Q_{20} is limited by Q_{18} and keeps Q_{20} from going very deeply into saturation.

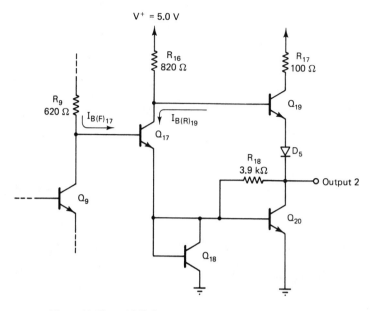

Figure 10.13 μA760 Output stage: high-to-low transition.

For the transition from the low state to the high state, Q_{17} must be rapidly brought out of saturation and into cutoff. This occurs when the input voltage is such that the current through Q_9 is greater than that through R_9, the difference between these two currents being the reverse base drive of Q_{17}, as indicated by $I_{B(R)17}$ in Figure 10.14. This reverse base current rapidly removes the stored charge in Q_{17} and quickly

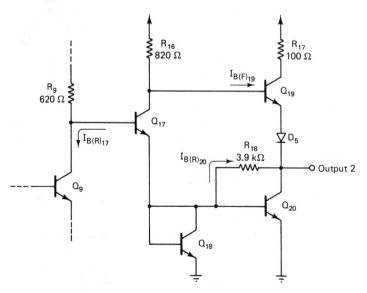

Figure 10.14 μA760 Output stage: low-to-high transition.

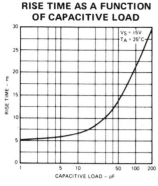

RISE TIME AS A FUNCTION
OF CAPACITIVE LOAD

(a)

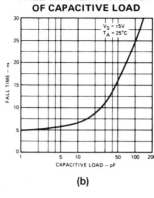

FALL TIME AS A FUNCTION
OF CAPACITIVE LOAD

(b)

Figure 10.15 Rise and fall times on the μA760 voltage comparator as a function of load capacitance: (a) rise time; (b) fall time (Fairchild).

brings it out of saturation and across the active region into cutoff. The rapidly rising voltage at the collector of Q_{17} turns Q_{19} and D_5 on. At the same time, Q_{18} and thus Q_{20} also are being turned off. The rapid turn-off of Q_{20} is aided by the flow of reverse base current $I_{B(R)20}$ through R_8 (3.9 kΩ).

With Q_{20} off and Q_{19} and D_5 on, the output voltage is now in the high state at a voltage level of $V_H = V^+ - V_{BE19} - V_{D5} = 5.0 - 0.7 - 0.7 = 3.4$ V. Since $V_L = V_{CE(SAT)20} \simeq 0.2$ to 0.4 V, we see that this comparator produces output voltages that are TTL compatible.

The input-voltage swing necessary to produce the output-voltage transition is approximately 1 mV, so the open-loop voltage gain is around 3000. This relatively small value of the open-loop gain as compared to the values commonly encountered for operational amplifiers is the result of purposely trading off a reduced gain for a faster response time.

The input offset voltage V_{OS} is 1.0 mV (typ.), 6.0 mV (max.), so we see that with the combination of the offset voltage and the input voltage swing of 1 mV that is required to produce the output transition, this device will typically switch at an input voltage level that is within about 2 mV of the reference voltage.

This output circuit has an active pull-up and pull-down configuration, with Q_{19} and Q_{23} used for the active pull-up to source currents into the load, and Q_{20} and Q_{24} used for the pull-down to sink load currents. This type of output circuit provides for the fast charging of load capacitances. Indeed, load capacitances of up to 10 pF will not produce any significant elongation of the response time. Large load capacitances, however, will start to produce a substantial lengthening of the response time, particularly with respect to the rise- and fall-time portions of the response time, as shown in Figure 10.15.

10.5.3 Comparator Example: LM160 and LM161

The use of a thinner, more lightly doped base region is an effective way of reducing the storage time. The thinner base width will result in a much smaller stored charge in the base region. The thinner and less heavily doped nature of the base region will also result in less injection of holes into the collector region, so the stored charge in

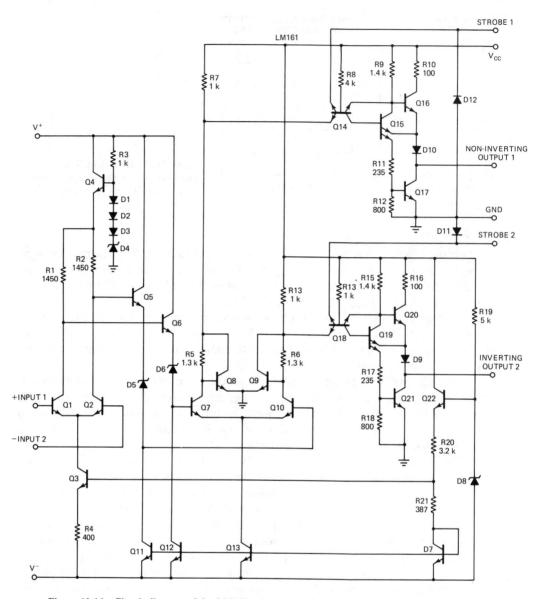

Figure 10.16 Circuit diagram of the LM161 voltage comparator (National Semiconductor).

the collector region will be reduced. The only serious drawback of the use of the thin, lightly doped base region is the reduction in the collector-to-emitter breakdown voltage, BV_{CEO}. This reduction in the breakdown voltage is due to the occurrence of the collector-to-emitter *punch-through* or *reach-through* phenomenon. This happens when the width of the depletion region on the base side of the collector–base junction is such that it extends all the way across the base region and reaches the

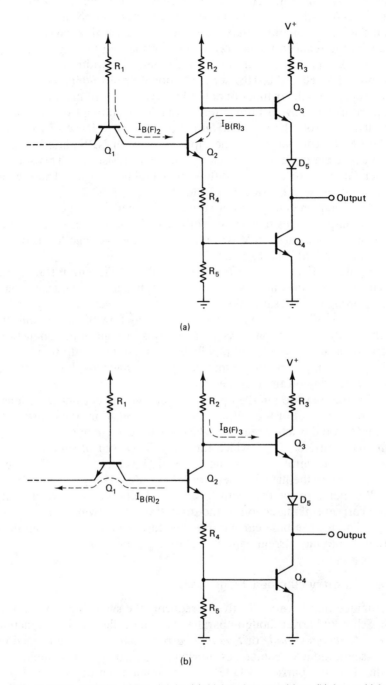

Figure 10.17 LM161 output circuit: (a) high-to-low transition; (b) low-to-high transition.

emitter–base junction. The effective base width is reduced to zero, and direct injection of electrons from the emitter to the collector can occur. The design of the IC and the specification of the maximum allowable applied voltages must be in conformance with the reduced breakdown voltage of the thin base transistors.

An example of comparators that use thin base width to achieve a fast response time is the LM160 and LM161 series (National Semiconductor). These comparators have propagation delay times of only 13 ns (typ.) for a 10-mV input overdrive. These comparators also have a storage time reduction resulting from the large reverse base current drawn from the saturated output transistors. Figure 10.16 is a circuit diagram of the LM161 comparator. The input stage is an NPN differential amplifier with current sink biasing and low-value passive load resistors. This stage is coupled by emitter followers to a second differential-amplifier stage. The second stage has a current-to-voltage converter type of load.

The complementary output voltages obtained from the second stage drive two identical output stages that are essentially high-speed TTL gates. These TTL gates are biased by a separate 5-V supply, so the output voltage levels of the comparator are at the standard TTL logic levels.

Figure 10.17a is a simplified diagram of the TTL output stage. In this diagram the current directions are shown for the output high-to-low transition for which Q_3 is being brought out of saturation. Note that with the emitter of Q_1 high the collector–base junction of Q_1 turns on. This results in the flow of current into the base of Q_2, which rapidly turns Q_2 on. As soon as Q_2 is brought into conduction, it sinks the reverse base current of Q_3 and rapidly brings Q_3 out of saturation. At the same time Q_4 is turned on, pulling the output down to the logic low level of $V_{CE(SAT)_4} \simeq 0.1$ to 0.3 V, depending on the load being driven.

For the transition in the opposite direction, the current directions are as shown in Figure 10.17b. For the output low-to-high transition, the emitter of Q_1 goes low, putting Q_1 into the active region. The collector current of Q_1 now results in a reversal of the base current of Q_2, which rapidly pulls Q_2 out of saturation and into cutoff. As Q_2 goes into cutoff, Q_4 is turned off and Q_3 is turned on. This causes the output voltage to rise to the high logic level of $V^+ - V_{BE_3} - V_{D_5} \simeq 5 - 0.7 - 0.7 = 3.6$ V.

We therefore see that both Q_2 and Q_3 can be rapidly brought out of saturation by the current-sinking action of the drive transistors, which for Q_2 is Q_1 and for Q_3 is Q_2. The reverse base current in either case is much larger than the forward base current due to the current gain of the driving transistor.

10.5.4 Schottky-clamped Transistors

A technique that is very effective in reducing the switching time of a transistor is to use a Schottky-barrier diode in parallel with the collector–base junction of the transistor. A *Schottky diode* or *Schottky-barrier diode* is a diode formed by a rectifying metal–semiconductor contact, as shown in Figure 10.18. The metals commonly used for the Schottky barrier include gold, aluminum, chromium, and platinum. The semiconductor is usually N-type silicon of moderate doping, generally in the range from 0.1 to 10 Ω-cm. The metal is a thin film typically of 1 μm thickness deposited by a vacuum evaporation or sputtering process.

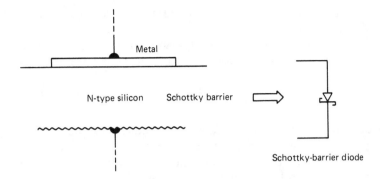

Figure 10.18 Schottky diode.

In Figure 10.19, the energy band (electric potential versus distance) diagram of a Schottky barrier formed between N-type silicon and a metal is shown. The height of the barrier (contact potential) ϕ_B at the metal/N-silicon junction is 0.56 V for chromium, 0.68 V for aluminum, 0.81 V for gold, and 0.90 V for platinum. This is the barrier for electron flow from the metal to the semiconductor and is high enough to prevent any significant flow of electrons in this direction.

The potential barrier for electron flow from the semiconductor to the metal is ϕ_S and is typically in the range of about 0.3 to 0.4 V. This is to be compared to the contact potential (or built-in voltage) of a silicon PN junction, which is generally in the range from 0.8 to 0.9 V. As a result of the much smaller barrier for electron flow in the Schottky diode, the forward-bias voltage that is required to produce a given level of forward current flow is about one-half of that for the PN junction diode.

Figure 10.19b is a diagram for forward-bias conditions. Under forward-bias conditions the applied voltage is such that the metal (the anode) is made positive with

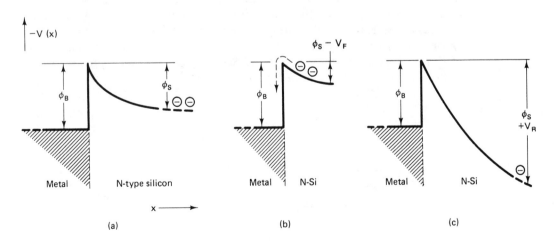

Figure 10.19 Energy-band diagram of Schottky barrier: (a) zero bias; (b) forward bias; (c) reverse bias.

respect to the semiconductor (the cathode). The voltage barrier for electron flow from the semiconductor to the metal is reduced from ϕ_S under zero-bias conditions to $\phi_S - V_F$, where V_F is the applied forward-bias voltage. The forward current I_F varies exponentially with the forward-bias voltage as given by the diode equation

$$I_F = I_O \left[\exp \left(\frac{V_F}{V_T} \right) - 1 \right] \approx I_O \exp \left(\frac{V_F}{V_T} \right) \tag{10.10}$$

The forward-bias voltage level required to produce currents in the milliampere range is only about 0.3 V, compared to voltages in the range from 0.6 to 0.7 V for the silicon PN junction. From the standpoint of the forward-bias voltage drop, the silicon Schottky-barrier diode is similar to the germanium PN junction diode.

In Figure 10.19c the Schottky barrier is shown under reverse-bias conditions. Now the potential barrier for electron flow from the semiconductor to the metal has been increased to $\phi_B + V_R$, where V_R is the reverse-bias voltage. This barrier is so high that there is negligible flow of electrons from the semiconductor to the metal. The high barrier ϕ_B for electron flow from the metal to the semiconductor also prevents any significant flow of electrons from the metal to the semiconductor. Therefore, under these reverse-bias conditions, the diode is "off." There is actually a very small reverse current or leakage current, due principally to electrons flowing from the metal to the semiconductor by climbing over the ϕ_B potential barrier.

Under forward-bias conditions the current that flows in the device is due to the injection of electrons from the semiconductor into the metal. Since electrons are the majority carriers on both sides of the junction, as soon as an electron leaves the N-type semiconductor its place is taken by an electron entering the semiconductor from the external circuit. Similarly, as soon as an electron enters the metal from the semiconductor, another electron leaves the metal by going out into the external circuit. As a result, charge neutrality is preserved in both the semiconductor and the metal and there are no excess electrons anywhere.

Since the concentration of holes in the N-type silicon is extremely small and there are no holes in the metal, there will be no injection or flow of holes in either direction. As a result, this is an entirely majority carrier device. There is no accumulation or storage of excess charge carriers on either side of the junction. As a result of this absence of charge storage, the Schottky-barrier diode proves to be a very fast device and exhibits no storage time.

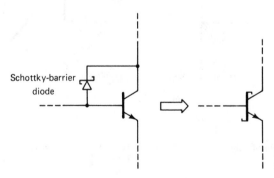

Schottky-barrier diode

Figure 10.20 Schottky-clamped transistor.

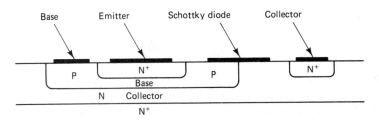

Figure 10.21 Integrated Schottky diode–transistor combination.

Figure 10.20 shows the combination of a transistor and a Schottky-barrier diode across the collector–base junction of the transistor. Usually, the Schottky diode is fabricated as an integral part of the transistor structure; the result is known as a *Schottky transistor* or a *Schottky-clamped transistor*. The symbol for this device is shown in Figure 10.20. Figure 10.21 is a cross-sectional view of an IC NPN transistor with an integral Schottky diode. The Schottky diode is formed by the overlap of the base contact metallization over the collector region. It should be noted that a Schottky-barrier diode is not formed at the collector contact, for here the metallization is in contact with the heavily doped N^+ region. For N^+ or P^+ regions with doping levels of about 1×10^{19} cm^{-3} or greater, a rectifying Schottky barrier is not produced, but rather a nonrectifying or *ohmic* contact will result.

When the collector–base junction is forward biased, the Schottky diode turns on at about 0.2 to 0.3 V and will be in full conduction at about 0.3 to 0.4 V. The Schottky diode bypasses current around the collector–base junction of the transistor. The forward-bias voltage across the collector–base junction is thus limited (or clamped) to a maximum value of about 0.3 to 0.4 V by the Schottky diode. Since the threshold for conduction of a silicon PN junction is about 0.5 V, we see that the clamping action of the Schottky diode prevents the collector–base junction from turning on.

As a result of the collector–base junction not being biased into conduction, there will be no injection of holes from the base into the collector, nor will there be any electrons emitted from the collector into the base. The only stored charge in the device is the electrons in transit across the base region, as given by $Q_{\text{stored}} = Q_{\text{base}} = I_C t_{\text{transit}}$. Since $I_{B(R)} t_{\text{storage}} \simeq Q_{\text{base}}$, the storage time can be expressed approximately as

$$t_{\text{storage}} \simeq \frac{I_{C(\text{SAT})}}{I_{B(R)}} t_{\text{transit}} \qquad (10.11)$$

where $I_{B(R)}$ is the reverse base current. Since the transit time is extremely short for the IC NPN transistors, typically in the range from 30 to 100 ps, we see that storage times of less than 1 ns can be achieved.

The switching speed of the transistor is now limited principally by the time required to change the voltage across the emitter–base junction capacitance, which is the fall time. Since

$$\Delta V_{BE} = \frac{\Delta Q_{BE}}{C_{BE}} \simeq \frac{I_{B(R)} t_{\text{fall}}}{C_{BE}} \qquad (10.12)$$

Sec. 10.5   Techniques to Reduce Response Time   **505**

we have that

$$t_{\text{fall}} \simeq \frac{\Delta V_{BE} C_{BE}}{I_{B(R)}}$$

Since $V_{BE(SAT)} \simeq 0.8$ V and $V_{BE(\text{cutoff})} \simeq 0.5$ V, we see that the change in the base-to-emitter voltage required to take the transistor from saturation to cutoff is only about 0.3 V. If $C_{BE} = 10$ pF and $I_{B(R)} = 10$ mA, this gives a fall time of $t_{\text{fall}} \simeq 0.3$ V $\times$ 10 pF/10 mA = 3 ns. We therefore see that with the Schottky-clamped transistor switching times down in the range of 3 ns can be obtained.

10.5.5 Comparator Using Schottky-clamped Transistors

An example of a comparator that uses Schottky-clamped transistors is the NE529 (Signetics). This device has the following propagation delay times for a 50-mV input overdrive:

$$\begin{aligned} t_{pd}(0) &= 10 \text{ ns (typ.)}, \ 20 \text{ ns (max.)} \\ t_{pd}(1) &= 12 \text{ ns (typ.)}, \ 20 \text{ ns (max.)} \end{aligned} \tag{10.13}$$

Figures 10.22 and 10.23 show the response-time characteristics for various input overdrives.

Figure 10.24 shows the circuit of this comparator. The circuit consists of two cascaded NPN differential amplifiers that drive two complementary TTL gates. The input stage is the Q_1–Q_2 differential amplifier biased by the Q_{27} current sink. Resistors R_2 and R_3 (1.5 kΩ each) serve as a low-resistance passive load for this stage to give a fast response time. The output voltage of the first stage is shifted downward in dc level by the emitter followers Q_4 and Q_5 and the zener diodes D_4 and D_5. This level-shift circuit is biased by the current sink transistors Q_6 and Q_7. The total dc shift is $V_{BE} + V_Z \simeq 7.0$ V.

The second differential-amplifier stage is Q_{10} and Q_{11} biased by the current sink transistor Q_8. The load for each half of the differential amplifier is a low-input-impedance current-to-voltage converter circuit consisting of Q_{12} and R_{11} (750 Ω) and R_{13} (1 kΩ) on one side and Q_{13}, R_{12}, and R_{14} on the other side.

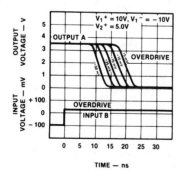

Figure 10.22 NE529 response times for various input overdrives (Signetics).

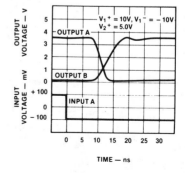

Figure 10.23 NE529 output propagation delays (Signetics).

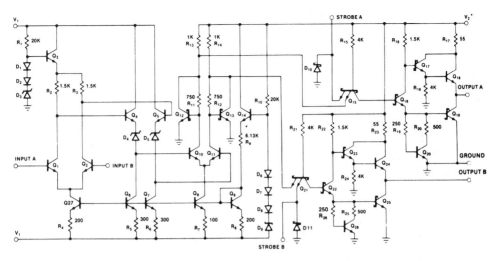

Figure 10.24 Circuit diagram of the NE529 voltage comparator (Signetics).

The output voltage produced by the second stage is used to drive the TTL gates in a complementary fashion. The TTL gates have active pull-up (Q_{17}, Q_{18} and Q_{23}, Q_{24}) and active pull-down (Q_{19}, Q_{20} and Q_{25}, Q_{26}) configurations for fast charging and discharging of the load capacitance.

Notice that most of the transistors in the circuit are of the Schottky-clamped type. The principal exceptions are those transistors that are not subject to being driven into saturation, such as Q_4 and Q_5, Q_{10} and Q_{11}, and Q_{18}. Also, the transistors used in the biasing circuitry do not need to be Schottky clamped.

The TTL output stages can be biased separately from the rest of the circuit by supplying a dc voltage of $+5.0$ V to the V_2^+ pin so as to give TTL-compatible logic outputs. The rest of the circuit can be biased with a substantially larger voltage, which will typically be $V_1^+ = +10$ V and $V_1^- = -10$ V so as to produce a faster response time and to obtain a larger common-mode input-voltage range.

10.5.6 Comparator Strobe Input

Some voltage comparators have a strobe input that allows the comparator to become deactivated by a digital input signal. The NE529 comparator that was just considered has two strobe inputs that are controlled by TTL-level digital signals, STROBE A and STROBE B. When a strobe input is low (0), the corresponding comparator output stays in the high (1) state independent of the input voltages that are applied to the comparator. When the strobe input goes high (1), the comparator becomes activated and responds to the input signals.

10.5.7 Additional High-speed Voltage Comparator Examples

Some other examples of high-speed voltage comparators and their response times are the 710 (TL710 Texas Instruments, LM710 National Semiconductor) at 40 ns, CMP-05 (Precision Monolithics, Inc.) at 37 ns, LM106 (National Semiconductor) at 28 ns,

Am686 (Advanced Micro Devices) at 9 ns, Am687 (Advanced Micro Devices) at 7 ns, and the Am685 (Advanced Micro Devices) at 5.5 ns.

10.6 CMOS INVERTER VOLTAGE GAIN

A CMOS inverter as shown in Figure 10.25 can be used as the basis of a voltage comparator. For a CMOS inverter that is operating in the high-gain transition region, both FETs will be on and operating in the active or saturated region. Each FET can be considered to operate as an active load for the other. For the two FETs we have that

$$I_{DS} = K(V_{GS} - V_t)^2 \qquad (10.14)$$

where the parameters K and V_t are the values that are appropriate for each transistor. The dynamic transfer conductance g_{fs} is given by

$$g_{fs} = \frac{dI_{DS}}{dV_{GS}} = 2K(V_{GS} - V_t) = \frac{2I_{DS}}{(V_{GS} - V_t)}$$

The active load resistance for each FET is the dynamic drain-to-source resistance r_{ds} of the other FET. This is given approximately by $r_{ds} = V_A/I_{DS}$, where $1/V_A$ is the channel-length modulation coefficient and V_A is typically in the range from 30 to 300 V. Considering the combined effects of the two FETs and the active loads, the CMOS voltage gain in the transition region for the simple case of $K_N = -K_P$, $V_{tN} = -V_{tP}$, and $V_{AN} = -V_{AP}$ is given by

$$A_V = 2g_{fs}\frac{r_{ds}}{2} = \frac{2I_{DS}}{V_{GS} - V_t}\frac{V_A}{I_{DS}} = \frac{2V_A}{V_{GS} - V_t} \qquad (10.15)$$

For the case under consideration, a quiescent operating point in the middle of the transition region is approximately halfway between the two supply voltages, V^+ and V^-, so $V_{GS(NMOS)} = -V_{GS(PMOS)} = (V^+ - V^-)/2$. Therefore, the voltage gain can be written approximately as

$$A_V = \frac{2|V_A|}{(V^+ - V^-)/2 - |V_t|} \qquad (10.16)$$

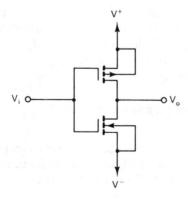

Figure 10.25 CMOS inverter.

Voltage Comparators   Chap. 10

As an example, if $|V_t| = 2.0$ V and $|V_A| = 60$ V for both FETs, and the total supply voltage $(V^+ - V^-)$ is 10 V, we obtain for the voltage gain the result $A_V = 2 \times 30$ V/$(5 - 2)$ V $= 40$. Note that this is under no-load conditions, and the voltage gain can be considerably lower if a substantial load is driven. If the supply voltage is reduced to just 5 V, the voltage gain in the transition region would reach a maximum of 240.

10.7 CMOS VOLTAGE COMPARATOR

A simple CMOS voltage comparator circuit is shown in Figure 10.26a, and the transfer curve is shown in Figure 10.26b. Transistors Q_1, Q_2, and Q_3 serve as analog switches, and the CMOS inverter can consist of one or several CMOS stages.

When ϕ_1 is high, transistors Q_1 and Q_2 are on, and for the CMOS inverter we have that $V_i = V_O = V_Q$. The voltage across capacitor C_1 is $V_{REF} - V_Q$. Then ϕ_1 goes low and the MOSFET switches Q_1 and Q_2 turn off. When ϕ_2 then goes high, transistor Q_3 turns on. The voltage across capacitor C_1 cannot change, however, since there is no current flow through the capacitor when Q_2 is off. Since the voltage

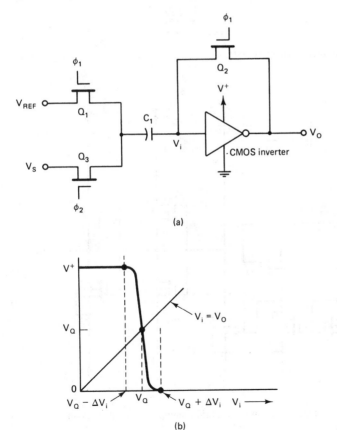

(a)

(b)

Figure 10.26 CMOS voltage comparator: (a) CMOS comparator circuit; (b) CMOS transfer curve.

at the left side of the capacitor changes from V_{REF} to V_S, the voltage at the CMOS inverter input is correspondingly changed by an amount $\Delta V_i = V_S - V_{REF}$ from V_Q to $V_Q + \Delta V_i$. As shown in Figure 10.26b, this can cause the output voltage V_O to switch to the low state if $V_S > V_{REF}$ ($\Delta V_i > 0$) or to the high state if $V_S < V_{REF}$ ($\Delta V_i < 0$).

The entire voltage comparator circuit can be implemented with just MOS transistors. As a result, this comparator requires very little chip area and can be especially useful for such applications as "flash" analog-to-digital converters that require very large numbers ($\gtrsim 100$) of comparators on one chip. The very low power drain of the CMOS inverters is also a very desirable feature for such applications.

10.7.1 Quad CMOS Voltage Comparator

An example of a quad CMOS voltage comparator is the MC14574 (Motorola). A simplified circuit diagram is shown in Figure 10.27. Transistors Q_1 and Q_2 constitute a PMOS differential amplifier with Q_3 and Q_4 serving as a current mirror active load. The differential amplifier is biased by the Q_5–Q_6 current mirror with a quiescent current given approximately by

$$I_Q \simeq I_{set} \simeq \frac{V^+ + V^- - 1\ V}{R_{set}}$$

where R_{set} is an external programming resistor.

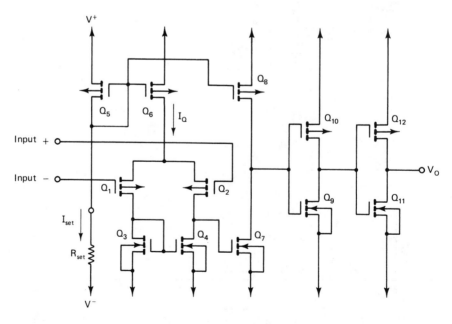

Figure 10.27 CMOS voltage comparator: MC14574 simplified circuit diagram. (Courtesy of Motorola Semiconductor, Inc.)

The second stage uses Q_7 as a common-source amplifier with Q_8 serving as a current-source active load. The third stage uses the Q_9-Q_{10} CMOS pair, which drives the $Q_{11}-Q_{12}$ CMOS output stage. When biased at $I_Q = I_{set} = 50$ μA, this comparator offers an open-loop gain, $A_{OL}(0)$, of 96 dB, but the propagation delay time is approximately 1 μs.

A second example of a quad CMOS voltage comparator is the TLC374 (Texas Instruments), which comes in a 14-pin DIP package. The differential-amplifier input stage uses a PMOS source-follower/NMOS common-gate configuration with a PMOS current-mirror active load. The second stage is a PMOS common-source amplifier with a current-source active load. This is followed by three cascaded CMOS inverter stages. The output stage is an NMOS common-source amplifier with an uncommitted drain such that an external pull-up resistor must be used, as in the case of the LM139. MOSFET op amps and voltage comparators often exhibit large offset voltage temperature coefficients and long-term drifts due to shifts in the threshold voltage caused by the migration of sodium ions through the gate oxide. This device uses a phosphorus-doped polycrystalline silicon gate technology that acts to immobilize sodium ions and as a result stabilize the MOSFET threshold voltage. For the TLC374, the offset voltage temperature coefficient is down to just 0.7 μV/°C and the long-term drift is 0.1 μV/month (typ.). Along with the low input bias current of 20 pA and offset current of 1 pA, this device offers the special feature of single supply operation with dc supply voltages as low as 2 V. The response time to a 100-mV step with a 5-mV overdrive is 650 ns, and for a TTL level input step the response time is down to 150 ns.

PROBLEMS

10.1. (*Pulse-width modulator*) Given a comparator with a triangular waveform (see Figure P10.1a) applied to the inverting input and an analog signal applied to the noninverting input terminal.

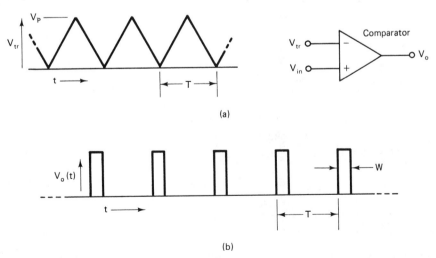

Figure P10.1

(a) Show that the output voltage waveform is as shown in Figure P10.1b with a pulse width given by $W = (V_{in}/V_P)T$ for $0 \leq V_{in} \leq V_P$.

(b) If $V_P = +10$ V and the voltage-to-pulse width conversion accuracy is to be within $\pm 1\%$ over a 1000 : 1 input voltage range, find the maximum value for V_{OS} if no offset voltage nulling is used.   [*Ans.:* ± 0.1 mV (max.)]

(c) If V_{OS} is nulled out at 25°C and the operating temperature range is to be from 0° to $+50°$C, find the maximum acceptable value for $TC_{V_{OS}}$.   (*Ans.:* ± 4 μV/°C)

(d) If $TC_{V_{OS}}$ is related to V_{OS} by $TC_{V_{OS}} = V_{OS}/T$, find the maximum acceptable value of V_{OS} for the above.   [*Ans.:* (± 1.2 mV (max.)]

(e) If $t_{pd(L-H)} = 50$ ns and $t_{pd(H-L)} = 40$ ns, find the maximum triangular wave frequency for the pulse width error due to the propagation delay times not to exceed 0.5% and the corresponding maximum signal frequency.   (*Ans.:* 500 kHz, 250 kHz)

10.2. (*Square-wave generator*)   Refer to Figure P10.2.   The high-state and low-state output-voltage levels of the comparator are V_{OH} and V_{OL}, respectively.

(a) Show that the frequency of oscillation f_{osc} is given by

$$f_{osc} = \frac{1}{2R_1 C_1 \ln 3}$$

(b) Show that the output voltage of the comparator is a square wave.

(c) If $R_1 = 1000$ Ω and $C_1 = 1$ nF, find the frequency of oscillation.   (*Ans.:* 455 kHz)

(d) What will limit the maximum frequency of oscillation available from this circuit?

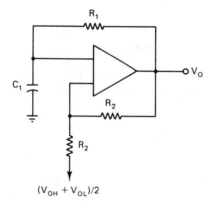

Figure P10.2

10.3. (*Square-wave generator*)   Refer to Figure P10.3.   The high-state and low-state output-voltage levels of the comparator are V_{OH} and V_{OL}, respectively.

(a) Show that the frequency of oscillation f_{osc} is given by

$$f_{osc} = \frac{1}{2R_1 C_1 \ln 2}$$

(b) Show that the output voltage of the comparator is a square wave.

(c) If $R_1 = 1000$ Ω and $C_1 = 1.0$ nF, find the frequency of oscillation.   (*Ans.:* 721 kHz)

(d) What will limit the maximum frequency of oscillation available from this circuit?

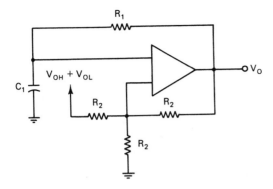

Figure P10.3

10.4. (*Pulse generator*)   Refer to Figure P10.4.

   (a) Find t_1.   (*Ans.: 6.9 μs*)
   (b) Find t_2.   (*Ans.: 69 μs*)
   (c) Find the frequency of oscillation.   (*Ans.: 13.1 kHz*)
   (d) Sketch V_O versus t for the case in which diodes D_1 and D_2 are interchanged.
   (e) By making the RC time constants (R_1C_1 and R_2C_1) smaller, the frequency can be increased.   What limits the maximum frequency of operation?
   (f) What is the function of the pull-up resistor?

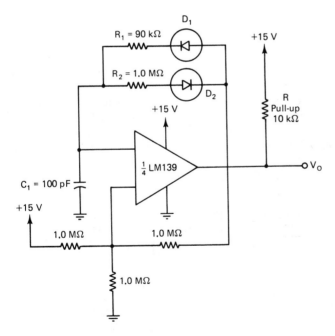

Figure P10.4

10.5. (*Time-delay generator*)   Refer to Figure P10.5.

   (a) Find the time delay of V_{O_1}, V_{O_2}, and V_{O_3}.   (*Ans.: 1.05, 2.23, 3.57 μs*)
   (b) Find the pulse lengths of V_{O_1}, V_{O_2}, and V_{O_3}.   (*Ans.: 8.95, 7.77, 6.43 μs*)

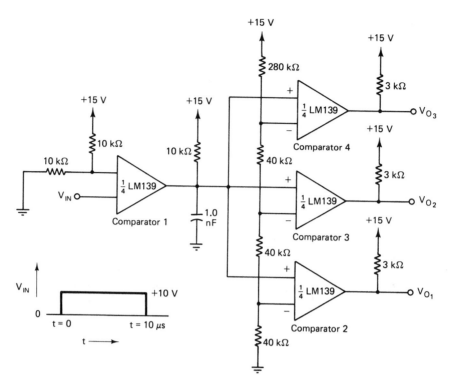

Figure P10.5

10.6. (*Thermostatic control system*)   Refer to Figure P10.6.   Given: The comparator C_1 has $V_{o(\text{high})} = +10$ V (heater on) and $V_{o(\text{low})} = 0$ (heater off).   The temperature sensor produces an output voltage given by $V_S = 10$ mV/°C $\times$ T, where T is the absolute temperature (K).   The temperature setpoint T_{SET} is to be adjustable from 20° to 30°C. The heating system is to turn on when the temperature drops below $T_{\text{SET}} - 1$°C and is to turn off when the temperature increases above $T_{\text{SET}} + 1$°C.   The reference voltage is $V_{\text{REF}} = +1.00$ V.

   (a) Find R_2, R_{F_1}, and R_{F_2}.   (*Ans.:* 2.0 kΩ, 1.93 kΩ, 100 Ω)

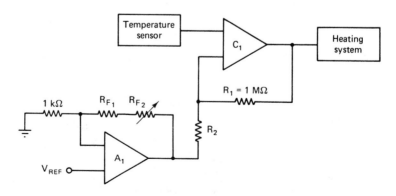

Figure P10.6

(b) Repeat part (a) for the case in which the reference voltage V_{REF} has a temperature coefficient of $+1000$ ppm/°C and is adjusted to $+1.00$ V at 20°C.   (*Ans.:* 1.4 kΩ, 1.93 kΩ, 70 Ω)

10.7. (*CMOS voltage comparator*)   A voltage comparator as shown in Figure 10.26 uses a three-stage CMOS inverter.   Each CMOS stage has a voltage gain of 20 in the transition region, and the total supply voltage is $V^+ = 15$ V.

(a) Find the approximate width of the transition region ΔV_i for the CMOS inverter. (*Ans.:* 1.9 mV)

(b) The leakage currents of MOSFET switches Q_1 and Q_2 are each 20 pA (min.)   and the minimum clock frequency is to be 10 kHz.   Find C_1 such that the voltage drift of V_i when ϕ_2 is high (ϕ_1 low) will not exceed 1.0 mV.   [*Ans.:* 2.0 pF (min.)]

(c) If $C_1 = 8.0$ pF and the net shunt capacitance at the input of the CMOS inverter is 4.0 pF, find ΔV_i in terms of $V_S - V_{REF}$ when ϕ_2 is high.   [*Ans.:* $\Delta V_i = \frac{2}{3}(V_S - V_{REF})$]

(d) If the inverter output resistance is 300 Ω, the MOSFET switch on resistance is 200 Ω, and the net capacitance driven by the inverter is 10 pF, find the time required for the CMOS inverter voltage to settle to within 10 mV of the quiescent level ($V_i = V_O = V_Q$) when ϕ_1 is high.   (*Ans.:* 31 ns)

(e) Find the maximum clock frequency subject to the conditions of part (d) and the requirement that the time that ϕ_2 is high should not be less than the time that ϕ_1 is high.   (*Ans.:* 15 MHz)

(f) For both the NMOS and PMOS transistors, $k = 10$ µA/V^2 and $V_{tN} = -V_{tP} = 2.5$ V.   Find V_Q, I_{DS} at $V_i = V_Q$ and the average power dissipation at $f = f_{MAX}$.   (*Ans.:* 7.5 V, 250 µA, 938 µW)

(g) Repeat part (d) for a total supply voltage of $V^+ = 10$ V.   (*Ans.:* 5.0 V, 63 µA, 156 µW)

(h) This CMOS voltage comparator is a sampled-data system.   Sketch the output voltage waveform for **(1)** $V_S > V_{REF}$ and **(2)** $V_S < V_{REF}$.

REFERENCES

ALLEN, P. E., and D. H. HOLBERG, *CMOS Analog Design*, Holt, Rinehart and Winston, New York, 1987.

CONNELLY, J. A., *Analog Integrated Circuits*, Wiley, New York, 1975.

GRAEME, J. G., G. E. TOBEY, and L. P. HUELSMAN, *Operational Amplifiers—Design and Applications*, McGraw-Hill, New York, 1971.

HAMILTON, D. J., and W. G. HOWARD, *Basic Integrated Circuit Engineering*, McGraw-Hill, New York, 1975.

LENK, J. D., *Handbook of Integrated Circuits*, Prentice-Hall, Englewood Cliffs, N.J., 1978.

————, *Manual for Integrated Circuit Users*, Prentice-Hall, Englewood Cliffs, N.J., 1973.

McMENAMIN, J. M., *Linear Integrated Circuits: Operation and Applications*, Prentice-Hall, Englewood Cliffs, N.J., 1985.

MILLMAN, J., *Microelectronics*, McGraw-Hill, New York, 1979.

MITRA, S. K., *An Introduction to Digital and Analog Integrated Circuits and Applications*, Harper & Row, New York, 1980.

MORLEY, M. S., *The Linear IC Handbook*, TAB Books, Blue Ridge Summit, Pa., 1986.

SEDRA, A. S., and K. C. SMITH, *Microelectronic Circuits*, Holt, Rinehart and Winston, New York, 1982.

11

Voltage Regulators

A voltage regulator is an electronic device that supplies a constant voltage to a circuit or load. The output voltage of the voltage regulator is regulated by the internal circuitry of the regulator to be relatively independent of the current drawn by the load, the supply or line voltage, and the ambient temperature. A voltage regulator may be part of some larger electronic circuit, but is often a separate unit or module, usually in the form of an integrated circuit.

A basic block diagram of a voltage regulator in its simplest form is shown in Figure 11.1. It is comprised of three basic parts:

1. A voltage reference circuit that produces a reference voltage that is independent of temperature and the supply voltage
2. An amplifier to compare the reference voltage with the fraction of the output voltage that is fed back from the voltage regulator output to the inverting input terminal of the amplifier
3. A series-pass transistor or combination of transistors to provide an adequate level of output current to the load being driven

The combination of the amplifier (often called an *error amplifier*) and the series-pass transistors, together with the resistive voltage divider to tap off a portion of the output voltage, constitutes a feedback amplifier.

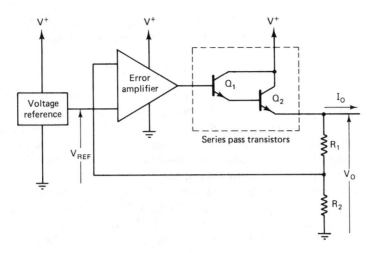

Figure 11.1 Voltage regulator: basic block diagram.

In the base circuit of Figure 11.1, the closed-loop amplifier configuration acts to maintain the fraction of the output voltage fed back to the amplifier inverting input terminal equal to the reference voltage that is applied to the noninverting input terminal. As a result we have that

$$V_O \frac{R_2}{R_1 + R_2} = V_{\text{REF}}$$

so

$$V_O = V_{\text{REF}} \frac{R_1 + R_2}{R_2} = V_{\text{REF}} \left(1 + \frac{R_1}{R_2} \right) \qquad (11.1)$$

11.1 OUTPUT RESISTANCE AND LOAD REGULATION

The ideal voltage regulator would be similar to an ideal voltage source in that the output voltage would be completely independent of the output current, or equivalently of the load impedance. In an actual voltage regulator, as is the case with an actual voltage source, there will be some variation of the output voltage with load or output current. The rate of change of the output voltage with output current is defined as the output resistance as given by

$$\text{dynamic output resistance} = r_o = -\frac{\Delta V_O}{\Delta I_O} \qquad (11.2)$$

The negative sign in the preceding expression is used because, for the reference direction of I_O that has been chosen, an increase in I_O leads to a decrease in V_O. Therefore, with the negative sign in front of $\Delta V_O / \Delta I_O$, the output resistance will turn out to be an algebraically positive quantity.

For a closed-loop amplifier, the following relationship exists between the closed-loop output impedance and the open-loop output impedance:

$$z_{o(CL)} = z_{o(OL)} \frac{A_{CL}}{A_{OL}} \qquad (11.3)$$

For the voltage regulator circuit shown in Figure 11.1, the amplifier input voltage is V_{REF} and the output voltage is V_O, so the closed-loop voltage gain is given by $A_{CL} = V_O/V_{REF}$. Note that the series-pass transistors (the Darlington configuration) are within the feedback loop and so can be considered to be part of the amplifier. Since the voltage gain of the Darlington emitter-follower stage is close to unity, the open-loop gain, A_{OL}, is essentially that of the error amplifier.

The open-loop output resistance is the dynamic resistance that is seen under open-loop conditions looking back into the voltage regulator circuit from the emitter of Q_2. This consists of two parts. One is the dynamic resistance of the Darlington transistors, as seen looking into the emitter of Q_2 back toward the base of Q_1. This dynamic resistance is $2V_T/I_{E_2} = 2V_T/I_O$.

Note that the dynamic resistance between the emitter and base of a single transistor, seen looking into the emitter, is V_T/I_E. Since the series-pass output stage under consideration here is a Darlington compound transistor configuration in which there are two emitter–base junctions in series, V_T in the preceding expression is replaced by $2V_T$.

The second part of the open-loop output resistance is the transformed value of the error amplifier output resistance, $r_{o(\text{amplifier})}$. The impedance transformation involved here is the result of the net current gain of the Darlington configuration from base to emitter, which is $(\beta_1 + 1)(\beta_2 + 1) \simeq \beta_1\beta_2$. The transformed value of the error amplifier output resistance is $r_{o(\text{amplifier})}/(\beta_1\beta_2)$. Since the transistor current gains are very large, generally on the order of 100, the product is on the order of 10,000, so the transformed value of the error amplifier output resistance will be very small indeed.

Combining the two resistance quantities, we now obtain for the open-loop output resistance the expression

$$r_{o(OL)} = \frac{2V_T}{I_O} + \frac{r_{o(\text{amplifier})}}{\beta_1\beta_2} \qquad (11.4)$$

As a result of the large value of the beta product, the second term is generally negligibly small, so we can write $r_{o(OL)} \simeq 2V_T/I_O$.

The closed-loop output resistance can now be written as

$$r_{o(CL)} = r_{o(OL)} \frac{A_{CL}}{A_{OL}} = \frac{(2V_T/I_O)(V_O/V_{REF})}{A_{OL}} \qquad (11.5)$$

where A_{OL} is the gain of the error amplifier.

Since $r_{o(CL)} = -\Delta V_O/\Delta I_O$, we now can write

$$\frac{\Delta V_O}{\Delta I_O} = -\frac{(2V_T/I_O)(V_O/V_{REF})}{A_{OL}} \qquad (11.6)$$

so

$$\frac{\Delta V_O}{V_O} = -\frac{\Delta I_O}{I_O} \frac{2V_T}{V_{REF}} \frac{1}{A_{OL}} \qquad (11.7)$$

Thus we now have a simple expression relating the fractional change in the voltage regulator output voltage to the fractional change in the output current. The *load regulation* is the change in the output voltage of a voltage regulator for a given change in the output current, generally from some no-load current condition to some specified full-load condition. Therefore, the equation above will give us some information relating to the voltage regulation, although it should be remembered that this equation is valid only for relatively small fractional changes in the voltage and current.

If we assume a reference voltage of 1.28 V (a band-gap voltage reference), we have

$$\frac{\Delta V_O}{V_O} = \frac{-\Delta I_O}{I_O} \frac{50 \text{ mV}}{1.28 \text{ V}} \frac{1}{A_{OL}} \cong \frac{-\Delta I_O}{I_O} \frac{1}{26 A_{OL}} \tag{11.8}$$

Therefore, if $A_{OL} = 1000$, for example, we obtain

$$\frac{\Delta V_O}{V_O} = \frac{-\Delta I_O}{I_O} \times 3.85 \times 10^{-5} \tag{11.9}$$

Thus a 10% change in the output current of the regulator results in only $3.85 \times 10^{-4}\%$ change in the output voltage. The load regulation for this regulator is thus very good indeed. The corresponding value of output resistance is given by

$$r_{o(CL)} = \frac{50 \text{ mV}}{I_O} \frac{10/1.28}{1000} = \frac{3.9 \times 10^{-4} \text{ V}}{I_O} \tag{11.10}$$

At an output current level of 100 mA, $r_{o(CL)}$ is 3.9 mΩ, and for a current level of 1.0 A the closed-loop output resistance is only 0.39 mΩ. These very low values of output resistance again attest to be the very good load regulation of this voltage regulator.

In looking at the expression for the regulator output resistance, we see that as the output current decreases the output resistance increases, and for very small values of output current, the load resistance can become unacceptably large. This can lead to a serious degradation of the load regulation characteristics of the voltage regulator. For this reason, in some voltage regulator circuits a means is provided to allow for a small output current to flow under no-load condition. To examine this in more detail, we can rewrite the expression for the fractional change in the output voltage in differential form as

$$\frac{dV_O}{V_O} = -\frac{dI_O}{I_O} \frac{2V_T}{V_{\text{REF}}} \frac{1}{A_{OL}} \tag{11.11}$$

If we now integrate both sides of this expression, we obtain

$$\frac{V_{O(\text{FL})}}{V_{O(\text{NL})}} = \left(\frac{I_{O(\text{FL})}}{I_{O(\text{NL})}}\right)^{-(2V_T/A_{OL}V_{\text{REF}})} \tag{11.12}$$

Again using the values of $V_{\text{REF}} = 1.28$ V and $A_{OL} = 1000$, the exponent is -3.9×10^{-5}. If the no-load current, $I_{O(\text{NL})}$, is chosen to be 1% of the full-load current,

$I_{O(FL)}$, we have that

$$\frac{V_{O(FL)}}{V_{O(NL)}} = (100)^{-3.9 \times 10^{-5}} = 0.99982 = 1 - 0.00018 \qquad (11.13)$$

Thus the decrease in the output voltage in going from the no-load current of 1% of the full-load current to the full-load current is only 0.018%. Therefore, for many applications, a no-load current that is only 1% of the full-load value will still give acceptable load regulation results.

11.2 REGULATOR WITHIN A REGULATOR DESIGN

In the basic voltage regulator system, we have seen that the output resistance is directly proportional to the closed-loop gain. If the closed-loop gain could be reduced to unity, the output resistance would be reduced to a minimum and the load regulation characteristics of the regulator can be improved.

A regulator configuration that achieves this objective is illustrated in Figure 11.2. This system is actually two regulators in one. The voltage reference, error amplifier A_1, and the R_1, R_2 resistance voltage divider constitute a voltage regulator that produces an output voltage given by $V_O = V_{REF}(1 + R_1/R_2)$. The output voltage of this regulator in turn serves as the input or reference voltage for the second regulator comprised of error amplifier A_2 and the series-pass transistors. Note that this second voltage regulator has unity feedback, so its closed-loop gain is unity.

The closed-loop output resistance of this voltage regulator is therefore given by

$$r_{o(CL)} = \frac{2V_T/I_O}{A_{OL}} \qquad (11.14)$$

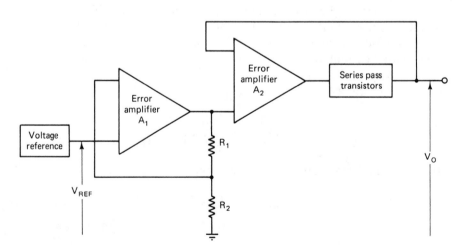

Figure 11.2 Voltage regulator using the regulator within a regulator design for improved load regulation.

Voltage Regulators   Chap. 11

where A_{OL} is the open-loop gain of the second error amplifier. The fractional change in the output voltage is related to the fractional change in load current by

$$\frac{\Delta V_O}{V_O} = -\frac{\Delta I_O}{I_O}\frac{2V_T/V_O}{A_{OL}} \qquad (11.15)$$

Note that the "reference voltage" for the output regulator is V_O since it is operated in a unity-gain condition.

If we again take $A_{OL} = 1000$ for an example, and $V_O = 10$ V, we obtain

$$\frac{\Delta V_O}{V_O} = \frac{-\Delta I_O}{I_O}\frac{50 \text{ mV}/10 \text{ V}}{1000} = -\frac{\Delta I_O}{I_O} \times 5 \times 10^{-6} \qquad (11.16)$$

Thus a 10% change in the load current results in only a $5 \times 10^{-5}\%$ change in V_O, or only 5 µV, a very small change indeed.

To evaluate the change in the output voltage in going from some no-load current value to a full-load current, we have the relationship

$$\frac{V_{O(FL)}}{V_{O(NL)}} = \left(\frac{I_{O(FL)}}{I_{O(NL)}}\right)^{-(2V_T/A_{OL}V_O)} \qquad (11.17)$$

For the example under consideration here, the exponent will have a value of 50 mV/$(1000 \times 10$ V$) = 5 \times 10^{-6}$.

If we take as a no-load current a current equal to 1% of the full-load current, we obtain

$$\frac{V_{O(FL)}}{V_{O(NL)}} = (100)^{-5 \times 10^{-6}} = 0.99998 = 1 - (2.3 \times 10^{-5}) \qquad (11.18)$$

so the decrease in the regulator output voltage in going from the no-load condition to the full-load condition is only a very minute 0.0023%. This represents very good load regulation indeed.

11.3 VOLTAGE REGULATOR PROTECTION CIRCUITRY

In addition to the basic elements of the voltage regulator circuit that have thus far been discussed, most voltage regulators also incorporate circuitry to protect the voltage regulator from excessive power dissipation. This excessive power dissipation can lead to overheating and permanent damage. For most voltage regulators, this internal protection circuit takes the form of a current-limiting circuit that senses the output current and acts to prevent the current from exceeding some preset safe limit.

11.3.1 Current Limiting

The power dissipated within the voltage regulator will be principally in the series-pass transistors, and mostly in the second transistor (Q_2) of the Darlington pair, which carries most of the current. The power dissipated in the voltage regulator is therefore given approximately by $P_d = V_{CE_2}I_{C_2} = (V_{IN} - V_O)I_O$, wherein V_{IN} is the unregulated dc supply voltage for the voltage regulator circuit (V^+).

The worst-case condition from the standpoint of power dissipation in the voltage regulator occurs when the output voltage is zero, for in this situation the entire dc supply voltage will appear across Q_2 and therefore the power dissipation will be given by $P_d = V_{IN}I_O$. This worst-case situation may occur, for example, when there is an accidental short-circuit condition existing across the load.

From the standpoint of a conservative design philosophy, it would be wise to design the regulator such that it will be safe from damage even under a sustained short-circuited load condition. If the maximum allowable power dissipation of the voltage regulator is indicated as $P_{d(max)}$, for short-circuit protection the protection circuitry should act to limit the current to a value given by $V_{IN}I_{CL} = P_{d(max)}$, so that $I_{CL} = P_{d(max)}/V_{IN}$, where I_{CL} is the value at which the output current is limited.

The circuit diagram of a simple current-limiting circuit is shown in Figure 11.3 together with the Darlington output stage of the regulator. The operation of this circuit is relatively simple. We will assume, however, that the current output of the error amplifier is limited to some maximum value by the circuit configuration of this amplifier. For a specific example, let us assume that this maximum current that the amplifier can supply is 50 mA. Let us also assume that the maximum power dissipation limit of Q_2 is 50 W and that the supply voltage for the regulator is no more than 20 V. Therefore, for short-circuit protection the value of I_{CL} should not exceed 50 W/20 V = 2.5 A. Let us assume that the two transistors that comprise the Darlington pair each have minimum current gains of 50, so the overall current gain of the Darlington configuration is 2500 (min).

If there were no current limiting, under short-circuit load conditions, the amplifier would be supplying its maximum available current output of 50 mA. This current,

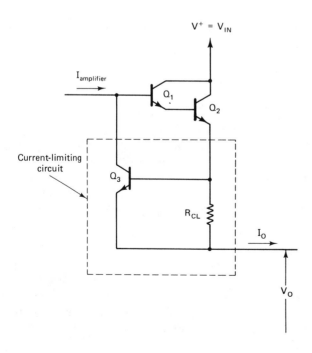

Figure 11.3 Current-limiting circuit.

when multiplied by the overall current gain of the Darlington stage, would result in a current of at least 50 mA $\times$ 2500 = 125 A! This high current level would most certainly irreparably damage the voltage regulator, so the use of current limiting is certainly indicated for this case.

At lower current levels such that the voltage drop across R_{CL} is less than about 500 mV, the base–emitter junction of Q_3 is "off" and Q_3 will not shunt any base current away from Q_1. Indeed, the behavior of the circuit is essentially the same as if Q_3 were not present. At higher current levels, the voltage drop across R_{CL} is such that Q_3 starts to turn on, thereby bypassing some current from the base to Q_1 directly to the load through Q_3. Note that this bypassed current is not afforded the high current gain of the Darlington pair. Due to the action of the feedback loop, however, the output current of the amplifier increases to make up for the loss of the current shunted through Q_3, and thereby maintain the output voltage at its regulated value of V_O as the load current increases. Nevertheless, since the current shunted through Q_3 increases exponentially with the base-to-emitter voltage of Q_3, which is $I_O R_{CL}$, as the load current increases a point will soon be reached where the current shunted through Q_3 approaches the maximum current available from the amplifier. When this point is reached, the base current of Q_1 can no longer increase and as a result the output current reaches a limiting value known as I_{CL}.

To continue with the example, we have seen that for protection of this regulator circuit a current limit of 2.5 A is necessary. The base drive required at the base of Q_1 to supply this current is only 2.5 A$/\beta_1\beta_2$ = 2.5 A/2500 (min.) = 1.0 mA (max.). Since the amplifier is capable of supplying 50 mA, this means that in the current-

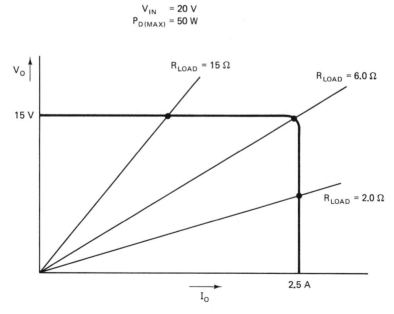

Figure 11.4 Output characteristics of a voltage regulator with simple current limiting.

limiting condition transistor Q_3 must be shunting almost 50 mA. If we assume that the base-to-emitter voltage of Q_3 is 650 mV at $I_C = 1.0$ mA, the V_{BE} at 50 mA is given by $V_{BE} = 650$ mV $+ V_T \ln 50 = 650$ mV $+ 98$ mV $= 748$ mV. Therefore, the voltage drop across R_{CL} should be 748 mV at $I_O = I_{CL} = 2.5$ A. The value of R_{CL} is therefore 748 mV/2.5 A $= 0.30$ Ω.

Figure 11.4 gives the output characteristics of this voltage regulator. These characteristics are for a regulated output voltage of 15 V. Load lines corresponding to several different load resistance values are shown. Note that for load resistances greater than 6.0 Ω the regulator maintains the output voltage at 15 V. However, when the load resistance drops below 6.0 Ω, the regulator becomes current limited at $I_{CL} = 2.5$ A.

11.3.2 Foldback Current Limiting

In the voltage regulator circuit just considered, the current limit was set such that under the worst-case short-circuit conditions the power dissipation $V_{IN}I_{CL} = 20$ V $\times$ 2.5 A $= 50$ W does not exceed the maximum power dissipation of the device. Note, however, that in the region where V_O is maintained at the regulated value of 15 V, the voltage drop across the output resistor (Q_2) is only 5 V, so under these conditions a maximum current of 50 W/5 V $= 10$ A could be allowed without exceeding the

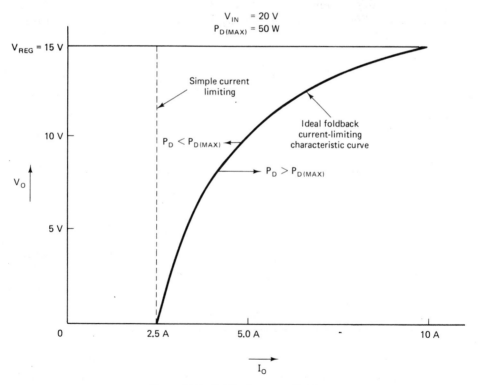

Figure 11.5 Foldback current limiting.

maximum power dissipation. As a result, we see that with the simple current limiting of the type considered the full potential of the voltage regulator is not utilized.

Since $P_D = (V_{IN} - V_O)I_O$, the output current limit that must be set to prevent the power dissipation from exceeding $P_{D(MAX)}$ is given by

$$I_{O(CL)} = \frac{P_{D(MAX)}}{V_{IN} - V_O} \tag{11.19}$$

Therefore, to ensure full protection of the device and at the same time to obtain the maximum output current possible under these conditions, a current-limiting characteristic of the type described by the equation above should be used. Note that the current limit is a function of the output voltage. As the output voltage decreases, the value of $I_{O(CL)}$ also decreases. This type of current-limiting characteristic is shown in Figure 11.5. This type of current limiting is called *foldback* current limiting, due to the shape of the regulator output characteristic curve.

For foldback current limiting, the limiting circuitry must sense not only the output current but the output voltage as well. The simplest type of foldback current limiting is the linear foldback type, in which the decrease in $I_{O(CL)}$ is a linear function of V_o. This is the easiest type of foldback limiting to implement in circuit form, and the characteristics of this type of current limiting are shown in Figure 11.6. The current limit when the output voltage is V_O is given by $I_{O(CL)} = P_{D(MAX)}/(V_{IN} - V_O)$. For full regulator protection, the regulator should not be allowed to operate in the

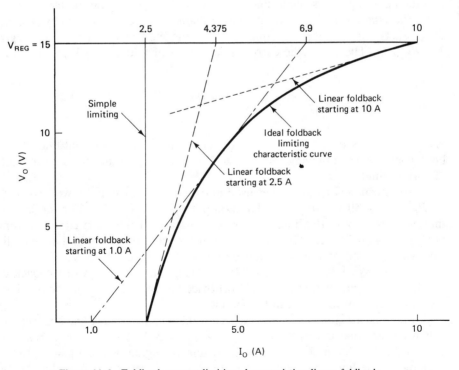

Figure 11.6 Foldback current limiting characteristics: linear foldback.

region beyond the foldback limiting curve described by this equation. For the linear foldback case, we must now determine the required short-circuit current limit.

Short-circuit current limit for linear foldback. Since $I_{O(CL)} = P_{D(MAX)}/(V_{IN} - V_O)$, we have that $V_O = V_{IN} - (P_{D(MAX)}/I_{O(CL)})$. The slope of the current limit curve is therefore given by $dV_O/dI_{O(CL)} = P_{D(MAX)}/(I_{O(CL)})^2$. The slope at $I_{O(CL)} = P_{D(MAX)}/(V_{IN} - V_O)$ is therefore $(V_{IN} - V_O)^2/P_{D(MAX)}$. The current axis intercept of the linear foldback line that starts at $V_O = V_{REG}$ is therefore at

$$I_{O(CL)}(V_O = 0) = \frac{P_{D(MAX)}}{V_{IN} - V_{REG}} - \frac{P_{D(MAX)} V_{REG}}{(V_{IN} - V_{REG})^2} \tag{11.20}$$

This last expression can be rewritten as

$$I_{O(CL)}(V_O = 0) = \frac{P_{D(MAX)}}{V_{IN} - V_{REG}} \left(1 - \frac{V_{REG}}{V_{IN} - V_{REG}} \right)$$

$$= \frac{P_{D(MAX)}}{V_{IN} - V_{REG}} \frac{V_{IN} - 2V_{REG}}{V_{IN} - V_{REG}} \tag{11.21}$$

From this last equation we note that if the regulator supply voltage is not more than twice the regulated output voltage V_{REG}, the required value for the short-circuit current limit is negative. Since this is not practical, for these cases a type of linear foldback that starts at a short-circuit current limit of $I_{O(CL)}(V_O = 0) = P_{D(MAX)}/V_{IN}$ and then goes up in a straight line as shown in Figure 11.6 can be used.

Since the slope of the current-limiting characteristic curve is given by $(V_{IN} - V_O)^2/P_{D(MAX)}$, the initial slope starting up from the current axis ($V_O = 0$) is given by $V_{IN}^2/P_{D(MAX)}$. Therefore, the current limit for linear foldback at $V_O = V_{REG}$ is given by

$$I_{O(CL)}(V_{REG}) = \frac{P_{D(MAX)}}{V_{IN}} + V_{REG} \frac{P_{D(MAX)}}{V_{IN}^2} = \frac{P_{D(MAX)}}{V_{IN}} \left(1 + \frac{V_{REG}}{V_{IN}} \right) \tag{11.22}$$

As a result, we see that for this type of current limiting the current limit the current limit when the output voltage is V_{REG} is $(1 + V_{REG}/V_{IN})$ times the short-circuit value of current limit.

To return now to the values used in the preceding example, we have $V_{IN} = 20$ V, $P_{D(MAX)} = 50$ W, and a regulated output voltage of 15 V. For this case we see that we must choose the linear foldback that starts at the short-circuit current limit of 50 W/20 V = 2.5 A. The current limit will start at this value and increase linearly with voltage, and at the regulated output voltage level of 15 V, it will be 2.5 A $(1 + 15/20) = 2.5$ A $\times$ 1.75 = 4.375 A. Thus, in this case with foldback current limiting, an output current that is almost twice the value that would be obtained with the simple current limiting can be obtained.

Looking at Figure 11.6, we note that if the linear foldback line begins at a short-circuit current that is less than $P_{D(MAX)}/V_{IN} = 2.5$ A, then an even greater output current can be obtained in the voltage-regulated region where $V_{REG} = 15$ V. For example, if the linear foldback starts at a current level of 1.0 A under short-circuit conditions, it can increase to 6.9 A at $V_{REG} = 15$ V without exceeding the maximum power dissipation rating of the device. For this case an output current that is almost three times that which would be obtained with simple current limiting (2.5 A) can be obtained.

Linear current foldback circuit. An example of a linear current foldback circuit is shown in Figure 11.7. The linear foldback characteristics of this circuit will be obtained by first obtaining an expression for the base-to-emitter voltage of Q_3, which is

$$V_{BE3} = (V_O + I_O R_{CL}) \frac{R_4}{R_4 + R_3} - V_O$$

$$= I_O \frac{R_{CL} R_4}{R_3 + R_4} - V_O \left(1 - \frac{R_4}{R_3 + R_4} \right) \tag{11.23}$$

$$= I_O \frac{R_{CL} R_4}{R_3 + R_4} - V_O \frac{R_3}{R_3 + R_4}$$

In this derivation the base current of Q_3 was assumed to be small compared to the current through R_3 and R_4. If we now solve for I_O, we obtain

$$I_O = \frac{R_3 + R_4}{R_{CL} R_4} \left(V_{BE} + V_O \frac{R_3}{R_3 + R_4} \right) \tag{11.24}$$

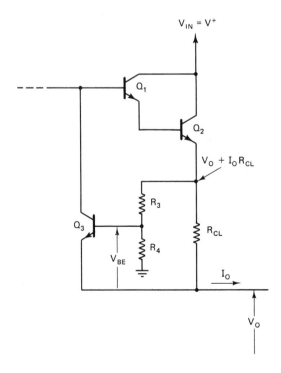

Figure 11.7 Linear foldback circuit.

Current limiting will occur when $V_{BE} = V_{BE(CL)}$, which is the value of the base-to-emitter voltage of Q_3 required to shunt a current approximately equal to the maximum output current of the amplifier. The value of $V_{BE(CL)}$ is generally in the range 0.6 to 0.7 V. We therefore have that

$$I_{O(CL)} = \frac{R_3 + R_4}{R_{CL}R_4} V_{BE(CL)} + V_O \frac{R_3}{R_{CL}R_4} \tag{11.25}$$

We see that this circuit will indeed produce linear foldback since $I_{O(CL)}$ increases linearly with V_O.

The short-circuit current limit $I_{O(CL)}(0)$ is given by

$$I_{O(CL)}(0) = I_{O(CL)}(V_O = 0) = \frac{R_3 + R_4}{R_{CL}R_4} V_{BE(CL)} \tag{11.26}$$

The change in $I_{O(CL)}$ in going from $I_{O(CL)}(0)$ to $I_{O(CL)}$ at V_O is given by

$$\Delta I_{O(CL)} = I_{O(CL)}(V_O) - I_{O(CL)}(0) = \frac{V_O R_3}{R_{CL}R_4} \tag{11.27}$$

Another useful design relationship is the ratio of $\Delta I_{O(CL)}$ to $I_{O(CL)}(0)$, which is

$$\frac{\Delta I_{O(CL)}}{I_{O(CL)}(0)} = \frac{V_O R_3}{(R_3 + R_4)V_{BE(CL)}} \tag{11.28}$$

Notice that R_{CL} cancels out of this relationship.

Design example. Let us consider the design of a linear foldback circuit to meet the following requirements based on the current and voltage values obtained from the previous discussion of linear foldback:

$$\begin{aligned} I_{O(CL)}(0) &= 1.0 \text{ A} \\ I_{O(CL)}(15 \text{ V}) &= 6.9 \text{ A} \end{aligned} \tag{11.29}$$

so we want to design the foldback circuit to limit the output current to a maximum of 6.9 A at the regulated output voltage of 15 V; but then as the voltage drops to zero, the current limit will correspondingly drop to 1.0 A. We will assume that $V_{BE(CL)} = 0.6$ V. We will also choose a current of 50 mA through R_3 and R_4 so that the base current of Q_3 can be neglected.

For $R_3 + R_4$ we therefore have $R_3 + R_4 \simeq 15$ V/50 mA = $\underline{300\ \Omega}$. We can now solve for R_3 using the following relationship:

$$\frac{\Delta I_{O(CL)}}{I_{O(CL)}(0)} = \frac{6.9 - 1}{1} = \frac{V_O R_3}{(R_3 + R_4)V_{BE(CL)}} = \frac{15 \text{ V} \times R_3}{300\ \Omega \times 0.6 \text{ V}} \tag{11.30}$$

so we obtain for R_3 the result

$$R_3 = 5.9 \times 300\ \Omega \times 0.6 \text{ V}/15 \text{ V}$$

$$= \underline{71\ \Omega}$$

TABLE 11.1 POWER
DISSIPATION VALUES
(V_{IN} = 20 V)

V_O (V)	$I_{O(CL)}$ (A)	P_D (W)
0	1.0	20
1.0	1.39	26.5
2.0	1.79	32.2
3.0	2.12	37.1
4.0	2.57	41.2
5.0	2.97	44.5
6.0	3.36	47.0
7.0	3.75	48.8
8.0	4.15	49.8
9.0	4.54	49.9
10.0	4.93	49.3
11.0	5.33	47.9
12.0	5.72	45.8
13.0	6.11	42.8
14.0	6.51	39.0
15.0	6.90	34.5

Therefore, $R_4 = 300 - 72 = \underline{229 \ \Omega}$. For R_{CL} we can now use the equation for $I_{O(CL)}(0)$ as

$$I_{O(CL)}(0) = 1.0 \ A = \frac{R_3 + R_4}{R_{CL}R_4} V_{BE(CL)} = \frac{300}{R_{CL}229} \times 0.6 \ V \qquad (11.31)$$

so $R_{CL} = \underline{0.79 \ \Omega}$. If we now substitute these resistance values back into the equation for $I_{O(CL)}$, we obtain

$$I_{O(CL)} = 1.0 + (0.393 \ V_O) \quad A \qquad (11.32)$$

Thus we can verify that with the resistance values chosen the current limit is 1.0 A under short-circuit conditions and 6.9 A when V_O is 15 V.

We can verify that this foldback circuit will indeed protect the voltage regulator from excessive power dissipation by looking at the power dissipation values given in Table 11.1.

Linear foldback in which I_{CL} is a function of $V_{IN} - V_O$. The power dissipation in the voltage regulator circuit is given by $P_D = V_{IN}I_Q + (V_{IN} - V_O)I_O$, where I_Q is the quiescent current of the regulator, generally only a few milliamperes. Since I_Q is so very much smaller than I_O in the region where we are concerned about the power dissipation limitation, the equation for P_D can be written approximately as $P_D = (V_{IN} - V_O)I_O$. To look at it in a slightly different way, the power dissipation due to I_Q is on the order of 5 mA × 20 V = 100 mW; so as long as the maximum power dissipation is much greater than this, we can neglect the quiescent power dissipation, $V_{IN}I_Q$, of the voltage regulator.

Since the power dissipation in the regulator is essentially proportional to $V_{IN} - V_O$, the input–output voltage differential, the foldback current limit should be related to this voltage difference rather than on just V_O, for the greatest flexibility. For the foldback current limiting of the preceding example, the design should be based on the maximum allowable value of the input voltage, V_{IN}. If foldback current limiting is based on a current limit set by $V_{IN} - V_O$, a greater maximum output current can be allowed when the input voltage is less than the maximum rated value.

An example of a linear foldback circuit in which the current limit is directly related to the input–output voltage differential is shown in Figure 11.8. Transistors Q_1 and Q_2 constitute the Darlington emitter-follower series pass output circuit, and Q_3 is the current-limit transistor. The base-to-emitter voltage of Q_3, assuming the base current to be small compared to the current through R_3 and R_4, is given by

$$V_{BE_3} = I_O R_{CL} + \frac{(V_{IN} - V_Z - I_O R_{CL} - V_O)R_4}{R_3 + R_4} \tag{11.33}$$

After rearrangement we obtain

$$V_{BE_3} = \frac{I_O R_{CL} R_3}{R_3 + R_4} + \frac{[(V_{IN} - V_O) - V_Z]R_4}{R_3 + R_4} \tag{11.34}$$

Current limiting occurs when the base-to-emitter voltage of Q_3 is high enough to cause Q_3 to shunt the base drive away from the Q_1, Q_2 Darlington pair. Designating this voltage as $V_{BE(CL)}$ (generally around 0.6 to 0.7 V) and the corresponding current-limited output current as $I_{O(CL)}$, we have after rearrangement of the equation above the result that

$$I_{O(CL)} = V_{BE(CL)} \frac{R_3 + R_4}{R_3 R_{CL}} - [(V_{IN} - V_O) - V_Z] \frac{R_4}{R_3 R_{CL}} \tag{11.35}$$

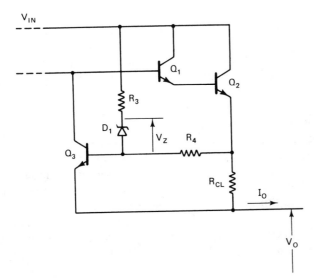

Figure 11.8 Foldback current limiting that is dependent on $V_{IN} - V_O$.

Voltage Regulators   Chap. 11

From this last equation we see that $I_{O(CL)}$ will indeed be a function of the input–output voltage differential and that, as this voltage differential increases, the current limit will decrease.

To consider a design example of this type of foldback circuit, we will choose a short-circuit current limit of 1.0 A, which increases to 6.0 A at $V_{IN} - V_O = 6.0$ V. The input voltage will be 20 V and we will choose $R_3 + R_4 = 400 \, \Omega$. We will assume a V_Z of 6.0 V. Upon substitution into the equation for $I_{O(CL)}$, we obtain the following resistance values:

$$R_4 = 14.3 \, \Omega$$

$$R_3 = 385.7 \, \Omega \tag{11.36}$$

$$R_{CL} = 0.104 \, \Omega$$

Substitution of these resistance values back into the equation for $I_{O(CL)}$ gives

$$I_{O(CL)} = 6.0 \text{ A} - [(V_{IN} - V_O) - 6 \text{ V}]0.357 \tag{11.37}$$

At this point it should be noted that the equations above are valid only for the case in which $V_{IN} - V_O > V_Z$. If the input–output differential is less than this, the zener diode, D_1, will be off (reversed biased, but not in the breakdown region). For this case the current through R_4 is negligible and we have that $V_{BE_3} = I_O R_{CL}$, so the current-limiting condition becomes simply $I_{O(CL)} = V_{BE(CL)}/R_{CL}$. For the example above, the simple current-limiting case occurs for input–output voltage differentials of 6.0 V or less, for which case the value of $I_{O(CL)}$ will be constant at 6.0 A.

A graph of the voltage regulator characteristics for the example under consideration is shown in Figure 11.9. At various points along the current-limit line, values of the output voltage, output current, and power dissipation are given. Note that all the power dissipation values lie safely below the $P_{D(MAX)}$ of 50 W. If just simple current limiting were to be used instead, a current limit of only 50 W/20 V = 2.5 A would have to be established.

The constant-current region that occurs for input–output differentials of less than V_Z (6 V) is useful in many cases to prevent excessive current that could damage the circuit due to such effects as electromigration of the metal-to-silicon contacts.

For a voltage regulator to operate properly, a minimum value of the input–output voltage differential is required, typically around 2 V. Therefore, for this voltage regulator, operation above around 18 V is not possible, as indicated by the shaded portion of the regulator characteristics.

Figure 11.10 shows the output characteristics of the same voltage regulator circuit again, except in this case the input voltage has been increased to 28 V. The same current-limiting circuit is that used for the preceding case with the same resistance values. Again note that, at various points along the current-limit line, values of output voltage, output current, and the power dissipation are given. We see that, in spite of the fact that the input voltage had been raised to 28 V, the power dissipation values are still safely below the maximum power dissipation rating of 50 W. This illustrates the versatility of this type of foldback current-limiting circuit. We may also note that if simple current limiting were to be used here, the current limit would have to be set at only 50 W/28 V = 1.8 A.

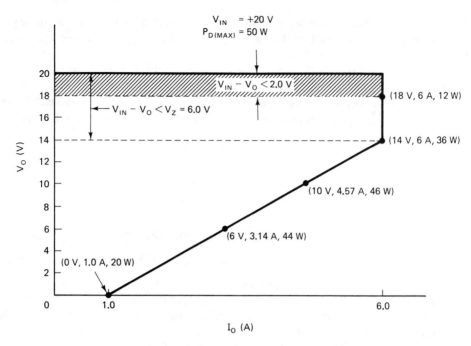

Figure 11.9 Foldback current-limiting characteristics.

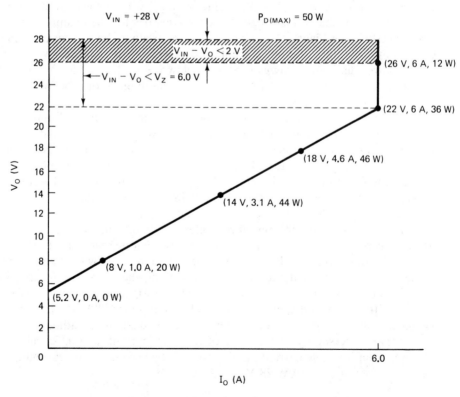

Figure 11.10 Foldback current-limiting characteristics.

11.4 THERMAL SHUTDOWN

In many IC voltage regulators and power amplifiers a *thermal shutdown circuit* is incorporated on the IC chip. The thermal shutdown circuit senses the chip temperature, and when this temperature exceeds some predetermined design limit, the circuit acts to shut down the device. Thus the thermal shutdown circuit is an additional protective circuit in addition to the current-limiting circuitry.

The basic circuitry of a thermal shutdown circuit is shown in Figure 11.11. Transistors Q_1 and Q_2 constitute the Darlington emitter-follower series pass output stage. Diode D_1 (actually a diode-connected transistor) is operated as a zener diode and biased into the reverse-bias breakdown region by current source I_Q. The voltage drop across this diode, V_Z, produces a bias voltage for transistors Q_4 and Q_5. The V_Z value for an IC zener diode is typically around 6.5 V, corresponding to the emitter-base junction breakdown voltage of the diode-connected transistor. The resistance ratio of R_6 and R_5 is such that at room temperature the base-to-emitter voltage of Q_5 is not large enough to bias it into conduction. However, as the temperature increases, the voltage across the zener diode increases due to the positive temperature coefficient of the zener voltage. At the same time, due to the negative temperature coefficient of the transistor base-to-emitter voltage, the voltage required to bias Q_5 into conduction decreases. Ultimately, a temperature is reached at which Q_5 is in full conduction such that it bypasses the base drive of the Darlington output stage (Q_1 and Q_2), thus preventing any further rise in temperature.

We will now obtain the basic design equations for this thermal shutdown circuit. For this analysis we will assume that the base current of Q_5 is small compared to the current through R_5 and R_6, such that R_5 and R_6 can be considered to act as a simple resistive voltage divider. As a result, we have that

$$V_{BE5} = \frac{R_6}{R_5 + R_6} (V_Z - V_{BE4}) \qquad (11.38)$$

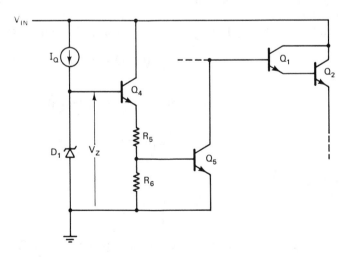

Figure 11.11 Thermal shutdown circuit.

Rearranging this equation yields

$$V_Z = V_{BE_4} + \frac{R_5 + R_6}{R_6} V_{BE_5} = V_{BE_4} + \left(1 + \frac{R_5}{R_6}\right)V_{BE_5} \qquad (11.39)$$

We must now consider the temperature dependence of V_Z and V_{BE}. Over the limited range of temperatures that we are interested in, we can assume that both V_Z and V_{BE} will vary linearly with temperature as a good approximation. We can therefore write

$$V_Z(T) = V_Z + \frac{dV_Z}{dT}(T - T_R)$$
$$V_{BE}(T) = V_{BE} + \frac{dV_{BE}}{dt}(T - T_R) \qquad (11.40)$$

where T_R is a reference temperature (generally 25°C) and V_Z and V_{BE} represent the values of these quantities at $T = T_R$. If we let $\Delta T = T - T_R$, we can now write

$$V_Z + \frac{dV_Z}{dT}\Delta T = V_{BE_4} + \left(1 + \frac{R_5}{R_6}\right)V_{BE_5} + \left(2 + \frac{R_5}{R_6}\right)\frac{dV_{BE}}{dT}\Delta T$$
$$= V_{BE_4} + V_{BE_5} + 2\left(\frac{dV_{BE}}{dT}\right)\Delta T + \frac{R_5}{R_6}\left[\left(V_{BE_5} + \frac{dV_{BE}}{dT}\Delta T\right)\right] \qquad (11.41)$$

We will now consider a design example. We will assume that $V_Z = 6.5$ V and $dV_Z/dT = +3.0$ mV/°C. For V_{BE} a temperature coefficient of -2.1 mV/°C will be used. We will assume that at $T = 25°C$, $V_{BE} = 650$ mV at $I_C = 10$ mA for Q_4 and Q_5. A quiescent current of 10 mA for Q_4 will be chosen. The maximum current available from the error amplifier will be 50 mA. The thermal shutdown circuit will be designed for complete shutdown of the regulator output when the temperature reaches 175°C, so $\Delta T = 150°C$. At 50 mA, we have that $V_{BE_5} = 690$ mV.

If we now substitute into the equation above we obtain

$$6.5 + (0.003 \times 150) = 0.650 + 0.690 + 2(-0.0021 \times 150) \qquad (11.42)$$
$$+ \frac{R_5}{R_6}(0.690 - 0.0021 \times 150)$$

so
$$6.95 = 1.34 - 0.630 + \frac{R_5}{R_6}(0.690 - 0.315) \qquad (11.43)$$
$$6.95 = 0.710 + \frac{R_5}{R_6}(0.375)$$

Solving for R_5/R_6, we obtain

$$\frac{R_5}{R_6} = \frac{6.24}{0.375} = \underline{16.64} \qquad (11.44)$$

The voltage across the resistive voltage divider, $R_5 + R_6$, is $V_Z - V_{BE_4} = 6.5 - 0.65 = 5.85$ V. Since the current through R_5 and R_6 has been specified as 10 mA, we have that $R_5 + R_6 = 585 \, \Omega$. Using the resistance ratio of $R_5/R_6 = 16.64$, we have that

$$1 + \frac{R_5}{R_6} = \frac{R_5 + R_6}{R_6} = \frac{585}{R_6} = 17.64 \tag{11.45}$$

so $R_6 = 33 \, \Omega$ and thus $R_5 = 585 - 33 = 552 \, \Omega$.

To illustrate the sharpness of the thermal limiting characteristics, we will now determine the temperature at which Q_5 shunts one-half of the base drive of Q_1 and then the temperature at which one-tenth of the base drive is shunted, the latter condition being considered as the threshold of the thermal limiting. We first rewrite the basic equation for thermal limiting in the form

$$\Delta T = \frac{V_{BE_4} + V_{BE_5}(1 + R_5/R_6) - V_Z}{dV_Z/dt - (2 + R_5/R_6) \, dV_{BE}/dT} \tag{11.46}$$

For the condition of Q_5 shunting one-half of the base drive, the current through Q_5 is 25 mA. The base-to-emitter voltage of Q_5 for this current at 25°C is $690 - 17 = 673$ mV. If we now substitute into the equation above, we obtain

$$\Delta T = \frac{0.650 + 0.673(17.64) - 6.5}{0.04214} = 143°C \tag{11.47}$$

so the temperature at which "one-half shutdown" occurs is 168°C.

For the one-tenth shutdown condition, the value of V_{BE} at which Q_5 carries 5 mA is 632 mV at 25°C. The corresponding value of ΔT is given by

$$\Delta T = \frac{0.650 + 0.632(17.64) - 6.5}{0.04214} = 126°C \tag{11.48}$$

so the temperature at which one-tenth shutdown occurs is 151°C. Thus we see that the thermal shutdown is initiated and comes to full completion over a rather narrow temperature span. This is a result of the exponential relationship between the transistor collector current and the base-to-emitter voltage.

11.5 ADJUSTABLE POSITIVE VOLTAGE REGULATOR EXAMPLE

Figure 11.12 shows the circuit diagram of an adjustable positive voltage regulator. This voltage regulator circuit is similar to that of the LM 376 and the LM 105/205/305/305A (National Semiconductor) series of voltage regulators. In Figure 11.12 the various functional blocks of this voltage regulator have been delineated and identified.

The current source portion of the circuit is comprised of transistors Q_6, Q_7, and Q_{13}. Transistor Q_{13} is a diode-connected field-effect transistor (JFET) and operates as a current regulator diode. For a JFET with $V_{GS} = 0$ and a drain-to-source voltage V_{DS} greater than the pinch-off voltage V_p, the drain current I_{DS} will essentially saturate at a value designated as I_{DSS}. The value of I_{DSS} is determined by the construction of the JFET, but will vary slightly with V_{DS}.

Figure 11.13a shows the terminal characteristics, I_{DS} versus V_{DS}, of a JFET used as a current regulator diode. Note that for values of V_{DS} above the pinch-off voltage,

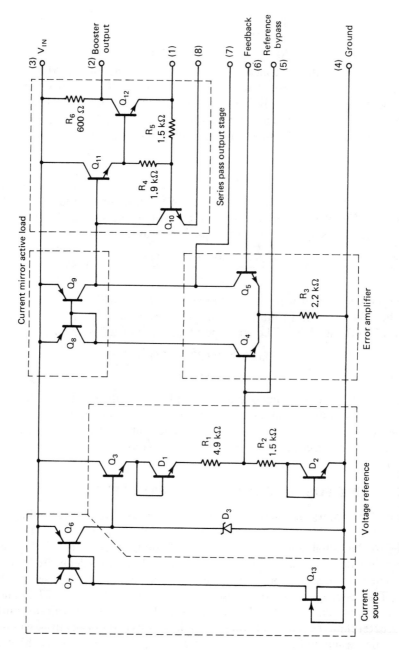

Figure 11.12 Circuit diagram of an adjustable positive voltage regulator: LM105 (National Semiconductor).

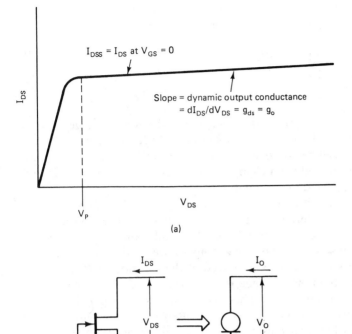

$I_{DSS} = I_{DS}$ at $V_{GS} = 0$

Slope = dynamic output conductance
= $dI_{DS}/dV_{DS} = g_{ds} = g_o$

V_{DS}

V_p

I_{DS}

(a)

I_{DS}

V_{DS}

I_o

V_o

(b)

Figure 11.13 Diode-connected field-effect transistor as a current regulator diode: (a) output characteristics of diode-connected JFET; (b) diode-connected JFET and the schematic symbol for a current regulator diode.

V_p, the drain current becomes relatively independent of V_{DS}, up to the breakdown voltage BV_{DSS}, so the voltage compliance range can be said to extend from V_p to BV_{DSS}. The value of V_p is generally in the range 3 to 30 V, with values of around 6 to 10 V being typical. The breakdown voltages are generally in the range from 30 to 300 V, with most devices being in the range from 50 to 100 V.

The slope of the current regulator diode characteristics within the compliance range is the dynamic output conductance as given by $g_o = g_{ds} = dI_{DS}/dV_{DS}$. The reciprocal quantity is the dynamic output resistance as given by $r_o = r_{ds} = 1/g_o = dV_{DS}/dI_{DS}$. The value of r_{ds} is generally approximately inversely proportional to current level and is typically in the range from 100 kΩ to 1.0 MΩ at a current of $I_{DSS} = 1.0$ mA. Figure 11.13b shows a schematic representation of the current diode.

The current through Q_{13} produces an approximately equal current through Q_6 by action of the Q_6–Q_7 current mirror. This in turn provides a bias current for the zener diode, D_3, which in reality is a reverse-biased diode-connected transistor with the zener voltage being the emitter–base breakdown voltage. The action of the current-source circuit is to produce a current through D_3 that is relatively independent of the input voltage V_{IN}. This, in turn, keeps the voltage drop across the zener diode relatively constant and almost completely independent of the supply voltage. For example, if the zener impedance is 10 Ω and the dynamic output resistance of the current regulator diode is 100 kΩ, the rate of change of V_Z with V_{IN} is given by

$$\frac{dV_Z}{dV_{IN}} = \frac{dV_Z}{dI_Z}\frac{dI_Z}{dI_{13}}\frac{dI_{13}}{dV_{IN}} = Z_z \times 1 \times \frac{1}{r_o}$$
$$= 10 \ \Omega \times (1/100 \ \text{k}\Omega) = 1 \times 10^{-4}$$

(11.49)

Therefore, a 1.0 V change in the input voltage will change the zener voltage drop by only 1×10^{-4} V = 0.1 mV. If the regulator output voltage is $V_O = 12$ V, the corresponding change in V_O will be approximately 0.1 mV(V_O/V_Z) = 0.2 mV. Thus the output voltage is well insulated from changes in the input or line voltage, so this regulator should exhibit very good line-regulation characteristics.

The voltage reference circuit consists of Q_3, D_1, D_2, and D_3. This is a temperature-compensated voltage reference. The negative temperature coefficient of the base-to-emitter voltage drop of transistor Q_3 and the diode-connected transistors, D_1 and D_2, is used to compensate for the positive temperature coefficient of the zener diode, D_3. The reference voltage that is produced at the junction of R_1 and R_2 is approximately 1.8 V. This reference voltage is internally connected to the base of Q_4, which is part of the error amplifier. There is also an external pin tied to this point (pin 5) labeled "reference bypass." This can be used for several purposes. A capacitor can be connected between pins 5 and 4 (ground) to bypass noise that is present at the output of the voltage reference circuit. This noise is produced principally in the zener diode as a result of statistical fluctuations in the avalanche multiplication process. The bypass capacitor can also help to bypass ripple and other fluctuations that may come through the reference circuit from the input voltage. Another use of pin 5 is in switching regulators, in which case a switching waveform can be fed directly into the error amplifier.

The error amplifier is a differential amplifier consisting of transistors Q_4 and Q_5. The biasing is provided by resistor R_3. Transistors Q_8 and Q_9 serve as a current-mirror active load for the differential amplifier. The quiescent current of the differential amplifier, I_Q, is determined by R_3 and the voltage drop across R_3, which is $V_{REF} - V_{BE}$. For $V_{REF} = 1.8$ and $V_{BE} = 0.7$, this becomes $I_Q = I_{R_3} = (1.8 - 0.7)$ V/2.2 kΩ = 1.1 V/2.2 kΩ = 0.5 mA.

The base of Q_4 is the noninverting input of the amplifier, and it is to this point that the reference voltage is applied. The base of Q_5 is the inverting input, and the feedback voltage that is obtained from a resistive voltage divider across the output is fed to this point via pin 6.

The series-pass output stage consists of transistors Q_{11} and Q_{12} connected as a Darlington emitter-follower configuration. Transistor Q_{10} is used to provide current limiting in conjunction with an externally connected resistor R_{CL}, as shown in Figure 11.14.

Figure 11.14a shows the basic external connections for this voltage regulator. Note that the output voltage of this regulator can be adjusted to the desired value by suitable choice of the ratio of R_2 to R_1. The output voltage is given by $V_{REF} = V_O R_2/(R_1 + R_2)$, so $V_O = V_{REF}(1 + R_1/R_2)$. The minimum value of V_O is thus $V_{REF} = 1.8$ V.

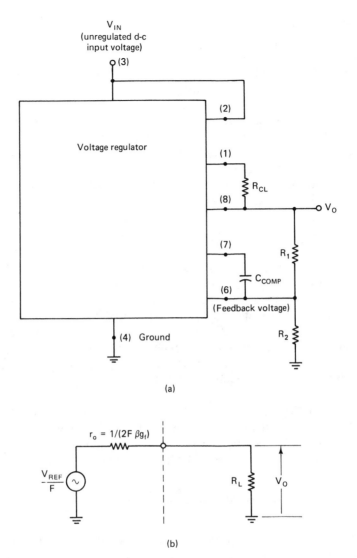

(a)

(b)

Figure 11.14 (a) Basic external connections for the adjustable positive voltage regulator; (b) equivalent circuit for voltage regulator.

The maximum value of V_O is determined by the minimum allowable input–output voltage differential for this regulator. Inspection of this circuit reveals that for this regulator to operate properly such that all transistors are operating in the active mode the minimum input–output differential is given by

$$(V_{IN} - V_O)_{MIN} = V_{CE(SAT)}(Q_9) + V_{BE}(Q_{11}) + V_{BE}(Q_{12}) \qquad (11.50)$$
$$= 0.2 + 0.6 + 0.7 = 1.5 \text{ V}$$

Thus the maximum output voltage attainable is limited to $V_{IN} - 1.5$ V.

Actually, this will be true only for low output current levels. At higher current levels the required voltage drops across the transistors increase somewhat, and in addition there will be a voltage drop across R_{CL} of as much as 0.7 V, so the required input–output differential may increase to about 2 to 2.5 V. Also, if no external pass transistors are to be used, resistor R_6 should be bypassed by a short-circuit between pins 2 and 3.

In the output stage, resistors R_4 and R_5 constitute a voltage divider across the base–emitter junction of Q_{12}. This voltage divider produces a small voltage to produce a prebias across the base-to-emitter junction of the current limit transistor Q_{10}; as a result the voltage drop that has to be produced across the current limit resistor, R_{CL}, does not have to be as large, and thus a smaller value of R_{CL} can be used.

The prebias voltage produced across R_5 is given by

$$V_{R_5} = \frac{V_{BE_{12}}R_5}{R_4 + R_5} = 0.7 \text{ V} \times \frac{1.5}{3.4} = 0.3 \text{ V} \tag{11.51}$$

Therefore, the voltage drop across R_{CL} that is required to bias Q_{10} into conduction is only about 0.4 V. To consider an example, let us assume that $P_{D(MAX)} = 500$ mW and $V_{IN} = 20$ V and $V_O = 10$ V, so a current limit of 50 mA is appropriate. For this current limit the value of R_{CL} that is required is 0.4 V/50 mA $= 8.0 \text{ }\Omega$. Without the prebias, an R_{CL} of 0.7 V/50 mA $= 14 \text{ }\Omega$ is required. The smaller value of R_{CL} is advantageous from several standpoints, including a lower output resistance and thus better load regulation, a greater output voltage range due to a lower voltage drop across R_{CL}, and a higher overall circuit efficiency due to a lower power loss in R_{CL}.

Note that the current limit of 50 mA chosen above will not provide for short-circuit protection of the regulator. For short-circuit protection, a current limit of 500 mW/20 V $= 25$ mA would have to be chosen if simple current limiting is used. With foldback current limiting, a considerably larger current limit could be obtained.

11.5.1 External Connections for Basic Positive Regulator Circuit

Figure 11.14a shows a diagram of the external connections that are needed for this adjustable positive regulator. Note the connection between pins 2 and 3, which places a short-circuit across the current boost resistor R_6. This resistor is needed only when external pass transistors are used, as will be seen a little later. Note also the current-limit resistor placed between pins 1 and 8. There is a resistive voltage divider placed between the output voltage pin (8) and ground to feed a fraction $R_2/(R_1 + R_2)$ of the output voltage back to the inverting input of the error amplifier (pin 6), where it is compared to the reference voltage.

A small compensation capacitor, C_{COMP}, is required between pins 6 and 7 to ensure stability from oscillations. The voltage regulator is basically a high-gain feedback amplifier, so it should not be surprising to find the requirement for a compensation capacitor for stability, just as in the case of an operational amplifier. C_{COMP} is a small capacitor, typically around 50 pF.

The design of the resistive voltage divider is based on two considerations. First and foremost is that the resistance ratio be what is needed to give the required output voltage, as indicated by the relationship $V_O = V_{REF}(1 + R_1/R_2)$. The second factor is that of minimizing the effects of the differential amplifier bias (base) current. To this end, the parallel combination of R_1 and R_2 should optimally be equal to the parallel combination of the R_1 and R_2 of the voltage reference circuit, which in this case is 1.15 kΩ. For example, if $V_{REF} = 1.8$ V and $V_O = 10$ V, we have that $1 + (R_1/R_2) = V_O/V_{REF} = 10/1.8 = 5.56$. Since $(R_2 + R_1)/R_1R_2 = 1/1.15$ kΩ,

$$R_1 = \frac{(1.15 \text{ k}\Omega)(R_2 + R_1)}{R_2} = 1.15 \text{ k}\Omega \times 5.56 = \underline{6.39 \text{ k}\Omega} \qquad (11.52)$$

Therefore, $R_1/R_2 = 4.56$, so $R_2 = \underline{1.40 \text{ k}\Omega}$.

11.5.2 Load Regulation

To evaluate the load regulation of this regulator circuit, we will first obtain an expression for the dynamic output resistance. As a result of the current-mirror active load, the output current of the error amplifier is given by $2g_f(V_{REF} - V_{FB}) = 2g_f(V_{REF} - FV_O)$, in which g_f is the dynamic transfer conductance of the differential amplifier as given by $g_f = I_Q/4V_T$, and F is the feedback factor, which is the fraction of the output voltage that is fed back to the differential amplifier, $F = V_{FB}/V_O$. The factor of 2 is the result of the current-doubling action of the current mirror.

The output current of the differential amplifier supplies the base drive of the Darlington output stage, in which there is an overall current gain of $\beta = \beta_{11}\beta_{12}$. The output current is therefore given by

$$I_O = 2\beta g_f(V_{REF} - FV_O) \qquad (11.53)$$

Since $I_O = V_O/R_L$, this can be rewritten as

$$\frac{V_O}{R_L} = 2\beta g_f(V_{REF} - FV_O) \qquad (11.54)$$

Collecting terms in V_O and solving for V_O gives

$$V_O = \frac{1}{F} V_{REF} \frac{R_L}{R_L + (1/2F\beta g_f)} \qquad (11.55a)$$

In Figure 11.14b, an equivalent circuit for the voltage-regulator output that is based on the equation above is presented. From this equivalent circuit and the equation that it represents, it is clear that the output resistance of the regulator is given by $r_o = 1/2F\beta g_f$. Since $g_f = I_Q/4V_T = I_Q/100$ mV and $1/F = V_O/V_{REF}$, this equation can be restated as

$$r_o = \frac{V_O}{V_{REF}} \frac{50 \text{ mV}}{\beta I_Q} \qquad (11.55b)$$

To consider a representative example, let us use the value of $I_Q = 0.5$ mA for the differential amplifier that was determined earlier. For the Darlington output stage, let us assume a typical current gain of around 70 for each transistor so that the overall current gain will be approximately 5000. We will use a reference voltage of 1.8 V and assume an output voltage of 10 V. As a result, we have that

$$r_o = \frac{10}{1.8} \left(\frac{50 \text{ mV}}{5000} \times 0.5 \text{ mA} \right) = \underline{0.11 \ \Omega} \tag{11.56}$$

For a no-load to full-load current change of 50 mA, the corresponding decrease in the output voltage is given by

$$\Delta V_O = \Delta I_O r_o = 50 \text{ mA} \times 0.11 \ \Omega = \underline{5.6 \text{ mV}} \tag{11.57}$$

This is the load regulation of the voltage regulator. On a percentage basis, the load regulation is

$$\frac{\Delta V_O}{V_O} \times 100\% = \frac{5.6 \text{ mV}}{10 \text{ V}} \times 100\% = \underline{0.056\%} \tag{11.58}$$

It is of interest to note that since the output resistance is directly proportional to V_O, the load regulation, on a percentage basis, is not a function of V_O.

11.5.3 Line Regulation

The line regulation is the ratio of the change in the output voltage to a given change in the regulator input voltage. We have seen from an earlier calculation that based on assuming a zener impedance of 10 Ω and a dynamic output impedance of 100 kΩ for the JFET current diode, the rate of change of the zener voltage with respect to the input voltage is $dV_Z/dV_{\text{IN}} = 10 \ \Omega/100 \text{ k}\Omega = 1 \times 10^{-4}$. Since the output voltage is derived from the reference voltage, which in turn is obtained from the zener voltage, the rate of change of the output voltage with respect to input voltage, which is the line regulation, is given by

$$\frac{dV_O}{dV_{\text{IN}}} = \frac{V_O}{V_Z} \frac{dV_Z}{dV_{\text{IN}}} = \frac{V_O}{V_Z} \times 1 \times 10^{-4} \tag{11.59}$$

For $V_O = 10$ V, this becomes

$$\frac{dV_O}{dV_{\text{IN}}} = 1.67 \times 10^{-4} \text{ V/V} = 0.167 \text{ mV/V} \tag{11.60}$$

Thus the output voltage increases by approximately 0.167 mV for every 1-V increase in the input voltage. On a percentage basis, this is 0.00167%/V for the line regulation.

11.5.4 Ripple Rejection

The input voltage of a voltage regulator is generally derived from a half-wave, or more commonly, full-wave rectifier circuit and associated filter. The unregulated input voltage of the regulator will accordingly not be a pure dc waveform but will have some ac ripple riding on it. We have just seen that the regulator acts to maintain a constant output voltage, almost completely independent of changes in the input voltage. This relative independence of the output voltage from the input voltage holds true for the variations in the output voltage that are a manifestation of the ac ripple voltage. As a result, the ac ripple voltage appearing on the output of the regulator will be a very much attenuated version of the input ripple.

The *ripple rejection* of a voltage regulator is the ratio of the ac ripple voltage on the output to the ac input ripple voltage and is usually expressed in decibels as

$$\text{ripple rejection} = 20 \log_{10} \frac{\text{output ripple voltage}}{\text{input ripple}} \tag{11.61}$$

Since the ratio of the output ripple to the input ripple is the same as the ratio of the change in the output voltage to the change in the input voltage, we can write the ripple rejection equation as

$$\text{ripple rejection} = 20 \log_{10} \frac{\Delta V_O}{\Delta V_{\text{IN}}} \tag{11.62}$$

so we see a very close relationship with the line regulation.

For the example under consideration above, the line regulation was 1.67×10^{-4} V/V, so the corresponding ripple rejection should be $20 \log_{10} (1.67 \times 10^{-4}) = -75$ dB. Therefore, the output ripple will be some 75 dB below the input ripple.

For example, if the input voltage is 20 V with a 1% ripple factor corresponding to a ripple voltage of 200 mV, the output ripple voltage will be 200 mV $\times$ 1.67 $\times$ $10^{-4} = 0.0334$ mV $= 33.4$ μV. The output ripple factor will consequently be 33.4 μV/10 V) $\times$ 100% = 0.000334%, so the output waveform will be very smooth indeed. As a result, we see that a voltage regulator with good line regulation can serve as a very useful adjunct to the filtering circuit of the rectifier in providing for a dc output voltage with a very low ripple factor.

11.5.5 Current Boost

The maximum current output of an IC voltage regulator is limited by the maximum power dissipation rating of the device. For some IC regulators the maximum output current may be limited to as little as 50 mA.

The maximum output current can, however, be greatly increased by the use of additional series-pass transistors external to the IC. The output current will now be limited only by the maximum power dissipation rating of the external transistors, so now output currents of several amperes, or even several tens of amperes, are possible.

An example of a *current boost* circuit is presented in Figure 11.15. The voltage regulator is the same as has been under previous analysis (Figure 11.12). The external series-pass transistors are Q_A and Q_B connected in a cascade arrangement, with Q_A driving Q_B.

The voltage drop across the current boost resistor R_6 of 600 Ω that is produced by the collector current of Q_{12} acts to bias Q_A on. Since the current through R_6 is limited by the V_{BE} of Q_A to about 700 mV/600 $\Omega \simeq 1$ mA, any current through Q_{12} in excess of this will be drawn from the base of Q_A. This current therefore becomes the base drive of Q_A.

The collector current of Q_A, in turn, serves to bias Q_B on via the voltage drop across the 68-Ω resistor. Again we see that the current through this resistor is limited

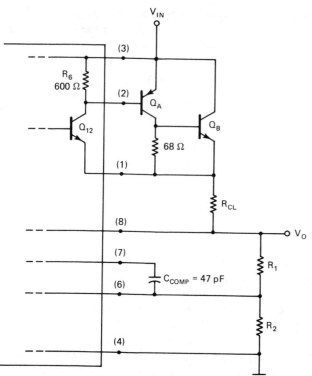

Figure 11.15 External connections for regulator with current boost.

by the V_{BE} of Q_B to about 700 mV/68 Ω = 10 mA. Any collector current of Q_A in excess of this becomes the base drive of Q_B. This base current, when multiplied by the current gain of this transistor, is the regulator circuit output current.

As a result of the base-to-collector current gain of Q_A, and the base-to-emitter gain of Q_B, the output current of the IC regulator can be multiplied by a very large factor. Using a representative current gain of 50 for Q_A and for Q_B, the overall current gain becomes 2500. As a result, an IC output current of 10 mA can be boosted up to 25 A by the external series-pass circuit. Nevertheless, the actual maximum output current is still limited by considerations of maximum allowable power dissipation.

Most of the power dissipation occurs in the final transistor stage, Q_B. If, for example, Q_B has a 50-W power rating, for an input voltage of V_{IN} = 20 V, an output current of 2.5 A is possible with short-circuit protection. If foldback current limiting were to be used, for an output voltage level of 15 V, an output current of 50 W/ (20 − 15) V = 10 A would be available.

The function of the 600-Ω resistor (R_6) and the 68-Ω resistor is to allow a small quiescent current to flow through Q_{12} and Q_A, respectively, under no-load conditions. This quiescent current serves to reduce the output resistance under no-load conditions and thus to improve the load regulation of the system, especially when the no-load current is very small.

Current boost for very large output currents. Depending on the amount of output current required, the external series-pass circuit can consist of one, two, or three transistors. Figure 11.16 shows an example of a current boost circuit that is comprised of three transistors. This circuit is similar to the current boost circuit just considered, except that a third transistor is added and foldback current limiting is used. Transistors Q_A, Q_B, and Q_C are connected in cascade so that the overall current gain is now extremely large.

The 600-, 68-, and 6.8-Ω resistors are used to provide a small no-load quiescent current for transistors Q_{12}, Q_A, and Q_B, respectively. The 1.32- and 1.86-kΩ resistors across the base–emitter junction of Q_{12} are used to provide a small (~ 0.3 V) prebias voltage across the base–emitter junction of the current-limit transistor, Q_{10}. This allows a smaller value of R_{CL} to be used than would otherwise be the case. The foldback limiting characteristic is determined by the 47- and 160-Ω resistors in conjunction with R_{CL}.

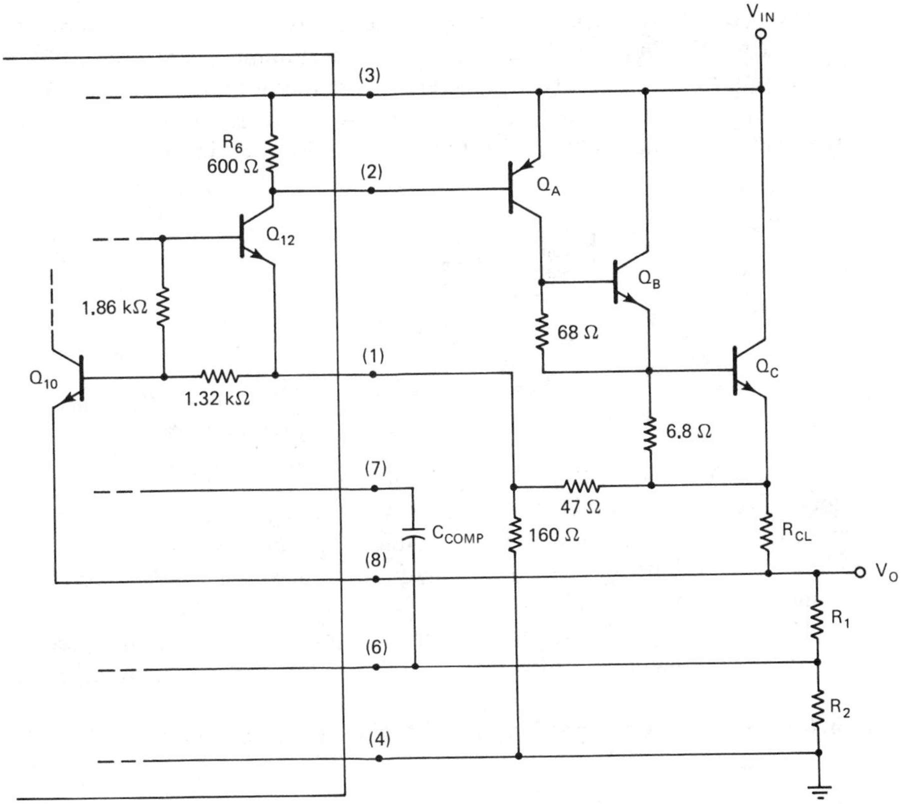

Figure 11.16 External connections for regulator with three-transistor current boost circuit and foldback current limiting.

11.6 THREE-TERMINAL REGULATORS (FIXED REGULATORS)

Three-terminal voltage regulators are voltage regulators in which the output voltage is set at some predetermined value. They therefore do not require any external feedback connections. As a result, only three terminals are required for this type of regulator: input (V_{IN}), output (V_O), and a ground terminal. Since these regulators operate at a preset output voltage, the current-limit resistor R_{CL} is also internal to the regulator.

The principal advantage of three-terminal regulation is the simplicity of connection to the external circuit, with a minimum of external components required. Indeed, in many applications no external components are required. In some applications the use of filter capacitors across the input and output terminals may be desirable. Figure 11.17 shows the basic circuit configuration of a three-terminal voltage regulator. The simplicity and ease of application is evident. The capacitor across the input terminals is required only when the voltage regulator is located more than about 5 cm from the power supply filter capacitor such that the lead inductance between the supply and the regulator may cause stability problems and high-frequency oscillations. This capacitor should be characterized by a very low effective series resistance (ESR). Acceptable values are generally 0.2 μF for a ceramic disk, 2 μF or greater for tantalum, or 25 μF or greater for aluminum electrolytic.

A capacitor is generally not needed across the output terminals. The use of a suitable capacitor will, however, improve the regulator response to transient changes in the load conditions and will also reduce the noise present at the regulator output.

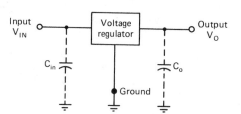

Figure 11.17 Basic three-terminal voltage regulator.

Although the three-terminal regulator offers only fixed output voltages, a wide variety of voltages is available, both positive and negative. The output voltages of commercially available three-terminal voltage regulators are 5, 5.2, 6, 8, 10, 12, 15, 18, and 24 V in both positive and negative output-voltage polarities. The output currents range from 100 mA to 3 A.

The three-terminal regulators generally offer a line regulation of about 0.005 to 0.02%/V, a load regulation of 0.1% to 1.0%, and a ripple rejection of 65 to 85 dB.

11.6.1 Adjustable Voltage Output with the Three-terminal Regulator

Although the three-terminal regulator is basically a fixed voltage regulator, with the simple addition of two resistors to the circuit an adjustable output voltage can be obtained. Figure 11.18 shows a three-terminal regulator connected as an adjustable voltage regulator.

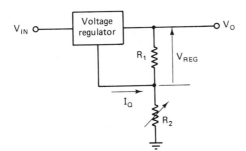

figure 11.18 Three-terminal voltage regulator connected as an adjustable regulator.

The voltage regulator acts to maintain a fixed voltage between its output and common or ground terminals. This voltage regulator output voltage is designated in Figure 11.18 as V_{REG}. The current through R_1 is therefore V_{REG}/R_1. The current through R_2 is equal to the current through R_1 plus the voltage regulator quiescent current, I_Q. The voltage drop across R_2 is therefore given by $V_{R_2} = (V_{\text{REG}}/R_1 + I_Q)R_2$. As a result, the output voltage of this circuit is

$$V_O = V_{\text{REG}} + V_{R_2} = V_{\text{REG}}(1 + R_2/R_1) + I_Q R_2 \qquad (11.63)$$

The voltage regulator quiescent current is defined as that part of the regulator input current that does not go out of the output terminal. This current will vary with changes in input voltage and in load voltage. For example, for the LM 7805 voltage regulator, which has a 5.0-V output voltage, the quiescent current is 7 mA typical, 10 mA maximum. The change in the quiescent current, ΔI_Q, for a change in the output current of from 5 mA to 1.5 A (the full-load current) is 0.5 mA maximum. The change in the quiescent current for a change in input voltage from 7 to 25 V is 1.3 mA maximum.

It is evident that the quiescent current, and in particular the changes in the quiescent current, degrade the performance of the regulator, particularly with respect to the line regulation and load regulation. The line regulation of this regulator configuration is given by

$$\text{line regulation} = \Delta V_O/\Delta V_{\text{IN}} = (1 + R_2/R_1)\Delta V_{\text{REG}}/\Delta V_{\text{IN}} + R_2(\Delta I_Q/\Delta V_{\text{IN}}) \qquad (11.64)$$

The degradation or increase in the line regulation factor due to the quiescent current is thus equal to $R_2(\Delta I_Q/\Delta V_{\text{IN}})$. Thus we see that for good line regulation both R_2 as well as ΔI_Q should be small.

The load regulation is given by

$$\Delta V_O/\Delta I_O = (1 + R_2/R_1)\Delta V_{\text{REG}}/\Delta I_O + R_2 \, \Delta I_Q/\Delta I_O \qquad (11.65)$$

The degradation (increase) in the load regulation is equal to $R_2(\Delta I_Q/\Delta I_O)$, and thus again we see the importance of using a small R_2 and, if possible, choosing a voltage regulator with a quiescent current that does not change much with load current.

The quiescent current will also change with temperature so that the temperature coefficient of the adjustable three-terminal voltage regulator is given by

$$\Delta V_O/\Delta T = (1 + R_2/R_1)\Delta V_{\text{REG}}/\Delta T + R_2 \, \Delta I_Q/\Delta T \qquad (11.66)$$

Again we see the importance of using a small-value resistor for R_2. Note, however, that as R_2 is decreased the value of R_1 will also have to be decreased for a given output voltage. As a result of the increased current drain through R_1 and R_2, the maximum output current available from the regulator will decrease.

To consider an example of the problem arising from the quiescent current, we will use the LM 7805 three-terminal voltage regulator in an adjustable voltage regulator configuration for a range of output voltages from 5 to 10 V. Since 5.0 V is the regulated output of the 7805, this means that resistor R_2 will have to vary from 0 to R_1.

The rated output current of the 7805 is 1.5 A, so a reasonable choice for the current through R_1 and R_2 would be around 0.2 A. Since the voltage across R_1 is 5.0 V, the value of R_1 is 5.0 V/0.2 A = 25 Ω. As a result, we see that R_2 ranges from 0 to 25 Ω. As mentioned above, the $I_Q R_2$ voltage drop will produce a degradation in the line regulation, load regulation, and temperature coefficient. For this example, we will turn our attention to the temperature coefficient.

The rate of change of quiescent current with temperature for the 78XX series of regulators is approximately -7 μA/°C. Therefore, the contribution of this temperature coefficient to the temperature coefficient of the output voltage is 7 μA/°C $\times$ 25 Ω = 175 μV/°C. This is, of course, assuming that R_2 does not change appreciably with temperature. The temperature coefficient of the regulator output voltage is typically 500 μV/°C, so we see that the effect of the quiescent current can be significant.

Figure 11.19 shows one very good approach to minimizing the effects of the quiescent current on the three-terminal adjustable regulator performance. An operational amplifier, connected in the voltage-follower feedback configuration, is interposed between the R_1–R_2 resistive voltage divider and the voltage regulator ground terminal. The voltage follower sinks the regulator quiescent current. The current flowing from the junction of R_1 and R_2 into the operational amplifier is the bias current of the operational amplifier, I_B. By suitable choice of the operational amplifier, this bias current can be so small as to be a negligible factor in the performance of the regulator circuit. At the same time, the output impedance of the voltage follower will be so small that changes in I_Q will not affect the voltage at the ground terminal of the regulator.

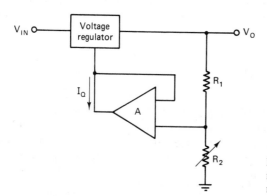

Figure 11.19 Three-terminal adjustable regulator with voltage follower to reduce effects of the quiescent current.

Voltage Regulators   Chap. 11

As a result of the just-described buffering action of the voltage follower, the output voltage for this circuit can be expressed to a high degree of precision by $V_O = V_{\text{REG}}(1 + R_2/R_1)$, and there will be no degradation of the voltage regulator characteristics.

11.6.2 Three-terminal Adjustable Regulators with Very Low Quiescent Currents

There is a series of three-terminal voltage regulators with very low quiescent currents. These voltage regulators achieve the very low quiescent currents by simply redesigning the internal circuitry of the regulator so that almost all the biasing currents of the device go out of the output terminal rather than the ground or common terminal. As a result, there is a minimum load current specification in order to assure proper functioning of the regulator. This minimum load current specification corresponds essentially to the total biasing currents of the device and is typically around a few milliamperes. Since the current going out of the common or ground terminal of the regulator is no longer the quiescent current of the device, this common terminal current is called the *adjustment pin current*, I_{ADJ}.

Examples of adjustable three-terminal voltage regulators of the type being discussed now are the following: LM 117, 217, and 317; LM 117HV, 217HV, and 317HV; LM 138, 238, and 338; LM 150, 250, and 350. The above mentioned voltage regulators all have adjustment pin current specifications of 50 μA typical, 100 μA maximum. The change in the adjustment pin current in going from a load current of 10 mA to the maximum rated current, and/or for a change in input voltage from 3 V to the maximum rated input voltage, is specified as 0.2 μA typical and 5 μA maximum for all these voltage regulators. The minimum load current is 3.5 mA. The LM 117 and 117HV series have rated output currents of 1.5 A. For the LM 138 series, there is a 5 A maximum current rating, and for the LM 150 series the current rating is 3 A. The regulated output voltage (reference voltage) of these regulators is 1.25 V.

As a result of the small values of adjustment pin current and, even more importantly, of the very small change in the adjustment pin current over the regulator operating range, precision adjustable voltage regulators can be constructed with a minimum of external components.

11.6.3 Current Regulator Based on the Three-terminal Voltage Regulator

With the three-terminal voltage regulator, it is a relatively simple matter to construct a current regulator. A current regulator is the circuit dual of a voltage regulator and is essentially a current source. With the current regulator, however, the emphasis is generally on much larger output currents than is the case with the conventional current source circuits.

Figure 11.20 shows a simple current regulator that is based on the use of a three-terminal voltage regulator. In this circuit, the voltage regulator acts to maintain a constant voltage, V_{REG}, across resistor R_1. As a result, the current through R_1 is similarly constrained to a constant value given by V_{REG}/R_1. The output current is the sum of this current and the current coming out of the common terminal of the

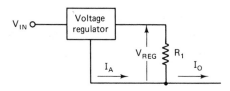

Figure 11.20 Use of a three-terminal voltage regulator as a current regulator.

regulator, which is the quiescent current, or the adjustment pin current, depending on the type of three terminal regulator that is used. In either case this current will be designated as I_A. Therefore, we have that

$$I_O = \frac{V_{REG}}{R_1} + I_A \qquad (11.67a)$$

One of the most important characteristics of a current regulator or current source is the dynamic output conductance, g_o, which in the ideal case will be zero. Since $g_o = -dI_O/dV_O$, we have

$$g_o = \frac{-dI_O}{dV_O} = -\left(\frac{1}{R_1}\frac{dV_{REG}}{dV_O} + \frac{dI_A}{dV_O}\right) \qquad (11.67b)$$

We must now note that the input voltage V_{IN}, as seen by the voltage regulator, is really the voltage between the input terminal and the common (ground) terminal of the regulator. For the circuit under consideration, the voltage at the regulator common terminal is V_O, so for a change in the output voltage of amount dV_O the change in the voltage between the regulator input and common terminals is $-dV_O$. Therefore, we can rewrite the equation for g_o as

$$g_o = \frac{1}{R_1}\frac{dV_{REG}}{dV_{IN}} + \frac{dI_A}{dV_{IN}} \qquad (11.68)$$

Thus we see again that the voltage regulator common pin current, I_A, will have an influence on the overall circuit characteristics.

An appropriate example at this point is a 5-A current regulator based on the use of the LM 138 (National Semiconductor) series of three-terminal voltage regulators. For these regulators, the change in the adjustment pin current is specified as 0.2 μA typical and 5 μA maximum for an input voltage range of 3 to 35 V. The rate of change of the adjustment pin current can therefore be estimated as

$$\frac{dI_A}{dV_{IN}} = \frac{200\ \text{nA}}{32\ \text{V}} \cong 6\ \text{nA/V (typ.)}$$

$$\frac{dI_A}{dV_{IN}} = \frac{5000\ \text{nA}}{32\ \text{V}} \cong 160\ \text{nA/V (max.)} \qquad (11.69)$$

The line regulation is specified as 0.02%/V (typ.). Therefore, we have that $dV_{REG}/dV_{IN} = 1.25\ \text{V} \times 0.0002/\text{V} = 250\ \text{μV/V}$, since the regulated output voltage is 1.25 V.

For a 5-A current regulator, the value of R_1 should be 1.25 V/5 A = 0.25 Ω. We consequently have for the dynamic output conductance the result

$$g_o = \frac{1}{0.25\ \Omega}\ (250\ \mu\text{V/V}) + \frac{dI_A}{dV_{\text{IN}}}$$

$$= 1.0\ \text{mA/V} + \frac{dI_A}{dV_{\text{IN}}}$$

(11.70)

Since dI_A/dV_{IN} will be no more than 160 nA/V, we see that the adjustment pin current will not be a significant factor in the output conductance. We see therefore that the dynamic output conductance of this current regulator is 1.0 mA/V = 1.0 mS. On a percentage basis, this is (1.0 mA/5 A) × 100%/V = 0.02%/V, so the output current changes by only 0.02% for every 1-V change in the load voltage. This represents very good current regulation indeed.

11.6.4 Current Boost for Three-terminal Voltage Regulators

The limited number of external pins notwithstanding, it is possible to have external transistors for the purpose of extending the output current range of the three-terminal voltage regulator. An example of a simple current boost circuit is shown in Figure 11.21.

For the circuit of Figure 11.21, we have that $I_1 R_1 + V_{EB_1} \simeq I_{\text{REG}} R_2 + V_{D_1}$, where I_{REG} is the regulator input current, and R_3 is assumed to be large enough such that it does not shunt a major amount of current from the emitter of Q_1. If diode D_1 is chosen such that its forward voltage drop is a reasonable match to the emitter–base voltage drop of Q_1, we have that $V_{D_1} \simeq V_{EB_1}$; as a result, $I_1 R_1 \simeq I_{\text{REG}} R_2$ and thus $I_1/I_{\text{REG}} = R_2/R_1$. The output current of the circuit is therefore given as

$$I_O \cong I_{\text{REG}} + I_1 = I_{\text{REG}} \left(1 + \frac{R_2}{R_1} \right), \qquad \text{since } I_Q \text{ is small} \qquad (11.71)$$

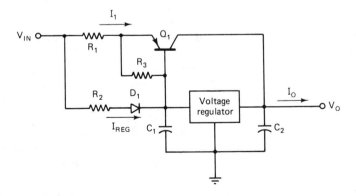

Figure 11.21 Current boost for three-terminal voltage regulator.

so we see that by this means the output current of the circuit has been increased by a factor of $1 + R_2/R_1$.

As an example of this technique, let us consider the design of a circuit to boost the current output of a 1.5-A regulator to 6 A. The resistor ratio is therefore $R_2/R_1 = 3$. To minimize the effects of the difference between the diode voltage drop and the emitter–base voltage drop, we want to choose resistance values large enough such that the I_1R_1 and the $I_{REG}R_2$ voltage drops are large compared to the uncertainty in the diode and the transistor voltage drops. At the same time, too large a value for R_1 and R_2 should certainly be avoided because this will reduce the input voltage at the voltage regulator input pin. A reasonable choice for the $I_{REG}R_2$ and I_1R_1 voltage drops is 0.5 V.

We will choose a target value of 1.0 A for the regulator current to allow a reasonable safety factor. The value of R_2 is therefore 0.5 V/1.0 A = 0.5 Ω. For R_1 we have that R_1 = 0.5 V/5.0 A = 0.1 Ω. Resistor R_3 should be chosen such that, at low output current levels, transistor Q_3 is not turned on and the voltage regulator supplies the entire output current. A reasonable current level for this is about 50 mA, so for R_3 we have R_3 = 700 mV/50 mA = 14 Ω.

In this circuit the current limiting of the three-terminal regulator will also act to limit the current through Q_1 and thus provide overall protection for the circuit. When the output current of the voltage regulator becomes current limited, the regulator input current will similarly be limited since the quiescent regulator current flow out the regulator common (ground) terminal is very small in comparison. Since I_{REG} thus becomes limited by the action of the voltage regulator current-limiting circuitry, the transistor current will also be limited, since the ratio of these two currents is determined by the R_2/R_1 resistor ratio.

11.6.5 Three-terminal Positive Voltage Regulator Example: LM78XX

An example of a three-terminal positive voltage regulator is shown in Figure 11.22. This is representative of the LM78XX (National Semiconductor) voltage regulator circuit and uses a band-gap type of voltage reference circuit.

The band-gap voltage reference circuit is comprised basically of transistors Q_1 through Q_8. The reference voltage generated by this circuit appears at the base of Q_6 and therefore appears across R_{18}. This reference voltage is also applied to the base of Q_1. Note, however, that the reference voltage is not available externally.

Transistor Q_{10} in the emitter-follower configuration in conjunction with resistor R_9 is used to increase the current output of the reference circuit. This current output provides the base drive for the series-pass output stage, consisting of transistors Q_{15} and Q_{16} in a Darlington emitter-follower configuration.

Transistor Q_{11} is a multiple-collector lateral PNP transistor and serves as a multiple current source. One of the collectors of Q_{11} serves as a current source active load for Q_{10}.

Foldback current limiting is provided by Q_{14} together with the current-limit resistor R_{16}. Resistors R_{13} and R_{14} and diode D_2 produce the foldback dependence on the input–output voltage differential.

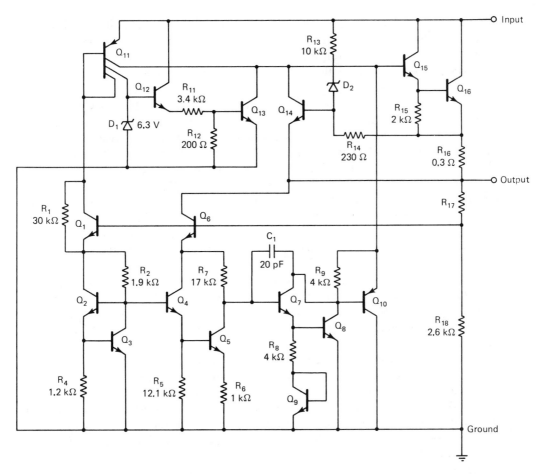

Figure 11.22 Positive voltage regulator with fixed output voltage: LM78XX (National Semiconductor).

Thermal shutdown is obtained from the circuitry of D_1, Q_{12}, Q_{13}, R_{11}, and R_{12}. One of the collectors of Q_{11} provides a current source biasing for D_1. When the thermal shutdown transistor Q_{13} is turned on, it bypasses the base drive away from the Q_{15}–Q_{16} output stage.

The output voltage is derived from the reference voltage that appears across R_{18}. Resistors R_{17} and R_{18} constitute a resistive voltage divider, and, neglecting the effects of the Q_6 and Q_1 base current, the following relationship between the output and reference voltage is obtained: $V_{REF} = V_O R_{18}/(R_{17} + R_{18})$. As a result, we have that $V_O = V_{REF}(1 + R_{17}/R_{18})$. Note again that the reference voltage is not available externally. Since the resistive voltage divider that sets the output voltage is internal to the regulator, the regulator output voltage that appears between the output and ground terminals is not adjustable.

The biasing voltage for the reference voltage circuitry is derived from the reference voltage itself via Q_1 and Q_6. An important consequence of this is that the reference voltage will be almost completely independent of the input voltage, thus leading to very good line regulation. Since, when the circuit is first turned on, the reference voltage will be zero, resistor R_1 is provided to bypass Q_1 to provide a biasing current for the voltage reference circuit.

Having now described the basic operation of this voltage regulator, we will briefly return to consider the operation of the voltage reference circuit. A band-gap voltage reference circuit produces a temperature-compensated reference voltage equal to $V_{REF} = (n + 1)(E_{GO} + 3V_T) = (n + 1)1.283$ V, where $n + 1$ is the number of base-to-emitter junction drops across which V_{REF} is developed. Inspection of the LM78XX circuit reveals that V_{REF} is developed across the base–emitter junctions of Q_6, Q_7, and Q_8, so $n + 1 = 3$, and therefore that $V_{REF} = 3.85$ V. Note that the reference voltage is given by $V_{REF} = V_{BE_6} + I_5R_7 + V_{BE_7} + V_{BE_8}$, and it is the positive temperature coefficient of the voltage drop across R_7 that compensates for the negative temperature coefficient of the base-to-emitter voltage drops.

The voltage drop across R_2 is $V_{REF} - V_{BE_1} - V_{BE_2} - V_{BE_3}$. Using a representative figure of 650 mV for the base-to-emitter voltage drops, the voltage drop across R_2 is $3.85 - 3(0.65) = 1.9$ V. Therefore, the current through R_2 is 1.0 mA. Neglecting base currents as being small, this also gives the current through Q_3 as $I_3 = I_{R_2} = 1.0$ mA. The current through R_4 is $I_{R_4} = 0.65$ V/1.2 kΩ = 0.5 mA. The voltage across R_5 is $V_{BE_2} + V_{BE_3} - V_{BE_4} = 0.65$ V, so the current through R_5 and therefore through Q_4 is given by $I_4 = I_{R_5} = 0.65$ V/12.1 kΩ = 0.05 mA = 50 μA.

We will now consider the current through Q_5, which is I_5, noting that it is this current that is used for the temperature compensation of the base-to-emitter negative temperature coefficients. Inspection of the circuit allows us to write that $V_{BE_2} + V_{BE_3} = V_{BE_4} + V_{BE_5} + I_5R_6$. We will assume identical transistors, all operating in the active region. We can therefore express the relationship between collector current and base-to-emitter voltage as $I_C = I_{CO} \exp(V_{BE}/V_T)$, so $V_{BE} = V_T \ln(I_C/I_{CO})$. Solving for I_5R_6 gives

$$I_5R_6 = V_{BE_2} + V_{BE_3} - V_{BE_4} - V_{BE_5} = V_T \ln \frac{I_2I_3}{I_4I_5} \qquad (11.72)$$

Since $I_2 = 0.5$ mA, $I_3 = 1.0$ mA, and $I_4 = 0.05$ mA, this becomes

$$I_5R_6 = V_T \ln \frac{10 \text{ mA}}{I_5} \qquad (11.73)$$

This is a transcendental equation for I_5 and can be solved by an iterative technique to give a value of 0.1 mA for I_5. Corresponding to this value of I_5, the I_5R_6 voltage drop is 25 mV ln (10 mA/0.1 mA) = 115 mV. The voltage drop across R_7 is therefore $I_5R_7 = (R_7/R_6)(I_5R_6) = 17 \times 115$ mV = 1.96 V. Since the reference voltage is equal to the voltage drop across R_7 plus the sum of three base-to-emitter voltage drops, we obtain for the reference voltage the value $V_{REF} = 1.96 + 3(0.65) = 3.9$ V, which is close to the expected value of $3(E_{GO} + 3V_T) = 3.85$ V.

In the reference voltage circuit, capacitor C_1 is used for frequency compensation to ensure stability. Resistor R_8 and diode-connected transistor Q_9 are used to prebias Q_8 to produce an acceptable quiescent current in that transistor.

The LM78XX series of three-terminal voltage regulators are available with output voltages of 5, 6, 8, 12, 15, 18, and 24 V, with the designation of the 5-V regulator being the 7805, the 6-V being the 7806, and so on. The voltage regulators of the 78XX series all have the same internal circuitry, except for different values for R_{17}, which determines the output voltage level.

11.6.6 Three-terminal Positive Voltage Regulator Example: LM78LXX

We will now consider another example of a three-terminal positive voltage regulator, the LM78LXX (National Semiconductor) as shown in Figure 11.23. In this circuit the voltage reference is provided by zener diode D_2 with Q_3, Q_2, and Q_1 used for temperature compensation. The temperature-compensated reference voltage appears at the junction of resistors R_1 and R_2 and is applied to the base of Q_7, which is part of the error amplifier. The zener diode D_2 is biased by Q_4, which acts as a current source.

The current-source transistor Q_4 is biased by Q_5, which is a multiple emitter and multiple collector PNP transistor. Transistor Q_5 is in turn biased by the current

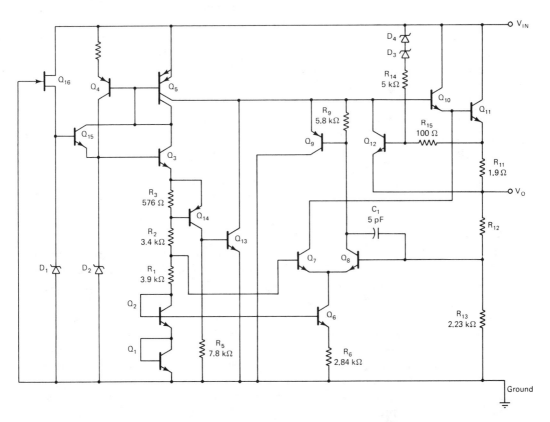

Figure 11.23 Three-terminal positive voltage regulator circuit: LM78XX (National Semiconductor).

through Q_3, which is controlled by the zener voltage drop and the base-to-emitter drops of Q_3, Q_2, and Q_1. As a result, the bias current of the voltage reference circuit, and in particular of D_2, will be almost completely independent of the input voltage so that, correspondingly, the reference voltage will be almost completely independent of the input voltage as well. This will result in very good line regulation for the regulator.

Transistors Q_{15} and Q_{16} and diode D_1 are used as a startup circuit to apply the initial bias to the voltage reference circuit. Once this has been achieved and the voltage across D_2 has risen to the zener voltage, transistor Q_{15} will be turned off, since its base-to-emitter voltage will now be zero, and the startup circuit will thus be disconnected from the voltage reference circuit.

The error amplifier is a differential amplifier comprised of Q_7 and Q_8 with Q_6 together with R_6 used for current sink biasing of the amplifier. The output of the differential amplifier is fed to the base of Q_9, which is in the emitter-follower configuration and drives the series-pass output stage. Capacitor C_1 connected across the collector–base junction of Q_8 is a compensation capacitor to ensure stability.

The series-pass output stage consists of Q_{10} and Q_{11} in a Darlington emitter-follower configuration. Foldback current limiting is provided by Q_{12} together with the current-limit resistor R_{11}. Resistors R_{15} and R_{14}, together with diodes D_3 and D_4, will produce a foldback characteristic that is dependent on the input–output voltage differential.

A portion of the output voltage is tapped off by the resistive voltage divider, consisting of R_{12} and R_{13}, and fed back to the base of Q_8, which is the inverting input of the error amplifier. Note that this voltage divider is internal to the IC regulator, so a fixed output voltage will result. Note also that the reference voltage is not accessible at any external terminal.

Thermal shutdown is provided by Q_{14}, Q_{13}, and R_3. The voltage across R_3 is given by

$$V_{R_3} = \frac{(V_Z - V_{BE_3} - V_{BE_2} - V_{BE_1})R_3}{R_1 + R_2 + R_3} = (V_Z - 3V_{BE}) \times 0.073 \quad (11.74)$$

With a zener voltage of 6.5 V and a base-to-emitter drop of 0.6 V, the voltage across R_3 is approximately 340 mV at 25°C. This is insufficient to turn Q_{14} on, and thus both Q_{14} and Q_{13} will be off. Due to the positive temperature coefficient of the zener voltage and the negative temperature coefficient of the base-to-emitter voltage drops, the voltage across R_3 will increase with temperature at a rate given by $dV_{R_3}/dT = 0.073 \times (dV_Z/dT - 3dV_{BE}/dT) = 0.72$ mV/°C. At the same time the voltage required to turn Q_{14} on decreases at a rate of about 2.1 mV/°C. Therefore, at some elevated temperature Q_{14} will indeed turn on and thereby turn Q_{13} on. This will cause the base drive of Q_{10} to be shunted through Q_{13} and thus produce the thermal shutdown.

We see that since the voltage across R_3 increases at a rate of about 0.7 mV/°C and the base-to-emitter voltage required to turn Q_{14} on decreases at a rate of 2.1 mV/°C, the temperature for thermal shutdown is determined by the relationship $\Delta T \simeq (650 \text{ mV} - 340 \text{ mV})/(2.8 \text{ mV/°C})$, where ΔT is the temperature rise for thermal shutdown. Solving for ΔT gives $\Delta T \simeq 110°C$, so thermal shutdown will occur at a chip temperature of about 135°C.

The LM78LXX series of three terminal voltage regulators is available with output voltages of 5, 8, 12, 15, 18, and 24 V, with the device designation being that the 78L05 is a 5-V regulator, and so on.

11.7 NEGATIVE VOLTAGE REGULATORS

In principle, it might be thought that negative voltage regulators could be constructed using the same circuit configuration as for positive voltage regulators, simply replacing the NPN transistors by PNP transistors, and vice versa. All the voltages and currents would be reversed, so the operation of the regulator would be completely equivalent to the complementary positive regulator. However, due to the fact that the performance of IC PNP transistors is in general much inferior to that of the NPN transistors, the design of the negative regulator is not the same as the positive regulator. Indeed, the design of the negative regulator is based on the preponderate use of NPN transistors, as is the case with the positive regulators. As a result of this, the circuit configuration will be somewhat different, although the basic principles of operation are still the same.

An example of a three-terminal negative regulator is shown in Figure 11.24.

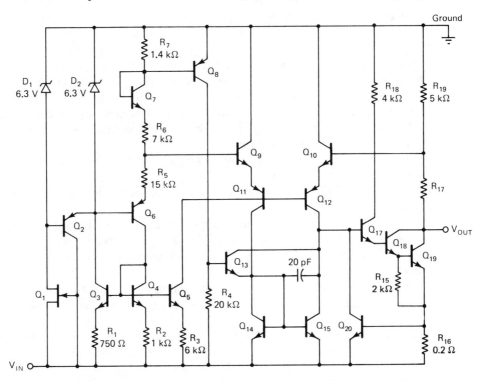

Figure 11.24 Three-terminal negative voltage regulator: LM79XX (National Semiconductor).

This voltage regulator circuit is similar to that of the LM120/145/320/79XX/79LXXAC (National Semiconductor) series of voltage regulators.

The voltage reference for this regulator circuit is of the temperature-compensated zener diode type. The zener diode for the voltage reference circuit is D_2 (6.3 V), and the positive temperature coefficient of this zener voltage is compensated by the negative temperature coefficients of Q_6, Q_7, and Q_8. The temperature-compensated reference voltage appears at the junction of R_5 and R_6.

The bias current for the zener diode, D_2, is supplied by Q_3, which acts as a current sink. Q_3 is in turn biased by Q_4, which is a diode-connected transistor. The current through Q_4 is supplied by the D_2, Q_6, Q_7, R_6, R_7 voltage reference circuit and is given by $I_4 = (V_Z - 3V_{BE})/(R_5 + R_6)$. We see, therefore, that the bias current of the zener diode is to a first approximation independent of the supply voltage, which thus leads to a very good line regulation for the voltage regulator. The startup circuit of D_1, Q_1, and Q_2 acts to supply the initial bias voltage across the voltage reference circuit. Once this has happened, the equilization of the voltage drops across D_1 and D_2 causes transistor Q_2 to turn off, thereby disconnecting the startup circuit from the voltage reference so that thereafter the voltage reference becomes self-biasing.

The reference voltage produced by the voltage reference circuit is supplied to the base of Q_9, which is the noninverting input of the error differential amplifier. This differential amplifier consists of transistors Q_9 through Q_{12}. Each half of the differential amplifier consists of a common-collector/common-base cascode direct-coupled compound transistor configuration. The bias current for this differential amplifier is the base current of Q_{11} and Q_{12} and is supplied by the current sink Q_5. Note that the current of Q_5 is determined by the current through Q_4, which is controlled by the self-biasing voltage reference. As a result, the quiescent current of the differential amplifier will exhibit little dependence on the input voltage, this again leading to very good line regulation. Transistors Q_{14} and Q_{15} constitute a current-mirror active load for the differential amplifier. The 20-pF capacitor from collector to base of Q_{15} is for frequency compensation.

The series-pass output stage consists of Q_{17}, Q_{18}, and Q_{19} in a Darlington emitter-follower configuration. Resistor R_{15} (2 kΩ) provides for an acceptable quiescent current for Q_{17} and Q_{18} and provides a return path for the leakage current of Q_{17} and Q_{18}.

A portion of the output voltage is tapped off by the resistive voltage divider of R_{17} and R_{19} and fed back to the base of Q_{10}, which is the inverting input of the error amplifier. Note that this is an internal feedback network, so the output voltage is not variable.

Current limiting is accomplished by the action of Q_{20} and resistor R_{16}. When Q_{20} is turned on, it diverts base drive away from the base of Q_{17} and produces current limiting.

Thermal shutdown is produced by Q_8, Q_{13}, R_7, and R_4. At 25°C the voltage drop across R_7 is given by $(V_Z - 2V_{BE})R_7/(R_5 + R_6 + R_7) = 300$ mV. This voltage is insufficient to turn Q_8 on, so Q_{13} will also not be turned on. The voltage across R_7 will, however, increase with temperature at a rate of about 0.4 mV/°C due to the temperature coefficient of the zener diode (D_2) and the two base–emitter junctions

$(Q_6$ and $Q_7)$. At the same time, the required base-to-emitter voltage needed to turn Q_8 on decreases at a rate of about 2.1 mV/°C. To turn Q_8 on, the combination of the change in the voltage across R_7 and the decrease in the required base-to-emitter voltage must add up to about 300 mV. This requires a temperature increase of 300 mV/(2.5 mV/°C) = 120°C. As a result, a thermal shutdown temperature of about 150°C is expected.

11.8 SWITCHING VOLTAGE REGULATORS

The voltage regulators discussed thus far have been of the dissipative type. In these regulators, the flow of current to the load is controlled by a series-pass circuit that produces a voltage drop equal to the required input–output voltage difference, $V_{IN} - V_O$. The input voltage is the unregulated dc voltage. In performing this function, the series-pass circuit must dissipate an amount of power equal to the difference between the input power $V_{IN}I_O$ and the output power V_OI_O. As a result, the efficiency of this type of voltage regulator is quite low, being given by

$$\text{efficiency} = \eta = \frac{P_O}{P_{IN}} = \frac{V_OI_O}{V_{IN}I_O} = \frac{V_O}{V_{IN}} \tag{11.75}$$

The fraction of the input power that is dissipated in the voltage regulator is therefore

$$\frac{P_{IN} - P_O}{P_{IN}} = 1 - \eta = \frac{V_{IN} - V_O}{V_{IN}} \tag{11.76}$$

Since the output voltage is typically in the range from 0.5 to 0.7 V_{IN}, we see that the circuit efficiency will typically be in the range from 50% to 70%. As a result, some 30% to 50% of the input power will have to be dissipated by the regulator.

The power dissipation in the regulator is not a serious problem for regulators supplying less than about 1 W. However, for output power levels much in excess of this, and in particular for levels above 10 W, the circuit inefficiency and in particular the power that must be dissipated by the series-pass circuit may pose serious problems. Due to the power transistor and the associated heat sinking requirements, the regulator cost, size, and weight will escalate rapidly with increasing output power requirements.

A useful alternative to the dissipative type of voltage regulator is one of the switching type. In the dissipative regulator, the series-pass transistors operate continuously in the active region wherein the combination of a substantial current flow and concurrently a major voltage drop results in a large power dissipation as given by $P_d = (V_{IN} - V_O)I_O$. In contrast, in the switching regulator the series-pass transistors are not operated in the active mode, but rather are switched back and forth between cutoff and saturation.

In the switching regulator the power flow to the load is controlled by the duty cycle of the series-pass transistors, that is, the fraction of time that these transistors are in the "on" state. A fraction of the output voltage is supplied to the control circuitry and therein compared to the reference voltage. The difference between the feedback voltage and the reference voltage is amplified and used to control the duty cycle of the series-pass transistors. If the output voltage is too low, the control circuitry acts to lengthen the duty cycle, thus causing the output voltage to increase.

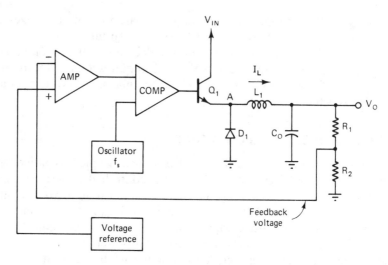

Figure 11.25 Basic diagram of a switching voltage regulator.

If the output voltage is too high, the duty cycle is diminished and the output voltage decreases. In either case the output voltage will be stabilized at a level determined by the resistance ratio of the feedback voltage divider and the reference voltage. Figure 11.25 shows a basic block diagram of a switching regulator.

The power dissipation of the switching transistor is relatively low. When the transistor is in the cutoff region of operation, the current through the transistor is negligibly small, so the power dissipation will similarly be negligible. When the transistor is turned on, it will rapidly be driven into saturation by the control circuitry. In the saturation mode, the voltage drop across the transistor will be relatively small, generally less than 1 V, and often down in the range of about 0.2 V. As a result of the small voltage drop across the transistor, the instantaneous power dissipation will be correspondingly small. The only time in which there is a substantial amount of power dissipated in the transistor is during the switching transition, when the transistor is passing through the active region in going from cutoff to saturation, or vice versa. The switching transition time, however, represents only a very small fraction of the total time, so the average power dissipation is small. Figure 11.26 shows the instantaneous power dissipation of the transistor as a function of time.

If we now look at Figure 11.25, we see that a fraction $R_2/(R_1 + R_2)$ of the output voltage is fed back to the inverting input of the error amplifier, where it is compared to the reference voltage that is applied to the noninverting input of the error amplifier. The difference between the reference voltage and the feedback voltage is thus amplified and thence applied to the inverting input of the comparator.

The oscillator generates a triangular waveform at a fixed repetition frequency, and this voltage is applied to the noninverting input of the comparator. The output of the comparator is thus in the high state during the time when the triangular voltage waveform is above the level of the error amplifier output, and it is during this time that the series-pass switching transistor will be turned on. During the remainder of the period, the comparator output will be low and transistor Q_1 will be switched off.

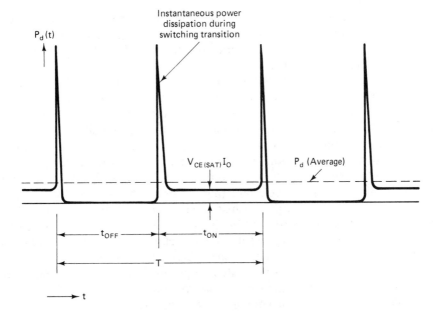

Figure 11.26 Instantaneous power dissipation of a switching transistor.

The output of the comparator is thus a pulse waveform, the period of which, T, is the same as the period of the oscillator output. The duty cycle, $\delta = t_{on}/T$, of the pulse waveform is controlled by the difference between the feedback voltage and the reference voltage.

When Q_1 is on, the voltage applied to the inductor L_1 (at point A) is $V_A = V_{IN} - V_{CE(SAT)} \cong V_{IN}$. This voltage acts to promote the flow of current through L_1. When Q_1 is turned off, the inductor L_1 acts to continue the current through itself and thus on to the load. Diode D_1 is used to provide a complete circuit during this period of time. When Q_1 is on, this diode is biased off. Capacitor C_O acts to smooth out the voltage so that the output voltage will be a relatively smooth dc voltage with very little ac ripple.

During the time that Q_1 is off, the voltage at point A is close to zero, actually one diode drop (0.7 V) below ground. The average voltage at point A is therefore given by

$$V_A(dc) = (V_{IN} - V_{CE(SAT)}) \frac{t_{on}}{T} + (-V_{D_1}) \frac{t_{off}}{T}$$

$$= V_{IN}\delta - V_{CE(SAT)}\delta - V_{D_1}(1 - \delta) \qquad (11.77)$$

$$\simeq V_{IN}\delta$$

Since L_1 and C_O will not affect the dc voltage, the dc voltage at the output will be the same as at point A, that is, $V_O = V_A \cong \delta V_{IN}$. Thus we see that the dc output voltage can be controlled or regulated by the control of the duty cycle by the feedback loop. The output voltage is slightly smaller than δV_{IN} due to $V_{CE(SAT)}$ and the voltage drop of D_1. Both voltage drops are generally in the neighborhood of about 0.7 V, so we see that V_O will typically be around 0.7 V less than $V_{IN}(\delta)$. There is also a small dc voltage drop due to the series resistance of L_1, although this probably will not be in excess of about 0.1 V.

We have seen that the output voltage is controlled by the duty cycle $\delta = t_{on}/T$. For example, if $V_{IN} = 15$ V and V_O is to be 10 V, the duty cycle is approximately given by $\delta = V_O/V_{IN} = 0.67$. Due to the transistor saturation voltage and the diode drop, the duty cycle under steady-state conditions will actually have to be somewhat larger than this, probably around 0.70, so Q_1 will be on for about 70% of the time (in saturation) and off (cutoff) for the remaining 30% of the time.

If the output voltage for some reason falls below 10 V, the feedback loop will act to increase the duty cycle and thereby bring the output voltage up to 10 V. Conversely, if for some reason, such as variations in load conditions or in line voltage, the output voltage goes above 10 V, the feedback loop will act to decrease the duty cycle and thereby bring the output voltage back down. The net result of the action of the feedback loop is to regulate V_O at the design value given by $V_O = V_{REF}(1 + R_1/R_2)$. The entire system acts basically as a pulse-width modulator (PWM) to control the duty cycle with which the series-pass switching transistor is driven.

The optimum switching frequency f_s is generally in the range from 10 to 100 kHz. A high switching frequency allows for relatively small values of L_1 and C_O to be used, thus reducing the size, weight, and cost of the system. On the other hand, as the switching frequency goes up, the power dissipated by the switching transistor increases as a result of the increase in the number of switching transitions per unit time. This increase in the transistor power dissipation is the result of the finite switching speed of the transistor. To maximize the circuit efficiency, therefore, a power transistor with a fast switching speed should be chosen. As a result of these considerations, a switching frequency in the range from 10 to 100 kHz is usually chosen, often about 20 to 50 kHz.

11.8.1 Current Waveforms

Now that we have been introduced to the basic principles of operation of the switching regulator, we will consider the current waveform through L_1 and the ac voltage ripple across C_O, and then conclude with a design example.

The inductance value of L_1 is chosen such that there is a continuity of current flow through the inductor throughout the entire switching cycle. Indeed, both L_1 and C_O will be large enough such that the current waveform of the inductor current I_L will have an approximately triangular waveform superimposed on a dc current level, which is equal to the load current. This is shown in Figure 11.27.

The relationship between the voltage across L_1 and the current through L_1 is given by $V_L = L_1(di_L/dt)$. As a result of the approximately linear slope of the current waveform, we can express this relationship during the "on" time as

$$V_L = V_{IN} - V_O = L\frac{\Delta I_L}{t_{on}}$$

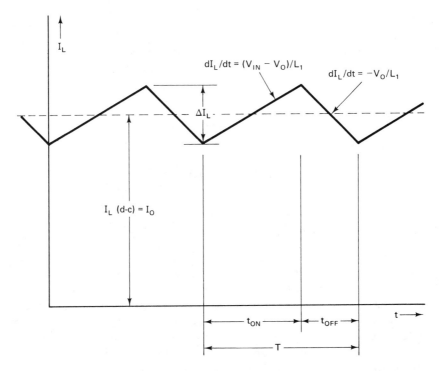

Figure 11.27 Inductor current waveform.

so the change in the current during the "on" time is given by

$$\Delta I_L = \frac{1}{L} (V_{IN} - V_O)t_{on} \tag{11.78}$$

During the "off" time, the voltage across L_1 is $-V_D - V_O \cong -V_O$, so the change in the current through L_1 during this time period is given by

$$\Delta I_L \cong \frac{1}{L_1} (-V_O)t_{off} \tag{11.79}$$

Since the current is a continuous function of time, the algebraic sum of these two changes in the current must be zero, so we have $(V_{IN} - V_O)t_{on} = V_O t_{off}$, giving $V_{IN}t_{on} = V_O(t_{on} + t_{off}) = V_O T$; thus $V_O/V_{IN} = t_{on}/T = \delta$, which is the same as the result obtained earlier. A somewhat more rigorous analysis in which the transistor saturation voltage and the diode drop are taken into account yields

$$V_O = V_{IN}\delta - V_{CE(SAT)}\delta - V_{D_1}(1 - \delta) \tag{11.80}$$

which is again consistent with the previous results.

For the proper operation of this switching regulator, in particular for a maximally smooth output waveform, it is important that the current through the inductor never drop to zero or even reverse direction. Therefore, we must ensure that ΔI_L never approach the dc inductor current level, which is the same as the dc output current I_O. To take into account variations in I_O with loading conditions, a suitable safety factor is to have $\Delta I_L \lesssim 0.4 \, I_O$. Since we can express ΔI_L as $\Delta I_L = (1/L)V_{ON}t_{off}$, we can solve for L_1 as

$$L_1 = \frac{V_o t_{off}}{|\Delta I_L|} \gtrsim \frac{V_o t_{off}}{0.4 \, I_O} \qquad (11.81)$$

Since $t_{off} = T(1 - \delta)$, this can be rewritten as

$$L_1 \geq \frac{(V_O/I_O)T(1 - \delta)}{0.4} = R_L T(1 - \delta)2.5$$

11.8.2 Output Ripple Factor

The voltage at point A is basically a repetitive pulse waveform with a duty cycle δ and an amplitude of approximately V_{IN}. The pulse waveform is applied to L_1 and C_O, which act as a low-pass filter to smooth out this voltage waveform and produce a dc voltage across C_O that is the average value of the input voltage, V_{IN}.

The capacitance value for C_O is such that the reactance of C_O at the ac frequencies of interest is small compared to the load resistance R_L and also small compared to $j\omega L_1$. As a result, L_1 and C_O can, to a good degree of approximation, be considered at the ac frequencies of interest to be a simple voltage divider. The ratio of the ac output voltage at frequency f to the ac input voltage at the same frequency is given by

$$\frac{v_o(f)}{v_A(f)} = \frac{(1/j\omega C_O)}{j\omega L_1 + 1/j\omega C_O} \simeq -\frac{1}{\omega^2 L_1 C_O} \qquad (11.82)$$

where $\omega = 2\pi f$ is the radian frequency. The negative sign for the voltage ratio is simply indicative of a 180° phase shift in the low-pass network.

To obtain an expression for the output ac ripple voltage, we will consider the simple representative case of a 50% duty cycle for which case the dc output voltage is $V_O \simeq \frac{1}{2}V_{IN}$, and the voltage applied to the low-pass filter (at point A) is a square wave of amplitude V_{IN} and period $T = 1/f_s$, where f_s is the oscillator frequency.

The Fourier series for this square wave is given by

$$v_A(t) = V_{IN}\left(\frac{1}{2} + \frac{2}{\pi}\cos \omega t - \frac{2}{3\pi}\cos 3\omega t + \frac{2}{5\pi}\cos 5\omega t - \cdots\right) \qquad (11.83)$$

where $\omega = 2\pi f_s$. We see that the input voltage to the low-pass filter consists of a dc component $V_{IN}/2$, a fundamental component at $f = f_s$, and various odd harmonic components at $f = nf_s$, where n is an odd integer. We further note that the amplitude of the various harmonic terms decreases as $1/n$.

The attenuation of the low-pass filter is proportional to f^2. This, together with the fact that the amplitudes of the various harmonic terms at the filter input decrease as $1/n$, means that the dominant component of the output ac ripple voltage will be the fundamental term at $f = f_s$, with the other harmonics being of relatively little importance.

The ac output ripple voltage is therefore given approximately by

$$v_o(\text{ac}) = v_o(f_s) = \frac{2}{\pi} V_{IN} \frac{1}{\omega_s^2 L_1 C_O} \tag{11.84}$$

Since $V_O \simeq V_{IN}/2$ and $\omega_s = 2\pi f_s$, this can be rewritten as

$$v_o(\text{ac}) = \frac{4}{\pi} V_O \frac{1}{4\pi^2 f_s^2 L_1 C_O} \simeq \frac{1}{30 f_s^2 L_1 C_O} V_O \tag{11.85}$$

The output ripple factor is the ratio of the peak-to-peak ripple voltage to the dc output voltage. From the preceding equation, we can thus directly write the ripple factor as

$$r = \text{ripple factor} = \frac{v_o(\text{ac})_{\text{peak-to-peak}}}{V_O} \simeq \frac{1}{15 f_s^2 L_1 C_O} \tag{11.86}$$

We thus see that the ripple factor is inversely proportional to the square of the switching frequency and to the $L_1 C_O$ product. With a high switching frequency, therefore, a moderately small value of inductance and capacitance can be used.

11.8.3 Design Example

For a design example, we will stipulate the following:

I_O = dc output (load) current = 1.0 A
V_O = dc output voltage = 10 V
V_{IN} = 20 V
f_s = switching frequency = 30 kHz
output ripple factor = 0.05%

For the inductor L_1, we have the following requirement:

$$L_1 \gtrsim \frac{R_L T(1 - \delta)}{0.4} = \frac{2.5\, R_L(1 - \delta)}{f_s} = \frac{2.5 \times 10\, \Omega \times 0.5}{30\ \text{kHz}} \tag{11.87}$$

so $L_1 \gtrsim 417\ \mu\text{H}$. A choice of 500 μH is therefore adequate. From the ripple factor condition, we have

$$5 \times 10^{-4} \geq \frac{1}{15 f_s^2 L_1 C_o} = \frac{1}{15(30\ \text{kHz})^2(500 \times 10^{-6}) C_O} \tag{11.88}$$

so

$$C_O \geq 296\ \mu\text{F} \tag{11.89}$$

Therefore, a choice of 300 to 500 μF for C_O should prove satisfactory. The 0.05% ripple factor corresponds to a peak-to-peak output ripple voltage of only 5 mV, so the dc output waveform of this regulator circuit will be quite smooth.

11.8.4 Step-up Switching Voltage Regulator

The basic circuit of the switching regulator that has been discussed thus far is shown in Figure 11.28a. This type of switching regulator is called a step-down regulator since the output voltage is less than the input voltage ($V_O < V_{IN}$), and V_O is related to V_{IN} approximately by $V_O = \delta V_{IN}$, where $\delta = t_{on}/T$ is the duty cycle.

With the switching regulator it is also possible to achieve a voltage step up so that V_O will be greater than V_{IN}. The basic circuit of the step-up regulator is shown in Figure 11.28b.

It might at first be thought that it is not possible to have the output voltage greater than the input voltage. It is, however, the voltage induced in the inductor adding to the input voltage that produces an output voltage that is greater than the input voltage. During the time that Q_1 is off, the reduction in the current through the inductor produces a voltage across the inductor given by $V_L = L(di_L/dt)$ that adds to V_{IN} to make V_O greater than V_{IN}. It should also be noted that even though $V_O > V_{IN}$ the output power $P_O = V_O I_O$ will not, of course, be any greater than the input power, $P_{IN} = I_{IN} V_{IN}$, since the input current will be larger than the output current.

We can now easily derive the relationship between the input and output voltages of the step-up switching regulator. For this derivation, we will assume that L_1 is large enough such that the current through L_1 will have a triangular waveform. Dur-

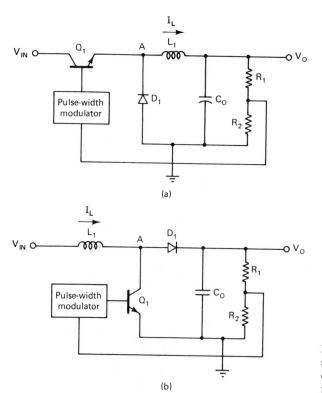

(a)

(b)

Figure 11.28 (a) Basic circuit of the step-down switching regulator; (b) basic circuit of the step-up switching regulator.

ing the time that Q_1 is off, the voltage across the inductor is given by $V_{IN} - V_D - V_O = L(dI_L/dt)$, where V_D is the forward-bias voltage drop of D_1 (about 0.7 V). Assuming a linear variation of current with time, we therefore have that

$$\Delta I_L = \frac{1}{L}(V_{IN} - V_D - V_O)t_{off} \tag{11.90}$$

During the time interval that Q_1 is on (in saturation), we have that

$$V_{IN} - V_{CE(SAT)} = L\frac{dI_L}{dt} \tag{11.91}$$

so

$$\Delta I_L = \frac{1}{L}(V_{IN} - V_{CE(SAT)})t_{on}$$

Since the current flow through L_1 must be a continuous function of time, the algebraic sum of these two current changes must be zero, so we have $\Delta I_L(t_{off}) + \Delta I_L(t_{on}) = 0$. Substituting in the equations above gives us

$$(V_{IN} - V_D - V_O)t_{off} + (V_{IN} - V_{CE(SAT)})t_{on} = 0 \tag{11.92}$$

Solving this for V_O gives

$$V_O = \frac{V_{IN}(t_{on} + t_{off}) - V_D t_{off} - V_{CE(SAT)}t_{on}}{t_{off}} \tag{11.93}$$

so

$$V_O = V_{IN}\left(1 + \frac{t_{on}}{t_{off}}\right) - V_{CE(SAT)}(t_{on}/t_{off}) - V_D \tag{11.94}$$

Since V_D is approximately 0.7 V and $V_{CE(SAT)}$ is less than 1 V, we see that it is quite possible to have V_O greater than V_{IN} if t_{off} is less than t_{on}. In terms of the duty cycle δ, the equation above can be rewritten as

$$V_O = \frac{V_{IN}}{1 - \delta} - V_{CE(SAT)}\frac{\delta}{1 - \delta} - V_D \tag{11.95}$$

For example, if we have that V_D and $V_{CE(SAT)}$ are both approximately 0.7 V, for a duty cycle of 0.7 and an input voltage of 10 V, we have

$$V_O = \frac{10}{0.3} - 0.7\left(\frac{0.7}{0.3}\right) - 0.7$$
$$= 31 \text{ V} \tag{11.96}$$

Due to losses in L_1 and in the transistor during the switching transitions, the actual output voltage may be closer to 30 V. In any case, we see that a 3:1 voltage step-up has been achieved.

When Q_1 is on, the voltage at point A is $V_{CE(SAT)}$. When Q_1 is off, the voltage at point A is equal to the input voltage plus the voltage induced in L_1 due to the falling current. This induced voltage is approximately $V_{IN}(t_{on}/t_{off})$, so the total voltage at point A during this time interval is approximately $V_{IN}(1 + t_{on}/t_{off})$. This relationship can be readily obtained by noting that when Q_1 is on, the increase in the current through L_1 is given by $\Delta I_L = (1/L)V_{IN}t_{on}$. When Q_1 is off, the change in the current is $\Delta I_L = (1/L)(V_{IN} - V_A)t_{off}$. The diode drop and the transistor saturation voltage have been neglected for purposes of simplicity. Since the net change in the inductor current over the complete switching cycle must be zero, we have that $V_{IN}t_{on} + (V_{IN} - V_A)t_{off} = 0$, so solving for V_A we have $V_A = V_{IN}(1 + t_{on}/t_{off})$.

Figure 11.29 is diagram of the voltage waveform at point A. Notice that this is a repetitive pulse waveform. The circuit to the right of point A, particularly the combination of the diode D_1 and the capacitor C_O, can be considered to be essentially a peak detector circuit giving the output waveform shown in Figure 11.29. During the pulses the capacitor will charge up via D_1 to the peak pulse amplitude $V_{IN}(1 + t_{on}/t_{off})$. Between pulses, however, the diode D_1 will be off, so the capacitor will discharge through the load. The rate of discharge of the capacitor is $dQ_c/dt = I_O = V_O/R_L$, where R_L is the load resistance. If the circuit is designed for a low-output ripple factor V_O, and therefore I_O will not change much on a percentage basis during the discharge time, the total charge lost by the capacitor during this time is given by $\Delta Q_C = I_O t_{on} = V_O t_{on}/R_L$.

The droop or sag in the output voltage at the end of the discharge time is given by

$$\Delta V_O = \frac{\Delta Q_C}{C_O} = \frac{I_O t_{on}}{C_O} = \frac{V_O t_{on}}{R_L C_O} \tag{11.97}$$

This drop in the output voltage is the peak-to-peak ripple voltage. The ripple factor

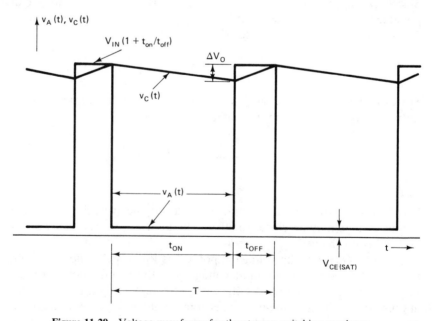

Figure 11.29 Voltage waveforms for the step-up switching regulator.

is therefore given by

$$\text{ripple factor} = r = \frac{\Delta V_O}{V_O} = \frac{t_{on}}{R_L C_O} = \frac{\delta T}{R_L C_O} = \frac{\delta}{f R_L C_O}$$

where f is the switching frequency.

For a representative example, let us take $\delta = 0.7$, $f = 30$ kHz, $V_O = 30$ V, and $I_O = 0.5$ A. For a ripple factor at the output not to exceed 0.1%, the required value for C_O is given by

$$C_O = \frac{\delta}{rfR_L} = \frac{0.7}{30 \text{ kHz} \times 60 \ \Omega \times 0.001}$$

$$= \underline{389 \ \mu F}$$

(11.98)

Thus a choice of a 500-μF, 50-V capacitor would prove to be satisfactory.

11.8.5 Effective Series Resistance

It is appropriate at this point to discuss another important factor in the capacitor selection for switching voltage regulators. This factor is the *effective series resistance* or ESR of the capacitor. Any real capacitor will exhibit, together with the desired capacitance effect, a parasitic series resistance, R_S, which is the effective series resistance. This series resistance will vary somewhat with frequency and with temperature.

To illustrate the importance of the ESR in the capacitor specification, we will return to a consideration of the preceding example. During the discharge interval, the current flowing out of the capacitor is I_O, so the additional drop in the capacitor voltage due to R_S is $I_O R_S$. During the charge interval (D_1 on, Q_1 off), the charge supplied to the capacitor must make up for the charge lost during the discharge time, so we have

$$I_C t_{off} = I_O t_{on}$$

(11.99a)

and thus

$$I_C = I_O \frac{t_{on}}{t_{off}}$$

(11.99b)

where I_C is the current into the capacitor during the charging time. During this time interval, there is an additional rise in the capacitor voltage due to R_S given by $\Delta V_C = \Delta V_O = I_O R_S(t_{on}/t_{off})$. Therefore, the increase in the peak-to-peak ripple voltage due to the effects of R_S is given by

$$\Delta V_{O(\text{peak-to-peak})} = I_O R_S\left(1 + \frac{t_{on}}{t_{off}}\right) = I_O R_S \frac{V_O}{V_{IN}}$$

(11.100)

For a 0.1% ripple factor and a 10-V output voltage, the corresponding peak-to-peak ripple voltage is 10 mV. For an output current of 0.5 A and a step-up voltage ratio of 3 : 1, the maximum allowable value of R_S is given by the condition that

$$10 \text{ mV} = 0.5 \text{ A} \times R_S \times 3$$

(11.101)

so

$$R_S = \frac{10 \text{ mV}}{1.5 \text{ A}} = 7 \text{ m}\Omega$$

This low value of effective series resistance may be very difficult to achieve in practice, so the 0.1% ripple factor may not be readily achievable. We see therefore that the ESR may be the dominant factor in determining the ripple factor and is therefore a very important factor in the capacitor selection.

11.8.6 Self-oscillating Switching Regulators

The switching regulators described thus far have been of the externally excited or driven type, in which an oscillator or square-wave generator external to the voltage regulator circuit is used to determine the switching frequency. The regulator circuit itself acts as a pulse-width modulator to control the duty cycle of the pulses supplied by the oscillator.

In the interest of greater overall circuit simplicity, it is possible to have a self-oscillating switching regulator in which a separate oscillator circuit is not needed. The self-excited condition of the switching regulator arises by means of a positive-feedback loop, as shown in Figure 11.30. The positive feedback is obtained via the R_3–R_4 voltage divider. A small fraction of the voltage (typically 0.01 to 0.001) at point A is fed back to the noninverting input of the error amplifier together with the reference voltage. There is at the same time negative feedback via the R_1–R_2 voltage divider in which a fraction of the output voltage is fed back to the inverting input of the amplifier.

To see how the self-oscillating switching regulator operates, let us start off when the input voltage is first applied so that V_O starts off at zero. The reference voltage applied to the amplifier causes Q_1 to turn on and go into saturation, and V_O starts to increase. This continues until the output voltage reaches a value such that the voltage fed back to the inverting input via the R_1–R_2 voltage divider equals the voltage applied to the noninverting input. These two voltages are $V_O R_2/(R_1 + R_2)$ and $[V_{REF}R_3/(R_3 + R_4)] + [(V_{IN} - V_{SAT})R_4/(R_3 + R_4)]$, respectively. When the voltage fed back via the negative-feedback loop approaches the voltage applied to the non-inverting input of the amplifier, the output voltage of the amplifier starts to drop. This in turn causes the voltage at point A to drop, and via the positive feedback loop,

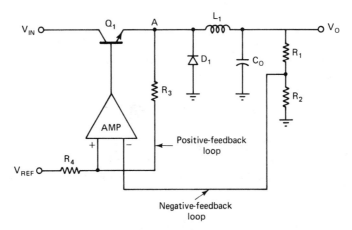

Figure 11.30 Basic diagram of the self-oscillating switching regulator.

the voltage at the noninverting input terminal also starts to drop. This drop in the voltage at the noninverting input accelerates the drop in the amplifier output voltage and thus in the drop in the voltage at point A. As a result of this positive-feedback condition, Q_1 is rapidly driven into cutoff.

With Q_1 in cutoff, the output voltage starts to fall. As soon as the voltage fed back by the negative-feedback loop approaches the voltage at the noninverting input, the output of the amplifier once again starts to go up. This causes Q_1 to turn on, raising the voltage at point A. This rise in voltage is fed back to the amplifier via the positive feedback loop and accelerates the turn-on of Q_1, with Q_1 rapidly being driven into saturation. The output voltage again starts to go up and the cycle is then repeated.

11.8.7 Switching Frequency

We will now derive an approximate expression for the switching frequency. We will first note that when Q_1 is on, the positive-feedback voltage is $(V_{IN} - V_{SAT})R_4/(R_3 + R_4)$, where $V_{SAT} = V_{CE(SAT)}$ is the saturation voltage of Q_1. When Q_1 is off, the positive-feedback voltage is $(-V_D)R_4/(R_3 + R_4)$, where V_D is the forward-bias voltage drop of D_1. The peak-to-peak swing in the positive-feedback voltage is therefore $(V_{IN} - V_{SAT} + V_D)R_4/(R_3 + R_4) \cong V_{IN}R_4/(R_3 + R_4)$, since both V_{SAT} and V_D have about the same value (0.7 V).

As a result of the switching action of Q_1, the output voltage will have some peak-to-peak ripple component. The switching of Q_1 via the amplifier occurs when the voltages at the two amplifier input terminals become equal to each other. Therefore, the peak-to-peak excursions in the positive-feedback voltage must be matched by an equal excursion in the negative-feedback voltage that is fed back to the inverting input of the amplifier via the R_1–R_2 voltage divider.

We will now turn our attention to the voltage fed back to the inverting input terminal via the negative-feedback loop. In addition to the dc component of this feedback voltage, there is a small ac component resulting from the ac ripple on the output. Assuming a duty cycle of around 50%, the input voltage can be considered to be a square wave with a peak amplitude of V_{IN}. The L_1–C_O network can be considered to be a simple voltage divider for the ac frequencies of interest, with a voltage-division ratio of

$$\frac{v_o}{v_A} = \frac{1/j\omega C_O}{j\omega L_1 + (1/j\omega C_O)} \simeq \frac{1}{\omega^2 L_1 C_O} \tag{11.102}$$

since $\omega L_1 \gg I/\omega C_O$ for the frequencies of interest.

The voltage at point A can be analyzed using a Fourier series to give a dc component, a fundamental term at the switching frequency, and various odd harmonic terms. Due to the rapidly increasing attenuation of the L_1–C_O network with increasing frequency, only the fundamental term in the output will be of any importance (except for the dc term, of course). The fundamental term in the Fourier series has an amplitude given by $(2/\pi)V_{IN}$ at point A. At the output of the L_1–C_O low-pass filter, the amplitude will be reduced by the factor $1/\omega^2 L_1 C_O$ to $v_o = (2/\pi)(1/\omega^2 L_1 C_O)V_{IN}$. The peak-to-peak voltage ripple at the output is therefore twice this.

The peak-to-peak swing in the voltage fed back to the inverting input terminal of the amplifier is therefore given by $[R_2/(R_1 + R_2)] (4/\pi)(1/\omega^2 L_1 C_O)V_{IN}$. If we now equate the peak-to-peak voltage excursions at the amplifier input terminals, we obtain

$$\frac{V_{IN}(4/\pi)}{\omega^2 L_1 C_O} \frac{R_2}{R_1 + R_2} = V_{IN} \frac{R_4}{R_3 + R_4} \tag{11.103}$$

so that, solving for ω, we obtain

$$\omega^2 = \frac{4/\pi}{L_1 C_O} \frac{R_2}{R_1 + R_2} \frac{R_3 + R_4}{R_4} \tag{11.104}$$

The switching frequency is therefore given by

$$(2\pi f)^2 \cong \frac{4/\pi}{L_1 C_O} \frac{V_{REF}}{V_O} \frac{R_3 + R_4}{R_4} \tag{11.105}$$

after noting that $V_O R_2/(R_1 + R_2) \cong V_{REF}$. We see that the switching frequency is a function of the $L_1 C_O$ product as well as the feedback factors for both the positive- and negative-feedback loops.

We will now consider a representative example and will use some of the design specifications of the switching regulator considered previously: $I_O = 1.0$ A, $V_O = 10$ V, $V_{IN} = 20$ V, $f_s = 30$ kHz, and a ripple factor of 0.05%. To meet the ripple factor specification, we obtained $L_1 = 500$ μH and $C_O = 300$ μF. We will now use these values to determine the values of R_3 and R_4 required to make this a self-oscillating regulator with a 50% duty cycle under equilibrium conditions. We will assume a reference voltage of 1.8 V.

If we solve the equation above for the positive-feedback factor, $R_4/(R_3 + R_4)$, we obtain

$$\frac{R_4}{R_3 + R_4} = \frac{4/\pi}{\omega^2 L_1 C_O} \frac{V_{REF}}{V_O} \tag{11.106a}$$

so

$$\frac{R_4}{R_3 + R_4} = 4.3 \times 10^{-5} \tag{11.106b}$$

Thus the positive-feedback factor required is extremely small. The small value of the positive-feedback factor is due in most part to the very small ripple factor. Indeed, the positive feedback factor can be obtained more directly by expressing the peak-to-peak negative-feedback excursion directly in terms of the ripple factor as $rV_O R_2/(R_1 + R_2)$, where r is the peak-to-peak ripple factor. Since $V_O R_2/(R_1 + R_2) = V_{REF}$, this can be rewritten as rV_{REF}. Thus the peak-to-peak excursion in the negative-feedback voltage is simply equal to the product of the ripple factor and the reference voltage. Since the peak-to-peak excursion in the positive-feedback voltage has previously been given as $V_{IN}R_4/(R_3 + R_4)$, we can now obtain the equation for the required positive-feedback factor by equating the two peak-to-peak excursions, giving $V_{IN}R_4/(R_3 + R_4) = rV_{REF}$, so $R_4/(R_3 + R_4) = rV_{REF}/V_{IN}$.

If we use a ripple factor of 0.05%, a reference voltage of 1.8 V, and an input voltage of 20 V, we obtain a positive-feedback factor of 4.5×10^{-5}, which is essentially the same as obtained previously. If R_4 is the source resistance of the voltage reference circuit of 1 kΩ, the value of $R_3 + R_4$ is 22 MΩ, so R_3 is essentially 22 MΩ.

In practice, a ripple factor as low as 0.05% may be very difficult to achieve in a switching regulator due to the effects of the capacitor series resistance (ESR). A ripple factor in the range of 0.1% to 1% may be the best that can be achieved. For a ripple factor of 1%, the calculations above would yield a positive-feedback factor of 9×10^{-4}, so R_3 would now be 1.1 MΩ.

One principal disadvantage of the switching type of voltage regulator in comparison to the linear or nonswitching type (the dissipative type) is the greater ripple factor present at the output of the switching regulator. With the linear regulator the output ripple is reduced below the input ripple level by the ripple rejection factor of the regulator. This factor is typically in the range from 60 to 80 dB. For example, if the input ripple is 1% and the ripple rejection factor is 60 dB, the output ripple is only 0.001%.

On the other hand, it must be understood that in the switching regulator the output ripple is due principally to the switching action of the regulator itself and is not due to the feedthrough of ripple from the input circuit. As a result of this, a smaller filter capacitor and inductor can be used in the rectifier power supply that provides the input voltage for the switching regulator. In addition, the output ripple that does result is at a relatively high frequency, typically in the range from 20 to 50 kHz, as compared to the predominant output ripple component of 120 Hz for the linear type of regulator. The higher ripple frequency of the switching regulator can much more easily be filtered than that of the linear regulator by the use of low-pass filters or bandpass filters at various critical points in the circuits that are supplied by the regulator. Furthermore, with the driven switching regulator it is possible to synchronize the switching rate with some other frequency in the system supplied by the regulator such that the ripple is less of a problem.

11.8.8 Examples of Switching Regulator Circuits

We will now consider some examples of switching regulator circuits, starting off with the externally driven regulators. The externally driven regulators offer the advantage of operating at a fixed switching frequency that is independent of line and load conditions. This fixed switching frequency can be chosen so as to optimize the performance of the regulator. The disadvantage of the externally driven regulator is, of course, the necessity of proving for a square-wave oscillator circuit to drive the regulator.

Figure 11.31 is a circuit diagram of an adjustable voltage regulator that is connected as an externally driven switching regulator. The voltage regulator itself can be any of the LM105 or LM376 type and is similar to that presented in Figure 11.12. The current boost circuitry consisting of Q_1, Q_2, and R_5 is the same as for the linear regulator circuit. Resistor R_4 is used for current limiting, and there is a feedback of a portion of the output voltage via the R_1–R_2 voltage divider to the inverting input of the error amplifier (pin 6).

The square-wave input is integrated by the R_3–C_3 network to become a triangular waveform that is applied to the noninverting input of the error amplifier (pin 5). The

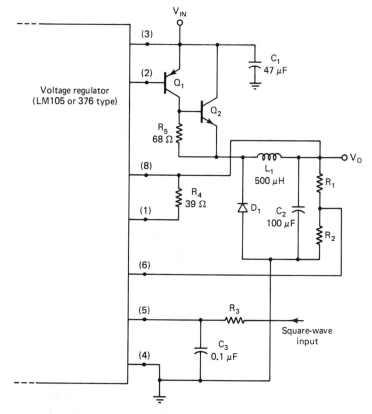

Figure 11.31 Switching regulator driven by external square-wave generator.

internally generated reference voltage is also applied to this amplifier input so that the total instantaneous voltage consists of a small-amplitude triangular waveform superimposed on the reference voltage as shown in Figure 11.32. The peak-to-peak excursion of the triangular waveform is typically about 40 mV for the best performance of the circuit.

With the input voltages just described, the error amplifier operates as a pulse-width modulator. The two input voltages of the error amplifier are shown in Figure 11.32. The output of the error amplifier is in the high state whenever $V_{REF} + v_{triangular}$ is greater than the feedback voltage, $V_{FB} = V_O R_2/(R_1 + R_2)$. When this happens, Q_1 and Q_2 are turned on (in saturation). Conversely, when $V_{REF} + v_{triangular}$ is less than V_{FB}, the error amplifier output is in the low state and Q_1 and Q_2 are off. We therefore see that the series-pass transistors will be turned on and off every cycle of the input square wave. The duty cycle, $\delta = t_{on}/T$, is a function of the value of V_{FB} relative to V_{REF}.

Under steady-state equilibrium conditions, the duty cycle is adjusted by the feedback loop at the value necessary to maintain the desired output voltage. For example, if $V_{IN} = 20$ V and V_O is to be 5 V, a duty cycle of approximately 25% is required, essentially as shown in Figure 11.32. Assuming a reference voltage of

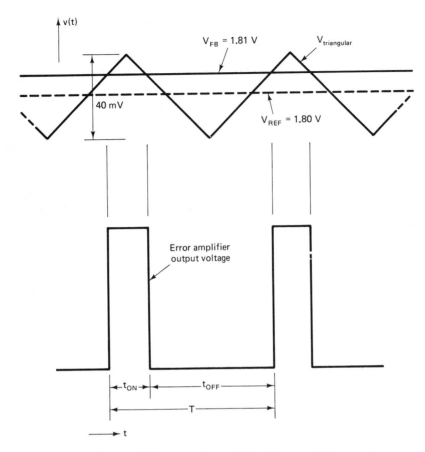

Figure 11.32 Voltage waveforms for the driven switching regulator.

$V_{REF} = 1.80$ V and a 40 mV peak-to-peak swing for the triangular waveform, the value of V_{FB} under equilibrium conditions is $V_{FB} = V_{REF} + 10$ mV $= 1.81$ V. For a 5.0-V output, the required resistance ratio is given by the equation 5.0 V$[1 + (R_1/R_2)] = V_{REF} + 10$ mV $= 1.81$ V.

Now, if for some reason, such as an increased load current or a drop in line voltage, the output voltage drops below 5.0 V, we see that the duty cycle will lengthen and become greater than 25%. The increased duty cycle will bring V_O up back toward 5.0 V. Conversely, if for some reason, such as a sudden drop in load current, V_O goes above 5.0 V, we see that the duty cycle will shorten to less than 25%. The shorter duty cycle will act to bring V_O back toward the 5.0-V equilibrium level.

We see from this example that the triangular waveform input produces, in effect, a small shift in the reference voltage. In the example above, with a 25% duty cycle, a 10-mV shift resulted. For this reason, and the fact that the amplitude of the triangular waveform may not be temperature compensated or well regulated, an excessively large peak-to-peak amplitude for the triangular waveform is to be avoided. On the other hand, too small an amplitude for the triangular waveform may allow the circuit to go into self-oscillation. The peak-to-peak amplitude of the triangular wave should usually be in the range from 10 to 100 mV, with 40 mV generally being about the optimum value.

Self-oscillating switching regulator. A diagram of a self-oscillating switching regulator is presented in Figure 11.33. The connections shown are made to a linear voltage regulator of the LM105 or LM376 type. Most of the circuitry is the same as for the externally driven regulator just discussed. The principal difference has to do with the signal fed to the noninverting input of the error amplifier (pin 5). In the externally driven regulator, this was a triangular waveform obtained by the integration of a square-wave input. In the self-oscillating regulator, it is obtained from the emitter of Q_2 via resistor R_3. It is this connection that provides the positive feedback that is necessary to sustain oscillations.

The resistance seen looking into pin 5 is essentially the source resistance of the voltage reference of about 1 kΩ. Therefore, the positive-feedback factor is approximately 1 kΩ/1 MΩ = 0.001.

From the analysis of the self-oscillating switching regulator, we have obtained the simple relationship between the output ripple factor r and the positive-feedback

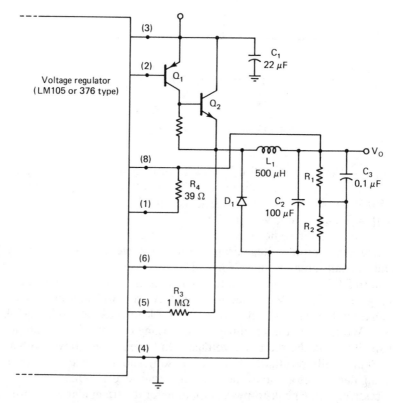

Figure 11.33 Self-oscillating switching regulator.

factor as given by

$$\frac{rV_{REF}}{V_{IN}} = \frac{R_4}{R_3 + R_4} = \text{positive-feedback factor} \tag{11.107}$$

Now if, for an example, we let $V_{IN} = 20$ V and $V_{REF} = 1.8$ V, and for the moment neglect the effect of C_3, we have that

$$r\left(\frac{1.8}{20}\right) = \frac{1 \text{ k}\Omega}{1 \text{ M}\Omega} = 0.001 \tag{11.108}$$

so the ripple factor is $r = 1.1\%$.

We will now consider the effect of C_3. Capacitor C_3 will, of course, have no influence on the dc output voltage. If, however, the reactance of C_3 is small compared to the resistance of R_2, then almost the full ac ripple voltage will appear at the noninverting amplifier input (pin 6). We must accordingly modify the relationship used above. Since the peak-to-peak excursion at pin 5 is $V_{IN}R_4/(R_3 + R_4)$ and that at pin 6 is now given by rV_O, we have that

$$r\frac{V_O}{V_{IN}} = \frac{R_4}{R_3 + R_4} \tag{11.109}$$

Assuming an output voltage of 10 V, the ripple factor becomes

$$r = 0.001 \times \frac{20 \text{ V}}{10 \text{ V}} = 0.002 \quad \text{or} \quad \underline{0.2\%} \tag{11.110}$$

In practice, this low a value for the ripple factor may be difficult to achieve due to the effects of the capacitor (C_2) effective series resistance.

With this type of regulator operating at an output voltage of 10 V and input voltages in the range from 13 to 40 V, efficiencies in the range from 85% to 90% are obtainable at an output current level of 1 A. As the output current increases, the efficiency decreases slowly as a result of the increasing I^2R losses in the inductor and in the transistor; but even at an output current of 5-A, the efficiency is still up around 72% with $V_{IN} = 28$ V.

With the component values as shown in Figure 11.32, the switching frequency ranges from about 20 kHz at an output current of 0.5 A to 40 kHz at 5.0 A. The change in the switching frequency is due in large part to the decrease in the inductance of the coil, with increasing current resulting from the effects of magnetic saturation of the inductor core.

11.8.9 Regulating Pulse-width Modulator

Another switching voltage regulator that is worthy of mention is the 1524/2524/3524 (National Semiconductor, Signetics, Silicon General, and others) type of regulating pulse-width modulator. This device is basically a switching voltage regulator with an on-chip oscillator circuit and so can be connected as a driven switching regulator without the need of any external oscillator. The frequency of oscillation is set by an external resistor and capacitor from as low as 1 kHz up to a maximum of 350 kHz.

Sec. 11.8   Switching Voltage Regulators

This regulator can be connected in a variety of configurations as a step-down or as a step-up switching regulator, and with current boosting using external series-pass transistors.

When used as a 1-A, 5-V, step-down regulator with V_{IN} = 10 V at a 20-kHz switching frequency, an efficiency of 80% is obtainable. The ripple is 10 mV peak-to-peak, corresponding to a ripple factor of 0.2%. The load regulation for a variation in the output current of from 200 mA to 1 A is 3 mV, corresponding to 0.06%/V. The line regulation is 6 mV for an input voltage change of from 10 to 20 V, corresponding to a line regulation factor of 0.12%/V.

A similar type of pulse-width modulator control circuit is the MC34060/35060 (Motorola). This IC contains an oscillator, 5.0-V reference, comparator, and error amplifier and can be used for step-down, step-up, and inverting voltage regulator circuits.

Some other examples of switching regulator controllers and their reference voltages are the TL497 (Texas Instruments), 1.2 V; LH1605 (National Semiconductor), 2.5 V; NE5561 (Signetics), 3.75 V; TL494/495/594/595 (Texas Instruments), 5 V; CA2524/3524 (RCA), 5 V; and MC3420 (Motorola), 7.8 V.

11.8.10 Monolithic Switching Regulator

An example of a monolithic switching regulator is the LAS6320P made by Lambda Semiconductors. This device comes in a 14-pin DIP package and contains a temperature-compensated voltage reference, sawtooth oscillator, pulse-width modulator, error amplifier, and a Darlington output transistor with a 2-A rating and internal current-limiting protection. The internal reference voltage is 2.25 V, the error amplifier has an open-loop gain of 70 dB, and the output section current limit is set at 2.8 A. Maximum ratings include an input voltage of 35 V and an oscillator frequency of 200 kHz. The junction-to-case and case-to-ambient (free air) thermal resistances are 13°C/W and 47°C/W, respectively. Typical performance characteristics include a line regulation of 0.015%/V and an output-voltage temperature coefficient of 0.01%/°C. Both step-down and step-up modes of operation are possible. When operated as a step-down regulator with V_{IN} = 24 V, V_O = 5 V, and an oscillator frequency of 50 kHz, the system conversion efficiency is 72% at an output current of 0.5 A and is 76% for an output current of 2 A. For an output voltage of 12 V and an output current of 2 A, the efficiency is at 85%.

11.9 VOLTAGE CONVERTERS

In many systems there is just a single positive supply voltage available, such as the +5 V that is used so often in digital systems. There are some ICs, however, that require a dual supply voltage, such as +5 V and -5 V, or supply voltages greater than the existing supply voltage. If the system basically operates with a single supply voltage and there is one device or just a few devices that require a negative supply voltage or a different supply voltage, and the amount of current required is just a few milliamperes, the most economical solution often is to use an IC voltage converter, rather than to install a separate power supply.

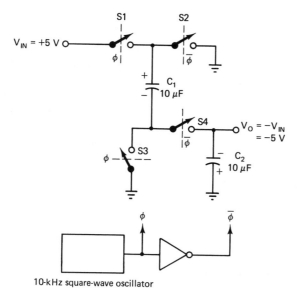

Figure 11.34 Idealized negative voltage converter.

10-kHz square-wave oscillator

To get an idea of what a voltage converter is and how it operates, let us first look at Figure 11.34, which shows an idealized negative voltage converter. This voltage converter has a 10-kHz square-wave oscillator that produces complementary, but nonoverlapping clock pulses of phases ϕ and $\bar{\phi}$. When ϕ is high, switches $S1$ and $S3$ are closed and switches $S2$ and $S4$ are open. During this time, capacitor C_1 charges up to approximately V_{IN} with the polarity indicated. Then, on the next clock phase, when ϕ is low, switches $S1$ and $S3$ open, and switches $S2$ and $S4$ are closed. This puts capacitors C_1 and C_2 in parallel, with the top side of C_1 grounded and the bottom, or negative side, of C_1 connected to the top side of C_2. Note that the bottom end of C_1 is at a voltage of $-V_{IN}$ with respect to ground. This voltage is transferred to the output capacitor C_2 via switch $S4$. Thus, under lightly loaded steady-state conditions, the output voltage is approximately equal to the negative of the input voltage.

Let us assume that $C_1 = C_2 = C$. When power is initially applied to this circuit and C_2 is uncharged, the voltage that appears across C_2 during the first cycle of operation is $-V_{IN}/2$. This is the result of the sharing of the charge by these two equal capacitors when they are connected in parallel. Before the capacitors are connected in parallel, we have that $Q = V_{IN}C$. After the capacitors are connected in parallel, charge Q is now shared equally between the two capacitors, so we now have $Q = -V_O(1)(2C)$, where $V_O(1)$ is the output voltage at the end of the first cycle of operation. Thus we have that $V_O(1) = -V_{IN}/2$. During subsequent cycles of operation, the output voltage increases as capacitor C_2 charges up and, under no-load conditions, asymptotically approaches $-V_{IN}$. We have that $V_O(1) = -0.5 V_{IN}$, $V_O(2) = -0.75 V_{IN}$, $V_O(3) = -0.875 V_{IN}$, $V_O(4) = -0.9375 V_{IN}$, and so on, reaching a value of $-0.999 V_{IN}$ after 10 cycles under no-load conditions. Thus after just a few milliseconds after start-up the output voltage reaches its steady-state value.

In actual practice, the switches are implemented as CMOS analog switches, as shown in Figure 11.35, which is a simplified diagram of a CMOS voltage converter. The IC voltage converters are fully self-contained, requiring capacitors C_1 and C_2 as the only two external components. An example of a CMOS voltage converter is the ICL7660S. This device can convert a +5-V supply voltage to a −5-V output. The voltage conversion ratio or efficiency is $-V_O/V_{IN}$, and under no-load conditions this is specified as 99.9% (typ.), 99% (min.). The effective output resistance is 60 Ω (typ.), 100 Ω (max.) for currents from 0 to 40 mA, so at an output current of 1.0 mA the output voltage will be down to −4.94 V (typ.), and at a load current of 10 mA the output voltage will be −4.40 V (typ.). The power conversion efficiency is the ratio of the output power to the input power, P_O/P_{IN}. For a load of 5 kΩ ($I_O = 1$ mA), the power conversion efficiency is 98% (typ.), 96% (min.), so we see that for applications that require a negative supply voltage of no more than a few milliamperes, these voltage converters are very efficient.

The ICL7660S can perform supply voltage conversions over an input-voltage range of from +1.5 up to +12 V, producing output voltages of −1.5 to −12 V. This device can be connected as a voltage doubler to produce output voltages up to +18.6 V, with a +10-V input. These devices can also be cascaded for larger output voltages and operated in parallel for larger output currents. This voltage converter operates at a nominal oscillator frequency of 10 kHz, which can be raised or lowered by means of external control. The ICL7662 (Intersil) is similar to the ICL7660S, but it can operate over an input-voltage range of from +4.5 to +20 V and thereby produce output voltages of −4.5 to −20 V.

Figure 11.36 is the connection diagram for a simple negative voltage converter. Figure 11.37 shows connections for paralleling voltage converters for a larger output-current capability. In this circuit, N voltage converters are connected in parallel. Each voltage converter has a separate *pump* capacitor C_1, whereas a common output capacitor C_2 serves all N voltage converters. The net output resistance is equal to

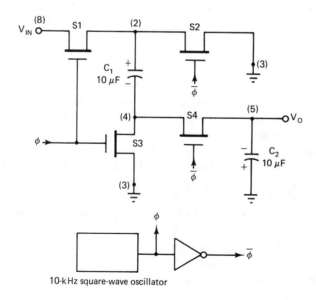

10-kHz square-wave oscillator

Figure 11.35 Simplified diagram of a CMOS voltage converter.

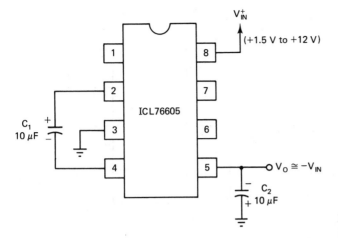

Figure 11.36 Simple negative converter (Intersil).

the output resistance of a single voltage converter divided by N. Voltage converters can be cascaded to produce a greater output voltage, as shown in Figure 11.38, where N voltage converters are cascaded. The output of each voltage converter (pin 5) is connected to the common input terminal (pin 3) of the next voltage converter, and the supply voltage input (pin 8) of every voltage converter, except the first, is connected to ground. As a result, for the first voltage converter, we have that $V_{O1} = -V_{IN}$. Let us now consider the situation with respect to the second voltage converter. The pump capacitor of the second voltage converter is charged up to a voltage given by

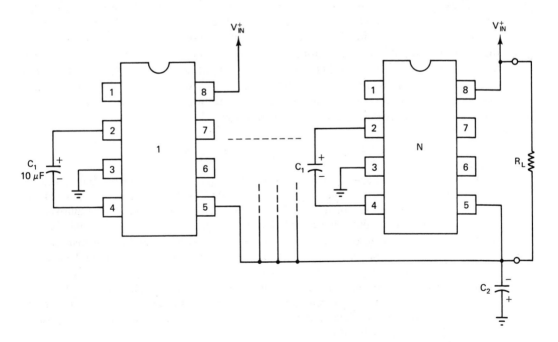

Figure 11.37 Paralleling voltage converters for greater output current (Intersil).

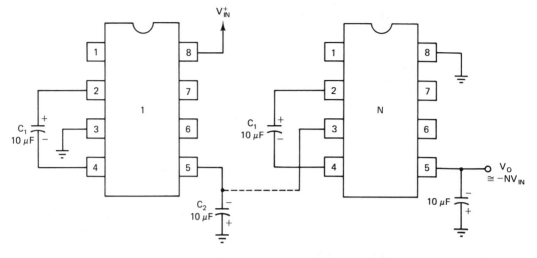

Figure 11.38 Cascading devices for increased output voltage.

$V_{C1} = V_8 - V_3 = 0 - V_{O1} = V_{IN}$. During the second phase of the clock cycle, the bottom end of this capacitor (pin 4) is connected to the top side of the output capacitor C_2 as before. However, in this case, since pin 3 no longer goes to ground, but rather is connected to the output voltage of the first voltage converter, the voltage at the bottom end of capacitor C_1 (at pin 4) is $V_{C1} + V_3 = -V_{IN} - V_{IN} = -2V_{IN}$. Under steady-state conditions, this will be the voltage that capacitor C_2 charges up to. Thus, for the second voltage converter and similarly for every following voltage converter, the output voltage is greater than the input voltage by $-V_{IN}$. For a cascaded chain of N voltage converters, the output voltage is $-NV_{IN}$, where V_{IN} is the input voltage of the first voltage converter. The effective output resistance of the cascaded chain of N voltage converters is approximately N times the output resistance of an individual voltage converter, so the number of voltage converters that can be cascaded depends on the loading conditions, with up to 10 voltage converters for lightly loaded ($I_O < 1$ mA) conditions.

PROBLEMS

11.1. The reference voltage of a voltage regulator is $V_{REF} = 2.0$ V and has a supply voltage sensitivity of 2 mV/V. The supply voltage is $+15$ V with a 1% ripple factor at 120 Hz. The regulated output voltage is $+10$ V. Find the line regulation, the output-voltage ripple, and the output-voltage ripple factor. (*Ans.*: 10 mV/V, 1.5 mV, 0.015%)

11.2. The output voltage of a voltage regulator decreases by 10 mV as the load current increases from a no-load value of zero to a full-load value of 1.0 A. Find the average dc output resistance. (*Ans.*: 10 mΩ)

11.3. A voltage regulator has a reference voltage of 2.0 V with a temperature coefficient of $+100$ ppm/°C. The regulated output voltage is $+10$ V. Find the temperature coefficient of the output voltage expressed in terms of ppm/°C, %/°C, and mV/°C. (*Ans.*: $+100$ ppm/°C, $+0.01$%/°C, $+1.0$ mV/°C)

11.4. A voltage regulator has a reference voltage of 2.0 V and a regulated output voltage of $+10$ V. The dc output resistance of the regulator is 0.05 Ω. The error amplifier has a dc gain of 10,000 and a unity-gain frequency of 1.0 MHz.

  (a) Find the output impedance at 100 kHz.

  (b) Find the value of bypass capacitor needed across the output terminals of the regulator to limit the output impedance to a maximum of 1.0 Ω over the frequency range 0 to 100 kHz. [*Ans.*: $+j50$ Ω, 1.6 μF (min.)]

11.5. A voltage regulator has a maximum supply voltage of $+20$ V and a maximum power dissipation rating of 10 W. The regulated output voltage is $+15$ V.

  (a) Find the required current limit I_{CL} at $V_O = +15$ V and at $V_O = 0$ for an ideal foldback current-limiting characteristic.   (*Ans.*: 2.0 A, 0.5 A)

  (b) For a linear foldback current-limiting characteristic with $I_{CL} = 0.5$ A at $V_O = 0$, find I_{CL} at $V_O = +15$ V.   (*Ans.*: 0.875 A)

  (c) Use a computer to generate the ideal foldback current-limiting characteristic curve needed for this problem. Then generate linear foldback lines that are subject to the requirement that $P_d \leq P_{d(MAX)}$ and that have the following current axis intercepts: 0.4 A, 0.3 A, 0.2 A, and 0.1 A. Find the corresponding values of I_{CL} at $V_O = +15$ V.

11.6. A 5.0-V three-terminal voltage regulator is to be used as a current regulator, as shown in Figure 11.20. The voltage regulator has $V_{IN} = 20$ V and a line regulation of 10 mV/V. The regulated output current is to be 1.0 A.

  (a) Find R_1.   (*Ans.*: 5.0 Ω)

  (b) Find the dynamic output conductance of this current source, neglecting the effect of the voltage regulator quiescent current.   (*Ans.*: 2.0 mS)

  (c) Find the percentage of change in the output current per volt of change in the output (load) voltage.   (*Ans.*: 0.2%/V)

  (d) The input–output voltage differential for this voltage regulator is 3.0 V (min.) and 15 V (max.). Find the voltage compliance range.   (*Ans.*: from 0 to $+12$ V)

  (e) Find the power dissipation rating needed for the voltage regulator if the output (load) voltage is to range from 0 to $+5$ V and the input voltage is 15 V.   (*Ans.*: 10 W)

11.7. (*Switching regulator*) A switching voltage regulator operates at a switching frequency of 50 kHz and is to supply a load current I_O of 10 A at a dc output voltage V_O of $+12$ V. The dc input voltage is $V_{IN} = +18$ V and the output (peak-to-peak) ripple factor is not to exceed 0.1%.

  (a) Find the value of the filter inductor L_1 such that the maximum change or ripple in the current through the inductor will not exceed 40% of the average or dc current.   [*Ans.*: 20 μH (min.)]

  (b) Find the value of the output filter capacitor C_o for $L_1 = 20$ μH and for $L_1 = 100$ μH.   (*Ans.*: 1300 μF, 267 μF)

11.8. Given: A voltage regulator circuit using an LM305 or LM376 type of voltage regulator together with the external pass transistors as shown in Figure P11.8. Assume that $V_{BE} = 0.7$ V for transistors in full conduction and $V_{BE} = 0.5$ V at threshold of conduction. Let $\beta = 100$ (min.) and $V_{REF} = 1.72$ V (LM305 typ. value).

  (a) Find V_{R9}.   (*Ans.*: 0.292 V)

  (b) Find I_{R9} and I_{C15}.   (*Ans.*: 0.22 mA, 1.0 mA)

  (c) Find R_2 for $V_O = +15.0$ V.   (*Ans.*: 719 Ω)

  (d) Show that $V_1 = V_O[R_6/(R_4 + R_6)] + (I_O R_3)[R_6/(R_4 + R_6)] + (1.22$ mA$)(R_4 \| R_6)$.

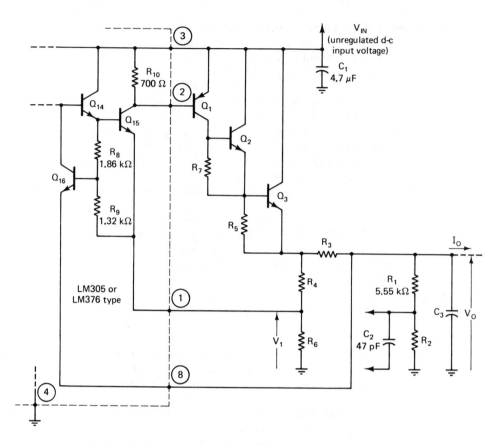

Figure P11.8

(e) Show that current limiting occurs when

$$I_O R_3 \frac{R_6}{R_4 + R_6} - V_O \frac{R_4}{R_4 + R_6} + (1.22 \text{ mA})(R_4 \| R_6) = V_{BE16} - V_{R9} = 0.21 \text{ V}$$

(at threshold) and 0.41 V when Q_{16} is in full conduction.

(f) If $P_{D(MAX)} = 50$ W for Q_3 and $V_{IN(MAX)} = +20$ V, find $I_{O(CL)}$ for the case of $V_O = 0$, and for $V_O = +15$ V.   (*Ans.:* 2.5 A, 10 A)

(g) If the parallel combination of R_4 and R_6 is set at 10 Ω and the required value of V_{BE16} is 0.542 V for current limiting, find the values of R_3, R_4, and R_6 for foldback current limiting to meet the conditions above for $I_{O(CL)}$.   (*Ans.:* 100 mΩ, 10.5 Ω, 210 Ω)

(h) (1) Sketch and dimension the V_O versus I_O foldback characteristics of this circuit.

(2) What will $I_{O(CL)}$ and P_D be when $V_O = +10$ V and $V_{IN} = +20$ V?   (*Ans.:* 7.5 A, 75 W)

(3) Is this foldback curve suitable for the full protection of Q_3?

(i) Find R_5 for $I_2 = 100$ mA.   (*Ans.:* 7.0 Ω)

(j) Find R_7 for $I_1 = 10$ mA.   (*Ans.:* 70 Ω)

(k) **(1)** Find the open-loop output resistance $r_{o(OL)}$ when $I_O = 10$ A.   (*Ans.:* $r_{o(OL)} = 104$ mΩ)

(2) If the voltage gain of the difference (error) amplifier is 500 (min.) including the gain of the current boost circuit, find $r_{o(CL)}$ for $V_O = +15$ V and $I_O = 10$ A.   [*Ans.:* $r_{o(CL)} = 1.74$ mΩ (max.)]

(3) If the difference amplifier has a unity-gain frequency of $f_u = 1.0$ MHz, find the dynamic output impedance $z_{o(CL)}$ at $f = 20$ kHz.   (*Ans.:* $z_{o(CL)} = j17.4$ mΩ)

(l) If $TC_{V_{REF}} = +50$ μV/°C, find TC_{V_O} when $V_O = +15$ V.   (*Ans.:* $TC_{V_O} = 436$ μV/°C or 29 ppm/°C or 0.0029%/°C)

(m) A temperature-compensated voltage reference circuit (Figure 11.12) produces V_{REF} with $Z_z = 10$ Ω and with the dynamic output resistance of the current source I_Q being $r_o = 100$ kΩ. Find the sensitivity of the reference voltage to changes in the supply voltage, dV_{REF}/dV_{supply}.   (*Ans.:* $dV_{REF}/dV_{supply} = 23$ μV/V $= 0.00133\%$/V)

(n) Find the *line regulation* of this voltage regulator, dV_O/dV_{supply}, for an output voltage of $+15$ V.   (*Ans.:* 200 μV/V $= 0.00133\%$/V)

(o) Find the value of C_3 needed to keep the voltage regulator output impedance less than 50 mΩ at 200 kHz.   (*Ans.:* $C_3 \geq 16$ μF)

(p) Find the *load regulation* for $I_{NL} = 1.0$ A and $I_{FL} = 10$ A if $A_{OL} = 500$. The regulated output voltage is $+15$ V.   (*Ans.:* $\Delta V_O = 2.01$ mV $= 0.0134\%$)

(q) Find the ripple factor reduction ratio. This is the ratio of the ac ripple voltage on the output voltage of the regulator to the ac ripple voltage on the supply voltage.   (*Ans.:* 202 μV/V)

(r) Find the value of the REF bypass capacitor required to reduce the ripple factor ratio to 20 μV/V at 120 Hz.   (*Ans.:* $C_{REF\ BYPASS} \geq 17.5$ μF)

(s) Find the maximum power dissipation (Q_3) that will occur for the foldback characteristics under consideration and the output voltage level at which this maximum power dissipation occurs.   (*Ans.:* 78.1 W at $V_O = 7.5$ V)

(t) Choosing a power dissipation rating for Q_3 to be 80 W to provide full protection for Q_3 under all output voltage and load conditions, what thermal impedance is required for the heat sink of Q_3? The junction-to-case thermal impedance Θ_{JC} of Q_3 (2N3772) is 1.17°C/W, and the maximum junction temperature $T_{J(MAX)}$ is 200°C.   (*Ans.:* heat sink thermal impedance $= \Theta_{CA} \leq 1.02$°C/W)

(u) With this value of $P_{D(MAX)} = 80$ W, what would the corresponding value of $I_{O(CL)}$ be for the case of straight current limiting (no foldback)?   (*Ans.:* $I_{O(CL)} = 4.0$ A as compared to 10 A available when foldback current limiting is used)

11.9. A voltage regulator circuit is designed with a linear foldback characteristic that goes to a current limit of 20 A at $V_O = +10$ V and has a short-circuit limit of 5.0 A. The unregulated input voltage has a maximum value of $+15$ V. What power dissipation rating should the output transistor have for full protection, and at what value of output voltage and current will the maximum power dissipation occur?   (*Ans.:* 126 W, 5.83 V, 13.75 A)

11.10. (*Latching foldback circuit*) **(a)** A latching foldback circuit is shown in Figure P11.10. Find R_{CL} and R_{FB} such that the current limiting circuit will be triggered at $I_{L(MAX)} = 10$ A and then will immediately reduce the load current to $I_{L(FB)} = 1.0$ A. Use $V_{BE} = 500$ mV at the threshold of conduction and $V_{BE} = 600$ mV when in full conduction.   (*Ans.:* $R_{CL} = 0.050$ Ω, $R_{FB} = 27.5$ Ω)

(b) Sketch the V_L versus I_L output curve for this circuit.

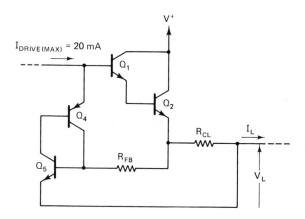

$I_{DRIVE\,(MAX)} = 20\text{ mA}$

Figure P11.10

11.11. (*Crowbar circuit for overvoltage protection*) Figure P11.11 shows a *crowbar circuit.* This circuit protects the load from excessive voltage due to various causes such as power-line voltage spikes or failure of the voltage regulator. Given $V_z = 6.5$ V for D_1 and when $V_{REG} = 20$ V, $I_z = 1.0$ mA, and the current through resistors R_1 and R_2 will be 1.0 mA. When V_{REG} exceeds 20 V, the silicon-controlled rectifier (SCR) is to turn on to protect the load from an overvoltage condition. Find:
  (a) R_1 and R_2.   (*Ans.:* 13.5 kΩ, 6.5 kΩ)
  (b) R_3.   (*Ans.:* 13.5 kΩ)

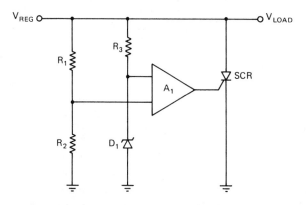

Figure P11.11

11.12. (*DC-to-DC inverter*) Figure P11.12 shows a dc-to-dc inverter circuit. If the square-wave generator produces a 50-kHz square-wave voltage that goes from zero to +10 V, find the output voltage of the circuit
  (a) Under no-load conditions.   (*Ans.:* −10 V)
  (b) Under lightly loaded conditions.   (*Ans.:* approx. −8.8 V).
  (c) Would there be any advantage to operating this circuit at a much lower frequency, say 100 Hz? Explain. Would there be any advantage to be gained in going to a much higher frequency? Explain.

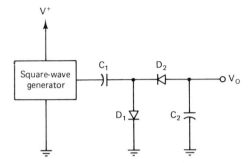

Figure P11.12

REFERENCES

COWLES, L. G., *Sourcebook of Modern Transistor Circuits*, Prentice Hall, Englewood Cliffs, N.J., 1976.

FRANCO S., *Design with Operational Amplifiers and Analog Integrated Circuits*, McGraw-Hill, New York, 1988.

GOTTLIEB, I., *Design and Operation of Regulated Power Supplies*, Howard W. Sams, Indianapolis, Ind., 1966.

_____, *Regulated Power Supplies*, 3rd ed. Howard W. Sams, Indianapolis, Ind., 1981.

_____, *Switching Regulators and Power Supplies*, TAB Books, Blue Ridge Summit, Pa., 1976.

GREBENE, A. B., *Analog Integrated Circuit Design*, Van Nostrand Reinhold, New York, 1972.

_____, *Bipolar and MOS Analog Integrated Circuit Design*, Wiley, New York, 1984.

HNATEK, E. R., *Design of Solid-State Power Supplies*, Van Nostrand Reinhold, New York, 1971.

_____, *A User's Handbook of Integrated Circuits*, Wiley, New York, 1973.

_____, *Design of Solid State Power Supplies*, 2nd ed., Van Nostrand Reinhold, New York, 1981.

LENK, J. D., *Handbook of Integrated Circuits*, Prentice-Hall, Englewood Cliffs, N.J., 1978.

_____, *Manual for Integrated Circuit Users*, Prentice-Hall, Englewood Cliffs, N.J., 1973.

McMENAMIN, J. M., *Linear Integrated Circuits: Operation and Applications*, Prentice-Hall, Englewood Cliffs, N.J., 1985.

MITCHELL, D. M., *DC–DC Switching Regulator Analysis*, McGraw-Hill, New York, 1988.

OXNER, E. S., *Power FETs and Their Applications*, Prentice-Hall, Englewood Cliffs, N.J., 1982.

PRESSMAN, A. I., *Switching and Linear Power Supply: Power Converter Design*, Hayden, Rochelle Park, N.J., 1977.

12

Power Amplifiers

12.1 BASIC CONSIDERATIONS

For amplifiers that are designed to deliver substantial amounts of power to a load, generally greater than 1 W, the question of the amplifier power conversion efficiency becomes one of major concern.   The ac output power that the amplifier delivers to the load is derived almost entirely from the dc power supply.   The input signal acts to control the conversion of the dc power to the ac output power, but the signal source that drives the amplifier generally supplies a negligible amount of input power.

The amplifier power conversion efficiency, η, is defined as

$$\eta = \frac{P_O(\text{ac})}{P_{\text{IN}}(\text{dc})} \tag{12.1}$$

The power that must be dissipated by the power amplifier is therefore given by $P_D = P_{\text{IN}} - P_O$, and therefore can be expressed in terms of the power conversion efficiency as

$$\text{amplifier power dissipation} = P_D = P_{\text{IN}} - P_O \tag{12.2}$$
$$= P_O \left(\frac{1}{\eta} - 1 \right)$$

This power dissipation occurs almost entirely in the last or output stage of the power amplifier.   From this expression we see the importance of the power conversion

588

efficiency in determining the amount of power that must be dissipated by the power amplifier.

12.2 CLASS A POWER AMPLIFIERS

In a class A amplifier, the output stage is biased in the active region such that there is an uninterrupted flow of current during the entire cycle, and at no time does the amplifier go into the cutoff or saturation region. Thus the amplifier operates in the active region during the entire 360° cycle of the input signal.

Consider the power amplifier shown in Figure 12.1 with a supply of voltage of $+V_S$. The maximum peak-to-peak swing of the output voltage cannot exceed V_S, and in most practical cases will be limited to 1 or 2 V less than V_S, and sometimes more, in part due to distortion considerations.

For a peak-to-peak output voltage swing of V_S, the peak swing is $V_S/2$ and the rms output voltage is therefore $V_S/(2\sqrt{2})$. If the quiescent current of the output stage is I_Q, the peak output current can be no greater than I_Q, so the rms output current is $I_Q/\sqrt{2}$. These conditions essentially correspond to the output transistor being biased in the middle of the active region, halfway between cutoff and saturation, the coordinates of the quiescent point being $I_C = I_{C(SAT)}/2$ and $V_C = (V_S - V_{CE(SAT)})/2 \simeq V_S/2$ for the case of a bipolar transistor.

The ac power output under conditions of maximum voltage and current swing is given by $P_O(ac) = v_o i_o = (V_S/2\sqrt{2})(I_Q/\sqrt{2}) = V_S I_Q/4$, where v_o and i_o are the rms output voltage and current, respectively. Since the dc input power is $P_{IN}(dc) = V_S I_Q$, the maximum power conversion efficiency η is given by

$$\eta = \frac{P_O}{P_{IN}} = \frac{1}{4} \quad \text{or} \quad 25\% \tag{12.3}$$

Thus, of the total dc input power, a maximum of 25% can be converted to ac power and delivered to the load, and the rest will be dissipated by the amplifier. For example, if the ac output power is to be 5 W, the amplifier power dissipation rating will have to be at least 15 W, and the total input power will have to be at least 20 W.

It is important to note that the 25% power conversion efficiency is approached only when the amplifier is driven hard enough such that the maximum output voltage and current swings are obtained. If the input drive conditions are such that this is not the case, the efficiency is correspondingly less. In any case, it is not really possible to achieve the 25% efficiency figure in practice due in part to various other circuit

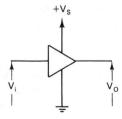

Figure 12.1 Class A power amplifier.

losses. Most important, however, because of the nonlinearities of the amplifier output stage, it is generally not desirable, based on distortion considerations, to drive the amplifier over the full span of the active region between cutoff and saturation. To limit distortion to an acceptable level, the output swing is often limited to no more than about 50% to 80% of the total active region voltage and current spans. As a result of this, the actual power conversion efficiency values of class A power amplifiers generally fall in the range from 10% to 20%, and many times even lower if a very low amount of distortion is required.

If the actual amplifier efficiency is 10%, for example, we have that $P_D = 9P_O$, so for a 5-W power output the dissipation rating of the output stage must be at least 45 W, and a total input power of at least 50 W will be required. We can thus see from these considerations the desirability of operating the power amplifier at as high a power conversion efficiency as possible, subject to the limitations imposed by distortion considerations.

12.3 CLASS B POWER AMPLIFIERS

In the class B mode of operation, the output transistor is biased at or very near cutoff, such that conduction occurs for only about one-half of the input waveform cycle, or 180° for a sinusoidal type of input signal. In this mode of operation the quiescent current is essentially zero, and as a result of this the class B mode of operation offers the possibility of a much higher power conversion efficiency than does the class A mode.

In virtually all practical applications of a class B type of amplifier, the output stage is comprised of two transistors whose outputs are combined in such a way as to reconstruct the full 360° waveform cycle. Each transistor operates in the class B mode and conducts during alternative half-cycles of the input signal so that by combining the two outputs an amplified replica of the input signal will be obtained. The two transistors of the class B output stage are usually connected in a push–pull arrangement, a simple example of which is shown in Figure 12.2a.

Transistors Q_1 and Q_2 constitute a complementary push–pull emitter-follower output stage operating in the class B mode. Transistor Q_1 conducts during the positive half-cycles of the input voltage applied to the Q_1–Q_2 output stage and will "source" or "push" current into the load. Transistor Q_2 conducts during the negative half-cycles and will "sink" or "pull" current from the load. Each transistor thus conducts for one-half of the entire cycle, and the two outputs are combined to give the full-cycle (360°) sinusoidal output current. Under quiescent condition ($v_o = 0$, $i_o = 0$), both transistors are off and the power dissipation is negligible.

Transistors Q_1 and Q_2 constitute a complementary pair in the sense that one is an NPN transistor and the other a PNP. Both bases are fed from the same point, so when the base voltage goes up in the positive direction Q_1 will be turned on and conduct while Q_2 will be off. Conversely, when the base voltage goes negative, Q_1 will be turned off and Q_2 will be biased into conduction. For greater current gains in the output stage, Q_1 and Q_2 can be Darlington compound transistors, as shown in Figure 12.2, and power MOSFETs can also be used.

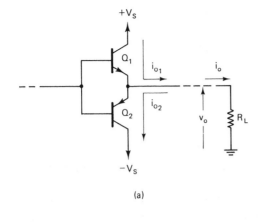

(a)

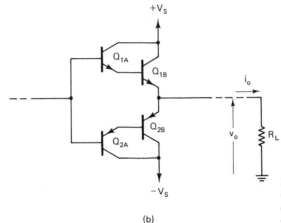

(b)

Figure 12.2 (a) Complementary emitter-follower class B push–pull output stage; (b) Darlington output stage.

12.3.1 Class B Power Output and Power Conversion Efficiency

The ac output power of the amplifier is $P_O \text{ (ac)} = v_o i_o = i_o^2 R_L$, where v_o and i_o are the rms output voltage and current, respectively, and R_L is the load resistance. The peak current for a sinusoidal waveform is related to the rms current by $i_{o(\text{peak})} = \sqrt{2}\, i_o$. During the positive half-cycles of the output current, the current is drawn from the positive supply terminal through Q_1 to the output. During the negative half-cycles of the output current, the current flow is from the load through Q_2 and then into the negative supply terminal. The average or dc current drawn from each power supply terminal is related to the peak and rms output current by $I_{\text{DC}} = i_{o(\text{peak})}/\pi = i_o \sqrt{2}/\pi$. The average of dc power drawn from each power supply terminal is therefore given by $V_S I_{\text{DC}} = V_S i_o \sqrt{2}/\pi$, where V_S is the dc supply voltage. The total power that is drawn from the dc power supply is therefore twice this, or

$$P_{\text{IN}} = 2 V_S I_{\text{DC}} = \frac{2 V_S i_o \sqrt{2}}{\pi} \tag{12.4}$$

The amplifier power dissipation is therefore

$$P_D = P_{IN} - P_O = \frac{2 V_S i_o \sqrt{2}}{\pi} - P_O \tag{12.5}$$

Since $i_o = \sqrt{P_O/R_L}$, this can be rewritten as

$$P_D = \frac{2 V_S}{\pi} \sqrt{\frac{2 P_O}{R_L}} - P_O \tag{12.6}$$

From this expression we see that the power dissipation at first increases with P_O at a rate proportional to $\sqrt{P_O}$. It then bends over and reaches a maximum and thereafter decreases with increasing P_O.

The maximum available ac power output is limited, however, by the peak output voltage swing, which cannot exceed the supply voltage, V_S. Since $v_{o(peak)} = V_S$, and $P_O = v_o i_o = v_o^2/R_L = v_{o(peak)}^2/2R_L$, we have that the ac output power is limited to a maximum given by

$$P_{O(MAX)} = \frac{V_S^2}{2 R_L} \tag{12.7}$$

In terms of the rms output current, this expression for the maximum output power can be expressed as

$$P_{O(MAX)} = v_o i_o = \frac{V_S i_o}{\sqrt{2}} \tag{12.8}$$

The power conversion efficiency is therefore given by

$$\eta = \frac{P_O}{P_{IN}} = \frac{V_S i_o / \sqrt{2}}{2 V_S i_o \sqrt{2} / \pi} = \frac{\pi}{4} = \underline{0.785 \text{ or } 78.5\%} \tag{12.9}$$

Since this is the power conversion efficiency under the conditions of maximum power output, this will be the maximum possible power conversion efficiency for a class B amplifier. The value of η that is actually obtained in practice is generally substantially less than this due to various other circuit losses, and most of all due to the fact that the peak output voltage swing is always less than the supply voltage, typically by about several volts. Part of this is due to the saturation voltage drop of the transistor, but it also results from distortion considerations. To limit distortion to an acceptable value, the output voltage swing is purposely limited to something considerably less than the full extent of the active region between cutoff and saturation.

The amplifier power dissipation is related to the power conversion efficiency by

$$P_D = P_{IN} - P_O = P_O \left(\frac{1}{\eta} - 1 \right) \tag{12.10}$$

For a class B amplifier operating at the maximum efficiency of 78.5%, the power dissipation is related to the ac power output by

$$P_D = P_O \left(\frac{4}{\pi} - 1 \right) = 0.273 P_O \tag{12.11}$$

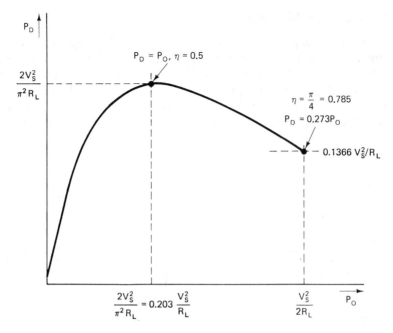

Figure 12.3 Power dissipation versus power output for a class B amplifier.

For example, for an ac power output of 5 W, the input power is $P_{IN} = P_O/\eta = 5$ W/0.785 = 6.4 W and the power dissipation is 1.4 W. This represents a very substantial improvement over the class A amplifier.

We will now return to the expression for power dissipation given earlier as

$$P_D = \frac{2V_S}{\pi}\sqrt{\frac{2P_O}{R_L}} - P_O \tag{12.12}$$

To determine the value of P_O at which P_D is a maximum, we will take the derivative of P_D with respect to P_O and set the result equal to zero, giving

$$\frac{dP_D}{dP_O} = \frac{V_S}{\pi}\sqrt{\frac{2}{P_O R_L}} - 1 = 0 \tag{12.13}$$

so

$$P_O = \frac{2V_S^2}{\pi^2 R_L} \tag{12.14}$$

at the point where P_D is a maximum. To find $P_{D(MAX)}$ under these circumstances, we will substitute P_O given above back into the expression for P_D and thereby obtain

$$P_{D(MAX)} = \frac{2V_S^2}{\pi^2 R_L} \tag{12.15}$$

Since P_O also has the value of $2V_S^2/\pi^2 R_L$ at this point, the power conversion efficiency is therefore 50% at this point. Figure 12.3 shows a generalized graph of P_D versus P_O.

12.3.2 Crossover Distortion

The ratio of output power to power dissipation is limited to a maximum value of unity for the class A amplifier, compared to a ratio of 3.66 for the class B amplifier, so from this standpoint the class B amplifier does offer a substantial advantage over the class A circuit. The push–pull class B amplifier does, however, suffer from an important type of distortion known as *crossover distortion*. This type of distortion is due to the very low or almost nonexistent gain of the transistors in the cutoff region.

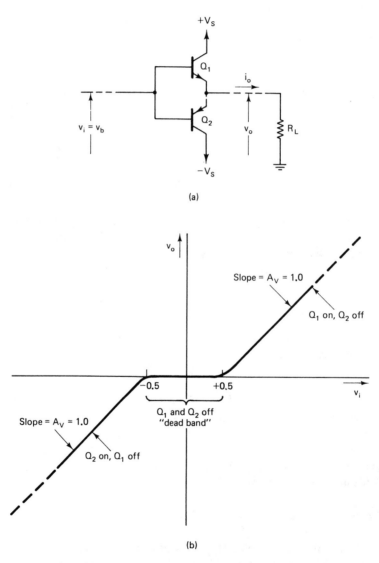

Figure 12.4 (a) Complementary push–pull emitter-follower output stage; (b) transfer characteristics.

The voltage gain of an emitter-follower stage is given as

$$A_V = \frac{g_m R_L}{1 + g_m R_L} = \frac{R_L}{R_L + 1/g_m} \tag{12.16}$$

Since $g_m = I_C/V_T = I_O/V_T$, this can be rewritten as

$$A_V = \frac{R_L}{R_L + V_T/I_O} = \frac{1}{1 + V_T/I_O R_L} = \frac{1}{1 + V_T/V_O} \tag{12.17}$$

We see from this expression that in the region near cutoff, where I_O becomes very small such that $I_O R_L$ becomes comparable to or less than V_T, the voltage gain of the emitter follower will decrease considerably below unity.

Figure 12.4a shows a single complementary push–pull emitter-follower output stage, and Figure 12.4b shows the v_o versus v_b transfer characteristic. We see that in the region where v_b is between -0.5 and $+0.5$ V there will be a "dead region" or "dead band" wherein both transistors will be in the cutoff region. In this region the voltage gain of both Q_1 and Q_2 is very small. To bring the voltage gain up toward unity for small input voltages, it is necessary to bias the transistors with a small quiescent current so as to bias the transistors at a point that is slightly into the active region.

This can be done by applying a small dc bias voltage between the bases of the two transistors, as shown in Figure 12.5a. The resulting transfer characteristics are shown in Figure 12.5b. To minimize the crossover distortion, the bias voltages, V_{B1} and V_{B2}, must be sufficiently large to bias Q_1 and Q_2 enough into the active region to ensure that the emitter-follower voltage gain is close to unity, even under quiescent condition ($v_i = 0$ and $v_o = 0$). The total bias voltage required between the two bases, $V_{B1} + V_{B2}$, is in the range of about 1.0 V.

For the complementary push–pull emitter-follower circuit, we will now derive an expression for the voltage gain under quiescent conditions. For this condition we have that $V_O = I_O = 0$, so the quiescent current in Q_1 and Q_2 is equal as given by $I_1 = I_2 = I_Q$. The dynamic transfer conductances of Q_1 and Q_2 are therefore equal, as given by $g_{m_1} = I_1/V_T = I_Q/V_T$ and $g_{m_2} = I_2/V_T = I_Q/V_T$; so $g_{m_1} = g_{m_2} = g_m = I_Q/V_T$. The ac small-signal output currents of the two transistors are given by $i_{o_1} = g_m v_{be_1} = g_m(v_{b_1} - v_o)$ and $i_{o_2} = g_m v_{be_2} = g_m(v_{b_2} - v_o)$. The ac voltage at the two bases is the same since the two bases are separated only by a dc voltage, V_B; so $v_{b_1} = v_{b_2} = v_b$, and we have $i_o = i_{o_1} + i_{o_2} = 2g_m(v_b - v_o)$. The ac small-signal output voltage is $v_o = i_o R_L$, so we now have that

$$v_o = 2g_m R_L(v_b - v_o)$$

and thus

$$v_o = \frac{2g_m R_L v_b}{1 + 2g_m R_L} \tag{12.18}$$

Thus the ac small-signal voltage gain of the output stage under quiescent conditions is given by

$$A_V = \frac{v_o}{v_b} = \frac{2g_m R_L}{1 + 2g_m R_L} \tag{12.19}$$

Sec. 12.3   Class B Power Amplifiers

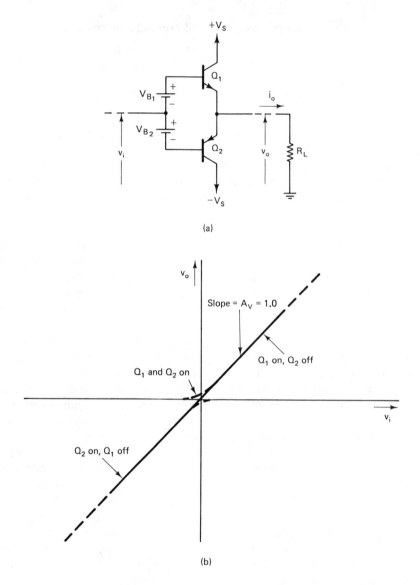

Figure 12.5 (a) Push–pull output stage biased to minimize crossover distortion; (b) transfer characteristics.

This equation can be rewritten as

$$A_V = \frac{R_L}{R_L + 1/2g_m} = \frac{R_L}{R_L + V_T/2I_Q} = \frac{1}{1 + V_T/2I_Q R_L} \quad (12.20)$$

Therefore, for example, for the voltage gain under quiescent conditions to be no more than 10% below the maximum gain of unity, we will have the condition that

$$V_T/2I_Q R_L = 1/0.9 - 1 \leq 0.11$$

so the required quiescent current flow through Q_1 and Q_2 is given by

$$I_Q \geq \frac{4.5V_T}{R_L} \qquad (12.21)$$

If the minimum expected value of load resistance R_L is 8 Ω, the required value of quiescent current is $I_Q \geq (4.5 \times 25 \text{ mV})/8$ Ω = 14.1 mA. The relatively large value of required quiescent current in this case is a direct consequence of the small value of load resistance. Furthermore, to put the quiescent current into the proper perspective, let us assume a representative supply voltage of ± 12 V for the amplifier so that the peak output-voltage swing will be about 10 V. The peak output current will consequently be 10 V/8 Ω = 1.25 A. Therefore, we now see that the quiescent current of 14 mA is only 1.1% of the peak output current swing, so the resulting small loss in the overall power conversion efficiency is a relatively small price to pay for the great reduction in crossover distortion that results from this mode of operation.

In general, we can say that to minimize crossover distortion we will require that the quiescent current satisfy the condition that $I_Q \geq 5V_T/R_L$. Since $V_{O(\text{peak})} = R_L I_{O(\text{peak})}$, this inequality can be rewritten as

$$I_Q/I_{O(\text{peak})} \geq 5V_T/V_{O(\text{peak})}$$

Since $V_{O(\text{peak})}$ is always somewhat less than the dc supply voltage V_S, we can write that

$$I_Q/I_{O(\text{peak})} \geq 5V_T/V_S = 0.125V/V_S$$

Since V_S is generally in the range from 10 to 30 V, we see again that the required quiescent current is but a very small fraction of the peak output-current swing.

Since Q_1 and Q_2 are now no longer biased in the cutoff region, but slightly into the active region, the conduction angle for each transistor will be greater than 180° (one-half of a cycle), although it still will be considerably less than 360° (a full cycle). Therefore, strictly speaking, the mode of operation is no longer class B, nor is it class A. The mode of operation is called class AB, and it combines the low-distortion attribute of class A operation with the high power conversion efficiency characteristics of class B operation. Since the mode of operation is generally considerably closer to class B than to class A, the high power conversion efficiency available from class B operation can, to a large extent, still be obtained.

Along with a high power conversion efficiency, a low-percentage distortion figure can be obtained. Indeed, with a well-balanced push–pull amplifier, the even harmonic distortion components will cancel out, leaving only the odd-numbered harmonics. Since the amplitude of the various harmonics decreases rapidly with increasing harmonic number, the cancellation of the second harmonic term, which will be the largest harmonic component, along with the other even harmonics, can lead to a very great reduction in the distortion.

Power amplifiers are usually operated under closed-loop conditions with either an internally or externally connected negative-feedback loop. Since the closed-loop gain is related to the open-loop gain by $A_{CL} = A_{OL}/(1 + FA_{OL})$, where F is the feedback factor, it can be easily shown that the fractional or percentage variation in the closed-loop gain is related to the fractional or percentage variation in the open-loop gain by $dA_{CL}/A_{CL} = (dA_{OL}/A_{OL})(A_{CL}/A_{OL})$. Since the closed-loop gain is less than the open-loop gain, and generally very much less, we see that any variations in the open-loop gain result in smaller variations in the closed-loop gain.

As an example of this, let us consider a variation in the open-loop gain of 10% that is due to crossover distortion. If we take as representative values an open-loop gain of 1000 and a closed-loop gain of 50, the variation of A_{CL} due to a 10% variation of A_{OL} is 10% $\times$ (50/1000) = 0.5%. The resulting contribution of this to the total amplifier distortion is considerably less than 0.5% since the crossover gain variation is present only over a relatively small part of the output voltage swing.

12.4 CLASS AB PUSH–PULL OUTPUT-STAGE EXAMPLES

Figure 12.6a shows an example of a simple class AB complementary push–pull emitter-follower output stage. Transistor Q_3 operates as a common-emitter gain stage with a current source active load of strength I_Q. The push–pull output stage is comprised of transistors Q_1 and Q_2. The voltage drop across diodes D_1 and D_2 is the bias voltage for the class AB operation of Q_1 and Q_2. In IC applications these two diodes will actually be diode-connected transistors. The active areas of these transistors are scaled so as to obtain the desired standby or quiescent current through Q_1 and Q_2. If, for example, D_1 and D_2 were diode-connected transistors, identical in construction to Q_1 and Q_2, respectively, but scaled in active area by a ratio of n with respect to Q_1 and Q_2, the standby current of Q_1 and Q_2 would be I_Q/n. The voltage drop across D_1 and D_2 decreases with temperature, but this temperature variation is exactly what is needed to compensate for the negative temperature coefficient of the base-to-emitter voltage of Q_1 and Q_2.

Figure 12.6b shows a Darlington configuration class AB push–pull stage. Transistors Q_{1A} and Q_{1B} constitute an NPN Darlington emitter follower for sourcing currents into the load. Transistors Q_{2A}, Q_{2B}, and Q_{2C} constitute a composite PNP Darlington that in effect acts like a PNP transistor with a very high current gain. The effective current gain of this compound transistor configuration is equal to the product of the individual current gains of the three transistors involved. The PNP Darlington is made up of three transistors, as compared to the NPN Darlington, which is composed of two transistors. This is done to make up for the relatively low current gain of IC PNP transistors.

12.5 SINGLE-SUPPLY AUDIO POWER AMPLIFIER EXAMPLE

A simple example of a single-supply audio power amplifier is shown in Figure 12.7. This amplifier consists of a Darlington differential-amplifier input stage with a current-mirror active load. The output stage is a class AB Darlington emitter-follower complementary push–pull configuration. The input stage is a Darlington differential amplifier comprised of Q_1 through Q_4 with the current-mirror active load using Q_5 and Q_6. The quiescent current of the differential amplifier is controlled by R_1 and the dc voltage applied to the base of Q_1 by the R_2–R_3 voltage divider. The output of this stage is fed directly to the emitter-follower output stage, so this amplifier has

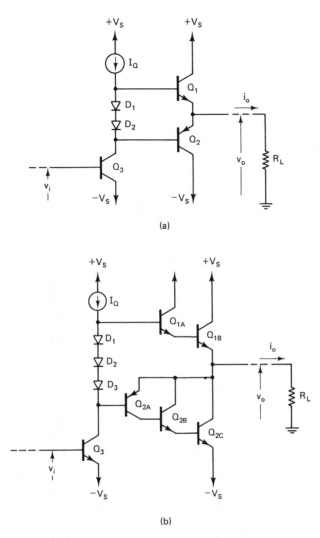

(a)

(b)

Figure 12.6 (a) Class AB push–pull output stage; (b) class AB Darlington output stage.

only one gain stage. Nevertheless, a very high open-loop gain is possible, as will be seen.

The output stage uses Q_7 and Q_8, acting as a Darlington emitter follower to source current into the load, and Q_9, Q_{10}, and Q_{11}, which are connected as a composite PNP–Darlington emitter-follower configuration to sink current from the load. To minimize crossover distortion, this output stage is operated in the class AB mode by means of the bias voltage produced across the series string of D_1, D_2, and D_3. Resistors R_4 and R_5 allow Q_7 and the Q_9–Q_{10} composite pair to operate at a higher quiescent current level than would otherwise be the case. These two resistors also provide a path for the more rapid removal of charge stored in the base region in case the above-mentioned transistors go into saturation and then have to be brought out of saturation.

The voltage divider, composed of R_6 and R_7, provides dc feedback from the

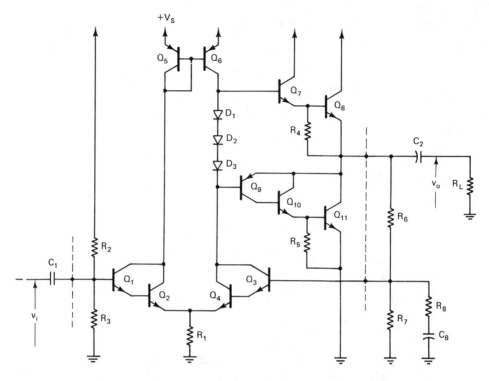

Figure 12.7 Single-supply audio power amplifier.

output back to the base of Q_3 so as to stabilize the quiescent output-voltage level at the desired value. Under quiescent conditions the differential amplifier will be in a balanced condition with $V_{B1} = V_{B3}$. If we assume that the base currents of Q_1 and Q_3 are small compared to the currents going through the R_2–R_3 and R_6–R_7 voltage dividers, respectively, the voltage at the base of Q_1 is given by $V_{B1} = V_S R_3/(R_2 + R_3)$, and that at the base of Q_3 is $V_{B3} = V_{O(Q)} R_7/(R_6 + R_7)$, where $V_{O(Q)}$ is the quiescent output voltage.

To allow this amplifier to develop the maximum possible undistorted peak-to-peak output voltage swing, the output quiescent voltage level should be midway between the positive supply voltage and ground so that $V_{O(Q)} = V_S/2$. Since $V_{B1} = V_{B3}$ under quiescent conditions, we can now express the condition for the voltage-divider ratios as $(V_S/2)R_7/(R_6 + R_7) = V_S R_3/(R_3 + R_2)$ and thus $R_7/(R_6 + R_7) = 2R_3/(R_3 + R_2)$, or $2(1 + R_6/R_7) = 1 + R_2/R_3$.

Ac feedback is also established by means of feedback to the base of Q_3. Assuming that C_8 is large enough such that the capacitive reactance of C_8 is small compared to R_8 over the ac frequency range of interest, the feedback factor is given by $F = (R_7 \| R_8)/[(R_7 \| R_8) + R_6]$. If the loop gain, FA_{OL}, is much greater than unity, the closed-loop gain is approximately equal to $1/F$, so we have $A_{CL} = 1/F = 1 + R_6/(R_7 \| R_8)$.

We will now consider a simple design example for this circuit. We will use the following quantities for this example: $V_S = +20$ V; $V_{O(Q)} = V_S/2 = +10$ V; $V_{B1} =$

+5.0 V; differential-amplifier quiescent current I_Q = 10 mA; R_L = 8 Ω; A_V = 50; f_1 (lower 3-dB frequency) = 20 Hz; β = 50 (min.) for all NPN transistors; and V_A = 200 V (Early voltage). For R_1 we have R_1 = $(V_{B1} - 2V_{BE})/I_Q$ = (5 − 1.3) V/ 10 mA = 370 Ω. The maximum expected base current of Q_1 and Q_3 is given by I_{B_1} = I_{B_3} = $(I_Q/2)/β_1β_2$ = 5.0 mA/(50 × 50) = 2.0 μA (max.). Therefore, the current through the R_2–R_3 and the R_6–R_7 voltage dividers should be much larger than 2 μA so as to minimize the effect of β variations on the quiescent levels. We will choose a current of 2 μA × 50 = 100 μA as satisfying this requirement, so we have that R_2 + R_3 = 20 V/0.1 mA = 200 kΩ. For R_6 + R_7 we have R_6 + R_7 = 10 V/0.1 mA = 100 KΩ. Since $R_3/(R_2 + R_3)$ = 5/20, we have that R_3 = (1/4)(200 kΩ) = 50 kΩ, and thus R_2 = 150 kΩ. Similarly, for the R_6–R_7 voltage divider, we have that $R_7/(R_6 + R_7)$ = 5/10, so R_7 = R_6 = 50 kΩ.

The open-loop voltage gain is approximately just the voltage gain of the differential amplifier stage since the voltage gain of the emitter-follower output stage is approximately unity. Accordingly, we have A_{OL} = $A_{V_1}A_{V(EF)}$ ≅ A_{V_1} ≅ $2g_f/g_{total}$, where g_f is the dynamic transfer conductance of the differential amplifier and g_{total} is the total dynamic conductance driven by the differential amplifier at the C_3–C_4–C_6– B_7–B_9 node. For g_f we have g_f = $I_Q/[4(2V_T)]$ = 10 mA/0.2 V = 50 mS. During the positive half-cycles the total conductance driven by the first stage is g_{total} = g_{o6} + g_{o3} + g_{o4} + g_{i7}, and during the negative half-cycles it is g_{total} = g_{o6} + g_{o3} + g_{o4} + g_{i9}. Since g_{i9} is considerably less than g_{i7} as a result of the greater overall current gain of the Q_9–Q_{10}–Q_{11} combination as compared to the Q_7–Q_8 combination, for a minimum-gain calculation we will use g_{i7}. This is given approximately by

$$g_{i7} = \frac{1}{β_7β_8R_L} = \frac{1}{50 × 50 × 8\ Ω} = 50\ μS\ (max.) \tag{12.22}$$

Since the current through Q_3 is much less than that through Q_4, we can neglect g_{o3}. The transistor output conductances are given approximately by g_o = I_C/V_A = 5 mA/ 200 V = 25 μS, so the total conductance driven by the first stage is g_{total} ≅ 25 μS + 25 μS + 50 μS (max.) = 100 μS (max.). The open-loop gain is therefore A_{OL} = $2g_f/g_{total}$ ≅ 2 × 50 mS/100 μS (max.) = 1000 (min.). We see therefore that a very large open-loop gain can be obtained with only one gain stage and with a load resistance of only 8 Ω.

Since the open-loop gain of 1000 (min.) is much greater than the closed-loop gain of 50, we have that $A_{OL} \gg A_{CL}$; thus the closed-loop gain is determined principally by the feedback factor F as given by A_{CL} = 1/F, so F = $1/A_{CL}$ = 1/50. The feedback factor is given in terms of the R_6–R_7–R_8 voltage divider as 1/F = 1 + $R_6/(R_7||R_8)$ = 50, so since R_6 = 50 kΩ we have that $(R_7||R_8)$ = 50 kΩ/49 = 1020 Ω. With R_7 = 50 kΩ, we obtain for R_8 a value of 1040 Ω.

For the required lower half-power frequency of 20 Hz, we have the condition for C_8 as $1/ωC_8$ = R_8 = 1040 Ω at 20 Hz. Solving this for C_8 gives the requirement that $C_8 ≥ 8.0\ μF$.

Diodes D_1 through D_3 should be designed to produce a no-load or standby current through the output transistors Q_8 and Q_{11} that is large enough to minimize the effects of crossover distortion. This current should satisfy the condition derived earlier that $I_Q ≥ 4.5V_T/R_L$, where I_Q represents the quiescent or standby current of the output transistors. For the case under consideration here, with $R_L = 8\,\Omega$, the standby current should be at least 14 mA.

12.6 IC AUDIO POWER AMPLIFIER EXAMPLES

12.6.1 The LM380

We will start off the consideration of examples of IC audio power amplifiers by looking at the LM380 (National Semiconductor). The LM380 is a fixed-gain (50) power amplifier capable of ac power outputs of up to 5 W. Figure 12.8a is a schematic diagram of this device.

The input stage is a Darlington compound differential amplifier comprised of Q_1 through Q_4 with a current-mirror active load composed of Q_5 and Q_6. This differential amplifier is biased by resistors R_{1A}, R_{1B}, and R_2. The quiescent current through R_1 comes from the positive supply voltage, while that through R_2 is a dc feedback current from the output terminal. Under quiescent conditions the currents through each half of the differential amplifier are approximately equal. We therefore have that $(V^+ - 3V_{BE})/(R_{1A} + R_{1B}) = (V_O - 2V_{BE})/R_2$, where V^+ is the dc supply voltage, V_{BE} is the base-to-emitter voltage drop of the diode connected transistor Q_{10}, and V_O represents the quiescent output voltage level. We therefore have that

$$
\begin{aligned}
V_O &= \frac{R_2}{R_{1A} + R_{1B}} (V^+ - 3V_{BE}) + 2V_{BE} \\
&= \frac{V^+}{2} - \frac{3V_{BE}}{2} + 2V_{BE} = \frac{V^+}{2} + \frac{V_{BE}}{2}
\end{aligned}
\tag{12.23}
$$

Thus the quiescent output voltage level will be centered at approximately one-half the supply voltage. This quiescent voltage level will allow for the maximum peak-to-peak output voltage swing and therefore the maximum ac output voltage.

In addition to this dc feedback loop, which acts to center the output voltage at one-half the supply voltage, there is also an ac feedback loop via the voltage divider action of R_2 and R_3. A fraction of the output voltage is fed back to the differential amplifier by this internal negative feedback loop to fix the overall voltage gain of the amplifier. If we split the differential amplifier down the axis of symmetry, we see that the feedback factor F is given by

$$
F = \frac{R_3/2}{R_2 + R_3/2} = \frac{500\,\Omega}{25\,\text{k}\Omega + 500\,\Omega} \cong \frac{1}{50}
\tag{12.24}
$$

Since the open-loop gain of the amplifier is much greater than $1/F \cong 50$, the closed-loop amplifier gain is fixed by the feedback factor to a value of $1/F \cong 50$, corresponding to 34 dB.

Note that the input-stage configuration of this power amplifier is similar to that of the LM124 operational amplifier and many other operational amplifiers that can be operated with a single supply voltage. This is a direct result of the fact that the input voltage range includes ground potential, and indeed extends to about 0.5 V below ground. As a result, this power amplifier can be operated with a single supply

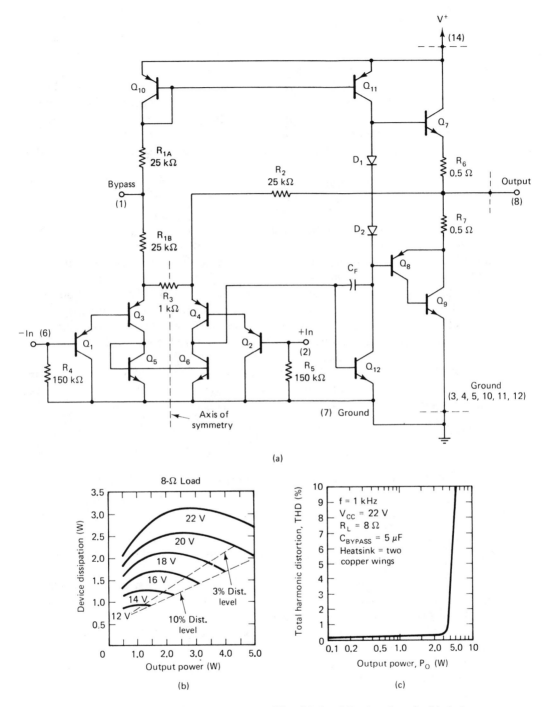

Figure 12.8 (a) LM380 IC audio power amplifier (National Semiconductor); (b) device dissipation versus output power for the LM380 (National Semiconductor); (c) total harmonic distortion versus output power for the LM380 (National Semiconductor).

Sec. 12.6   IC Audio Power Amplifier Examples

voltage, $+V_S$, and the ac input voltage can be fed into either input terminal without the need for any dc bias voltage superimposed on the ac input voltage. Resistors R_4 and R_5 (150 kΩ) provide a dc return path for the input bias (base) current, so this amplifier can be operated with either input terminal open.

The single-ended output voltage of the differential amplifier stage is fed to the second gain stage, which is comprised of Q_{12} in a common-emitter configuration with Q_{11} acting as a current-source active load. Capacitor C_F, which is connected as a feedback capacitor from the output back to the input of the second stage, is used for frequency compensation to stabilize this amplifier against any type of oscillatory response.

The single-ended output of the second gain stage drives the third, or output, stage. This is a class B, complementary push–pull emitter-follower output stage comprised of transistors Q_7, Q_8, and Q_9 and diodes D_1 and D_2. Transistor Q_7 is the NPN half of the push–pull output stage and is used to source currents into the load. Transistors Q_8 and Q_9 operate as a composite PNP–NPN transistor and are equivalent to a single PNP transistor with a current gain equal to the product of the current gains of Q_8 and Q_9. This composite configuration is used to compensate for the low current gain of the IC PNP transistors and is used as the PNP half of the push–pull output stage to sink currents from the load. Diodes D_1 and D_2 are used to develop a small prebias voltage across the base–emitter junctions of Q_7 and Q_8 so as to minimize the crossover distortion. Resistors R_6 and R_7 are used for current limiting.

Now that we have discussed the basic circuit configuration of this power amplifier, we will calculate some representative quiescent current values and voltage gains. Under quiescent conditions, the current through Q_3 is one-half of the total differential-amplifier quiescent current, I_Q, or $I_3 = I_Q/2$. This current is given by

$$I_3 = \frac{V^+ - 3V_{BE}}{R_{1A} + R_{1B}} = \frac{V^+ - 2\ \text{V}}{50\ \text{k}\Omega} \tag{12.25}$$

For this sample calculation, let us use a supply voltage of 20 V so that

$$I_Q = \frac{2(20 - 2)\ \text{V}}{50\ \text{k}\Omega} = \underline{0.72\ \text{mA}} \tag{12.26}$$

The dynamic transfer conductance of the differential amplifier is given by $g_f = I_Q/4(2V_T) = \underline{3.6\ \text{mS}}$.

By the symmetry of the circuit, we see by inspection that the base current of Q_{12} is equal to the sum of the base currents of Q_5 and Q_6. Therefore, on the assumption of equal current gains for all the NPN transistors, we have that the collector current of Q_{12} is equal to the sum of the collector currents of Q_3 and Q_4, this sum being equal to I_Q; so $I_{12} = I_Q = 0.72$ mA.

If we assume a representative current value of $\beta = 200$ and $n = 1.5$, the input conductance of Q_{12} is given by

$$g_{i12} = \frac{I_{12}}{n\beta V_T} = \frac{720\ \mu\text{A}}{1.5 \times 200 \times 25\ \text{mV}} = 96\ \mu\text{S} \tag{12.27}$$

The voltage gain of the differential amplifier stage is given by

$$A_{V_1} = \frac{2g_f}{g_o} = \frac{2g_f}{g_{o4} + g_{o6} + g_{i12}} \qquad (12.28)$$

where the factor of 2 comes from the current doubling action of the current-mirror active load. Since g_{o4} and g_{o6} are given essentially by $g_o = I_C/V_A = I_Q/2V_A$, where V_A is the Early voltage, and $g_{i12} = I_Q/n\beta V_T$, we note that since $2V_A = 500\ V \gg n\beta V_T = 7.5\ V$ we will have that $g_{i12} \gg g_{o4} + g_{o6}$. As a result, the voltage gain of the first stage can be written approximately as

$$A_{V_1} \simeq \frac{2g_f}{g_{i12}} = \frac{2I_Q/8V_T}{I_Q/n\beta V_T} = \frac{n\beta}{4} \qquad (12.29)$$

Since $\beta = 200$, the voltage gain of the first stage is 75. Note that this result is essentially independent of the quiescent current level.

The voltage gain of the common-emitter second stage is given by

$$A_{V_2} = -g_{m12}R_L' \qquad (12.30)$$

where g_{m12} is the dynamic transconductance of Q_{12} as given by $g_{m12} = I_{12}/V_T$, and R_L' is the transformed value of the load resistance, R_L. The impedance transformation occurs by means of the emitter-follower output stage consisting of Q_7 for the positive half-cycles and the Q_8–Q_9 composite transistor for the negative half-cycles. The transformed load resistance is $R_L' = \beta_7 R_L$ during the positive half-cycles and $R_L' = \beta_8\beta_9 R_L$ during the negative half-cycles. For a representative load resistance value of 8 Ω (a typical loudspeaker impedance level), the transformed value of load resistance R_L' is about 1600 Ω during the positive half-cycles and much larger than this during the negative half-cycles. Since $g_{m12} = I_{12}/V_T = 720\ \mu A/25\ mV = 29$ mS, the voltage gain of the second stage is given by $A_{V_2} = g_{m12}R_L' = 29\ mS \times 1600$ $\Omega = 46$ during the positive half-cycles and is much larger during the negative half-cycles.

The voltage gain of the emitter-follower output stage will be on the order of unity, so $A_{V(EF)} \cong 1$, and the overall open-loop gain of the amplifier is given by $A_{OL} = A_{V_1}A_{V_2}A_{V(EF)} = 75 \times 46 \times 1 = 3450$. The loop gain FA_{OL} is $(1/51)(3450) = 68$. Since the loop gain is considerably greater than unity, we see that the closed-loop gain of the amplifier is controlled principally by the feedback factor. If the open-loop gain has an actual value of 3450, as given above, the closed-loop gain is $A_{CL} = A_{OL}/(1 + FA_{OL}) = 3450/(1 + 68) = 50$, compared to a closed-loop gain of 51 that would be obtained if the open-loop gain went to infinity. The gain error would thus be 2%.

Note that the gain on the negative half-cycles of the output voltage swing is larger than during the positive half-cycles due to the greater overall current gain of the Q_8–Q_9 combination compared to that of just Q_7 alone. This imbalance in the push–pull amplifier results in the negative half-cycles being slightly larger in amplitude than the positive half-cycles. This discrepancy in the two halves of the output voltage swing contributes to the overall distortion of the amplifier, principally in terms of the second harmonic component. Nevertheless, in spite of this gain asymmetry, the total harmonic distortion of the output waveform can be kept to values of around 0.2% as long as the peak-to-peak output voltage swing is restricted to be at least about 8 V less than the supply voltage.

Power output characteristics. We will now look at the power output characteristics of the LM380 as shown in Figure 12.8b and c. We will consider a supply voltage of 22 V and an 8-Ω load resistance. With a 22-V supply voltage, the maximum peak-to-peak output voltage swing that can possibly be obtained is about 20 V, although there will be some degree of flattening of the output voltage peaks due to the output transistor being driven fairly close to saturation. The power output that is achieved with this 20-V peak-to-peak output voltage swing is given by $P_{O(MAX)} = V_{O(p-p)}^2/8R_L = 6.25$ W. At this point the power conversion efficiency should be close to the theoretical value of the maximum possible power conversion efficiency of 78.5%. Since the power dissipation at this point is 2.5 W, the actual efficiency is about 71%. The price to be paid for this large peak-to-peak output voltage swing will, of course, be a considerable amount of distortion, which will be in the range from 5% to 10%.

If the peak-to-peak voltage swing is now reduced to 18 V, the power output is reduced correspondingly to about 5.1 W, and the distortion is about 4%, which is still an unacceptably high value for many applications. A further reduction in the peak-to-peak output swing to 16 V reduces the output power to about 4.0 W, but now reduces the total distortion down to about 0.5%. The power dissipation under these conditions is about 3.0 W, so the power conversion efficiency is now about 57%.

To reduce the distortion to a minimum, the peak-to-peak output voltage swing has to be restricted to values less than about 14 V, for which case the distortion can be reduced to levels of about 0.2%. The very rapid increase in distortion that occurs for peak-to-peak output voltage swings in excess of about 16 V ($P_o = 4.0$ W) is due to the onset of the flattening or clipping of the voltage peaks due to the output transistors entering into the transition zone between the active and saturation regions.

The equation for the point at which the power dissipation is a maximum has been given as $P_D = P_O = 2V_S^2/\pi^2R_L$. For a total supply voltage of $2V_S = 22$ V, this gives a result of $P_D = P_O = 3.06$ W. The actual values of P_O and P_D obtained from the graph of the device dissipation versus output power for this device are very close to this.

12.6.2 The LM384

The LM384 (National Semiconductor) is identical to the LM380 except that it has a maximum voltage rating of 28 V, compared to 22 V for the LM380. With this higher supply voltage, a larger peak-to-peak output voltage swing is possible, and hence a substantially larger output power is available. For a supply voltage of 26 V and an 8-Ω load resistance, a peak-to-peak output voltage swing of 22 V produces an output power of 7.6 W, but the total harmonic distortion is up at about 5%. A reduction in the peak-to-peak swing to 20 V reduces the power output to 6.25 W, but brings the distortion all the way down to about 0.5%. A further reduction in the output voltage

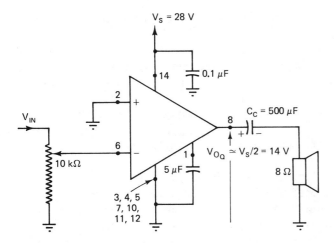

Figure 12.9 A 5-watt audio amplifier.

swing to 18 V reduces the output power to 5.1 W, but the distortion is now only about 0.2%.

Figure 12.9 shows the external circuitry for a 5-W audio amplifier using the LM384 power amplifier. The 500-μF coupling capacitor provides the ac drive to the loudspeaker while at the same time maintaining the quiescent voltage at the output at one-half the supply voltage. The dc voltage across the coupling capacitor is therefore $V_{O(Q)} \simeq V_S/2 = 14$ V. This allows the power amplifier to operate with a single supply and still be able to both source and sink output currents.

This device comes in a 14-pin, plastic, dual-in-line package (DIP). The center three pins on either side (3, 4, 5, and 10, 11, 12) are ground pins. This availability of a multiplicity of ground pins is used to maximize the heat flow from the package to the heat sink and thus to decrease the thermal impedance. The pins are of copper for good thermal conductivity and are usually soldered to a large copper foil area on a printed circuit board.

The free-air thermal impedance of this device is 82°C/W. The maximum allowable junction temperature is 150°C, so the maximum allowable power dissipation under free-air conditions is only 1.5 W. With the six ground pins soldered to a total copper foil area of 6 in.^{2} on a printed circuit board, the thermal impedance is reduced to 35°C/W. This will increase the maximum power dissipation rating to 3.6 W. With an "infinite heat sink" the thermal impedance is reduced to 12°C/W, which corresponds to the junction-to-case thermal impedance. Under these conditions, the maximum power dissipation rating is 10.4 W.

12.6.3 Audio Power Amplifiers with External Power Transistors

The power output available from an IC power amplifier is limited by the power dissipation rating of the IC itself to a maximum of generally about 10 W. Greater ac power output levels than this can, however, be achieved by the use of external power transistors that are driven by the IC power amplifier. This is analogous to the use of current boost transistors in conjunction with IC voltage regulators for higher output currents.

An example of an IC power amplifier that is specifically designed for this type of application is the LM391 audio power driver (National Semiconductor). With this device driving suitable external power transistors, output power levels of up to 100 W are available.

Figure 12.10 shows a somewhat simplified circuit diagram of this IC, and Figure 12.11 shows the typical external circuitry. The input stage is a PNP differential amplifier (Q_1 and Q_2) with a current-mirror active load (Q_3 and Q_4). The differential amplifier is biased by a current source (Q_5).

The second gain stage is transistor Q_6 in a common-emitter configuration with Q_{11} acting as a current-source active load. Transistor Q_6 via common-base transistors Q_9 and Q_{14} drives the emitter-follower output transistor (Q_7), which is used for sourcing currents into the load, via externally connected transistors (Q_{20} and Q_{21}). For sinking output currents, transistor Q_6 is used directly and drives the externally connected transistors, Q_{30} and Q_{31}.

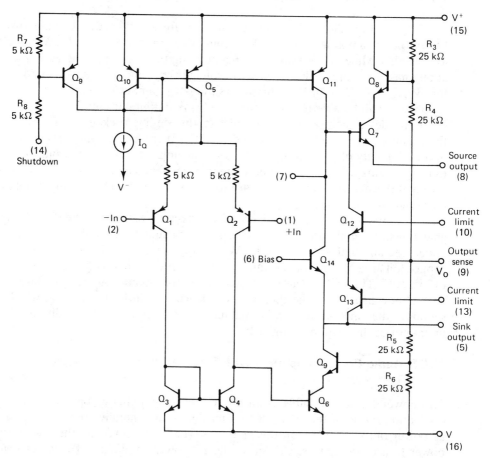

Figure 12.10 Audio power driver: LM391 (National Semiconductor).

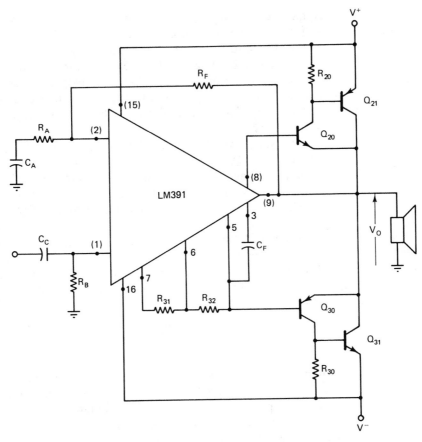

Figure 12.11 External connections for the audio power driver.

Transistor Q_{14} in conjunction with the externally connected resistors R_{31} and R_{32} produces a dc voltage drop that biases the push–pull output stage in the class AB mode of operation. Resistors R_{31} and R_{32} act as a voltage divider, with R_{32} being across the base–emitter junction of Q_{14} (pins 6 and 5). The voltage developed across Q_{14}, from collector (pin 7) to emitter (pin 5), is given by the relationship

$$V_{CE14} \frac{R_{32}}{R_{31} + R_{32}} = V_{BE14} \tag{12.31}$$

so
$$V_{CE14} = V_{BE14} \left(1 + \frac{R_{31}}{R_{32}}\right) \tag{12.32}$$

This voltage appears in series across the base–emitter junction of Q_7, Q_{20}, and Q_{30} as

$$V_{CE14} = V_{BE7} + V_{BE20} + V_{EB30} \tag{12.33}$$

If V_{CE14} is designed to be of suitable magnitude (about 2.0 V), class AB operation with an acceptably low amount of crossover distortion will be achieved.

For any given quiescent or standby current for the push–pull output stage, the corresponding base-to-emitter voltages of Q_7, Q_{20}, and Q_{30} will decrease with temperature. If these three transistors share the same heat sink and are in good thermal contact, all three base-to-emitter voltages will decrease with temperature at approximately the same rate. Since the bias voltage that is developed across Q_{14} is proportional to the base-to-emitter voltage drop of Q_{14}, we see that this bias voltage decreases at essentially the rate that is required to keep the standby current reasonably constant with increasing temperature.

This IC has a total supply voltage rating of 100 V (or ± 50 V). To limit the collector-to-emitter voltage of Q_6 and Q_7 to voltages that are no more than 50 V, emitter-follower transistors Q_8 and Q_9 are used as voltage sources in conjunction with the R_3–R_4 and R_5–R_6 voltage dividers, respectively. The total voltage drop across the R_3–R_4 voltage divider is $V^+ - V_O$, so the voltage at the base of Q_8 will be one-half of this voltage drop plus V_O or $\frac{1}{2}(V^+ - V_O) + V_O = \frac{1}{2}(V^+ + V_O)$. The voltage at the emitter of Q_8, which is the same as the collector voltage of Q_7 is V_{BE} less than this, so we have $V_{C_7} = V_{E_8} = V_{B_8} - V_{BE} = \frac{1}{2}(V^+ + V_O) - V_{BE}$. The voltage at the emitter of Q_7 is equal to the output voltage, V_O, plus the base-to-emitter voltage drop of Q_{20} or $V_O + V_{BE}$. Therefore, the collector-to-emitter voltage that appears across Q_7 is given by

$$V_{CE_7} = V_{C_7} - V_{E_7} = \frac{1}{2}(V^+ + V_O) - (V_{BE} + V_O) - V_{BE}$$
$$= \frac{1}{2}(V^+ - V_O) - 2V_{BE} \tag{12.34}$$

Let us now consider the operation of this circuit with the maximum rated supply voltages given by $V^+ = +50$ V and $V^- = -50$ V. Under these conditions the output voltage cannot drop below -50 V, so the maximum voltage to which Q_7 is subject is $V_{CE_{7(MAX)}} = \frac{1}{2}(50 + 50) - 2(0.7) \simeq 48$ V.

There is a similar situation for Q_6, for which $V_{CE_6} = V_{C_6} - V_{E_6} = V_{C_6} - V^- = V_{B_9} - V_{BE} - V^- \simeq \frac{1}{2}(V_O - V^- - V_{BE})$. Since V_O cannot go above V^+, which is $+50$ V, the maximum voltage to which Q_6 is subject is approximately 49 V.

Thus, by means of the R_3–R_4 and R_5–R_6 voltage dividers in association with Q_8 and Q_9, the maximum voltage that appears across the source and sink output transistors (Q_7 and Q_6) of the IC is limited to no more than one-half of the total supply voltage. This makes the IC implementation of these transistors much easier, especially from the standpoint of the safe operating area (SOA) protection of these transistors based on second breakdown considerations.

12.6.4 Class B Audio Driver

Another example of an audio power driver is the MC3320P/MC3321P (Motorola), and Figure 12.12 shows a circuit schematic of this IC. Figure 12.13 shows a typical application circuit. With a 30-V dc supply and the external power transistors, this IC is capable of delivering up to 10 W to an 8-Ω load. The feedback loop using the 100-kΩ resistor (R_4) and the 1-kΩ resistor establishes a closed-loop gain of 101 for this circuit. The quiescent output level at the emitters of the external power output transistors, Q_1 and Q_2, is set by the feedback through R_4 to pin 6 to a value closely equal to the voltage at pin 5. The dc voltage level at pin 5 is set by the 820-kΩ and

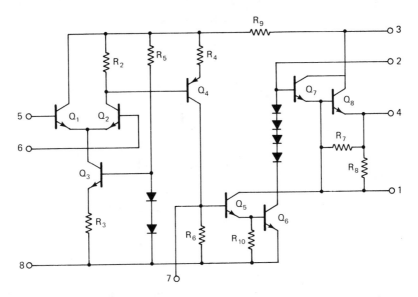

Figure 12.12 Circuit schematic of the MC3320P Audio Driver (Motorola Semiconductor Products).

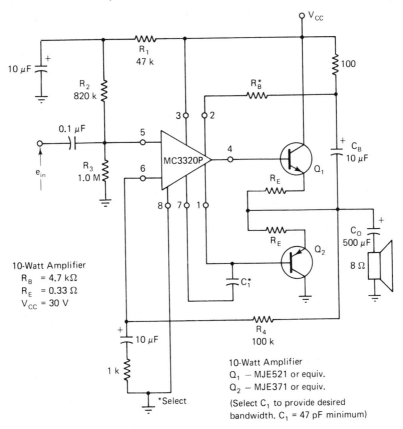

Figure 12.13 Application circuit for the MC3320P (Motorola Semiconductor Products).

Sec. 12.6   **IC Audio Power Amplifier Examples**   **611**

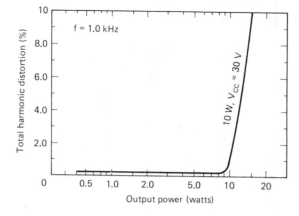

Figure 12.14 Total harmonic distortion versus output power (Motorola Semiconductor Products).

1.0-MΩ voltage divider to a value of 0.55 V_{CC}. As a result, the quiescent output voltage level is approximately at the mid-point between the positive supply voltage and ground. This permits the maximum symmetrical undistorted output-voltage swing and therefore the maximum output power. Figure 12.14 is a graph of the total harmonic distortion versus output power for this circuit. An output power of up to 8 W is available with a total harmonic distortion not exceeding 0.2%, and at a power output of 10 W the distortion level is just reaching 1%. Beyond this, however, the distortion increases very rapidly due to the clipping of the output voltage. At a power output of 10 W, the peak-to-peak voltage swing is 25.3 V. If we add to this the base-to-emitter voltage drops of the four IC output transistors plus those of Q_1 and Q_2, which total about 4.5 V, we obtain a total voltage very close to the 30-V dc supply voltage, so it should be expected that the distortion should increase rapidly beyond this output power level.

12.7 DISTORTION IN PUSH–PULL CLASS AB POWER AMPLIFIERS

There are two principal causes of distortion in class AB power amplifiers. One is the gain mismatch between the NPN and PNP halves of the push–pull output stage. This gain mismatch results in an asymmetrical output waveform with the amplitude of the positive half-cycles not being the same as that of the negative half-cycles. The other possible cause of distortion is the flattening or clipping of the peaks of the output voltage waveform that will occur if the output transistors are driven from the active region into the transition zone between the active and saturation regions, or all the way into the saturation region itself.

12.7.1 Gain Mismatch

We will first look at the distortion that results from the push–pull gain asymmetry. We will use the Fourier series representation for positive half-cycles taken separately, then that for the negative half-cycles, and then we will combine the two Fourier series expressions to obtain that which will correspond to the entire output waveform. The

Fourier series representation for the half-wave rectified sine wave will be used for each set of half-cycles. For a half-wave rectified sine wave of peak amplitude A, the Fourier series is given by

$$v(t) = \frac{2A}{\pi}\left(\frac{1}{2} + \frac{\pi}{4}\cos \omega t + \frac{1}{3}\cos 2\omega t - \frac{1}{15}\cos 4\omega t + \cdots\right) \qquad (12.35)$$

The peak amplitude of the positive half-cycles of the push–pull output–voltage waveform is represented by A_1, so the positive half-cycles can be represented as

$$v_+(t) = \frac{2A_1}{\pi}\left(\frac{1}{2} + \frac{\pi}{4}\cos \omega t + \frac{1}{3}\cos 2\omega t - \frac{1}{15}\cos 4\omega t + \cdots\right) \qquad (12.36)$$

For the negative half-cycles, the peak amplitude is $-A_2$, and the negative half-cycle waveform is shifted by π radians (180°) with respect to the positive half-cycles. Since $\cos (\theta + n\pi) = \cos \theta$ for n being an even integer, and $\cos (\theta + n\pi) = -\cos \theta$ for n being an odd integer, we have for the negative half-cycles the Fourier series given by

$$v_-(t) = -\frac{2A_2}{\pi}\left(\frac{1}{2} - \frac{\pi}{4}\cos \omega t + \frac{1}{3}\cos 2\omega t - \frac{1}{15}\cos 4\omega t + \cdots\right) \qquad (12.37)$$

The total output-voltage waveform can be obtained by taking the algebraic sum of v_+ and v_-, giving

$$\begin{aligned}
v_o(t) &= v_+(t) + v_-(t) \\
&= \frac{A_1 - A_2}{\pi} + \frac{A_1 + A_2}{2}\cos \omega t + \frac{2(A_1 - A_2)}{3\pi}\cos 2\omega t \qquad (12.38) \\
&\quad + \frac{2(A_1 - A_2)}{15\pi}\cos 4\omega t + \cdots
\end{aligned}$$

From this expression we see that the distortion components are all even harmonics, and the dominant distortion component is the second harmonic term. As a result, the total harmonic distortion (THD) can be expressed as

$$\% \text{ THD} \simeq \% \text{ 2HD} = \frac{\text{amplitude of second harmonic term}}{\text{amplitude of fundamental term}} \times 100\% \qquad (12.39)$$

Since the amplitude of the fundamental term is $(A_1 + A_2)/2$ and that of the second harmonic term is $2(A_1 - A_2)/3\pi$, the percentage of total harmonic distortion is given by

$$\% \text{ THD} \simeq \% \text{ 2HD} = \frac{4}{3\pi}\frac{A_1 - A_2}{A_1 + A_2} \times 100\% \qquad (12.40)$$

For the usual cases of interest, the difference between the two amplitude values, $A_1 - A_2$, is small compared to magnitude of either A_1 or A_2; so if we let $A_1 = A + \Delta A$, and $A_2 = A$, for the usual case of $\Delta A \ll 1$ we have that $A_1 + A_2 = 2A + \Delta A$

$\simeq 2A$. Using this approximation, the expression for the percentage of distortion can be written as

$$\% \text{ THD} \simeq \frac{4 \, \Delta A}{3\pi(2A)} \times 100\% = \frac{2}{3\pi} \frac{\Delta A}{A} \times 100\%$$

$$\simeq 0.2\left(\frac{\Delta A}{A}\right) \times 100\% = 20\left(\frac{\Delta A}{A}\right)\%$$

(12.41)

To consider an example, let us take a power amplifier for which $A_{OL} = 1000$ and $A_{CL} = 50$, and assume that there is a 20% open-loop gain difference between the positive and negative half-cycle output-voltage swings. The closed-loop gain difference is given approximately by the relationship $dA_{CL}/A_{CL} = (dA_{OL}/A_{OL})(A_{CL}/A_{OL})$, so a 20% change in the open-loop gain will result in a change in the closed-loop gain given by $dA_{CL}/A_{CL} \simeq 20\% \times (50/1000) = 1\%$. This 1% gain difference will produce a corresponding difference of 1% in the amplitudes of the positive and negative half-cycles, so $\Delta A/A = 0.01$. As a result, the percentage of distortion is approximately 0.2%. We see in this example the beneficial action of the negative feedback in stabilizing the closed-loop gain against variations in the open-loop gain and thereby reducing the distortion produced by the open-loop gain asymmetry.

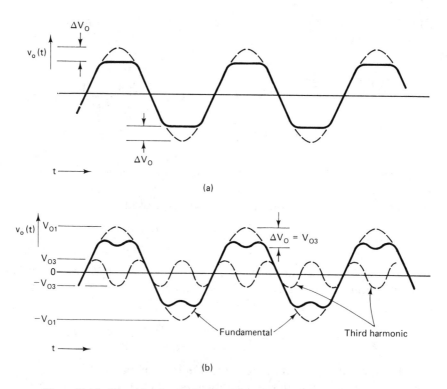

Figure 12.15 Distortion due to clipping: (a) output voltage waveform; (b) approximation of clipped output waveform using the fundamental and third harmonic sinusoidal components.

12.7.2 Distortion Due to Clipping

We will now consider the distortion that results from a symmetrical flattening or clipping of the output waveform. If we look at Figure 12.15, we see that the waveform that results from this type of clipping can be approximated by combining a sine wave at the fundamental frequency and a sine wave at the third harmonic frequency. As we see from Figure 12.15b, the third harmonic waveform causes the net waveform to be flattened or clipped by an amount given by the amplitude of the third harmonic term, V_{O_3}. If the flattening that has been incurred by the output voltage waveform is ΔV_O, we have that $V_{O_3} = \Delta V_O$, and therefore the harmonic distortion is given by

$$\% \text{ THD} \simeq \% \text{ 3HD} = \frac{V_{O_3}}{V_{O_1}} \times 100\% = \frac{\Delta V_O}{V_O} \times 100\% \qquad (12.42)$$

Thus, even a small amount of clipping of the output voltage can result in a very significant increase in distortion. The use of negative feedback provides only a limited amount of help in this situation since the clipping level will not be affected much by the presence of feedback. We see, therefore, that for a minimum amount of distortion the output transistors should be kept out of the saturation region, and even well out of the transition zone between the active and saturation regions.

12.8 POWER OPERATIONAL AMPLIFIERS

Operational amplifiers can exhibit some very good performance characteristics, such as a very high open-loop gain (100,000 to several million), a very high input impedance (up into the 10^9-Ω range for some devices), and a very low input bias current. Also available are op amps with very low offset voltages (V_{OS}) and very low offset voltage temperature coefficients ($TC_{V_{OS}}$). Op amps are, however, generally limited in the amount of ac output power available.

The current limit for most op amps is generally in the range of about 20 or 25 mA. The maximum supply voltage rating is usually around 36 V with a single supply and ± 18 V if a split supply is used. For an op amp using a ± 18-V supply, the peak output voltage swing obtainable is about ± 15 V. For a peak output-current swing of ± 20 mA and a peak output-voltage swing of ± 15 V, the ac power delivered to the load is given by $P_O = V_{O(\text{peak})}I_{O(\text{peak})}/2 = 15 \text{ V} \times 20 \text{ mA}/2 = 150 \text{ mW}$. While this power output level is satisfactory for many applications, for some applications a considerably larger power output is required.

Larger ac power outputs can be obtained by adding external *current-boost* power transistors to the circuit in a fashion similar to that of the IC voltage regulators with external current-boost transistors for greater current outputs.

Figure 12.16a shows a basic circuit of an op amp with external current-boost power transistors. Transistors Q_3 and Q_5 are the current-boost transistors and are driven as a complementary push–pull output stage. Transistors Q_4 and Q_6 together with the R_{CL} resistors perform the current-limiting function to protect Q_3 and Q_5 against excessive current flow.

When the op amp is sourcing current to the load, a voltage drop is produced across R_1 equal to $(I_Q + I_{O_1})R_1$, where I_Q is the quiescent or biasing current of the

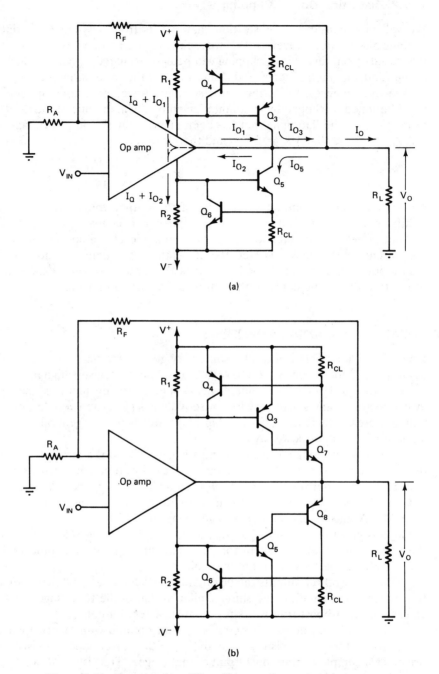

Figure 12.16 Operational amplifier with external power transistors.

op amp and I_{O_1} is the current sourced by the op amp. When the voltage drop across R_1 rises above 500 mV, Q_3 starts to turn on, and when the voltage drop reaches the range from 600 to 700 mV, Q_3 is in full conduction. When this happens, Q_3 supplies additional current to the load, I_{O_3}. A similar situation holds true for Q_5 for sinking current from the load when the voltage drop across R_2 has risen above 500 mV. This causes Q_5 to turn on and to sink load current.

If $I_Q = 5$ mA and the op amp output current limit is 25 mA, for example, a suitable choice for R_1 and for R_2 would be a value that will result in a voltage drop of about 600 mV when the op amp output current is 15 mA. We therefore have that $R_2 = 600$ mV/(5 + 15) mA = 30 Ω, and R_1 will have the same value as R_2.

With this value for R_1 and R_2, the output stage of the op amp will supply the load current in its entirety up to a level of about 15 mA. Above that current level, the external transistors, Q_3 and Q_5, turn on and supply any additional current required by the load. For a current gain of 50 (min.) for Q_3 and Q_5, the maximum output current available from these two transistors is $50 \times I_{B(MAX)}$ where $I_{B(MAX)}$ is the maximum base drive that is available. To bias the transistors on, the current flow through R_1 or R_2 is 20 mA. Since the maximum output current of the op amp is limited to 25 mA, the maximum current left over to serve as the base drive for the external current boost transistors is 10 mA; so an output current of 525 mA (min.) is available, with the op amp supplying 25 mA and the current boost transistor supplying 500 mA (min.). Note that Q_3 and Q_5 operate in essentially a class B push–pull fashion, but that the operational amplifier itself supplies the output current during the transition interval between the conduction period of Q_3 and the conduction period of Q_5.

For even greater output currents than that available with the circuit just described, a Darlington configuration can be used as shown if Figure 12.16b. In this figure, the Q_3–Q_7 combination is a direct-coupled PNP–NPN combination used for sourcing currents into the load. This combination has an overall current gain equal to the product of the current gains of Q_3 and Q_7. In a similar fashion, the Q_5–Q_8 NPN–PNP combination is used to sink load currents and has an overall current gain equal to the product of the individual transistor current gains. With this arrangement, output current levels up into the range of several amperes are now available. The maximum current that may be safely drawn from this circuit will still, of course, be limited by considerations of maximum power dissipation and of the safe operating area of Q_7 and Q_8.

12.8.1 Hybrid IC Power Operational Amplifiers

A hybrid integrated circuit that is the combination of a conventional op amp chip and current boost power transistors, all in one package, is the LH0021 and the LH0041 (National Semiconductor). A schematic diagram of this device is shown in Figure 12.17. The circuit is basically similar to that of a 741-type op amp with the addition of the current boost transistors Q_{13} and Q_{14}.

The differential amplifier input stage consists of Q_1 through Q_4 connected as a CC–CB compound differential amplifier. The differential amplifier is biased by Q_{17}, which sinks the combined base currents of Q_3 and Q_4. Transistors Q_5 and Q_6 serve as a current-mirror active load for the differential amplifier. The second stage consists

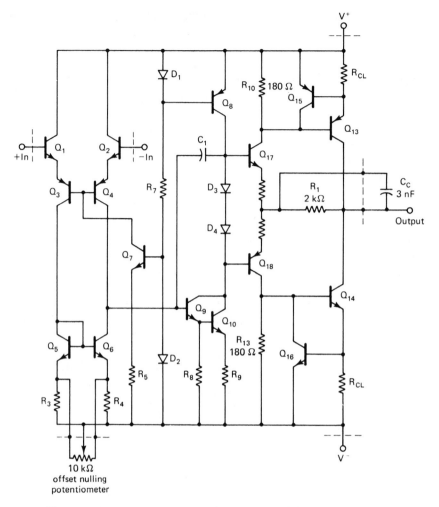

Figure 12.17 Power operational amplifier: LH0021 (National Semiconductor).

of Q_9 and Q_{10} connected as a Darlington common-emitter gain stage with Q_8 serving as a current-source active load for this stage.   Capacitor C_1 is a compensation capacitor that provides feedback around the second stage.

Transistors Q_{17} and Q_{18} serve as a complementary emitter-follower push–pull output stage, with the voltage drop provided by diodes D_3 and D_4 biasing this stage in the class AB mode to minimize crossover distortion.

Whenever the voltage drop across R_{10} or R_{13} exceeds approximately 500 mV, the current-boost transistor, Q_{13} or Q_{14}, respectively, is turned on.   The value of R_{10} and R_{13} is 180 Ω, so for output-current levels in excess of ± 500 mV/180 Ω $=$ ± 2.8 mA, either Q_{13} or Q_{14} will be turned on and contribute additional current to the load. The actual turn-on threshold is somewhat less than the value just calculated due to the standby current of Q_{17} and Q_{18} that will result from the class AB biasing condition resulting from the voltage drop across D_3 and D_4.

Resistor $R_1(2 \text{ k}\Omega)$ limits the current flow through Q_{17} and Q_{18} to a maximum of about $\pm 18 \text{ V}/2 \text{ k}\Omega = 6 \text{ mA}$ and so provides protection for these two transistors. Protection of the current boost transistors Q_{13} and Q_{14} is provided by the action of Q_{15} and Q_{16} and the voltage drop produced across the associated current-limit resistor, R_{CL}.

Capacitor C_C is an externally connected bypass capacitor (3 nF) connected across R_1. If C_C were not present, and if this device is driving a load with a substantial capacitive component, the combination of R_1 and the load capacitance would constitute an RC low-pass network. As a result of the rather large size of R_1 (2 kΩ), a large phase shift could be produced, which would have a destabilizing effect on the amplifier and could result in such effects as gain peaking or even oscillations. To eliminate this problem, a capacitor of suitably large size (~ 3 nF) is connected across R_1. Since it is not possible to have such a large capacitor on the chip, an externally connected capacitor is used.

The LH0021 comes in an 8-pin TO-3 type of package and, with adequate heat sinking, is capable of supplying a peak current of ± 1.2 A of output current, with an output-voltage swing of ± 12 V across a 10-Ω load.

Figure 12.18 shows a 10-W *bridge* amplifier circuit using two LH0021 power op amps. The input signal is applied to the noninverting input of one amplifier and to the inverting input of the other, so the two output voltages are 180° out of phase with respect to each other. For each amplifier the closed-loop gain is $1 + (91 \text{ k}\Omega/10 \text{ k}\Omega)$ $\cong 10$, so the two output voltages are $+10 V_{\text{IN}}$ and $-10 V_{\text{IN}}$, respectively.

With a ± 15-V split supply, the LH0021 is capable of a peak output voltage swing of about 12 V into a 10-Ω load, but to limit the distortion to an acceptable value the output voltage swing should be reduced to a maximum of about 10 V.

With each amplifier supplying a peak voltage of 10 V across the load, the peak voltage across the load is 20 V since the two amplifier output voltages are 180° out of phase with respect to each other. Note that the peak voltage that appears across the load is *twice* the peak voltage that is supplied by each amplifier. The peak load current is therefore 20 V/20 Ω = 1.0 A.

The average (rms) ac power delivered to the load is

$$P_O = \frac{V_{O(\text{peak})}^2}{2R_L} = \frac{I_{O(\text{peak})}^2 R_L}{2} = \frac{(20 \text{ V}^2)}{2 \times 20 \text{ }\Omega} = 10 \text{ W} \tag{12.43}$$

Note that if a single amplifier were used, rather than the bridge arrangement shown here, the peak voltage across the load would be only 10 V, and the power delivered to the load would be limited to 2.5 W, or only one-fourth of that obtained with the bridge amplifier configuration. The 0.5-Ω resistors are used in the current-limiting circuit (R_{CL}) and set the current limit at a value of about 600 mV/0.5 Ω = 1.2 A.

The LH0021 has a full-power bandwidth (FPBW) of 15 kHz (when driving a 10-Ω load), so this circuit is capable of supplying the 10 W of output power over the entire audio-frequency range. The distortion is only about 0.2% for frequencies up to 2 kHz and then increases to about 1.0% at 7 kHz and 1.6% at 10 kHz. Above

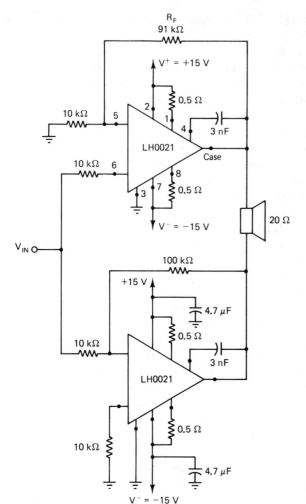

Figure 12.18 A 10-watt bridge audio amplifier.

10 kHz the distortion increases very rapidly as the condition of slew-rate limiting is approached at a frequency of 15 kHz. Note, however, that the distortion harmonic components produced for fundamental frequencies above 7 kHz fall outside the audio spectrum and thus are generally not noticeable.

12.8.2 The OPA501

An example of a power operational amplifier with a very high output power capability is the OPA501 (Burr-Brown). This hybrid op amp comes in a TO-3 package and can deliver as much as 260 W continuous (26 V at 10 A) to a load. It can operate with supply voltages as high as ± 40 V and has an output current capability of ± 10 A. The small-signal bandwidth of this op amp is 1.0 MHz, and the full-power bandwidth is 16 kHz when delivering 40 V peak-to-peak to an 8-Ω load.

12.8.3 PA04 Power Operational Amplifier

The PA04 (Apex Microtechnology Corporation) is an example of a high-voltage, high-current hybrid IC power operational amplifier. It can operate with supply voltages up to 200 V and is capable of a peak output-current swing of 20 A. It has a dc open-loop gain of 102 dB (typ.) and 94 dB (min.). It offers a gain–bandwidth product of 2 MHz and a full-power bandwidth of 90 kHz when producing an output-voltage swing of 180 V peak-to-peak. This is a FET input operational amplifier, so along with these very impressive power output characteristics is an input bias current of 10 pA (typ.), 50 pA (max.) and an input impedance of 10^{11} Ω (typ.) in parallel with 13 pF (typ.).

Circuit description. A simplified schematic of the PA04 power operational amplifier is shown in Figure 12.19. The input stage consists of the Q_1–Q_2 PMOS differential-amplifier stage that is biased by a current source comprised of Q_5, R_5,

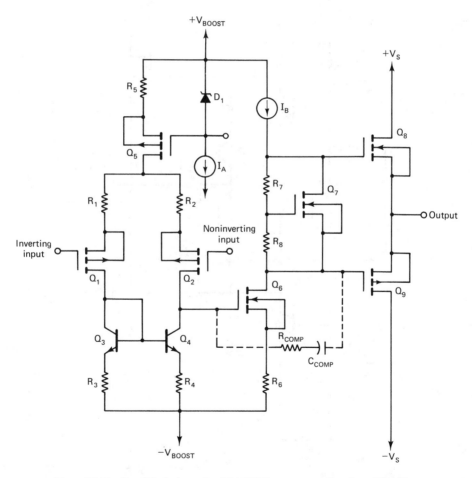

Figure 12.19 Simplified schematic of the PA04 power operational amplifier (Apex Microtechnology).

and D_1. The differential amplifier drives an NPN current-mirror active load that is comprised of Q_3 and Q_4. The second stage uses Q_6 as a common-source amplifier stage with a current source active load. The last stage uses Q_8 and Q_9 in the form of a complementary NMOS–PMOS class AB source-follower push–pull output stage. Transistor Q_7, in conjunction with resistors R_7 and R_8, generates a voltage drop that is used to bias the Q_8–Q_9 push–pull output stage for class AB operation to minimize crossover distortion.

There are two separate sets of positive and negative supply voltage connections, $\pm V_{\text{BOOST}}$ and $\pm V_S$. This operational amplifier can be operated with $+V_{\text{BOOST}} =$

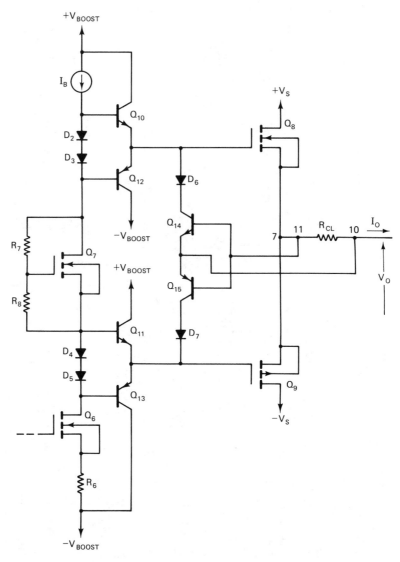

Figure 12.20 PA04 output stage (Courtesy of Apex Microtechnology).

Power Amplifiers   Chap. 12

$+V_S$ and $-V_{\text{BOOST}} = -V_S$. Under these conditions, the peak output-voltage swing is about 5 V less than the supply voltage under no-load conditions and about 10 V less than the supply voltage for a load current of 20 A. For a larger output-voltage swing and a greater power conversion efficiency, the boost voltage can be raised as much as 20 V above the supply voltage. A boost voltage that is 5 V above the supply voltage is enough to cause the output transistor to be driven into saturation and results in a substantial improvement in the output-voltage swing and in the power conversion efficiency. With this value of boost voltage, the peak output-voltage swing is now only 2 V less than the supply voltage under no-load conditions, and 5 V less for a 20-A load current.

A more detailed view of the output stage is presented in Figure 12.20. Current limiting is provided by transistors Q_{14} and Q_{15} together with an external resistor R_{CL}. Note that pins 10 and 11 are for the connection of the current-limiting resistor R_{CL} to the operational amplifier and that pin 11 is separate from the operational-amplifier output terminal, pin 7. It might seem that there is no need to have both pins 7 and 11. However, at output-current levels up in the ampere range, there can be a significant voltage drop in the connections at leads between the operational-amplifier output terminal (pin 7) and the R_{CL} connection at pin 11. For example, for a current limit of 20 A, the value of R_{CL} is given by $R_{CL} \cong 700$ mV/20 A $= 35$ m$\Omega = 0.035$ Ω. A resistance of only 0.001 Ω at a load current of 20 A will produce a voltage drop of 20 mV. If this 20-mV drop were to be sensed along with R_{CL}, it would produce a very substantial change in the current limit.

Transistors Q_{10} and Q_{12}, together with biasing diodes D_2 and D_3, constitute an emitter-follower push–pull driving circuit for the NMOS output transistor Q_8. Transistors Q_{11} and Q_{13}, together with biasing diodes D_4 and D_5, perform a similar function for the PMOS output transistor Q_9. Since Q_8 and Q_9 are MOSFETS, the dc gate current drive required for these transistors is negligible. However, these power MOSFETs have an appreciable input capacitance, and these bipolar transistor push–pull pairs (Q_{10}–Q_{12} and Q_{11}–Q_{13}) are useful for the rapid charging and discharging of the MOSFET gate input capacitance under transient or ac conditions. This improves the full-power bandwidth and the slewing rate characteristics of this power operational amplifier.

This OA has a *sleep mode* in which the amplifier quiescent current can be reduced from the normal value of 70 mA (typ.) to a sleep value of 3 mA. This is done by connecting the SLEEP pin (pin 12) to the V_{BOOST} terminal (pin 9).

Power output characteristics. For any power device, the thermal characteristics are an important part of the device specifications. The PA04 has a junction-to-case thermal resistance of 0.6°C/W (max.) and a junction-to-air (that is, no heat sink, free air conditions) thermal resistance of $R_{TH(JC)} = 12$°C/W (typ.). The maximum allowable junction temperature is 150°C. Let us consider two extreme cases, assuming an ambient temperature of 25°C. First, under free air conditions (no heat sink), the maximum allowable power dissipation for this operational amplifier is given by $P_{D(\text{MAX})} = \Delta T_{\text{MAX}}/R_{TH(JA)} = (T_{J(\text{MAX})} - T_A)/R_{TH(JA)} = (150°\text{C} - 25°\text{C})/12°\text{C/W}$ $= 10.4$ W. For the second extreme case, we will assume that the IC is placed in contact with an infinite heat sink such that the case-to-ambient thermal resistance

$R_{TH(CA)}$ is reduced to zero. For this case, we have that $R_{TH(JA)} = R_{TH(JC)} + R_{TH(CA)}$ = 0.6°C/W. The corresponding value of the maximum power dissipation is $P_{D(MAX)}$ = $\Delta T_{(MAX)}/R_{TH(JA)}$ = 125°C/(0.6°C/W) = 208 W. Thus, depending on the heat sinking conditions, the device maximum power dissipation rating is somewhere between 10 W with no heat sink at all and 208 W with an infinite heat sink.

An ideal class B amplifier has a power conversion efficiency of 78.5% and the corresponding ratio of output power to device dissipation is 3.66. Thus, ideally, this IC would have an output power capability of around 760 W. In actual practice, output powers of up to 400 W are obtainable.

Other power operational amplifiers from Apex Microtechnology Corporation include the PA03, which offers a peak output current of ±30 A and a peak output voltage of ±68 V. The gain–bandwidth product is 1 MHz, the full-power bandwidth is 30 kHz for an output-voltage swing of 88 V peak-to-peak, and the slewing rate is 8 V/ms. The PA85 offers a peak output-voltage swing of ±215 V and a peak output current of ±0.2 A. The gain–bandwidth product is 20 MHz and the full-power bandwidth is 500 kHz. The slewing rate is an impressive 1000 V/μs. These are FET-input operational amplifiers with input bias currents of 5 pA (typ.) and input resistances of 10^{11} Ω.

The PB50 (Apex Microtechnology) is a power-booster amplifier that can be driven by an operational amplifier and placed within the overall feedback loop. It can operate with total supply voltages of up to 200 V and can supply output currents of up to 2 A. It provides a full-power bandwidth of 320 kHz (typ.) for a 100-V peak-to-peak output-voltage swing and a slewing rate of 100 V/μs. The gain of this power booster can be set from a minimum of 3 V/V to a maximum of 25 V/V by a single external gain setting resistor.

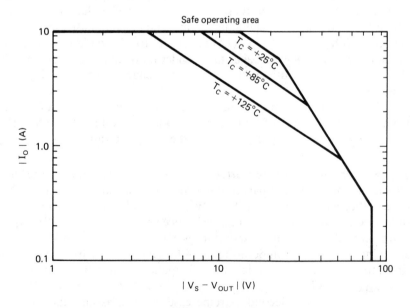

Figure 12.21 Safe operating area (Burr-Brown).

The OPA541 (Burr-Brown) is a FET-input, bipolar transistor output power operational amplifier that can be operated with ± 40-V supplies and can deliver a peak load current of 10 A. The safe operating area (SOA) of this operational amplifier is shown in Figure 12.21. In this graph, $|V_S - V_{OUT}|$ is the voltage across the conducting output transistor. For output currents of less than 0.3 A, the maximum difference between the supply voltage and the output voltage is the full supply voltage of 80 V. At a case temperature of 125°C and an output current of 1.0 A, the maximum voltage difference between the supply and V_O output is 40 V, so for a ± 40-V supply a peak output-voltage swing that is equal to the supply can be allowed. This will correspond to a power of 20 W delivered to the load.

The OPA541M comes in a TO-3 package, and the thermal resistances are $R_{TH(JC)}$ = 1.9°C/W (max.), and under free air conditions (no heat sink), $R_{TH(JA)}$ = 30°C/W (typ.). The full-power bandwidth is 45 kHz (min.) for a 20-V rms output, and the slewing rate is 10 V/μs (typ.). A dual version of the OPA541 is available as the OPA2541, and it also comes in an 8-pin TO-3 package.

12.9 HIGH-VOLTAGE INTEGRATED CIRCUITS

Most ICs have a maximum rated supply voltage of about 36 V (or ± 18 V if a dual supply is used). This limits the peak output-voltage swing that is available to about ± 16 V or 32 V peak-to-peak. The maximum supply voltage rating of ICs is based on the junction breakdown voltages. These breakdown voltages are usually around 50 V, so to keep well away from this critical voltage and to allow for a sufficient safety margin a maximum supply voltage rating of about 36 to 40 V is common, although many ICs have a lower maximum voltage rating.

The breakdown voltage of a planar PN junction is determined by two principal factors: (1) the doping levels on the two sides of the junction, and (2) the junction curvature. The breakdown voltage of a flat (no curvature) abrupt PN junction is given by the approximate relationship $BV \simeq 2.7 \times 10^{12}/N^{2/3}$, where BV is the breakdown voltage (volts) and N is the net doping level as given by $1/N = (1/N_A) + (1/N_D)$, with N_A being the net doping on the P-type side and N_D being the net doping on the N-type side of the junction. For the usual case of a PN junction in which the doping level is much greater on one side than it is on the other, the breakdown voltage is determined principally by the doping level on the more lightly doped side, and the net doping N given by $N \simeq N_D$ for a P$^+$N junction and $N \simeq N_A$ for an N$^+$P junction.

The junction curvature that is inherent in the planar device structure produces a reduction in the breakdown voltage as a result of the increase in the electric field strength in the curved portion of the junction, as shown in Figure 12.22. The junction curvature can have a very major effect on the breakdown voltage. For example, if we consider a one-sided N$^+$P or P$^+$N junction with a doping level of 1×10^{15} on the more lightly doped side of the junction, the breakdown voltage for a flat junction is about 300 V. If the junction depth is 10 μm, the breakdown voltage will be reduced to about 200 V due to the effect of the junction curvature. For a 3-μm junction depth, the breakdown voltage will be reduced to about 100 V, and if the junction depth is 1 μm, the breakdown voltage will be down to about 40 V.

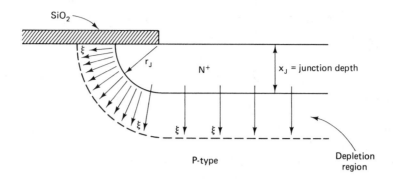

Figure 12.22 Effect of junction curvature on electric field intensity.

To obtain high junction breakdown voltages, it is necessary to have a light doping level and a deep junction. For the conventional junction-isolated monolithic IC, the light doping levels for the collector and substrate regions can cause problems due to the increased parasitic series resistance, especially for high-frequency operation. To have a high breakdown voltage, it is also necessary to increase the thickness of the N-type epitaxial layer so as to accommodate the larger depletion-layer thickness of the collector–base junction. This requirement for an increased epitaxial-layer thickness together with the requirements for greater junction depths will lead to a substantial increase in the chip area needed for the IC.

One approach that is taken for high-voltage ICs is to use a hybrid IC in which the high-voltage parts of the circuit are located on separate chips from the lower-voltage portion of the circuit. Another approach is to use a dielectric isolation IC structure. In the case of the dielectrically isolated ICs, there is an SiO_2 layer between the various IC components and the common substrate so that the collector-to-substrate breakdown voltage is no longer a limiting factor in the operation of the device. The collector–base breakdown voltage will, however, still be a limitation for high-voltage operation.

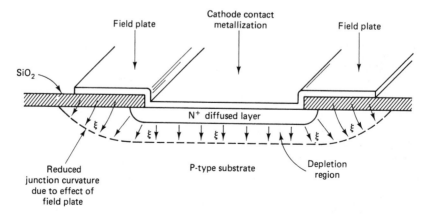

Figure 12.23 Use of a field plate to increase the breakdown voltage of a PN junction.

One method of increasing the breakdown voltage of a planar junction is to use a *field plate*, as shown in Figure 12.23. The electric field produced by the field plate acts to extend the depletion region laterally and acts to reduce the curvature (produce a larger radius of curvature) of the peripheral region of the junction. This can result in a very substantial increase in the breakdown voltage.

12.9.1 Voltage-sharing Circuits

In addition to the device fabrication techniques for increasing the voltage ratings of ICs, there are some circuit design techniques that can also be applied. One interesting method for increasing the output-voltage swing available from a circuit is to use the *voltage-sharing* technique of Figure 12.24. In this example the total supply voltage is 200 V. The resistive voltage divider acts to divide the output voltage V_O in equal parts across the bases of Q_2 through Q_4. Transistor Q_1 operates in the common-emitter configuration, and Q_2 through Q_4 operate as common-base transistors. This circuit can be thought of as a modified or compound cascode circuit. The ac voltage gain is given approximately by $A_V = -g_m[R_L \| 4R_1] \simeq -g_m R_L$ if $4R_1 \gg R_L$ and $g_m = I_C/V_T$.

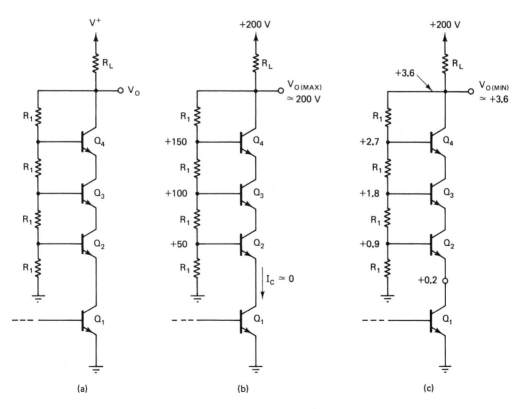

Figure 12.24 High-voltage output circuit using voltage sharing: (a) circuit diagram; (b) voltage levels for $V_{O(MAX)}$; (c) voltage levels for $V_{O(MIN)}$.

The positive peak or maximum output voltage is given by $V_{O(MAX)} = V^+ \times 4R_1/(4R_1 + R_L) \simeq V^+ = 200$ V if $4R_1 >> R_L$, as shown in Figure 12.24b. We see that $V_{CB_1} \simeq V_{CB_2} \simeq V_{CB_3} \simeq 50 - 0.7$ V $= 49.3$ V, and $V_{CB_4} \simeq 50$ V, so the maximum collector-to-base voltage of any transistor in the circuit will not exceed 50 V.

At the other extreme, the minimum output-voltage level is limited by the collector-to-emitter saturation voltage and the base-to-emitter forward bias voltages, as shown in Figure 12.24c. The voltage at the base of Q_2 is limited to a minimum value of $V_{B_2} = V_{BE_2} + V_{CE(SAT)_1} \simeq 0.7 + 0.2 = 0.9$ V. Since the voltage drops across all the resistors in the voltage divider are approximately equal, this will result in a minimum output-voltage level of $V_{O(MIN)} \simeq 3.6$ V. As a result, we see that with a 200-V supply a peak-to-peak output-voltage swing of about 195 V is possible. At the same time, the maximum voltage drop that must be sustained by any transistor will not exceed 50 V.

Figure 12.25 shows a voltage-sharing circuit with an emitter-follower output.

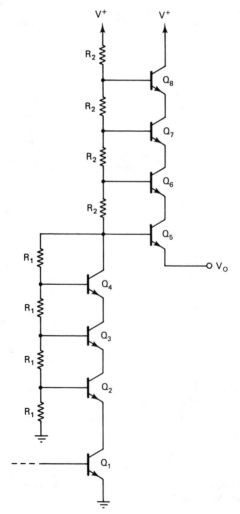

Figure 12.25 High-voltage output stage with an emitter follower.

The lower part of the circuit, consisting of Q_1 through Q_4 and the associated R_1 voltage divider, operates in the same fashion as the compound cascode circuit shown in Figure 12.24. The emitter-follower part of the circuit uses Q_5 through Q_8 with the associated R_2 voltage divider. With a supply voltage of $V^+ = 200$ V and with Q_1 off, the output voltage is approximately given by $V_{O(MAX)} \simeq V^+ \times 4R_1/(4R_1 + 4R_2) = V^+ \times R_1/(R_1 + R_2)$. If $R_1 >> R_2$, then $V_{O(MAX)} \simeq V^+ = 200$ V. At the other extreme, when Q_1 is in saturation, the voltage at the collector of Q_4 can drop to as low as about $+3.6$ V, so $V_{O(MIN)}$ is approximately $3.6 - 0.7 = 2.9$ V.

12.10 IC POWER TRANSISTORS WITH INTERNAL PROTECTION CIRCUITRY

Although power transistors can be implemented as simple discrete devices, better performance and protection from excessive power dissipation can be obtained by a monolithic IC power transistor. An example of such a device is the LM195 (National Semiconductor), and a simplified schematic diagram is given in Figure 12.26. This device can be thought of as a triple-transistor Darlington circuit, and it has an overall current gain of about 1×10^6, requiring a base drive of only 3 μA to produce the maximum output collector current of 1.8 A.

This circuit incorporates a thermal limiting circuit that senses the chip temperature and limits the device power dissipation when the chip temperature reaches 165°C. Current limiting is also provided. This current limiting limits the maximum collector current to 1.8 A for V_{CE} values of less than 13 V. As V_{CE} increases above this level, the current limit decreases and so provides safe operating area (SOA) protection for the device. The current-limiting characteristic is shown in Figure 12.27. The maximum power dissipation rating for this device is 35 W. Note that the current-limiting

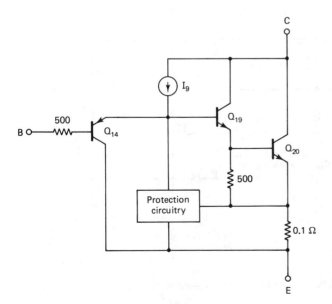

Figure 12.26 Simplified schematic diagram of a monolithic IC power transistor: LM195 (National Semiconductor).

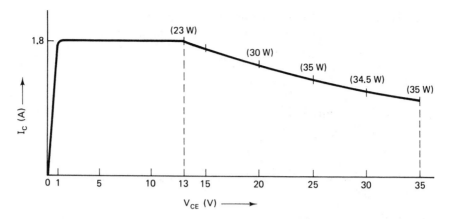

Figure 12.27 LM195 current-limiting characteristic.

characteristic keeps the power dissipation from exceeding 35 W. The maximum current is also limited to 1.8 A to protect against the fusing of the aluminum metallization and to also provide protection against the second breakdown effect, to be described later.

The use of a PNP transistor Q_{14} as the input stage offers two advantages. First, the net base-to-emitter voltage of this device is given by $V_{BE} = V_{BE20} + V_{BE19} + V_{BE14} = +0.7 + 0.7 - 0.7 = +0.7$ V, instead of about $+2.1$ V, which would be the case if the first transistor were an NPN transistor. The second advantage is the high emitter-to-base breakdown voltage BV_{EB} of greater than 40 V, which is inherent in the PNP transistor structure.

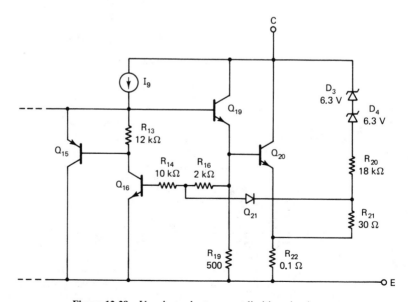

Figure 12.28 V_{CE}-dependent current-limiting circuit.

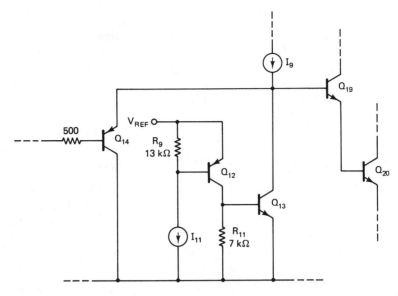

Figure 12.29 Thermal limiting circuit.

Figure 12.28 shows a detail of the V_{CE}-dependent current-limiting circuit. When the voltage drop across the current-limit resistor R_{22} (0.1 Ω) becomes large enough, transistor Q_{16} turns on, which in turn will turn Q_{15} on, which will divert base drive away from the Q_{19}–Q_{20} Darlington pair. Note that diode Q_{21}, in conjunction with resistors R_{14} and R_{16}, provide a prebias for Q_{16}, so the voltage drop across R_{22} that is needed to turn Q_{16} on is only 180 mV, corresponding to a current limit of 1.8 A.

The current-limiting threshold is lowered when V_{CE} goes above 12.6 V, for then diodes D_3 and D_4 go into breakdown. The voltage divider, comprised of R_{20} (18 kΩ) and R_{21} (30 Ω), then adds a small fraction of V_{CE} to the voltage drop across R_{22} to reduce the current through R_{22} that is required to produce current limiting.

A simplified diagram of the thermal limiting or shutdown circuit is given in Figure 12.29. The base-to-emitter voltage supplied to Q_{12} is the $I_{11}R_9$ voltage drop that is developed across R_9. As the temperature increases, the current through Q_{12} increases rapidly. This in turn activates Q_{13}. The current flow through both Q_{12} and Q_{13} increases very rapidly with increasing temperature due to the negative base-to-emitter voltage temperature coefficients ($TC_{V_{BE}}$) of the two transistors.

12.10.1 Second Breakdown

At high current levels in a transistor, a localized current concentration can lead to the following sequence of events: localized current density increase → increased temperature in the localized region (formation of a hot spot) → increased current density as a result of the temperature increase → and so on. This is an electrothermal positive feedback loop, which results in the formation of current filaments of very small diameter (~10 μm) and extremely high current density (a *microplasma*). The temperature rise in the filamentary region can reach very high values, causing the alloying or fusing of the overlaying metallization and even the localized melting of the silicon. This will result in the irreversible destruction of the device and can occur at power dissipation levels that are well below the maximum power dissipation rating of the device, especially in the region where the collector current is relatively large. The region of the I_C versus V_{CE} characteristics of the device within which it is relatively safe from damage from the second breakdown phenomenon just described is called the *safe operating area* (SOA).

12.11 TRANSISTOR GEOMETRY AND CURRENT CROWDING

The transistor top surface geometry is an important consideration for power transistors. At high collector current levels, the lateral voltage drop produced by the flow of base current through the internal base resistance $r_{bb'}$ results in the base-to-emitter voltage drop being a function of position. The V_{BE} is higher in the peripheral regions of the junction than in the central region, as shown in Figure 12.30b, c, and d. This results in a shift in the current distribution of the emitter-to-collector current flow from the uniform distribution of Figure 12.30a to the *current-crowding* situation of Figure 12.30d, in which most of the current flow is concentrated near the edges of the junction. An excessive amount of current crowding is bad because it leads to a decrease in the transistor current gain and, more important, increases the power dissipation density in the current-crowding regions. This localized increase in the power density results in hot spots, which can result in the second breakdown effect just described.

To minimize the current-crowding effect, an interdigitated transistor geometry is often used for power transistors, and this geometry is used for Q_{20} in the LM195. The lateral geometry of Q_{20} is shown in Figure 12.31. The total emitter area is seen to be subdivided into many separate emitter "fingers" in order to increase the total emitter periphery. The constricted N^+ regions between each emitter region and the emitter contact strip form small-value emitter ballast resistors to help equalize the emitter current, as illustrated in Figure 12.30e.

To minimize the lateral base voltage drop, a multiplicity of base contact fingers is used and is interdigitated with the emitter fingers. In a monolithic IC, the collector contact must be taken out of the top surface. To minimize the collector series resistance, the base region has been subdivided into 10 regions with an N^+ contact strip between the base regions. Each base region has 6 N^+ emitter areas, so that the transistor has a total of 60 separate emitter regions.

12.12 POWER MOSFETS

A device that is increasingly competing with bipolar power transistors is the MOSFET power transistor. The MOSFET power transistors are either of the vertical double-diffused type (V-DMOS) or the V-groove double-diffused (VMOS) construction. Power MOSFETs that can handle currents as high as 25 A at voltages up to 500 V are available. Drain-to-source "on" resistances as low as 0.018 Ω have been obtained. While not strictly speaking an IC, these devices generally consist of a very large

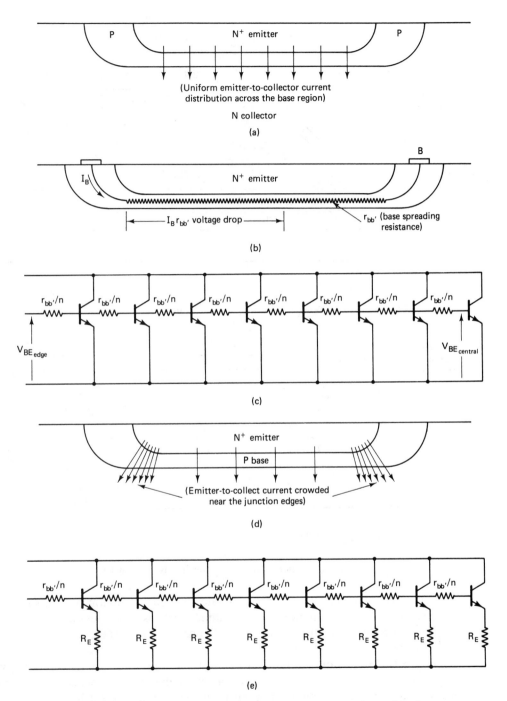

Figure 12.30 Current crowding in transistors: (a) transistor at low and moderate collector current levels; (b) lateral voltage drop in the base region; (c) equivalent representation of transistor; (d) transistor current distribution at high collector current levels; (e) use of emitter ballast resistors to reduce current crowding.

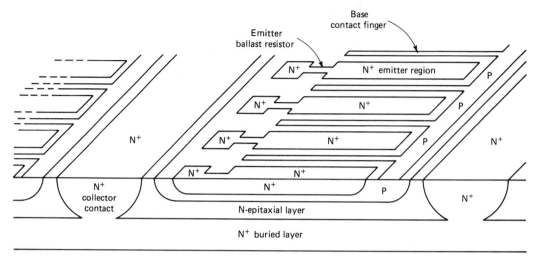

Detail of emitter-base contact interdigitation

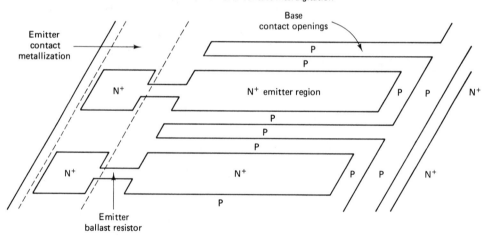

Figure 12.31 Interdigitated IC power transistor geometry.

number of parallel-connected source and gate cells on a common N-type substrate, which serves as the drain region. The channel length is generally on the order of 1 μm, but as a result of the thousands of parallel-connected cells the net channel width can be up in the range of 1 m.

Power MOSFETs offer a number of substantial advantages over bipolar power transistors. First, power MOSFETs have a very high input impedance, so the input-current requirements are minimal, even when the output-current level is up in the range of tens of amperes. MOSFETs are a majority carrier device, so there is no minority carrier storage time. As a result, very fast switching of large currents is possible. Also related to the fact that the MOSFET is a majority carrier device is the negative temperature coefficient of the drain current of typically about $-0.5\%/°C$. This negative temperature coefficient is the result of the decrease of electron mobility in the conducting channel with increasing temperature. The decrease of the drain current with increasing temperature results in thermal stability for the device, and there will be no formation of hot spots and the resulting second breakdown phenomenon, as is the case for bipolar transistors. This negative temperature coefficient also makes it easy to operate MOSFET power transistors in parallel to increase the net current-handling capability. Any transistor mismatch that tends to make one transistor carry more of the current will result in an increase in the temperature of that transistor. This temperature increase will result in a decrease of the drain current of that transistor, which acts to make the currents through the various transistors more equal. Finally, power MOSFETs have an approximately linear transfer characteristic. This can be an important feature since power MOSFETs are generally the output stage of an amplifier and operated under large-signal conditions. As a result, the linear transfer characteristic results in lower distortion.

While the power MOSFET is basically a discrete device, it can be combined with bipolar transistors and with CMOS devices on a single IC substrate. Such ICs have been made with these devices on a common P-type substrate and using the conventional junction isolation process with a P+ isolation diffusion. Power MOS-FETs can be of the N-channel vertical DMOS type with an N-type drain region and an N+ buried layer, together with a deep N+ drain diffusion to minimize the drain series resistance. Lateral DMOS transistors can also be used, in which case the N+ buried layer is not needed. The IC power MOSFET devices offer the convenience of the high-output-current capability of the power MOSFET together with all the necessary driving and signal-processing circuitry on one chip.

A variety of power MOSFETs is now being made by a number of manufacturers, including General Electric, International Rectifier, Motorola, RCA, Siemens, and Siliconix. As an example of power MOSFET capabilities, the Motorola MTE100N06 has a continous current rating of 100 A and a drain-to-source "on" resistance of only $0.018\ \Omega$ at a drain current of 50 A. The drain-to-source breakdown voltage rating is 60 V minimum. Another Motorola device, the MTP1N100, has a drain-to-source breakdown voltage of 1000 V minimum, combined with a continuous current rating of 1.0 A and a drain-to-source "on" resistance of $10\ \Omega$ maximum at 0.5 A.

Some other examples of the very impressive high voltage and current capabilities of DMOS power MOSFETs are from the IXYS Corporation. The IXFN100N10 is an N-channel device with a voltage rating of 100 V and a current rating of 100 A continuous, and it can handle a power dissipation of up to 300 W. The drain-to-source "on" resistance is only $0.013\ \Omega$. The IXTM12N100 has a voltage rating of 1000 V, a current rating of 8 A, and a power dissipation rating of 300 W. The drain-to-source "on" resistance of this MOSFET is $1.0\ \Omega$. From Advanced Power Technology, there is the APT6040AN with a rating of 600 V, 15.5 A, and a drain-to-source "on" resistance of just $0.40\ \Omega$. The APT4020 has ratings of 400 V, 22.5 A, and a drain-to-source "on" resistance of only $0.20\ \Omega$. Siliconix makes the SMM70N10, which is a 100-V device with a 70-A continuous current rating and a drain-to-source "on" resistance of a mere $0.025\ \Omega$. The use of these and other power MOSFETs, together with various ICs for signal processing and applications, is a valuable combination for many applications ranging from audio systems to the control of large electrical motors and other electrical machinery.

Power MOSFETs that combine a very high voltage rating with a very high current rating and a very low "on" resistance are available and are basically integrated combinations of an N-channel power vertical double-diffused MOSFET and a PNP transistor. The P-type body region of the MOSFET is also the collector of the PNP transistor, and the lightly doped N-type drain region is the base of the transistor. The P+ substrate is the emitter of the PNP transistor. The PNP transistor is off when the MOSFET is off, but when the MOSFET is turned on, the drain current of the MOSFET results in a base current for the PNP transistor. The PNP transistor then turns on with a heavy emission of holes from the P+ substrate region into the N-type base–drain region. The high concentration of holes injected into the N-type base–drain region results in the drawing in of an approximately equal number of electrons to maintain overall charge neutrality. The net increase in the electron and hole population in this region can result in a very substantial increase in the conductivity of this region. This effect is known as *conductivity modulation*. As a result of this conductivity modulation effect, a very lightly doped drain region can be used to obtain a very high breakdown voltage rating for the device, and at the same time a very low value for the drain-to-source "on" resistance and a correspondingly high current rating can be obtained. With this type of conductivity modulated MOSFET voltage ratings as high as 500 V together with current ratings of 25 A are available.

PROBLEMS

12.1. (*Class B audio power amplifier*) Given: A class B audio power amplifier has a positive supply voltage of $+V_S = +15$ V and a negative supply voltage of $-V_S = -15$ V. This amplifier has a closed-loop gain of 50 and is to deliver 10 W of power into an 8.0-Ω load.
  (a) Find the peak output-voltage swing, $V_{O(peak)}$. (*Ans.*: ± 12.65 V)
  (b) Find the peak output-current swing, $I_{O(peak)}$. (*Ans.*: ± 1.58 A)
  (c) Find the input signal required (rms). [*Ans.*: 179 mV (rms)]
  (d) Find the total power from power supply. (*Ans.*: 15.1 W)
  (e) Find the power dissipated in the amplifier. (*Ans.*: 5.1 W)
  (f) Find the power conversion efficiency. (*Ans.*: 66.2%)
  (g) Find the power output level at which the amplifier power dissipation will be a maximum and the maximum amplifier power dissipation. (*Ans.*: 5.70 W, 5.70 W)
  (h) If the ambient temperature T_a is 25°C and the maximum allowable device temperature $T_{J(MAX)}$ is 125°C, find the maximum allowable value for the case-to-ambient thermal resistance $R_{TH(CA)}$ given that the junction-to-case thermal resistance $R_{TH(JC)}$ is 10°C/W. (*Ans.*: 7.54°C/W)

Power Amplifiers: Thermal Shutdown Circuits

12.2 Refer to Figure P12.2. Given $V_{BE} = 0.70$ V, $V_Z = 6.0$ V, $T_O = 25°C$, $TCV_{BE} = -2.1$ mV/°C, and $TC_{V_Z} = +3.0$ mV/°C, assume that the base-to-emitter voltage drop of all transistors at 25°C will be 700 mV when biased into full conduction and that thermal shutdown will occur when Q_8 is biased into full conduction.
  (a) Show that for thermal shutdown to occur at temperature T_{SD} the required resistance

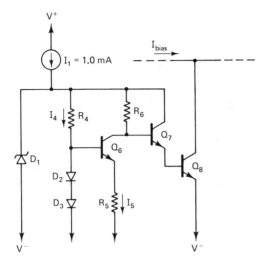

Figure P12.2

ratio is given by $(R_6/R_5) + 2 = V_Z(T_{SD})/V_{BE}(T_{SD})$, where $V_Z(T_{SD})$ and $V_{BE}(T_{SD})$ are the values of V_Z and V_{BE} at the shutdown temperature T_{SD}, respectively.

(b) Find the resistance ratio for thermal shutdown at $T_{SD} = 175°C$. (*Ans.*: 14.75)

(c) Find R_4, R_5, and R_6 for $I_4 = I_5 = 0.5$ mA (at 25°C). (*Ans.*: 9.2, 1.4, 20.7 kΩ)

12.3. Refer to Figure P12.3. Given $V_{REF} = 1.220$ V, $V_{BE} = 700$ mV at 1.0 mA and 25°C, and $\beta = 100$, $TC_{V_{BE}} = -2.0$ mV/°C for all transistors over the range from 25 to 175°C.

    (a) Find the temperature at which complete shutdown occurs (I_{bias} is reduced to zero). (*Ans.*: 123.7°C)

    (b) Find the temperature at which I_{bias} has been reduced by one-half to 50 μA. (*Ans.*: 115°C)

    (c) Find the temperature at the threshold of shutdown for which I_{bias} has undergone a 10% reduction to 90 μA. (*Ans.*: 95°C)

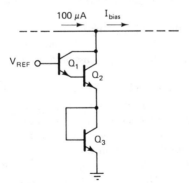

Figure P12.3

12.4. (*Operational amplifier with external power MOSFETs for current boost*) An operational amplifier uses external power MOSFETs for current boosting as shown in Figure P12.4. The threshold voltages are +3 V for the NMOS transistor (Q_1) and −4 V for the PMOS transistor (Q_2). The supply voltages are $V^+ = +16$ V and $V^- = -16$ V, and the quiescent supply current of the operational amplifier is 1 mA. The operational amplifier

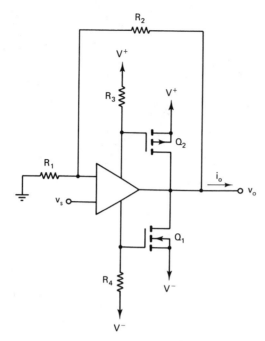

Figure P12.4

has a current limit of ± 20 mA. The peak output-voltage swing is to be ± 12 V across a 4-Ω load. Resistor R_1 is 1000 Ω.

(a) Find R_3 and R_4 for transistor Q_2 to turn on when the output current reaches $+10$ mA and for Q_1 to turn on when the output current reaches -10 mA. *(Ans.: R_3 = 364 Ω, R_4 = 273 Ω)*

(b) Find the peak output currents through Q_1 and Q_2. *(Ans.: I_{PEAK} = 3 A)*

(c) Find the average power dissipation in transistors Q_1 and Q_2. *(Ans.: 6.28 W for each transistor)*

(d) If $T_{J(\text{MAX})} = 150°C$, the ambient temperature is $T_A = 25°C$, and the junction-to-case thermal resistance is $R_{TH(JC)} = 1°C/W$, find the maximum allowable value for the heat sink thermal resistance, $R_{TH(CA)}$, assuming the conditions of part (c). *(Ans.: $R_{TH(CA)} = 18.9°C/W$)*

(e) Find R_2 for a voltage gain of 12. *(Ans.: R_2 = 11 kΩ)*

12.5. *(Power operational amplifier)* Given a PA04 power operational amplifier operating with a supply voltage of ± 200 V and a boost voltage of ± 220 V. The operational amplifier is driven to produce an output voltage of 200 V sin ωt across a 200-Ω load resistance.

(a) Find the average power dissipation in the operational amplifier. *(Ans.: 40 W)*

(b) If $T_{J(\text{MAX})} = 150°C$, $T_A = 25°C$, and $R_{TH(JC)} = 0.5°C/W$, find the maximum allowable heat sink thermal resistance, $R_{TH(CA)}$. *(Ans.: $R_{TH(CA)} = 2.625°C/W$)*

(c) Find the output power delivered to the load and the circuit power conversion efficiency, η, and compare the result to the maximum possible efficiency for a class B amplifier. *(Ans.: P_{LOAD} = 100 W, η = 71.4% as compared to η_{MAX} = 78.5% for class B)*

(d) Find the current limit resistance R_{CL} for a current limit of 1.2 A. *(Ans.: R_{CL} = 0.58 Ω)*

(e) If two PA04 operational amplifiers are connected in a bridge configuration, find the power delivered to the load, assuming that each operational amplifier delivers a peak output voltage of 200 V and a peak output current of 1 A.   (*Ans.*: 200 W)

REFERENCES

ANTOGNETTI, P., *Power Integrated Circuits: Physics, Design, and Applications*, McGraw-Hill, New York, 1986.

COWLES, L. G., *Sourcebook of Modern Transistor Circuits*, Chapter 8, Prentice-Hall, Englewood Cliffs, N.J., 1976.

EVANS, A. D., *Designing with Field-Effect Transistors*, McGraw-Hill, New York, 1981.

GHANDI, S. K., *Semiconductor Power Devices*, Wiley, New York, 1977.

GREBENE, A. B., *Analog Integrated Circuit Design*, Van Nostrand Reinhold, New York, 1972.

_____, *Bipolar and MOS Analog Integrated Circuit Design*, Wiley, New York, 1984.

HAFT, R. G., *Semiconductor Power Electronics*, Van Nostrand Reinhold, New York, 1986.

KRAUSS, H. L., C. W. BOSTIAN, and F. H. RAAB, *Solid State Radio Engineering*, Wiley, New York, 1980.

LENK, J. D., *Handbook of Integrated Circuits*, Prentice-Hall, Englewood Cliffs, N.J., 1978.

OXNER, E. S., *Power FETs and Their Applications*, Chapter 7, Prentice-Hall, Englewood Cliffs, N.J., 1982.

_____, *FET Technology and Applications*, Marcel Dekker, New York, 1989.

RASHID, M. H., *Power Electronics*, Prentice-Hall, Englewood Cliffs, N.J., 1988.

RIPS, E. M., *Discrete and Integrated Electronics*, Prentice-Hall, Englewood Cliffs, N.J., 1986.

RISTENBATT, M. P., *Semiconductor Circuits*, Chapter 7, Prentice-Hall, Englewood Cliffs, N.J., 1975.

13

Wide-bandwidth (Video) Amplifiers

For operational amplifiers the emphasis is on a high open-loop voltage gain, typically on the order of 100,000 (100 dB) to 1,000,000 (120 dB) at low frequencies. Operational amplifiers are almost always operated in a closed-loop configuration, and the high open-loop gain values are needed to obtain a high loop gain. The high loop gain is desirable to minimize the gain error, to produce a large input impedance, and to obtain a very small output impedance.

To obtain the high voltage gain, sacrifices are made in the frequency response. Indeed, to stabilize the amplifier against an oscillatory response, the open-loop frequency response is usually made to start rolling off at a very low frequency, often at only about 10 Hz. The resulting unity-gain frequencies are generally in the range from 1 to 3 MHz, with some operational amplifiers extending up to about 10 MHz.

Video, or wideband, amplifiers are designed to give relatively flat gain versus frequency-response characteristics over the frequency range that is generally required to transmit video information. This frequency range is from low frequencies, generally down in the range of about 20 Hz, up to several megahertz. For television applications, bandwidths generally in the range from 4 to 6 MHz are required. For some other applications, bandwidths as high as 50 MHz may be needed.

In contrast with this, the bandwidths required for audio amplifiers extend only over the frequency range corresponding to that which the human ear is sensitive to, from about 50 Hz to 15 kHz. For many applications, such as telephony or AM radio,

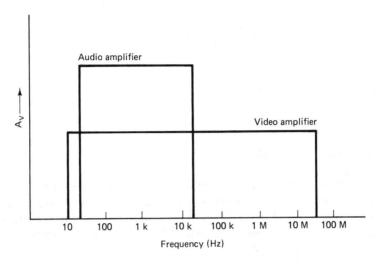

Figure 13.1 Comparison of audio and video amplifier bandwidths.

a considerably narrower bandwidth is used, typically from 100 Hz to 5 kHz. In Figure 13.1, the frequency-response characteristics of video and audio amplifiers are compared.

The principal technique that is used to obtain the large bandwidths that are required for video amplifiers is the trading off of gain for increased bandwidth. This trade-off is accomplished by use of reduced load resistance for the various gain stages of the amplifier and by the use of negative feedback. In many, if not most, video amplifiers, the techniques of reduced load resistance and negative feedback are both employed at the same time.

13.1 TRANSISTOR FREQUENCY RESPONSE

Before we look at some representative examples of IC video amplifiers, we will first engage in a brief review of the frequency-response characteristics of a single common-emitter transmitter stage. In a later section of this chapter, high-frequency FET amplifiers will be discussed. The frequency response of a transistor is controlled by the interelectrode or junction capacitances, in conjunction with the various resistances in the circuit. The two transistor capacitances that are involved here are $C_{b'e}$, the emitter–base junction capacitance, and $C_{cb'}$, the collector–base capacitance. Also of interest and importance is the base spreading resistance, $r_{bb'}$, which is due to the bulk series resistance of the base region. In the discussion to follow, we will analyze the frequency response of the transistor common-emitter stage to see what factors are of importance in the design of a video amplifier.

Figure 13.2 shows a diagram of a basic common-emitter amplifier stage, with the dc biasing details being omitted for purposes of clarity. The transistor junction capacitances and the base spreading resistance, which are actually internal to the transistor, are shown as external elements for purposes of analysis. Also shown in

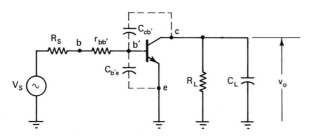

Figure 13.2 Common-emitter amplifier stage.

this circuit are the source resistance R_S, the load resistance R_L, and the load capacitance C_L.

Figure 13.3 shows a small-signal ac equivalent circuit for the common-emitter stage of Figure 13.2. The signal source v_s and the series resistance of $R_S + r_{bb'}$ has been transformed into the equivalent current source–parallel resistance combination for greater ease in analysis. The parallel resistance is represented as a conductance G_1, with $G_1 = 1/(R_S + r_{bb'})$.

For the analysis to follow, the node voltage equations will be written for the two nodes of interest, b' and c, by taking the algebraic sum of the currents going into and out of these two nodes.

If we now equate the current going out of node b' to that coming into this node, we obtain

$$v_{b'}[G_1 + g_{b'e} + j\omega(C_{b'e} + C_{cb'})] = i_s + j\omega C_{cb'}v_c \tag{13.1}$$

and for node c we similarly obtain

$$v_c[G_L + j\omega(C_{cb'} + C_L)] + g_m v_{b'} = j\omega C_{cb'}v_b' \tag{13.2}$$

Solving the node c equation for v_c gives

$$v_c = \frac{-(g_m - j\omega C_{cb'})v_{b'}}{G_L + j\omega(C_{cb'} + C_L)} \tag{13.3}$$

Substitution of this expression back into the b' node equation and rearranging yield

$$v_{b'}\left[G_1 + g_{b'e} + j\omega(C_{b'e} + C_{cb'}) + \frac{(j\omega C_{cb'})(g_m - j\omega C_{cb'})}{G_L + j\omega(C_{cb'} + C_L)}\right] = i_s \tag{13.4}$$

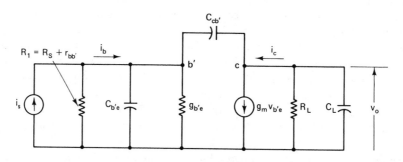

Figure 13.3 AC small-signal equivalent circuit.

In these equations, $g_{b'e}$ is the dynamic input conductance of the transistor looking into the base as given by

$$g_{b'e} = \frac{1}{r_{b'e}} = \frac{dI_B}{dV_{BE}} = \frac{I_B}{nV_T}$$

The constant n is a dimensionless factor between 1.0 and 2.0, typically about 1.5.

If we now solve the b' node equation for $v_{b'}$, we obtain

$$v_{b'} = \frac{i_s}{G_1 + g_{b'e} + j\omega(C_{b'e} + C_{cb'}) + \dfrac{(j\omega C_{cb'})(g_m - j\omega C_{cb'})}{G_L + j\omega(C_{cb'} + C_L)}} \tag{13.5}$$

Going back to the c node equation with this result allows us now to solve for v_c as

$$v_c = \frac{-(g_m - j\omega C_{cb'})}{G_L + j\omega(C_{cb'} + C_L)} \tag{13.6}$$

$$\cdot \frac{i_s}{G_1 + g_{b'e} + j\omega(C_{b'e} + C_{cb'}) + \dfrac{(j\omega C_{cb'})(g_m - j\omega C_{cb'})}{G_L + j\omega(C_{cb'} + C_L)}}$$

Since $i_s = v_s/(R_s + r_{bb'}) = v_s G_1$, we can now obtain the voltage gain as

$$A_V = \frac{v_c}{v_s}$$

$$= \frac{-(g_m - j\omega C_{cb'})G_1}{[G_L + j\omega(C_{cb'} + C_L)]\left[G_1 + g_{b'e} + j\omega(C_{b'e} + C_{cb'} + \dfrac{(j\omega C_{cb'})(g_m - j\omega C_{cb'})}{G_L + j\omega(C_{cb'} + C_L)}\right]} \tag{13.7}$$

This is a rather complicated expression for the frequency dependence of the voltage gain, and we will now seek to simplify it by using the following suitable approximations:

1. In the frequency range of interest, it will virtually always be true that $g_m \gg \omega C_{cb'}$, so we can replace $g_m - j\omega C_{cb'}$ by just g_m alone.
2. The input conductance $g_{b'e}$ will usually be sufficiently small compared to $G_1 = 1/(R_S + r_{bb'})$ such that we can replace the sum of $G_1 + g_{b'e}$ by just G_1 alone.
3. In the frequency range of interest, we will usually have that $G_L \gg \omega(C_{cb'} + C_L)$ and $g_m \gg \omega C_{cb'}$ such that the expression

$$\frac{g_m - j\omega C_{cb'}}{G_L + j\omega(C_{cb'} + C_L)} \tag{13.8}$$

can be replaced by simply $g_m/G_L = g_m R_L$.

Using these approximations, we can now write the expression for A_V as

$$A_V \simeq \frac{-g_m G_1}{[G_L + j\omega(C_{cb'} + C_L)][G_1 + j\omega(C_{b'e} + C_{cb'} + g_m R_L C_{cb'})]} \tag{13.9}$$

We will now define two breakpoint radian frequencies as

$$\omega_1 = \frac{G_1}{C_{b'e} + C_{cb'}(1 + g_m R_L)} = \frac{1}{(R_S + r_{bb'})[C_{b'e} + C_{cb'}(1 + g_m R_L)]}$$

$$= \frac{1}{(R_S + r_{bb'})C_i} \tag{13.10}$$

$$\omega_2 = \frac{G_L}{C_{cb'} + C_L} = \frac{1}{R_L(C_{cb'} + C_L)}$$

Note that in the net input capacitance C_i the collector–base capacitance C_{cb} is multiplied by $1 + g_m R_L = 1 - A_v$. This is known as the Miller effect.

The foregoing equation for the voltage gain can now be rewritten in terms of these two breakpoint frequencies as

$$A_V \simeq \frac{-g_m/G_L}{[1 + j(\omega/\omega_1)][1 + j(\omega/\omega_2)]} = \frac{-g_m R_L}{[1 + j(f/f_1)][1 + j(f/f_2)]} \tag{13.11}$$

where $f_1 = \omega_1/2\pi$ and $f_2 = \omega_2/2\pi$.

We see that the frequency response will be characterized by two breakpoint frequencies, f_1 and f_2, of which f_1 will generally be the smaller and thereby be the principal factor in determining the bandwidth of the amplifier stage. For an IC transistor, there will be an additional capacitance C_{CS} between collector and substrate (ac ground). This can be taken into account by replacing C_L by $C_L + C_{CS}$.

13.2 TRANSISTOR UNITY-GAIN FREQUENCY

The unity-gain frequency of a transistor is defined as the frequency at which the common-emitter short-circuit current gain has dropped to unity and is denoted by the symbol f_T. The current gain is defined as $A_i = i_o/i_{in} = i_c/i_b$, and the term "short-circuit" in the definition above indicates that there is an ac short circuit from collector to ground.

If we now consider the circuit of Figure 13.3 for the case of $R_L = 0$, we see that the ac output or collector current is given simply by $i_o = i_c = g_m v_{b'}$. The voltage from point b' to ground is related to the input or base current by

$$v_{b'} = \frac{i_b}{g_{b'e} + j\omega(C_{b'e} + C_{cb'})} \tag{13.12}$$

As a result, the short-circuit current gain can be written as

$$A_{i(sc)} = \frac{i_o}{i_b} = \frac{g_m}{g_{b'e} + j\omega(C_{b'e} + C_{cb'})} = \frac{g_m/g_{b'e}}{1 + j\omega[(C_{b'e} + C_{cb'})/g_{b'e}]} \tag{13.13a}$$

Since $g_m = dI_C/dV_{BE}$ and $g_{b'e} = dI_B/dV_{BE}$, the ratio of g_m to $g_{b'e}$ is given by $g_m/g_{b'e}$ $= dI_C/dI_B = \beta(\text{ac}) = h_{fe}$ in terms of the hybrid parameters. We can therefore rewrite the current-gain expression as

$$A_{i(\text{sc})} = \frac{\beta}{1 + j\omega[(C_{b'e} + C_{cb'})/(g_m/\beta)]} \tag{13.13b}$$

Using the definition of the unity-gain frequency, we have that, at $\omega = \omega_T$, $A_{i(\text{sc})} = 1$. Since $\beta \gg 1$, for this condition we have

$$\frac{\beta}{[\omega_T(C_{b'e} + C_{cb'})]/(g_m/\beta)} = 1$$

and thus

$$\boxed{\omega_T = \frac{g_m}{C_{b'e} + C_{cb'}} = \frac{I_C/V_T}{C_{b'e} + C_{cb'}}} \tag{13.14}$$

Note that the unity-gain frequency is independent of β and, as a result of the β independence of ω_T, the unit-to-unit variation in ω_T among transistors of the same type designation will be relatively small, generally within a $\pm 10\%$ range. As a result, we can understand why ω_T is a very often used parameter for the characterization of transistors.

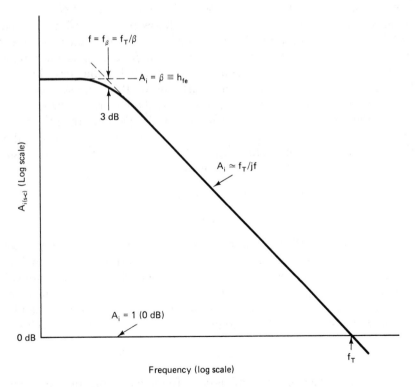

Figure 13.4 Common-emitter short-circuit current gain versus frequency.

In terms of ω_T, the short-circuit current gain of the transistor can be written as

$$A_{i(\text{sc})} = \frac{\beta}{1 + j(\beta\omega/\omega_T)} = \frac{\beta}{1 + j(\beta f/f_T)} = \frac{\beta}{1 + j(f/f_\beta)} \qquad (13.15)$$

where $f_\beta = f_T/\beta$ is the half-power (or -3 dB) frequency of the short-circuit current gain. For frequencies that are more than one-half a decade above f_β ($f \gtrsim 3f_\beta$), we can express the current gain approximately as

$$A_{i(\text{sc})} \simeq \frac{\beta}{j(f/f_\beta)} = \frac{\beta f_\beta}{jf} = \frac{f_T}{jf} \qquad (13.16)$$

as in Figure 13.4. This relationship is often used for the experimental determination of f_T without the need for going all of the way up in measurement frequency to f_T. For example, if $A_{i(\text{sc})} = 100$ at 10 kHz, and at 50 MHz the current gain is $A_{i(\text{s-dc})} = 10$, we can conclude that 50 MHz is much geater than f_β since the current gain at 50 MHz is much less than that at 10 kHz. We can therefore say that $f_T/50$ MHz $= 10$, so $f_T = 500$ MHz. Furthermore, from the current-gain value at 10 kHz, we can say that $\beta = 100$, so f_β is given by $f_\beta = f_T/\beta = 500$ MHz/100 $= 5$ MHz.

13.3 MULTISTAGE AMPLIFIER BANDWIDTH AND RISE TIME

A multistage amplifier can usually be modeled as a series of cascaded, but noninteracting simple RC low-pass networks. Each low-pass network is characterized by an RC time constant and a breakpoint frequency (or *pole*) that is related to the corresponding time constant by $f_n = 1/(2 \pi \tau_n)$. If f_1 represents the lowest breakpoint frequency and it is at least about one decade below the next highest breakpoint frequency f_2 ($f_1 < f_2/10$), then the 3-dB bandwidth of the system is approximately equal to (but always less than) the lowest breakpoint frequency f_1.

If the breakpoint frequencies are not well separated, then the situation is much more complicated and there is no exact simple relationship between the 3-dB bandwidth and the breakpoint frequencies. There is, however, a relatively simple approximate relationship that can be used as given by

$$\frac{1}{BW^2} \simeq \frac{1}{f_1^2} + \frac{1}{f_2^2} + \frac{1}{f_3^2} + \cdots \qquad (13.17)$$

where BW is the 3-dB bandwidth of the system and $f_1, f_2, \ldots$, represent the various breakpoint frequencies in the system. Note that if the first breakpoint frequency f_1 is much smaller than the other breakpoint frequencies the bandwidth will be approximately equal to f_1. In the special case of a system with n coincident breakpoint frequencies, there is an exact expression that can be used for the bandwidth. For a system with n coincident breakpoint frequencies, the 3-dB bandwidth is given by

$$BW = f_1 \sqrt{2^{1/n} - 1} \qquad (13.18)$$

For the case of two coincident breakpoint frequencies at f_1, this gives $BW = 0.6436f_1$, for $n = 3$ the result is $BW = 0.5098f_1$, and for $n = 4$ the bandwidth is down to BW

$= 0.4350f_1$, so we see that there will be a considerable bandwidth shrinkage as the number of stages increases.

The 10% to 90% rise time, t_{rise}, of a system that is characterized by a single time constant τ is given by $t_{rise} = 2.2\tau$, and in terms of the 3-dB bandwidth, BW, is given by $t_{rise} = 0.35/BW$. For a system that is characterized by several time constants (a multiple-pole system), the foregoing relationship between rise time and bandwidth will not be exactly true, though in almost all cases it can serve as a very good approximation. Another good approximate relationship that can be used for a multiple-pole system is that the rise time will be approximately 2.2 times the largest time constant if the largest time constant is at least twice the next largest time constant.

13.4 DESIGN CONSIDERATIONS FOR VIDEO AMPLIFIERS

For a video or wideband amplifier, it is necessary that the two characteristic or breakpoint frequencies, f_1 and f_2, be sufficiently large, since in no case can the bandwidth be any greater than either of these two frequencies. Since

$$\omega_1 = \frac{1}{(R_S + r_{bb'})(C_{b'e} + C_{cb'} + C_{cb'}g_m R_L)} \quad \text{and} \quad \omega_2 = \frac{1}{R_L(C_{cb'} + C_L)} \quad (13.19)$$

we see that to achieve a wide bandwidth we should do the following:

1. Trade off gain for increased bandwidth by decreasing R_L.
2. Make the source resistance R_S as small as possible consistent with other circuit requirements. In some cases the use of an emitter-follower in the input circuit for impedance transformation may be useful.
3. Make the net load capacitance as small as possible consistent with the other circuit requirements. In this case again the use of an emitter follower on the output side of the circuit for impedance transformation may be very useful.
4. Choose a transistor with a small value of $C_{cb'}$ (a "high-frequency" transistor).
5. Choose a transistor with a small value for the base spreading resistance, $r_{bb'}$. Sometimes, transistors are characterized in terms of the $r_{bb'}C_{cb'}$ time constant, in which case a small value of this constant is the desirable feature.
6. Choose a transistor with a high unity-gain frequency f_T. Since $\omega_T = g_m/(C_{b'e} + C_{cb'})$, this means that $C_{b'e}$ will be small.
7. Use one or more negative-feedback loops, if necessary to trade off reduced gain for increased bandwidth. The negative feedback can be done in terms of a local feedback loop around or within each gain stage and an overall feedback loop around the entire amplifier from the output to the input.

For video amplifiers, the use of local feedback is often a better choice than an overall feedback loop because of the large phase shifts that will be encountered at high frequencies with multistage amplifiers. There will consequently be stability problems with feedback loops that encompass several gain stages. In many cases, both types of feedback loops are used simultaneously.

13.5 VIDEO-AMPLIFIER COMMON-EMITTER STAGE EXAMPLE

For consideration of a simple example of a single common-emitter amplifier stage frequency response, we will use the circuit of Figure 13.2. The biasing details have been omitted for clarity. For this example we will use the following representative circuit and device parameters:

$$r_{bb'} = 60 \ \Omega, \qquad R_S = 40 \ \Omega$$

$$C_{cb'} = 1.5 \ \text{pF}, \qquad C_L = 1.0 \ \text{pF} \tag{13.20}$$

$$f_T = 1.6 \ \text{GHz at a quiescent current of 2.5 mA}$$

Since $f_T = 1.6$ GHz, we have for ω_T the value of $\omega_T = 1 \times 10^{10} \ \text{s}^{-1}$. At the quiescent current level of 2.5 mA, g_m is given by $g_m = I_C/V_T = 2.5 \ \text{mA}/25 \ \text{mV} = 0.1 \ \text{S}$. We can now solve for $C_{b'e}$ as

$$C_{b'e} + C_{cb'} = \frac{g_m}{\omega_T} = \frac{0.1 \ \text{S}}{10^{10} \ \text{s}^{-1}} = 1 \times 10^{-11} \ \text{F} = 10 \ \text{pF} \tag{13.21}$$

We can now write the equations for f_1, f_2, and the midfrequency gain as

$$f_1 = \frac{1}{(2\pi)(R_S + r_{bb'})(C_{b'e} + C_{cb'} + g_m R_L C_{cb'})} \tag{13.22}$$

$$= \frac{1}{(2\pi)(100 \ \Omega)(10 \ \text{pF} + 0.1 \ \text{S} \times R_L \times 1.5 \ \text{pF})} \tag{13.23}$$

$$f_2 = \frac{1}{(2\pi)R_L(C_{cb'} + C_L)} = \frac{1}{(2\pi)R_L(2.5 \ \text{pF})}$$

and

$$A_{V(\text{MID})} = -g_m R_L = (0.1 \ \text{S})R_L \tag{13.24}$$

For the values of load resistance, R_L, listed in Table 13.1, we obtain the corresponding values of f_1, f_2, bandwidth, $A_{V(\text{MID})}$, and gain–bandwidth product.

TABLE 13.1

R_L (Ω)	f_1 (MHz)	f_2 (MHz)	BW (MHz)	$A_{V(\text{MID})}$	$A_{V(\text{MID})} \times BW$ (MHz)
30	110	2122	110	3	330
100	64	637	64	10	640
300	29	212	29	30	868
1,000	10	64	10	100	1,000
3,000	3.5	21	3.5	300	1,038
10,000	1.05	6.4	1.04	1000	1,040

We see from this table that the f_1 breakpoint frequency is much lower than the f_2 breakpoint frequency for all the load resistance values listed. Therefore, f_1 is the principal factor in determining the 3-dB bandwidth of the amplifier stage in this example.

Wide-bandwidth (Video) Amplifiers   Chap. 13

13.6 DIFFERENTIAL VIDEO AMPLIFIER ANALYSIS: TYPE 733

The first type of video IC that we will look at is the 733 type (LM733 by National Semiconductor, MCI733 from Motorola, and others), of video amplifier. This video amplifier has a differential input and output, so it can accept both single-ended and balanced input signals, and either single-ended or balanced outputs can be obtained.

The circuit diagram for this device is shown in Figure 13.5. The input stage is a differential amplifier comprised of Q_1 and Q_2 with current sink biasing provided by Q_7, and with R_1 and R_2 (2.4 kΩ each) being used as the load resistors. Resistors R_3 through R_6 are used to provide local negative feedback (emitter degeneration) within the first gain stage.

A balanced output is taken from the first gain stage and drives the second gain

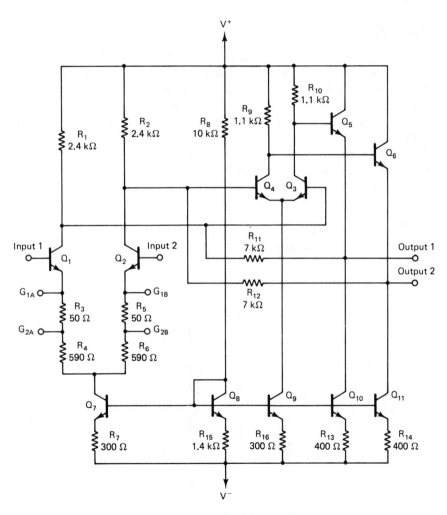

Figure 13.5 Type 733 video amplifier.

stage, which is another differential amplifier consisting of Q_3 and Q_4 with current sink biasing provided by Q_9. Resistors R_9 and R_{10} (1.1 kΩ each) are the load resistors for the second gain stage. The second gain stage provides a balanced output to drive a pair of NPN emitter-follower output stage transistors, Q_5 and Q_6. These emitter-follower output transistors are biased by current sink transistors Q_{10} and Q_{11}.

Resistors R_{11} and R_{12} provide negative feedback from the two output terminals back to the input terminals of the second stage. There are therefore two negative-feedback loops, one within the first stage (using R_3 through R_6), and the other around the second stage using R_{11} and R_{12}.

The overall biasing for this IC is provided by the diode-connected transistor Q_8 in conjunction with resistors R_8 and R_{15}. This biasing circuit drives the current sink transistors, Q_7, Q_9, Q_{10}, and Q_{11}, and provides current sink biasing for all the stages in the amplifier.

13.6.1 Biasing Analysis

For the biasing analysis of this device, we will consider the circuit shown in Figure 13.6 showing the biasing network, and we will assume representative supply voltages of ±6.0 V. We will have for I_8 a dc current given by $I_8 = (12 - 0.6)V/(10 + 1.4)$ kΩ = 11.4 V/11.4 kΩ = 1.0 mA. The voltage drop across R_{15} will therefore be 1.4 V. Since the base-to-emitter voltage drops of all the transistors involved in the biasing network are approximately equal (to within ±0.05 V), the voltage drops across all the emitter resistors (R_{13}, R_{14}, R_{15}, R_{16}, and R_7) are approximately the same. Consequently, we have that $I_7 = I_9 = 1.4$ V/300 Ω = 4.7 mA, and $I_{13} = I_{14} = 1.4$ V/ 400 Ω = 3.5 mA.

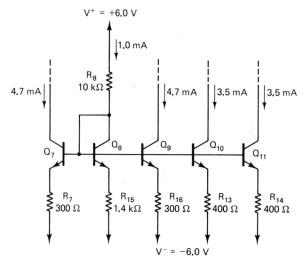

Figure 13.6 Biasing circuitry for the 733 video amplifier.

13.6.2 AC Gain Analysis

We will perform the ac gain analysis of this amplifier by splitting each stage down the axis of symmetry. Doing this gives the ac circuit of Figure 13.7. The dynamic transfer conductance for each half of the first-stage differential amplifier (Q_1 and Q_2) is given by

$$g_{f_1} = \frac{i_{c_1}}{v_{b_1}} = \frac{g_m}{1 + g_m R_E} = \frac{1}{(1/g_m) + R_E} = \frac{1}{(V_T/I_C) + R_E} \qquad (13.25)$$

Note that when $R_E = 0$ this reduces to simply $g_{f_1} = g_m = I_C/V_T$. Since v_{b_1} is one-half of the total (base-to-base) input voltage V_{IN}, the ac collector current produced by each half of the differential amplifier is $i_{c_1} = g_{f_1}v_{b_1} = g_{f_1}(V_{IN}/2)$.

The open-loop voltage gain produced by each half of the second-stage differential amplifier Q_3 and Q_4 is

$$A_{OL_2} = -g_{f_2}R_{L_2} = \frac{-I_C}{V_T}(1.1 \text{ k}\Omega) = \frac{4.7 \text{ mA/2}}{25 \text{ mV}} \times 1.1 \text{ k}\Omega = -103$$

The actual gain of the second stage is slightly lower than this due to the finite input impedance presented by the emitter-follower output stage.

The voltage gain produced by each half of the emitter-follower output stage (Q_5 and Q_6) is given by $A_{V_3} = R_L/[R_L + (V_T/I_C)]$, where I_C is the quiescent current of

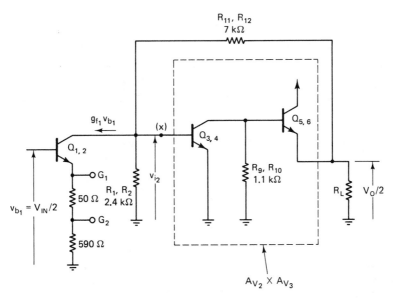

Figure 13.7 AC circuit for the 733 video amplifier obtained by splitting the amplifier along the axis of symmetry.

the emitter-follower transistors, 3.5 mA. We therefore have that

$$A_{V_3} = \frac{R_L}{R_L + (25 \text{ mV}/3.5 \text{ mA})} = \frac{R_L}{R_L + 7.1\ \Omega} \tag{13.26}$$

From this we see that for any reasonable value of load resistance the voltage gain of the emitter-follower output stage is close to unity. Therefore, the combined open-loop gain of the second and third stages, $A_{V_2} \times A_{V_3}$, is approximately 100.

We will now write a node-voltage equation for node x in Figure 13.7 to obtain

$$-g_{f_1}\frac{V_{\text{IN}}}{2} + \frac{V_{\text{OUT}}/2}{7\ \text{k}\Omega} - V_X\left(\frac{1}{7\ \text{k}\Omega} + \frac{1}{2.4\ \text{k}\Omega}\right) = 0 \tag{13.27}$$

Noting that $V_{\text{OUT}}/2 = A_{V_2}A_{V_3}V_X = -100V_X$, we have that $V_X = -(V_{\text{OUT}}/2)/100$, and thus the node equation can be rewritten as

$$-g_{f_1}\frac{V_{\text{IN}}}{2} + \frac{V_{\text{OUT}}/2}{7\ \text{k}\Omega} + \frac{V_{\text{OUT}}/2}{100}\left(\frac{1}{7\ \text{k}\Omega} + \frac{1}{2.4\ \text{k}\Omega}\right) = 0 \tag{13.28}$$

Collecting terms, this gives

$$V_{\text{OUT}}\left[\frac{1}{7\ \text{k}\Omega} + \frac{1}{100}\left(\frac{1}{7\ \text{k}\Omega} + \frac{1}{2.4\ \text{k}\Omega}\right)\right] = g_{f_1}V_{\text{IN}} \tag{13.29}$$

and thus

$$A_V = \frac{V_{\text{OUT}}}{V_{\text{IN}}} = \frac{g_{f_1}}{\dfrac{1}{7\ \text{k}\Omega} + \dfrac{1}{100}\left(\dfrac{1}{7\ \text{k}\Omega} + \dfrac{1}{2.4\ \text{k}\Omega}\right)} = \frac{g_{f_1}(7\ \text{k}\Omega)}{1.04} \simeq g_{f_1}(7\ \text{k}\Omega)$$

We see from this result that changes in the voltage gain of the second and third stages have relatively little effect on the overall gain of the amplifier. Indeed, if the combined gain of the second and third stages were to go from a minimum of 100 all the way up to infinity, the net gain of the amplifier would change by only about 4%.

For g_{f_1}, we have

$$g_{f_1} = \frac{1}{(V_T/I_C) + R_E} = \frac{1}{(25 \text{ mV}/2.35 \text{ mA}) + R_E} = \frac{1}{11\ \Omega + R_E} \tag{13.30}$$

so the overall gain of the amplifier is

$$A_V = g_{f_1}(7\ \text{k}\Omega) = \frac{7\ \text{k}\Omega}{11\ \Omega + R_E} \tag{13.31}$$

TABLE 13.2

Gain option	A_V (calc.)	A_V (spec.)	BW (MHz)	A_V BW (MHz)
Gain 1	640	400 (typ.)	40 (typ.)	16,000
Gain 2	115	100 (typ.)	90 (typ.)	9,000
Gain 3	10.8	10 (typ.)	120 (typ.)	1,200

The value of R_E is determined by which, if any, of the gain pins are shorted together. If all the gain select pins are left open, R_E will attain its maximum value of 590 + 50 = 640 Ω. If G_{2A} and G_{2B} are connected together, R_E will be 50 Ω, and if pins G_{1A} and G_{1B} are shorted together, R_E will be zero. The fixed gain values corresponding to these three conditions are therefore

Gain 1 ($R_E = 0$): $A_V = $ 7 kΩ/11 $\Omega = \underline{640}$
Gain 2 ($R_E = 50\ \Omega$): $A_V = $ 7 kΩ/11 + 50 $\Omega = \underline{115}$
Gain 3 ($R_E = 640\ \Omega$): $A_V = $ 7 kΩ/(11 + 640) $\Omega = \underline{10.8}$

Values of gain that are intermediate between the maximum and minimum values given above can be obtained by connecting a fixed or variable resistor, R_{ADJ}, between the G_{1A} and G_{1B} gain select pins, for which case R_E is given by $R_E = $ 640 $\Omega \parallel (R_{\text{ADJ}}/2)$. Note that $R_{\text{ADJ}}/2$ is used for R_E since, when the amplifier is split down the axis of symmetry, the gain adjust resistor will be cut in half.

The 3-dB bandwidth values (for a 50-Ω source resistance) that are obtained for the three gain-select options are listed in Table 13.2 using values obtained from the specification sheet for this device.

From these values we do note that as the gain is decreased by using more feedback (increasing R_E) the bandwidth does increase. However, the increase in the bandwidth is by a far smaller factor than the decrease in the gain. This is further evidenced by the decrease in the gain–bandwidth product. Thus the decrease in the gain from a value of 400 (gain 1) to 100 (gain 2) produces an increase in the bandwidth from 40 MHz up to 90 MHz, or an increase by a factor of 2.25. The decrease in the gain from 100 (gain 2) to 10 (gain 1) results in a further increase in the bandwidth of from 90 to 120 MHz, a factor of only 1.3.

The basic reason that the decrease in the voltage gain that is produced by an increase in the negative feedback does not produce an increase in the bandwidth by the same factor that the voltage gain is decreased is the result of the multiplicity of breakpoint frequencies involved in the high-frequency region of interest (about 40 to 120 MHz). If we look at the graph of the amplifier phase shift as a function of frequency for the 733, as shown in Figure 13.8, we see that, in the region from 40 to

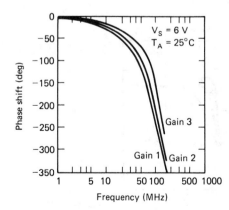

Figure 13.8 Phase shift versus frequency for the LM733 video amplifier (National Semiconductor).

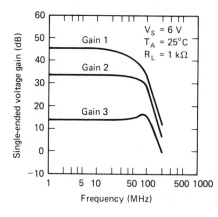

Figure 13.9 Voltage gain versus frequency for the LM733 video amplifier (National Semiconductor).

120 MHz, phase shifts ranging from about 100° to 330° will be encountered. These large phase-shift values indicate that this frequency range extends from beyond the first breakpoint frequency at 40 MHz to beyond the fourth breakpoint frequency at 120 MHz. As a result, the gain versus frequency curve decreases quite rapidly in this frequency range, especially at the upper end of the range, as shown in Figure 13.9. For example, if the slope is 60 dB/decade, corresponding to the effect of three breakpoint frequencies, then a sacrifice of 20 dB in gain (a 10 : 1 ratio) will produce an improvement in the bandwidth by only about 20 dB/(60 dB/decade) $= \frac{1}{3}$ decade, or a ratio of $(10)^{1/3} = 2.15$. We therefore see why a decrease in the gain does not lead to an increase in the bandwidth by the same factor as illustrated in Figure 13.9.

The large phase shifts encountered at high frequencies also lead to some gain peaking. This is shown by the voltage gain versus frequency curve for gain 3, where there is about 15% gain peaking at about 100 MHz. We note that at this frequency the phase shift is approximately 180°, so this indeed corresponds to the condition in which what was negative feedback at lower frequencies has turned around to become positive feedback near 100 MHz.

If we look at the pulse response for the gain 3 condition in Figure 13.10, we see that the gain peaking in the frequency domain is manifested in the time domain as

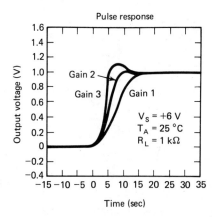

Figure 13.10 LM733 pulse response (National Semiconductor).

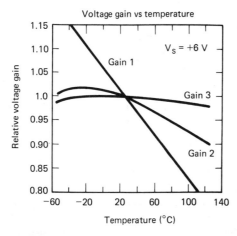

Figure 13.11 LM733 voltage gain versus temperature (National Semiconductor).

overshoot in the pulse response. Indeed, not only is there about 10% overshoot, but there is also a slight amount of *ringing*, or a heavily damped oscillatory response.

The voltage gain as a function of temperature for this device is also of interest, as shown in Figure 13.11. The voltage gain for the gain 1 option ($R_E = 0$) decreases rapidly with temperature. The voltage gain for the gain 2 condition decreases at a much slower rate with increasing temperature, whereas the voltage gain for the gain 3 conditions hardly changes at all with temperature. This behavior is indeed consistent with that expected on the basis of the expression for the voltage gain as given by $A_V = g_{f_1}$ (7 kΩ) = (7 kΩ)/[$(V_T/I_C) + R_E$]. For the gain 1 condition, $R_E = 0$, so we should expect that the voltage gain will be inversely proportional to temperature, since $A_V = (I_C/V_T)(7 \text{ k}\Omega)$ for this case and $V_T = kT/q$. For the gain 3 condition, on the other hand, $R_E = 640 \ \Omega$, so $A_V = (7 \text{ k}\Omega)/[(V_T/2.35 \text{ mA}) + 640 \ \Omega]$; so the gain should indeed exhibit very little temperature dependence.

As a result of the two cascaded differential-amplifier configuration of this device and the balanced nature of the signal flow, we should expect to obtain a very good common-mode rejection ratio. This is indeed the case, and for gain 2 for which the voltage gain is 100 (typ.) or 40 dB the common-mode rejection ratio (CMRR) is 86 dB (typ.) for frequencies below 50 kHz, as shown in Figure 13.12. The 86-dB CMRR values correspond to a common-mode voltage gain of 40 dB − 86 dB = − 46 dB, or only 0.005, compared to the difference-mode gain of + 40 dB or 100. Looking at the graph of the CMRR versus frequency, we see that the CMRR does decrease with frequency above about 50 kHz. Part of this decrease in the CMRR with frequency is due to the falloff of the difference-mode gain, but perhaps to a larger extent it is due to a greater capacitative feedthrough of the common-mode signal and the decrease in the output impedance of the current sinks (Q_7 and Q_9) of the first and second differential-amplifier stages.

Closely related to the 733 type of video amplifier is the NE/SE592 (Signetics), which has a circuit similar to that of the 733 type except that there is just one pair of pin-selectable gain-setting resistors, as compared to two pairs for the 733. The NE/SE592 offers fixed gains of 100 and 400 and adjustable gains from 0 to 400 with a single external resistor. The small-signal bandwidths are 90 MHz at a gain of 100

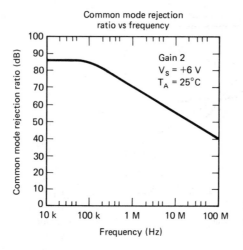

Figure 13.12 LM733 common-mode rejection ratio versus frequency (National Semiconductor).

and 40 MHz at a gain of 400. The NE5592 (Signetics) is a dual video amplifier. It is similar to the NE592, but offers just a single fixed gain of 400 and adjustable gains ranging from 0 to 400. The small-signal bandwidth is 25 MHz at a gain of 400.

13.7 CASCODE-AMPLIFIER CONFIGURATION

The cascode configuration is an amplifier stage composed of a direct-coupled common-emitter/common-base combination as shown in Figure 13.13a. This configuration offers the possibility of very large bandwidths and is very often used for video and radio-frequency (RF) amplifiers.

Figure 13.13b shows the ac small-signal equivalent circuit for the cascode stage. The net load admittance driven by the collector of the common-emitter transistor Q_1 is the input admittance of the common-base stage Q_2 plus the collector–base capacitance of Q_1, so this will be

$$Y_{L(CE)} = g_{eb'2} + j\omega(C_{b'e} + C_{cb'}) = \frac{I_Q}{V_T} + j\omega(C_{b'e} + C_{cb'}) \qquad (13.32a)$$

where I_Q is the quiescent current of Q_1 and Q_2. Since $I_Q/V_T = g_m$, this can be rewritten as

$$Y_{L(CE)} = g_m + j\omega(C_{b'e} + C_{cb'}) = g_m\left[1 + \frac{j\omega(C_{b'e} + C_{cb'})}{g_m}\right] \qquad (13.32b)$$

$$= g_m\left(1 + \frac{j\omega}{\omega_T}\right) = g_m\left(1 + \frac{jf}{f_T}\right) \qquad (13.32c)$$

Over the frequency range of interest, we have that $f \ll f_T$, so $Y_{L(CE)} \simeq g_m$. The voltage gain of Q_1 is therefore given by

$$A_{V_1} = \frac{-g_m}{Y_{L(CE)}} \simeq \frac{-g_m}{g_m} = -1 \qquad (13.32d)$$

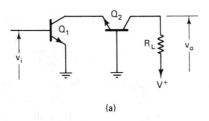

(a)

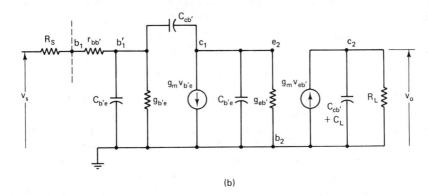

(b)

Figure 13.13 Cascode configuration: (a) basic cascode configuration (common emitter/common base) (biasing details omitted); (b) ac small-signal equivalent circuit.

Since the voltage gain of Q_1 is essentially unity, the equation for the first breakpoint (radian) frequency ω_1 is accordingly

$$\omega_1 = \frac{G_1}{C_{b'e} + C_{cb'} + C_{cb'}g_m R_{L(CE)}} = \frac{G_1}{C_{b'e} + C_{cb'} + C_{cb'}(g_m/g_m)} \tag{13.33a}$$

$$= \frac{1}{(R_S + r_{bb'})(C_{b'e} + 2C_{cb'})}$$

Since $C_{b'e}$ is usually considerably larger than $C_{cb'}$, we can say that $C_{b'e} + 2C_{cb'} \simeq C_{b'e} + C_{cb'} = g_m/\omega_T$. This approximation allows us to rewrite the expression for ω_1 in the form

$$\omega_1 \simeq \frac{1}{(R_S + r_{bb'})(C_{b'e} + C_{cb'})} = \frac{\omega_T}{g_m(R_S + r_{bb'})} \tag{13.33b}$$

and thus

$$f_1 \simeq \frac{f_T}{g_m(R_S + r_{bb'})} \tag{13.34}$$

The second breakpoint frequency is determined by the $g_{eb'}/(C_{b'e} + 2C_{cb'})$ time constant. Noting that $g_{eb'} = g_m$, we have for ω_2 the expression

$$\omega_2 = \frac{g_{eb'}}{C_{b'e} + 2C_{cb'}} = \frac{g_m}{C_{b'e} + 2C_{cb'}} \simeq \omega_T \tag{13.35}$$

so

$$f_2 \simeq f_T \tag{13.36}$$

The third breakpoint frequency is determined by the $R_L(C_{cb'} + C_L)$ time constant as

$$\omega_3 = \frac{G_L}{C_{cb'} + C_L} = \frac{1}{R_L(C_{cb'} + C_L)}$$

We should note at this point that the second breakpoint frequency f_2 is equal to the unity-gain frequency f_T, so in virtually every case of practical interest this breakpoint frequency will be much greater than the other two breakpoint frequencies. As a result, the second breakpoint frequency will not have any significant effect in determining the overall bandwidth of the circuit.

The voltage gain of the common-base transistor Q_2 is given by $A_{V_2} = g_m/G_L = g_mR_L$, and thus the overall voltage gain of the cascode combination is

$$A_{V(\text{MID})} = A_{V_1}A_{V_2} = (-1)(g_mR_L) = -g_mR_L$$

Looking at the expression for the first breakpoint frequency f_1, we see that it is no longer a function of the load resistance R_L and therefore is independent of the midfrequency voltage gain. This independence of f_1 from R_L is due to the isolation provided by the common-base transistor in that there is no longer any direct feedback from output to input via the collector–base capacitance. This isolation can be a very important consideration for RF amplifiers, for which case the stability of the amplifier from oscillations is of concern. The base of Q_2 in effect acts as an *electrostatic shield* between the input and output circuits and thus provides very good isolation from capacitative feedback effects.

We will now consider an example of the bandwidth and midfrequency gain available with the cascode configuration, and we will use the same transistor and circuit parameters as were used in the previous example of the common-emitter stage. Since $R_S + r_{bb'} = 100 \ \Omega$, $f_T = 1.6$ GHz at the quiescent current of 2.5 mA, and $g_m = 0.1$ S, for f_1 we have the value

$$f_1 = \frac{f_T}{g_m(R_S + r_{bb'})} = \frac{1600 \text{ MHz}}{(0.1 \text{ S})(100 \ \Omega)} = \underline{160 \text{ MHz}} \qquad (13.37)$$

Using $C_{cb'}$ and 1.5 pF and $C_L = 1.0$ pF, for f_3 we have the equation $f_3 = 1/[2\pi R_L(2.5 \text{ pF})]$, and $A_{V(\text{MID})} = -g_mR_L = -(0.1 \text{ S})R_L$. For the values of load resistance R_L listed in Table 13.3, we obtain the corresponding values of f_1, f_2, f_3, 3-dB bandwidth, and midfrequency gain, $A_{V(\text{MID})}$. From these values we note the large bandwidths and gain–bandwidth products that can be obtained with the cascode circuit. Indeed, if we compare these values with those obtained for the common-emitter stage con-

TABLE 13.3

R_L (Ω)	f_1 (MHz)	f_2 (MHz)	f_3 (MHz)	BW (MHz)	$A_{V(\text{MID})}$	$A_{V(\text{MID})} \times BW$ (MHz)
30	160	1600	2122	160	3	480
100	160	1600	637	160	10	1600
300	160	1600	212	128	30	3831
1,000	160	1600	64	59	100	5900
3,000	160	1600	21	21	300	6305
10,000	160	1600	6.4	6.4	1000	6400

sidered earlier, we see a large improvement, especially for load resistances in excess of about 300 Ω. For example, for a load resistance of 1000 Ω, for which the mid-frequency gain is 100 in both cases, the bandwidth for the common-emitter amplifier stage is 10 MHz, compared to a bandwidth of 59 MHz for the cascode-amplifier stage.

To be completely fair, however, we could compare the bandwidth available with the cascode stage with that obtain from two cascaded common-emitter stages with the same overall voltage gain. If we do this for the case of two cascaded common-emitter stages, each stage having a gain of 10 for an overall midfrequency voltage gain of 100 using $R_L = 100$ Ω for each stage, the bandwidth of each stage is 64 MHz. The net bandwidth of the cascaded combination of the two stages is given by BW_{NET} $= BW(2^{1/2} - 1)^{1/2} = 64$ MHz $\times$ 0.6436 $= 41$ MHz. Since the bandwidth of the cascode configuration under the same voltage-gain condition of $A_{V(MID)} = 100$ is 59 MHz, we see that the cascode amplifier still compares very favorably with the common-emitter amplifier.

13.8 USE OF EMITTER-FOLLOWER STAGES IN WIDEBAND AMPLIFIERS

We have seen the important role played by the source resistance R_S and the load resistance R_L in limiting the bandwidth of video amplifiers, both for the simple common-emitter case and for the cascode configuration. In many cases the impedance transformation characteristics of an emitter-follower stage can be used to great advantage on both the source as well as the load ends of the amplifier. The emitter-follower stage interposed between the signal source and the first gain stage of the amplifier can be used to transform the source resistance R_S down to a much smaller value to be called R_S'. An emitter-follower stage interposed between the last gain stage of the amplifier and the load can be used to transform the load capacitance C_L down to a much smaller value.

Figure 13.14 shows a cascode-amplifier configuration with the addition of emitter-follower stages for impedance transformation on both the input and output sides

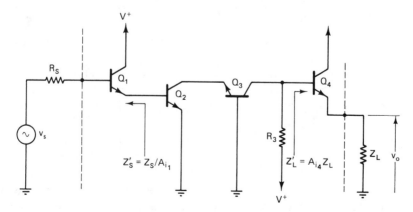

Figure 13.14 Cascode configuration with emitter-follower input and output stages (biasing details omitted).

of the amplifier. At the source end, the transformed value of the source resistance as seen by the cascode stage (at the base of Q_1) is $R_S' = R_S/A_{i1}$, where A_{i1} is the ac current gain of Q_1. This is given by the equation $A_i = \beta/(1 + jf/f_\beta)$. For frequencies more than half of a decade above f_β, this can be written approximately as $A_i \simeq \beta f_\beta/jf = f_T/jf$. Therefore, as long as the frequency range of interest is considerably below f_T, which it will be in virtually every case, a large impedance transformation ratio can be achieved.

At the load end, the emitter follower, Q_2, acts to transform the load impedance Z_L, so the effective value of the load impedance driven by Q_3 as seen looking into the base of Q_4 is given by $Z_L' = A_{i4}Z_L$, where A_{i4} is the ac current gain of Q_4 and is given by the same relationships as those used above for Q_1.

The use of emitter-follower input and output stages can prove to be especially useful when large source resistances and load capacitances are present. To illustrate the efficacy of the emitter-follower stages for this purpose, we will consider a cascode configuration similar to the one-considered earlier, except that the source resistance will be increased by a substantial amount from 40 to 1000 Ω. The load capacitance will be changed from 1.0 pF up to 50 pF. We will consider the design of the circuit for a 50-MHz bandwidth.

Without the use of an emitter-follower stage, the first breakpoint frequency f_1 is given by $f_1 = f_T/g_m(R_S + r_{bb'}) = 1600\ \text{MHz}/0.1(1060) = 15.1\ \text{MHz}$. We see that we will fall far short of the bandwidth requirement with this value for f_1.

We will now interpose an emitter-follower input stage, Q_1, between the source resistance and the input base terminal (B_2) of the cascode stage, using a transistor with characteristics similar to those of the cascode configuration, so that $f_T = 1.6$ GHz at a quiescent current of 2.5 mA. At a frequency of 50 MHz, the current gain of Q_1 is $A_{i1} = f_T/jf = 1600\ \text{MHz}/j50\ \text{MHz} = -j32$. The 1000-ω source resistance is therefore transformed to $Z_S' = Z_S/A_{i1} = 1000\ \Omega/(-j32) = j31\ \Omega$ at 50 MHz.

With a quiescent current of 2.5 mA, the dynamic emitter-to-base resistance $r_{eb'}$ of Q_1 is given by $r_{eb'} = V_T/I_{CQ} = 25\ \text{mV}/2.5\ \text{mA} = 10\ \Omega$. The capacitative reactance present at the base of Q_2 is given by

$$X_{C2} = \frac{1}{j\omega(C_{b'e} + 2C_{cb'})} \simeq \frac{1}{j\omega(C_{b'e} + C_{cb'})} = \frac{\omega_T}{j\omega g_m} = \frac{f_T}{jf g_m} \qquad (13.38)$$

For $g_m = 0.1\ \text{S}$, $f_T = 1600\ \text{MHz}$, and $f = 50\ \text{MHz}$, we obtain a capacitative reactance of $X_{C2} = 1600\ \text{MHz}/(j50\ \text{MHz} \times 0.1\ \text{S}) = -j320\ \Omega$.

We can now represent the input circuit for this cascode amplifier by the equivalent circuit shown in Figure 13.15. The voltage-gain factor $v_{b'e2}/v_s$ can be obtained by considering this circuit to be a simple voltage divider with a voltage-division ratio given by

$$\left|\frac{v_{b'e2}}{v_s}\right| = \left|\frac{-j320}{+j31 + 10 + 60) - j320}\right| = \left|\frac{-j320}{70 - j289}\right| = \frac{320}{297} = 1.08 \qquad (13.39)$$

Since the $v_{b'e2}/v_s$ factor is greater than 0.7071, we see that the bandwidth of the input circuit is above 50 MHz. Indeed, in this case it appears that f_1 will be considerably above 50 MHz since the $v_{b'e}/v_s$ ratio is actually slightly greater than unity. This is the result of the impedance transformation of Q_1, which has resulted in the source

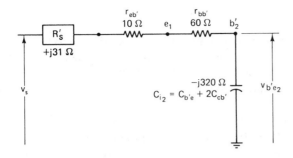

Figure 13.15 Input equivalent circuit showing transformed source resistance.

resistance of 1000 Ω in being transformed into a reactance of $+j31\ \Omega$. Since the algebraic sign of this reactance is opposite to that of the capacitative reactance at the base of Q_2, the net reactance around the loop is reduced as a result, as seen from the denominator of the $v_{b'e_2}/v_s$ ratio expression.

The capacitative reactance of the load capacitance of 50 pF at a frequency of 50 MHz is given by $X_{CL} = 1/j\omega C_L = -j64\ \Omega$. This capacitative reactance is transformed by Q_4 to an impedance given by

$$Z'_L = Z_L A_{i4} = Z_L \frac{f_T}{jf} = (-j64\ \Omega) \frac{1600\ \text{MHz}}{j50\ \text{MHz}} = -2040\ \Omega \qquad (13.40)$$

Thus the capacitative load C_L has been transformed by Q_4 into a negative resistance of $-2040\ \Omega$.

The net capacitance driven by Q_3 is the sum of the collector–base capacitances of Q_3 and Q_4, or $C_{cb'3} + C_{cb'4} = 3.0$ pF. The net conductance driven by Q_3 is the algebraic sum of G_3 and the conductance seen looking into the base of Q_4. The breakpoint frequency f_3 obtained for this part of the circuit is related to the net resistance R_{NET} and capacitance of $f_3 = 1/2\pi R_{\text{NET}} C_{\text{NET}}$, where $C_{\text{NET}} = 3.0$ pF. Since f_1 is considerably greater than the required bandwidth of 50 MHz, as is $f_2 = f_T = 1600$ MHz, we require that $f_3 = 50$ MHz. The corresponding value of R_{NET} is given by $R_{\text{NET}} = 1/2\pi (BW) C_{\text{NET}} = 1/2\pi\ (50\ \text{MHz}) \times 3.0$ pF $= 1061\ \Omega$. Since we have that $G_{\text{NET}} = G_3 + G_{i4}$, we obtain $1/1061 = (1/R_3) + (1/-2040)$, so $R_3 = 698\ \Omega$. The midfrequency voltage gain obtained from this circuit is given by $A_{V(\text{MID})} = -g_m R_{\text{NET}} = -0.1\ \text{S} \times 1061\ \Omega = -106$.

If there were no emitter-follower stage on the output side, the value of R_{NET} that would have to be used in order to obtain a 50-MHz bandwidth is given by

$$R_{\text{NET}} = \frac{1}{2\pi(BW)(C_L + 2C_{cb'})} = \frac{1}{2\pi(50\ \text{MHz}) \times 51.5\ \text{pF}} = 60\ \Omega \qquad (13.41)$$

With this very small value for $R_{\text{NET}} \simeq R_3$, the midfrequency voltage gain would be given by $A_{V(\text{MID})} = -g_m R_{\text{NET}} = -(0.1\ \text{S})(60\ \Omega) = -6.0$, compared to the gain of -106 that can be obtained if an emitter-follower stage is used. Thus we see that a very substantial improvement in the gain–bandwidth product can be achieved by the use of an emitter-follower stage for the transformation of the load impedance, and we have already seen the benefits that accrue from the use of an emitter follower on the input side for source resistance transformation.

Sec. 13.8   Use of Emitter-follower Stages in Wideband Amplifiers   **661**

13.9 DIFFERENTIAL VIDEO AMPLIFIER EXAMPLE: CA3040

We will now consider another IC video amplifier, the CA3040 (RCA), which has a differential cascode-amplifier configuration. The circuit diagram for this device is shown in Figure 13.16. This is a differential video amplifier and can accept single-ended or balanced input signals, and single-ended as well as balanced output voltages can be obtained.

The input stage consists of transistors Q_1 and Q_2 in a emitter-follower configuration driving the cascode (CE–CB) differential amplifier with Q_3 and Q_4 serving as the common-emitter part of the cascode combination and Q_5 and Q_6 serving as the common-base part. This differential amplifier has current sink biasing provided by Q_9. Resistors R_5 and R_6 (1.32 kΩ each) serve as the load resistors for this stage. Transistors Q_7 and Q_8 constitute a differential emitter-follower output stage with biasing provided by resistors R_7 and R_8.

Diodes D_1 and D_2 together with resistor R_{10} (4.5 kΩ) provide a dc bias voltage

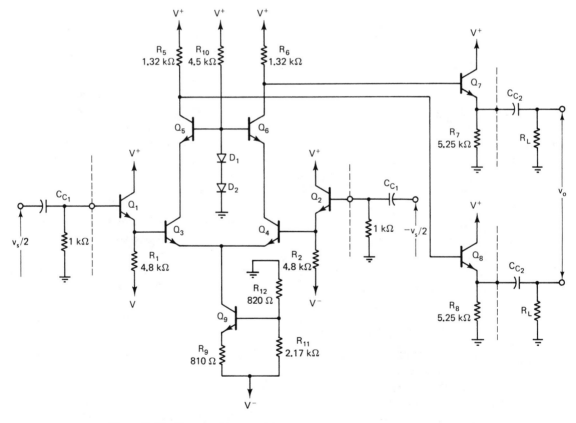

Figure 13.16 Cascode video amplifier integrated circuit: CA3040 (RCA Corporation).

Wide-bandwidth (Video) Amplifiers   Chap. 13

of about 1.4 V for the proper biasing of Q_3, Q_4, Q_5, and Q_6. Since the bases of Q_1 and Q_2 are essentially at ground potential (0 V), the bases of Q_3 and Q_4 are at about -0.7 V and the emitters are at about -1.4 V. Since the emitters of Q_5 and Q_6 are at a voltage of $V_{D1} + V_{D2} - V_{BE(5,6)} = +0.7$ V, the collector-to-emitter voltage of Q_3 and of Q_4 is approximately $+2.1$ V. This value of V_{CE} should prove adequate to keep Q_3 and Q_4 well out of saturation and to ensure that these two transistors will be operating under bias conditions that will give good high-frequency performance.

The bias voltage for the current sink transistor Q_9 is supplied by the voltage divider comprised of R_{11} and R_{12}. This bias voltage together with resistor R_9 (810 Ω) determines the collector current of Q_9 and hence the quiescent current of the cascode differential amplifier.

13.9.1 DC Biasing Calculations

For the dc biasing calculations, we will use a representative split supply voltage of ± 6.0 V. The current through the series circuit of R_{11} and R_{12} is 6.0 V/2.99 k$\Omega \simeq$ 2.0 mA, so the voltage at the base of Q_9 is -2.0 mA $\times$ 0.82 kΩ = -1.64 V. Assuming a V_{BE} of 0.7 V, the voltage at the emitter of Q_9 is -2.34 V, so the voltage drop across R_9 is $-2.34 - (-6.0) = 3.66$ V. The current through Q_9 is therefore $I_9 = 3.66$ V/0.81 kΩ = __4.5 mA__, and this is the quiescent current of the differential amplifier.

The quiescent current of Q_1 and of Q_2 is determined by R_1 and R_2 as $-0.7 - (-6.0)/4.8$ kΩ = __1.1 mA__. The quiescent voltage at the collectors of Q_5 and Q_6 and therefore at the bases of Q_7 and Q_8 is given by $+6.0 - (4.5$ mA/2)(1.32 kΩ) = $+3.0$ V, so the emitter voltage of Q_7 and Q_8 is 2.3 V. The quiescent current of Q_7 and Q_8 is as a result 2.3 V/5.25 kΩ = __0.44 mA__.

13.9.2 AC Voltage Gain Analysis

For the ac gain analysis, we will split the amplifier along the axis of symmetry producing the result shown in Figure 13.17. We see that in terms of the ac signal flow the amplifier is basically a CC–CE–CB–CC configuration, and thus uses a cascode gain stage with emitter-follower input and output stages for impedance transformation.

The midfrequency voltage gain of the cascode stage is given by

$$A_{V(\text{MID})} = -g_L R_L = -\frac{I_C R_L}{V_T} = -\frac{2.25 \text{ mA}}{25 \text{ mV}} (1.32 \text{ k}\Omega) = \underline{-119} \qquad (13.42)$$

Using a representative current gain value of 100, the input resistance seen looking into the base of Q_3 and Q_4 is $R_{i(3,4)} = n\beta V_T/I_{(3,4)} = 1.5 \times 100 \times 25$ mV/2.25 mA = 1.67 kΩ. The voltage gain of the emitter-follower input stage is therefore

$$A_{v1,2} = \frac{R_{L,(1,2)}}{(V_T(I_{C12}) + R_{L(1,2)})} = \frac{(4.8 \text{ k}\Omega \| 1.67 \text{ k}\Omega)}{(25 \text{ mV}/1.1 \text{ mA}) + (4.8 \text{ k}\Omega \| 1.67 \text{ k}\Omega)} \qquad (13.43)$$

$$= \frac{1239 \ \Omega}{(23 + 1237)\Omega} = 0.982$$

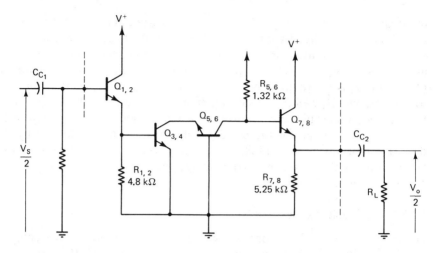

Figure 13.17 AC circuit of the CA3040-type video amplifier after splitting the circuit along the axis of symmetry.

The voltage gain of the emitter-follower output stage of Q_7 and Q_8 for a load resistance of 1000 Ω is in a similar fashion given by

$$A_{V7,8} = \frac{R_{L(7,8)}}{(V_T/I_{C(7,8)}) + R_{L(7,8)}} = \frac{(1\,\text{k}\Omega\|5.25\,\text{k}\Omega)}{(25\,\text{mV}/0.44\,\text{mA}) + (1.0\,\text{k}\Omega\|5.25\,\text{k}\Omega)}$$

$$= \frac{840}{57 + 840} = \underline{0.937}$$

(13.44)

The overall midfrequency voltage gain of this amplifier circuit is therefore given by $A_{V(\text{MID})} = -0.982 \times 119 \times 0.937 = -109$ or 40.8 dB. This is the voltage gain that will be obtained by taking a balanced output voltage, since for a given input voltage applied as either a single-ended or balanced input, the voltage V_S will be split evenly between the two halves of the circuit. The voltage applied to each half-circuit is therefore $V_S/2$ and the resultant output is $V_O/2$. By taking a balanced output, an output voltage of V_O is obtained. If a single-ended output is taken, the output voltage is only $V_O/2$, and the voltage gain is only half as much as for a balanced output, or 55, corresponding to 34.8 dB.

The amplifier output impedance is the impedance seen looking into the emitter of the emitter-follower output stage (Q_7 and Q_8) and is given by $Z_O = (V_T/I_{7,8}) + R_{L5,6}/\beta = (25\,\text{mV}/0.44\,\text{mA}) + (1.32\,\text{k}\Omega/100) = 57\,\Omega + 13\,\Omega = 70\,\Omega$. As a result, load resistances on the order of 1 kΩ or more will not have much effect on the performance of this amplifier.

13.9.3 Bandwidth Calculations

This amplifier with its cascode gain-stage configuration and its emitter-follower input and output stages should exhibit a very large gain–bandwidth product. Since the overall gain is relatively low [$A_{V(\text{MID})} = 109$] as a result of the small value of $R_{5,6}$ (1.32 kΩ), we should expect to see a relatively large bandwidth resulting.

To perform a sample calculation of the bandwidth to be expected from this circuit, we will assume a representative value for the f_T of Q_3 and Q_4 of 1.0 GHz at the quiescent current level of 2.25 mA. We will also use a base spreading resistance of $r_{bb'} = 60\ \Omega$, and we will assume that the amplifier is driven from a 50-Ω source.

The effective value of the source impedance as seen looking back from the base of $Q_{3,4}$ is

$$R'_S = R_{1,2} \| \left[\frac{V_T}{I_{1,2}} + \frac{R_S}{A_{i1,2}} \right]$$

$$= 4.8\ \text{k}\Omega \| \left(23\ \Omega + \frac{50\ \Omega}{A_{i1,2}} \right) \simeq 23\ \Omega$$

since $A_{i1,2}$ is much larger than unity in the frequency range of interest. For the first breakpoint frequency, f_1, we have

$$f_1 = \frac{f_T}{g_m(R'_S + r_{bb'})} = \frac{1000\ \text{MHz}}{(2.25\ \text{mA}/25\ \text{mV})(23 + 60)\ \Omega} \tag{13.45}$$

$$= \frac{1000\ \text{MHz}}{(0.09)(83)} = \underline{134\ \text{MHz}}$$

For the collector–base capacitance, we will choose a representative value of $C_{cb'} = 1.0$ pF. Since the impedance looking into the base of $Q_{7,8}$, $R_{i7,8}$, is much greater than $R_{5,6}$, the third breakpoint frequency f_3 is

$$f_3 = \frac{1}{2\pi R_{5,6}(C_{cb5,8} + C_{cb7,8})} = \frac{1}{2\pi(1.3\ \text{k}\Omega)(2.0\ \text{pF})} = \underline{61\ \text{MHz}}$$

Since $f_2 = f_T = 1000$ MHz and $f_1 = 134$ MHz, we see that the bandwidth will be controlled principally by f_3 and will be approximately $\underline{56\ \text{MHz}}$. The corresponding rise time (10% to 90%) is $t_{\text{rise}} = 0.35/BW = \underline{6.3\ \text{ns}}$. The actual bandwidth (3 dB) for this video amplifier is 55 MHz (typ.) [40 MHz (min.)], so this sample calculation has yielded a bandwidth value that agrees quite well with the actual results.

13.10 WIDE-BANDWIDTH AND FAST-SLEWING OPERATIONAL AMPLIFIERS

Most op amps have unity-gain frequencies (or gain–bandwidth products) in the range from 1 to 3 MHz with slewing rates in the range from 0.5 to 5 V/μs. There are many op amps that can be characterized as very wide bandwidths and fast-slewing op amps. A good example is the HA2539 (Harris Semiconductor) with a gain–bandwidth product of 600 MHz and a slewing rate of 600 V/μs. This op amp will be discussed in more detail in the next section. The NE5539 (Signetics) is a very wide bandwidth monolithic bipolar operational amplifier with a gain–bandwidth product of 1200 MHz at a gain of 7 and a slewing rate of 600 V/μs. The full-power bandwidth is 20 MHz with a peak-to-peak output of 3 V into a 150-Ω load. A very wide bandwidth hybrid op amp is the 3554 (Burr-Brown) with a gain–bandwidth product of 1700 MHz (at a gain of 1000) and a slewing rate of 1000 V/μs. The small-signal bandwidth is 22.5 MHz at a gain of 10, 7.25 MHz at a gain of 100, and 1.7 MHz at a gain of 1000. The full-power bandwidth is 19 MHz when driving a 100-Ω load with a 20-V peak-to-peak output voltage. Another wide-bandwidth hybrid op amp is the LH0032 (National Semiconductor). This JFET-input op amp has a unity-gain frequency of 70 MHz and a slewing rate of 500 V/μs, along with an input impedance of 10^{12} Ω and an input bias current of just 10 pA. A series of very wide bandwidth op amps is made by the Comlinear Corporation. The CLC200 features a 100 MHz 3-dB bandwidth for gains of from 1 to 50. The rise time is 3.6 ns and the settling time for the output voltage to settle to within 0.02% of the final value is 25 ns. The CLC210 combines a wide bandwidth with a large output-voltage swing capability. It has a 50-MHz small-signal bandwidth. It can supply 50 mA to a load with a 60-V peak-to-peak output-voltage swing and a full-power bandwidth of 5 MHz minimum. The CLC220 has a 200-MHz bandwidth with rise and fall times of only 1.6 ns. The slewing rate of this op amp is a very impressive 8000 V/μs. The settling time for the output voltage to settle to within 0.02% of the final value is 12 ns.

VTC, Inc., offers a series of video op amps that can be used in a variety of video and RF applications. They are all-bipolar devices with a complementary class AB push–pull output stage that is capable of supplying up to 50 mA to the load. The low output impedance and high output current capability of this IC are especially useful for RF and video applications where load impedances down in the 50- to 75-Ω range are often used. These op amps use NPN transistors that have a unity current gain frequency of $f_T = 6$ GHz and vertical PNP transistors with $f_T = 1.5$ GHz. The impressive high-frequency performance is obtained by using transistors with very narrow base widths. This does, however, result in lower breakdown voltages, and these op amps normally operate with supply voltages of just ±5 V.

The VTC op amps include the VA705 with a gain–bandwidth product of 25 MHz, a rise time of 7 ns, and a slewing rate of 36 V/μs. The open-loop gain is 5000 V/V, and this op amp is internally compensated for stability down to unity gain. This circuit is also available in the dual op amp VA2705 and the quad op amp 4705 versions. The VA708 has a gain–bandwidth product of 100 MHz, a rise time of 7 ns, and a slewing rate of 90 V/μs. The open-loop gain is 10,000 V/V, and this op amp is internally compensated down to a minimum closed-loop gain of 3. Dual and quad versions are available as the VA2708 and VA4708, respectively. The VA707 offers a gain–bandwidth product of 300 MHz, a rise time of 9 ns, and a slewing rate of 105 V/μs. The open-loop gain is 10,000 V/V, and this op amp is internally compensated down to a minimum closed-loop gain of 12. The dual version of this op amp is the VA2707 and the quad version is the 4707.

13.11 VERY WIDE BANDWIDTH OPERATIONAL AMPLIFIER: HA-2539

An example of an operational amplifier that has a very wide bandwidth and a very high slewing rate is the HA-2539 (Harris Semiconductor). The device can be used as a video or RF amplifier in addition to the usual operational-amplifier applications.

This IC has a gain–bandwidth product of 600 MHz and a slewing rate of 600 V/μs. When driving a 1000-Ω load with a ±10-V peak output-voltage swing, the full-power bandwidth (FPBW) is 9.5 MHz. This good high-frequency performance is, however, at the expense of the open-loop gain $A_{OL}(0)$, which is 10 V/mV.

A simplified circuit diagram of this IC is shown in Figure 13.18. The input stage consists of a complementary pair of differential amplifiers (Q_1–Q_2 and Q_1'–Q_2').

This complementary pair of differential-amplifier configurations provides for some degree of base current cancellation to produce a lower net input bias current.

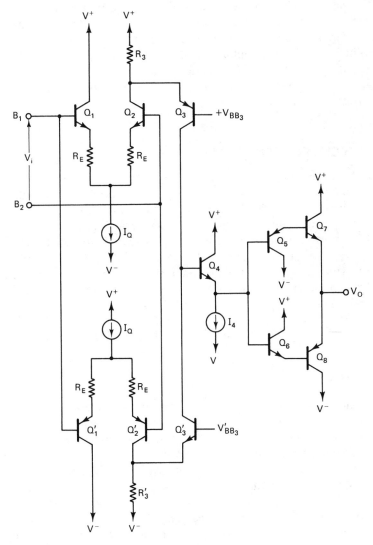

Figure 13.18 Very wide bandwidth operational amplifier simplified circuit diagram: HA-2539 (Harris Semiconductor).

This allows for the biasing current I_Q of this stage to be set at a much higher value than would otherwise be the case.

The complementary differential amplifiers drive a pair of common-base transistors (Q_3 and $Q_{3'}$). The combination of the differential amplifiers and common-base transistors essentially forms a pair of CE–CB cascode stages, which can be considered to constitute the single gain stage of the amplifier and which minimizes the total amplifier phase shift.

The output stage is a complementary push–pull Darlington emitter follower using transistors Q_4 through Q_8. Transistors Q_5 and Q_6 are used as emitter followers in the output stage and also act to minimize the crossover distortion.

The very wide bandwidth and high slewing rate of this device are the result of the following design features: (1) local negative feedback (emitter degeneration) in the differential amplifiers produced by resistors R_E and $R_{E'}$; (2) a high quiescent current I_Q for the differential-amplifier stage; (3) the single gain-stage configuration of the differential-amplifier/common-base cascode combination; and (4) the use of dielectric isolation to reduce the parasitic capacitances, especially the collector-to-substrate capacitance.

13.12 FET HIGH-FREQUENCY AMPLIFIERS

Most high-frequency amplifiers use bipolar transistors, but short-channel MOSFETs and GaAs MESFETs can offer very good performance for high-speed analog and digital circuits. The analysis of the frequency- and time-domain response of FET amplifier circuits follows along the same general lines as for bipolar circuits. Figure 13.19a shows a common-source FET amplifier and Figure 13.19b shows the equivalent-circuit representation. This small-signal model can be used for both MOSFETs as well as JFETs. The equivalent circuit is seen to be of the form of two cascaded RC low-pass networks with time constants $\tau_i = R_S C_i$ and $\tau_o = R_L (C_o + C_L)$. Using Miller's theorem, capacitances C_i and C_o are given by

$$
\begin{aligned}
C_i &= C_{gs} + C_{gd}(1 - A_v) \simeq C_{gs} + (1 + g_{fs}R_L)C_{gd} \\
C_o &= C_{ds} + C_{gd}\left(1 - \frac{1}{A_v}\right) \simeq C_{ds} + C_{gd}, \qquad \text{since usually } A_v \gg 1
\end{aligned}
\tag{13.46}
$$

The corresponding breakpoint frequencies are given by $f_i = 1/(2\pi\tau_i)$ and $f_o = 1/(2\pi\tau_o)$.

In most wide-bandwidth amplifiers, the source resistance R_S is small such that τ_i is substantially smaller than τ_o; so τ_o will be the dominant time constant. For this case, the 3-dB bandwidth is given approximately by $BW \simeq f_o = 1/[2\pi R_L(C_{ds} + C_{gs} + C_L)]$ and the rise time is $t_{\text{rise}} \simeq 2.2\,\tau_o = 2.2 R_L(C_{ds} + C_{gs} + C_L)$. Since the midfrequency voltage gain is $A_v = -g_{fs}R_L$, the gain–bandwidth product is given approximately by $|A_v \times BW| \simeq g_{fs}/[2\pi(C_{ds} + C_{gs} + C_L)]$.

For short-channel MOSFETs, a very simple relationship for the gain–bandwidth product can be obtained. The drain-to-source current can be expressed as $I_{DS} =$

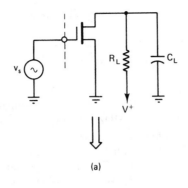

(a)

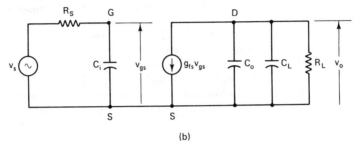

(b)

Figure 13.19 FET amplifier circuit: (a) FET common-source amplifier stage; (b) ac small-signal equivalent circuit.

$Q_{\text{channel}}/t_{\text{transit}}$. Since $Q_{\text{channel}} = C_{gs}(V_{gs} - V_{Th})$ and $t_{\text{transit}} = Lv_{\text{sat}}$, this can be rewritten as $I_{DS} = C_{gs}(V_{gs} - V_{Th})v_{\text{sat}}/L$. The dynamic transfer conductance g_{fs} is given by

$$g_{fs} = \frac{dI_{DS}}{dV_{gs}} = \frac{C_{gs}v_{\text{sat}}}{L} \tag{13.47}$$

Since C_{ds} is usually small compared to C_{gs}, and if we also assume C_L to be small compared to C_{gs}, the maximum gain–bandwidth product capability of the device can be expressed very simply by

$$(A_v \times BW)_{\text{MAX}} = \frac{g_{fs}}{2\pi C_{gs}} = \frac{v_{\text{sat}}}{2\pi L} \tag{13.48}$$

The saturation electron velocity v_{sat} in silicon is approximately 8×10^6 cm/s and in GaAs it is approximately 2×10^7 cm/s. For a silicon MOSFET with a channel length of $L = 1.0$ μm, the maximum gain–bandwidth product is about 12 GHz. For a GaAs MESFET with a 1-μm channel length, the maximum gain–bandwidth product is up to 32 GHz. Indeed, very high speed digital ICs have been developed using GaAs that offer substantial speed advantages over silicon ICs. We see therefore that short-channel field-effect transistors can indeed offer very good high-frequency performance.

PROBLEMS

13.1. (*Transistor frequency response*) A transistor has a common-emitter short-circuit current gain of 200 at $f = 10$ kHz and of 20 at $f = 50$ MHz. Find β, f_β, and f_T. (*Ans.*: 200, 5.0 MHz, 1.0 GHz)

13.2. (*Multistage amplifier frequency response*) A multistage amplifier has breakpoint frequencies at 20, 30, and 40 MHz. Find the approximate 3-dB bandwidth and rise time of this system. (*Ans.*: 15.4 MHz, 22.8 ns)

13.3. (*Common-emitter amplifier frequency response*) Refer to Figure P13.3. Given: Common-emitter amplifier stage with $I_C = 2.5$ mA, $r_{bb'} = 100$ Ω, $f_T = 1.59$ GHz, $C_{cb'} = 2.0$ pF, $\beta = 100$, and $n = 1$. (For the voltage gain, take into account the effect of $r_{bb'}$ by considering the $r_{bb'} - r_{b'e}$ voltage-division ratio.) Find the midfrequency voltage gain, 3-dB bandwidth, and gain–bandwidth product for the following values of load resistance:

(a) $R_L = 100$ Ω.   (*Ans.*: 9.1, 53 MHz, 482 MHz)
(b) $R_L = 300$ Ω.   (*Ans.*: 27, 23 MHz, 621 MHz)
(c) $R_L = 1.0$ kΩ.   (*Ans.*: 91, 7.6 MHz, 691 MHz)
(d) $R_L = 5.0$ kΩ.   (*Ans.*: 455, 1.6 MHz, 727 MHz)
(e) Show that the gain–bandwidth product can never be greater than $A_{V(MID)} \times BW = 1/(2\pi r_{bb'}C_{cb'})$, and calculate this limiting value.   (*Ans.*: 796 MHz)

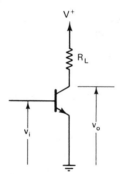

Figure P13.3

13.4. (*Cascode video amplifier*) Given: Cascode amplifier using same transistors and quiescent conditions as in Problem 13.3 and as shown in Figure P13.4.

(a) Find $A_{V(MID)}$ and BW for $R_L = 100$ Ω and $C_L = 5$ pF. Also calculate the gain–bandwidth product.   (*Ans.*: 9.1, 130 MHz, 1183 MHz)
(b) Repeat for $R_L = 1.0$ kΩ.   (*Ans.*: 91, 22 MHz, 2002 MHz)
(c) Repeat for $R_L = 100$ Ω and a source resistance of 50 Ω.   (*Ans.*: 8.7, 96 MHz, 835 MHz)

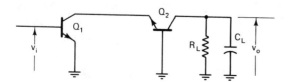

Figure P13.4

13.5. (*Video integrated circuit*) Given: Integrated circuit as shown in Figure P13.5 with V^+ = +6.0 V and V^- = −6.0 V. For all transistors, f_T = 700 MHz, $r_{bb'}$ = 40 Ω, C_{cb} = 1.0 pF, β = 100 (typ.), and V_{BE} = 0.7 V. The circuit is driven from a 50-Ω source and drives a load resistance R_L of 1.0 kΩ. Find:

(a) $I_{C1,2}$.   (*Ans.*: 1.1 mA)

(b) $V_{E7,8}$ and $I_{C7,8}$.   (*Ans.*: +2.33 V, 0.44 mA)

(c) I_{C9}.   (*Ans.*: 4.5 mA)

(d) Total quiescent supply current.   (*Ans.*: 10.6 mA)

(e) Voltage gain (midfrequency) of differential-amplifier stage with a balanced output.   (*Ans.*: 117)

(f) Voltage gain of the emitter-follower input transistors Q_1 and Q_2.   (*Ans.*: 0.975)

(g) Voltage gain of the emitter-follower output stage, Q_7 and Q_8.   (*Ans.*: 0.937)

(h) Overall voltage gain of the amplifier, v_o/v_s.   (*Ans.*: 107 or 40.6 dB)

(i) Dynamic output resistance, r_o.   (*Ans.*: 69 Ω)

(j) 3-dB bandwidth and rise time.   (*Ans.*: 54 MHz, 6.5 ns)

(k) Coupling capacitor values C_{C_1} and C_{C_2} for a lower 3-dB frequency of 10 Hz.   [*Ans.*: C_{C_1} and C_{C_2} = 15 μF (min.) when considering the response due to C_{C_1} and C_{C_2} separately, and ~23 μF (min.) when considering the net effects of both coupling capacitors]

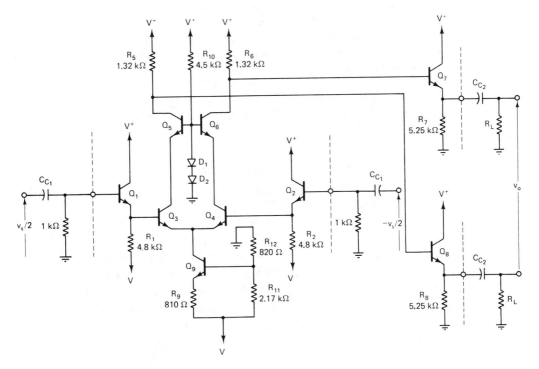

Figure P13.5

13.6. (*Wide-bandwidth operational amplifier*) Given: The very wide bandwidth, high-slewing-rate operational-amplifier circuit (HA-2539, Harris Semiconductor) shown in Figure 13.18.

(a) Show that the open-loop gain can be expressed approximately by

$$A_{OL} \simeq \frac{g_m}{(1 + g_m R_E)(j\omega C_{\text{NET}})} \simeq \frac{1}{j\omega R_E C_{\text{NET}}}, \qquad \text{if} \quad g_m R_E \gg 1$$

where $g_m = (I_Q/2)/V_T$ and $C_{\text{NET}} = C_{CB_3} + C_{CB_{3'}} + C_{CB_4}$.

(b) Show that the gain–bandwidth product (or unity-gain frequency f_u) is given approximately by

$$f_u \simeq \frac{g_m}{2\pi(1 + g_m R_E)C_{\text{NET}}} \simeq \frac{1}{2\pi R_E C_{\text{NET}}}, \qquad \text{if} \quad g_m R_E \gg 1$$

(c) If $I_Q = 2.0$ mA, $R_E = R_{E'} = 50\ \Omega$, and $C_{CB} = 1.2$ pF for all transistors, find the gain–bandwidth product (f_u).   (*Ans.*: 590 MHz)

(d) Find the slewing rate using the parameters above.   (*Ans.*: 555 V/μs)

(e) Find the full-power bandwidth for a ± 10-V swing.   (*Ans.*: 8.83 MHz)

(f) If $\beta = 100$ and $n = 1.5$ for all transistors, find the difference-mode input resistance.   (*Ans.*: 8800 Ω)

(g) Find I_{BIAS} if $\beta_{\text{NPN}} = 120$ and $\beta_{\text{PNP}} = 80$.   (*Ans.*: -4.2 μA)

(h) Describe the advantage that is gained by the use of the complementary differential-amplifier configuration of this circuit.

(i) Find the values of f_u, SR, and FPBW for the preceding problems if I_Q is reduced to only 20 μA.   (*Ans.*: 17 MHz, 5.6 V/μs, 88 kHz)

(j) Describe the relative advantages and disadvantages of the operation of this circuit with a large value of I_Q.

13.7. (*Multistage video amplifier*)   A multistage video amplifier is to amplify an input signal of 100 μV rms up to an output-voltage level of 2.0 V rms with a 30-MHz 3-dB bandwidth. This amplifier is to use wide-bandwidth operational amplifiers that have a gain–bandwidth product of 600 MHz and a slewing rate of 600 V/μs.   Determine the number of cascaded op amps that are needed and the closed-loop gain needed for each op amp.   (*Ans.*: five operational amplifiers, 7.25)

REFERENCES

Cowles, L. G., *Sourcebook of Modern Transistor Circuits*, Chapter 12, Prentice-Hall, Englewood Cliffs, N.J., 1976.

Fitchen, F. C., *Electronic Integrated Circuits and Systems*, Van Nostrand Reinhold, New York, 1970.

Grebene, A. B., *Analog Integrated Circuit Design*, Van Nostrand Reinhold, New York, 1972.

Hughes, R. S., *Logarithmic Video Amplification*, Artech House, Dedham, Mass., 1986.

Lenk, J. D., *Handbook of Integrated Circuits*, Prentice-Hall, Englewood Cliffs, N.J., 1978.

Manasse, F., *Semiconductor Electronics Design*, Prentice-Hall, Englewood Cliffs, N.J., 1977.

Millman, J., *Microelectronics*, McGraw-Hill, New York, 1979.

Rips, E. M., *Discrete and Integrated Electronics*, Prentice-Hall, Englewood Cliffs, N.J., 1986.

14

Modulators, Demodulators, and Phase Detectors

14.1 BALANCED MODULATOR–DEMODULATOR ANALYSIS

The balanced modulator–demodulator IC can be used to perform a number of functions. These include use as a suppressed carrier double-sideband modulator, suppressed carrier single-sideband modulator, single-sideband demodulator, double-sideband demodulator, frequency modulation demodulator, frequency doubler, and phase detector.

We will first analyze the basic balanced modulator–demodulator circuit shown in Figure 14.1. For this analysis, we will neglect the base currents as being very small compared to the emitter and collector currents and consider all transistors to be identical.

For the currents through Q_5 and Q_6, we have that $I_5 = I_Q + I_R$ and $I_6 = I_Q - I_R$. We also have that $I_5 = I_{CO} \exp(V_{BE5}/V_T)$ and $I_6 = I_{CO} \exp(V_{BE6}/V_T)$. Therefore,

$$V_{BE5} - V_{BE6} = (V_{B5} - V_{E5}) - (V_{B6} - V_{E6})$$

$$= V_T \left[\ln \left(1 + \frac{I_R}{I_Q} \right) - \ln \left(1 - \frac{I_R}{I_Q} \right) \right]$$

(14.1)

673

Since $V_{B5} - V_{B6} = v_m$ and $V_{E5} - V_{E6} = I_R R_M$, we have that

$$v_m - I_R R_M = V_T \left[\ln \left(1 + \frac{I_R}{I_Q} \right) - \ln \left(1 - \frac{I_R}{I_Q} \right) \right] \tag{14.2}$$

The power series expansion for $\ln (1 + x)$ is $\ln (1 + x) = x - x^2/2 + x^3/3 - \cdots$. Using this power series expansion, the equation above can be rewritten as

$$v_m - I_R R_M = V_T \left\{ \left[\frac{I_R}{I_Q} - \frac{(I_R/I_Q)^2}{2} + \cdots \right] - \left[-\frac{I_R}{I_Q} - \frac{(I_R/I_Q)^2}{2} - \cdots \right] \right\}$$

$$\approx 2V_T \frac{I_R}{I_Q}, \quad \text{for} \left(\frac{I_R}{I_Q} \right)^3 \ll 1 \tag{14.3}$$

Solving for I_R, we obtain $I_R [R_M + (2V_T/I_Q)] \approx v_m$, so $I_R \approx v_m/[R_M + (2V_T/I_Q)]$.

If v_c is of large enough amplitude, the transistor pairs Q_1 and Q_2 and Q_3 and Q_4 will be switched from near cutoff to near full conduction. That is, when v_c is positive, Q_2 and Q_3 will be near cutoff, and Q_1 and Q_4 will carry almost all the current of the differential-amplifier pair. This is I_5 for the Q_1–Q_2 pair and I_6 for the Q_3–Q_4 pair. Therefore, $I_1 \approx I_5$ and $I_4 \approx I_6$ under these conditions. When v_c is negative,

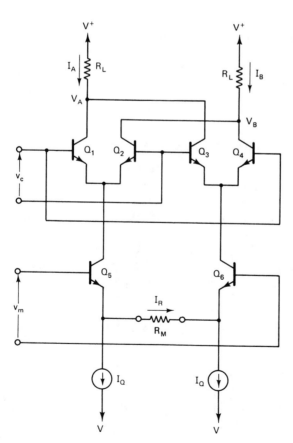

Figure 14.1 Balanced modulator–demodulator.

Modulators, Demodulators, and Phase Detectors   Chap. 14

we have that $I_2 \simeq I_5$ and $I_3 \simeq I_6$. The switching action of the two differential-amplifier transistor pairs is essentially the case for signal amplitudes in excess of 200 mV, although even down to 100 mV it will be a good approximation.

14.1.1 Switching Function

The switching function $S(t)$ is shown in Figure 14.2a. It is a square wave with an amplitude of 1.0 and a period of T_C. The period of the switching function is related to the frequency of the voltage v_c applied to the carrier input of the modulator as $T_C = 1/f_c$. We will assume that v_c is of the form $v_c = V_C \sin \omega_c t$.

Since $S(t)$ is a square wave of unity amplitude and frequency f_c, it can be represented in terms of the following Fourier series:

$$S(t) = \frac{1}{2} + \sum_{n=1}^{\infty} \frac{2}{n\pi} \sin n\omega_c t, \qquad \text{where } n \text{ is an odd integer} \qquad (14.4)$$

Shifting $S(t)$ by one-half of the period along the time axis produces the function $S(t - T_C/2)$, as shown in Figure 14.2b.

If we assume a sufficiently large signal amplitude for v_c, the collector currents of Q_1, Q_2, Q_3, and Q_4 can be expressed as

$$
\begin{aligned}
I_1 &= I_5 S(t), & I_3 &= I_6 S\left(t - \frac{T_C}{2} \right) \\
I_2 &= I_5 S\left(t - \frac{T_C}{2} \right), & I_4 &= I_6 S(t)
\end{aligned}
\qquad (14.5)
$$

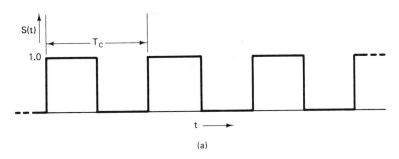

(a)

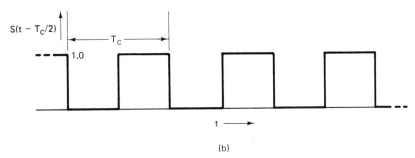

(b)

Figure 14.2 Switching functions: (a) switching function $S(t)$; (b) switching function $S(t - T_c)$.

The currents I_A and I_B are $I_A = I_1 + I_3 = I_5 S(t) + I_6 S(t - T_C/2)$ and $I_B = I_2 + I_4 = I_5 S(t - T_C/2) + I_6 S(t)$. If we now express I_5 and I_6 as $I_5 = I_Q + I_R$ and $I_6 = I_Q - I_R$, we obtain

$$I_A = (I_Q + I_R)S(t) + (I_Q - I_R)S\left(t - \frac{T_C}{2}\right)$$

$$= I_Q\left[S(t) + S\left(t - \frac{T_C}{2}\right)\right] + I_R\left[S(t) - S\left(t - \frac{T_C}{2}\right)\right] \tag{14.6}$$

and

$$I_B = (I_Q + I_R)S\left(t - \frac{T_C}{2}\right) + (I_Q - I_R)S(t)$$

$$= I_Q\left[S(t) + S\left(t - \frac{T_C}{2}\right)\right] - I_R\left[S(t) - S\left(t - \frac{T_C}{2}\right)\right] \tag{14.7}$$

We now note that $S(t) + S(t - T_C/2) = 1$, and thus we have that

$$S(t) - S\left(t - \frac{T_C}{2}\right) = S(t) - [1 - S(t)] = 2S(t) - 1 = \sum_{n=1}^{\infty} \frac{4}{n\pi} \sin n\omega_c t \tag{14.8}$$

where n is an odd integer. We can now write the expressions for I_A and I_B as

$$I_A = I_Q + I_R \sum_{n=1}^{\infty} \frac{4}{n\pi} \sin n\omega_c t \tag{14.9}$$

and

$$I_B = I_Q - I_R \sum_{n=1}^{\infty} \frac{4}{n\pi} \sin n\omega_c t \tag{14.10}$$

If $v_m = V_M \sin \omega_m t$, then I_R is

$$I_R = \frac{V_M \sin \omega_m t}{R_M + (2V_T/I_Q)} = \frac{V_M \sin \omega_m t}{R_M + 2r_{eb}} \tag{14.10a}$$

where $r_{eb} = V_T/I_Q$. If we substitute this expression for I_R into the equations for I_A and I_B above, we obtain

$$I_A = I_Q + \sum_{n=1}^{\infty} \frac{4}{n\pi} \frac{V_M \sin \omega_m t \sin n\omega_c t}{R_M + 2r_{eb}} \tag{14.11}$$

and

$$I_B = I_Q - \sum_{n=1}^{\infty} \frac{4}{n\pi} \frac{V_M \sin \omega_m t \sin n\omega_c t}{R_M + 2r_{eb}} \tag{14.12}$$

If we now use the trigonometric identity that

$$\sin x \sin y = \frac{\cos (x - y) - \cos (x + y)}{2}$$

in the equation above, we obtain

$$I_A = I_Q + \sum_{n=1}^{\infty} \frac{2}{n\pi} \frac{V_M}{R_M + 2r_{eb}} [\cos (n\omega_c - \omega_m)t - \cos (n\omega_c + \omega_m)t] \tag{14.13}$$

and

$$I_B = I_Q - \sum_{n=1}^{\infty} \frac{2}{n\pi} \frac{V_M}{R_M + 2r_{eb}} [\cos(n\omega_c - \omega_m)t - \cos(n\omega_c + \omega_m)t] \quad (14.14)$$

Since $V_A = V^+ - I_A R_L$ and $V_B = V^+ - I_B R_L$, $V_O = V_B - V_A = (I_A - I_B)R_L$. Using the expressions for I_A and I_B above, we obtain for V_O the equation

$$
\boxed{
\begin{aligned}
V_O &= V_B - V_A \\
&= \sum_{n=1}^{\infty} \frac{4V_M R_L}{(n\pi)(R_M + 2r_{eb})} [\cos(n\omega_c - \omega_m)t - \cos(n\omega_c + \omega_m)t]
\end{aligned}
} \quad (14.15)
$$

where again we note that n is an odd integer.

14.1.2 Modulator Operation with Small Carrier Amplitudes

We will now consider the operation of this modulator–demodulator circuit for the case in which the carrier signal amplitude is small such that the Q_1–Q_2 and Q_3–Q_4 differential-amplifier transistor pairs will be operating in the approximately linear region of the transfer characteristics. For this to be the case, the amplitude of the signal applied to the carrier input port v_c should not be in excess of about 25 mV.

For this case, we have that the ac collector current components of Q_1 and Q_2 are given by $i_1 = g_f v_c$ and $i_2 = -g_f v_c = -i_1$, where g_f is the dynamic forward transfer conductance of the Q_1–Q_2 differential amplifier. Since the quiescent current of the Q_1–Q_2 pair is I_5, g_f is given by $g_f = I_5/4V_T$, and $I_5 = I_Q + v_m/(R_m + 2r_{eb})$. We therefore have for i_1 the equation

$$
\begin{aligned}
i_1 = g_f v_c &= \frac{I_Q + v_m/(R_M + 2r_{eb})}{4V_T} v_c \\
&= \frac{I_Q v_c}{4V_T} + \frac{v_m v_c}{4V_T(R_M + 2r_{eb})}
\end{aligned}
\quad (14.16)
$$

For the Q_3–Q_4 pair, we similarly have

$$i_4 = \frac{I_Q v_c}{4V_T} - \frac{v_m v_c}{4V_T(R_M + 2r_{eb})} \quad (14.17)$$

We again note that $i_2 = -i_1$ and also that $i_3 = -i_4$. The ac components of $I_A = I_1 + I_3$ and $I_B = I_2 + I_4$ are therefore

$$i_A = i_1 + i_3 = \frac{v_m v_c}{2V_T(R_M + 2r_{eb})} \quad \text{and} \quad i_B = i_2 + i_4 = \frac{-v_m v_c}{2V_T(R_M + 2r_{eb})} \quad (14.18)$$

Note here the cancellation of the $I_Q v_c/4V_T$ term in i_A and i_B.

The ac balanced output voltage v_o of this circuit is

$$v_o = v_b - v_a = (i_a - i_b)R_L = \frac{v_m v_c R_L}{V_T(R_M + 2r_{eb})} \qquad (14.19)$$

If $v_c = V_C \sin \omega_c t$ and $v_m = V_M \sin \omega_m t$, we obtain

$$v_o = \frac{V_C V_M R_L}{2V_T(R_M + 2r_{eb})} [\cos(\omega_c - \omega_m)t - \cos(\omega_c + \omega_m)t] \qquad (14.20)$$

14.2 CONVERSION TRANSFER CONDUCTANCE AND GAIN

We will define the *dynamic sideband conversion transfer conductance* y_c as the ratio of the ac output current at the sideband frequency $nf_c \pm f_m$ to the modulation input voltage v_m at frequency f_m. We accordingly have that $y_c = i_o(nf_c \pm f_m)/v_m(f_m)$, where i_o is the ac component of I_A or I_B. By inspection of the equations for I_A and I_B for the large carrier amplitude case ($V_C \gtrsim 50$ mV), the conversion transfer conductance is

$$y_c = \frac{2}{(n\pi)(R_M + 2r_{eb})} \qquad (14.21)$$

with respect to the sidebands at frequency $nf_c \pm f_m$. This is the conversion transfer conductance with respect to I_A. For the transfer conductance with respect to I_B, the value is the same except for a negative sign.

We note that the expressions for I_A and I_B and therefore for V_O do not contain any term at the carrier frequency f_c. The only terms present are those of the various sidebands at $nf_c \pm f_m$.

If the output of this modulator is passed through a low-pass or bandpass filter such that only the $f_c + f_m$ and $f_c - f_m$ components of the total signal are passed through, the ac output voltage is

$$v_o = \frac{4V_m R_L}{\pi(R_M + 2r_{eb})} [\cos(\omega_c - \omega_m)t - \cos(\omega_c + \omega_m)t] \qquad (14.22)$$

The term at frequency $f_c - f_m$ is the lower sideband and the term at $f_c + f_m$ is the upper sideband. We again note the absence of the term at the carrier frequency. This type of signal voltage waveform is called a *double-sideband suppressed carrier* (DSB/SC) *amplitude-modulated* (AM) signal.

The dynamic sideband conversion transfer conductance for the small carrier case ($V_C \lesssim 25$ mV) can be obtained from inspection of the equations for I_A and I_B as

$$y_c = \frac{V_C}{4V_T(R_M + 2r_{eb})} \qquad (14.23)$$

We note that in this case the transfer conductance is proportional to the signal level at the carrier input port.

Inspection of the equation for v_o for the case of a small carrier amplitude at the carrier input port shows that the only frequency components present are the upper sideband frequency at $f_c + f_m$ and the lower sideband frequency at $f_c - f_m$. We therefore again have the production of a DSB/SC AM signal.

The *sideband conversion gain* is the ratio of the output-voltage component at the sideband frequency to the modulation input voltage. In terms of the dynamic sideband conversion transfer conductance, this gain can be expressed as

$$\text{conversion gain} = \frac{v_o(nf_c \pm f_m)}{v_m(f_m)} = 2y_c R_L \tag{14.24}$$

for a balanced or double-ended output. For a single-ended (or unbalanced) output from either side of the modulator (V_A or V_B), the gain is one-half of this or just $y_c R_L$.

14.3 AMPLITUDE MODULATION

The amplitude-modulated waveform that is produced by the modulation of the amplitude or envelope of a radio-frequency (RF) carrier can be described mathematically by $v(t) = A(t) \sin \omega_c t$, where $A(t)$ is the instantaneous amplitude of the waveform. The amplitude $A(t)$ will usually be a linear function of the modulating voltage $v_m(t)$.

If the modulating voltage is of the form $v_m(t) = V_M \sin \omega_m t$, then the amplitude function $A(t)$ can in general be written as

$$A(t) = V_C + kV_M \sin \omega_m t = V_C(1 + m \sin \omega_m t) \tag{14.25}$$

where $m = kV_M/V_C$ is the *modulation index* and V_C is the amplitude of the unmodulated carrier.

As long as the modulation index m is less than unity, the envelope of the AM waveform will be a replica of the original modulating signal, and this signal can be detected or demodulated at the receiver by a simple *envelope detector*. An envelope detector is a circuit that produces a voltage that is equal to or proportional to the envelope or amplitude variations of the amplitude-modulated signal. An envelope detector can be a peak detector circuit or a rectifier circuit with a suitable low-pass filter to remove the carrier frequency components of the rectified voltage.

This ease of demodulation or detection is the principal advantage of this type of amplitude-modulated waveform with $m \leq 1$. This type of AM waveform is called *double-sideband/large carrier* (DSB/LC). The principal disadvantage of the DSB/LC signal is the inefficient use of the transmitted power.

The information in an amplitude-modulated signal is contained entirely in the sidebands. The DSB/LC voltage is given by

$$v(t) = A(t) \sin \omega_c t = V_C(1 + m \sin \omega_m t) \sin \omega_c t \tag{14.26}$$

$$= V_C\left[\sin \omega_c t + \frac{m}{2} \cos (\omega_c - \omega_m)t - \frac{m}{2} \cos (\omega_c + \omega_m)t\right]$$

The spectrum of this AM signal consists of the carrier of amplitude V_C at f_c and the upper and lower sidebands of amplitude $mV_C/2$ at $f_c + f_m$ and $f_c - f_m$, respectively.

The amount of power contained in each spectral component of the signal is proportional to the square of the amplitude. Therefore, the ratio of the power in each sideband to the carrier power is $m^2/4$ and the ratio of the total sideband power to the carrier power is $m^2/2$.

Note that the carrier component is of fixed amplitude and thus carries no information. The ratio of the power in the information-carrying sidebands to the total transmitted power is thus $(m^2/2)/[1 + (m^2/2)] = m^2/(2 + m^2)$. The maximum value that this ratio has for the DSB/LC case occurs when $m = 1$ and is 0.33. Thus only 33% of the total transmitted power is in the information-carrying sidebands and the remaining 67% is in the carrier, which contains no information. In practice, the modulation index is often limited to a maximum value of around 0.8 to prevent any possibility of overmodulation. Overmodulation occurs whenever m exceeds unity and leads to severe distortion of the detected signal when an envelope type of detector is used. If the maximum value of m is set at 0.8, the average value is 0.4. For this case the sideband power is only 7.4% of the total transmitted power. We therefore see that the system is operated very inefficiently indeed.

In the case of the double-sideband/suppressed carrier (DSB/SC) signal, all the power is in the sidebands, so the efficiency can approach 100% for the case of complete carrier suppression. The DSB/SC system does, however, have the disadvantage of not being able to utilize the simple envelope-detection method of demodulation.

In both the DSB/SC and the DSB/LC systems, the same information is contained in each of the two sidebands; that is, whatever information is contained in the lower sideband is exactly repeated in the upper sideband. As a result, the total bandwidth involved in the double-sideband systems is twice as much as is really needed for the transmission of the information. A system that uses only one sideband is called a *single-sideband suppressed carrier* (SSB/SC) system. This type of system represents the ultimate in efficiency with respect to both power and bandwidth.

14.3.1 Modulation and Demodulation Circuits for DSB/SC AM Signals

Figure 14.3 shows a modulation circuit for DSB/SC AM. The bandpass filter has a passband that includes the frequency range from $f_c - f_m$ to $f_c + f_m$, but does not include the sidebands of the carrier harmonic frequencies, such as $3f_c \pm f_m$, $5f_c \pm f_m$, and so on.

If the modulator is well balanced, there will be very little carrier feedthrough and there will thus be almost complete suppression of the carrier and the output will just be at the carrier sideband frequencies $f_c - f_m$ and $f_c + f_m$. The demodulator is very similar to the modulator, as shown in Figure 14.4, and indeed the same type of IC can be used.

The DSB/SC signal is applied to the modulation input port and a locally generated large amplitude (≥ 100 mV) signal at a frequency f_c', which is nominally equal to the carrier frequency is injected into the carrier input port. The nonlinear mixing action that occurs in the modulator–demodulator results in an output voltage that includes the following frequency components: $f_c' - (f_c \pm f_m) = f_c' - f_c + f_m$ and

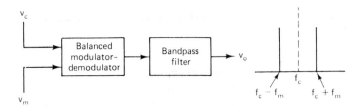

Figure 14.3 Balanced modulator circuit to produce DSB/SC AM.

$f'_c - f_c - f_m$, and $f'_c + (f_c \pm f_m) = f'_c + f_c + f_m$ and $f'_c + f_c - f_m$. In addition, there will also be components involving various odd multiples of f'_c.

If this output voltage now goes through a low-pass filter with a cutoff frequency f_1 such that $f_m < f_1 < 3f_c - f_m$, we obtain at the output only $f_m \pm (f'_c - f_c)$. We see that for an exact recovery of the original modulating waveform the reinjected carrier frequency must be exactly equal to the original suppressed carrier frequency f_c. If there is any difference between these two frequencies, the detected signal is given by

$$v(t) = V_M \cos(\omega_m t + \Delta\omega_c t) + V_M \cos(\omega_m t - \Delta\omega_c t)$$

$$= 2V_M \cos\omega_m t \cos\Delta\omega_c t \qquad (14.27)$$

where $\Delta\omega_c = \omega'_c - \omega_c$. Thus the detected signal is amplitude modulated at the difference frequency $\Delta f_c = f'_c - f_c$. If there is exact frequency synchronism, but there is still a phase angle difference θ between the original carrier and the reinjected carrier, we have that the detected signal is $v(t) = V_M \cos\omega_m t \cos\theta$. This is not a serious problem as long as θ is constant and is not very large. However, if θ were to approach 90°, this would result in a severe reduction in the amplitude of the detected signal.

To facilitate the exact frequency and phase matching of the reinjected carrier at the receiver with the original carrier at the transmitter, a small-amplitude *pilot carrier* is often transmitted either at the original carrier frequency or at some integer multiple thereof. This pilot carrier serves as a reference for the locally generated carrier in order that it exactly match the original carrier in frequency and also be very close to the phase of the original carrier.

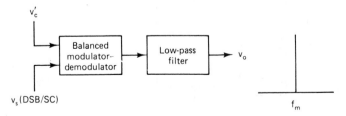

Figure 14.4 Circuit for the demodulation of a DSB/AM signal.

14.3.2 Modulation and Demodulation Circuits for SSB/SC AM Signals

Figure 14.5 shows a modulation circuit for the generation of a SSB/SC AM signal. It might at first be thought that the simplest way of producing a SSB/SC signal would be to just filter out one of the sidebands using a bandpass filter. In many cases this is indeed done, especially in cases where the sidebands extend over a relatively wide frequency range.

In the case of television broadcasting this method of filtering out one of the sidebands is indeed done. Since it is not possible to have a wide-bandwidth filter that will sharply cut off at the carrier frequency, part of one sideband (the lower sideband) is transmitted along with the entire upper sideband. The video bandwidth is 4.5 MHz, so the total bandwidth that would be required if both sidebands were to be transmitted would be 9.0 MHz. The bandpass characteristic at the transmitter is such that the entire 4.5-MHz upper sideband is transmitted, but only 1.25 MHz of the lower sideband is transmitted, so the total bandwidth required for the video portion of the transmission is now only 5.75 MHz. This technique of transmitting all of one sideband and a small part of the other sideband is called *vestigial sideband transmission*. The actual bandwidth allotted per TV channel is 6.0 MHz to allow room for the audio portion of the signal.

In the circuit of Figure 14.5, the removal of one of the sidebands occurs not by the action of a bandpass filter, but rather by the canceling effect that results from the 90° phase shift given to both the carrier and signal input voltages supplied to one of the modulators. In many cases, this method is chosen because it would be very difficult to filter out one of the sidebands without severely affecting the other sideband.

The output voltage of modulator 1 can be expressed as

$$v_{o1} = \sum_{n=1}^{\infty} \frac{K}{n} \left[\cos\,(n\omega_c - \omega_m)t - \cos\,(n\omega_c + \omega_m)t \right] \tag{14.28}$$

where $K = 4V_M R_L / \pi(R_M + 2r_{eb})$ and n is an odd integer.

The output voltage of modulator 2 is the same as that given above except for the 90° phase shift given to both the carrier and modulation input signals. Noting that $\cos\,(x \pm 180°) = -\cos x$, we have that

$$v_{o2} = \sum_{n=1}^{\infty} \frac{K}{n} \left[\cos\,(n\omega_c - \omega_m)t + \cos\,(n\omega_c + \omega_m)t \right] \tag{14.29}$$

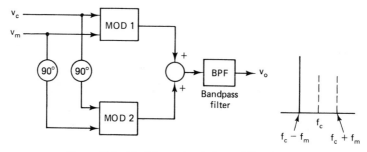

Figure 14.5 Modulator circuit for an SSB/SC signal.

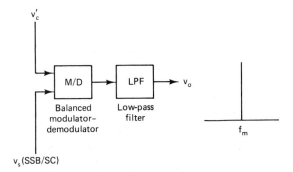

Figure 14.6 Demodulator circuit for a SSB/SC signal.

When these two voltages are added together, we obtain

$$v_o = v_{o_1} + v_{o_2} = 2 \sum_{n=1}^{\infty} \frac{K}{n} [\cos (n\omega_c - \omega_m)t] \qquad (14.30)$$

We therefore see that the upper sideband has canceled out, so we obtain an SSB/SC signal with just the lower sideband.

In the single-sideband modulator circuit just discussed, the 90° phase shift given to the carrier and modulating input signals of modulator 2 is of the same algebraic sign for both inputs. If the 90° phase shifts of these two signals are of opposite algebraic sign, we obtain for v_{o_2} the expression

$$\begin{aligned} v_{o_2} &= \sum_{n=1}^{\infty} \frac{K}{n} [\cos [(n\omega_c - \omega_m)t + 180°] - \cos (n\omega_c + \omega_m)t] \\ &= \sum_{n=1}^{\infty} \frac{K}{n} [-\cos (n\omega_c - \omega_m)t - \cos (n\omega_c + \omega_m)t] \end{aligned} \qquad (14.31)$$

When this voltage is added to v_{o_1}, we obtain

$$v_o = v_{o_1} + v_{o_2} = -2 \sum_{n=1}^{\infty} \frac{K}{n} [\cos (n\omega_c + \omega_m)t] \qquad (14.32)$$

so now we have an SSB/SC signal with just the upper sideband present.

Figure 14.6 shows a demodulator circuit for an SSB/SC signal. The reinjected carrier signal is at frequency f_c', which is nominally the same as the carrier frequency f_c. The low-pass filter has a passband that cuts off somewhere beyond f_m, but well below f_c.

If the carrier and signal input voltages are $v_c'(t) = V_C' \sin \omega_c' t$ and $v_s(t) = V_S \sin (\omega_c - \omega_m)t$, respectively, at the output of the low-pass filter we have

$$v_o(t) = \frac{4V_S R_L}{\pi(R_M + 2r_{eb})} \cos (\omega_m t + \Delta\omega_c t) \qquad (14.33)$$

where $\Delta\omega_c = \omega_c' - \omega_c$. Thus, to obtain an exact replica of the original modulating signal, it is necessary to have the reinjected carrier frequency exactly match the original carrier frequency. Any difference in the two frequencies will result in a frequency shift in the demodulated output by an amount $\Delta f = f_c' - f_c$.

14.4 FREQUENCY MODULATION

For a frequency-modulated (FM) waveform, the instantaneous frequency of a carrier signal is varied or modulated by a modulating signal. The modulation is usually done in a way such that the variation or shift in the carrier frequency is a linear function of the modulating voltage.

If the modulating signal is of the form $v_m(t) = V_M \sin \omega_m t$, the instantaneous carrier frequency is $f(t) = f_c + kV_M \sin \omega_m t = f_c + \Delta f \sin \omega_m t$, where f_c is the unmodulated carrier frequency and Δf is the *peak frequency deviation*.

The radian frequency ω of a sinusoidal waveform is related to the phase angle θ by $\omega = d\theta/dt$, so that $\theta = \int_0^t \omega \, dt$. Applying this to the modulated carrier frequency, we obtain

$$\theta = \int_0^t \omega \, dt = \int_0^t (\omega_c + \Delta\omega \sin \omega_m t) \, dt = \omega_c t + \frac{\Delta\omega}{\omega_m} \cos \omega_m t$$

$$= \omega_c t + \beta \cos \omega_m t \qquad (14.34)$$

where $\beta = \Delta\omega/\omega_m = \Delta f/f_m = $ *modulation index*. The frequency-modulated signal can be written as

$$v(t) = V_C \sin \theta(t) = V_C \sin (\omega_c t + \beta \cos \omega_m t) \qquad (14.35)$$

If we now apply the trigonometric identity that $\sin (x + y) = \sin x \cos y + \cos x \sin y$, the equation for $v(t)$ can be written as

$$v(t) = V_C[\sin \omega_c t \cos (\beta \cos \omega_m t) + \cos \omega_c t \sin (\beta \cos \omega_m t)] \qquad (14.36)$$

For small values of the modulation index such that $\beta \leq 0.3$, we can use the approximation that $\cos (\beta \cos \omega_m t) \simeq 1$ and $\sin (\beta \cos \omega_m t) \simeq \beta \cos \omega_m t$. We can therefore express $v(t)$ approximately as

$$v(t) \simeq V_C(\sin \omega_c t + \beta \cos \omega_m t \cos \omega_c t)$$

$$\simeq V_C\left[\sin \omega_c t + \frac{\beta}{2} \cos (\omega_c + \omega_m)t + \frac{\beta}{2} \cos (\omega_c - \omega_m)t \right] \qquad (14.37)$$

This type of frequency modulation with small values of modulation index ($\beta \leq 0.3$) is called *narrow-bandwidth frequency modulation* (NBFM).

We will now compare the expression just obtained for the NBFM signal with that obtained earlier for the DSB/LC AM signal, which can be expressed as

$$v(t) = V_C(1 + m \cos \omega_m t) \cos \omega_c t$$

$$= V_C\left[\cos \omega_c t + \frac{m}{2} \cos (\omega_c - \omega_m)t + \frac{m}{2} \cos (\omega_c + \omega_m)t \right] \qquad (14.38)$$

after noting that $\cos x \cos y = \frac{1}{2} \cos (x - y) + \frac{1}{2} \cos (x + y)$. We see that the spectra of the two cases are the same with a carrier at f_c, an upper sideband at $f_c + f_m$, and a lower sideband at $f_c - f_m$. The principal difference in the two signals is that in the NBFM case the carrier is in phase quadrature with respect to the sidebands;

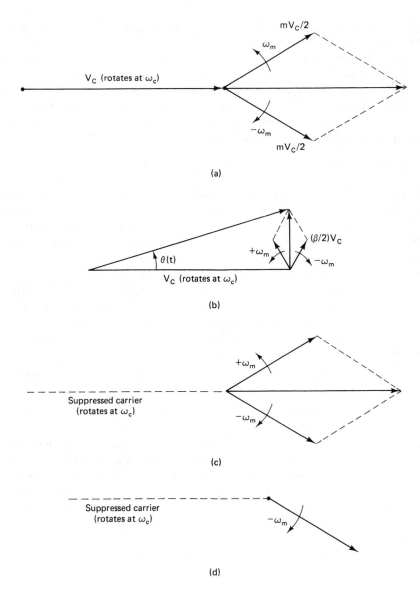

Figure 14.7 Phasor diagram for various types of modulation: (a) DSB/AM; (b) NBFM; (c) DSB/AM; (d) SSB/SC AM.

that is, at a reference time $t = 0$ there is a 90° phase angle between the carrier and the sideband phasors.

In Figure 14.7 a number of phasor diagrams are presented showing the relationships between DSB/LC AM, DSB/SC AM, SSB/SC AM, and NBFM. For these diagrams the phasor representing the carrier is used as the reference, and the phasor rotational angular velocities are taken with respect to the carrier phasor, which itself rotates at an angular velocity of ω_c. Notice in particular the relationship between DSB/LC AM and NBFM. For the FM case, note that the resultant phasor has an angular velocity that is not constant with time, but rather will fluctuate about an average value of ω_c at a rate determined by the modulating frequency ω_m.

From the phasor diagram, we see that the instantaneous phase of the resultant signal with respect to the unmodulated carrier $\theta(t)$ will vary with time as $\theta(t) \simeq \beta \cos \omega_m t$. We therefore see again that there is a close interrelationship between *phase modulation* and *frequency modulation*.

14.4.1 NBFM Modulator Circuit

Narrow-bandwidth FM can be generated using a balanced modulator circuit, as shown in Figure 14.8. The modulating signal $v(t) = V_M \sin \omega_m t$ first goes through an integrator circuit to produce $\int v(t)\,dt = (V_M/\omega_m) \cos(\omega_m t)$. This modulating voltage then mixes with the carrier in the balanced modulator. The carrier signal is $v_c(t) = V_C \sin \omega_c t$, but after passing through the 90° phase shift circuit becomes $V_C \cos \omega_c t$ and is then applied to the carrier input port of the modulator.

The output voltage of the modulator then goes through a bandpass filter to remove the sidebands of the carrier harmonic frequencies, such as $3f_c \pm f_m$, $5f_c \pm f_m$, and so on. At the output of the bandpass filter, we have $v_1(t) = V_0[\cos(\omega_c - \omega_m)t + \cos(\omega_c + \omega_m)t]$. This voltage is then added to the carrier to produce the signal given by

$$v_o(t) = V_C \sin \omega_c t + V_0[\cos(\omega_c + \omega_m)t + \cos(\omega_c - \omega_m)t]$$

$$= V_C\left[\sin \omega_c t + \frac{\beta}{2}\cos(\omega_c + \omega_m)t + \frac{\beta}{2}\cos(\omega_c - \omega_m)t\right]$$

(14.39)

where $\beta = 2V_0/V_C =$ FM modulation index. Comparison of this signal with that obtained earlier for NBFM shows that it is identical to the NBFM expression.

The circuit for NBFM can also be used as the basis for generating wide-bandwidth FM. This can be accomplished by taking the NBFM signal and feeding it into a frequency-multiplier circuit. If the signal supplied to the frequency multiplier is $v_i(t) = V_C \sin(\omega_c t + \beta \cos \omega_m t)$, the output voltage is of the form $v_o(t) = V'_C \sin(n\omega_c t + n\beta \cos \omega_m t)$. We see that the modulation index is now $\beta' = n\beta$. The peak frequency deviation Δf is now given by $\Delta f' = n\,\Delta f = n\beta f_m$.

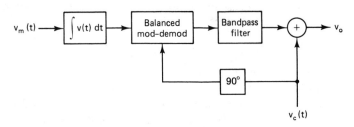

Figure 14.8 NBFM modulator circuit.

14.4.2 FM Demodulator

The objective of an FM demodulator or detector is to produce an output voltage that is directly proportional to the instantaneous frequency deviation away from the unmodulated carrier frequency. An FM demodulator circuit is shown in Figure 14.9. The FM signal first goes through a limiter to remove any amplitude modulation that may be present. This amplitude modulation may be the result of noise, interference, variations in incoming signal strength, or variations in amplifier gain. The amplitude modulation may also be present intentionally, such as in the case of television broadcasting wherein the video signal is carried by the amplitude modulation and the sound by frequency modulation.

After the signal passes through the limiter circuit it is split, part of it going directly to the carrier input port of the balanced modulator-demodulator. The other part of the signal goes through the *quadrature circuit* and then is fed to the modulation port.

The quadrature circuit consists of a small capacitor C_2 and a parallel resonant circuit represented here by L_1, C_1, and G_1. The admittance of the resonant circuit can be written as $Y = G_1(1 + j2Q_o\delta)$, where δ is the fractional detuning as given by $\delta = (f - f_o)/f_o = \Delta f/f_o$ and $Q_o = \omega_o C_1/G_1 = 1/\omega_o L_1 G_1$. The voltage transfer ratio of the quadrature circuit is

$$T = \frac{v_2}{v_1} = \frac{1/Y}{(1/Y) + (1/j\omega C_2)} = \frac{j\omega C_2}{j\omega C_2 + Y} = \frac{j\omega C_2}{G_1(1 + j2Q_o\delta) + j\omega C_2} \qquad (14.40)$$

If $(\omega C_2)^2 \ll G_1^2$, the transfer ratio can be written approximately as

$$T = \frac{v_2}{v_1} \simeq \frac{j\omega C_2}{G_1(1 + j2Q_o\delta)} \qquad (14.41)$$

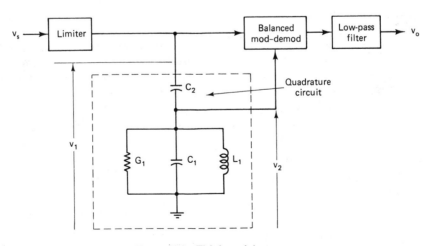

Figure 14.9 FM demodulator.

The angle of the transfer ratio $\angle T = \phi$ will be $\phi = (\pi/2) - \tan^{-1}(2Q_o\delta)$. Since the 3-dB bandwidth of the tuned circuit is related to the Q-factor as $BW = f_o/Q_o$, we have that $2Q_o\delta = 2Q_o\,\Delta f/f_o = 2\,\Delta f/BW$, where $\Delta f = f - f_o$ is the frequency deviation from the resonant frequency of the tuned circuit.

The equation for the phase shift ϕ can now be expressed as $\phi = (\pi/2) - \tan^{-1}(2\,\Delta f/BW)$. For $(\Delta f/BW)^2 \ll 1$, we can rewrite this expression as $\phi = (\pi/2) - 2\,\Delta f/BW$. We see that the phase angle ϕ will be $90°$ when $f = f_o$ and will vary linearly from that value as the frequency increases or decreases.

The input signals to the balanced modulator–demodulator can be expressed as $v_1 = V_1 \sin(\omega_c + \Delta\omega)t$ and $v_2 = V_2 \sin(\omega_c t + \Delta\omega t + \phi)$. We will assume that the resonant frequency of the tuned circuit f_o is equal to the unmodulated carrier frequency f_c. The output voltage of the modulator–demodulator then passes through a low-pass filter. The filter has a cutoff frequency such that the modulation frequency is passed through, but the carrier frequency and harmonics thereof are rejected.

At the output of the filter we have only the lowest-frequency difference term as given by

$$v_o = V_o \cos\left[(\omega_c + \Delta\omega)t - (\omega_c + \Delta\omega)t - \phi\right]$$

$$= V_o \cos\phi = V_o \cos\left(\frac{\pi}{2} - \frac{2\,\Delta f}{BW}\right) \qquad (14.42)$$

$$= V_o \sin\frac{2\,\Delta f}{BW}$$

Since $\sin x \simeq x$ for small x, we have that $v_o \simeq V_o (2\,\Delta f/BW)$ for $(\Delta f/BW)^2 \ll 1$. Thus we see that the objective of the FM detector has been achieved in that an output voltage is produced that is a linear function of the frequency deviation of the FM signal away from the carrier frequency.

14.5 FREQUENCY DOUBLER

The balanced modulator–demodulator can also be used as a frequency doubler. A frequency-doubler circuit is shown in Figure 14.10. The input signal represented by $v_s = V_S \sin\omega_s t$ is supplied to both the carrier and the modulation input ports. Since $f_c = f_m = f_s$ in this case, the output signal (for the large signal case) is given by

$$v_o = \sum_{n=1}^{\infty} \frac{4V_S R_L}{(n\pi)(R_M + 2r_{eb})} \left[\cos(n\omega_s - \omega_s)t - \cos(n\omega_s + \omega_s)t\right] \qquad (14.43)$$

where n is an odd integer. There are therefore frequency components at dc, $2f_s$, $4f_s$, $6f_s$, and so on.

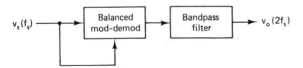

Figure 14.10 Frequency-doubler circuit.

If the output is now passed through a bandpass filter that passes the $2f_s$ component of the modulator output but rejects the higher harmonics, we obtain

$$v_o(t) = \frac{-4V_S R_L}{\pi(R_M + 2r_{eb})} (\cos 2\omega_s t)\left(1 - \frac{1}{3}\right)$$

$$= \frac{-8V_S R_L}{3\pi(R_M + 2r_{eb})} \cos 2\omega_s t$$

(14.44)

The voltage-doubler conversion gain is given as

$$\text{doubler conversion gain} = \frac{v_o(2f_s)}{v_s(f_s)} = \frac{8R_L}{3\pi(R_M + 2r_{eb})}$$

(14.45)

14.6 PHASE DETECTOR USING THE BALANCED MODULATOR–DEMODULATOR

The balanced modulator–demodulator circuit can be used as a phase detector. The basic phase detector circuit is shown in Figure 14.11. The two signals whose phase is to be compared are represented by $v_c = V_C \sin \omega t$ supplied to the carrier input port and $v_m = V_M \sin(\omega t + \phi)$ supplied to the modulation input port.

We will assume that the amplitude of the voltage applied to the carrier input port is large enough ($\gtrsim 100$ mV) such that we have the same relationships for the currents through Q_1, Q_2, Q_3, and Q_4 as given before in terms of the switching functions $S(t)$ and $S(t - T/2)$. These relationships are $I_1 = I_5 S(t)$, $I_2 = I_5 S(t - T/2)$, $I_3 = I_6 S(t - T/2)$, and $I_4 = I_6 S(t)$, where T is the period of the input voltage.

For this analysis we will consider the case in which there is a short-circuit connection between the emitters of Q_5 and Q_6 ($R_M = 0$). If the voltage applied to the modulation input port is of sufficiently large amplitude such that $V_M \gtrsim 100$ mV, the transistor pair of Q_5 and Q_6 will be operating essentially in a switching mode with I_5 and I_6 being given by $I_5 = 2I_Q S(t + \Delta t)$ and $I_6 = 2I_Q S(t + \Delta t - T/2)$, where $\Delta t = \phi/\omega$. Note that $v_m = V_M \sin(\omega t + \phi)$ can be reexpressed as $v_m = V_M \sin \omega(t + \phi/\omega) = V_M \sin \omega(t + \Delta t)$, so a phase shift ϕ corresponds to a time delay of $\Delta t = \phi/\omega$.

If we now insert the equations for I_5 and I_6 into the equations for I_1 through I_4 given above, we obtain

$$I_1 = 2I_Q S(t + \Delta t)S(t) \qquad\qquad I_2 = 2I_Q S(t + \Delta t)S\left(t - \frac{T}{2}\right)$$

$$I_3 = 2I_Q S\left(t + \Delta t - \frac{T}{2}\right)S\left(t - \frac{T}{2}\right), \qquad I_4 = 2I_Q S\left(t + \Delta t - \frac{T}{2}\right)S(t)$$

(14.46)

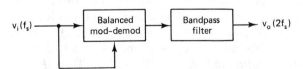

Figure 14.11 Phase detector.

Combining these four currents to obtain I_A and I_B gives

$$I_A = I_1 + I_3 = 2I_Q\left[S(t)S(t + \Delta t) + S\left(t - \frac{T}{2}\right)S\left(t + \Delta t - \frac{T}{2}\right)\right] \qquad (14.47)$$

and

$$I_B = I_2 + I_4 = 2I_Q\left[S\left(t - \frac{T}{2}\right)S(t + \Delta t) + S(t)S\left(t + \Delta t - \frac{T}{2}\right)\right] \qquad (14.48)$$

By inspection of the various switching functions, it is easy to verify that the average values (averaged over one or an integer number of periods) of the switching function products are given by

$$\overline{S(t)S(t + \Delta t)} = \frac{1}{2} - \frac{|\Delta t|}{T}$$

$$\overline{S(t - T/2)S\left(t + \Delta t - \frac{T}{2}\right)} = \frac{1}{2} - \frac{|\Delta t|}{T}, \qquad \text{for } |\Delta t| < \frac{T}{2}$$

$$\qquad\qquad\qquad (14.49)$$

$$\overline{S(t + \Delta t)S\left(t - \frac{T}{2}\right)} = \frac{|\Delta t|}{T}$$

$$\overline{S\left(t + \Delta t - \frac{T}{2}\right)S(t)} = \frac{|\Delta t|}{T}$$

As a result, we can write the average or dc values of I_A and I_B as

$$\overline{I_A} = 2I_Q\left(1 - \frac{2|\Delta t|}{T}\right) \quad \text{and} \quad \overline{I_B} = 2I_Q\frac{2|\Delta t|}{T} \qquad (14.50)$$

Since $\Delta t = \phi/\omega$ and $\omega = 2\pi f = 2\pi/T$, we have that $\Delta t/T = \phi/2\pi$. We can therefore express the average values of I_A and I_B in terms of the phase angle ϕ as

$$\overline{I_A} = 2I_Q\left(1 - \frac{|\phi|}{\pi}\right) \quad \text{and} \quad \overline{I_B} = \frac{2I_Q|\phi|}{\pi} \qquad (14.51)$$

In these equations the phase angle ϕ is restricted to the range of $-\pi \leq \phi \leq +\pi$. Outside this range the correct result can be obtained by adding or subtracting $2\pi n$ to the phase angle to put the result back into the range $-\pi$ to $+\pi$, with n being an integer.

In Figure 14.12a, a graph is presented showing the variation of $\overline{I_A}$ and $\overline{I_B}$ as a function of the phase angle ϕ. If a balanced or doubled-ended output is taken such that $\overline{V_o} = \overline{V_A} - \overline{V_B} = (\overline{I_B} - \overline{I_A})R_L$, we obtain $\overline{V_O} = 2I_QR_L[(2|\phi|/\pi) - 1]$ for $-\pi < \phi < +\pi$. In Figure 14.12b, the variation of $\overline{V_O}$ as a function of ϕ is shown. The *phase angle-to-voltage conversion coefficient* is $K_\phi = |dV_O/d\phi| = (4/\pi)I_QR_L$ (V/rad).

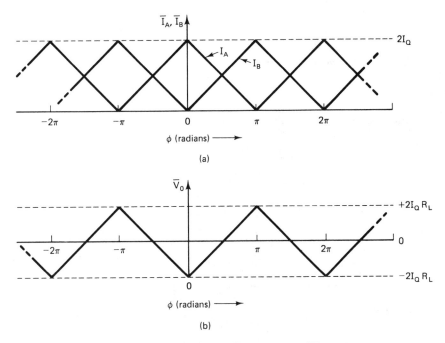

Figure 14.12 (a) $\overline{I}_A$ and $\overline{I}_B$ versus ϕ; (b) $\overline{V}_o$ versus ϕ.

14.7 1496/1596 BALANCED MODULATOR–DEMODULATOR

A very popular type of balanced modulator–demodulator is the 1496/1596 series that is available from several sources including National Semiconductor (LM1496/1596). The basic circuit configuration of this IC is shown in Figure 14.1 and a more detailed diagram is given in Figure P14.1a. A typical application of this IC as a balanced modulator is shown in Figure P14.1b. This device can be used at carrier frequencies up to 100 MHz. When used as a balanced modulator the carrier suppression is 65 dB (typ.) at 500 kHz and 50 dB at 10 MHz.

14.8 COMMUTATING AMPLIFIER BALANCED MODULATOR–DEMODULATOR

Another example of a very versatile balanced modulator–demodulator circuit is of the commutating amplifier type such as the AD630 (Analog Devices). A simplified block diagram of this IC is shown in Figure 14.13. The input signal v_M is fed to two identical amplifiers $A1$ and $A2$. The output of these two amplifiers goes to a third amplifier $A3$ via an electronic commutating switch S. The position of switch S is controlled by the comparator $C1$ and is a function of the comparator input voltage v_c such that when v_c changes polarity the switch changes position. Amplifiers $A1$ and $A2$ share a common feedback resistor R_F. When switch S is in the upper position, the feedback loop is closed via amplifier $A1$. The closed-loop gain in this case is given by $A_{CL} = v_o/v_M = -R_F/R_1$. When switch S is in the lower position, the

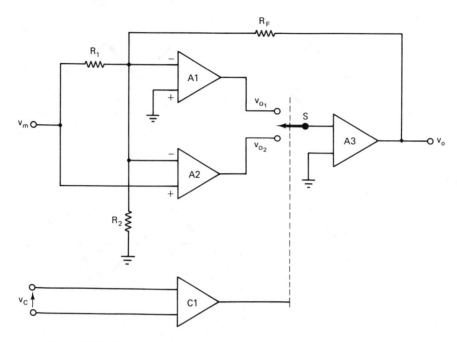

Figure 14.13 Block diagram of a balanced modulator–demodulator: AD630 (Analog Devices).

feedback loop is closed via amplifier $A2$. The closed-loop gain in this case is given by $A_{CL} = v_o/v_M = 1 + (R_F/R_2)$. We note that the gains in these two cases are of opposite algebraic sign. In most applications it is required that the two gains have equal magnitude. To satisfy this requirement, we have $R_F/R_1 = 1 + (R_F/R_2)$, which can be rewritten as $(1/R_1) = (1/R_F) + (1/R_2)$, and thus R_1 should be equal to the parallel combination of R_F and R_2.

We will now analyze the operation of this system and we will assume that the closed-loop gain values that are obtained for the two different switch positions are equal in magnitude, but opposite in algebraic sign. The voltage applied to $A1$ and $A2$ at the modulation input port v_M will be assumed to be given by $v_M = V_M \sin(\omega_M t)$, and the voltage applied to the comparator $C1$ at the carrier input port v_c will be given by $v_c = V_C \sin(\omega_c t)$. We will assume that the switch is in the lower position when the voltage applied to $C1$ via the carrier input port v_c is positive, so $v_o = (R_F/R_1) v_M$. When v_c is negative, the switch will be in the upper position and the output voltage will be $v_o = -(R_F/R_1)v_M$. We will again make use of the switching function $S(t)$, but this time $S(t)$ will be defined as $S(t) = +1$ when $v_c > 0$ and $S(t) = -1$ when $v_c < 0$; so $S(t)$ will be a symmetrical square wave of amplitude ± 1. In terms of the switching function, the output voltage can now be expressed as $v_o = [S(t)] (R_F/R_1)V_M = [S(t)] (R_F/R_1)V_M \sin(\omega_M t)$. The switching function can be expressed in the form of a Fourier series as

$$S(t) = \sum_{n=1}^{\infty} (4/n\pi) \sin(n\omega_c t) \tag{14.52}$$

where n is an odd integer. Thus the expression for v_o can be written as

$$v_o(t) = V_M(R_F/R_1) \sum_{n=1}^{\infty} (4/n\pi) \sin (\omega_M t) \sin (n\omega_c t) \qquad (14.53)$$

Using the trigonometric identities for the $\sin x \cdot \sin y$ product, the expression becomes

$$v_o(t) = V_M(R_F/R_1) \sum_{n=1}^{\infty} (2/n\pi) [\cos (n\omega_c - \omega_M)t - \cos (n\omega_c + \omega_M)t] \qquad (14.54)$$

We note that this result is similar to that obtained in Equation (14.15) for the balanced modulator–demodulator of Figure 14.1; as a result, we see that these balanced modulator–demodulators can be used for the same basic applications. This type of balanced modulator–demodulator, however, has the feature that the conversion gain can be set very accurately by the feedback network. With the modulator circuit of Figure 14.13, the sideband conversion gain is given by $A_{V(CONV)} = V_O(nf_c \pm f_M)/V_M$ $= (R_F/R_1)2/(n\pi)$. For the case of the AD630, resistors R_F (10 kΩ), R_1 (5 kΩ), and R_2 (10 kΩ) are available as laser-trimmed on-chip resistors so that excellent gain accuracy and stability can be expected. With these resistors as well as other on-chip resistors, closed-loop gains of ± 1 and ± 2 can be obtained with an accuracy of 0.1% (typ.) and a temperature coefficient of 2 ppm/°C (typ.). With other on-chip resistors, as well as with external resistors, other gain values can be obtained. In addition to the conventional balanced modulator–demodulator applications, the AD630 can be used for other applications, such as a lock-in amplifier or a two-channel multiplexer. When used as a two-channel multiplexer, the AD630 can be operated with fixed gains of 1, 2, 3, or 4, and the channel separation is 100 dB (typ.) at 10 kHz.

Internal circuit. Figure 14.14 shows a simplified schematic of the AD630 balanced modulator–demodulator. The comparator part of the circuit uses transistors Q_1 through Q_4. Transistors $Q1$ and $Q2$ constitute a simple PNP differential amplifier biased by current source I_{Q1} and driving a flip-flop comprised of transistors $Q3$ and $Q4$. When the voltage applied to the $+C_{IN}$ input is more than 1.5 mV above the voltage at the $-C_{IN}$ input, the increased current flow on the $Q2$ side of the differential amplifier causes the base voltage of $Q3$ to increase above that of $Q4$, and the flip-flop is forced into the state wherein the voltage at the collector of $Q3$ is low and the voltage at the collector of $Q4$ is high. This is a very rapid switching acting due to the positive-feedback loop that results from the collector-to-base cross coupling of $Q3$ and $Q4$. Similarly, when the voltage at the $+C_{IN}$ is more than 1.5 mV below that at the other input terminal of the comparator, the increased current through $Q1$ causes the flip-flop to change state to where the $Q3$ output is high and the $Q4$ output is low. It takes an input voltage of only ± 1.5 mV to produce the rapid switching action of the comparator. Thus, if the signal level at the carrier input port has an amplitude of greater than about 10 mV, the comparator will essentially operate in the switching mode, and the preceding analysis of the balanced modulator–demodulator operation will be valid.

Amplifiers $A1$ and $A2$ have separate input differential-amplifier stages, but share a second gain stage and an emitter-follower output stage. Amplifier $A1$ uses tran-

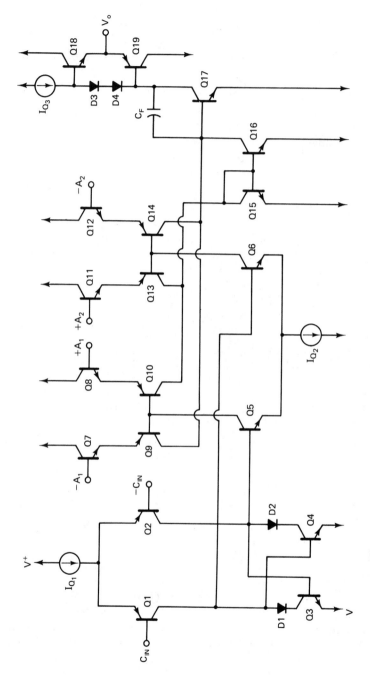

Figure 14.14 Simplified schematic diagram of a balanced modulator–demodulator: AD630 (Analog Devices).

sistors $Q7$ through $Q10$, and amplifier $A2$ uses transistors $Q11$ through $Q14$. Both amplifiers are biased by current source I_{Q2} via transistor $Q5$ for $A1$ and via $Q6$ for $A2$. The comparator output voltage goes to the bases of transistor $Q5$ and $Q6$. If the comparator output is such that the voltage at the base of $Q5$ is high with respect to the voltage at the base of $Q6$, then the current of I_{Q2} is supplied to amplifier $A1$ and no bias current is supplied to amplifier $A2$. Thus $A1$ is active and $A2$ is turned off. Conversely, when the comparator output is in the opposite state, we have that $A1$ is inactive and $A2$ is active. As a result of the switching action of the comparator, only one amplifier is turned on at any time.

Each amplifier is in the form of a common-collector, common-base differential amplifier, and both share the same current-mirror active load comprised of transistors $Q15$ and $Q16$. Since the amplifier that has been deactivated is not supplied with any bias current, it looks essentially like an open circuit and thus has no effect on the operation of the other amplifier. The output voltage of the $Q15$–$Q16$ current mirror is supplied to the base of $Q17$, which acts as a common-emitter second stage for both $A1$ and $A2$. This stage has a current-source active load and drives the last stage comprised of $Q18$ and $Q19$. This is in the form of a complementary push–pull emitter-follower output stage, with diodes $D3$ and $D4$ being used to minimize crossover distortion.

PROBLEMS

14.1 Given an LM1596 suppressed carrier modulator circuit (Figure P14.1); $R_{LOAD} = 3.9$ kΩ.
    (a) Find the quiescent currents for $V^+ = +12$ V and $V^- = -8$ V.   (*Ans.*: I_5 through $I_9 = 1.0$ mA, I_1 through $I_4 = 0.5$ mA)
    (b) $R_M = R_{23} = 1.0$ kΩ and $V_7 - V_8 = \pm 500$ mV:
        (1) Find the signal port transfer admittance at low frequencies.   (*Ans.*: 0.952 mS)
        (2) Find the sideband conversion transfer admittance at low frequencies and a carrier amplitude of 200 mV or greater.   (*Ans.*: 0.606 mS)
    (c) Repeat part (b) for $R_M = 100$ Ω.   (*Ans.*: (1) 6.67 mS; (2) 4.24 mS]
    (d) Find the conversion gain for a single-ended output with $R_M = 1.0$ kΩ and $f = 10$ kHz.   (*Ans.*: $A_{V\ CONV} = 2.365$ for each sideband)
    (e) If $R_{LOAD} = 3.9$ kΩ and $C_{TOTAL} = 2.0$ pF, repeat part (d) for a frequency of 10 MHz.   (*Ans.*: $A_{V\ CONV} = 2.12$ for each sideband)
    (f) If $R_M = 0$ and $v_c = v_s$ is a small signal (less than 30 mV), find the coefficient K in the relationship for the frequency doubler as given by $V_O(2f_i) = K[V_i(f_i)]^2$, where V_O and V_i are the peak values (amplitudes) of the output and input voltages, respectively. The load resistance R_{LOAD} is 3.9 kΩ.   (*Ans.*: $K = 780$ V^{-1}, single-ended output)
    (g) For the phase detector application with $R_M = 0$, $v_c(t) = 100$ mV sin ωt and $v_s(t) = 100$ mV sin ωt, find R_L such that $dV_O/d\phi = 1.0$ V/rad, where $V_O = V_B - V_A = (I_A - I_B)R_L$.   (*Ans.*: 785 Ω)
    (h) Find V_O for ϕ values of $0°$, $45°$, $90°$, $135°$, and $180°$ for the circuit of part (g).   (*Ans.*: -1.571, -0.785, 0, $+0.785$, $+1.571$ V)
    (i) Repeat part (g) for a phase-to-voltage conversion coefficient of 20 mV/deg.   (*Ans.*: $R_L = 900$ Ω)

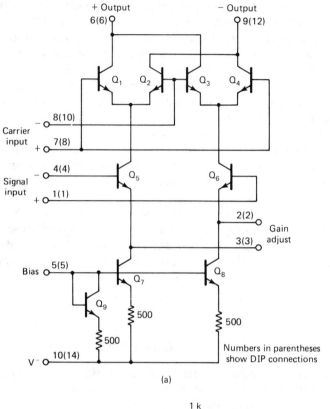

(a)

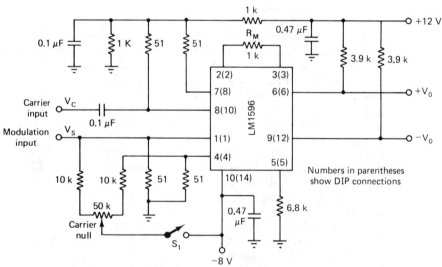

NOTE: S_1 is closed for "adjusted" measurements

(b)

Figure P14.1

14.2 Given, at 20 MHz, the sideband conversion transfer admittance $= y_f = 2.0 + j2.0$ mS, and the output admittance $= y_o = 100 + j1000$ μS (see Figure P14.2). The modulation input is a small signal with frequencies ranging up to 250 kHz. Find the sideband conversion gain, $v_{o(\text{sideband})}/v_{\text{mod}}$. *(Ans.:* $1.414 \angle 45°$)

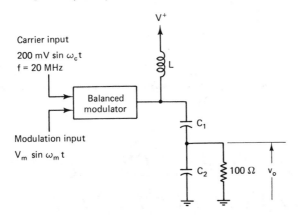

Figure P14.2

14.3. *(Phase detector)* Given a phase detector circuit. Assume all transistors identical unless otherwise indicated. Use Figure 14.1 with $R_M = 0$ for the phase detector.

(a) If the dc voltage drop across the load resistor R_L under quiescent conditions (no signal input) is 2.0 V, find the value of the constant in the equation $V_{O(\text{dc})} = -(\text{constant})\ V_M \cos \phi$, assuming that V_M is small (30 mV or less). *(Ans.:* constant $= 50.9$)

(b) If V_M is 10 mV rms, find the relationship between $V_{O(\text{dc})}$ and the phase angle ϕ. *(Ans.:* $V_{O(\text{dc})} = -0.720$ V cos ϕ)

(c) If the dc voltage drop across the load resistance R_L under quiescent conditions is 2.0 V, and V_M is $2|\phi|/\pi$ a large signal (100 mV or more), show that $V_{O(\text{dc})}$ is given by $V_{O(\text{dc})} = 4.0$ V $[2|\phi|/\pi - 1]$.

(d) Sketch a curve of $V_{O(\text{dc})}$ versus ϕ for the conditions of part (c). Dimension the graph showing the positions of the maxima, minima, and so on, and the values of the peak excursions of $V_{O(\text{dc})}$. *(Ans.:* peak excursion $= \pm 4.0$ V)

(e) Find the value of the phase angle-to-voltage transfer coefficient $K_{\phi-V}$ for the case of part (c). *(Ans.:* $K_{\phi-V} = 2.55$ V/rad)

REFERENCES

GRAEME, J. G., G. E. TOBEY, and L. P. HUELSMAN, *Operational Amplifiers—Design and Applications,* McGraw-Hill, New York, 1971.

GREBENE, A. B., *Analog Integrated Circuit Design,* Van Nostrand Reinhold, New York, 1972.

———, *Bipolar and MOS Analog Integrated Circuit Design,* Wiley, New York, 1984.

KRAUSS, H. L., C. W. BOSTIAN, and F. H. RAAB, *Solid State Radio Engineering,* Wiley, New York, 1980.

LENK, J. D., *Handbook of Integrated Circuits,* Prentice-Hall, Englewood Cliffs, N.J., 1978.

———, *Manual for Integrated Circuit Users,* Prentice-Hall, Englewood Cliffs, N.J., 1973.

MITRA, S. K., *An Introduction to Digital and Analog Integrated Circuits and Applications,* Harper & Row, New York, 1980.

15

Voltage-controlled Oscillators and Waveform Generators

15.1 VOLTAGE-CONTROLLED OSCILLATORS

A *voltage-controlled oscillator* (VCO) is an oscillator circuit in which the frequency of oscillation can be controlled by an externally applied voltage. One feature that is often required for VCOs is a linear relationship between the oscillation frequency and the control voltage.

A basic block diagram of the type of VCO circuit that is used in most integrated-circuit VCOs is shown in Figure 15.1. Voltage-controlled current sources are used to charge and to discharge the timing capacitor C_t. The charging and discharging time periods are controlled by the action of a Schmitt trigger circuit, the input voltage of which is the voltage across C_t. The Schmitt trigger has two threshold voltage switching levels, V_L and V_H. The width of the Schmitt trigger hysteresis curve V_W is given by $V_W = V_H - V_L$.

Since C_t is charged and discharged by current sources, the voltage across C_t will vary linearly with time. If the charging and discharging currents are of equal magnitude, the voltage waveform across C_t is a symmetrical triangular waveform. The timing capacitor C_t is charged from the top current source up to a voltage equal to the V_H triggering level. At this point the Schmitt trigger is activated and causes the current sources to be switched such that the top current source is disconnected from C_t and the bottom current source is connected to C_t. The capacitor will now be discharged (or equivalently, charged in the opposite direction) by the bottom current

698

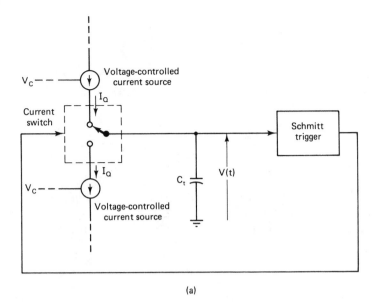

(a)

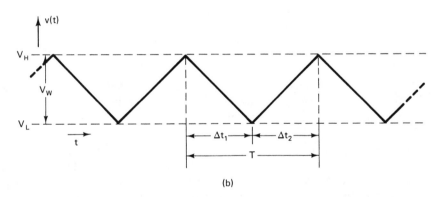

(b)

Figure 15.1 Voltage-controlled oscillator: (a) basic block diagram; (b) voltage waveform across capacitor.

source down to lower trigger level V_L. At this point the Schmitt trigger is activated again to cause the current sources to be switched again, and the entire cycle is repeated. The voltage across C_t is therefore a symmetrical triangular waveform extending between the limits of V_L to V_H for a total peak-to-peak excursion of $V_W = V_H - V_L$, as shown in Figure 15.1b.

Since the slope of the voltage across C_t is $dv(t)/dt = d(Q/C_t)/dt = (1/C_t)(dQ/dt) = \pm I_Q/C_t$, the time required to change the capacitor voltage from V_L to V_H, and vice versa, is given by

$$\frac{V_H - V_L}{\Delta t_2} = \frac{I_Q}{C_t} \quad \text{and} \quad \frac{V_L - V_H}{\Delta t_1} = \frac{I_Q}{C_t}$$

Therefore, we have that

$$\Delta t_1 = \Delta t_2 = \frac{(V_H - V_L)C_t}{I_Q} = \frac{V_W C_t}{I_Q}$$

so the period of oscillation T is given by $T = \Delta t_1 + \Delta t_2 = 2V_W C_t/I_Q$. The frequency of oscillation f_o is therefore

$$\boxed{f_o = \frac{1}{T} = \frac{I_Q}{2V_W C_t}} \tag{15.1a}$$

If the voltage-controlled current sources have a linear voltage-to-current transfer relationship as given by $I_Q = G_c(V_C + V_o)$, where G_c is the transfer conductance of the current source, V_c is the control voltage and V_o is a constant, we have that

$$f_o = \frac{I_Q}{2V_W C_t} = \frac{G_c(V_C + V_o)}{2V_W C_t} \tag{15.1b}$$

We therefore see that the oscillation frequency f_o is a linear function of the control voltage V_C. The *voltage-to-frequency transfer coefficient* K_V is given by

$$\boxed{K_V = \frac{df_o}{dV_C} = \frac{G_c}{2V_W C_t}} \tag{15.2}$$

A circuit implementation of a voltage-controlled current source is shown in Figure 15.2. Also shown in this diagram is the circuitry that produces charging and discharging currents of equal magnitude.

The voltage at the emitter of Q_2 is $V_{E2} = V_{EB_2} + V_{EB_1} + V_C = V_C - V_{BE_1} + V_{EB_2}$. Since Q_1 is an NPN transistor and Q_2 is a PNP transistor, we have that $V_{BE_1} \simeq V_{EB_2}$ to within something in the range from 10 to 50 mV. Therefore, we have that $V_{E2} \simeq V_C$. As a result, the current through R_1 is $I_{R_1} = (V^+ - V_{E2})/R_1 \simeq (V^+ - V_C)/R_1$ and the current source output current is $I_Q = \alpha_{pnp} I_{R_1} \simeq \alpha_{pnp}(V^+ - V_C)/R_1 \simeq (V^+ - V_C)/R_1$ if $\alpha_{pnp} \simeq 1$. We thus obtain a linear relationship between the current source output current I_Q and the control voltage V_C.

Transistor Q_5 is controlled by the Schmitt trigger. When Q_5 is off, C_t is charged with the output current of Q_2, which is I_Q via diode D_2, and Q_3 and Q_4 are off. When Q_5 is turned on (in saturation), Q_3, Q_4, and D_1 are on. The voltage at the anode of D_1 is $V_{CE(SAT)5} + V_{BE_3} + V_{D_1} \cong 0.2 + 0.65 + 0.65 = 1.5$ V. The voltage at the cathode of D_2 is the voltage across C_t, which is $v(t)$. Since $v(t)$ is larger than 1.0 V, we see that diode D_2 is off.

The current through Q_4 is equal to the current through Q_3, which is I_Q. Since D_2 is off, we see that C_t is discharged through Q_4 with a current equal to I_Q. Therefore, the charging and discharging currents are equal in magnitude and are linearly dependent on the control voltage V_C.

A diagram of a Schmitt trigger circuit is shown in Figure 15.3. To determine the high and low triggering levels, V_H and V_L, we will first examine the circuit for

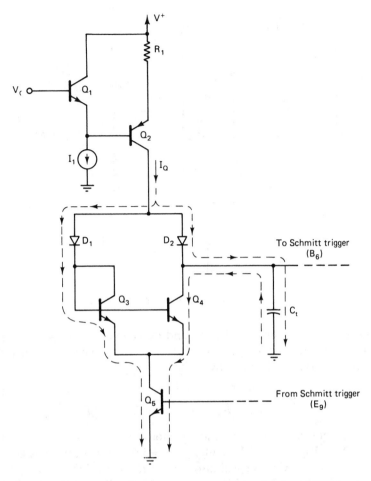

Figure 15.2 Voltage-controlled current sources for the charging and discharging of the timing capacitor.

the case in which the Q_6–D_3–Q_7 combination is off. With Q_7 off, Q_8 is turned on and driven into saturation. With Q_8 in saturation, Q_9 is turned off, and as a result Q_5 is also turned off.

When Q_8 is on (in saturation), the voltage across R_3 is $V_{R_3} = V_{E_7} = V_{E_8} = (V^+ - V_{CE(SAT)})R_3/(R_3 + R_4) \cong V^+ R_3/(R_3 + R_4)$. The trigger or threshold voltage V_H required to switch the Schmitt trigger from this state to the other state is therefore $V_H = V_{E_7} + V_{BE_7} + V_{D_3} + V_{BE_6} \cong 3V_{BE} + V^+ R_3/(R_3 + R_4)$.

If the voltage at the base of Q_6 goes above V_H, the Q_6–D_3–Q_7 combination is turned on and the Schmitt trigger will rapidly go into the other state. With Q_7 on (in saturation), Q_8 is off and as a result Q_9 is driven into saturation. This, in turn, drives Q_5 into saturation.

With Q_8 off, the voltage across R_3 drops to a value given by $V_{R_3} = V_{E_7} = (V^+ - V_{CE(SAT)7})R_3/(R_3 + R_2) \cong V^+ R_3/(R_3 + R_2)$. The input voltage $v(t)$ at the base

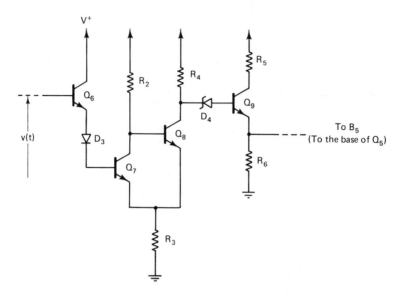

Figure 15.3 Schmitt trigger circuit.

of Q_6 required to turn on Q_6, D_3, and Q_7 is given by $V_L = V_{BE_6} + V_{D_3} + V_{BE_7} + V^+R_3/(R_3 + R_2) = 3V_{BE} + V^+R_3/(R_3 + R_2)$. The width of the Schmitt trigger hysteresis curve V_W is given by

$$V_W = V_H - V_L = \frac{V^+R_3}{R_3 + R_4} - \frac{V^+R_3}{R_3 + R_2} = V^+R_3\left(\frac{1}{R_3 + R_4} - \frac{1}{R_3 + R_2}\right) \qquad (15.3)$$

To consider an example, let us choose the following representative values: $V^+ = +12$ V, $V_H = +8.0$ V, and $V_L = +3.0$ V for a total hysteresis span of $V_W = V_H - V_L = 5.0$ V. We will also choose $R_3 = 1.0$ kΩ.

We have that $V_H = 3V_{BE} + V^+R_3/(R_3 + \overline{R_4)} = 1.8 + 12R_3/(R_3 + R_4) = 8.0$, so $12R_3 = 6.2(R_3 + R_4)$ and thus $6.2R_4 = 5.8R_3 = 5.8$ kΩ, giving $R_4 = 935$ Ω. For R_2 we have that $V_L = 3V_{BE} + V^+R_3/(R_3 + R_2) = 1.8 + 12R_3/(R_3 + \overline{R_2)} = 3.0$, so $12R_3 = 1.2(R_3 + R_2)$, giving $1.2R_2 = 10.8R_3 = 10.8$ kΩ and thus $R_2 = \underline{10.8 \text{ kΩ}/1.2 = 9.0 \text{ kΩ}}$.

When Q_8 is in saturation, we have that $V_{C_8} \cong V^+R_3/(R_3 + R_4) = 12 \times 1 \text{ kΩ}/1.935 \text{ kΩ} = 6.2$ V. Since the zener voltage of D_4 is 6.2 V, the voltage at the base of Q_9 is zero, and thus Q_9 and Q_5 are off.

When Q_8 is off, Q_9 and Q_5 are on. The base drive available to turn Q_9 on is $I_{B_9} = (V^+ - V_Z - V_{BE_9} - V_{BE_5})/R_4 = (12 - 6.2 - 0.7 - 0.7)/935 \text{ Ω} = 4.4 \text{ V}/935 \text{ Ω} = 4.7$ mA. This base current is more than sufficient to drive Q_9 and Q_5 into full saturation.

Another type of Schmitt trigger is shown in Figure 15.4. This Schmitt trigger features a differential amplifier comprised of transistors Q_{10} and Q_{11} biased by a current sink I_Q. If the voltage at the base of Q_{10} drops below the voltage at the base of Q_{11}, the action of the positive-feedback loop composed of R_7 and R_8 will be to rapidly raise the base voltage of Q_{11}. The difference in the base voltages of Q_{10} and

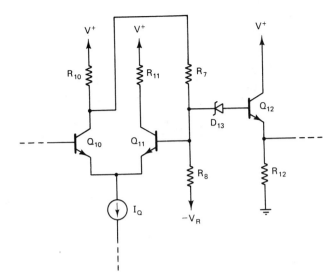

Figure 15.4 Schmitt trigger circuit using a differential amplifier.

Q_{11} that is established will be such that almost all of current I_Q will now flow through Q_{11} and very little will go through Q_{10}.

The voltage at the base of Q_{11} is now given by $V_{B_{11}} = (V^+ + V_R)R_8/(R_7 + R_8 + R_{10}) - V_R$. This is the high-level trigger voltage V_H.

If the voltage at the base of Q_{10} now rises above $V_{B_{11}} = V_H$, the action of the feedback loop will be to very rapidly drop the voltage at the base of Q_{11} such that almost all the current will now flow through Q_{10} and the current through Q_{11} will essentially be cut off. The voltage level at the base of Q_{11} is now given by

$$V_L = V_{B_{11}} = \frac{(V^+ - I_Q R_{10} + V_R)R_8}{R_8 + R_7 + R_{10}} - V_R \qquad (15.4)$$

and this is the low-level trigger voltage. The difference between these two trigger voltage levels is the hysteresis width of the Schmitt trigger, as given by $V_W = V_H - V_L = I_Q R_{10} R_8/(R_7 + R_8 + R_{10})$.

Transistors Q_{10} and Q_{11} can be prevented from going into saturation by suitable choice of I_Q, R_{10}, and R_{11} such that the collector voltages never drop by more than 0.5 V below the base voltages. This, combined with the all-NPN nature of this circuit, means that this Schmitt trigger will have the capability of very high-frequency switching.

Another Schmitt trigger circuit closely related to the first one discussed is shown in Figure 15.5. In this circuit, transistors Q_6 through Q_9 operate in the same fashion as in the circuit of Figure 15.3. When Q_6 and Q_7 are on, Q_8 is off and the collector of Q_8 is up near V^+, thus raising the emitter of Q_9 and the base of Q_{10} close to V^+. This will essentially turn off the Q_{10} side of the differential amplifier and Q_{11} will carry the full current. The collector current of Q_{11} is the base drive of Q_5, turning that transistor on (to saturation). When Q_6 and Q_7 are off, Q_8 is on. This drops the voltage at the collector of Q_8 and therefore that at the emitter of Q_9 and the base of Q_{10} to a low value. This causes the current in the Q_{10}–Q_{11} differential amplifier

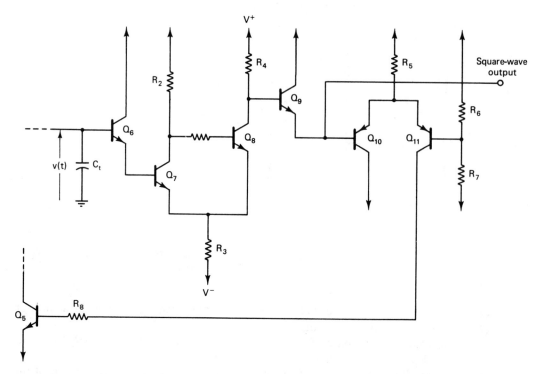

Figure 15.5 Schmitt trigger using a differential-amplifier output circuit.

to shift to the Q_{10} side, with Q_{11} being essentially turned off. This, in turn, turns Q_5 off.

Note that with all these VCO circuits a square-wave output is available from the Schmitt trigger circuit. There will also be a triangular-wave output available from across the timing capacitor.

An example of a VCO of the type just described is the NE/SE566 (Signetics), also available as the LM566 (National Semiconductor). This VCO produces square- and triangular-wave outputs up to a frequency of 1 MHz. The frequency can be swept over a 10 : 1 range by the modulation input (control) voltage.

15.1.1 Emitter-coupled Voltage-controlled Oscillator

An emitter-coupled VCO is shown in Figure 15.6. Transistors Q_5 and Q_6 are voltage-controlled current sources, each producing a current I_Q. Transistors Q_3 and Q_4 are the switching transistors and drive loads are comprised of R_1, R_2, Q_1, and Q_2. The load configuration is designed to produce a voltage drop V_D that will be relatively independent of the current level.

The voltage drop V_D can be obtained from the relationship $V_{BE} = V_D R_2/(R_1 + R_2)$. Solving for V_D, we obtain $V_D = V_{BE}(1 + R_1/R_2)$ when current is flowing through the load. Otherwise, the voltage drop is zero.

If we now consider the condition wherein Q_3 is assumed to be on and Q_4 is off, the voltage at the base of Q_3 is $(V^+ - V_{BE})R_6/(R_5 + R_6)$ and the voltage at the base

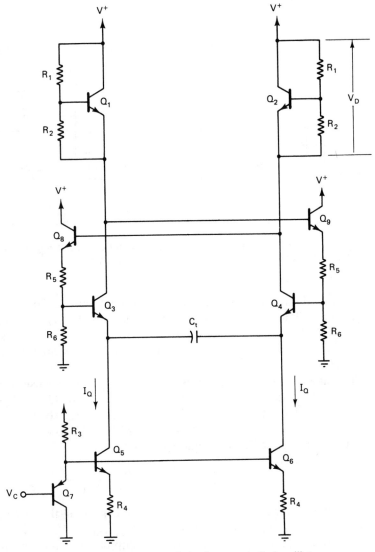

Figure 15.6 Emitter-coupled voltage-controlled oscillator.

of Q_4 is $(V^+ - V_{BE} - V_D)R_6/(R_5 + R_6)$. The difference in the voltages between the two bases is $V_{B_3} - V_{B_4} = V_D R_6/(R_5 + R_6) = V'_D$.

The voltage across the timing capacitor C_t is $V_{C_t} = V_{E_3} - V_{E_4}$ and thus the difference in the base-to-emitter voltages of Q_3 and Q_4 is $V_{BE_3} - V_{BE_4} = (V_{B_3} - V_{B_4}) - (V_{E_3} - V_{E_4}) = V_D R_6/(R_5 + R_6) - V_{C_t}$. As long as V_{C_t} is less than V'_D, we have that $V_{BE_3} > V_{BE_4}$ and Q_3 will remain on and Q_4 off.

As illustrated in Figure 15.7a, capacitor C_t is charged by current I_Q and the voltage across C_t will increase at a rate given by $dV_{C_t}/dt = I_Q/C_t$. As soon as V_{C_t} rises above V'_D, transistor Q_3 turns off and Q_4 turns on. This results in a change in

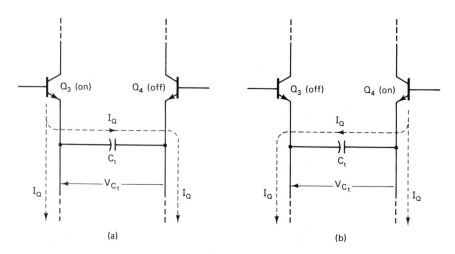

Figure 15.7 Charging and discharging cycles for C_t: (a) charging cycle with Q_3 on and Q_4 off; (b) discharge cycle with Q_3 off and Q_4 on.

the direction of the current through C_t as shown in Figure 15.7b. This circuit remains in this state as long as $V_{BE_3} < V_{BE_4}$, which requires that $V_{C_t} > -V'_D$. The voltage across C_t now decreases at a rate given by $-dV_{C_t}/dt = -I_Q/C_t$. As soon as V_{C_t} drops below $-V'_D$, transistor Q_3 is switched back on and Q_4 is switched off and the cycle is completed.

The change in the voltage across C_t that is required to cause Q_3 and Q_4 to switch from one state to the other is $\Delta V_{C_t} = \pm 2V'_D = \pm 2V_D R_6/(R_5 + R_6)$. Since $dV_{C_t}/dt = \pm I_Q/C_t$, we have that $\Delta V_{C_t} = \pm 2V'_D = \pm I_Q \Delta t/C_t$, so $\Delta t = 2V'_D C_t/I_Q$. The oscillation period T is therefore $T = 2\Delta t = 4V'_D C_t/I_Q = 4V_D C_t R_6/I_Q(R_5 + R_6)$. The corresponding oscillation frequency f_o is

$$f_o = \frac{1}{T} = \frac{I_Q}{4V'_D C_t} = \frac{I_Q(1 + R_5/R_6)}{4V_D C_t} \tag{15.5a}$$

For the voltage-controlled current source part of the circuit, we note that $V_{E_5} = V_{E_6} = V_C + V_{EB_7} - V_{BE_5} \simeq V_C$ since $V_{EB_7(PNP)} \simeq V_{BE_5(NPN)}$. As a result, $I_Q = V_{E_5}/R_4 \simeq V_C/R_4$ and there is a linear relationship between the current source current I_Q and the control voltage V_C. The oscillation frequency f_o can now be expressed in terms of the control voltage as

$$f_o = \frac{V_C(1 + R_5/R_6)}{4V_D R_4 C_t} \tag{15.5b}$$

and the voltage-to-frequency transfer coefficient is

$$K_V = \frac{df_o}{dV_C} = \frac{1 + R_5/R_6}{4V_D R_4 C_t} \tag{15.6}$$

This type of VCO has the capability of operation at relatively high frequencies ($\gtrsim 50$ MHz). This is the result of the fact that the switching transistors Q_3 and Q_4 can be prevented from going into saturation. The required condition for this is that the base voltage never go more than 0.5 V above the collector voltage. Thus we should require that $(V^+ - V_{BE})R_6/(R_5 + R_6) < (V^+ - V_D) + 0.5$ V.

15.2 WAVEFORM GENERATORS

A waveform generator is a device that generates the following three types of voltage waveforms: square waves, triangular waves, and sinusoidal waves. The square and triangular waves can be generated by the same types of circuits as those used for voltage-controlled oscillators. These circuits involve the charging and discharging of a timing capacitor from a current source and the use of a Schmitt trigger or comparator to switch the current source between the charging and discharging modes of operation. If the charging and discharging currents are equal in magnitude, the duty cycle will be 50%; that is, the capacitor charges for one-half of the total period and discharges for the other half.

A square-wave output is available from the Schmitt trigger or comparator circuit and a triangular wave is available across the timing capacitor. In the cases in which the timing capacitor has one side connected to ground, a triangular wave will appear at the high (ungrounded) side of the capacitor. In the case of the emitter-coupled VCO type of circuit, neither side of the timing capacitor is grounded. To obtain a single-ended triangular-wave output, the voltage across the timing capacitor must be supplied as a balanced input to a differential amplifier. The differential amplifier will then convert this balanced or double-ended input to an unbalanced or single-ended output at the collectors of the differential amplifier.

15.2.1 Sine-wave Generation

For the generation of a sinusoidal waveform, there are two basic techniques that can be used. One approach is to use a feedback amplifier circuit using an LC tuned circuit or RC phase-shift network in the feedback loop, as shown in Figure 15.8a. The other approach is to use a nonlinear wave-shaping circuit to convert the triangular waveform into a sinusoidal waveform. The feedback oscillator circuit will be discussed first.

For a closed-loop system of the type shown in Figure 15.8a, the condition for oscillations to occur in that the magnitude of the loop gain FA_{OL} be greater than unity at the frequency at which the net phase shift around the feedback loop is $0°$ or, equivalently, an integer multiple of $360°$. Since the feedback in Figure 15.8a goes to the inverting input terminal of the amplifier, there must be an additional phase shift of $180°$ in the feedback loop in order to satisfy this condition. This condition can be written as $|FA_{OL}| > 1$ at $f_{180°}$, where $f_{180°}$ is the frequency at which $\underline{/FA_{OL}} = 180°$. The frequency of oscillation f_o is determined by the phase-shift condition and is equal to $f_{180°}$.

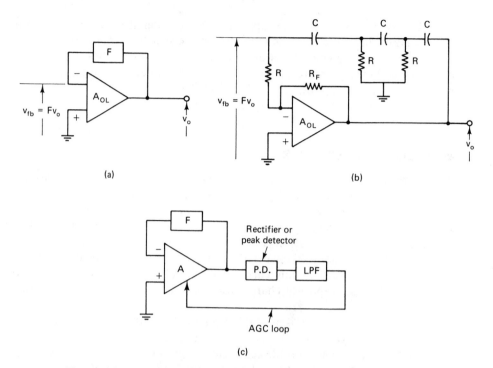

Figure 15.8 Feedback sinusoidal oscillators: (a) basic diagram of a feedback oscillator; (b) RC phase-shift oscillator; (c) low-distortion sinusoidal oscillator using AGC.

15.2.2 *RC* Phase-shift Oscillator

An example of a feedback oscillator that uses an RC phase-shift network in the feedback loop is the RC phase-shift oscillator shown in Figure 15.8b. The transfer function of the RC phase-shift network is given by

$$F = \frac{1}{1 - 5\alpha^2 - j(6\alpha - \alpha^3)}, \qquad \text{where } \alpha = \frac{1}{\omega RC} \qquad (15.7a)$$

The phase shift θ of the feedback network is given by

$$\theta = \angle F = \tan^{-1} \frac{6\alpha - \alpha^3}{1 - 5\alpha^2} \qquad (15.7b)$$

Assuming that the amplifier phase shift is zero, we have that $\angle FA_{OL} = \angle F = \theta$. For a 180° phase shift, the argument of the inverse tangent function is zero, so $6\alpha - \alpha^3 = 0$. After noting that the $\alpha = 0$ case corresponds to a 0° phase shift, we see that for a 180° phase shift $\alpha^2 = 6$. Therefore, the 180° phase-shift frequency, which is also the frequency of oscillation f_o, is given by the condition that $\alpha = 1/\omega_o RC = \sqrt{6}$, so $\omega_o = 1/(\sqrt{6} RC)$ and $f_o = 1/(2\pi \sqrt{6} RC)$.

When $\alpha^2 = 6$, the value of the feedback factor F is $F = 1/(1 - 5\alpha^2) = 1/(1 - 30) = -1/29 = (1/29) \angle 180°$. For oscillations to occur, the magnitude of the loop gain must be greater than unity at $f = f_{180°} = f_o$. Therefore, we require that $|F(f_{180°})|$ $\times (R_F/R) > 1$, so $R_F/R > 29$.

If the loop gain is substantially in excess of the minimum value required for oscillations, however, the amplitude of the oscillations is limited only by the amplifier going into saturation (or cutoff) and there will be clipping of the peaks of the output voltage. The voltage waveform produced will therefore be a clipped sine wave and a large amount of distortion will be present. To produce a relatively undistorted sine wave, the loop gain should be adjusted or controlled to a value that is just slightly greater than unity at f_o. The closer that the loop gain is to unity, the less distortion there will be.

For the phase-shift oscillator just considered, a relatively undistorted sine wave can be produced by adjusting R_F to a value that is just slightly larger than that needed to initiate oscillations. This approach, however, suffers the possible drawback that changes in operating conditions such as the temperature or supply voltage may change the amplifier gain so that the optimum value of R_F may drift with time, and R_F may require periodic readjustment.

Another approach for obtaining an undistorted sine-wave output is shown in Figure 15.8c. In this case an automatic gain control (AGC) feedback loop is used to control the gain of the amplifier to the value corresponding to an output voltage that has a peak value that is substantially less than that which would produce clipping. The cutoff frequency of the low-pass filter should be much less than f_o. With this circuit a sine-wave output of very low distortion can be produced.

15.2.3 Sinusoidal VCO Output: Triangle to Sine-wave Converter

The feedback oscillator can produce a very low distortion sine wave, but it is difficult to modulate the oscillation frequency by means of a control voltage, especially with respect to a large-frequency sweep range. The VCO, on the other hand, is capable of a frequency sweep ratio as large as 100 : 1 with very good linearity between the frequency and the control voltage. A sine wave can be obtained from a VCO by using a wave-shaping network to convert the triangular-wave output to a sine wave.

A triangular wave-to-sine wave converter uses a nonlinear circuit to convert a triangular wave to a sine wave, as shown in Figure 15.9. In this figure a seven-segment piecewise linear transfer characteristic that can be used for this purpose is shown together with the corresponding sine-wave output, as shown in Figure 15.9b.

The Fourier series representation of a triangular wave is given by

$$v_T(t) = \frac{8}{\pi^2} V_{PT}\left(\sin \omega t - \frac{1}{9} \sin 3\omega t + \frac{1}{25} \sin 5\omega t - \frac{1}{49} \sin 7\omega t + \cdots \right) \qquad (15.8)$$

We see that the triangular wave has a large component at the fundamental frequency, and can be considered to be a very distorted sine wave with about 12% total harmonic distortion (THD). By the use of a nonlinear wave-shaping network to modify the shape of the triangular wave, particularly by rounding off of the sharp peaks of the triangular wave, a sine wave with a THD of less than 1% can easily be obtained.

Figure 15.10 shows a simple example of a waveform converter circuit. This circuit produces a nine-segment piecewise-linear transfer characteristic. The slope

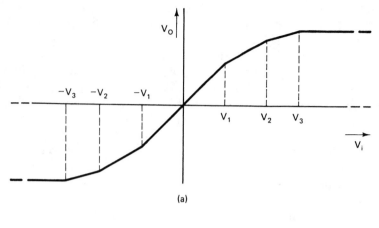

(a)

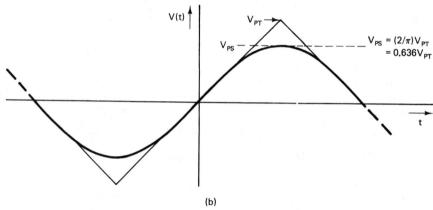

(b)

Figure 15.9 Transfer characteristics for triangular waveform-to-sine wave converter; (a) seven-segment converter; (b) comparison of triangular wave and sine wave.

dV_o/dV_i of the transfer curve for various values of V_i is as follows:

V_i	$Slope = dV_o/dV_i$
(1) $-V_1 < V_i < +V_1$	1
(2) $+V_1 < V_i < +V_2,$ or $-V_2 < V_i < -V_1$	$\dfrac{R_1}{R_1 + R_A}$
(3) $+V_2 < V_i < +V_3,$ or $-V_3 < V_i < -V_2$	$\dfrac{R_1 \| R_2}{(R_1 \| R_2) + R_A}$
(4) $+V_3 < V_i < +V_4,$ or $-V_4 < V_i < -V_3$	$\dfrac{R_1 \| R_2 \| R_3}{(R_1 \| R_2 \| R_3) + R_A}$
(5) $+V_4 < V_i,$ or $V_i < -V_4$	≈ 0

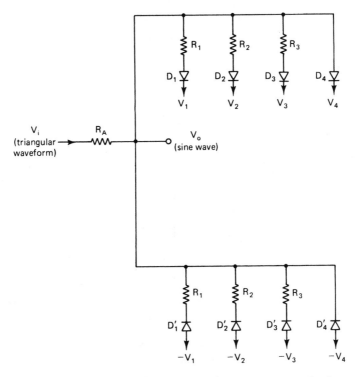

Figure 15.10 Triangular waveform-to-sine wave converter circuit.

The last case should occur at a voltage level of approximately $V_4 = V_{PS} = 0.636 \, V_{PT}$ to produce the final flattening of the triangular waveform to produce the rounded peak of the sine wave.

15.2.4 Intersil 8038

Figure 15.11 shows a practical waveform converter circuit (Intersil 8038). The reference voltages V_1, V_2, V_3, and so on, are produced by a voltage divider connected between the positive and negative supply voltages as shown in Figure 15.12. The PNP–NPN transistor combinations are used to transform the impedance of the voltage divider down to a very low value that is small compared to R_1, R_2, and R_3. The voltage level that appears at the output emitters of the PNP–NPN pairs is approximately equal to the reference voltages $(V_1, V_2, \ldots)$ since the base-to-emitter voltage drops of the PNP and NPN transistors are of opposite algebraic sign.

The effective source resistance due to the voltage divider is reduced by a factor equal to the product of the PNP and NPN current gains and thus is reduced to a negligibly small value ($\lesssim 1 \, \Omega$). We note, however, that due to the finite voltage change needed to bias the various transistor pairs into full conduction there will not be a sharp transition between the linear segments of the transfer characteristic, but rather somewhat of a rounded-off transition. The rounded-off or smooth transition

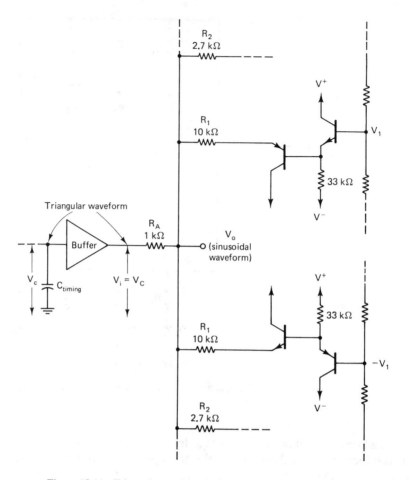

Figure 15.11 Triangular-to-sinusoidal waveform converter circuit.

between the piecewise-linear segments is actually beneficial in producing a less dis-torted sine wave.

For the circuit of Figure 15.11, the value of R_3 (not shown) is 800 Ω and $R_4 = 0$. The complete voltage-divider circuit is shown in Figure 15.12. These reference voltage levels are suitable for the conversion of a symmetrical triangular wave with a peak-to-peak amplitude $2V_{PT}$ of $\frac{1}{3}(V^+ - V^-) = 20/3 = 6.7$ V, or $V_{PT} = 3.33$ V. A sine wave with an amplitude of $V_{PS} = V_4 = 0.247\ V^+ = 2.47$ V is produced.

Figure 15.13 shows a buffer amplifier circuit for this converter. Note that in taking the algebraic summation of all the base-to-emitter voltage drops going from C_{timing} to the output of the buffer circuit, we obtain the result that the output voltage of the buffer is approximately equal to the voltage across the timing capacitor. The principal function of the buffer is to act as a unity-gain, low-output-impedance ampli-fier for driving the waveform converter circuit.

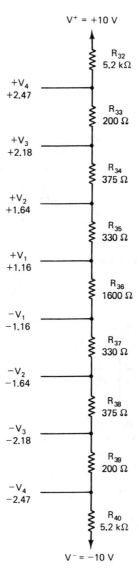

$V^+ = +10$ V

R_{32}
5.2 kΩ

$+V_4$
$+2.47$

R_{33}
200 Ω

$+V_3$
$+2.18$

R_{34}
375 Ω

$+V_2$
$+1.64$

R_{35}
330 Ω

$+V_1$
$+1.16$

R_{36}
1600 Ω

$-V_1$
-1.16

R_{37}
330 Ω

$-V_2$
-1.64

R_{38}
375 Ω

$-V_3$
-2.18

R_{39}
200 Ω

$-V_4$
-2.47

R_{40}
5.2 kΩ

$V^- = -10$ V

Figure 15.12 Voltage divider for the breakpoint voltages.

15.2.5 ICL8038 (INTERSIL)

The ICL8038 is a very versatile VCO and waveform generator that is capable of generating sine, square, pulse, and triangular waveforms over a frequency range from 0.001 Hz up to 300 kHz. The sine, square, and triangular waveforms are available simultaneously. A functional diagram of this IC is given in Figure 15.14.

As a result of the $R_1-R_2-R_3$ voltage divider that is comprised of three equal 5-kΩ resistors, comparator 1 has a threshold or reference voltage of $\frac{2}{3}V^+$ and comparator 2 has a reference voltage of $\frac{1}{3}V^+$. If we start with switch S in the open position, the

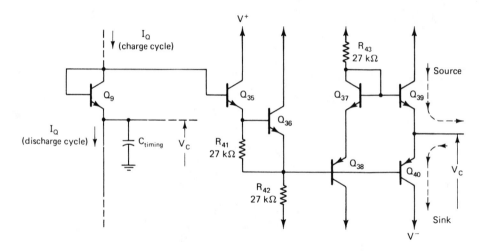

Figure 15.13 Buffer amplifier for waveform converter.

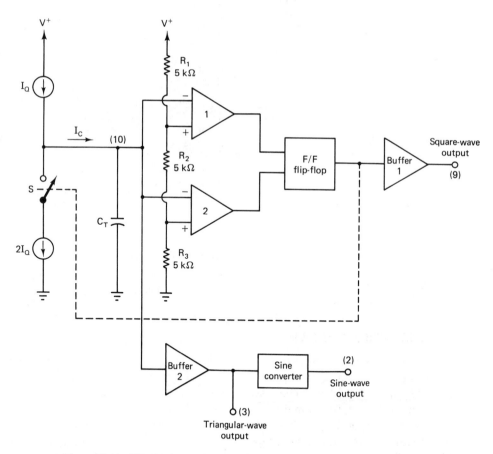

Figure 15.14 ICL8038 precision waveform generator/voltage-controlled oscillator (Intersil).

timing capacitor C_T will be charged with constant current $I_C = I_Q$, so its voltage increases linearly with time at a rate given by $dV_C/dt = I_C/C_T = I_Q/C_T$. When $V_C = \frac{2}{3}V^+$, the output of comparator 1 goes low. This causes the output of the flip-flop to change state, which results in the closing of switch S. After this happens, capacitor C_T will have a net charging current of $I_C = I_Q - 2I_Q = -I_Q$, so the voltage across C_T now decreases linearly with time at a rate of $dV_C/dt = -I_Q/C_T$. When the voltage reaches $\frac{1}{3}V^+$, the output of comparator 2 switches to the high state. This causes the flip-flop to change state again, and now switch S opens and the cycle is repeated. Since the capacitor is charging and discharging between the limits of $\frac{1}{3}V^+$ and $\frac{2}{3}V^+$, the peak-to-peak swing in the capacitor voltage V_C is $\frac{1}{3}V^+$. Noting that the capacitor voltage ramps up and down at a constant rate of $dV_C/dt = I_Q/C_T$, the time required for the capacitor voltage to go between the lower and upper limits is given by $(I_Q/C_T)t = \frac{1}{3}V^+$, so $t = V^+C_T/(3I_Q)$. This time is one-half of the period T, so $T = 2V^+C_T/(3I_Q)$, and the repetition rate or frequency of oscillation is $f_{OSC} = 1/T = 3I_Q/(2V^+C_T)$.

A simplified diagram of the voltage-controlled current source circuit is shown in Figure 15.15. It bears some similarity to the circuit of Figure 15.2 that was considered earlier. The two currents I_3 and I_4 are controlled by the two external resistors R_A and R_B and the control voltage V_C. We note that $V_{E2} = V_{E3} = V_C$, so $I_2 = (V^+ - V_C)/R_B$ and $I_3 = (V^+ - V_C)/R_A$. When the flip-flop output is high, transistor Q_8 is turned on and driven into saturation. This causes Q_4 to be turned off; as a result, Q_5, Q_6, and Q_7 are also off. As a result, $I_4 = 0$ and thus the capacitor charging current is $I_C = I_3 = (V^+ - V_C)/R_A$. When the flip-flop output then goes low, transistor Q_8 turns off, and now the Q_4, Q_5, Q_6, and Q_7 combination acts as a Wilson current mirror. Assuming identical transistors, we have that $Q_5 = Q_6 = Q_7$. Since $I_5 = I_2$, and $I_4 = I_6 + I_7 = 2I_5 = 2I_2$, we have that $I_4 = 2I_2 = 2(V^+ - V_C)/R_B$. The capacitor discharging current is now given by $I_C = I_3 - I_4 = (V^+ - V_C)/R_A - 2(V^+ - V_C)/R_B$.

If $R_A = R_B$, then this current is given by $I_C = -(V^+ - V_C)/R_A$, and thus the capacitor currents during this time period are equal in magnitude, but opposite in direction, to the current during the previous time period. As a result, the capacitor charging time is equal to the capacitor discharge time, so the output voltage is a square wave and the voltage across the capacitor is a symmetrical triangular waveform. Since $f_{OSC} = 3I_Q/(2V^+C_T)$ and $I_Q = (V^+ - V_C)/R_A$, we have that $f_{OSC} = 3(V^+ - V_C)/(2V^+R_AC_T)$. Therefore, a linear voltage-to-frequency transfer relationship is obtained. The triangular waveform across the capacitor goes to a triangular-wave to sine-wave converter via buffer 2. This converter rounds off the peaks of the triangular waveform and converts it into a low-distortion sine wave. If pins 7 and 8 are connected together, a constant voltage of $V_C = (39\text{ k}\Omega/50\text{ k}\Omega)V^+ = 0.78\,V^+$ is obtained, and as a result the oscillation frequency is now given as $f_{OSC} = 3(0.22V^+)/(2V^+R_AC_T) = 1/(3R_AC_T)$. Note that under these conditions the oscillation frequency is independent of the supply voltage.

Now let us consider the case in which the two external resistors are not equal. Then the capacitor charging and discharging currents will no longer be equal. The charging current will be $I_C = I_3 = (V^+ - V_C)/R_A$, and the discharging current will be $I_C = I_3 - I_4 = (V^+ - V_C)/R_A - 2(V^+ - V_C)/R_B = -(V^+ - V_C)[(2/R_B) - $

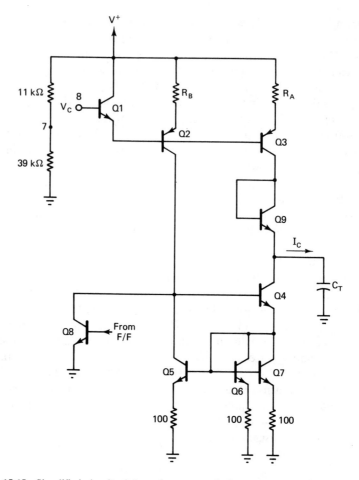

Figure 15.15 Simplified circuit of the voltage-controlled current sources for the ICL8038 (Intersil).

$(1/R_A)]$. The ratio of the charging to discharging currents is therefore I_C (charging)/I_C (discharging) $= (1/R_A)/[(2/R_B) - (1/R_A)] = 1/[(2R_A/R_B) - 1]$. The output voltage of the flip-flop will now no longer be a square wave, but a pulse waveform with a duty cycle given by

$$\delta = \frac{t_{CHARGE}}{t_{CHARGE} + t_{DISCHARGE}} = \frac{1}{1 + (t_{DISCHARGE}/t_{CHARGE})} = 1 - \frac{R_B}{(2R_A)}$$

and the frequency is given by $f_{OSC} = [3(V^+ - V_C)/(V^+ R_A C_T)] [1 - R_B/(2R_A)]$. By suitable choice of the R_B/R_A resistance ratio, duty cycles from 2% to 98% can be obtained.

15.2.6 IC Timer Example: 555 Type

The 555 type of IC timer is a very versatile and popular device. This device is also available as a dual timer (556). It can be used as a monostable multivibrator (a *one-shot*) for the generation of pulses ranging in duration from 1 μs to 100 s. It can also be used as an astable multivibrator with oscillation frequencies ranging from 0.01 Hz to 1 MHz.

The basic block diagram of this device is shown in Figure 15.16. It consists of two comparators that drive the set (S) and reset (R) terminals of a flip-flop, which in turn controls the on and off cycles of the discharge transistor Q_{14}. The comparator reference voltages are fixed at $\frac{2}{3}V^+$ for comparator 1 and $\frac{1}{3}V^+$ for comparator 2 by means of the R_3, R_4, and R_5 (5 kΩ each) voltage divider.

In the operation of this circuit, an external timing capacitor C is charged through an external resistor from the positive supply voltage. When the voltage across the capacitor reaches the threshold voltage level of comparator 1, which is $\frac{2}{3}V^+$, the comparator output goes *high*. This causes the flip-flop to reset so that $\overline{Q}$ is latched

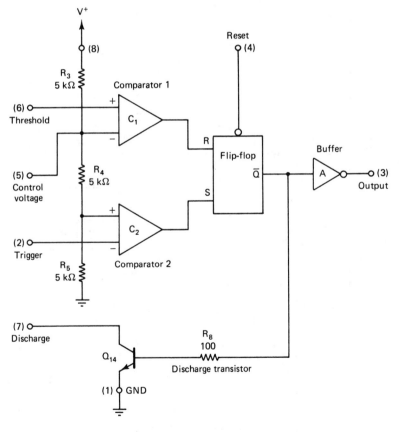

Figure 15.16 Basic block diagram of the 555-type timer.

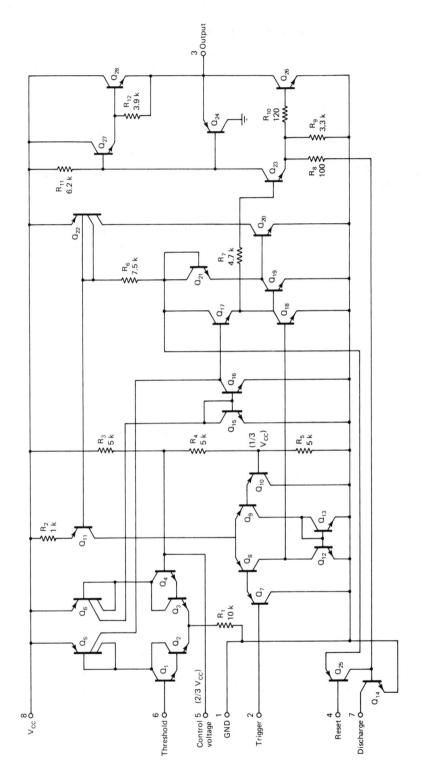

Figure 15.17 Circuit diagram of the LM555 timer (National Semiconductor).

into the *high* state, which turns the discharge transistor Q_{14} *on*. This results in the discharge of the capacitor. In the monostable mode of operation, the capacitor discharges very rapidly through Q_{14} and this terminates the action of the circuit until the next triggering pulse comes along. In the astable mode, the capacitor is connected to both comparators so that when the capacitor voltage drops below the trigger level of comparator 2, which is $\frac{1}{3}V^+$, the comparator output goes *high*. This causes the flip-flop output $\overline{Q}$ to go *low*, turning off the discharge transistor, thus terminating the discharge cycle. The capacitor then once again starts charging, completing the full cycle.

The capacitor charges from the trigger level of $\frac{1}{3}V^+$ up toward V^+, but the charging cycle is interrupted at the threshold voltage level of $\frac{2}{3}V^+$. At this point the capacitor starts discharging toward ground potential. When the capacitor voltage reaches $\frac{1}{3}V^+$ the discharge cycle is interrupted and the capacitor starts charging again.

Figure 15.17 shows the full schematic diagram of the 555-type timer. Comparator 1 is comprised of transistors Q_1 through Q_6, with Q_{15} and Q_{16} in conjunction with Q_5 and Q_6 serving as an active load. Comparator 2 is composed of transistors Q_7 through Q_{10} biased by current source transistor Q_{11}, and Q_{12} and Q_{13} serve as the active load for this comparator.

The flip-flop is comprised of Q_{19} and Q_{20}, which act as two cascaded inverters, with positive feedback from the collector of Q_{20} back to the base of Q_{19} being provided by R_7 (4.7 kΩ). The reset voltage R for the flip-flop is obtained from comparator 1 via its active load Q_{15} and Q_{16} and buffered by Q_{17}. When the voltage applied to the base of Q_{17} goes *high*, Q_{19} goes *on* and Q_{20} goes *off* and the output of the flip-flop $\overline{Q}$ goes *high*.

The set voltage S for the flip-flop is obtained from comparator 2 and is applied to the base of Q_{18}. When the voltage at the base of Q_{18} goes *high*, Q_{18} turns *on* and goes into saturation. This turns Q_{19} *off* and Q_{20} *on* (in saturation), so the flip-flop output voltage $\overline{Q}$ is now *low*.

The flip-flop drives Q_{23}, which in conjunction with R_{11} (6.2 kΩ) and R_9 (3.3 kΩ) acts as a phase splitter that drives the push–pull (or totem-pole) output stage of the Q_{27}–Q_{28} Darlington pair for sourcing currents into the load and Q_{24} and Q_{26} for sinking load currents.

The discharge transistor Q_{14} is driven by the emitter of Q_{23} via R_8 (100 Ω). When the flip-flop output $\overline{Q}$ (at the collector of Q_{20}) is *high*, Q_{14} is turned *on* and discharges the timing capacitor. When $\overline{Q}$ is *low*, Q_{23} is *off* and therefore Q_{14} is also *off*. This condition allows the timing capacitor to charge up.

This circuit may be reset or cleared at any time by means of the reset transistor Q_{25}. With the base of Q_{25} (pin 4) *high* at V^+ or open, Q_{25} is *off* and the circuit operates in the normal fashion. When the base of Q_{25} goes *low* (<0.4 V), Q_{25} is turned on, which then drives Q_{14} on; this causes the discharge of the timing capacitor and also causes the output voltage of the circuit (pin 3) to go *low* (~0 V) for as long as the reset voltage is held low.

Monostable multivibrator. The external circuit connections for the 555 timer operating as a monostable or one-shot multivibrator are shown in Figure 15.18. With the voltage at the trigger input pin 2 *high* (at V^+), the output of comparator 2 is *low*,

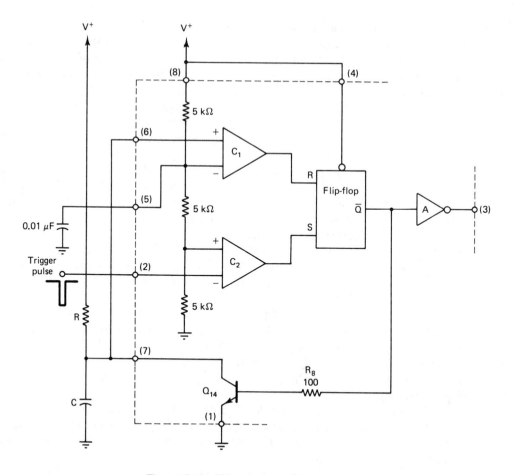

Figure 15.18 555-type monostable multivibrator.

causing the flip-flop output $\overline{Q}$ to be *high*. The discharge transistor Q_{14} will be *on* and the voltage across the timing capacitor will be essentially zero. The output voltage (pin 3) will be *low* (~0 V) and this will be the quiescent state of the device.

When a negative polarity trigger pulse is applied to the trigger input pin 2 such that the voltage at this pin drops below $\frac{1}{3}V^+$, the output of comparator 2 will go *high*. This causes the flip-flop to switch to the opposite state with $\overline{Q}$ now *low* and the discharge transistor will be turned off. Note that after the termination of the trigger pulse the flip-flop remains in the $\overline{Q}$ low state. The timing capacitor can now charge up toward V^+ via resistor R. The voltage across the capacitor is given by $v(t) = V^+[1 - \exp(-t/RC)]$. When $v(t)$ reaches the threshold voltage level of $\frac{2}{3}V^+$, however, comparator 1 switches states and its output voltage now goes *high*. This causes the flip-flop to reset, so $\overline{Q}$ goes *high*. This turns on the discharge transistor Q_{14}, and the voltage across the capacitor rapidly drops to zero. The circuit will be back in its quiescent state with the output voltage *low* (~0 V).

The end of the output pulse occurs at time $t = T$, at which point $v(t) = \frac{2}{3}V^+$, so $\frac{2}{3}V^+ = V^+[1 - \exp(-T/RC)]$, where T is the pulse duration. Solving for T gives

$$\boxed{T = RC \ln 3 = 1.1\,RC} \tag{15.10}$$

During the pulse the output voltage is *high* at $V_{O(\text{HIGH})} = V^+ - V_{BE28} - V_{BE27} - I_{11}R_{11} \simeq V^+ - 1.7$ V. Note that the pulse duration T is independent of the supply voltage V^+. Note also that the trigger pulse must be shorter in duration than T.

Astable multivibrator. Figure 15.19 shows the external connections for the operation of the 555 timer as an astable multivibrator. In this mode of operation, the timing capacitor charges up toward V^+ through $R_A + R_B$ until the voltage across the capacitor reaches the threshold level of $\frac{2}{3}V^+$. At this point, comparator 1 switches states, causing the flip-flop output $\overline{Q}$ to go *high*. This turns *on* the discharge transistor

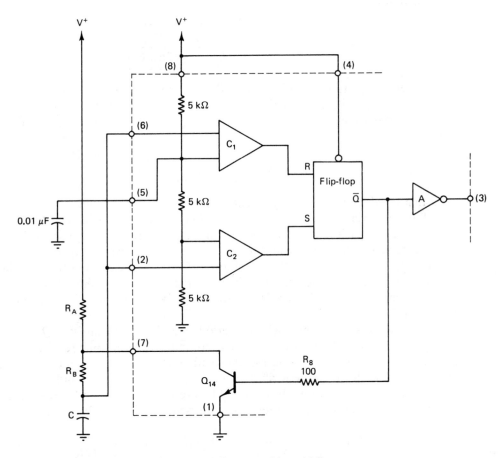

Figure 15.19 555-type astable multivibrator.

Q_{14} and the timing capacitor then discharges through R_B and Q_{14} (pin 7). This discharge continues until the capacitor voltage drops to $\frac{1}{3}V^+$, at which point comparator 2 switches states, causing the flip-flop output $\overline{Q}$ to go *low*, turning off the discharge transistor Q_{14}. At this point the capacitor starts to charge again, thus completing the cycle.

During the charging period, $0 \leq t \leq T_C$, the voltage across the capacitor is given by $v(t) = \frac{2}{3}V^+\{1 - \exp[-t/(R_A + R_B)C]\} + \frac{1}{3}V^+$. At time $t = T_C$ the capacitor voltage reaches the threshold level of $\frac{2}{3}V^+$, so $\frac{2}{3}V^+ = \frac{2}{3}V^+[1 - \exp(-T_C/(R_A + R_B)C)] + \frac{1}{3}V^+$. Solving for the charging time T_C gives $T_C = (R_A + R_B)C \ln 2 = 0.693(R_A + R_B)C$.

During the discharge interval $0 < t' < T_D$, we have that $v(t) = \frac{2}{3}V^+ \exp(-t'/R_BC)$. At time $t' = T_D$, the voltage across the capacitor reaches the trigger level of $\frac{1}{3}V^+$, so we have that $v(t = T_D) = \frac{1}{3}V^+ = \frac{2}{3}V^+ \exp(-T_D/R_BC)$. Solving for T_D, we obtain $T_D = R_BC \ln 2 = 0.693R_BC$, where T_D is the discharge time.

The total period T is $T = T_C + T_D = 0.693(R_A + 2R_B)C$, and the frequency of oscillation is

$$f = \frac{1}{T} = \frac{1}{0.693(R_A + 2R_B)C} \tag{15.11}$$

The output voltage (pin 3) is high during the charging time T_C at a value of $V_{O(HIGH)} = V^+ - V_{BE28} - V_{BE27} - I_{11}R_{11} \simeq V^+ - 1.7$ V. During the discharge interval T_D, the output voltage is low at $V_{O(LOW)} \simeq 0$. The duty cycle of the output pulse waveform is given by

$$\text{duty cycle} = \frac{T_C}{T} = \frac{R_A + R_B}{R_A + 2R_B} \tag{15.12}$$

Note that the duty cycle is always greater than 50% for this circuit.

For duty cycles of 50% and less, the circuit of Figure 15.20 can be used. In this circuit the capacitor C is charged only through R_A. During the charging interval, the voltage across the timing capacitor is $v(t) = \frac{2}{3}V^+[1 - \exp(-t/R_AC)] + \frac{1}{3}V^+$ for $0 < t < T_C$. At $t = T_C$ the threshold level of $\frac{2}{3}V^+$ is reached and the charging of the capacitor ceases. The charging time can therefore be obtained from the relationship that $\frac{2}{3}V^+ = \frac{2}{3}V^+[1 - \exp(-T_C/R_AC)] + \frac{1}{3}V^+$. Solving this for T_C gives $T_C = R_AC \ln 2 = 0.693R_AC$.

During the subsequent discharge period, the discharge transistor Q_{14} is on (in saturation), so the voltage at pin 7 is near ground potential. Figure 15.21 shows a diagram for the discharge period together with a Thévenin equivalent-circuit representation. The discharge cycle starts with $v(t') = \frac{2}{3}V^+$ at $t' = 0$ and ends at $t' = T_D$ with $v(T_D) = \frac{1}{3}V^+$. From the circuit of Figure 15.21, the voltage across the timing capacitor during the discharge interval can be written as

$$v(t') = \left(\frac{2}{3}V^+ - \frac{V^+R_B}{R_A + R_B}\right)\exp\left(\frac{-t'}{\tau_d}\right) + \frac{V^+R_B}{R_A + R_B} \tag{15.13}$$

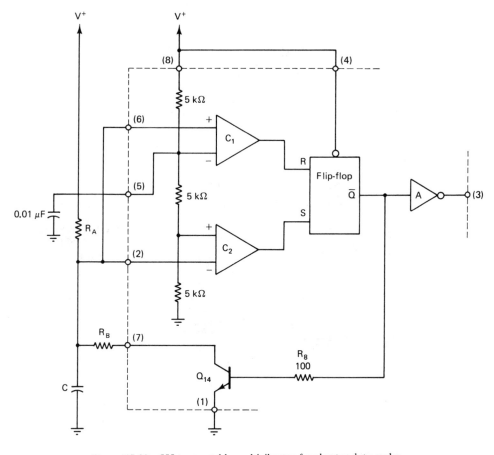

Figure 15.20 555-type astable multivibrator for shorter duty cycles.

where $\tau_d = (R_A \| R_B)C$ = discharge time constant. At $t' = T_D$, $v(T_D) = \frac{1}{3}V^+$, so solving for T_D gives

$$T_D = (R_A \| R_B)C \ln \frac{2R_A - R_B}{R_A - 2R_B} \qquad (15.14)$$

The ratio of T_C to T_D is

$$\frac{T_C}{T_D} = \frac{0.693(1 + R_A/R_B)}{\ln \dfrac{2(R_A/R_B) - 1}{(R_A/R_B) - 2}} \qquad (15.15)$$

For a 50% duty cycle, $T_C/T_D = 1$. The corresponding resistance ratio, R_A/R_B, can

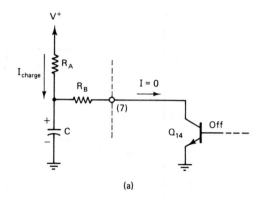

(a)

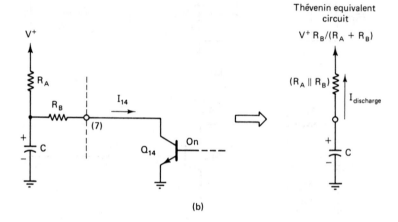

(b)

Figure 15.21 Charging and discharging cycles of the astable multivibrator: (a) charging cycle; (b) discharging cycle.

be obtained by an iterative solution of the equation above to give

$$\frac{R_A}{R_B} = 2.362 \tag{15.16}$$

For the case of a 50% duty cycle, the total period is $T = 2T_C = 2R_A C \ln 2 = 1.386R_A C$ and the corresponding oscillation frequency is

$$f = \frac{1}{T} = \frac{0.721}{R_A C} \tag{15.17}$$

The lowest value that the voltage across the timing capacitor can drop to during the discharge period is $V^+ R_B/(R_A + R_B)$, as seen from the Thévenin circuit representation. To reach the trigger level of comparator 2 and so to end the discharge cycle, the capacitor voltage must drop to $\frac{1}{3}V^+$. Therefore, we see that R_B should not be any greater than $R_A/2$ in order to satisfy this requirement, so oscillations will occur.

Modulation capability. The voltage at pin 5 is nominally $\frac{2}{3}V^+$ and is the voltage applied to the inverting input of comparator 1. This voltage can, however, be varied by means of an externally applied voltage at this pin. Since the upper limit of the voltage across the timing capacitor (pin 6) is equal to the threshold voltage at pin 5, the charging time can be varied by means of a control or modulation voltage applied to pin 5. Thus, for monostable operation, if this device is triggered by a continuous pulse train, the pulse width of the output voltage can be varied or modulated to produce a *pulse-width-modulated* (PWM) signal.

The voltage across the timing capacitor is given as before as $v(t) = V^+[1 - \exp(-t/RC)]$ during the charging time $0 < t < T_C$. The output pulse ends when the voltage across the capacitor reaches the voltage level at pin 5, which is V_M, so at time $t = T$ we have that $v(T) = V_M = V^+[1 - \exp(-T/RC)]$. Solving this equation for the pulse duration T gives $T = RC \ln[V^+/(V^+ - V_M)]$. Thus by varying V_M the pulse width can be modulated, although the relationship between pulse width and modulating voltage will not be linear. Note also that, for triggering the voltage level at the trigger input, pin 2 must drop to less than $V_M/2$.

If this device is operated as an astable multivibrator as shown in Figure 15.19, the modulation voltage applied to pin 5 will again act to vary the upper limit of the capacitor voltage and thereby modulate the charging time. The capacitor now charges from a lower limit of $\frac{1}{2}V_M$ up toward V^+, but reaches the upper limit of the charging cycle when $v(t) = V_M$. Note that if the voltage at pin 5 is V_M, the trigger level voltage is $\frac{1}{2}V_M$.

The voltage across the timing capacitor during the charging cycle of $0 < t < T_C$ is

$$v(t) = \left(V^+ - \frac{V_M}{2}\right)\left[1 - \exp\left(\frac{-t}{(R_A + R_B)C}\right)\right] + \frac{V_M}{2} \qquad (15.18)$$

At $t = T_C$, $v(T_C) = V_M$, so solving for T_C gives $T_C = (R_A + R_B)C \ln[V^+ - (V_M/2)/(V^+ - V_M)]$. Note that when $V_M = \frac{2}{3}V^+$ we obtain $T_C = (R_A + R_B)C \ln 2$, as before.

For the discharge period of $0 < t' < T_D$, we have that $v(t) = V_M \exp(-t/R_BC)$. At time $t' = T_D$, we have that $v(T_D) = \frac{1}{2}V_M$, so solving for T_D gives $T_D = R_BC \ln 2 = 0.693R_BC$.

Note that the discharge time T_D is not a function of the modulation voltage. We see that the width of the pulse, and therefore the position of the pulse trailing edge (the negative-going transition) with respect to the leading edge can be varied. This device therefore can form the basis of a pulse-position modulator (PPM).

15.2.6 Linear Ramp Waveform Generator

In both the monostable and astable modes of operation, the voltage across the timing capacitor will not be a linear function of time, but rather will be characterized by an

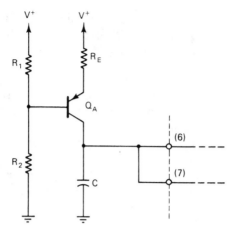

Figure 15.22 Current source circuit for the linear ramp generator.

exponential type of charging and discharging curve. For the monostable mode, we have obtained that the charging time or pulse width is $T = 1.1RC$. Since the pulse width is equal to $1.1RC$ time constants, we see that the voltage waveform across the capacitor will be quite nonlinear.

For the monostable (one-shot) multivibrator, a linear ramp can be obtained by charging the timing capacitor by a constant-current source rather than by a resistor. A linear ramp-generating one-shot is shown in Figure 15.22. In this circuit, transistor Q_A in conjunction with resistors R_E, R_1, and R_2 acts as a constant-current source of strength $I_S = [V^+ R_1/(R_1 + R_2) - V_{EB}]/R_E$.

The voltage across the timing capacitor during the charging period is given by $v(t) = I_S t/C$, and at $t = T$ we will have the end of the charging period at which time $v(T) = \frac{2}{3}V^+$. This is the threshold voltage for comparator 1, and at this point the discharge transistor is turned on due to the action of comparator 1 and the flip-flop. This action rapidly discharges the capacitor and ends the charging period. Setting $v(T) = \frac{2}{3}V^+$ equal to $I_S T/C$ and solving for T gives

$$T = \frac{\frac{2}{3}V^+ C}{I_S} = \frac{\frac{2}{3}V^+ R_E C}{[V^+ R_1/(R_1 + R_2)] - V_{EB}} \qquad (15.19)$$

PROBLEMS

15.1. (*Emitter-coupled voltage-controlled oscillator*)   Given an emitter-coupled VCO with R_1 = 18.6 kΩ, R_2 = 10.0 kΩ, R_3 = 50 kΩ, and $R_5 = R_6 = 100$ kΩ. The circuit is shown in Figure 15.6 with $V^+ = 12$ V.
   (a) Find V_D.   (*Ans.:* $V_D = 2.0$ V)
   (b) Find the control voltage range (V_C) over which the VCO will operate properly.   [*Ans.:* V_C from ∼ 0 V (min.) to +3.75 V (max.)]
   (c) Find the VCO frequency range for $R_4 = 1.0$ kΩ and $C_t = 0.01$ μF.   (*Ans.:* f_{MIN} = 0 and $f_{MAX} = 93.75$ kHz)
   (d) If $TC_{V_{BE}} = -2.2$ mV/°C and $TC_R = +2000$ ppm/°C, find the temperature coefficient of the VCO frequency, TC_f.   (*Ans.:* $TC_f = 1.143 \times 10^{-3}$/°C $= 0.1143\%$/°C $= 1143$ ppm/°C)

(e) If the matching tolerance of the IC resistors is $\pm 2\%$, find the effect on the square-wave output symmetry.   (*Ans.*: the square-wave duty cycle will be $50 \pm 1\%$)

(f) For small values of V_C (less than about 50 mV), the square-wave symmetry becomes progressively worse as V_C gets smaller due to the offset voltage between Q_5 and Q_6. Explain. Find the resulting square-wave symmetry if $V_{os}(Q_5, Q_6) = \pm 2.0$ mV and $V_C = 10$ mV.   [*Ans.*: the square-wave symmetry (duty cycle) will be 45.5% (min.), 54.5% (max.)]

(g) Sketch and dimension the voltage versus time waveforms at the collectors of Q_3 and Q_4.

(h) Sketch and dimension the waveform across the timing capacitor C_6.

15.2. (*Voltage-controlled current source for a VCO*)   Given a current source circuit for a VCO (Figure P15.2). The VCO frequency will be directly proportional to I_Q. Current source I_1 is set to give $V_{EB_1} = V_{BE_2} = V_{BE_3}$ at $I_Q = 10$ mA. The VCO frequency is $f_{MAX} = 1.0$ MHz when $V_C = V_{C(MAX)} = 10$ V.

(a) Find the voltage-to-frequency transfer coefficient, K_V.   (*Ans.*: 100 kHz/V)

(b) Find the VCO frequency f at the following V_C values and the percentage departure from linearity. (*Hint*: Use the iterative technique.)

    (1) $V_C = 1.0$ V   (*Ans.*: 106 kHz, 6%)
    (2) $V_C = 0.50$ V   (*Ans.*: 57.1 kHz, 14.2%)
    (3) $V_C = 0.20$ V   (*Ans.*: 28.9 kHz, 44.4%)

(c) If $V_{BE_2} = V_{BE_3} = 680$ mV and $V_{EB_1} = 700$ mV, all at a current level of 10 mA, find the required value of I_1 to satisfy the condition that $V_{EB_1} = V_{BE_2} = V_{BE_3}$ at $I_Q = 10$ mA.   (*Ans.*: 4.5 mA)

(d) What is the advantage of the use of a current source I_1 over the use of a resistor in its place in this application?

(e) The voltage-controlled current source under consideration here is an open-loop system. For improved linearity, a closed-loop system can be used. Show a voltage-controlled current source that will operate as a closed-loop system and exhibit very good linearity down to very low current levels.

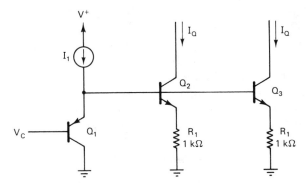

Figure P15.2

15.3. (*Low distortion Wien bridge oscillator*)   In the Wien bridge oscillator circuit of Figure P15.3, a ± 15 V supply is used, $R_1 = R_2$, and $C_1 = C_2$. A small thermistor is used for R_3 to automatically adjust the negative feedback and thereby produce a low distortion sinusoidal output. The thermistor has a resistance of 4000 Ω at 25°C and a thermal resistance of 1°C/mW. The resistance of the thermistor decreases exponentially with

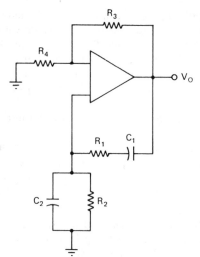

Figure P15.3

temperature, and at 50°C has decreased by a ratio of 8:1. The thermistor temperature is to stabilize at 50°C due to self-hearing when stable oscillations are established.

(a) Find R_4 for stable oscillations.   (*Ans.*: 250 Ω)

(b) Find the rms and peak output voltage.   (*Ans.*: 5.3 V rms, 7.5 V peak)

(c) Using a thermistor similar to the one described, find the thermistor resistance (at 25°C) and the value of R_4 for a 6-V peak output swing.   (*Ans.*: 2560 Ω, 160 Ω)

REFERENCES

FRANCO, S., *Design with Operational Amplifiers and Analog Integrated Circuits*, McGraw-Hill, New York, 1988.

GRAEME, J. G., G. E. TOBEY, and L. P. HUELSMAN, *Operational Amplifiers—Design and Applications*, McGraw-Hill, New York, 1971.

GREBENE, A. B., *Bipolar and MOS Analog Integrated Circuit Design*, Wiley, New York, 1984.

HORN, D. T., *How to Use Special-Purpose ICs*, TAB Books, Blue Ridge Summit, Pa., 1986.

MILLMAN, J., *Microelectronics*, McGraw-Hill, New York, 1979.

MORLEY, M. S., *The Linear IC Handbook*, TAB Books, Blue Ridge Summit, Pa., 1986.

RIPS, E. M., *Discrete and Integrated Electronics*, Prentice-Hall, Englewood Cliffs, N.J., 1986.

STOUT, D. F., *Handbook of Microcircuit Design and Applications*, McGraw-Hill, New York, 1980.

United Technical Publications, *Modern Applications of Integrated Circuits*, TAB Books, Blue Ridge Summit, Pa., 1974.

16

Phase-locked Loops

A *phase-locked loop* (PLL) is a feedback loop comprised of a phase detector, low-pass filter, amplifier, and voltage-controlled oscillator (VCO). The phase detector or phase comparator compares the phase angle of the signal to that of the VCO output voltage, as shown in Figure 16.1, and produces an output voltage that is related to this phase angle difference.

If the phase difference between the signal and the VCO voltage is ϕ radians, the output voltage of the phase detector is given by

$$V_o = 2I_Q R_L \left(\frac{2\,|\phi|}{\pi} - 1 \right) = \frac{4I_Q R_L}{\pi} \left(|\phi| - \frac{\pi}{2} \right) = K_\phi \left(|\phi| - \frac{\pi}{2} \right) \qquad (16.1)$$

where K_ϕ is the *phase-angle-to-voltage transfer coefficient* of the phase detector and the phase angle ϕ is reduced to the range from $-\pi$ to $+\pi$ radians by addition or subtraction of integer multiples of 2π radians.

The output voltage of the phase detector is filtered by the low-pass filter (LPF) to remove the high-frequency components such as those at the signal and VCO frequencies and harmonics thereof. The output of the LPF is then amplified and applied as the input or control voltage V_C of the VCO as given by $V_C = K_\phi A(|\phi| - \pi/2)$, where A is the voltage gain of the amplifier.

This control voltage will result in a shift in the VCO frequency from its *free-running frequency* f_o to a frequency f given by $f = f_o + K_V V_C$, where K_V is the *voltage-to-frequency transfer coefficient* of the VCO.

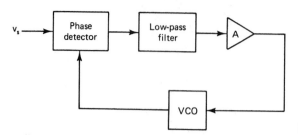

Figure 16.1 Phase-locked loop.

16.1 LOCK-IN RANGE

When the PLL is *locked in* to the signal frequency f_s, we have that $f = f_s = f_o + K_V V_C$ and ϕ will be in the range from 0 to π radians. Since $V_C = (f_s - f_o)/K_V = K_\phi A(\phi - \pi/2)$, we obtain that $\phi - \pi/2 = (f_s - f_o)/(K_V K_\phi A)$. Thus, when the PLL is locked into the signal, a phase angle difference ϕ is established between the signal voltage and the VCO output voltage, given by

$$\phi = \frac{\pi}{2} + \frac{f_s - f_o}{K_V K_\phi A} \tag{16.2}$$

and the two frequencies will be in *exact synchronism*.

The maximum output voltage magnitude available from the phase detector occurs for $\phi = \pi$ and 0 rad and is $V_{O(MAX)} = \pm 2 I_Q R_L = \pm K_\phi(\pi/2)$. The corresponding value of the maximum control voltage available to drive the VCO is $V_{C(MAX)} = \pm(\pi/2)K_\phi A$. The maximum VCO frequency swing that can be obtained is $(f - f_o)_{MAX} = K_V K_{C(MAX)} = \pm K_V K_\phi(\pi/2)A$. Therefore, the maximum range of signal frequencies over which the PLL can remained locked in is $f_s = f_o \pm K_V K_\phi(\pi/2)A = f_o \pm \Delta f_L$, where $2 \Delta f_L$ is the *lock-in frequency range*, given by

$$\boxed{\text{lock-in range} = 2 \Delta f_L = K_V K_\phi A \pi} \tag{16.3}$$

Note that the lock-in range is symmetrically located with respect to the VCO free-running frequency f_o.

Figure 16.2 shows a graph of the VCO control voltage V_C versus the signal frequency f_s. Outside the lock-in range, the VCO frequency cannot be brought into synchronism with the signal frequency. The resulting phase angle difference is $\phi = (\omega_s t + \theta_s) - (\omega_o t + \theta_o) = (\omega_s - \omega_o)t + (\theta_s - \theta_o)$ and will thus change rapidly with time. The rate of change of ϕ with time is $d\phi/dt = \omega_s - \omega_o$. The phase detector output voltage will therefore vary rapidly with time and be heavily attenuated by the low-pass filter. As a result, there will be very little voltage available to drive the VCO and the VCO frequency will revert back to essentially the free-running value of f_o. Thus we see that outside the lock-in range the VCO control voltage will essentially drop to zero.

When the VCO is locked in to the signal, we have that

$$\phi = \frac{\pi}{2} + \frac{f_s - f_o}{K_\phi K_V A} \tag{16.4}$$

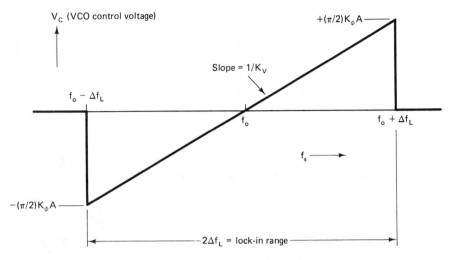

Figure 16.2 PLL lock-in range.

Note that when $f_s = f_o$ the VCO voltage is in *phase quadrature* (90° phase difference) with respect to the signal voltage. As f_s goes above f_o, the phase angle increases from 90° toward a maximum of 180° at the upper end of the lock-in range. As f_s goes below f_o, the phase angle drops from 90° toward 0° at the lower end of the lock-in range.

16.2 CAPTURE RANGE

The discussion of the lock-in range was based on the assumption that the PLL was initially locked in on the signal. We will now investigate the situation wherein this condition is not initially present to determine the range of frequencies over which the PLL can lock in on a signal. This range is called the *capture range* or *acquisition range*.

When the PLL is *not* initially locked into the signal, the frequency of the VCO is the free-running frequency f_o. The phase angle difference between the signal and the VCO voltage is $\phi = (\omega_s t + \theta_s) - (\omega_o t + \theta_o) = (\omega_s - \omega_o)t + \Delta\theta$ and thus is not constant, but changes with time at a rate given by $d\phi/dt = \omega_s - \omega_o$. The phase detector output voltage will therefore not have a dc component, but rather will produce an ac voltage with a triangular waveform of peak amplitude $K_\phi(\pi/2)$ and a fundamental frequency of $f_s - f_o$.

If the low-pass filter is a simple RC low-pass network, it has a transfer function given by

$$T(\omega) = \frac{1}{1 + j\omega\tau} = \frac{1}{1 + j(\omega/\omega_1)} = \frac{1}{1 + j(f/f_1)} \tag{16.5}$$

where $\tau = RC$ and $\omega_1 = 1/RC$, so the breakpoint frequency f_1 is $f_1 = 1/2\pi RC$. For the condition that $(f/f_1)^2 \gg 1$, the transfer function can be expressed approximately as $T(f) \simeq 1/j(f/f_1) = f_1/jf$.

The fundamental input frequency term supplied to the low-pass filter by the phase detector is at the difference frequency $\Delta f = f_s - f_o$. If $\Delta f > 3f_1$, the LPF transfer function is approximately given by $|T(\Delta f)| \simeq f_1/\Delta f = f_1/(f_s - f_o)$. The voltage available to drive the VCO is given by $V_C = V_{o(\text{phase detector})} \times T(f) \times A$ and has a maximum value of $V_{C(\text{MAX})} = \pm K_\phi(\pi/2)(f_1/\Delta f)A$. The corresponding value of the maximum VCO frequency shift is given by

$$(f - f_o)_{\text{MAX}} = K_V V_{C(\text{MAX})} \simeq \pm K_V K_\phi \frac{\pi}{2} A \frac{f_1}{\Delta f} \qquad (16.6a)$$

For the acquisition of the signal frequency f_s, we must have that $f = f_s$, so the maximum signal frequency range that can be acquired by the PLL is

$$(f_s - f_o)_{\text{MAX}} = \Delta f_c \simeq \pm K_V K_\phi \frac{\pi}{2} A \frac{f_1}{\Delta f_c} \qquad (16.6b)$$

where $\Delta f_c = (f_s - f_o)_{\text{MAX}}$. We therefore have that $(\Delta f_c)^2 \simeq K_V K_\phi(\pi/2)Af_1$. Since $\Delta f_L = K_V K_\phi(\pi/2)A$, we can express Δf_c as $(\Delta f_c)^2 \simeq f_1 \Delta f_L$ and thus $\Delta f_c \simeq \pm\sqrt{f_1 \Delta f_L}$.

This gives the range of frequencies that can be captured by the PLL. The total *capture range* is given by

$$\boxed{\text{capture range} = 2 \Delta f_c \simeq 2\sqrt{f_1 \Delta f_L}} \qquad (16.7)$$

for the usual case of $\Delta f_L \gg f_1$. Note that the capture range will be located symmetrically with respect to the VCO free-running frequency f_o.

In Figure 16.3 a graph of the VCO control voltage V_C versus the signal frequency is given again, but this time showing both the capture range and the lock-in range. The PLL cannot acquire a signal outside of the capture range, but once the PLL captures the signal, it will hold on to it until the signal frequency goes beyond the limits of the lock-in range.

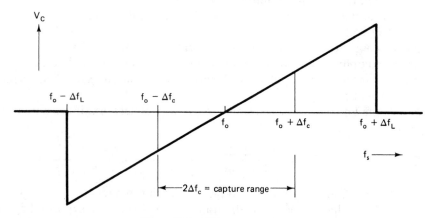

Figure 16.3 PLL capture range.

Phase-locked Loops   Chap. 16

A large capture range is desirable from the standpoint of the ability of the lock-in on a signal. However, a large capture range makes the PLL more susceptible to interference by undesired signals and noise. For the maximum amount of rejection of interference and noise, a small capture range is desirable. In many cases a suitable compromise is reached between these two opposing requirements for the capture range.

In some cases where a suitable compromise cannot be reached, the low-pass filter bandwidth is first set to a large value for the initial acquisition of the signal. Once the signal has been captured and the PLL is locked in on the signal, the band-width of the low-pass filter will be reduced substantially. This will minimize the effects of interfering signals and noise. Indeed, one key feature of a PLL is the ability to remain locked in on a signal even under very adverse noise conditions in which the signal-to-noise ratio may be less than unity. The PLL is used very often in very low level signal applications.

16.3 THE PLL AS AN FM DETECTOR

The PLL can be very easily used as an FM detector or demodulator. A diagram of a PLL FM detector is shown in Figure 16.4.

When the PLL is locked in on the FM signal, we have that the VCO frequency $f_o + K_V V_C$ is equal to the instantaneous frequency of the FM signal f_s, so $f_s = f_o + K_V V_C$ and thus the VCO control voltage voltage V_C is given by $V_C = (f_s - f_o)/K_V$. If the instantaneous frequency of the FM signal is given by $f_s(t) = f_c + \Delta f \sin \omega_m t$, where f_c is the (unmodulated) carrier frequency, Δf is the peak frequency deviation, and ω_m is the radian frequency of the modulating signal, we have that

$$V_C(t) = \frac{f_s(t) - f_o}{K_V} = \frac{f_c - f_o + \Delta f \sin \omega_m t}{K_V} \tag{16.8}$$

The ac component of $V_C(t)$ is $v_c(t) = \Delta f \sin \omega_m t/K_V$, so we see that this will represent a true replica of the modulating voltage that is applied to the FM carrier at the

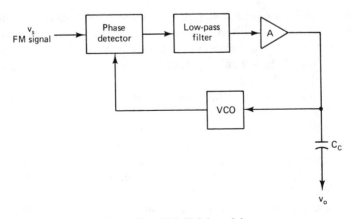

Figure 16.4 PLL FM demodulator.

transmitter. We see that the VCO control voltage is a linear function of the instantaneous frequency deviation, so the FM signal will be demodulated with little or no distortion.

For a maximum symmetrical lock-in range, the VCO free-running frequency f_o should be set as close as possible to the FM carrier frequency f_c.

The maximum VCO control voltage that will be available to drive the VCO is given by $V_{C(MAX)} = \pm K_\phi (\pi/2) A |T(f_m)|$, where $T(f_m)$ is the low-pass filter transfer function at the modulation frequency f_m. For values of f_m such that $f_m > 3f_1$, we have that $|T(f_m)| \simeq f_1/f_m$, where f_1 is the breakpoint frequency of the low-pass filter. Note that when the PLL is locked in on the FM signal the VCO control voltage will vary with time as $v_C(t) = \Delta f \sin \omega_m t / K_V$. This is the voltage that is supplied to the VCO input from the low-pass filter.

For the case of $f_m > 3f_1$, we have that $V_{C(MAX)} \simeq \pm K_\phi (\pi/2) A (f_1/f_m)$. If $f_0 \simeq f_c$, the limiting combination of FM frequency deviation and modulation frequency that the PLL can remain locked in to is given by $V_{C(MAX)} = \Delta f / K_V \simeq K_\phi (\pi/2) A (f_1/f_m)$, so $\Delta f \simeq K_V K_\phi (\pi/2) A (f_1/f_m)$. We therefore have that

$$(f_m \, \Delta f)_{MAX} \simeq K_V K_\phi \frac{\pi}{2} A f_1$$
$$\simeq \Delta f_L f_1 = (\Delta f_c)^2 \tag{16.9}$$

If the product of the modulation frequency f_m and the frequency deviation Δf exceeds this value, the VCO will not be able to follow the instantaneous frequency variations of the FM signal. The output voltage of this FM detector will no longer be an exact replica of the FM modulation signal and there will be distortion. The amount of distortion is very high whenever there is a combination of a large modulation signal amplitude that produces a large Δf and a large modulating frequency f_m. Nevertheless, under conditions in which the PLL can remain locked in on the FM signal, a very linear demodulation characteristic is obtained and there is very little distortion of the demodulated signal.

16.3.1 Frequency-shift Keying

Frequency-shift keying (FSK) is a type of frequency modulation in which the frequency of the FM signal is varied between two fixed levels. This can be considered to be a binary digital type of information transmission system in which one frequency f_1 represents a 0 and the other frequency level f_2 represents a 1, the 0-to-1 frequency deviation Δf being $\Delta f = f_2 - f_1$.

A PLL can be used as an FSK demodulator, as shown in Figure 16.5. It is similar to the PLL demodulator for analog FM signals except for the addition of a comparator to produce a reconstructed digital output signal.

If the PLL remains locked in to the FSK signal at both f_1 and f_2, the VCO control voltage that is also supplied to the comparator is given by $V_{C_1} = (f_1 - f_0)/K_V$ and $V_{C_2} = (f_2 - f_0)/K_V$, respectively. The difference between the two control voltage levels is $\Delta V_C = (f_2 - f_1)/K_V$.

The reference voltage for the comparator is obtained by passing the VCO control voltage through an additional low-pass filter (LPF-2). This low-pass filter has a very

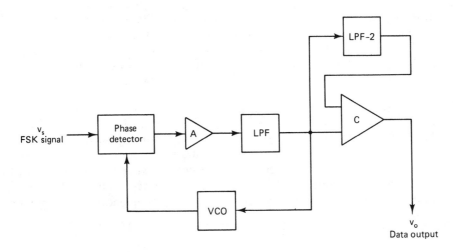

Figure 16.5 Frequency shift keying demodulator.

long time constant compared to the FSK pulse period such that an essentially dc voltage is obtained. This dc voltage has a level that is midway between V_{C_1} and V_{C_2} and therefore is at the optimum value to produce a minimum bit error rate.

16.3.2 PLL Response Time

For an FSK demodulator, the response time of the PLL to a step-function type of frequency change is of importance. For an approximate analysis of the PLL response time for a step-function frequency change from f_1 to $f_2 = f_1 + \Delta f$, we will assume that both f_1 and f_2 fall within the PLL capture range.

For the VCO frequency to change from f_1 to f_2, the required change in the VCO control voltage is $\Delta V_C = \Delta f / K_V$. The VCO control voltage is obtained from across the capacitor of the RC low-pass filter. The time required for the voltage across this capacitor to change by an amount ΔV_C is given approximately by $\Delta V_C = \Delta Q/C = I \, \Delta t/C \simeq [(K_\phi(\pi/2)A/R)](\Delta t/C)$. We therefore have that

$$\Delta t \simeq \frac{RC \, \Delta V_C}{K_\phi(\pi/2)A} \simeq \frac{RC \, \Delta f}{K_V K_\phi(\pi/2)A} \tag{16.10}$$

Since $K_V K_\phi(\pi/2)A = \Delta f_L$ and $RC = \tau$ is the RC time constant of the low-pass network, we have that $\Delta t = \tau \Delta f / \Delta f_L$. The breakpoint frequency of the low-pass network is related to the time constant τ by $f_1 = 1/(2\pi\tau)$, so $\Delta t \simeq \tau \Delta f/\Delta f_L = \Delta f/(2\pi f_1 \Delta f_L)$. After noting that $(\Delta f_c)^2 = f_1 \Delta f_L$, we can also express Δt as $\Delta t \simeq \Delta f/2\pi(\Delta f_c)^2$. The ratio $\Delta f/\Delta t \simeq 2\pi(\Delta f_c)^2 = 2\pi f_1 \Delta f_L$ can be considered to be essentially the slewing rate of the PLL in response to a step-function frequency change and represents the maximum rate at which the VCO frequency can change with time. Note that the PLL slewing rate is a function of RC time constant of the low-pass network and the loop gain $K_V K_\phi A$ around the feedback loop.

The same PLL that is used for the demodulator of the FSK signal can also be used for the production of an FSK signal. This is done by just using the VCO portion of the circuit and driving the VCO by a binary digital input voltage so that the VCO is switched between two fixed frequency levels. The PLL can thus serve very conveniently as a *modem* (or modulator–demodulator) for data communications systems using FSK, such as for the transmission of digital data over telephone lines.

16.4 PLL FREQUENCY SYNTHESIZER

The PLL can be used as the basis for frequency synthesizers that can produce a precise series of frequencies that are derived from a stable crystal-controlled oscillator. A basic diagram of a PLL frequency synthesizer is shown in Figure 16.6. The frequency of the crystal-controlled oscillator is divided by an integer factor M by a counter circuit to produce a frequency f_{osc}/M, where f_{osc} is the frequency of the crystal-controlled oscillator.

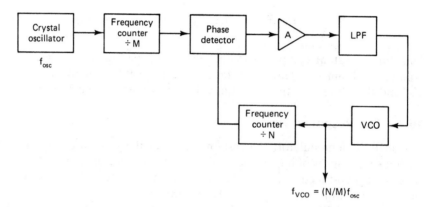

Figure 16.6 PLL frequency synthesizer.

The VCO frequency f_{VCO} is similarly divided by a counter circuit by an integer factor N to become f_{VCO}/N. When the PLL is locked in on the divided-down oscillator frequency, we have that $f_{osc}/M = f_{VCO}/N$, so $f_{VCO} = (N/M)f_{osc}$.

The frequency counters or dividers can be programmed to produce a number of different frequency-division factors so that a large number of frequencies can be produced, all derived from crystal-controlled oscillator.

16.5 THE PLL AS AN AM DETECTOR

The use of a PLL for AM detection is shown in Figure 16.7. When the PLL is locked in on the AM signal, the VCO frequency is equal to the AM carrier frequency f_c. The phase angle of the VCO output voltage with respect to the AM carrier is given by

$$\phi = \left(\frac{\pi}{2} + \frac{f_c - f_o}{K_V K_\phi A}\right) = \frac{\pi}{2}\left(1 + \frac{f_c - f_o}{\Delta f_L}\right) \tag{16.11}$$

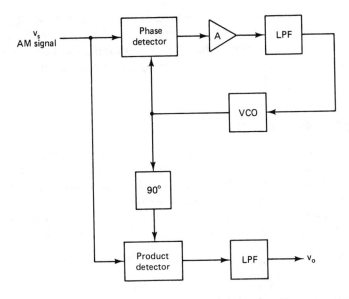

Figure 16.7 PLL amplitude-modulation detector.

where f_o is the VCO free-running frequency and Δf_L is one-half of the lock-in range. If the PLL free-running frequency f_o is close to the AM carrier frequency such that $|f_c - f_o| \ll \Delta f_L$, the VCO output voltage is approximately in phase quadrature (90° out of phase) with the carrier of the AM signal.

The VCO output voltage is passed through a 90° phase-shift network to produce a large-level signal that is in frequency synchronism with the original carrier and is also approximately in phase with the carrier. It is then supplied to one input port of a product detector, such as a balanced modulator–demodulator. The other input port is supplied with the AM signal. The output of the demodulator is then filtered by a low-pass filter to remove all the higher-frequency components such as those at $nf_c \pm f_m$, and passing through only the lowest-frequency difference terms that are obtained from the AM signal sidebands being heterodyned with the VCO frequency at f_c. For the upper sideband at $f_c + f_m$, this gives $(f_c + f_m) - f_c = f_m$, with the same result being obtained for the lower sideband.

Note that for this AM demodulation system the carrier does not have to be transmitted at the full amplitude that would be characteristic of the DSB/LC signal. For the demodulation of a DSB/LC signal with an envelope or peak detector the carrier amplitude must be such that the modulation index will never go above unity, for if it does, severe distortion will result. As a result of the large carrier amplitude, a substantial fraction of the total transmitted power is used in transmitting just the carrier.

For the system under consideration, the PLL can regenerate a carrier that has been transmitted at a much reduced level and produce a large-amplitude signal in frequency synchronism with the original carrier. It will not work, however, in most cases with a DSB/SC or SSB/SC signal in which there has been complete suppression of the carrier. The required amplitude of the carrier for the PLL to the lock in on is a function of the PLL feedback loop parameters. In general, the smaller the PLL capture range and the closer the VCO free-running frequency is to the carrier frequency, the lower the minimum carrier signal amplitude can be for accurate carrier regeneration by the PLL.

16.6 STEREO DEMODULATION

An interesting application in which a PLL is used to regenerate a carrier is the PLL stereo demodulator as shown in Figure 16.8. For FM stereo broadcasting, a total modulation bandwidth of 75 kHz is available and is divided up as shown in Figure 16.9. The range 0 to 15 kHz is used for the L + R (left + right) audio signal for monaural (nonstereo) reception. The L − R difference signal is sent as a double-sideband/suppressed carrier (DSB/SC) signal using a 38-kHz suppressed carrier. A

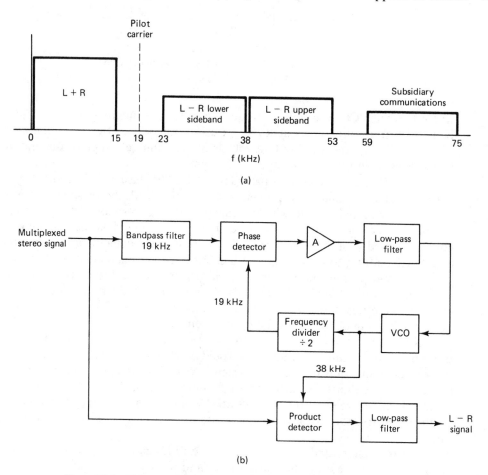

Figure 16.8 PLL stereo demodulator: (a) stereo modulation spectrum; (b) PLL demodulator for the L−R signal.

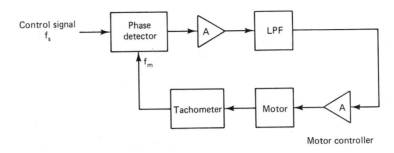

Figure 16.9 Motor speed controller.

small-amplitude pilot carrier at exactly one-half of the carrier frequency (19 kHz) is also transmitted.

At the receiver, a PLL operates on the 19-kHz pilot carrier to produce a 38-kHz signal. This 38-kHz signal is in frequency synchronism with the original 38-kHz carrier and is combined with the multiplexed stereo signal in a product detector. The output of the product detector then passes through a low-pass filter to produce the demodulated L − R signal, the heterodyned L + R signal being rejected by the low-pass filter. The L − R signal can then be combined with the L + R signal to produce the L and R signals for stereo sound.

16.7 MOTOR SPEED CONTROL

Figure 16.9 shows a PLL motor speed control circuit. In this case the PLL is an electromechanical feedback loop, and the tachometer output is a voltage waveform with a frequency that is proportional to the motor speed. Under lock-in conditions, the motor speed is exactly proportional to the signal frequency f_s.

16.8 IC PLL EXAMPLE: LM565

An example of an IC PLL is the LM565 (National Semiconductor). Figure 16.10 is a diagram of this PLL. The phase detector is comprised of the Q_1–Q_2, Q_3–Q_4, and Q_5–Q_6 differential amplifier pairs with Q_{37} together with R_3 (200 Ω) serving as a current sink bias source. Resistors R_1 and R_2 (7.2 kΩ) serve as the load for the phase detector. The output voltage of the phase detector is limited by the diode-connected transistors Q_7 and Q_8 to a maximum of ±0.7 V. This limiting action helps to minimize the effect of high-amplitude noise pulses and other transient effects on the operation of the PLL.

A balanced output is taken from the phase detector and supplied to the Q_{10}–Q_{11} differential-amplifier pair, which is biased by the Q_{39} current sink. The Q_{10}–Q_{11} differential amplifier serves as an amplifier stage in amplifying the output voltage of the phase detector. A single-ended output is taken from this stage from across the load resistor R_{12} (3.6 kΩ) and connected internally to the VCO. Resistor R_{12} serves as part of the low-pass filter. Connection of an external capacitor between pin 7 and ground produces a simple low-pass (lag) network. A capacitor C_2 and a

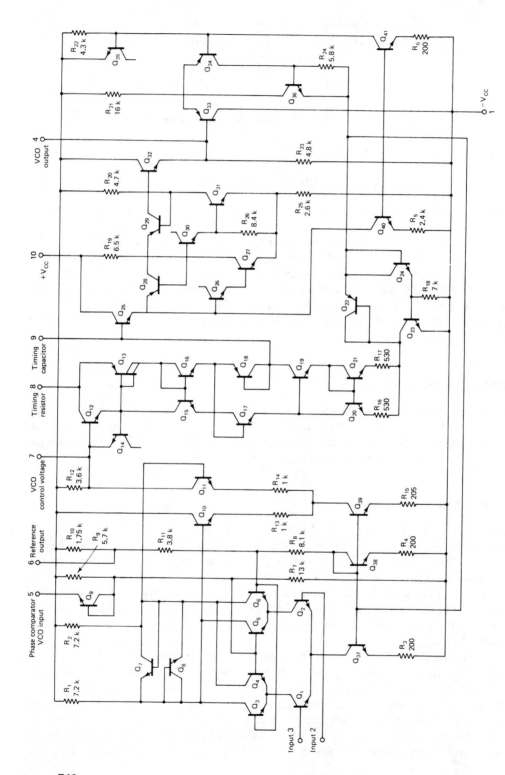

Figure 16.10 Circuit diagram of the LM565 phase-locked loop (National Semiconductor).

resistor R_2 connected in series between pin 7 and ground result in a lag–lead network with a transfer function given by $T(s) = (1 + s\tau_2)/1 + s(\tau_1 + \tau_2)$, where $\tau_1 = R_{12}C_2$ and $\tau_2 = R_2C_2$.

The VCO consists of a voltage-controlled current source (Q_{12} through Q_{23}) that supplies equal-magnitude charging and discharging currents to an externally connected (pin 9) timing capacitor C_t. A timing resistor R_t is connected between pin 8 and the positive supply V^+.

The rest of the VCO circuit is a Schmitt trigger (Q_{25} through Q_{36}) with a differential-amplifier output circuit (Q_{33} and Q_{34}). This controls the turn-on and turn-off of Q_{23} and Q_{24} for the switching action of the current source for the charging and discharging cycles.

At a supply voltage level of ± 6 V, the phase detector has a phase angle-to-voltage transfer coefficient (including the gain of the Q_{10}–Q_{11} differential amplifier) of $K_\phi = 0.68$ V/rad. The VCO has a voltage-to-frequency transfer coefficient or sensitivity of $K_V = 0.65f_o$ Hz/V or $4.1f_o$ rad/s-V, where f_o is the VCO free-running frequency.

The lock-in range is given by $\Delta f_L = K_V K_\phi A(\pi/2) = 0.694f_o$, so the lock-in range extends from $0.31f_o$ to $1.69f_o$, for a total extent of $2\,\Delta f_L = 1.39f_o$.

The loop gain $\omega_L = 2\pi K_\phi A K_V$ is $\omega_L = 2.8f_o$ (s^{-1}). The VCO in this PLL has a maximum free-running frequency of 500 kHz. If $f_o = 100$ kHz, for example, the loop gain is $\omega_L = 2.8 \times 10^5$ s^{-1} and the lock-in range extends from 31 to 169 kHz.

16.9 PLL DESIGN EXAMPLE

We will now consider an example of a 565 type of PLL used as an FSK demodulator for telephone line data transmission. We will consider a data channel for which $f_L = 1070$ Hz represents a 0 and $f_H = 1270$ Hz represents a 1, for a total frequency deviation of 200 Hz between the low and high states and a center frequency of 1170 Hz. We will assume a keying rate (bit rate) of 150 Hz.

The slewing rate of a PLL in response to a step-function frequency change of amount Δf has been given as approximately $\Delta f_{VCO}/\Delta t = 2\pi f_1 \Delta f_L$, where f_1 is the breakpoint frequency of the low-pass filter and Δf_L is the lock-in range. For the 565 PLL, we have that $\Delta f_L = 0.694f_o$, where f_o is the free-running frequency of the VCO. If f_o is set to the FSK center frequency of 1170 Hz, we obtain $\Delta f_L = 812$ Hz, which is more than adequate for this application.

For a keying rate of 150 Hz (150 bits/s), let us set the maximum value of the frequency transition time Δt at $1/(4f_{keying}) = 1/(4 \times 150/s) = 1.7 \times 10^{-3}$ s. We therefore have that $\Delta f_{VCO}/\Delta t = 2\pi f_1 \times 812$ Hz $= 200$ Hz/1.7×10^{-3} s $= 1.2 \times 10^5$ Hz^2. Solving for f_1 gives $f_1 = 24$ Hz. Since $f_1 = 1/2\pi\tau_1 = 1/2\pi R_{12}C_2$, where $R_{12} = 3.6$ kΩ, we obtain that for $f_1 = 24$ Hz the corresponding requirement that $C_2 \leqq 1.9$ μF.

16.9.1 Capture Range

The capture range is given approximately by $\Delta f_c = \sqrt{f_1 \Delta f_L} = \sqrt{\Delta f_L/2\pi\tau_1}$. For the case at hand, we have that $\Delta f_L = 812$ Hz and $f_1 = 24$ Hz, so $\Delta f_c =$

$\sqrt{24 \text{ Hz} \times 812 \text{ Hz}} = 140 \text{ Hz}$. The total capture range thus extends from $1170 - 140 = 1030$ Hz to $1170 + 140 = 1310$ Hz. Since this total capture range encompasses both the low- and high-state frequencies of the FSK signal, we see that the FSK signal can indeed be acquired and, if subsequently lost, can be reacquired.

16.10 MICROPOWER CMOS PHASE-LOCKED LOOP

An example of a micropower CMOS PLL is the CD4046B (National Semiconductor). Another version of this same device is the MC14046B (Motorola). A block diagram of this IC is shown in Figure 16.11. This PLL has two phase detectors that can be used. One phase detector PD1 uses exclusive OR gates and, if used, results in a 90° phase difference between the signal and VCO voltages when in a locked-in condition with the signal frequency f_s equal to the VCO free-running frequency f_o. The other phase detector operates by sensing the leading edge of the signal and VCO waveforms, and when it is selected for use, there is a 0° phase difference between the signal and VCO voltages when in a locked-in condition with the signal frequency equal to the VCO free-running frequency.

This IC operates over the usual CMOS IC supply voltage range from 3 to 15 V dc. This is truly a micropower device with a quiescent device current for the CD4046 of only 5 nA (typ.), 5 μA (max.) with a 5-V supply, increasing to 15 nA (typ.), 20 μA (max.) with a 15-V supply. As is characteristic of CMOS devices, the power dissipation increases at approximately a linear rate with increasing operating frequency. At a frequency of 10 kHz, the power dissipation is 70 μW for a 5-V supply, increasing to 600 μW with a 10-V supply, and 2.4 mW with a 15-V supply.

The maximum VCO frequency is 0.8 MHz when operated with a 5-V supply, increasing to 1.2 MHz with a 10-V supply and 1.6 MHz with a 15-V supply, with similar results obtained for the MC14046. The VCO voltage-to-frequency transfer characteristic has a linearity of 1% (typ.).

The phase comparators have a common signal input with a self-biasing circuit that sets the quiescent level of the input circuit near the middle of the CMOS transition or linear region. As a result, this system can operate with small signal levels, but capacitative coupling must be used so that the biasing conditions are not disturbed. Larger CMOS logic level signals can, however, be fed directly into this system. Large signals can be fed directly into the phase comparator, but small signals must go in via a coupling capacitor. The small-signal input voltage sensitivity with ac coupling using a 1-nF capacitor at 50 kHz is 200 mV (typ.) with a 5-V supply, increasing to 700 mV (typ.) with a 15-V supply.

An interesting feature of this PLL is the very wide lock-in range with respect to the center frequency f_0. The lock-in range $2 \Delta f_L$ extends from the VCO maximum frequency f_{MAX} to the VCO minimum frequency f_{MIN}, where f_{MAX} is given approximately by $f_{MAX} = 1/[R_1(C_1 + 32 \text{ pF})]$ and f_{MIN} is given approximately by $f_{MIN} = 1/[R_2(C_1 + 32 \text{ pF})]$. The allowable range for both R_1 and R_2 is from a minimum of 10 kΩ up to a maximum of 1 MΩ, and C_1 can range from 100 pF up to 10 nF. Since the ratio of R_2 to R_1 can be as large as 100 : 1, the ratio of f_{MAX} to f_{MIN} can be equally large. For example, if $f_{MAX} = 100$ kHz, f_{MIN} can be as small as 1 kHz, so with a

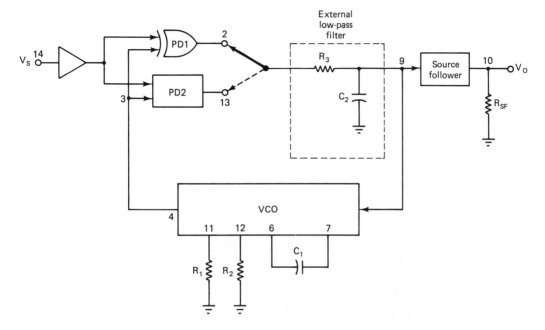

Figure 16.11 CMOS micropower phase-locked loop: CD4046 (National Semiconductor).

center frequency of $f_0 \cong 50$ kHz, the lock-in range $2\,\Delta f_L$ will extend from 1 to 100 kHz. The capture range will be determined by the low-pass filter. If a simple lag filter of the type shown in Figure 16.11 is used, the capture range $2\,\Delta f_c$ is given by $2\,\Delta f_c = \sqrt{2\,\Delta f_L}/(\pi R_3 C_2)$.

This PLL includes a source follower to minimize loading on the phase detector output and should be used with an external load resistor of at least 10 kΩ. This being a CMOS device, a large rail-to-rail output voltage swing can be obtained under lightly loaded conditions.

This device can be used for all the usual PLL applications, such as FM and FSK modulation and demodulation, frequency synthesis, data synchronization, voltage-to-frequency conversion, and motor speed control.

PROBLEMS

16.1. If the frequency of a VCO varies with the VCO input voltage, V_i, as $f =$ (constant) $(V_i + V_o)$, show that the normalized voltage-to-frequency conversion factor is given by $(1/f)(df/dV_i) = 1/(V_o + V_i)$, where V_o is a constant voltage.

16.2. If a VCO has a free-running frequency f_o of 50 kHz, and $V_o = 2.0$ V in the equation in Problem 16.1, find the VCO voltage-to-frequency conversion coefficient for the relationship $\Delta f = K_V\,\Delta V_i$. (*Ans.*: $K_V = 25$ kHz/V)

16.3. Given that a phase-locked loop has a VCO (Figure P16.3) with $K_V = 25$ kHz/V and $f_o = 50$ kHz. The amplifier gain is $A = 2.0$, and the phase detector has a maximum output

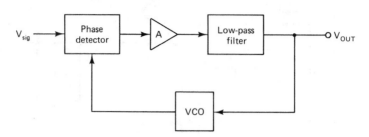

Amplifier

Figure P16.3

voltage swing of ± 0.70 V. Find the *hold-in range* of the PLL. (*Ans.*: $\Delta f_{\text{MAX}} = \pm 35$ kHz, so the hold-in range is from 15 to 85 kHz)

16.4. Find the approximate *capture range* with respect to f_0 of the PLL of Problem 16.3 if the loop low-pass filter time constant is $\tau = RC$, where R is 3.6 kΩ for the following values of filter capacitors. (Note that C is an external element.)

(a) 0.01 μF   (*Ans.*: ± 12.4 kHz)
(b) 0.03 μF   (*Ans.*: ± 7.2 kHz)
(c) 0.10 μF   (*Ans.*: ± 3.9 kHz)
(d) 0.30 μF   (*Ans.*: ± 2.3 kHz)
(e) 1.0 μF   (*Ans.*: ± 1.24 kHz)

16.5. Under small-signal input conditions ($V_{\text{signal}} = 25$ mV or less), the phase detector output voltage will become proportional to the signal voltage amplitude, so that we will have $V_{o(\text{dc})\text{phase detector}} = -(\text{constant})V_{\text{signal}} \cos \phi$. If for a signal voltage of 1.0 mV rms the phase detector output voltage is given by $V_{O(\text{dc})} = -50$ mV $\cos \phi$, find the hold-in range using the same PLL as for Problem 16.3. (*Ans.*: $\Delta f_{\text{max}} = \pm 2.5$ kHz, so the hold-in range runs from 47.5 to 52.5 kHz)

16.6. If the input signal to the PLL is a frequency-modulated (FM) signal, and the PLL is in lock, find the frequency modulation to output-voltage conversion coefficient ($\Delta V_O/\Delta f_{\text{signal}}$) if the VCO characteristics are given by $f = (\text{constant})(V_i + V_o)$, where $V_o = 2.0$ V and the free-running frequency of the VCO is 400 kHz. (*Ans.*: FM conversion coefficient $= 5$ mV/kHz)

16.7. Given that the input signal to the PLL is an FM signal of frequency $f_s = f_o(1 + m \sin \omega_m t)$, show that the maximum modulation frequency for which the PLL remains in lock is given approximately by $f_{m(\text{MAX})} = (\Delta f_L/mf_o)(1/2\pi\tau)$, where $\pm \Delta f_L$ is the lock-in range, f_o is the VCO free-running frequency, and τ is the low-pass filter RC time constant.

16.8. A PLL has a VCO free-running frequency of $f_o = 100$ kHz and a hold-in range of $\Delta f_H = \pm 50$ kHz. A frequency-modulated (FM) signal has a center frequency of $f_c = 100$ kHz (carrier frequency) and a frequency deviation of ± 10 kHz. Find the maximum modulating frequency for which the PLL will stay in lock for the following values of low-pass filter capacitance ($R_L = 3.6$ kΩ).

(a) $C = 0.01$ μF   (*Ans.*: 22.1 kHz)
(b) $C = 0.03$ μF   (*Ans.*: 7.4 kHz)
(c) $C = 0.10$ μF   (*Ans.*: 2.2 kHz)
(d) $C = 0.3$ μF   (*Ans.*: 737 Hz)
(e) $C = 1.0$ μF   (*Ans.*: 221 Hz)

16.9. A signal that is applied to a PLL occurs in short bursts, between which the signal is absent. The signal frequency is $f_s = 1.0$ MHz and the signal bursts occur at intervals of 1.0 ms and have a duration of 10 μs. The PLL has a crystal-controlled VCO with a free-running frequency f_o that is within 0.1% of the signal frequency. It is required that the PLL remain continuously locked in on the signal and that the phase angle of the VCO output voltage not drift by more than 0.1 rad in the interval between the signal bursts.

(a) Find the maximum allowable drift in the VCO frequency between the signal bursts. (*Ans.*: 15.9 Hz)

(b) A simple lag low-pass filter is used in this PLL. Find the minimum value of the filter time constant. (*Ans.*: 62.3 ms)

(c) If the net resistance of the filter is 4.0 kΩ, find the minimum value that the filter capacitor can have. (*Ans.*: 15.6 μF)

REFERENCES

BERLIN, H. M., *Design of Phase-Locked Loop Circuits with Experiments*, Howard M. Sams, Indianapolis, Ind., 1978.

BLANCHARD, A., *Phase-Locked Loops*, Wiley, New York, 1976.

CONNELLY, J. A., *Analog Integrated Circuits*, Wiley, New York, 1975.

EGAN, W. F., *Frequency Synthesis by Phase Lock*, Wiley, New York, 1981.

GARDNER, F. M., *Phaselock Techniques*, Wiley, New York, 1979.

GLASER, A. B., and G. E. SUBAK-SHARPE, *Integrated Circuit Engineering*, Addison-Wesley, Reading, Mass., 1977.

GREBENE, A. B., *Analog Integrated Circuit Design*, Van Nostrand Reinhold, New York, 1972.

――――, *Bipolar and MOS Analog Integrated Circuit Design*, Wiley, New York, 1984.

HORN, D. T., *How to Use Special-Purpose ICs*, TAB Books, Blue Ridge Summit, Pa., 1986.

IRVINE, R. G., *Operational Amplifiers—Characteristics and Applications*, 2nd ed., Prentice-Hall, Englewood Cliffs, N.J., 1987.

KLAPPER, J., and J. T. FRANKLE, *Phase-Locked and Frequency Feedback Systems*, Academic Press, New York, 1972.

KRAUSS, H. L., C. W. BOSTIAN, and F. H. RAAB, *Solid State Radio Engineering*, Wiley, New York, 1980.

LENK, J. D., *Handbook of Integrated Circuits*, Prentice-Hall, Englewood Cliffs, N.J., 1978.

MEADE, M. L., *Lock-in Amplifiers: Principles and Applications*, Peter Peregrinus Ltd., London, 1983.

MCMENAMIN, J. M., *Linear Integrated Circuits: Operation and Applications*, Prentice-Hall, Englewood Cliffs, N.J., 1985.

STOUT, D. F., *Handbook of Microcircuit Design and Applications*, McGraw-Hill, New York, 1980.

United Technical Publications, *Modern Applications of Integrated Circuits*, TAB Books, Blue Ridge Summit, Pa., 1974.

17

Digital-to-analog and Analog-to-digital Converters

17.1 DIGITAL-TO-ANALOG CONVERTERS

A *digital-to-analog converter* (D/A or DAC) is a device that converts a digital input signal to an analog output voltage (or current) that is proportional to the digital signal. A simple example of an N-bit DAC for a binary digital input is shown in Figure 17.1. The output voltage is given by

$$V_O = -R_F \left(\frac{b_1 V_{\text{REF}}}{R} + \frac{b_2 V_{\text{REF}}}{2R} + \frac{b_3 V_{\text{REF}}}{4R} + \cdots + \frac{b_N V_{\text{REF}}}{2^{N-1} R} \right)$$

$$= -2 \frac{R_F}{R} V_{\text{REF}} \left(\frac{b_1}{2} + \frac{b_2}{4} + \frac{b_3}{8} + \cdots + \frac{b_N}{2^N} \right) \tag{17.1}$$

where b_1 is the most significant bit (MSB), b_2 is the next significant bit, . . . , and b_N is the least significant bit (LSB).

The output voltage corresponding to the LSB input (. . . 000001) is $V_{O(\text{LSB})} = 2(R_F/R)V_{\text{REF}}(1/2^N) = (R_F/R)V_{\text{REF}}(1/2^{N-1})$. The output voltage corresponding to a MSB digital input (1000 . . .) is $V_{O(\text{MSB})} = V_{O(\text{LSB})}2^{N-1}$.

The maximum output voltage occurs when all bits are 1 (1111 . . .) and is $V_{O(\text{MAX})} = 2V_{O(\text{MSB})} - V_{O(\text{LSB})} = (2^N - 1)V_{O(\text{LSB})} = 2V_{O(\text{MSB})}(1 - 1/2^N)$. The *nominal full-scale output voltage* $V_{O(\text{FS})}$ is defined as $V_{O(\text{FS})} = 2V_{O(\text{MSB})} = 2^N V_{O(\text{LSB})}$, and is thus greater than the maximum output voltage by an amount equal to $V_{O(\text{LSB})}$.

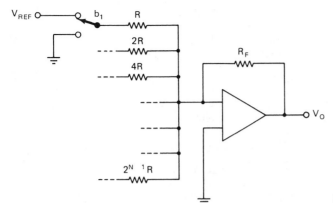

Figure 17.1 *N*-bit digital-to-analog converter.

Many DACs are set to have a full-scale output voltage of 10.000 V. For a 4-bit ($N = 4$) DAC, $V_{O(LSB)} = 10.00 \text{ V}/2^N = 10.00/2^4 = 0.625 \text{ V}$. The output voltage for a MSB input (1000) is $V_{O(MSB)} = V_{O(FS)}/2 = 5.00 \text{ V}$. The output voltage for a digital input of 1111 is $V_{O(MAX)} = V_{O(FS)} - V_{O(LSB)} = 10.00 - 0.625 = 9.375 \text{ V}$.

Similarly, for the case of an 8-bit DAC with a full-scale output voltage of 10.000 volts, we have

$$V_{O(LSB)} = \frac{10.000}{2^8} = 0.039 \text{ V}$$

$$V_{O(MSB)} = 5.000 \text{ V} \tag{17.2}$$

$$V_{O(MAX)} = 9.961 \text{ V}$$

For a 12-bit converter, we have $V_{O(LSB)} = 0.00244 \text{ V} = 2.44 \text{ mV}$, $V_{O(MSB)} = 5.00$ V, and $V_{O(MAX)} = 9.9976 \text{ V}$.

Another commonly used value for the full-scale output voltage for DACs is 10.24 V. With this choice of $V_{O(FS)}$, the LSB output voltage and therefore all the voltage increments in the output voltage will be integer multiples or submultiples of 10 mV. Table 17.1 gives some values for $V_{O(LSB)}$ and $V_{O(MAX)}$ for various values of *N*. For all cases, the output voltage corresponding to a digital MSB input (1000 . . .) is 5.12 V.

Figure 17.2 is a graph of the ideal transfer characteristic of an *N*-bit DAC for the case of a 4-bit DAC. For the ideal DAC, the envelope of the transfer characteristic is a straight line passing through the origin ($V_O = 0$ for a digital input of 0000) and through the point $V_{O(FS)} - V_{O(LSB)}$ at a digital input of 1111.

Figure 17.3 shows the envelope of the transfer characteristic of an actual or nonideal DAC and compares it with that of the ideal DAC. The accuracy of a DAC is measured by the maximum error from the ideal characteristic, usually expressed as a fraction of the LSB voltage increment, $V_{O(LSB)}$. A typical accuracy specification for a DAC is $\pm\frac{1}{2}$ LSB. As long as the maximum error of the DAC is less than $\pm\frac{1}{2}$ LSB, the DAC will have a monotonic transfer characteristic; that is, as the digital input increases the analog voltage output will also increase.

TABLE 17.1 $V_{O(FS)} = 10.24$ V

N	$V_{O(LSB)}$ (mV)	$V_{O(MAX)}$ (V)
3	1280	8.96
4	640	9.60
5	320	9.92
6	160	10.08
7	80	10.16
8	40	10.20
9	20	10.22
10	10	10.23
11	5.0	10.235
12	2.5	10.2375
13	1.25	10.23875
14	0.625	10.239375
15	0.3125	10.239688
16	0.15625	10.239844

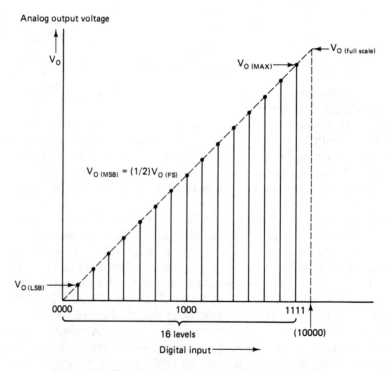

Figure 17.2 Transfer characteristic of an ideal 4-bit DAC.

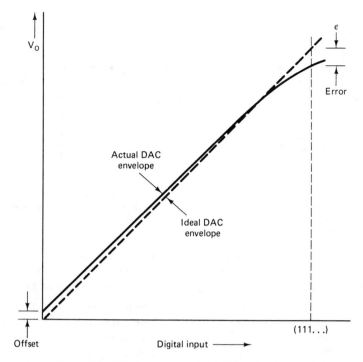

Figure 17.3 Ideal versus actual DAC transfer characteristics.

The ratio of the maximum resistance to the minimum resistance of the simple resistor ladder DAC of Figure 17.1 is 2^{N-1}, where N is the number of bits. For large values of N ($N \gtrsim 6$), the resistance ratio becomes very large, and it becomes very difficult to establish accurate resistance ratios over this large a resistance ratio.

This problem can be solved by the use of cascaded resistor ladder networks as shown in Figure 17.4. This is an 8-bit DAC using two cascaded resistor ladder quads. The resistor R_A placed between the quads for current division must produce a current division of $1:16$, so only $\frac{1}{16}$ of the current produced by the second quad will pass through R_A and appear at the operational amplifier input.

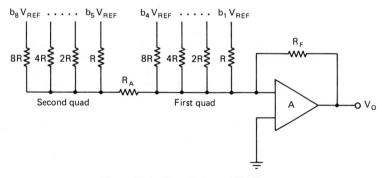

Figure 17.4 Cascaded quad DAC.

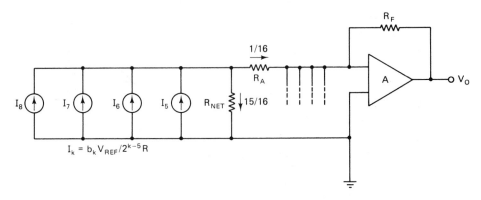

Figure 17.5 Equivalent current source–parallel resistance representation of the second quad.

If we redraw the second quad in the form of a current source (Norton) equivalent circuit, we obtain the circuit of Figure 17.5. The resistance R_{NET} is the parallel combination of R, $2R$, $4R$, and $8R$ and is given by $R_{\text{NET}} = 0.5333333R$. For a 16:1 current attenuation, we must have that $R_A/R_{\text{NET}} = 15$, so $R_A = 8.00R$.

If the cascaded quad DAC is driven by current sources as shown in Figure 17.6, a current divider comprised of R_B and R_C is required. For a 15:1 current ratio between R_C and R_B, we require that $R_B = 15R_C$.

To carry the cascaded quad DAC one step further, let us consider a 12-bit DAC using weighted current sources as shown in Figure 17.7. For this circuit we have two current ratio conditions to satisfy and four resistance values (R_D, R_E, R_F, and R_G). Therefore, we can arbitrarily set two conditions for the resistance values. For a representative example, let us set $R_D = 1.00 \text{ k}\Omega$. For the b_5–b_8 quad, we have the requirement that $1/(R_F + R_G) + 1/R_D = 15/R_E$. Let us now set $R_F + R_G$ equal to R_E, so we now have that $1/R_D = 15/R_E - 1/R_E = 14/R_E$, so $R_E = \underline{14.00 \text{ k}\Omega}$.

For the b_9–b_{12} quad, we have the requirement that

$$\left[\frac{R_F}{R_F + R_G + (R_E \parallel R_D)}\right]\left[\frac{R_D}{R_D + R_E}\right] = \frac{1}{16 \times 16} \tag{17.3}$$

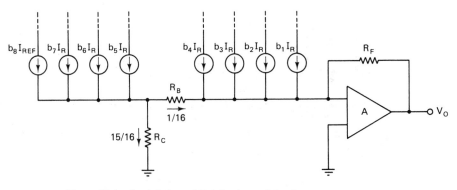

Figure 17.6 Cascaded quad DAC using weighted current sources.

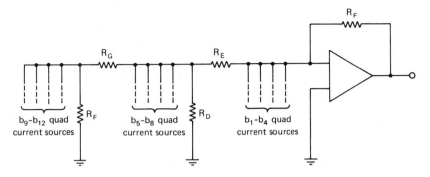

Figure 17.7 Twelve-bit cascaded quad DAC using weighted current sources.

Since $R_D/(R_D + R_E) = \frac{1}{15}$, $R_F + R_G = R_E = 14.00$ kΩ, and $R_E \parallel R_D = 1$ k$\Omega \parallel$ 14 k$\Omega = 0.9333$ kΩ, we obtain $R_F/(14$ k$\Omega + 0.9333)$ k$\Omega = 15/(16)^2$; so $R_F = \underline{0.875}$ $\underline{\text{k}\Omega}$ and therefore $R_G = \underline{13.125 \text{ k}\Omega}$.

17.1.1 *R–2R* Resistor Ladder Networks

The ultimate case of a cascaded resistor ladder is the $R–2R$ ladder network shown in Figure 17.8a. This type of ladder network has the distinct feature and advantage of requiring only two resistance values, R and $2R$. For a small resistance ratio such as this, the resistance ratio tolerance of IC resistors will be much better than for the case of large resistance ratios.

At every node $(1, 2, 3, \ldots, N)$ along the $R–2R$ ladder network, the resistance looking toward the left $(2R)$ is equal to the resistance seen looking toward the right $(2R)$. Figure 17.8b is a view of three nodes in the $R–2R$ ladder network. The current coming into node m from the previous node $(m + 1)$ will split into two equal parts at node m. One half of the current will go down through the $2R$ resistor and the other half will be supplied through resistor R to the next node $(m - 1)$. We see that the current propagating through the ladder will be split in half at every node. The current from every current source will be split into three equal parts at the node at which it is injected. Thus current source $b_m I_R$ will produce a current $b_m I_R/3$ at node m traveling to the right. This current will then be split in half at every node, with half going down through resistor $2R$ and half continuing on to the right through resistor R. The output current I_O is therefore given by $I_O = \frac{2}{3}I_R (b_1/2 + b_2/4 + b_3/8 + \cdots + b_N/2^N)$.

Figure 17.9 shows the $R–2R$ ladder network connected to an operational amplifier for current-to-voltage conversion. The output voltage of this DAC is given by

$$V_O = -R_F I_O = -\frac{2}{3}R_F I_R \left(\frac{b_1}{2} + \frac{b_2}{4} + \frac{b_3}{8} + \cdots + \frac{b_N}{2^N} \right) \qquad (17.4)$$

The resistor of $3R \parallel R_F$ connected between the noninverting input of the operational amplifier and ground is for the purpose of bias current cancellation.

Another $R–2R$ ladder DAC is shown in Figure 17.10. The bottom end of the $2R$ resistors will be at essentially ground potential for either position of the b_1, b_2,

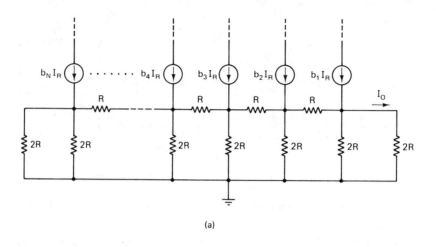

(a)

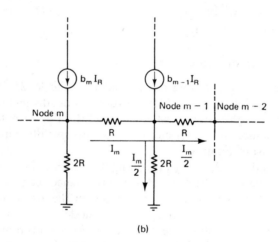

(b)

Figure 17.8 $R-2R$ ladder with current source inputs: (a) basic circuit; (b) two nodes.

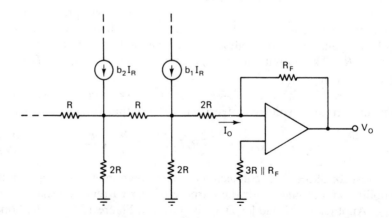

Figure 17.9 $R-2R$ ladder DAC.

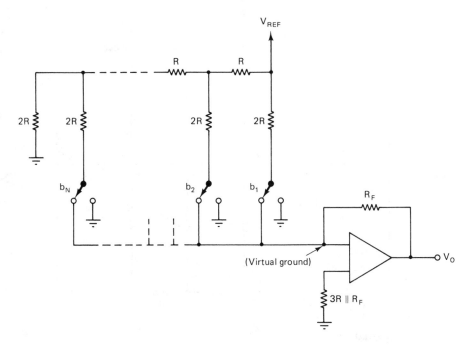

Figure 17.10 Inverted $R-2R$ ladder DAC.

$b_3, \ldots, b_N$ switches since the inverting input terminal of the operational amplifier is a virtual ground. The current division at each node of the $R-2R$ ladder is the same as that considered before, and we obtain for the output current I_O of the ladder the expression $I_O = (V_{REF}/R)(b_1/2 + b_2/4 + b_3/8 + \cdots + b_N/2^N)$. This *inverted ladder* DAC circuit has the advantage over the first $R-2R$ ladder circuit that the voltage level at the switching nodes remains at essentially ground potential for either position of the $b_1, b_2, b_3, \ldots, b_N$ switches. In the other $R-2R$ ladder circuit, the current switching node voltages change depending on the values of $b_1, b_2, b_3, \ldots, b_N$. As a result, the inverted ladder DAC is capable of much faster operation since the voltage across the switching node capacitances remains at essentially ground potential.

An implementation of the inverted ladder $R-2R$ network using current source transistors is shown in Figure 17.11a. Let us first assume that all transistors are identical except that they are scaled in area such that the current densities are equal in all transistors. As a result, the base-to-emitter voltage drops are all equal, and therefore the voltages at the top end (the emitter end) of all the $2R$ resistors are the same. We therefore see that the current I_1 is equal to the current I_O through transistor Q_O (1.0 mA), current I_2 is one-half of I_1, and so on. The last two transistors Q_N and Q_{N+1} will, however, carry equal currents. This is necessary in order to have the $R-2R$ ladder terminated at the proper resistance level.

For large numbers of bits $(N > 4)$, it becomes impractical to scale the transistor areas for equal current densities since the largest transistors will become very large and take up an excessive amount of area on the IC chip. Therefore, only the four

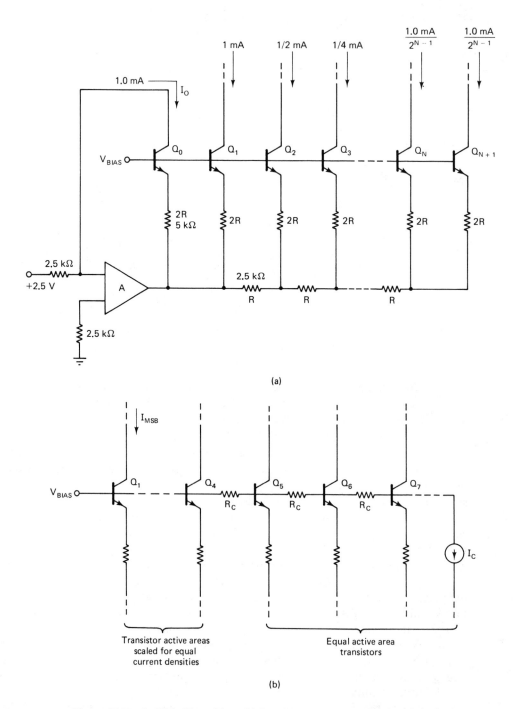

(a)

(b)

Figure 17.11 *R–2R* ladder with multiple current source transistors: (a) basic circuit; (b) V_{BE} compensation.

most significant bit transistors are scaled in area, and the remaining transistors are of equal active area. A method for compensating for the nonequal base-to-emitter voltage drops of the lesser significant bit transistors is shown in Figure 17.11b.

Since the current carried by every successive equal-area transistor ($Q_4 - Q_{N-1}$) decreases by a 2:1 ratio, the base-to-emitter voltage drop will correspondingly decrease by $V_T \ln 2 = 18$ mV. To compensate for this 18-mV voltage difference between transistors, the IR drop across the interbase resistors R_C is set to be 18 mV. For example, if $I_C = 200$ μA, the value of R_C should be 18 mV/0.20 mA $= 90 \ \Omega$.

The decrease in the V_{BE} drop for transistors Q_5 through Q_N is $\Delta V_{BE} = V_T \ln 2$, and it will therefore have a temperature coefficient of $d(\Delta V_{BE})/dT = (k/q) \ln 2 = \Delta V_{BE}/T = 18{,}000 \ \mu V/300 \ K = 60 \ \mu V/°C$. This is a rather small temperature coefficient and, since it affects only the lesser significant bit transistors, it will generally not be a problem.

Figure 17.12 shows a diagram of a current switching cell for an inverted R–$2R$ ladder. This current switching cell is comprised of transistors Q_A and Q_B acting as a differential amplifier, which in turn drives the Q_C–Q_D differential amplifier. Transistor Q_N acts as a current sink for biasing the Q_A–Q_B differential amplifier. When the logic voltage level at the base of Q_B is high such that the voltage at the base of

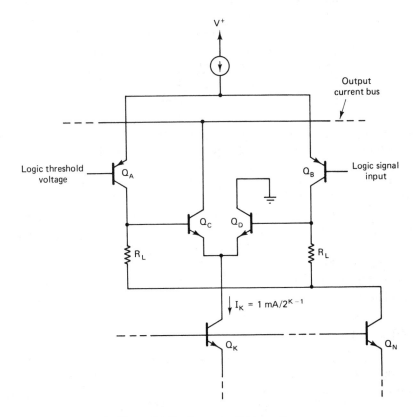

Figure 17.12 Current switching cell.

Q_B is higher than the logic reference voltage at the base of Q_A, the current in the Q_A–Q_B differential-amplifier pair is shifted to the Q_A side. This causes the current in the Q_C–Q_D differential-amplifier pair to shift to the Q_C side. The current from transistor Q_K is thus delivered to the DAC output current bus.

If now the input logic voltage level is low (below the logic threshold voltage level), the current in the Q_C–Q_D differential amplifier will be shifted to the Q_D side and thus will not flow into the DAC output current bus, but rather will be shunted to ground. Since the DAC output current bus is connected to the inverting input terminal of an op amp, it will remain at essentially ground potential, so the voltages at the collectors of Q_C and Q_D will be approximately the same. As a result of this and the fact that the total current through the Q_C–Q_D differential-amplifier pair remains constant at I_K and that none of the transistors involved is driven into saturation, this current switching circuit provides for a very fast (about 3 ns) switching action.

Example of 10-bit DAC: AD561. An example of a 10-bit DAC of the type just discussed is the AD561 (Analog Devices); a circuit diagram is shown in Figure 17.13. An internally generated -7.5 V reference voltage is developed on the chip by a current-source-biased zener diode CR_1. This is a temperature-compensated diode with a net temperature coefficient in the range from 0 to ± 15 ppm/°C. This reference voltage is then transformed to a $+2.5$-V reference by amplifier A_1. This 2.5-V reference voltage is applied to the 2.5-kΩ resistor connected to the inverting input terminal of amplifier A_2 and thereby produces a current flow of 1.0 mA, which sets the MSB current level through Q_2 at 1.0 mA.

Figure 17.14a shows the basic circuit of a buffered unipolar 0- to $+10$-V DAC. The output current I_O is given by $I_O = 1.0 \text{ mA}(b_1 + b_2/2 + b_3/4 + \cdots + b_N/2^{N-1})$, and the corresponding output voltage is $V_O = +5.0 \text{ V}(b_1 + b_2/2 + b_3/4 + \cdots + b_N/2^{N-1})$ for a nominal full-scale output voltage of $+10$ V. When all bits are zero (0000000000), the output voltage is zero. For an MSB digital input (1000000000), the output voltage is $+5.0$ V, and for a digital input of all ones (1111111111), the output voltage is $+9.990$ V. The LSB voltage increment for the output voltage is $10 \text{ V}/2^{10} = 9.766$ mV.

Figure 17.14b shows a buffered 10-bit DAC with a bipolar output voltage range. The output voltage V_O is given by $V_O = (I_O - 1.0 \text{ mA})R_F = +1.0 \text{ mA } R_F(b_1 + b_2/2 + b_3/4 + \cdots + b_N/2^{N-1} - 1)$. For a ± 5.0-V output voltage range, R_F should be 5.0 kΩ. The output voltage is -5.0 V for a digital input of 0000000000, 0 V for a digital input of 1000000000, and $+5.0 \text{ V}(1 - \frac{1}{2^9}) = +4.99$ V for a digital input of 1111111111.

17.1.2 CMOS Current Switches

Figure 17.15a shows an example of a CMOS current switch for a DAC. The CMOS transistor pairs typically have a switching threshold voltage of $+1.4$ V, which makes this circuit compatible with any of the commonly used TTL or CMOS logic families.

When the digital input voltage is in the low state (<1.4 V), Q_{1N} is *off* and Q_{1P} is *on* as shown in Figure 17.15b. The output voltage of the Q_{1N}–Q_{1P} CMOS inverter

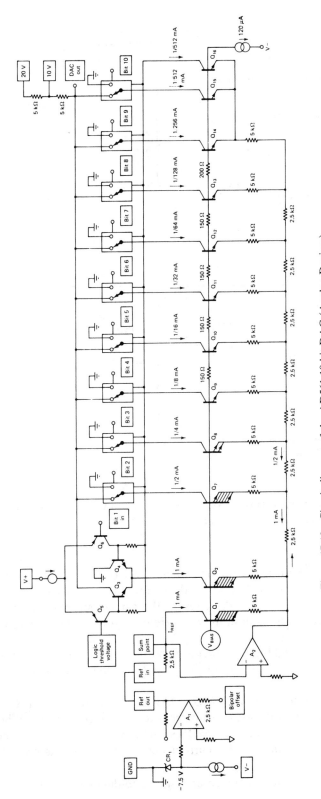

Figure 17.13 Circuit diagram of the AD561 10-bit DAC (Analog Devices).

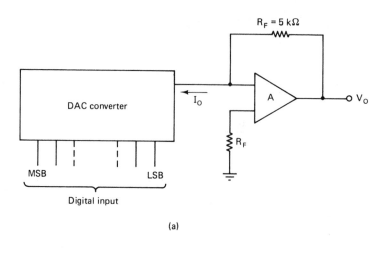

(a)

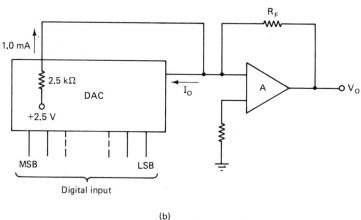

(b)

Figure 17.14 Buffered DACs: (a) with unipolar 0- to +10-V output voltage range; (b) with bipolar output voltage range.

is *high* ($= +V_{DD}$). This causes Q_{2N} to be *on* and Q_{2P} to be *off*, so the output voltage of the Q_{2N}–Q_{2P} inverter is *low* (~0 V), and similarly the output voltage of the Q_{3N}–Q_{3P} inverter is *high* ($\approx +V_{DD}$). As a result, the current switching transistor Q_5 is *on* and Q_4 is *off*. This will result in the bit current I_N from the ladder network being delivered to the I_{O_2} current bus.

In the case of a digital bit input that is high (>1.4 V), all of the transistors switch states. For this case, Q_4 is *on* and Q_5 is *off* and the current I_N is delivered to the I_{O_1} current bus.

Figure 17.16 is a diagram of a CMOS/DAC converter that uses a R–$2R$ ladder network, CMOS current switches, and a current-to-voltage converter. When the bit input (b_1, b_2, b_3, ...) is high, the CMOS switches (S_1, S_2, S_3, ...) send the bit currents (I_1, I_2, I_3, ...) into the I_{O_1} current bus. This current is given by $I_{O_1} =$

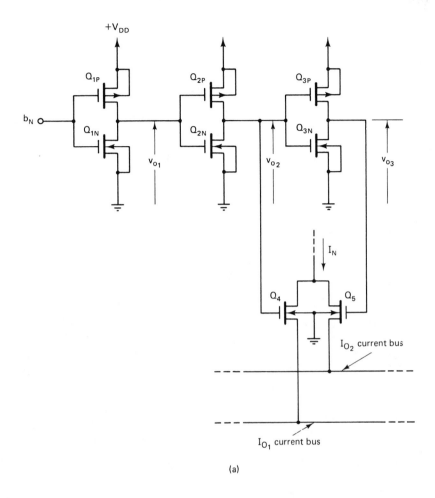

Figure 17.15 CMOS current switch: (a) basic circuit; (b) digital input low.

$(V_{\text{REF}}/10 \text{ k}\Omega) \, (b_1/2 + b_2/4 + b_3/8 + \cdots + b_N/2^N)$, and the output voltage V_O is given by $V_O = -I_{O_1}R_F = -V_{\text{REF}}(R_F/10 \text{ k}\Omega) \, (b_1/2 + b_2/4 + b_3/8 + \cdots + b_N/2^N)$. For a full-scale output voltage of $+10$ V with $V_{\text{REF}} = -10$ V, the feedback resistor R_F should be 10 kΩ.

This circuit can be easily modified for a bipolar output voltage range as shown in Figure 17.17. In this circuit, both output current buses, I_{O_1} and I_{O_2}, are used. The current I_{O_2} is converted by the A_2 op-amp circuit to a voltage $-R_F I_{O_2}$, which in turn will produce a current flow of $-R_F I_{O_2}/R_F = -I_{O_2}$ into the inverting input node of the A_1 operational-amplifier circuit. The net current flowing into this circuit node is therefore $I_{\text{NET}} = I_{O_1} - I_{O_2}$. The expression for I_{O_2} is the same as that given earlier for I_{O_1} except for the complementation of the bits as given by $I_{O_2} = (V_{\text{REF}}/10 \text{ k}\Omega)(\overline{b}_1/2 + \overline{b}_2/4 + \overline{b}_3/8 + \cdots + \overline{b}_N/2^N)$, where $\overline{b}_n$ is the complement of b_N. The

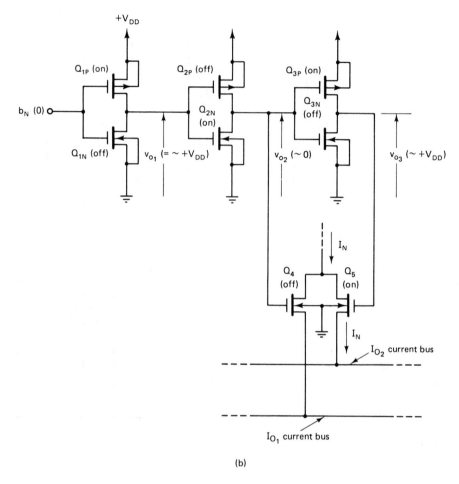

(b)

Figure 17.15 (continued)

output voltage of this circuit is therefore given by

$$V_O = -V_{\text{REF}} \frac{R_F}{10 \text{ k}\Omega} \left(\frac{b_1 - \bar{b}_1}{2} + \frac{b_2 - \bar{b}_2}{4} + \cdots + \frac{b_N - \bar{b}_N}{2^N} \right) \quad (17.5)$$

For $V_{\text{REF}} = -10$ V and a full-scale output voltage range of ± 10 V, the feedback resistor R_F should again be 10 kΩ. For a digital input of 000 . . . 000, the output voltage is given by $V_O = -[V_{O(FS)} - V_{O(LSB)}] = -10(1 - \frac{1}{2}N)$ V. For a digital input of 111 . . . 111, the output voltage will have its positive maximum value of $V_O = V_{O(FS)} - V_{O(LSB)} = +10(1 - \frac{1}{2}N)$ V. For a digital input of 1000 . . . 000, we have that $b_1 = 1, \bar{b}_1 = 0, b_2 = 0, \bar{b}_2 = 1, b_3 = 0, \bar{b}_3 = 1$, and so on. The analog output voltage V_O is therefore given by $V_O = V_{O(MSB)} - (V_{O(MAX)} - V_{O(MSB)}) =$

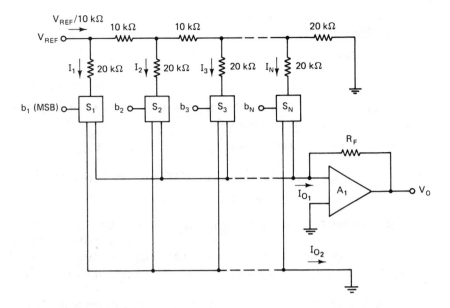

Figure 17.16   CMOS/DAC.

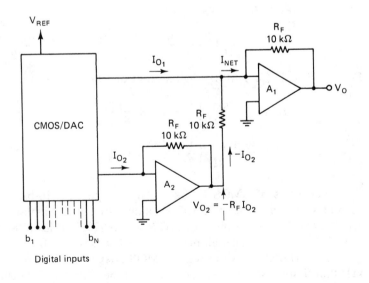

Figure 17.17   CMOS/DAC with bipolar output voltage range.

$2V_{O(\text{MSB})} - V_{O(\text{MAX})}$. Since $V_{O(\text{FS})} = 2V_{O(\text{MSB})}$ and $V_{O(\text{MAX})} = V_{O(\text{FS})} - V_{O(\text{LSB})}$, we have that

$$V_O(100\ldots000) = V_{O(\text{FS})} - [V_{O(\text{FS})} - V_{O(\text{LSB})}]$$

$$= +V_{O(\text{LSB})} = +10/2^N \text{ V} \qquad (17.6)$$

For a digital input of 1 LSB less than the previously considered input of $100\ldots000$, which is $011\ldots111$, the output voltage is

$$V_O(011\ldots111) = -V_{O(\text{MSB})} + V_{O(\text{MAX})} - V_{O(\text{MSB})}$$

$$= V_{O(\text{MAX})} - 2V_{O(\text{MSB})} = V_{O(\text{MAX})} - V_{O(\text{FS})} \qquad (17.7)$$

$$= -V_{O(\text{LSB})} = -10/2^N \text{ V}$$

We therefore see that there is no digital input for which the analog output voltage is zero. The output voltage is $-10/2^N$ V for a digital input of $011\ldots111$ and increases to $+10/2^N$ V for a digital input that is 1 LSB higher or $100\ldots000$. The output voltage increase for every 1 LSB increase in the digital input is $2 \times 10/2^N$ V $= 10/2^{N-1}$ V.

The transfer characteristic of the bipolar DAC can easily be shifted to produce an analog output voltage of zero for a "half-scale" digital input of $100\ldots000$ by supplying an offset current of magnitude $I_{\text{OFFSET}} = (V_{\text{REF}}/10 \text{ k}\Omega)(1/2^N) = -1.0$ mA/2^N to the inverting input node of the A_2 operational amplifier as shown in Figure 17.17. This is converted to a voltage of $(-V_{\text{REF}})(1/2^N)$ by this operational amplifier and produces a current flow into the inverting input node of the A_1 operational amplifier of $(-V_{\text{REF}}/10 \text{ k}\Omega)(1/2^N) = -I_{\text{OFFSET}}$. The net current coming into this node is now

$$I_{\text{NET}} = I_{O_1} - I_{O_2} - I_{\text{OFFSET}}$$

$$= \frac{V_{\text{REF}}}{10 \text{ k}\Omega}\left(\frac{b_1 - \bar{b}_1}{2} + \frac{b_2 - \bar{b}_2}{4} + \cdots + \frac{b_N - \bar{b}_N}{2^N} - \frac{1}{2^N}\right) \qquad (17.8)$$

For a half-scale (or 1 MSB) digital input of $100\ldots000$, this gives a current of

$$I_{\text{NET}} = \frac{V_{\text{REF}}}{10 \text{ k}\Omega}\left[\frac{1}{2} - \left(\frac{1}{4} + \frac{1}{8} + \frac{1}{16} + \cdots + \frac{1}{2^N} + \frac{1}{2^N}\right)\right]$$

$$= \frac{V_{\text{REF}}}{10 \text{ k}\Omega}\left(\frac{1}{2} - \frac{1}{2}\right) = 0 \qquad (17.9)$$

Examples of CMOS DACs.   An example of a 10-bit DAC that uses CMOS current switches is the AD7520 (Analog Devices). Similar CMOS/DACs are the AD7523/7524 series (8-bit), the AD7530/7533/7522 series (10-bit), and the AD7521/7531/7541 series (12-bit), all made by Analog Devices, Inc.

The AD7520 and the other CMOS/DACs listed above use $R-2R$ (10 and 20 kΩ) thin-film resistor ladders deposited on the silicon CMOS chip. The resistors are silicon–chromium alloy thin-film resistors with a sheet resistance of 2 kΩ/square, so the 10-kΩ resistors have a length-to-width ratio of 5:1 and the 20-kΩ resistors have

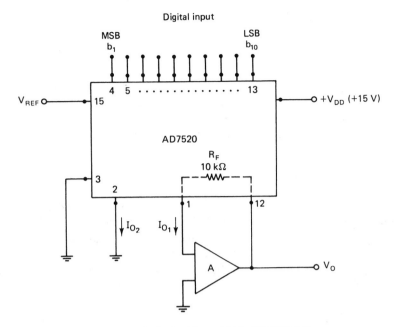

Figure 17.18 Unipolar binary 10-bit CMOS/DAC.

a 10:1 ratio. As a result of the moderate values of the length-to-width ratio, the resistors can be made fairly wide, which will lead to a very good absolute-value tolerance and, even more important, very close resistor ratio matching (a very small resistance ratio tolerance). These resistors have a temperature coefficient of 150 ppm/°C, and because of the very close thermal coupling between the resistors on the silicon chip, the temperature coefficient of the resistance ratio is less than 1 ppm/°C. The 10-kΩ feedback resistor R_F for the operational amplifier for the current-to-voltage conversion is also provided on the chip, so the R–$2R$ resistors and the feedback resistor track each other with respect to resistance changes with temperature. As a result, the temperature coefficient of the DAC scale factor is less than 10 ppm/°C.

To compensate for the "on" resistance of the CMOS switching transistors, Q_4 and Q_5, the gate geometry (channel width to channel length ratio) of the six most significant bit switches are designed so as to produce equal voltage drops of 10 mV across each switch. For a reference voltage of 10 V, the MSB current is 0.5 mA, so this corresponds to a channel resistance of 10 mV/0.5 mA = 20 Ω for the MSB transistor, 40 Ω for the next significant bit (b_2) transistor, 80 Ω for the b_3 transistor, and so on. To compensate for the 10-mV drop across these switching transistors, the actual reference voltage is set at a value of 10 V + 10 mV = 10.010 V.

The four least significant bit transistors (b_7 through b_{10}) are not scaled, but this will not have a significant effect on the accuracy of the DAC due to the very small level of current through these transistors.

Figure 17.18 shows the application of this 10-bit CMOS/DAC for a unipolar binary DAC with a buffered output. The analog output voltages for a number of digital input codes are:

Digital input	Analog output
1111111111	$-V_{REF}(1 - 2^{-10})$
1000000001	$-V_{REF}(\frac{1}{2} + 2^{-10})$
1000000000	$-V_{REF}(\frac{1}{2})$
0111111111	$-V_{REF}(\frac{1}{2} - 2^{-10})$
0000000001	$-V_{REF}(2^{-10})$
0000000000	0

This device can be operated as a *two-quadrant multiplying DAC* since the output voltage is proportional to the product of the reference voltage V_{REF} and the analog value of the digital input code. By varying V_{REF}, the gain or scale factor of the DAC can similarly be varied. Note that, although V_{REF} can have both positive and negative values, the digital code is unipolar; as a result, this multiplying DAC can operate in only two quadrants.

Another interesting DAC circuit is that of the DAC0806/7/8 series (National Semiconductor), which is an 8-bit DAC. A simplified schematic diagram of this device is shown in Figure 17.19a. It consists of two cascaded $R-2R$ ladder networks. The transistors in the first $R-2R$ section have their active areas scaled for equal current densities, and therefore the base-to-emitter voltage drops of these transistors are equal. The second $R-2R$ ladder network is driven by a current sink, which is the last transistor of the first ladder network. This transistor sinks a current equal to $\frac{1}{16}$ of the full-scale current.

The first transistor of the second $R-2R$ ladder network sinks a current of one-half of this, or $\frac{1}{32}$ of the full-scale current. The transistors of the second ladder network are also scaled for equal current densities.

The various currents produced by the ladder network are switched by a differential-amplifier type of current-mode switch, so the individual currents produced by the various branches of the $R-2R$ ladder network are either delivered to the output current bus or are shunted to ground.

The reference voltage for the current-mode switches is established by the voltage drop across the two series-connected diodes and is approximately $+1.4$ V, which makes the bit input compatible with the TTL and CMOS logic families.

Figure 17.19b is a block diagram of this DAC, and Figure 17.19c shows a circuit for a typical application for a unipolar DAC with a $+10$-V full-scale output-voltage range.

17.1.3 16-Bit Monolithic DAC

An example of a high-performance DAC is the DAC701/703 series (Burr-Brown). This is a complete monolithic 16-bit DAC that includes a low-noise buried zener voltage reference, $R-2R$ ladder resistor network, current switches, and a low-noise, fast-settling current-to-voltage converter op amp on a single chip. It features a linearity error of only 0.0015%. The gain error of 0.15% maximum can be adjusted to zero with an external potentiometer. The gain drift is just 10 ppm/°C, and the total full scale drift is also just 10 ppm/°C. The output voltage slewing rate is 10 V/μs with a 2-kΩ load, and the full-scale settling time (10 V for DAC701 and 20 V for DAC703) to within 0.003% is 4 μs.

17.1.4 DAC Speed and Settling Time

The speed or response time of a DAC is an important characteristic of the device. It is usually specified in terms of the *settling time*, which is the time required for the output voltage of the DAC to settle within a specified error band of the steady-state output voltage for a specified digital input change. The digital input change is usually specified as the worst-case condition of a change from 000 . . . 000 to 111 . . . 111, or vice versa. The corresponding output voltage change is $\Delta V_O = V_{O(MAX)} = V_{O(FS)} - V_{O(LSB)}$, and the error band is usually given as $\pm \frac{1}{2} V_{O(LSB)}$. For a 10-bit DAC with a full-scale output voltage of $V_{O(FS)} = 10.24$ V, we have that $V_{O(LSB)} = 10.24$ V/2^{10} = 10 mV, so $V_{O(MAX)} = 10.23$ V and $\frac{1}{2} V_{O(LSB)} = 5$ mV or 0.05% of the full-scale output voltage.

Figure 17.20 is a graph of the output voltage versus time for a 10-bit DAC in which the settling time is indicated. Figure 17.21 is an equivalent circuit of a DAC in which a load resistance R_L is used for the current-to-voltage conversion. If the time-domain response of this system is controlled by the single RC time constant $\tau = R_L C_{NET}$, we have that $v_o(t) = V_{O(MAX)} [1 - \exp(-t/\tau)]$. The $\frac{1}{2}$-LSB settling time can be obtained by setting $v_o(t)$ at $t = T_{SETTLING}$ equal to $V_{O(MAX)} - \frac{1}{2} V_{O(LSB)}$. Since $V_{O(MAX)} = V_{O(FS)} - V_{O(LSB)}$, this can be written as $v_o (t = T_{SETTLING}) = V_{O(FS)} - \frac{3}{2} V_{O(LSB)}$. We therefore have that

$$V_{O(FS)} - \frac{3}{2} V_{O(LSB)} = V_{O(FS)} \left[1 - \left(\frac{3}{2} \right) \left(\frac{1}{2^N} \right) \right]$$

$$= V_{O(MAX)} \left[1 - \exp \left(\frac{-t}{\tau} \right) \right] \qquad (17.10)$$

$$= V_{O(FS)} \left(1 - \frac{1}{2^N} \right) \left[1 - \exp \left(\frac{-t}{\tau} \right) \right]$$

Solving this for $t = T_{SETTLING}$ gives

$$T_{SETTLING} = \tau \ln [2(2^N - 1)] \simeq (N + 1)\tau \ln 2 = 0.693(N + 1)\tau \qquad (17.11)$$

Given below are some values of $T_{SETTLING}$ for various values of N (number of bits) and the corresponding ratio of the settling time (to $\pm \frac{1}{2}$ LSB) to the 10% to 90% rise time.

N	$T_{SETTLING}$	$T_{SETTLING}/t_{rise}$
8	6.234τ	2.834
10	7.624τ	3.465
12	9.011τ	4.096
14	10.397τ	4.72

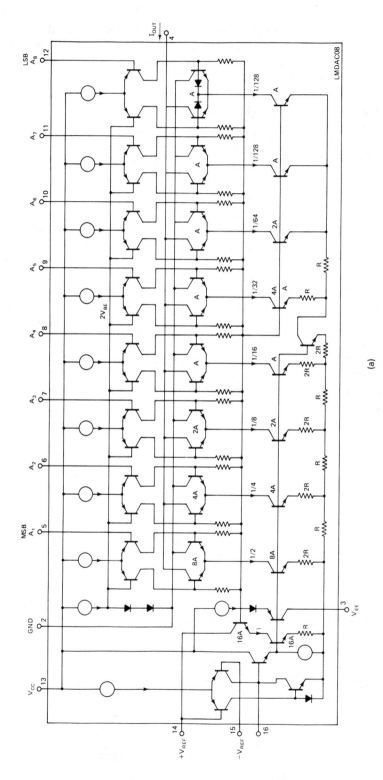

Figure 17.19 DAC0806 8-bit DAC: (a) circuit diagram; (b) block diagram; (c) unipolar DAC with +10-V full-scale output (National Semiconductor).

(a)

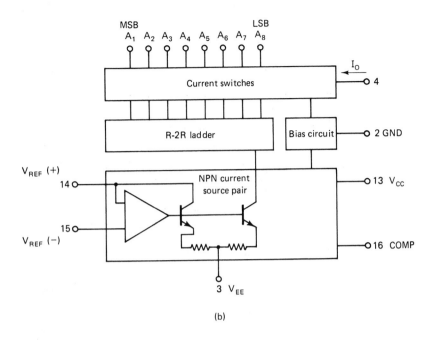

(b)

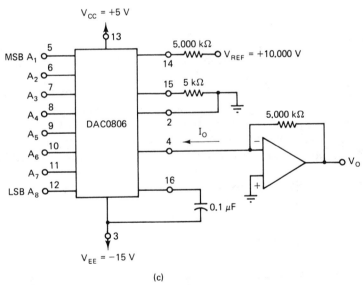

(c)

Figure 17.19 (continued)

We see that even in the case of a simple single-pole response the settling time is appreciably longer than the 10% to 90% rise time.

Let us consider a 10-bit DAC that uses an inverted $R-2R$ ladder network with current switches and assume that the output capacitance (collector–base capacitance, C'_{cb}) of each switch is 2.5 pF so that the total output capacitance will be 25 pF. If

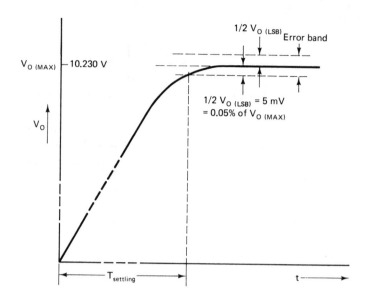

Figure 17.20 Settling time for a 10-bit DAC.

$I_{O(\text{MSB})} = 1.0$ mA, and thus $I_{O(\text{FS})} = 2.0$ mA, and if the full-scale output voltage is 4.0 V, the value of load resistance R_L will be 2 kΩ. With this value of load resistance, the time constant τ is $\tau = 2$ kΩ $\times$ 25 pF $= 50$ ns, and the corresponding settling time is $T_{\text{SETTLING}} = 7.62 \times 50$ ns $= 381$ ns. If the load resistance is now reduced to 1.0 kΩ, which will give a full-scale output voltage of 2.0 V, the time constant becomes 25 ns and the settling time is reduced to 191 ns.

If an operational amplifier is used for the current-to-voltage conversion, the response-time characteristics of the system will usually be limited by the slewing rate of the operational amplifier. For example, if the full-scale output voltage is 10.24 V and the operational amplifier has a slewing rate of 1.0 V/μs, the settling time is about 10 μs.

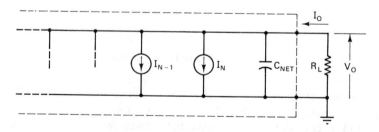

Figure 17.21 DAC using load resistance for the current-to-voltage conversion.

17.1.5 Monotonicity and Voltage-sharing DACs

In an ideal DAC the analog output voltage increases by a constant amount for every 1 LSB increase in the digital input. In the case of real DACs, this is, of course, not the case, and the increments in the analog output voltage are not all equal. This produces what is called the *differential nonlinearity* of the DAC transfer characteristic. Indeed, in some cases, especially for DACs with digital inputs of 10 bits or greater, the differential nonlinearity is such that the analog output voltage may actually drop between some of the steps of the increasing digital input. As a result, the DAC transfer characteristic is no longer monotonic, but is said to exhibit *nonmonotonicity*. The differential nonlinearity is due principally to deviations in resistor ratios from the design center values. For example, consider the simple N-bit DAC of Figure 17.1. For a digital input of 0111 . . . , switch b_1 will be open and the rest of the switches b_2 through b_N will be closed. When the digital input then increases by 1 LSB to 1000 . . . . . , switch b_1 will close and switches b_2 through b_N will open. If the MSB resistor R that is associated with switch b_1 is too large by a factor of more than 1 LSB (as compared to all the other resistors), the output voltage actually decreases as the digital input increases by 1 LSB from 0111 . . . to 1000 . . . , thus resulting in nonmonotonic behavior. A 1-LSB deviation in the resistor value corresponds to a 0.1% error for a 10-bit DAC, 0.02% for a 12-bit DAC, 0.006% for a 14-bit DAC, and only 0.0015% for a 16-bit DAC, so we see that a differential nonlinearity leading to nonmonotonicity is quite probable for DACs of 10 bits or greater. The nonmonotonicity can be a major problem in many systems, especially when the DAC is part of the feedback loop of a closed-loop system. When the DAC is operating in a nonmonotonic region, the output voltage decreases as the input increases. This produces a negative slope in the transfer characteristics of the system and can result in the negative feedback changing over to positive feedback, leading to instability and even oscillation in the system.

A voltage-scaling type of DAC is one that will inherently exhibit monotonic behavior. A simple diagram of a 3-bit voltage-scaling DAC is shown in Figure 17.22. The switches are generally implemented as CMOS analog switches. This system is inherently monotonic since, as the digital input increases, the output voltage is taken at higher taps on the resistive voltage divider so that even large errors in resistor values will not result in a decrease in the output voltage at any step, although there still can be some degree of differential nonlinearity. This type of DAC does, however, have the disadvantage of a high component count and a relatively slow speed, especially for DACs in excess of 8 bits. For high-resolution monotonic DACs, a cascaded voltage-scaling system can be employed as shown in Figure 17.23.

In this system, the first voltage-scaling DAC converts the N MSBs of the digital input to an analog output that is used as an input to the second voltage-sharing DAC. This second DAC also has as its input the M LSBs of the digital input and adds the analog voltage produced by the M LSB input to the voltage produced by the first DAC. This results in an analog output voltage that is representative of the $(N + M)$-bit digital input. An example of this type of DAC is the AD569 (Analog Devices), which is a monotonic 16-bit DAC. This IC contains two cascaded 8-bit voltage-sharing DACs, each of which has a resistor string comprised of 256 resistor segments.

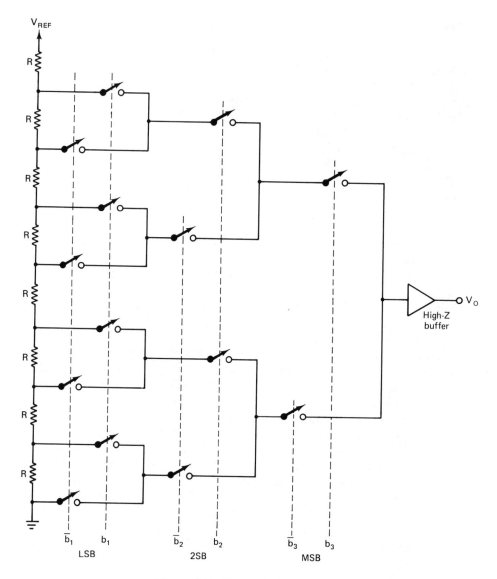

Figure 17.22 Voltage-scaling DAC.

This DAC has a settling time of 5 μs (max.) to ±0.001% of the full-scale range (corresponding to approximately ±1 LSB). The AD7536 (Analog Devices) uses a combination of a voltage-scaling circuit and an $R-2R$ circuit to produce a DAC with 14-bit monotonicity. The settling time is 1.5 μs (max.) to ±0.003% of the full-scale range (corresponding to approximately +1/2 LSB).

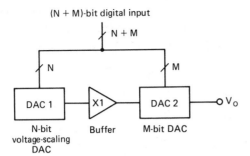

Figure 17.23 Cascaded voltage-scaling DAC.

N-bit
voltage-scaling
DAC

Buffer

M-bit DAC

17.2 ANALOG-TO-DIGITAL CONVERTERS

17.2.1 ADC Transfer Characteristic and Quantization Error

Figure 17.24a shows the transfer characteristic of an ADC. We note that since the digital output code can take on only certain discrete values, there is an inherent error in the system called the *quantization error*. The quantization error is defined as the difference between the analog voltage equivalent of the digital output code and the actual analog input voltage. A graph of the quantization error corresponding to the transfer characteristic of Figure 17.24a is presented in Figure 17.24b. Note that the maximum value of the quantization error is 1 LSB.

The maximum quantization error can be reduced to $\pm\frac{1}{2}$ LSB by adding a $-\frac{1}{2}$-LSB voltage offset to the analog signal or, equivalently, adding $\frac{1}{2}$ LSB to the digital reference voltage to which the analog signal is compared. The transfer characteristic obtained with this $\frac{1}{2}$-LSB offset is shown in Figure 17.25a, and the corresponding quantization error is shown in Figure 17.25b. Note that the maximum quantization error is the same for every range of input voltage and is equal to $\pm\frac{1}{2}$ LSB.

A simple example of the effect of the $\frac{1}{2}$-LSB offset on the quantization error can be seen for the case of a 3-bit parallel comparator type of ADC. Let us assume an LSB voltage of 1.0 V and a full-scale input voltage of 8.0 V. Without the $\frac{1}{2}$-LSB offset, the seven comparator reference voltages are 1.0, 2.0, 3.0, 4.0, . . . , and 7.0 V. The digital output codes obtained are as follows.

Analog input voltage	Digital output	Analog equivalent of output
$0 < V_A < 1.0$	000	0
$1.0 < V_A < 2.0$	001	1.0
$2.0 < V_A < 3.0$	010	2.0
$3.0 < V_A < 4.0$	011	3.0
$4.0 < V_A < 5.0$	100	4.0
$5.0 < V_A < 6.0$	101	5.0
$6.0 < V_A < 7.0$	110	6.0
$7.0 < V_A < 8.0$	111	7.0

The maximum digital output code is 111 and the maximum quantization error is 1 LSB in every 1.0 V (1 LSB) input voltage range. If the reference voltages are now shifted by $-\frac{1}{2}$ LSB to become 0.5, 1.5, 2.5, . . . , 6.5 V, the following transfer relationship is obtained.

Analog input voltage	Digital output	Analog equivalent of output
$0 \ < V_A < 0.5$	000	0
$0.5 < V_A < 1.5$	001	1.0
$1.5 < V_A < 2.5$	010	2.0
$2.5 < V_A < 3.5$	011	3.0
$3.5 < V_A < 4.5$	100	4.0
$4.5 < V_A < 5.5$	101	5.0
$5.5 < V_A < 6.5$	110	6.0
$6.5 < V_A < 7.5$	111	7.0

The maximum quantization error is $\pm\frac{1}{2}$ LSB in every range.

17.2.2 Parallel Comparator ADC

The first type of analog-to-digital converter (ADC) that will be considered is the *parallel comparator ADC*, also known as a *simultaneous* or *flash ADC*. An example of a 3-bit ADC of this type is shown in Figure 17.26. For this ADC, seven comparators are required. In general, for an *N*-bit ADC of the parallel conversion type, the number of comparators required is $2^N - 1$. Although this type of ADC has the advantage of being a very fast ADC, the number of comparators required increases very rapidly with the number of bits *N* and usually becomes an uneconomical approach for more than 3 or 4 bits.

The reference voltage divider sets up the following reference levels for the comparators: $V_{R_1} = (R/2)/7RV_{REF} = \frac{1}{14}V_{REF}$, $V_{R_2} = \frac{3}{14}V_{REF}$, $V_{R_3} = \frac{5}{14}V_{REF}$, $V_{R_4} = \frac{7}{14}V_{REF}$, $V_{R_5} = \frac{9}{14}V_{REF}$, $V_{R_6} = \frac{11}{14}V_{REF}$, and $V_{R_7} = \frac{13}{14}V_{REF}$. The ADC transfer characteristic is as shown in Figure 17.25 with the normalized analog input voltage given by $V_A/(V_{REF}/7)$.

The *quantization error* of an ADC is the range of analog input voltages that will produce a given digital output code, measured with respect to the midpoint value of the analog input. For the ADC of Figure 17.26, the choice of comparator reference voltage levels is such that the quantization error is uniformly distributed so that the quantization error is the same for every digital output level and is $\frac{1}{14}V_{REF}$. Since the total span of the analog input voltage corresponding to a 1-LSB change in the digital output is $\frac{2}{14}V_{REF}$, the quantization error of $\frac{1}{14}V_{REF}$ can be expressed in terms of the digital output as $\pm\frac{1}{2}$LSB.

Fast CMOS ADC. For flash converters of 6 bits and up, a very large number of comparators and logic gates are required. The use of CMOS ICs can be very advantageous for these flash ADCs from the standpoint of power dissipation. An

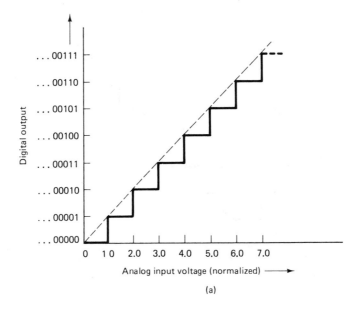

(a)

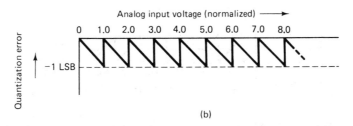

(b)

Figure 17.24 ADC transfer characteristic and quantization error: (a) transfer characteristic; (b) quantization error.

interesting example of a fast monolithic 6-bit flash CMOS ADC is the CA3300D (RCA). This device uses an array of 64 CMOS comparators and can operate at a sampling rate of 15 MHz, which corresponds to a conversion time of 66 ns. In addition to 64 comparators, it contains the encoding circuits, output latches, and tristate output gates, all on a single 2.4 mm × 3.3 mm chip. The reference voltages for the comparators are developed by a polysilicon resistor ladder with a resistance of 20 Ω between taps and a total resistance of 1250 Ω. It also produces a latched overflow bit. The overflow bit and the tristate outputs make it possible to easily connect two of these ADCs together to produce a 7-bit ADC. In addition, the parallel connection of two of these ADCs is possible. By connecting the chip enable inputs (CE and $\overline{CE}$) to the clock input, the two ADCs can be multiplexed such that each ADC samples the input signal on alternate half-cycles of the clock cycle. This arrangement will approximately double the sampling rate to 30 MHz. This device can operate with

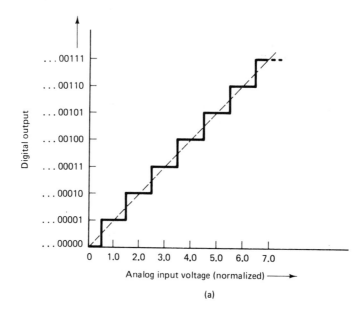

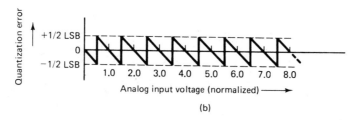

Figure 17.25 ADC transfer characteristic and quantization error with $\frac{1}{2}$-LSB offset of analog input voltage: (a) transfer characteristic: (b) quantization error.

single supply voltages from 10 V down to as low as 3 V. When operated with a 5-V supply and at an 11-MHz clock rate, the total power dissipation is less than 50 mW.

Subranging parallel comparator ADC. The hardware needed for a parallel comparator ADC in terms of the number of comparators and logic gates for encoding the comparator outputs increases exponentially with the number of bits. For an N-bit converter, $2^N - 1$ comparators are needed so that most parallel comparators or flash converters are limited to about 4 or 6 bits. There have been monolithic IC flash ADCs developed with 9-bit resolution and containing 511 comparators and associated circuitry on one IC chip. This flash converter has a conversion time of only 40 ns and can operate at a 25-MHz conversion rate.

For a fast high-resolution ADC, a *subranging* or *cascaded* flash ADC system as shown in Figure 17.27 can be used. The first ADC produces an N_1-bit digital code that is supplied to the output as the N_1 MSBs of the digital output code. This digital signal is also supplied to a DAC, which produces an analog voltage corresponding to the N_1 MSBs of the digital output code. This analog voltage is then subtracted from

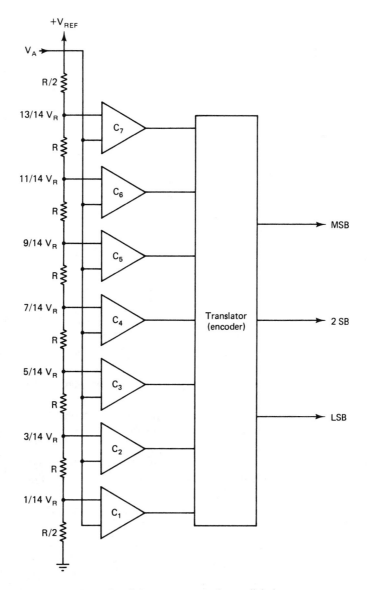

Figure 17.26 Parallel comparator analog-to-digital converter.

the analog input voltage in the difference amplifier A to produce a difference or error voltage. This difference voltage then goes to an N_2-bit ADC. The digital output of this second ADC becomes the N_2 LSBs of the digital output code, which will now have the full complement of $N_1 + N_2$ bits. Thus, if two 4-bit flash ADCs are used, an 8-bit digital output can be obtained.

Although the two ADCs in the preceding example have a 4-bit output for a total of 8 bits in the digital output code, for an 8-bit accuracy the accuracy of each ADC must be such that the conversion error in each must not exceed $V_{FS}/2^8$, where

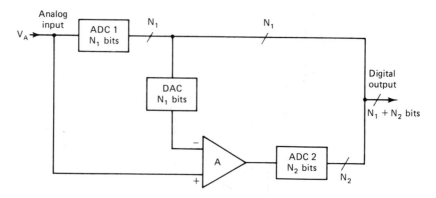

Figure 17.27 Cascaded (subranging) parallel comparator ADC.

V_{FS} is the full-scale analog input voltage range. This requirement will generally be no problem for the second ADC since this corresponds to a maximum error of $1/2^4$ of its full-scale voltage. It can, however, be a problem for the first comparator for which the maximum error must not exceed $1/2^8$ times its full-scale voltage. Another somewhat related problem results from the fact that the conversion process in the second ADC takes place at a slightly later time than it does in the first ADC due to the propagation delays in the first ADC, the DAC, and in the difference amplifier. This can be a significant problem for a rapidly changing analog input voltage.

Due to the effects mentioned above, the difference voltage that is supplied to the second ADC may fall outside the full-scale conversion range of the second ADC and errors will result. A means of minimizing this problem and of easing the accuracy requirements of the first ADC is shown in Figure 17.28. The range of the second ADC is increased from N_2 bits to $N_2 + M$ bits. The N_2 LSBs of this ADC still become the N_2 LSBs of the total digital output code. The M MSBs of this ADC are, however, combined with the N_1 bits of the first ADC in the digital adder to produce the N_1 MSBs of the digital output code, which will still have a total of $N_1 + N_2$ bits.

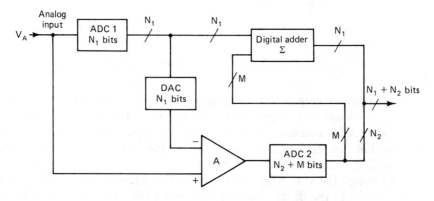

Figure 17.28 Subranging ADC with M-bit error correction.

If no errors were made in the first ADC or in the DAC and the analog input voltage is the same for both conversion cycles, the M MSBs of the second ADC will all be zeros, and no correction will be made to the output of the first ADC in the digital adder. If this is not the case, the M bits generated by the second ADC can be used to correct the output of the first ADC.

Subranging ADC examples. An example of a fast subranging flash ADC system is the MOD-1020 (Analog Devices), which offers a 10-bit digital output at a 20-MHz conversion rate. This system consists of several ICs and other components on a single PC board. The configuration is basically similar to that shown in Figure 17.28 with the addition of a delay line placed between the analog input and the difference amplifier. This compensates for the conversion delay of the N_1 bits through the first ADC and the DAC. The first conversion yields 5 bits (N_1), which are then combined with the 6 bits $(N_2 + M)$ of the second conversion to produce a 10-bit output, with the one extra bit being used for error correction. Another subranging flash ADC system is the MOD-1205 (Analog Devices), which offers a 12-bit output at a 5-MHz conversion rate.

17.2.3 Counting Type

The counting type of analog-to-digital converter is one of the simplest types of ADC; a basic block diagram of this type of converter is shown in Figure 17.29a. The digital output of a modulus N binary counter goes to an N-bit digital-to-analog converter (DAC), which produces the staircase type of waveform shown in Figure 17.29b. The DAC output voltage goes up in increments of 1 V_{LSB} up to a full-scale maximum of $2^N V_{LSB}$. This voltage is compared to the analog input voltage by the comparator. As long as the DAC voltage stays below the analog input voltage V_A, the comparator output voltage will stay in the *high* state. This keeps the clock gate G_1 enabled such that the clock pulses pass through to the counter so that the count will continue to increase and the DAC output voltage will continue with its incremental increases. As soon as the DAC voltage rises above the analog input voltage level, the comparator output goes *low*. This disables the clock gate G_1, so the count will stop at this point.

Some time after this, the HOLD voltage goes *high*, which enables the N-bit digital output gate, so the binary count of the counter now becomes available as the digital output of the ADC. The HOLD pulse then terminates and the RESET pulse then resets the counter back to zero, and the counter starts to count up again, thus starting another conversion cycle.

For an N-bit counting converter, a total conversion time of 2^N clock cycles is required for a full-scale input voltage. As a result, this type of ADC will be a relatively slow converter. For a 14-bit ADC of this type, the conversion time is $2^{14}T_{CLOCK} = 16,384T_{CLOCK}$. If $T_{CLOCK} = 100$ ns is chosen as a representative value, the resulting conversion time is 1.64 ms and the corresponding rate is 610 conversions per second.

The average conversion time can be reduced substantially if an up–down counter is substituted for the simple binary up counter and if the conversion cycles are terminated shortly after the counting stops. This type of counting ADC is called a *tracking* or *servo ADC*. In this type of converter, the counter is not reset to zero at the beginning of every conversion cycle, but rather is given a command to either

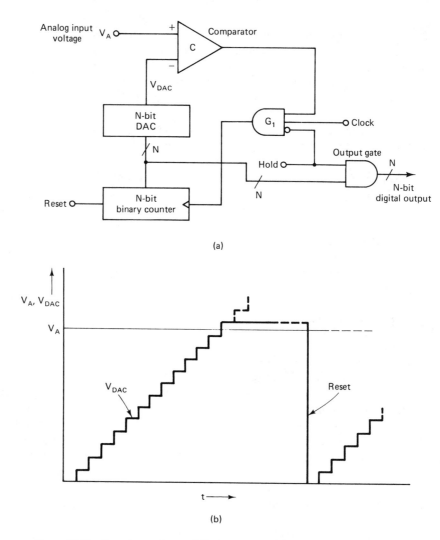

Figure 17.29 Counting analog-to-digital converter: (a) block diagram; (b) DAC output waveform.

continue to count up or to count down from the previous count, this depending on whether the analog input voltage is above or below the DAC output voltage at the beginning of the conversion cycle, respectively.

For a slowly changing analog voltage, the change in the count needed to match the DAC voltage with the analog input voltage will be small, so the conversion time will be very short.   For a full-scale analog voltage change occurring in a short period of time, the required conversion time will become equal to that of the simple counting ADC.

The accuracy of the counting ADC is a function of the offset voltage and voltage gain of the comparator and of the accuracy of the DAC.   It often is the DAC that is the limiting factor in the overall accuracy of the ADC.

This type of ADC is a feedback ADC because of the presence of a feedback loop around the comparator. This feedback loop consists of the clock gate G_1, the binary counter, and the DAC. Another type of feedback ADC that has a DAC in the feedback loop is the successive approximation converter (SAR), and it will be discussed next. The SAR converter offers the same basic accuracy as the counting ADC, but it offers a much shorter conversion time and is a very widely used ADC.

17.2.4 Successive Approximation ADC

The *successive approximation register* (SAR) type of ADC is one of the most popular types of ADC because it offers the combination of high accuracy and high conversion speed. Figure 17.30 shows a basic block diagram of an SAR ADC. The successive approximation register for an N-bit converter contains N flip-flops (FFs) that are set to the high state ($b = 1$), one at a time, to produce the bit inputs to the digital-to-analog converter (DAC). The output voltage of the DAC is compared to the analog input voltage.

The MSB FF is set to the high state ($b_1 = 1$) first, so the analog input voltage is first compared to a DAC output voltage that corresponds to the MSB voltage level of $V_{\text{MSB}} = V_{\text{FS}}/2$. If the analog voltage is greater than V_{MSB}, the MSB FF remains set ($b_1 = 1$) for the rest of the conversion period. If the analog input voltage, however, is below V_{MSB}, the MSB FF is reset to zero ($b_1 = 0$) and remains in that state for the rest of the conversion process.

The next significant bit (second bit) FF is set next ($b_2 = 1$) and the DAC output voltage now corresponds to $V_{\text{FS}}(b_1/2 + b_2/4)$ with $b_2 = 1$. If this voltage is less than the analog input voltage V_A, the second FF remains set at $b_2 = 1$. If this DAC voltage is more than the analog input voltage, the second FF is reset to give $b_2 = 0$. This process then continues for each successive FF for N clock cycles to complete the conversion process. In each case if the analog input voltage is greater than the DAC voltage, the FFs retain their present state. If the analog voltage is below the DAC voltage, the last FF to be set ($b = 1$) is reset to zero ($b = 0$).

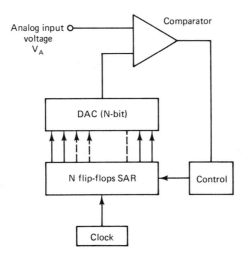

Figure 17.30 Block diagram of a successive approximation register analog-to-digital converter (SAR ADC).

The DAC voltage and the corresponding digital code are thus a successive approximation to the analog input voltage with each bit being tested at a time beginning with the most significant bit (b_1) and continuing one bit at a time down to the least significant bit (b_N) at the end of the conversion period.

Let us now consider the example of a 10-bit SAR ADC with a full-scale DAC output voltage of $V_{FS} = 10.24$ V, so $V_{LSB} = 10$ mV. We will also assume that a $\frac{1}{2}$-LSB offset voltage is added to the DAC output voltage in order to minimize the conversion quantization error.

For an analog input voltage of $V_A = 7.500$ V, the SAR bit outputs and the corresponding DAC output voltage values are as given in Table 17.2. After 10 clock cycles the SAR conversion process is completed and the final status of the SAR FFs is 1011101101, which produces a DAC output voltage of 7.495 V. The resulting quantization error is 5 mV, which corresponds to $\frac{1}{2}$ LSB. If the $\frac{1}{2}$-LSB offset were not present, the DAC output voltage would be 7.490 V, which would result in a quantization error of 10 mV or 1 LSB.

For an analog input voltage of less than 5 mV, the digital output is 0000000000 and the DAC output voltage (including the $\frac{1}{2}$-LSB offset voltage) is 5 mV. For an analog input voltage between 5 and 15 mV, the digital output code is 0000000001. The maximum DAC output voltage (including the $\frac{1}{2}$-LSB offset) corresponding to a digital output code of 1111111111 is $V_{MAX} = 10.24 - 10$ mV $+ 5$ mV $= 10.235$ V. The maximum analog voltage that can be converted with a quantization error not exceeding $\frac{1}{2}$ LSB (5 mV) is 10.240 V, which corresponds to the DAC full-scale output voltage.

The SAR ADC has the digital code output available in parallel form at the end of the conversion period. It is also possible to obtain a data output in serial form during the conversion period. As each data bit becomes valid at the end of the various clock cycle times during which the DAC output voltage is compared to the analog signal and the FFs are reset if necessary, the bits of the digital output code become available, starting with the MSB and ending with the LSB at the end of the conversion process.

TABLE 17.2 SAR DAC EXAMPLE

Clock cycle	Register FF states b_1(MSB) . . . b_N(LSB)	DAC output voltage ($+\frac{1}{2}$-LSB offset) (V)
1	1000000000	5.125
2	1100000000	7.685 (b_2 FF now reset to $b_2 = 0$)
3	1010000000	6.405
4	1011000000	7.045
5	1011100000	7.365
6	1011110000	7.525 (b_6 FF now reset to $b_6 = 0$)
7	1011101000	7.445
8	1011101100	7.485
9	1011101110	7.525 (b_9 FF now reset to $b_9 = 0$)
10	1011101101	7.495

The total conversion time for an N-bit SAR ADC is approximately $(N + 2)T_{\text{CLOCK}}$, where T_{CLOCK} is the clock period, which is typically on the order of 1 μs. Therefore, for a 12-bit SAR ADC, a total conversion time of around 14 μs would be required. This is to be compared to a conversion time of $2^N T_{\text{CLOCK}}$ for a digital ramp or counting type of ADC, so for a 12-bit digital ramp ADC with a 1-μs clock cycle time a total conversion time of around 4 ms is required. Of course, a parallel comparator ADC can produce conversion times of well under 1 μs, but the hardware requirements for a 12-bit ADC (or even for an 8-bit ADC) would be quite excessive.

The two principal factors that limit the accuracy of any of the DAC feedback type of ADCs such as the SAR ADC and the digital ramp ADCs are the precision of the DAC and the comparator offset voltage. For a 10-bit ADC with an accuracy of better than $\frac{1}{2}$ LSB and a full-scale voltage of 10 (or 10.24) V, the total conversion error from all sources (but not including the inherent quantization error) must not exceed ± 5 mV. Therefore, the comparator offset voltage V_{OS} should be substantially less than 5 mV, which is not a difficult requirement to meet. However, for a 12-bit converter the total error must not exceed 1.25 mV, and for a 14-bit converter the error must not exceed 0.313 mV, which becomes a much more difficult requirement to satisfy.

Let us consider a high-resolution 16-bit ADC. For a ± 1-LSB accuracy the total conversion error with a full-scale output voltage of 10 V must not exceed 10 V/2^{16} = 152 μV over the operating temperature range. For an operating temperature range from 0° to 50°C and with the ADC calibrated at 25°C, the maximum net temperature coefficient of the reference voltage and the comparator offset voltage must not exceed 152 μV/(± 25°C) = ± 6 μV/°C. This is indeed a very difficult condition to satisfy.

Decision tree. Further insight into the operation of the successive approximation ADC can be obtained by looking at the decision tree shown in Figure 17.31 for the simple case of a 4-bit ADC with a 16-V full-scale range. Figure 17.32 shows the decision tree path for the case of an analog input voltage of 10.7 V as an example. The following sequence of decisions is made in the ADC:

1. MSB = 1 since $V_A > 8.0$ V
2. 2SB = 0 since $V_A < 12.0$ V
3. 3SB = 1 since $V_A > 10.0$ V
4. 4SB = LSB = 0 since $V_A < 11.0$ V

For accurate analog-to-digital conversion, the analog input voltage V_A should be held constant during the conversion cycle. If the analog input voltage changes by more than $\pm \frac{1}{2}$ LSB, an error in the digital output code can result. The successive approximation ADC is especially susceptible to this problem.

To illustrate the effect of a changing analog input voltage on the conversion process, let us consider a worst-case situation of a successive approximation ADC with an analog input voltage that is nominally zero, but there happens to be a large amplitude noise or interference voltage spike occurring at the time that the first (MSB) decision is being made, as shown in Figure 17.33. At this decision point, the MSB

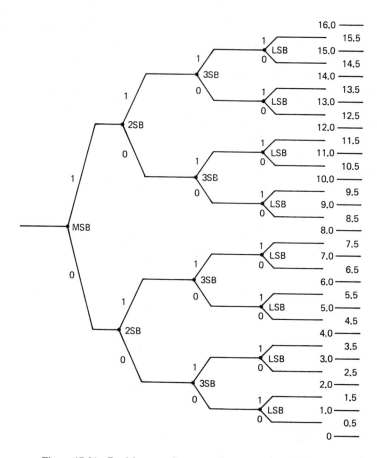

Figure 17.31 Decision tree for successive approximation ADC.

is set at 1 due to the presence of the voltage spike. At the remainder of the decision points, the rest of the bits are set at 0 since the voltage spike is gone and the input voltage has settled back to near zero. The digital output code will be 1000 as compared to the correct value of 0000. This error in the digital output code corresponds to an input voltage error of one-half of the full-scale voltage. To minimize the occurrence of these errors, *sample-and-hold* (S/H) or *tracking-and-hold* (T/H) amplifiers are often used between the analog input voltage and the ADC. These circuits sample or track the analog input voltage before the conversion cycles begin, and then at the beginning of the conversion period hold the value of the analog input voltage constant during the conversion process.

17.2.5 Integrating ADCs

In the integrating type of analog-to-digital converter, the analog input voltage or a fixed reference voltage, or both, are integrated and the result is used to clock or gate a binary counter to obtain a digital output that represents the analog input voltage.

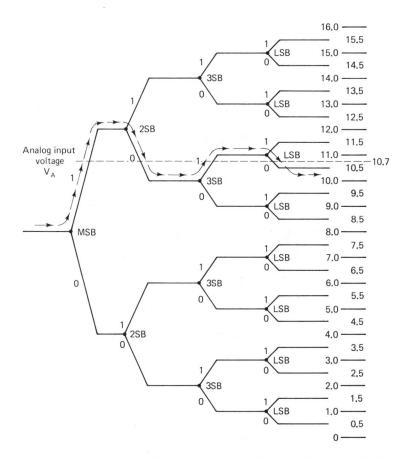

Figure 17.32 Decision tree for successive approximation ADC; example for $V_A = 10.7$ V.

Three basic types of integrating ADCs will be considered: the voltage-to-frequency (V/F) converter, the voltage-to-time (V/T) converter, and the dual-slope converter.

In the *voltage-to-frequency converter*, the analog input voltage is integrated and the result is compared to a fixed reference voltage to produce a pulse train whose frequency is proportional to the analog input voltage. This pulse train is used to clock a binary counter over a fixed gate period to obtain a binary digital output that represents the analog input voltage.

In the *voltage-to-time converter*, it is the reference voltage that is integrated and compared to the analog input voltage to produce a gating pulse whose length is proportional to the analog voltage. This pulse is used to gate a binary counter that is driven by a fixed-frequency clock. This produces a binary digital output that is based on the number of clock cycles in the gating period and will thus be a representation of the analog input voltage.

In the *dual-slope converter*, the analog input voltage is first integrated for a fixed period of time. Then a fixed reference voltage of polarity opposite to that of the analog voltage is applied and integrated for the period of time required to bring the

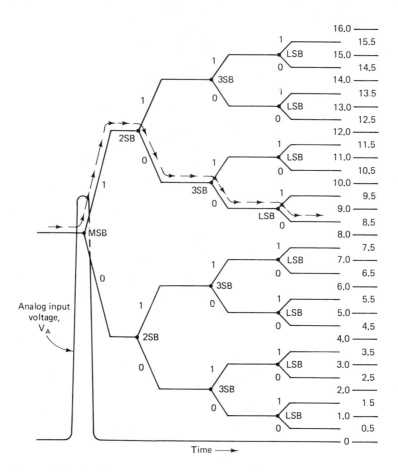

Figure 17.33 Decision tree for successive approximation ADC: effect of a large voltage spike.

output voltage of the integrator circuit back down to zero. This period of time is used to gate a binary counter, and the count so produced will be a binary digital output equivalent to the analog input voltage.

Integrating converters have the advantages of offering very high resolution (up to 14 bits) and very good noise and power frequency rejection, but have the disadvantage of a very low conversion rate. We will now consider the three basic types of integrating analog-to-digital converters in somewhat more detail.

Voltage-to-frequency ADC. In the voltage-to-frequency converter, the analog input voltage is used to control the frequency of a voltage-controlled oscillator (VCO) that has a very linear voltage to frequency transfer characteristic. The output pulse train of the VCO is then counted by a binary counter over a fixed sampling or gate time. The count obtained is then made available as the digital output at the end of the gate time.

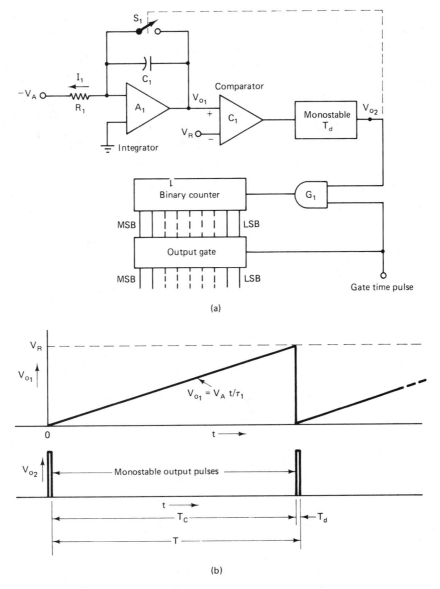

Figure 17.34 Voltage-to-frequency analog-to-digital converter: (a) basic circuit; (b) integrator and monostable output voltage.

A basic voltage-to-frequency ADC circuit is shown in Figure 17.34. The analog input voltage V_A produces a current $I_1 = V_A/R_1$ through the timing resistor R_1. This current is then used to charge up the timing capacitor C_1, producing an output voltage given by

$$V_{o_1} = \frac{1}{C} \int_0^t I_1 \, dt = \frac{1}{R_1 C_1} \int_0^t V_A \, dt = \frac{V_A t}{\tau_1} \tag{17.12}$$

where $\tau_1 = R_1 C_1$ and V_A is assumed to be constant during the integration period. When V_{o_1} goes above V_R, the comparator output switches to the high state, which activates the monostable multivibrator to produce a short pulse of length T_d. This causes switch S_1 to close, thus terminating the charging period. The charging period terminates at time $t = T_C$ as obtained from $V_A T_C / \tau_1 = V_R$, so $T_C = (V_R/V_A)\tau_1$. The switch remains closed for the time period T_d to allow the capacitor to discharge completely. At the end of this short period, the switch opens and the charging cycle starts again. The voltage waveforms of V_{o_1} and V_{o_2} are shown in Figure 17.34. The total period is $T = T_C + T_d$, and the corresponding frequency is $f = 1/T = 1/(T_C + T_d)$. If $T_d \ll T_C$, then the frequency is approximately given by $f = 1/T_C = (V_A/V_R)(1/\tau_1)$.

The pulse train produced by the monostable multivibrator becomes the clock to drive the binary counter, which will count the pulses during the fixed gate period T_G. At the end of the gate period, the count stops and the output of the counter is made available as the digital output of the ADC. The total number of pulses counted at the end of the gate period is $N = T_G \times f = T_G/T = (T_G/\tau_1)(V_A/V_R)$. The digital output is this count in binary form.

Although this is a relatively simple ADC, its accuracy and speed are limited by the requirement that $T_C = (V_R/V_A)\tau_1$ be much greater than the discharge time T_d, so the frequency is a linear function of the analog input voltage. The T_d pulse length is limited by the speed of the comparator or the minimum pulse available from the monostable multivibrator, as well as the capacitor discharge time constant.

The total conversion time for an M-bit ADC is approximately $NT_{C(\text{MIN})}$, where $N = 2^M$ and $T_{C(\text{MIN})} = (V_R/V_{A(\text{MAX})})\tau_1$. Since $T_{C(\text{MIN})}$ is limited by the requirement that $T_{C(\text{MIN})} \gg T_d$, this can result in a long conversion time. For example, if $C_1 = 1$ nF and the switch "on" resistance is 25 Ω, the discharge time constant will be 25 ns. For the full discharge of C_1, a discharge time of at least 10 time constants will generally be needed so that $T_d \geq 10 \times 25$ ns $= 250$ ns. For a 6-bit ADC with an accuracy of $\pm\frac{1}{2}$ LSB, we should require that $T_{C(\text{MIN})} = 2 \times 2^6 T_d = 128 T_d = 32$ μs. The total conversion time is therefore approximately $NT_{C(\text{MIN})} = 2^6 \times 32$ μs $= 2.05$ ms. The corresponding conversion rate is 488 conversions per second. For an 8-bit V/F-type ADC, the total conversion time is $2 \times 2^8 \times 2^8 \times 250$ ns $= 33$ ms and the corresponding conversion rate is only 30 conversions per second.

Voltage-to-time ADC. Another type of integrating ADC is the voltage-to-time converter. In this type of converter the reference voltage is integrated for a period of time until its integrated value reaches the analog input voltage. This time period is used to gate a binary counter that is driven by a fixed-frequency clock. The number of clock pulses counted by the counter is proportional to the analog input voltage, and the digital output is this number of clock pulses expressed in binary form.

Figure 17.35 is a diagram of this type of ADC, together with voltage waveforms. At time $t = 0$ the switch S_1 is opened, which initiates the integration period. At the same time the AND gate G_1 is enabled so that the clock pulses can now drive the binary counter. The output voltage of the integrator circuit A_1 is

$$V_{o_1} = \frac{-1}{R_1 C_1} \int_0^t -V_R \, dt = V_R t / \tau_1 \tag{17.13}$$

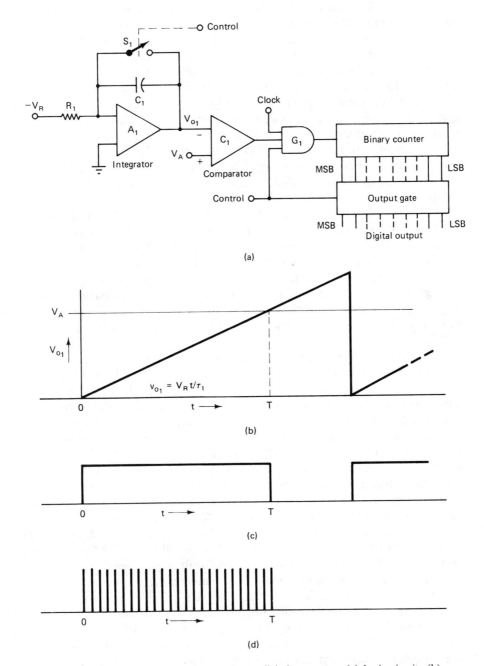

Figure 17.35 Voltage-to-time analog-to-digital converter: (a) basic circuit; (b) integrator output voltage; (c) comparator output voltage; (d) clock pulse input to counter.

where $\tau_1 = R_1 C_1$. At time $t = T$, the integral of the reference voltage is equal to the analog input voltage, so $V_R T / \tau_1 = V_A$ and thus $T = (V_A / V_R)\tau_1$.

At time $t = T$, the comparator output voltage goes low, which disables the gate G_1, so the counter stops counting and the binary counter outputs are gated out as the digital output of the converter. The control voltage next goes high, which resets the integrator to zero by closing switch S_1. At the same time the output gate G_2 is disabled and the clock input gate G_1 is enabled, allowing the counter to start counting again. The count will start from zero, however, since the counter will have been reset to zero shortly before switch S_1 is opened.

The digital output that is made available at the end of the integration time period T is a binary number corresponding to N clock pulses, where $N = T/T_{\text{CLOCK}} = (V_A / V_R)(\tau_1 / T_{\text{CLOCK}})$. Thus the digital output code is proportional to the analog input voltage.

For an M-bit V/T type of ADC, the total number of clock pulses that must be available for counting in a conversion cycle is $N_{\text{MAX}} = 2^M$, so the conversion time is approximately $2^M T_{\text{CLOCK}}$. Thus, for a 10-bit ADC with a representative clock period of $T_{\text{CLOCK}} = 50$ ns, the conversion time is approximately $2^{10} \times 50$ ns $= \underline{51\ \mu s}$ and the corresponding conversion rate is $\underline{20,000\ \text{conversions per second}}$. For a 14-bit converter the conversion time is up to $\underline{820\ \mu s}$ and the conversion rate is $\underline{1220\ \text{con-versions per second}}$. Thus the V/T ADC is a relatively slow converter, although it generally will be much faster than the voltage-to-frequency (V/F) type of integrating ADC.

In both the V/T and the V/F converters, the conversion accuracy is a direct function of the accuracy and stability of the timing resistor R_1 and capacitor C_1 that establish the integrator time constant τ_1. There will usually be provision for a small trimming or adjustment resistor to be connected externally to the IC that can be used to adjust the net value of R_1 to produce the correct ADC scale factor at some reference temperature, such as 25°C. Let us consider as an example a 10-bit ADC with an accuracy of ± 1 LSB over the temperature range from 0° to 50°C, and for which the net value of R_1 has been adjusted at 25°C to produce the correct ADC scale factor. Since the integrator time constant, and thus the ADC scale factor, will be directly proportional to R_1, the maximum fractional deviation of R_1 over the temperature range from 0° to 50°C is given by $(\Delta R_1 / R_1)_{\text{MAX}} = 1/2^{10} = 1/1024 = 0.001$. The corresponding maximum allowable value for the temperature coefficient of the resistance is given by $TC_{R_{\text{MAX}}} = (1/R_1)(dR_1/dT_{\text{MAX}}) = 0.001/\pm 25°C = \underline{4 \times 10^{-5}/°C} = \underline{40\ \text{ppm}/°C} = \underline{0.004\%/°C}$. This can be a difficult requirement to meet.

The comparator offset voltage V_{OS} can also be a factor in the accuracy of the ADC. If we assume that the offset voltage is initially nulled or compensated for at 25°C, the maximum allowable change in V_{OS} over the operating temperature range from 0° to 50°C is given by $\Delta V_{OS}/V_A = 1/2^{10} = 0.001$. For a full-scale analog input voltage of 10 V, this gives $\Delta V_{OS(\text{MAX})} = 10$ V $\times 0.001 = 10$ mV. The corresponding maximum allowable temperature coefficient of the offset voltage $TC_{V_{OS}}$ is $TC_{V_{OS(\text{MAX})}} = dV_{OS}/dT = 10$ mV$/\pm 25°C = \pm 0.4$ mV/°C $= \underline{\pm 400\ \mu V/°C}$. This is not a difficult requirement.

If we now consider an example of a 14-bit ADC of the V/F or V/T type with a ± 1-LSB accuracy over an operating temperature range of 0° to 50°C as in the previous example, the requirement of the temperature coefficient of resistance for R_1 becomes $TC_{R_{MAX}} = 2.5$ ppm/°C. This is indeed a very difficult requirement to meet. For the offset voltage, the maximum allowable temperature coefficient for this case is given by $TC_{V_{OS(MAX)}} = \pm 25$ μV/°C. This is still not a very difficult requirement, so the main problem is with respect to the temperature coefficient of the timing resistor R_1. Other possible problem areas are the temperature coefficient of the timing capacitor C_1 and the variation with temperature of the counter gating pulse in the case of the V/F converter and the clock frequency in the case of the V/T converter. In addition, the temperature coefficient of the reference voltage is important. For the 10-bit ADC the reference voltage, V_R must have a temperature coefficient of no more than 1000 ppm/°C, and for the 14-bit converter the temperature coefficient of V_R must not exceed 61 ppm/°C.

In the next section we will investigate the dual-slope type of integrating ADC. In the dual-slope converter, the only temperature coefficient that will be a factor in the accuracy of the ADC is that of the reference voltage, so converter resolutions of up to 14 bits become readily available.

Dual-slope ADC. The third type of integrating ADC that we will consider is the dual-slope ADC, and it will be seen that this type of converter offers some significant advantages over the V/F and V/T converters in that the conversion scale factor is independent of the integrator time constant and the clock frequency. A basic diagram of the dual-slope converter is shown in Figure 17.36, together with the integrator voltage versus time waveform.

At time $t = 0$ switch S_1 is connected to the analog input voltage V_A and switch S_2 is opened, so the analog input voltage integration begins. The output voltage of the integrator is

$$V_{o_1} = \frac{-1}{R_1 C_1} \int_0^t V_A \, dt = \frac{-V_A t}{R_1 C_1} = \frac{-V_A t}{\tau_1} \tag{17.14}$$

where $\tau_1 = R_1 C_1$ is the integrator time constant, and it is assumed that V_A remains constant over the integration time period. At the end of 2^N clock periods at time $t = 2^N T_{CLOCK}$, the output of the flip-flop, Q_N, goes high ($Q_N = 1$), which causes switch S_1 to be switched from V_A to $-V_R$. At this very same time the binary counter has gone through its entire count sequence and has cycled back to the 000 . . . state.

The integrator output voltage is now given by $v_{o_1} = -V_A(2^N T_{CLOCK})/\tau_1 + V_R t'/\tau_1$, where $t' = t - 2^N T_{CLOCK}$ and is the time measured from the instant of the change of switch S_1 at the end of the binary counter cycle. This output voltage reaches zero at $t' = \Delta t$, given by $V_R \Delta t/\tau_1 = V_A(2^N T_{CLOCK})/\tau_1$, so $\Delta t = (V_A/V_R)2^N T_{CLOCK}$. At this time the comparator output voltage goes low, which disables the clock AND gate G_1. This stops the clock pulse from reaching the counter, so that the counter will be stopped at a count corresponding to the number of clock pulses in time Δt as given by $n = \Delta t/T_{CLOCK} = (V_A/V_R)2^N$. Note that the clock period cancels out of this expression. The binary digital output of the counter corresponds to this count and therefore is directly proportional to the analog input voltage.

Note that the digital conversion scale factor is independent of the integrator time constant, which cancels out as a result of the two successive integrations of V_A

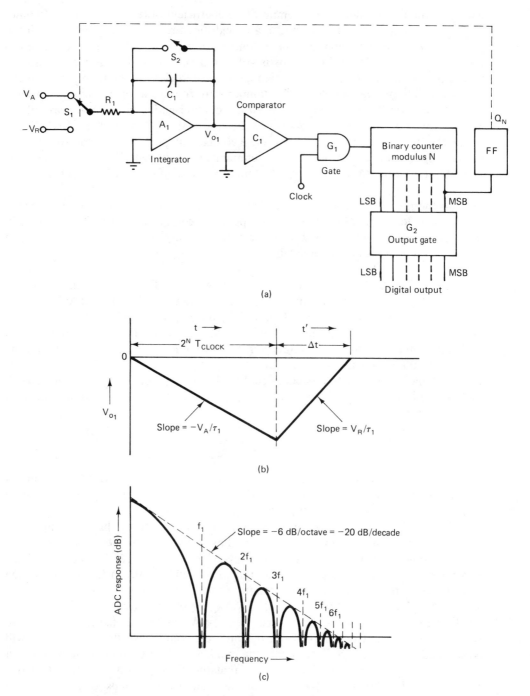

Figure 17.36 Dual-slope analog-to-digital converter: (a) basic circuit; (b) integrator output voltage; (c) frequency-response characteristic.

and V_R. Note also that the scale factor is not a function of the clock period, again because of the two successive integrations. Indeed, the principal factor that controls the conversion accuracy is the temperature coefficient of the reference voltage.

To consider a representative example, let us look at a 14-bit dual-slope ADC with an accuracy of ± 1 LSB over an operating temperature range of $0°$ to $50°C$. We will assume that the reference voltage has been adjusted to give the correct conversion scale factor at $25°C$. The maximum allowable temperature coefficient for the reference voltage is given by $(1/V_R)(dV_R/dT) = (1/2^{14})/25°C = \underline{2.44 \times 10^{-6}/°C} = \underline{2.44}$ ppm/°C. For a 10-V reference voltage, the maximum allowable temperature coefficient is therefore $\pm 24 \, \mu V/°C$. This is a difficult requirement and requires a well-compensated temperature reference circuit.

A very important feature of the dual-slope ADC is its rejection of frequency components that are integer multiples of $f_1 = 1/(2^N T_{CLOCK})$. Since the analog voltage integration time is fixed at $2^N T_{CLOCK}$, these frequency components will have periods that are integer submultiples of the analog signal integration time and so will execute an integer number of complete waveform cycles in this time. The contribution of these components of the analog input voltage to the integral of the analog voltage will therefore be zero.

Generally, the most troublesome interference frequency component is the 60-Hz power-line frequency and harmonics thereof. Therefore, if the analog voltage integration period $2^N T_{CLOCK}$ is made equal to 1/60 s, there will be a very good rejection of the power-line frequency and harmonics, often with rejections of 70 dB or more for these frequencies. The total conversion time for this case is 1/30 s for a conversion rate of 30 conversions per second. For a 14-bit converter the corresponding clock period is $T_{CLOCK} = (1/2^{14}) \times (1/60)$ s $= \underline{1.02 \, \mu s}$ and the corresponding clock frequency is $\underline{983 \text{ kHz}}$.

Figure 17.36c shows a normalized frequency response graph of the dual-slope ADC. Note that, in addition to the strong rejection of the frequency component at integer multiples of $f_1 = 1/(2^N T_{CLOCK})$, there will also be an attenuation of the higher-frequency components. This characteristic is useful for the rejection of noise and other interference.

The dual-slope ADC can also be used for voltage ratio conversions and measurements. The number of clock cycles counted to produce the digital output code is $n = (V_A/V_R)2^N$ and therefore is directly proportional to the ratio of two voltages V_A and V_R. If a second analog input voltage V_B is used in place of the fixed reference voltage V_R, the count will be proportional to the V_A/V_B ratio, so the digital output code will correspond precisely to the voltage ratio.

The dual-slope ADC is used as the basic element of most digital voltmeters (DVMs) and digital multimeters (DMMs). For these applications a high accuracy is needed and slow conversion rates are acceptable. A four-digit DVM or DMM with a full-scale range of 0 to 9.999 requires a resolution of 14 binary bits, so high-resolution ADCs are indeed required for many DVM and DMM applications.

An example of a dual-slope ADC with a very high resolution is the ADC100 (Thaler Corporation), which has a resolution of 22 bits and is fully monotonic. The maximum linearity error is 3 ppm. This ADC has an input-voltage range of ± 10.48 V, with an input of -10.485760 V producing a digital output of $000 \ldots 000$. An

TABLE 17.3 HIGH-PERFORMANCE ANALOG-TO-DIGITAL
CONVERTERS

ADC type	Resolution (bits)	Conversion rate
Parallel comparator (flash)		
CA3300D (RCA)	6	15 MHz
AD6020KD (AD)	6	50 MHz
AD9000SD (AD)	6	75 MHz
TDC1029 (TRW)	6	100 MHz
MC10315L/10317L (MOT)	7	15 MHz
TM1070 (Telmos)	7	15 MHz
TDC1025 (TRW)	8	75 MHz
TKAD10C (Tektronix)	8	500 MHz
TDC1019J (TRW)	9	25 MHz
Subranging parallel compar-ator systems		
MOD-1020 (AD)	10	20 MHz
MOD-1205 (AD)	12	5 MHz
THC1200 (TRW)	12	10 MHz
SP9550 (Sipex)	12	5 MHz
Successive approximation		
ADC1103 (AD)	8	1 MHz
ADC60 (BB)	8	1.1 MHz
ADC1103 (AD)	10	670 kHz
ADC60 (BB)	10	530 kHz
ADC1103 (AD)	12	300 kHz
ADC60 (BB)	12	300 kHz
HS9516-6 (HS)	16	10 kHz
ADC72 (AD)	16	20 kHz
Integrating (dual-slope)		
ADC141 (AD)	14	25 Hz
ADB1200/LF13300 (NS)	12	28 Hz
TSC850 (Teledyne)	16	40 Hz
Voltage-to-frequency		
ADC100 (BB)	12	80 Hz
ADC100 (BB)	14	20 Hz
ADC100 (BB)	16	5 Hz

Manufacturer: AD, Analog Devices; BB, Burr-Brown; HS, Hybrid Systems Corp.; MOT, Motorola; NS, National Semiconductor.

input of 0.0 V produces the digital output of 100 . . . 000 (200000 in hexadecimal), and an input of $+10.485755$ V produces the maximum digital output of 111 . . . . 111 (3FFFFF in hexadecimal). A 1-LSB change in the digital output is produced by a 5-μV change in the analog input voltage. The conversion time is 320 ms (max).

17.2.6 High-Performance Analog-to-Digital Converters

It is of interest to see what is currently available in terms of high-performance ADCs. Table 17.3 lists the resolution and conversion rate of some ADCs.

PROBLEMS

17.1. (*D/A converter*) Given a D/A converter using *binary-weighted quad current sources* (see Figure P17.1). Assume that the bit current transistors have their active areas scaled such that the *current density* is the same for all of the bit current transistors.

  (a) Given that $R_2 + R_3 = R_1$, find R_1, R_2, and R_3 for this 12-bit D/A converter. (*Ans.:* $R_1 = 14.000$ kΩ, $R_2 = 13.13$ kΩ, $R_3 = 875$ Ω)

  (b) Given that D_1 and D_2 constitute a temperature-compensated zener diode with $V_Z = 6.3$ V and $V_F = 0.7$ V, find R_8 such that $V_{\text{REF}} = 10.24$ V. (*Ans.:* $R_8 = 4.63$ kΩ)

  (c) For a minimum value of the temperature coefficient of V_{REF}, it is necessary to bias the zener diode D_1 and diode D_2 at a current level of about 7.5 mA. Find the required value of R_9. (*Ans.:* $R_9 = 432$ Ω)

  (d) Find the value of R_4 needed to produce an MSB current of 1.00 mA. (*Ans.:* $R_4 = 81.92$ kΩ)

  (e) Find the value of R_5 needed to minimize the error due to I_{BIAS} (assume that $R_6 \gg R_5$). (*Ans.:* $R_5 = 82$ kΩ)

  (f) For the case of unipolar operation (pins 1, 2, and 3 connected together), find the value of R_F needed to give a nominal full-scale output voltage of $+10.24$ V. (*Ans.:* $R_F = 10.24$ kΩ)

  (g) Find R_{10} for minimum error due to I_{BIAS}. (*Ans.:* $R_{10} = 3.82$ kΩ)

  (h) Find V_O for the following digital inputs (positive logic, unipolar operation): **(1)** 000 . . . 000; **(2)** 100 . . . 000; **(3)** 111 . . . 111. (*Ans.:* $+10.2375$ V; $+5.1175$ V; $+0.0000$ V)

  (i) Find the change in V_O for a 1-LSB change in the digital input. (*Ans.:* $\Delta V_o = 2.50$ mV)

  (j) Find the value of R_6 required for a ± 50-mV adjustment range for the full-scale voltage. (*Ans.:* $R_6 = 25$ MΩ)

  (k) Find the value of R_{11} needed for a ± 10-mV offset adjustment range (unipolar case). (*Ans.:* $R_{11} = 7.7$ MΩ)

  (l) **(1)** Find the emitter voltage swing on the MSB transistor such that $I_{\text{OFF}} \leq I_{\text{LSB}}/10$. (*Ans.:* $\Delta V_E = 248$ mV)
   (2) Find I' such that $\Delta V_E = 1.0$ V. (*Ans.:* $I' = 1.1$ mA)
   (3) Find R_{12} for $I' = 1.1$ mA (MSB input "high"). (*Ans.:* $R_{12} = 11.8$ kΩ)

  (m) For an accuracy, referred to the output voltage, of ± 1 LSB over the specified temperature range, what is the maximum allowable output error voltage, and the corresponding percentage of error (with respect to the full-scale output voltage)? (*Ans.:* ± 2.50 mV, $\pm 0.0244\%$)

  For the following problems, assume that the D/A converter is trimmed to give zero error at 25°C, and the specified operating temperature range is $-25°$ to $+75°$C.

  (n) **(1)** If the contribution of V_{OS} to the total maximum error is limited to one-fifth of the total, what is the maximum allowable value of $TC_{V_{OS}}$? (*Ans.:* $TC_{V_{OS}} = 10$ μV/°C)
   (2) If $TC_{V_{OS}} = V_{OS}/T$, where V_{OS} is the offset voltage before trimming, what is the maximum allowable value of V_{OS}? [*Ans.:* $V_{OS} = 3.0$ mV (max.)]

  (o) If I_{OS} is to contribute no more than one-fifth of the total error, find the maximum allowable value for $TC_{I_{OS}}$. [*Ans.:* $TC_{I_{OS}} = 2.0$ nA/°C (max.)]

  (p) If the error due to the transistor leakage current is not to exceed one-fifth of the total error, find the maximum allowable value for the transistor leakage current,

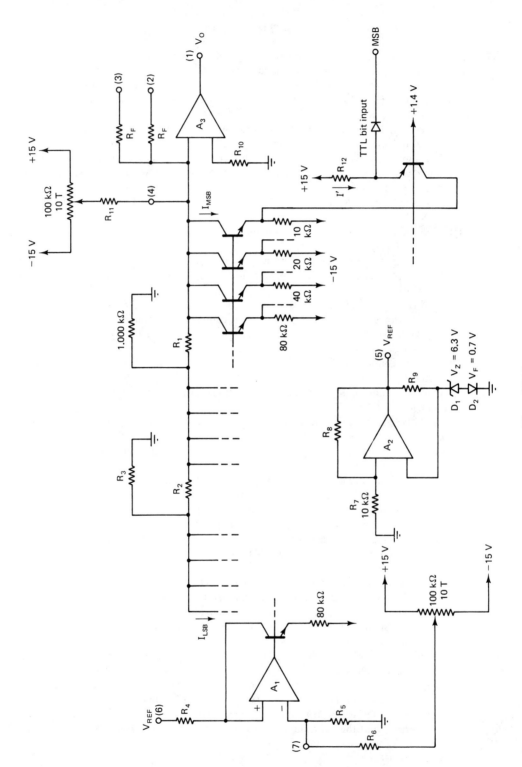

Figure P17.1

$I_{C(OFF)}$. (Note that the leakage current doubles for approximately every 10°C temperature increment.)   [*Ans.*: $I_{C(OFF)}$ = 715 pA (max.) at 25°C (corresponding to 23 nA at 75°C)]

(q) Find the maximum TCR tracking error if this tracking error is to contribute no more than one-half of the total error. Assume that the resistor network layout is such that there are no net temperature differentials between resistors.   (*Ans.*: ΔTC_R = 4.88 ppm/°C)

(r) Find the resistor ratio tolerance required if this is not to contribute more than one-half of the total error. Note that the ratio tolerance of the resistors associated with the MSB and the next significant bit will be the most important, since the other bits, going down to the LSB, will contribute successively smaller amounts of current.   (*Ans.*: 0.0244%)

(s) The offset voltage between the bit current transistors can produce a contribution to the total error, especially with respect to the matching between the MSB and the next significant bit (the second bit) transistor. Find the maximum allowable offset voltage if this is to contribute no more than one-fifth of the total error. Note that the transistor active areas are scaled to produce equal current densities.   (*Ans.*: ΔV_{BE} = 0.977 mV)

(t) If the $TC_{V_{REF}}$ is to contribute no more than one-fifth of the total error, find the maximum allowable value for $TC_{V_{REF}}$ and for TC_{V_Z}.   (*Ans.*: 1.0 ppm/°C for both $TC_{V_{REF}}$ and TC_{V_Z})

(u) If pin 2 is connected to pin 1 as before, but pin 3 is now connected to V_{REF} (pins 5, 6, and 3 connected together):
  (1) Show that bipolar operation is achieved with V_o = − 10.24 for a digital input of 1111 . . . 111 and V_O = + 10.235 for a digital input of 00000 . . . 000.
  (2) Find ΔV_o for a 1-LSB change in the digital input.   (*Ans.*: ΔV_o = 5.0 mV)

(v) If a $\frac{1}{2}$-LSB settling time of 1.0 μs is required for a full-scale output voltage of 10 V, what is the required slewing rate of the operational amplifier A_3?   [*Ans.*: $SR \simeq$ 10 V/μs (min.)]

17.2. (*Quantization noise*) (a) Show that the quantization error of an N-bit ADC (with a $\frac{1}{2}$-LSB offset) produces a quantization noise that has an rms value of $V_{O(FS)}/(2^N\sqrt{12})$. (b) A 10-bit ADC has a full-scale voltage of 10.24 V. Find the rms quantization noise.   (*Ans.*: 2.9 mV)

17.3. (*Effect of reference voltage temperature coefficient on ADC accuracy*) A 10-bit ADC has a reference voltage of 10.24 V. The maximum error of this ADC is not to exceed $\pm\frac{1}{2}$ LSB over the operating temperature range from 0° to 50°C, and the error is adjusted to zero at 25°C. Find the maximum allowable value for $TC_{V_{REF}}$. Express the result in terms of μV/°C and ppm/°C.   (*Ans.*: $\pm$200 μV/°C, $\pm$19.5 ppm/°C)

17.4. (*Flash ADC encoder logic*)
  (a) Show that a general $2^N - 1: N$ encoder to go from a thermometer code to a binary code can be based on the following logic gate configuration:
$$Q_{LSB} = (Q_1 \oplus Q_2) + (Q_3 \oplus Q_4) + (Q_5 \oplus Q_6) + (Q_7 \oplus Q_8) + \cdots$$
$$Q_{2SB} = (Q_2 \oplus Q_4) + (Q_6 \oplus Q_8) + (Q_{10} \oplus Q_{12}) + \cdots$$
$$Q_{3SB} = (Q_4 \oplus Q_8) + (Q_{12} \oplus Q_{16}) + \cdots$$
  .
  .
  .

  (b) Show that for a 4-bit flash ADC encoder this will reduce to
$$Q_{LSB} = (Q_1 \oplus Q_2) + (Q_3 \oplus Q_4) + (Q_5 \oplus Q_6) + (Q_7 \oplus Q_8) + (Q_9 \oplus Q_{10}) + (Q_{11} \oplus Q_{12}) + (Q_{13} \oplus Q_{14}) + Q_{15}$$

$$Q_{2SB} = (Q_2 \oplus Q_4) + (Q_6 \oplus Q_8) + (Q_{10} \oplus Q_{12}) + Q_{14}$$
$$Q_{3SB} = (Q_4 \oplus Q_8) + Q_{12}$$
$$Q_{MSB} = Q_8$$

17.5. (*Flash ADC encoder logic*) An N-bit parallel comparator ADC uses type D flip-flops with complementary outputs to latch the comparator output voltage levels. Show that the following logic gate configuration using AND and OR gates can be used to convert the thermometer code to a binary code.
$$Q_{LSB} = (Q_1 \cdot \overline{Q}_2) + (Q_3 \cdot \overline{Q}_4) + (Q_5 \cdot \overline{Q}_6) + \cdots$$
$$Q_{2SB} = (Q_2 \cdot \overline{Q}_4) + (Q_6 \cdot \overline{Q}_8) + (Q_{10} \cdot \overline{Q}_{12}) + \cdots$$
$$Q_{3SB} = (Q_4 \cdot \overline{Q}_8) + (Q_{12} \cdot \overline{Q}_{16}) + \cdots$$
.
.
.

17.6. (*Flash ADC encoder that uses only NOR gates*) Develop a general thermometer code to binary code encoder that uses only NOR gates. This will be for an N-bit parallel comparator ADC with the comparator output levels latched by D flip-flops with complementary outputs.

17.7. (*3-Bit flash ADC encoder*) Show a logic gate diagram of an NOR gate encoder for a 3-bit parallel comparator ADC.

17.8. (*Flash ADC design*)

(a) Show that the number of comparators needed for an N-bit flash ADC is $2^N - 1$.

(b) Show that the reference voltage levels needed for the $2^N - 1$ comparators are given by $V_{REF} = (M - \frac{1}{2})V_{FS}/(2^N - 1)$, where M takes on integer values from 1 to $2^N - 1$. This is for the condition that the quantization will not exceed $\pm\frac{1}{2}$ LSB at any point.

(c) A 4-bit flash ADC is to have a full-scale analog input voltage of $+15.0$ V. Find the reference voltage levels needed for the comparators such that the quantization error will not exceed $\pm\frac{1}{2}$ LSB at any point.

(d) For the comparator of part (c), find the maximum allowable net error due to the comparator reference voltages and the drift in the comparator offset voltage such that the total conversion error will not exceed ± 1 LSB at any point in the 0- to $+15.0$-V analog input voltage range.   (*Ans.*: ± 0.50 V)

17.9. (*Combining ADCs to increase the number of bits*) Show how two N-bit parallel comparator ADCs can be combined to produce an $N + 1$ bit ADC. Each ADC has a $-V_{REF}$ and a $+V_{REF}$ input terminal for the two ends of the reference voltage divider. Each ADC has D_O through D_{N-1} active high tristate outputs plus an active high overange (OVR) output. There is also an active high chip enable (CE) input that puts the D_O through D_{N-1} outputs into the high-Z state. The total conversion range is to be from -5.12 to $+5.12$ V.

17.10. [*8-Bit cascaded (subranging) flash ADC*] An 8-bit subranging flash ADC uses two 4-bit ADCs. Each of the two ADCs has a full-scale input voltage range of $+15.9375$ V. The DAC produces a maximum output voltage of $+15.0$ V.

(a) Design the difference amplifier and specify its gain.   (*Ans.*: $A_v = 16$)

(b) Find the number of comparators required for each of the two ADCs and the reference voltage levels needed for each comparator. The reference voltage levels should be such that the quantization error will not exceed $\pm\frac{1}{2}$ LSB at any point.

(c) Find the 8-bit digital output code and the quantization error both in absolute terms and as a fraction of an LSB voltage change for the following analog input voltages (volts):

(1) 0.000   (*Ans.*: 0000 0000, 0 V or 0 LSB)
(2) 0.1000   (*Ans.*: 0000 0010, 0.025 V or 0.4 LSB)
(3) 2.603   (*Ans.*: 0010 1010, 0.022 V or 0.352 LSB)
(4) 5.324   (*Ans.*: 0101 0101, -0.0115 V or -0.184 LSB)
(5) 7.276   (*Ans.*: 0111 0100, -0.026 V or -0.416 LSB)
(6) 12.870   (*Ans.*: 1100 1110, 0.005 V or 0.08 LSB)
(7) 14.220   (*Ans.*: 1110 0100, 0.030 V or 0.48 LSB)
(8) 15.375   (*Ans.*: 1111 0110, 0 V or 0 LSB)
(9) 15.772   (*Ans.*: 1111 1100, -0.022 V or -0.352 LSB)
(10) 15.9530 (*Ans.*: 1111 1111, -0.0155 V or -0.248 LSB)
(11) 15.9375 (*Ans.*: 1111 1111, 0 V or 0 LSB)

(d) For a $\pm\frac{1}{2}$-LSB accuracy (not including the quantization error), find the required tolerance limits for the reference voltage levels of the comparators of the LSB ADC and the MSB ADC.   (*Ans.*: ± 500 mV, ± 31.25 mV)

REFERENCES

ALLEN, P. E., and D. H. HOLBERG, *CMOS Analog Design,* Holt, Rinehart and Winston, New York, 1987.

ANALOG DEVICES, Inc., *Analog–Digital Conversion Notes,* Analog Devices, Inc., P.O. Box 796, Norwood, Mass. 02062.

CONNELLY, J. A., *Analog Integrated Circuits,* Wiley, New York, 1975.

FRANCO, S., *Design with Operational Amplifiers and Analog Integrated Circuits,* McGraw-Hill, New York, 1988.

GARRETT, P. H., *Analog I/O Design,* Prentice-Hall, Englewood Cliffs, N.J., 1981.

GRAEME, J. G., G. E. TOBEY, and L. P. HUELSMAN, *Operational Amplifiers—Design and Applications,* McGraw-Hill, New York, 1971.

GREBENE, A. B., *Analog Integrated Circuit Design,* Van Nostrand Reinhold, New York, 1972.

———, *Bipolar and MOS Analog Integrated Circuit Design,* Wiley, New York, 1984.

GRINICH, V. H., and H. G. JACKSON, *Introduction to Integrated Circuits,* McGraw-Hill, New York, 1975.

HOESCHEL, D. F., *Analog-to-Digital/Digital-to-Analog Conversion Techniques,* Wiley, New York, 1968.

McMENAMIN, J. M., *Linear Integrated Circuits: Operation and Applications,* Prentice-Hall, Englewood Cliffs, N.J., 1985.

MILLMAN, J., *Microelectronics,* McGraw-Hill, New York, 1979.

MITRA, S. K., *An Introduction to Digital and Analog Integrated Circuits and Applications,* Harper & Row, New York, 1980.

MORLEY, M. S., *The Linear IC Handbook,* TAB Books, Blue Ridge Summit, Pa., 1986.

SHEINGOLD, D. H. (ed.), *Analog–Digital Conversion Handbook,* 3rd ed., Prentice-Hall, Englewood Cliffs, N.J., 1986.

STOUT, D. F., *Handbook of Microcircuit Design and Applications,* McGraw-Hill, New York, 1980.

TAUB, H., and D. SCHILLING, *Digital Integrated Electronics,* Chapter 14, McGraw-Hill, New York, 1977.

WAIT, J. V., *Introduction to Operational Amplifier Theory and Applications,* McGraw-Hill, New York, 1975.

Appendix A

Physical Constants, Conversion Factors, and Parameters

Angstrom unit	Å	$1 \text{ Å} = 10^{-4} \text{ μm} = 10^{-8} \text{ cm} = 10^{-10} \text{ m}$
Boltzmann constant	k	$k = 1.380 \times 10^{-23} \text{ J/K} = 8.62 \times 10^{-5} \text{ eV/K}$
Electron charge	q	$q = 1.602 \times 10^{-19} \text{ C}$
Micron	μm	$1 \text{ μm} = 10^{-6} \text{ m} = 10^{-4} \text{ cm} = 10{,}000 \text{ Å}$
Mil		$1 \text{ mil} = 0.001 \text{ in.} = 25.4 \text{ μm}$
Nanometer	nm	$1 \text{ nm} = 10^{-9} \text{ m} = 10^{-3} \text{ μm} = 10 \text{ Å}$
Permittivity of free space	ε_0	$\varepsilon_0 = 8.854 \times 10^{-14} \text{ F/cm}$
Permeability of free space	μ_0	$\mu_0 = 4\pi \times 10^{-9} \text{ H/cm}$
Planck's constant	h	$h = 6.625 \times 10^{-34} \text{ J-s}$
Thermal voltage	V_T	$V_T = kT/q = 25.85 \text{ mV at 300 K (27°C)}$
Velocity of light in free space	c	$c = 2.998 \times 10^{10} \text{ cm/s}$

	Silicon	Germanium	GaAs	SiO_2
Dielectric constant, ε_r (relative permittivity)	11.8	16.0	10.9	3.8
Energy gap, E_G (eV at 300 K)	1.11	0.67	1.43	
Electron drift mobility, μ_n (cm²/V-s at 300 K)	1350	3900	8500	
Hole drift mobility, μ_p (cm²/V-s at 300 K)	500	1900	400	
Intrinsic carrier concentration, n_i (cm⁻³ at 300 K)	1.4×10^{10}	2.1×10^{12}	1.1×10^{7}	
Electron saturation velocity, v_{sat} (cm/s)	0.8×10^{7}	0.6×10^{7}	2×10^{7}	
Saturation electric field, E_{sat} (V/cm)	2×10^{4}	3×10^{3}	2×10^{3}	

Appendix B

Transistor Conductances

B.1 BIPOLAR TRANSISTOR CONDUCTANCES

For a bipolar transistor operating in the active mode (emitter–base junction "on" and collector–base junction "off"), we have the following basic relationships for the transistor currents:

1. $I_C = I_{CO} \exp (V_{BE}/V_T)$
2. $I_E = -I_C$
3. $I_B = I_{BO} \exp (V_{BE}/nV_T)$, where n is a dimensionless constant between 1 and 2 and usually near 1.5
4. By definition $\beta = I_C/I_B$, where $\beta = h_{FE}$ is the transistor current gain

The dynamic transfer and input conductances can readily be obtained from these simple exponential relationships, giving:

1. *Dynamic forward transfer conductance:* $g_m = dI_C/dV_{BE} = I_C/V_T$
2. *Dynamic emitter input conductance:* $g_{eb} = dI_E/dV_{EB} \simeq dI_C/dV_{BE} = g_m$
3. *Dynamic base input conductance:* $g_{be} = dI_B/dV_{BE} = I_B/nV_T$

The equation for I_C given above shows that there is an exponential dependence of the collector current on the base-to-emitter voltage V_{BE}, but there is no explicit dependence of I_C on the collector voltage. However, as will be seen, as a result of the *base-width modulation* or *Early effect* there will indeed be a slight dependence of I_C on the collector voltage.

The preexponential constant I_{CO} is inversely proportional to the *effective* or *electrical* base width W_B' of the transistor as given by $I_{CO} \propto 1/W_B'$. Figure B.1 shows both the actual base width W_B and the effective base width W_B'. The difference between these two is due to the incursion of the emitter–base and collector–base depletion regions into the base region, so the effective base width can become substantially smaller than the actual base width. The collector–base junction depletion region width will increase with increasing collector voltage, so as the collector voltage increases the effective base width W_B' decreases, and as a result I_{CO} increases. This in turn causes I_C to increase. It is by this chain of events that I_C becomes a function of the collector voltage, showing a slight increase as the collector voltage increases.

The *dynamic collector-to-emitter conductance* g_{ce} is defined as $g_{ce} = dI_C/dV_{CE}$ and we can write that

$$
\begin{aligned}
g_{ce} &= \frac{dI_C}{dV_{CE}} = \frac{dI_C}{dI_{CO}} \frac{dI_{CO}}{dW_B'} \frac{dW_B'}{dV_{CB}} \frac{dV_{CB}}{dV_{CE}} \\
&\simeq \exp\left(\frac{V_{BE}}{V_T}\right)\left(-\frac{I_{CO}}{W_B'}\right)\frac{dW_B'}{dV_{CB}} \times 1
\end{aligned}
\tag{B.1}
$$

We will now define the *base-width modulation coefficient* or *Early voltage* V_A by the equation

$$
\frac{1}{V_A} = \frac{1}{W_B'}\frac{dW_B'}{dV_{CB}}
\tag{B.2}
$$

Using V_A in the foregoing equation for g_{ce} gives

$$
\boxed{g_{ce} = I_{CO}\exp\left(\frac{V_{BE}}{V_T}\right)\frac{1}{V_A} = \frac{I_C}{V_A}}
\tag{B.3}
$$

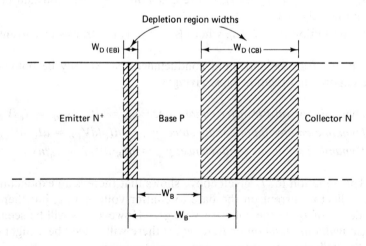

Figure B.1 Base-width modulation.

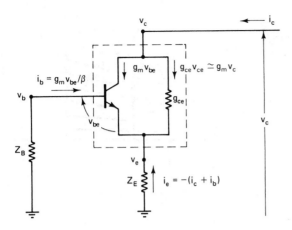

Figure B.2 Circuit for the calculation of collector conductance.

The value of the Early voltage V_A for most transistors is in the range from 100 to 300 V. For example, if $V_A = 200$ V, at a current of $I_C = 1.0$ mA the conductance g_{ce} is given by $g_{ce} = I_C/V_A = 1.0$ mA/200 V $= 1000$ μA/200 V $= 5$ μS. The reciprocal quantity of $r_{ce} = 1/g_{ce}$ is the *dynamic collector-to-emitter resistance* and for this example is $r_{ce} = 1/g_{ce} = V_A/I_C = 200$ V/1.0 mA $= 200$ kΩ.

The dynamic conductance looking into the collector of a transistor g_c is a function of g_{ce} and the impedances in series with the emitter and base of the transistor. For the analysis of the *dynamic collector conductance* g_c, we will use the circuit of Figure B.2. The transistor is represented as an ideal transistor ($g_{ce} = 0$) for which $i_c = g_m v_{be}$, in parallel with a conductance of value g_{ce} to represent the collector-to-emitter conductance that is actually present in the transistor. The total ac collector current of the transistor thus becomes $i_c = g_m v_{be} + g_{ce} v_{ce} \simeq g_m v_{be} + g_{ce} v_c$, since $v_c \simeq v_{ce}$.

Since $v_{be} = v_b - v_e$, $v_b = -i_b Z_B$, and $v_e = -i_e Z_E = (i_b + i_c) Z_E$, we have that $v_{be} = -i_b Z_B - (i_b + i_c) Z_E = -i_c Z_E - i_b(Z_E + Z_B)$. Since $i_b = g_m v_{be}/\beta$, we have that $v_{be} = -i_c Z_E - (g_m v_{be}/\beta)(Z_E + Z_B)$. If we now solve this last expression for v_{be}, we obtain $v_{be}[1 + (g_m/\beta)(Z_E + Z_B)] = -i_c Z_E$, so $v_{be} = -i_c Z_E/[1 + (g_m/\beta)(Z_E + Z_B)]$. If we now substitute this expression for v_{be} into the equation for i_c given earlier, we obtain

$$i_c = g_m v_{be} + g_{ce} v_c = -\frac{g_m(i_c Z_E)}{1 + (g_m/\beta)(Z_E + Z_B)} + g_{ce} v_c \qquad (B.4)$$

Combining all terms involving i_c on the left side now gives

$$i_c\left[1 + \frac{g_m Z_E}{1 + (g_m/\beta)(Z_E + Z_B)}\right] = g_{ce} v_c \qquad (B.5)$$

Now solving for the ratio $g_c = i_c/v_c$, we obtain

$$g_c = \frac{i_c}{v_c} = \frac{g_{ce}}{1 + g_m Z_E/[1 + g_m(Z_E + Z_B)/\beta]}$$

$$= \frac{g_{ce}[1 + g_m(Z_E + Z_B)/\beta]}{1 + g_m Z_E + g_m(Z_E + Z_B)/\beta} \qquad (B.6)$$

Appendix B: Transistor Conductances

801

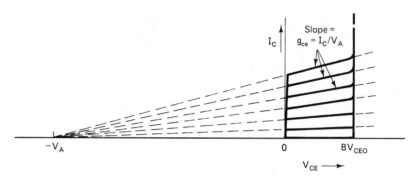

Figure B.3 Transistor output (collector) characteristics.

Since $g_m = I_C/V_T$ and $g_{ce} = I_C/V_A$, the expression for g_c can be written as

$$g_c = \frac{I_C}{V_A} \frac{1 + (I_C/V_T)(Z_E + Z_B)/\beta}{1 + (I_C/V_T)[Z_E + (Z_E + Z_B)/\beta]} \qquad (B.7)$$

From this last expression we note that, if $Z_E = 0$, then $g_c = I_C/V_A = g_{ce}$. This result is independent of the value of Z_B. At the other extreme, for large values of Z_E such that $(I_C/V_T)Z_E = g_m Z_E \gg \beta$ the collector conductance will approach a limiting value given by $g_c = I_C/\beta V_A$ and the corresponding dynamic collector resistance will become $r_c = \beta V_A/I_C$. This condition can result in a very small conductance. For example, if $V_A = 200$ V and $I_C = 1.0$ mA as in the preceding example, and if $\beta = 100$, then $g_c = I_C/\beta V_A = 1$ mA/100 $\times$ 200 V = $\underline{50 \text{ nS}}$, and the corresponding resistance is $r_c = 1/g_c = \underline{20 \text{ M}\Omega}$.

In Figure B.3, the common-emitter output or collector characteristics are shown, with curves of I_C versus V_{CE} for various constant values of base current I_B. The slope of the various I_C versus V_{CE} curves is the dynamic collector-to-emitter conductance g_{ce} as given by $g_{ce} = dI_C/dV_{CE}$ and is equal to $g_{ce} = I_C/V_A$. If the curves are extrapolated back from the active region to the voltage axis intercept, they all intersect the voltage axis at about the same place, this being at $-V_A$.

Finally, it is of interest to compare the effectiveness of the base-to-emitter voltage V_{BE} to the collector voltage (V_{CE} or V_{CB}) in controlling the collector current. We have that $i_c = g_m v_{be} + g_{ce} v_{ce}$. Since both conductances are proportional to the quiescent collector current as given by $g_m = I_C/V_T$ and $g_{ce} = I_C/V_A$, the ratio of g_m to g_{ce} is independent of I_C and is $g_m/g_{ce} = V_A/V_T$. For $V_A \sim 200$ V and $V_T = 25$ mV, this gives $g_m/g_{ce} = V_A/V_T \sim 200$ V/25 mV = $\underline{8000}$. Thus the input voltage V_{BE} is very much more effective in controlling the collector current than is the output voltage (V_{CE}). It is due to this that very large voltage gains are possible with just a single transistor stage, especially when active loads are used.

B.2 FIELD-EFFECT TRANSISTOR TRANSFER CONDUCTANCES

For a JFET in the saturated (active) region for which $|V_{DS} - V_{GS}| > |V_P|$, the transfer relationship between the drain-to-source current I_{DS} and the gate-to-source voltage V_{GS} is given approximately by $I_{DS} = I_{DSS}(1 - V_{GS}/V_P)^2$ for $0 \leq |V_{GS}| \leq |V_P|$, and where V_P is the pinch-off voltage and I_{DSS} is the value of I_{DS} when $V_{GS} = 0$. The dynamic transfer conductance, g_{fs} or g_m, is given by

$$g_{fs} \equiv g_m = dI_{DS}/dV_{GS} = (2I_{DSS}/ - V_P)(1 - V_{GS}/V_P) \qquad \text{(B.8)}$$

This can be rewritten as

$$g_{fs} \equiv g_m = \frac{2I_{DSS}}{-V_P}\sqrt{\frac{I_{DS}}{I_{DSS}}} = 2\frac{\sqrt{I_{DSS}I_{DS}}}{-V_P} = g_{fsO}\sqrt{\frac{I_{DS}}{I_{DSS}}} \qquad \text{(B.9)}$$

where $g_{fsO} = g_{mO}$ and $2I_{DSS}/ - V_P$ is the transfer conductance when $V_{GS} = 0$. All the preceding equations are applicable to both N- and P-channel JFETs, and the transfer conductances will be positive quantities in both cases.

For a MOSFET in the saturated (active) region for which $|V_{DS}| > |V_{GS} - V_t|$ and $|V_{GS}| > |V_t|$, the transfer relationship between I_{DS} and V_{GS} is given approximately by $I_{DS} = K(V_{GS} - V_t)^2$, where K is a constant having units of A/V^2 and V_t is the threshold voltage for channel formation (not to be confused with the thermal voltage). The dynamic transfer conductance, g_{fs} or g_m, is given by

$$g_{fs} \equiv g_m = \frac{dI_{DS}}{dV_{GS}} = 2K(V_{GS} - V_t) = \frac{2I_{DS}}{V_{GS} - V_t} = 2\sqrt{KI_{DS}} \qquad \text{(B.10)}$$

Again, all the preceding equations are equally applicable to both N- and P-channel MOSFETs and the transfer conductances will be positive in both cases.

The parameter K for the MOSFET is given by

$$K = \tfrac{1}{2}\mu\, C_{ox}\left(\frac{W}{L}\right) = \tfrac{1}{2}\mu\left(\frac{\epsilon_{ox}}{t_{ox}}\right)(W/L) \qquad \text{(B.11)}$$

where μ = mobility of the charge carriers in the surface inversion layer
$C_{ox} = \epsilon_{ox}/t_{ox}$ = capacitance per unit area of the MOS capacitor formed by the gate, the gate oxide, and the surface inversion layer
ϵ_{ox} = permittivity of the gate oxide (SiO_2) $\simeq 3.1 \times 10^{-13}$ F/cm
t_{ox} = thickness of the gate oxide
W/L = channel width to length ratio

The charge carrier mobility in the surface inversion layer channel is substantially less than the bulk mobility due to the scattering of the charge carriers in the very thin inversion layer at the Si–SiO$_2$ interface. The channel mobility is generally in the range of about one-half of the corresponding bulk mobility, so for an NMOS transistor $\mu_{channel} \approx \mu_n/2 \approx 500$ cm^2/V-s, and for PMOS, $\mu_{channel} \approx \mu_p/2 \approx 200$ cm^2/V-s.

For a representative example, let us consider an NMOS transistor for which t_{ox} = 1000 Å = 0.1 μm and we assume a channel mobility of 500 cm²/V-s. For the parameter K, we obtain

$$K = \frac{1}{2}\left[500 \text{ cm}^2/\text{V-s} \times \frac{3.1 \times 10^{-13}\text{F/cm}}{1 \times 10^{-5} \text{ cm}} \times \frac{W}{L} \right]$$

$$= 775 \times 10^{-8} \frac{W}{L} \text{ A/V}^2 = 7.75 \frac{W}{L} \text{ μA/V}^2$$

If $L = 10$ μm and $W = 100$ μm, then $K = 77.5$ μA/V². At $V_{GS} - V_t = +5.0$ V, the drain current is $I_{DS} = 77.5$ μA/V² × (5 V)² = 1.94 mA. The transfer conductance at this current level is $g_{fs} = 2I_{DS}/(V_{GS} - V_t) = 2 \times 1.94$ mA/5.0 V = 0.776 mS. For a PMOS transistor of similar construction, the value of K is 31 μA/V². At $V_{GS} - V_t = 5.0$ V, we obtain $I_{DS} = 776$ μA and $g_{fs} = 310$ μS.

B.3 FIELD-EFFECT TRANSISTOR OUTPUT CONDUCTANCE

The variation of the drain current I_{DS} of a field-effect transistor, JFET or MOSFET, will respect to the drain-to-source voltage V_{DS} in the saturated (active) region is due to the *channel length modulation effect* that is similar to the base-width modulation effect or Early effect in a bipolar transistor. The dynamic drain-to-source conductance g_{ds} of a FET in the saturated region can be expressed as

$$\boxed{g_{ds} = \frac{dI_{DS}}{dV_{DS}} = \frac{I_{DS}}{V_A}} \tag{B.12}$$

where the parameter $1/V_A$ is the channel length modulation coefficient. The parameter V_A has units of volts and is analogous to the Early voltage of a bipolar transistor. It is generally in the range of 20 to 200 V for most FETs.

We will now make a simple analysis of the channel length modulation coefficient for MOSFETs. Similar results will apply for JFETs. The drain current of a MOSFET in the saturated region is given by $I_{DS} = K(V_{GS} - V_t)^2$, where $K = \frac{1}{2}\mu C_{ox}(W/L)$. The rate of change of the drain current with respect to the drain voltage is due to the variation in the channel length L as a result of the expansion of the drain-body junction depletion region as shown in Figure B.4 and is given by

$$g_{ds} = dI_{DS}/dV_{DS} = \frac{dI_{Ds}}{dL} \times \frac{dL}{dV_{DS}} = \left(\frac{I_{DS}}{L}\right)\frac{dL}{dV_{DS}} = I_{DS}\left(-\frac{1}{L}\frac{dL}{dV_{DS}}\right) \tag{B.13}$$

We will define the channel length modulation coefficient as

$$\textit{channel length modulation coefficient} = \frac{1}{V_A} = -\left(\frac{1}{L}\right)\frac{dL}{dV_{DS}} \tag{B.14}$$

The effective or electrical channel length L is given by $L = L_0 - \Delta L$. L_0 is the channel length when $V_{DS} = 0$ and ΔL is the depletion layer width at the drain-body junction. This depletion layer width is given by $\Delta L = \sqrt{2\epsilon V_J/qN} =$

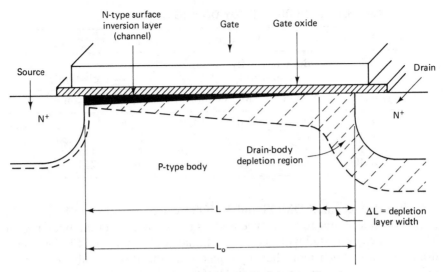

Figure B.4 Channel length modulation effect.

$\sqrt{2\epsilon(V_{DS} + \phi)/qN}$, where N is the body region doping (assuming heavily doped N^+ or P^+ source and drain regions) and ϕ is the contact potential of the junction and is typically about 0.8 V. For dL/dV_{DS}, we now have $dL/dV_{DS} = dL/d(\Delta L) \times d(\Delta L)/dV_{DS} = -1 \times \frac{1}{2}\Delta L/(V_{DS} + \phi)$, so

$$\frac{1}{V_A} = -\frac{1}{L}\frac{dL}{dV_{DS}} = \frac{\Delta L/L}{2(V_{DS} + \phi)}$$

and thus

$$V_A = 2(V_{DS} + \phi)\left(\frac{L}{\Delta L}\right) = 2(V_{DS} + \phi)\frac{L_0}{\Delta L} - 1 \qquad (B.15)$$

In Table B.1, some representative values of V_A are given based on this equation and for the case of $V_{DS} = 10$ V, $\phi = 0.8$ V, and for channel lengths of 5.0, 7.5, and 10 μm. From Table B.1 we note that V_A increases substantially with increases in the body doping level. However, increases in the body doping level also result in increases in the threshold voltage V_t, which in turn result in lower values for the transfer conductance. We also see the important role of the channel length L_0 in that V_A increases rapidly with channel length. However, K and therefore g_{fs} as well as the source-to-drain transit time will be inversely proportional to the channel length.

Short-channel MOSFETs generally exhibit very low values of V_A and also have relatively low values of the drain-to-source breakdown voltage BV_{DS} due to the drain-to-source punch-through. This can be partially compensated for by using a larger body doping. A very effective technique, however, is the use of an N/N^+ drain region, as is the case in DMOS and VMOS transistors.

	$V_{DS} = 10$ V	$\phi = 0.8$ V				
	Body resistivity (ohm-cm)		ΔL	V_A (volts) for $L =$		
N (cm^{-3})	NMOS	PMOS	(μm)	5 μm	7.5 μm	10 μm
1×10^{15}	13	5.0	3.75	7.2	21.6	36.0
3×10^{15}	4.5	1.7	2.17	28.3	53.2	78.1
1×10^{16}	1.5	0.6	1.19	69.4	115	160
3×10^{16}	0.65	0.23	0.68	136	215	294

In the N-channel DMOS and VMOS devices, the doping level of the N-type drain region adjacent to the P-type body is much lower than the body region doping. This is to be contrasted with the conventional N-channel MOSFET in which the N$^+$ drain region is much more heavily doped than the P-type body. As a result of the lightly doped N-type drain region in DMOS and VMOS transistors, the drain region absorbs most of the depletion region thickness, so the incursion of the depletion region into the P-type body, and therefore the decrease of the effective channel length L, will be relatively small. The result of this is a much larger value of V_A and drain-to-source breakdown voltage than would otherwise be the case for a short-channel device. Short-channel DMOS and VMOS transistors with channel lengths of less than 1 μm have very large values for the K parameter and therefore for the transfer conductance g_{fs}. The heavily doped N$^+$ portion of the drain region ensures a low drain series resistance. The short channel length also leads to a very good frequency response due to the short source-to-drain transit time. At the same time, these devices exhibit large values for V_A and therefore a relatively small value for g_{ds}, as well as a high value for the drain-to-source breakdown voltage.

B.3.1 FET Output Conductance with Impedance in Series with the Source

Let us now consider the circuit of Figure B.5a in which there is an impedance Z_S in series with the source. The FET can be either a JFET or a MOSFET, and in the equivalent circuit representation of Figure B.5b the drain to source conductance g_{ds} of the FET is represented as a conductance external to the transistor.

The total FET ac small-signal current i_d is given by $i_d = g_{fs}v_{gs} + g_{ds}v_{ds}$. Since the condition that $r_{ds} = 1/g_{ds} \gg Z_S$ will almost always hold true, we have that $v_{ds} \simeq v_d$, so i_d can be written as $i_d \simeq g_{fs}v_{gs} + g_{ds}v_d$. For v_{gs} we have that $v_{gs} = -i_dZ_S$, so $i_d \simeq -g_{fs}Z_Si_d + g_{ds}v_d$. Solving this last equation for i_d gives $i_d(1 + g_{fs}Z_S) \simeq g_{ds}v_d$ and thus $i_d \simeq [g_{ds}/(1 + g_{fs}Z_S)]v_d$.

The output or drain dynamic conductance of the FET circuit is given by $g_o = i_d/v_d$. Solving for g_o from the preceding equations gives

$$g_o \simeq \frac{g_{ds}}{1 + g_{fs}Z_S} \tag{B.16}$$

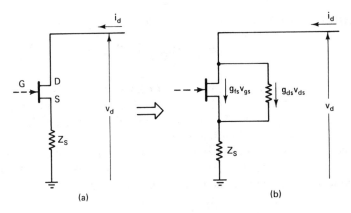

Figure B.5 Field-effect transistor output conductance.

B.4 SUBSTRATE BIAS EFFECT (BODY EFFECT)

The MOSFET is actually a four-electrode device with the gate, source, drain, and body (or substrate) electrodes. In most applications the body (or substrate) of the MOSFET is internally connected to the source so that the MOSFET appears as a three-electrode or triode device. In many IC applications, however, this is not possible since the MOSFET body is the same as the IC substrate that is common to many transistors on the chip.

A voltage difference between the body and the source of a MOSFET can affect the net charge in the source-to-drain channel and the threshold voltage of the device. The body-to-source voltage V_{BS} is, however, usually much less influential than the gate-to-source voltage V_{GS} in controlling the channel conductance. This is due to the much greater thickness of the body–channel junction depletion region as compared to the thickness of the gate oxide. The relative influence of the body voltage compared to that of the gate voltage over the net charge in the channel is proportional to the ratio of the body–channel junction capacitance C_J to the gate-to-channel (or oxide) capacitance C_{ox}. The effect of the body voltage on the MOSFET characteristics can most conveniently be described in terms of a shift in the threshold voltage ΔV_t that results from a change in the body-to-source voltage ΔV_{BS}.

The shift in the threshold voltage produced by a change in the body voltage is given by

$$K_{BS} = \frac{-\Delta V_t}{\Delta V_{BS}} = \frac{C_J}{C_{ox}} = \frac{\sqrt{(q\epsilon_{si}N_B)/(2|V_{BS} + \phi|)}}{\epsilon_{ox}/t_{ox}} \tag{B.17}$$

where N_B is the substrate (or body) doping and ϕ is the junction contact potential (~ 0.8 V). To obtain a normalized value of the K_{BS} coefficient, we will use $N_B = 1 \times 10^{16}$ cm^{-3}, $V_{BS} + \phi = 1.0$ V, and $t_{ox} = 1000$ Å, for which we obtain the following result:

$$K_{BS} = \frac{\sqrt{(1.602 \times 10^{-19} \text{ C} \times 1.04043 \times 10^{-12} \text{ F/cm} \times 10^{16} \text{cm}^{-3})/(2 \times 1.0 \text{ V})}}{(3.8 \times 10^{-13} \text{ F/cm})/(1 \times 10^{-5} \text{ cm})} = \underline{0.860}$$

$$\tag{B.18}$$

For a general relationship for K_{BS}, we have

$$K_{BS} = 0.860 \frac{t_{ox}}{1000 \text{ Å}} \sqrt{\frac{N_B}{10^{16} \text{ cm}^{-3}} \times \frac{1.0 \text{ V}}{V_{BS} + \phi}}$$ (B.19)

For a representative example, we will let $t_{ox} = 800$ Å, $N_B = 1 \times 10^{15}$ cm^{-3}, and $V_{BS} + \phi = 10$ V, for which we obtain $K_{BS} = 0.0688$; so we see that the body voltage can produce a significant effect on the MOSFET characteristics, although it is considerably less influential than the gate voltage.

The influence of the body voltage on the drain current can be expressed in terms of a body-to-drain current dynamic transfer conductance, $g_{fb} = dI_{DS}/dV_{BS}$. This transfer conductance can be obtained from the relationship

$$g_{fb} = \frac{dI_{DS}}{dV_{BS}} = \frac{dI_{DS}}{dV_t} \times \frac{dV_t}{dV_{BS}} = g_{fs}K_{BS}$$ (B.20)

where g_{fs} is the gate-to-source dynamic transfer conductance.

A principal consequence of the body effect is an often substantial increase in the dynamic drain-to-source conductance of the MOSFET. This can result in a reduced voltage gain for MOSFET amplifiers using active loads. Another consequence can be an increased dynamic conductance for MOSFET current source circuits.

To evaluate the contribution of the body effect to the dynamic output conductance of a MOSFET, let us consider the circuit of Figure B.6. In this circuit an N-channel depletion-mode MOSFET is used as a current source of strength $I_O = I_{DSS}$. The dynamic output conductance g_o is given by

$$\begin{aligned} g_o &= -\frac{dI_O}{dV_O} = -\left(\frac{dI_O}{dV_{DS}} \times \frac{dV_{DS}}{dV_O}\right) - \left(\frac{dI_O}{dV_{BS}} \times \frac{dV_{BS}}{dV_O}\right) \\ &= -(g_{ds} \times -1) - (g_{fs}K_{BS} \times -1) \\ &= g_{ds} + g_{fs}K_{BS} \\ &= \frac{I_{DSS}}{V_A} + \frac{2I_{DSS}}{-V_P} K_{BS} \end{aligned}$$ (B.21)

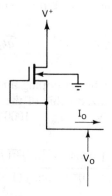

V^+

I_O

V_O

Figure B.6 Contribution of the body effect to the dynamic output conductance.

The fractional change in I_O per volt change in V_O is therefore given by

$$\frac{-1}{I_O}\frac{dI_O}{dV_O} = \frac{1}{V_A} + \frac{2K_{BS}}{-V_P} \tag{B.22}$$

Note that in the circuit of Figure B.6 the body of the MOSFET is not connected to the source so that a body-to-source voltage of $V_{BS} = -V_O$ will be present. This will result in a substantial increase in the output conductance. If the circuit construction were to allow the body to be connected to the source, then we would have $V_{BS} = 0$, and the output conductance would reduce to just $g_o = g_{ds} = I_{DSS}/V_A$.

B.5 COMPARISON OF BIPOLAR TRANSISTOR AND FET MODES OF OPERATION

It is of interest to compare the various regions or modes of operation of bipolar and field-effect transistors. In the cutoff region for a bipolar transistor, both junctions (EB and CB) are off and all transistor currents are reduced to zero, except for the small amount of junction leakage current. This corresponds to the cutoff region of a JFET (or a depletion-mode MOSFET) in which the gate-to-source voltage is beyond the pinch-off value, which fully depletes the channel all the way between the source and drain regions. The channel is thus pinched off at both the source and drain ends and the transistor currents are reduced to zero, except for the very small leakage currents. For the enhancement-mode MOSFET, the corresponding situation is that of the gate-to-source voltage being below the threshold value such that no conducting channel is formed between the source and drain regions.

In the active region for a bipolar transistor, the emitter–base junction is on and the collector–base junction is off. For the case of an NPN transistor, this correponds to the situation where the emitter is emitting electrons into the base region, and these electrons, after traversing the base region, are then collected by the collector. In this mode of operation, the collector current is relatively independent of the collector voltage, but exponentially dependent on the base-to-emitter voltage. For field-effect transistors (JFETs and MOSFETs), this corresponds to the situation of the channel being open at the source end, but being pinched off for a very short distance at the drain end. This produces the approximate saturation of the drain current such that the drain current increases very slowly with increasing drain voltage (but is very much dependent on the gate voltage). For this reason, this active region for FETs is often called the *saturated region* (not to be confused with the saturation region of bipolar transistors). The FETs operation in the active (saturated) region is obtained when the drain-to-gate voltage is such that the channel will be pinched off at the drain end, but at the same time the gate-to-source voltage will result in an open channel at the source end.

The saturation region for bipolar transistors occurs when both junctions (EB and CB) are on. In this region, the collector current is a strong function of the collector voltage, and the collector-to-emitter voltage drop across the transistor is small, generally less than 0.2 V. For field-effect transistors, this corresponds to the channel being open (or "on") at both the source and drain ends, and as a result the drain current is very much dependent on the drain voltage. This mode or region of operation for FETs is often called the *triode* or *nonsaturated* region. For small values of drain voltage, the drain current is an approximately linear function of the drain voltage; but as the drain voltage increases, the channel starts to narrow at the drain end, which produces a bending over of the drain current versus drain voltage curve. Finally, as the drain voltage increases further, the channel becomes closed (pinched off) at the drain end and the transistor then enters the saturated (active) region, where the drain current becomes relatively independent of drain voltage.

Appendix C

Transistor Small-signal Circuit Models

Although both bipolar and field-effect transistors are basically nonlinear devices, under ac small-signal conditions these devices can be represented in the form of a linear equivalent-circuit model. In Figure C.1, a general transistor amplifier stage is represented, and in Figures C.2 and C.3 are some approximate small-signal ac models for various bipolar and field-effect transistor configurations. These are simplified models but will be suitable for virtually all practical transistor amplifier applications.

The following is a summary of some of the basic parameters and the mid-frequency voltage gain equations for the various transistor circuits. For the parameter values given, $V_T = kT/q$ = thermal voltage ($\sim$25 mV) and V_A is the Early voltage or the reciprocal of the base-width modulation coefficient for bipolar transistors, and for field-effect transistors it is the reciprocal of the channel-length modulation coefficient. For both types of transistors, V_A is typically in the range from 30 to 300 V. The dimensionless factor n is between 1 and 2 and usually is close to 1.5. For the voltage-gain expressions, $A_{V(\text{MID})}$ is the mid-frequency ac voltage gain as given by $A_{V(\text{MID})} = v_o/v_i$, and $R_{L(\text{NET})}$ is the net load resistance driven by the transistor. The input and output resistances, r_i and r_o, are the mid-frequency values. At higher frequencies the capacitative reactances must be considered. For the bipolar transistor circuits some of the relationships with the hybrid or h parameters are shown. The parameter h_{fe} is the common-emitter (short-circuit) ac current gain, sometimes called β, h_{fb} is the common-base current gain, and $-h_{fb} = \alpha \simeq 1$.

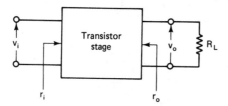

Figure C.1 Transistor amplifier stage.

Common emitter (Figure C.2a)

1. $r_i = h_{ie} = r_{be} = r_{bb'} + r_{b'e}$, where $r_{bb'}$ is the base spreading resistance and $r_{b'e} = nV_T/|I_B|$

2. $g_m = h_{fe}/h_{ie} = |I_C|/V_T$ = dynamic transfer conductance

3. $r_o = r_{ce} = 1/h_{oe} = V_A/|I_C|$

4. $A_{V(\text{MID})} \simeq -g_m R_{L(\text{NET})} = -h_{fe} R_{L(\text{NET})}/h_{ie}$

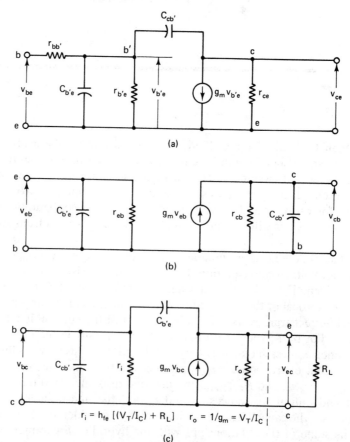

Figure C.2 Small-signal equivalent circuit models for bipolar transistors: (a) common emitter; (b) common base; (c) common collector (emitter follower).

Appendix C: Transistor Small-signal Circuit Models

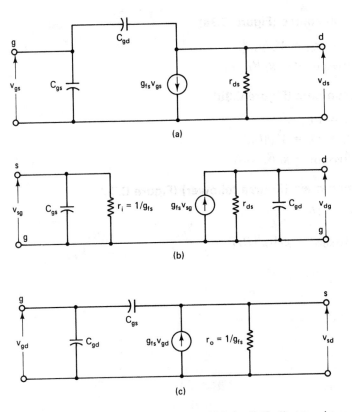

Figure C.3 Small-signal equivalent circuit models for field-effect transistors: (a) common source; (b) common gate; (c) common drain (source follower).

Common base (Figure C.2b)

1. $r_i = h_{ib} = V_T/|I_E| \simeq V_T/|I_C| = 1/g_m$
2. $g_m = h_{fe}/h_{ie} = -h_{fb}/h_{ib} = |I_C|/V_T$
3. $r_o = r_{cb} = 1/h_{ob} = h_{fe}/h_{oe} = h_{fe}(V_A/|I_C|)$
4. $A_{V(MID)} = g_m R_{L(NET)} = -h_{fb} R_{L(NET)}/h_{ib}$

Common collector (Emitter follower) (Figure C.2c)

1. $r_i = h_{fe}[(V_T/|I_C|) + R_L] = h_{fe}[(1/g_m) + R_L]$
2. $g_m = |I_C|/V_T$
3. $r_o = 1/g_m = V_T/|I_C|$
4. $A_{V(MID)} = g_m[r_o \| R_L] = \dfrac{g_m}{g_m + 1/R_L} = \dfrac{g_m R_L}{1 + g_m R_L} = \dfrac{R_L}{(1/g_m) + R_L}$

$$= \dfrac{R_L}{R_L + V_T/|I_C|}$$

(Note that, for the usual case of $R_L \gg V_T/I_C$, $A_{V(MID)} \simeq 1$.)

Common source (Figure C.3a)

1. $r_o = r_{ds} = V_A/|I_{DS}|$
2. $A_{V(\text{MID})} = -g_{fs}R_{L(\text{NET})}$

Common gate (Figure C.3b)

1. $r_i = 1/g_{fs}$
2. $r_o = r_{ds} = V_A/|I_{DS}|$
3. $A_{V(\text{MID})} = g_{fs}R_{L(\text{NET})}$

Common drain (Source follower) (Figure C.3c)

1. $r_o = 1/g_{fs}$
2.
$$A_{V(\text{MID})} = g_{fs}[r_o \| R_L] = \frac{g_{fs}}{g_{fs} + 1/R_L} = \frac{g_{fs}R_L}{1 + g_{fs}R_L}$$

Index